Coffee
Production, Quality and Chemistry

Coffee
Production, Quality and Chemistry

Edited by

Adriana Farah
Universidade Federal do Rio de Janeiro, Brazil
Email: afarah@nutricao.ufrj.br

ROYAL SOCIETY OF CHEMISTRY

Print ISBN: 978-1-78262-004-4
Two-volume set print ISBN: 978-1-78262-106-5
PDF ISBN: 978-1-78262-243-7
EPUB ISBN: 978-1-78801-658-2

A catalogue record for this book is available from the British Library

Reprinted 2019, 2023

The Royal Society of Chemistry is a charity, registered in England and Wales, Number 207890, and a company incorporated in England by Royal Charter (Registered No. RC000524), registered office: Burlington House, Piccadilly, London W1J 0BA, UK. Telephone: +44 (0) 20 7437 8656.

For further information see our website at www.rsc.org

For general enquiries, please contact books@rsc.org

For EU product safety enquiries, please email books@rsc.org or contact Royal Society of Chemistry Worldwide (Germany) GmbH, Römischer Hof, Unter den Linden 10, 10117 Berlin.

Preface

Since coffee started its Journey from Africa through the world more than thousand years ago so much knowledge has evolved around it. The unveiling of its chemical composition, the development of new agricultural and industrial technologies, the study of its physiological effects and so forth came to reveal coffee's enormous hidden potential both for flavor and health. In spite of currently being one of the most studied and consumed beverages in the world, it keeps surprising us with new flavor novelties and health properties.

These books are far from containing all that is known about coffee, which would be an impossible task, but they contains a good compilation of the most important technological and scientific data produced to date involving production, chemistry, quality and health implications. The handpicked authors are experienced scientists in their respective fields, with their post graduate students, and industry/market professionals. I would like to take the opportunity to thank all of them immensely for their precious contribution to making good quality scientific and technical knowledge available to academics and the general public. We tried to deliver this complex knowledge in a way that anyone can understand or at least have a good idea of the coffee world.

Adriana Farah

Coffee: Production, Quality and Chemistry
Edited by Adriana Farah

Published by the Royal Society of Chemistry, www.rsc.org

Contents

Coffee: Production, Quality and Chemistry

Part I: Coffee Production

Coffee: Production, Quality and Chemistry
Edited by Adriana Farah

Published by the Royal Society of Chemistry, www.rsc.org

Part III: Coffee Chemistry

Section I: Natural Coffee Compounds and Derivatives

Coffee: Consumption and Health Implications

Part I

Coffee Production

CHAPTER 1

Introduction to Coffee Plant and Genetics

THIAGO FERREIRA[a], JOEL SHULER[b], RUBENS GUIMARÃES[b] AND ADRIANA FARAH*[a]

[a]Núcleo de Pesquisa em Café Prof. Luiz Carlos Trugo, Laboratório de Química e Bioatividade de Alimentos, Instituto de Nutrição, Universidade Federal do Rio de Janeiro, 21941-902, Brazil; [b]Universidade Federal de Lavras/Departamento de Engenharia Agrícola - Cx. Postal 3037 Lavras, MG, 37200-000, Brazil
*E-mail: afarah@nutricao.ufrj.br

1.1 Introduction

The coffee beverage treasured by millions of people around the world results from roasted seeds of trees belonging to the botanical family Rubiaceae, genus *Coffea*. Coffee plants were discovered in Africa and eventually disseminated to countries throughout the world. Along this journey, a number of new cultivars have been created from selected varieties to fulfil the need for plants with higher productivity, resistance to diseases and superior cup quality, and over time, new wild varieties have been discovered as well. Currently, over 100 species within the genus *Coffea* are catalogued.[1–3] Despite this diversity, only two species are actually of great importance in the world market, *C. arabica* L. and *C. canephora* Pierre. Knowing the genetic origin of coffee varieties and cultivars within these two species is important to understand the main differences and similarities in their chemical composition and flavour.

Coffee: Production, Quality and Chemistry
Edited by Adriana Farah

Published by the Royal Society of Chemistry, www.rsc.org

Since its discovery, coffee has attracted the attention of explorers and botanists from all over the world, especially in the second half of the 19th century, when many new species were discovered. Because of the great variation in the types of coffee plants and seeds, botanists have failed to agree on a precise, single system to classify them or even to designate some plants as true members of the *Coffea* genus.[4]

Knowledge of the coffee plant and its characteristics is fundamental for understanding practical coffee growing topics, as well as topics related to interaction with the environment and its reactions to biotic and abiotic stresses.

In this chapter, we introduce the coffee plant, discuss its origin and genetic aspects of the two main species, and explain how they migrated from Africa to other continents, becoming the most commercially important coffee species in the world.

1.2 The Genus *Coffea*

The coffee tree is part of the sub-kingdom of plants known scientifically as the Angiosperm, or Angiospermae, meaning that the plant reproduces by seeds enclosed in a box-like compartment, the ovary, at the base of the flower. It belongs to the botanical family Rubiaceae, which has some 500 genera and over 6000 species, subfamily Ixoroideae. The current classification of the *Coffea* genus results from recent fusions of several subgenera and genera.[4,5] According to Leroy[6] and Bridson,[7] two genera existed in this subfamily, *Coffea* L. and *Psilanthus* Hook.f. (an Australasian genus), with the *Coffea* genus being split into two subgenera, *Coffea* and *Baracoffea*. After morphological and molecular studies by Davis *et al.*[8] and Maurin *et al.*,[9] respectively, the group concluded that a sister relationship between both subgenera was actually highly unlikely and untenable.[10,11] Later, subgenus *Coffea* and genus *Psilanthus* were merged according to additional phylogeny analysis (using molecular and morphological data), leading to the current *Coffea* genus,[12] which is by far the most economically important member of the Rubiaceae family.[4,13] The botanical classification of coffee is shown in Figure 1.1.

The various species of subgenus *Coffea* are largely present in the African continent, though they are mostly restricted to tropical zones when growing in the wild. There are 41 species from continental Africa (from Guinea to Tanzania and from Ethiopia to Mozambique), 59 from Madagascar and 4 from nearby islands (1 from Grand Comore and 3 from the Mascarenes Islands Mauritius and Réunion), each area having 100% endemicity for its species.[1,2,14] Considering the merge between subgenus *Coffea* and genus *Psilanthus*, located in Asia and in Australasia, currently there are at least 125 species in the genus *Coffea*.[1,5,10]

From all catalogued species under the genus *Coffea*, only three have commercial importance: *Coffea arabica*, *Coffea canephora* and, to a much lower degree, *Coffea liberica*, with the first being the most cultivated crop.[4] *C. arabica* is a tetraploid species ($2n = 4x = 44$) originating from a natural hybridization

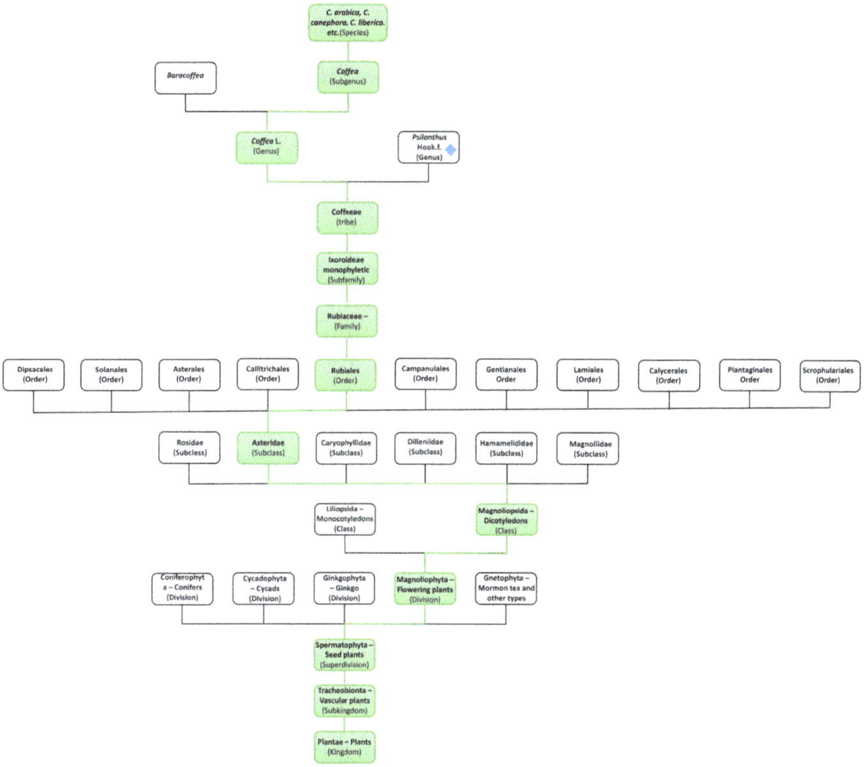

Figure 1.1 Botanical classification of the coffee plant according to Anthony *et al.*[14] and Natural Resources Conservation Service (USDA).[15] For further information on coffee specimens, access the website of the Royal Botanic Gardens, Kew.[16]

between either *C. canephora* and *C. eugenioides* or ecotypes related to these two diploid ($2n = 2x = 22$) species.[17–19] It is the species with highest cup quality compared to other known species, but the plant is not as strong and resistant as *C. canephora* species. Triploid hybrids, originating from crosses between *C. arabica* and diploid species, have been reported. They tend to be robust plants but are almost completely sterile.[4,17] *C. arabica* is self-compatible (self-fertile nature), which so far has only been reported in two other coffee species: *C. heterocalyx* Stoff. and *C. anthonyi* Stoff. & F. Anthony, ined. Despite its inferior cup quality, *C. canephora* maintains heterozygosity due to its cross-pollinating (self-incompatible) nature.[4,9] *Coffea liberica* Hiern is a diploid species cultivated to a minor extent, mainly because of its sensitivity to diseases, especially *Fusarium xylarioides*. Its seeds tend to have a better cup quality compared to *C. canephora* species, but still inferior compared to *C. arabica*.[20] Despite the close phylogenetic relationship between *C. liberica* and *C. canephora*, these species differ substantially in their morphological characteristics. C. *liberica* could thus be of interest for interspecific breeding programs.[20]

Owing to the richness of coffee species and varieties, and to the popularity of the coffee beverage, when referring to the main coffee species, some confusion has been observed regarding nomenclature, and the authors found it useful to clarify some misconceptions. For example, '*Coffea canephora*' has been described as '*Coffea robusta*', when 'robusta' is actually mostly reported as being a variety or subvariety of the *Coffea canephora* species. In the same way, the word 'robusta' has been popularly used for commercial and other purposes as a synonym of 'Kouilouensis' (also called 'Kouillon' or 'Conilon'), which is a different variety of *Coffea canephora*, widely cultivated in Brazil and with different chemical and sensory characteristics. Another misunderstanding sometimes occurs with the term '*Coffea dewevrei*', which has been used to refer to a separate species in some instances, and, in other instances, as a synonym for *Coffea liberica*. In fact, 'liberica' and 'dewevrei' (the latter also called 'excelsa' coffee) are different varieties within the *Coffea liberica* species. In addition, coffee varieties (wild genotypes) have been confused with cultivar names (plants selected by humans for cultivation).[4] As science advances and studies go deeper into unveiling the genetic, chemical and sensorial differences among coffee species and varieties/cultivars, knowledge of coffee genetics and nomenclature becomes ever more important for interpretation and dissemination of correct information in scientific reports.

1.3 Origin and Distribution of Subgenus *Coffea* in Africa

Coffea species have colonized many types of forests throughout a wide elevational distribution in the African continent. Up to 70% of species in *Coffea* subgenus are present in humid and evergreen forests, and at least 13% are adapted to seasonally dry forests in continental Africa. The other 17% of the species are adapted to various other types of forest, including humid evergreen forests, gallery forests, seasonally dry (evergreen to deciduous) forests, savannah woodlands and shrublands.[14,21]

In Madagascar, 67% of the species grow only in humid evergreen forests, 17% grow only in seasonally dry forests and the remaining species grow in both types of forests.[1,21,22]

Coffee trees are naturally found from sea level up to 2500 m, but no species grow throughout this entire range.[22] Species presenting the broadest elevational range of growth are: *C. eugenioides* (300–2200 m); *C. brevipes* (80–1450 m); *C. canephora* (50–500 m); *C. liberica* (80–1800 m); *C. mongensis* (400–200 m); *C. munfindiensis* (950–2300 m); *C. salvatrix* (400–1850 m); *C. dubardii* Jum., *C. homollei* J.-F. Leroy and *C. perrieri* (50–1200 m).[1,22] The largest number of endemic species in Africa is present between 200 and 1000 m above sea level, including *C. canephora* and *C. liberica* sub sp Dewevrei.[22] This broad range is mainly caused by variations in latitude. For example, in Uganda, an equatorial country where the minimum temperatures are warm and relatively stable, *C. canephora* grows above 1000 m. The altitude range for

C. arabica optimum growth is 1200–1950 m, with average growth occurring at 1575 m. It is worth noting that this elevation range is observed both on the continent and on islands, though the number of species that grow over 1000 m above sea level is higher in continental Africa than in Madagascar.[21,22] Figure 1.2 presents the average elevational distribution and type of forest colonized in Africa by important species of subgenus *Coffea*. The broadest elevational range species presented above are not in the pyramid. Throughout the rest of the world the presence or absence of species is largely defined by minimal temperatures, which is in most cases determined by elevation and latitude.[21,22]

The natural evolutionary history of coffee probably occurred between 150 000 and 350 000 years ago in the African continent.[21] Biogeographic analysis had indicated that the centre of origin of subgenus *Coffea* was in Kenya.[21] However, new DNA analysis and floristic records suggest that Lower Guinea in west equatorial Africa could be the centre of origin and speciation of *Coffea* subgenus *Coffea* as well as the richest sub-centre of endemism in

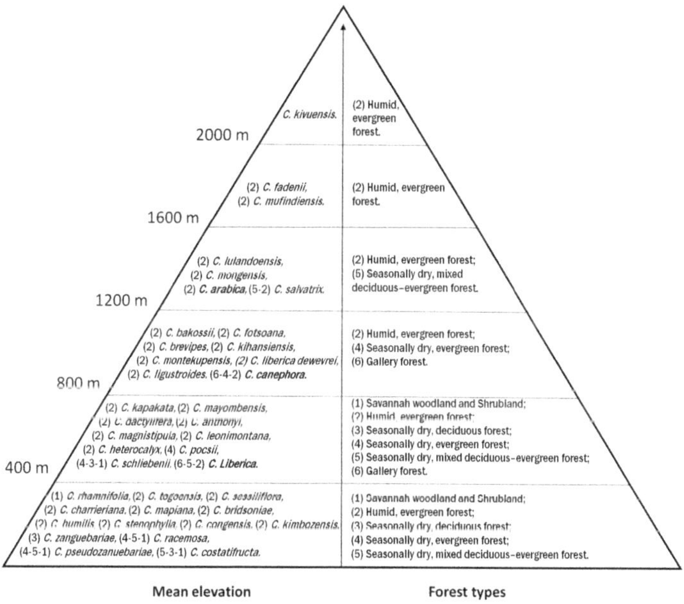

Figure 1.2 Elevational distribution (in mean) and types of forest colonized in Africa by *Coffea* species. Some species are not included in the pyramid because they have a wide range of elevational distribution (>1000 m), *i.e.*, *C. brevipes* (80–1450 m), *C. canephora* (50–1500 m), *C. eugenioides* (300–2200 m), *C. liberica* (80–1800 m), *C. mongensis* (400–2000 m), *C. mufindiensis* (950–2300 m) and *C. salvatrix* (400–1850 m). *C. eugenioides* is also naturally found in humid, evergreen forests, gallery forests, seasonally dry evergreen forests, savannah woodlands and shrublands. (Adapted with permission from ref. 21, Copyright 2011 Springer Nature, and ref. 22, Copyright 2015 Springer Nature.)

the Guineo-Congolese region. Diversity in subgenus *Coffea* has, therefore, been underestimated for a long time.[21,23] This region likely played the role of refuge for coffee trees during the last arid maximum (18 000 years before Pangea: B.P.) and previous arid phases. In Central Africa, a chain of small refuges has been located near the Atlantic Ocean: in west and south Cameroon, in the Crystal and Chaillu Mountains in Gabon and in the Mayombe Mountains in Congo. These areas, rich in coffee species, are known to be hotspots of biodiversity.[1,11,14] Figure 1.3 shows the original distribution of the current genus *Coffea* L., including subgenus *Coffea* in Africa and the additional Australasian *Psilanthus* spp.[12]

The *C. arabica* species has its primary centre of diversity in the southwestern Ethiopian highlands (in altitudes between 1000 and 2000 metres), the Boma Plateau of Sudan and Mount Marsabit of Kenya.[19,24] Its strict natural localization is due to the way that *C. arabica* speciation processes have occurred, as explained above. On the other hand, *C. canephora* has colonized various regions in Central Africa, stretching from West Africa through Cameroon, Central African Republic, Congo, the Democratic Republic of Congo, Uganda and northern Tanzania down to northern Angola.[25,26] In general, *C. liberica* habitats are localized to the same regions where *C. canephora* grow.[21,22]

The history of coffee cultivation is incompletely documented with regard to the domestication of the coffee plant in Africa and its dispersion throughout the world by humans (Figure 1.4).[27] Welman[28] reported in 1961 that the

Figure 1.3 Original distribution of the species included in the current classification of genus *Coffea* L. Grey colour area: distribution of the *Coffea* subgenus *Coffea* in the African continent.[12] Dark green colour area: additional areas of distribution of current *Coffea* genus, after the inclusion of Asian and Australasian *Psilanthus* spp.[12] Red circle: probable place of origin of *Coffea* subgenus *Coffea* in West-central Africa (Lower Guinea) before Pangea, considered to be a hotspot of *Coffea* biodiversity.[14]

cultivation of *C. arabica* varieties began when wild coffee was introduced from Ethiopia to Yemen as early as 575 AD or AC (Anno Domini, or After Christ), although other authors have reported possible cultivation even before that.[29] However, such data have been based on myths and legends, not scientific texts. Based on historical and scientific data, *C. arabica* diverged into two genetic bases, which have been described as two distinct botanical varieties: *Coffea arabica* var. arabica (usually called *Coffea arabica* var. Typica Cramer) and *Coffea arabica* var. Bourbon (B. Rodr.) Choussy.[17] These have subsequently led to most of the commercial *C. arabica* cultivars grown worldwide.[19] Bourbon-derived cultivars are characterized by a more compact and upright growth habit, higher yield and better cup quality (sensorial quality) than Typica-derived cultivars.[24]

Historical data indicate that the Typica variety originated from a single plant that was taken from Yemen to India.[30–32] Subsequent generations from this plant were taken to the island of Java in 1690 and then Amsterdam in 1706 or 1710, where plants were cultivated in the botanical gardens.[19,27] From Amsterdam, coffee was introduced to the Americas when seedlings were taken to Suriname in 1718. From there, an arabica coffee tree was introduced in the West Indies (Martinique) in 1720 or 1723.[33] In 1727, seeds were taken to the state of Pará in northern Brazil, apparently from French Guiana. Seeds

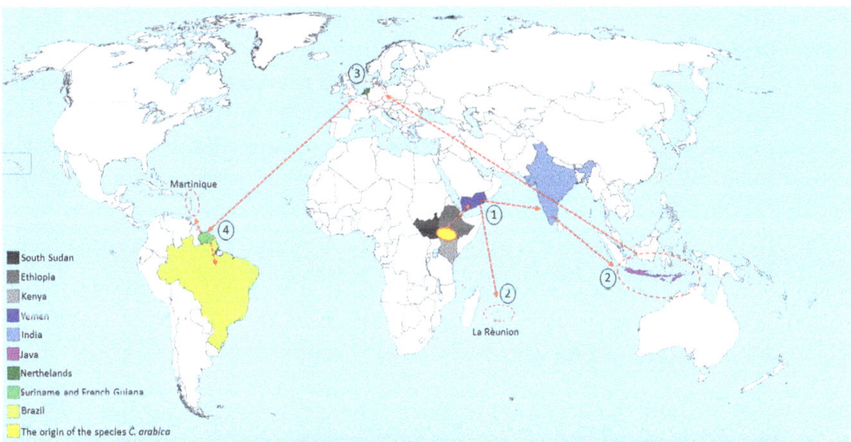

Figure 1.4 Origin and dissemination throughout the world of the most important coffee species, *Coffea arabica* L. Yellow circle: origin of cultivated *C. arabica* L. (mainly southwestern Ethiopia but also in the Boma Plateau of South Sudan and Mount Marsabit of Kenya). (1) *C. arabica* introduction into Yemen as early as 575 AD (after Christ).[19] (2) Coffee plant distribution to Réunion islands and taken from India to Java (Indonesia).[30,31] (3) From Java, coffee was introduced in Europe (Amsterdam) in 1710.[19,27] (4) From Europe, coffee was taken to South America (Suriname) in 1718. From there it was introduced in Martinique island (1720 or 1723) and Brazil *via* French Guiana (1727).[27,33,35] From South America the coffee was spread around the world. Note: colours indicate only the countries and not specific coffee growing regions within the countries.

from Suriname also became the parent of numerous self-progenies, which were further disseminated around the Americas (Jamaica, Puerto Rico, Haiti, Cuba, Central America, the Guianas, *etc.*).[27,34,35]

The Bourbon variety originated with the re-introduction of coffee trees to Bourbon Island (now Réunion, one of the Mascarenes Islands) with plants from Mocha, a city on the Yemeni coast (1715–1718). From there, Bourbon plants were possibly taken to Mauritius Island and later to various coffee growing origins worldwide.[18,19]

The spread of *C. canephora* from Central Africa throughout the world is more recent. It was initially taken to Indonesia in the 20th century as a solution to the coffee leaf rust that was attacking coffee plantations since it had presented resistance to this disease.[30] There are many varieties of *C. canephora* in Africa. However, only two have been commercially disseminated throughout the world: *C. canephora* from Guinea, and *C. canephora* from Congo.[26] *C. canephora* cultivars such as Laurenti (originated in the Belgian Congo), Apoã and Guarani (produced by the Agronomic Institute of Campinas, IAC) are less important economically.[25,26]

All of the places that grow *C. canephora* species, as well as hybrids with *C. arabica*, report its introduction due to the presence of coffee leaf rust and the need for breeding programs. Additionally, *C. canephora* thrives in warmer regions where *C. arabica* varieties are not well adapted.[25,26]

Currently, coffee is cultivated in the belt between the two tropics, being widely found in the tropical regions of South America (Brazil and Colombia), Asia, Oceania, Africa, Central America and Mexico.[36] *C. arabica* species prefer annual average temperatures between 18 °C and 22 °C and tend to grow in highlands. The closer this species gets to the equator, the higher the altitude needed for optimum growth. Therefore, the optimum altitude for growth and production to achieve a quality beverage will vary according to the country or growing region. *C. canephora* is more suitable for intertropical lowlands and can withstand higher temperatures than *C. arabica*.[22]

1.4 The Coffee Plant

This section covers the anatomy of the coffee plant, including the root system and aerial parts of the plant, and provides an overview of the flowering process and coffee fruit development.

1.4.1 Root System

Coffee plants are perennial, and the establishment of an adequate root system is fundamental to the health of the tree and its subsequent production throughout its lifetime. The root system (Figure 1.5) plays several key roles for the plant. Though often overlooked, it serves the basic function of fixing the plant in the soil or substrate. Perhaps the most widely known role is providing water to the plant. Apart from being a major constituent of plants, water acts as a solvent that serves to transport gases, minerals and other

Figure 1.5 Root system of *C. arabica* L. plant.

solutes from cell to cell and organ to organ; is a reactant in important processes such as photosynthesis; and maintains turgor, which is essential for cell enlargement and growth.[37] The root system also serves as a reserve for carbohydrates, and produces and accumulates key phytohormones such as auxins, abscisic acid and cytokines.[38,39]

It is impossible to succinctly define the root structure pattern of coffee plants since, as with all plants, it is patterned postembryonically, adapting its structure to optimize resources and respond to biotic and abiotic signals.[40] Many factors may affect the pattern of the root system and the size of the roots, including species and cultivar; physiological factors such as fruit load; vigour of the aerial part of the plant; plant reserves; pest and disease attacks; plant spacing; prunings; the chemical, physical and biological conditions of the soil; and the soil water content, among others.[39,41–44]

The aerial and root systems of the plant are directly related. Any alteration in the aerial part of the plant, such as pruning, excess fruit loads, pest attacks and diseases can lead to depletion of the root system, potentially causing root death, especially of roots with smaller diameters.[41,45] Similarly, the root system may, depending on conditions, either provide assimilates to the aerial parts of the plant, or it may act as a relatively important sink, such as during dry seasons, draining assimilates from non-fruiting and sometimes fruiting branches.[41]

Despite this variance, there are common features such as the presence of tap roots, axial roots, lateral roots, feeder roots and root hairs. In coffee, as in other dicotyledonous plants, the first root axis arises from the radicle and is called the tap root.[46] Though long lived, tap roots in coffee are generally not prominent, usually terminating at a depth no greater than 0.5 m.[39,44,47] Plants may also contain more than one tap root.[44,48] If the tap root becomes bent or twisted upon planting, this may result in a twisted or contorted condition, which may negatively affect the plant throughout its lifetime.[39,48] Because of this, many growers have adopted the practice of cutting the bottom few centimetres of the tap root before transplanting in an effort to avoid a twisted tap root. This results in removing the apical dominance of the tap root and triggers more lateral ramification.[39]

Ramifications from the tap root can be divided into two types, depending on the direction of their growth. Axial roots grow vertically below the plant, generally reaching depths of around 2–3 m. Lateral superficial roots, on the other hand, grow parallel to the soil surface and usually reach depths no greater than 2 m. Lateral roots tend to concentrate under the plant skirt, but can extend outward, often interweaving with neighbouring tree roots in densely planted fields. Feeder roots of various lengths are distributed on the axial and lateral roots. The root hairs that grow on these feeders are the main providers of mineral nutrition for the plant.[48]

1.4.2 Orthotropic and Plagiotropic Branches

Above the ground, coffee plants exhibit a dimorphic branching behaviour (Figure 1.6), in which orthotropic (vertical) stems produce plagiotropic (horizontal) branches, which in turn produce more plagiotropic branches and coffee fruit.[30,42,44,45,49]

The principal plant stem, or trunk, is orthotropic. There can be one or several main orthotropic stems per plant, depending on the desired plant stand. Orthotropic stems always grow vertically, or perpendicular to the soil. The apical meristem gives rise to two types of vegetative buds: serial buds and head of series buds. Serial buds on orthotropic stems form other orthotropic stems, called suckers. Head of series buds on orthotropic stems produce primary plagiotropic shoots, or branches. Each head of series bud is capable of producing only a single branch. Therefore, should the branches die (from frost, hail, over-shading, drought or other factors), it is necessary to stump the tree back, inciting the growth of new orthotropic stems, which will have new head of series buds capable of forming more primary plagiotropic branches.

Plagiotropic branches are the lateral branches, with primary plagiotropic branches originating from the orthotropic stems, and secondary and tertiary plagiotropic branches originating from other plagiotropic branches of respective orders. As with orthotropic stems, plagiotropic branches have serial buds and head of series buds. Serial buds, contained in the leaf axils, may form either fruit or more plagiotropic branches. Head of series buds only form other plagiotropic branches. Since plagiotropic branches cannot generate orthotropic stems, cuttings that will be used for plantings must originate from orthotropic stems in order to generate a normal, vertically growing tree.

The development and growth of the plant is dependent on species, variety and the environmental conditions in which the plant is situated. With *C. arabica*, within one year the plant typically develops six to ten levels of plagiotropic branches. After two years the orthotropic stem is usually 1.2–2 m in height, and the first flowers appear. After three years, the plant reaches maturity and usually begins to yield commercial crops.[30,48]

Figure 1.6 (A) *C. arabica* L. with one orthotropic stem and various fruit-bearing plagiotropic branches. (B) *C. canephora* Pierre with various orthotropic stems (photo courtesy of Pedro Malta Campos). (C) Fruit-bearing plagiotropic branches of *C. canephora* Pierre (photo courtesy of Dr Aymbiré Fonseca).

1.4.3 The Leaves

The foliar surface of adult coffee trees varies according to species, state of health, irradiance levels and many other factors.[48,50] In the principal commercial varieties, *C. arabica* and *C. canephora*, leaves are generally thin, shiny and waxed, elliptical in form and conspicuously veined. They typically grow in pairs that are opposite to each other on the branch. Between these

(A)

(B)

Figure 1.7 Coffee leaves of (A) *C. arabica* L. and (B) *C. canephora* Pierre.

two species, the main difference is that *Coffea arabica* leaves are smaller, with a glossy dark upper surface, while *Coffea canephora* leaves are often lighter in colour, less waxy, larger and slightly undulating (Figure 1.7).[30]

Leaf colour varies between species and variety. For example, younger leaves of *C. arabica* are either light green or bronze, depending on whether the plant is of Bourbon or Typica variety in origin, respectively (Figure 1.8). The bronze colour of Typica plants fades with age.[48] Leaf coloration is generally lighter on the abaxial (lower) leaf surface compared to the adaxial (upper) leaf surface, resulting from different cutin compositions (Figure 1.9).[39]

Leaves contain domatia, small cavities found in the lower epidermis. Although there is not a consensus regarding their exact function, it is possible that they play a positive role by harbouring mutually beneficial predators such as mites.[51,52] They can be used to distinguish *Coffea* species by

(A)

(B)

Figure 1.8 Young coffee leaves of (A) a *Coffee arabica* var. Bourbon plant with light green leaves and (B) a *Coffea arabica* var. Typica plant with bronze coloration in emerging leaves.

comparing their size, shape, placement and the presence or absence of stomata on the outermost cell layer of the domatia.

Stomata are apertures in the epidermis, facilitating the gas exchange of the plant with the external medium. Stomatal density is a function of both the number of stomata and the size of the epidermal cells, and it varies between species and even between leaves on the same plant. Stomata are typically composed of two stomatal cells, or 'guard cells', with an aperture between them called the ostiole. Through this pore, the internal atmosphere within the intercellular spaces communicates with the exterior. Like other epidermal cells, stomatal cells are lined with a cuticle, which spreads down into the ostiole and lines the external wall of the substomatal chamber.

The cuticle is a waxy substance that covers the leaf and is largely impervious to liquids and gases. It is made mainly of cutin, a fatty substance that becomes oxidized and polymerized on the outer cell surface through a process known as cuticularization.[53] The cuticle protects the leaf against abiotic damage and provides a barrier to water evaporation. In fact, it has been

(A)

(B)

Figure 1.9 *C. arabica* L. leaf. (A) Adaxial (upper) surface. (B) Abaxial (lower) surface.

estimated that only about 5% of the water lost from leaves escapes through the cuticle. Almost all of the water lost from leaves is lost by diffusion of water through the stomata.[54]

The lifecycle of coffee leaves varies between species. *C. arabica*, under greenhouse (phytotron) conditions, reaches full leaf expansion after 30–35 days and maximum dry weight after 50–60 days.[54,55] The lifecycle can be divided into four stages: quiescent buds, in which the apical meristem and paired leaf primordia are covered by two firm stipules (leaf-like appendages); the emergence of the bud, where the leaves emerge by pushing apart the stipules, although they remain tightly associated to each other; lamina expansion and mechanical strengthening of the leaf; and finally senescence.[30,56]

1.4.4 Flowering

While in equatorial regions, such as Colombia, the coffee flowering and fruit cycle may occur at various times throughout the year, in non-equatorial regions, which represent the majority of worldwide coffee production, coffee plants follow a single annual cycle of growth and fruiting.[42]

Figure 1.10 *C. canephora* Pierre inflorescence.

Coffee plant flowering consists of two distinct processes: flower bud initiation and flower opening, or anthesis (Figure 1.10). Flower bud initiation occurs when the serial buds of plagiotropic branches are induced to differentiate into flower buds. Buds grow to 4–6 mm and then enter a dormancy period, which in most growing regions coincides with a dry season.[48,57]

The dry period is necessary to break the dormancy of the floral buds. An extended dry season affects phytohormone levels in the plant. It also leads to low internal water potential which increases the unusually low hydraulic conductivity of the coffee roots, predisposing the trees to rapid rehydration following the first rains.[42]

During the first 3–4 days after a water stimulus, meiosis occurs and there is an increase in the levels of endogenous, active, gibberellic acid in the flower buds.[42] Inflorescences of both *C. arabica* and *C. canephora* are of the glomerular type, and flowers on *C. canephora* plants are generally more abundant and larger. The flowers are ephemeral, generally only lasting for two days. Several blossoming events can occur in each flowering season, and the greater their number and longer the spaces between them, the less uniform the coffee fruit will be upon the harvest.

1.4.5 The Fruit

The fruit of the coffee plant is typically described as a drupe: a fleshy, indehiscent fruit with a pericarp that is clearly differentiated into an exocarp, mesocarp and endocarp (Figure 1.11).[58,59] These layers surround the coffee seed, which comprises an embryo, endosperm and perisperm. How these layers develop, and their interaction during development and later post-harvest, will ultimately determine the quality and flavour profile of the coffee beverage. This development, as well as the anatomical components of a mature coffee fruit, are discussed in this section.

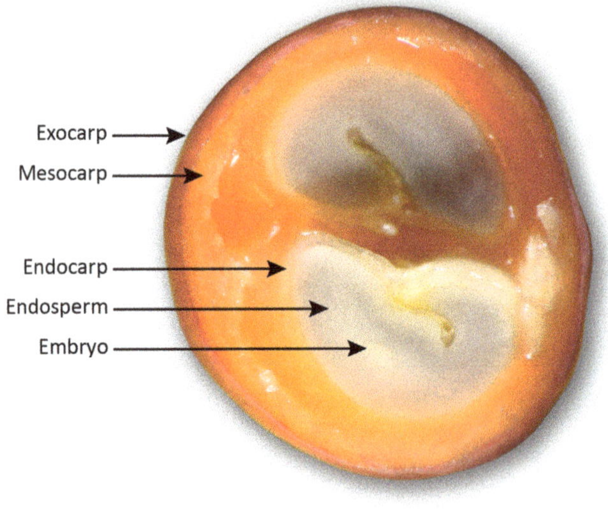

Figure 1.11 Transverse cut of a coffee fruit. Coffee is considered a drupe, having a clearly differentiated exocarp, mesocarp and endocarp. Photo courtesy of Thompson Owen, Sweet Maria's Coffee.

1.4.5.1 Stages of Fruit Growth

The time from flowering to the completion of fruit maturation varies greatly between species and is dependent on factors such as genotype, climate and cultivation practices. In general, the maturation times for several species are around 80–90 days for *C. racemosa*, 220 days for *C. arabica*, 300 days for *C. canephora* and 360 days for *C. dewevrei* and *C. liberica*.[60]

Despite these differences in maturation times, key steps in fruit development among commercial species appear to be identical and can be divided into five stages.[45,59] The first stage generally occurs for the first six to ten weeks after flowering in *C. arabica*, although fruits may enter into a latent state for up to 60 days after pollination.[61] This stage is one of limited fruit growth and is commonly referred to as the 'pinhead' stage (Figure 1.12).[42,45,48] The growth that occurs in this stage is mainly through cell division, not cell expansion.

The second stage, generally lasting from 6 to 16 weeks after flowering in arabica, is the rapid swelling stage, characterized by a rapid increase in volume and dry weight, mostly due to pericarp growth. Unlike the first stage, this second stage is dominated by rapid cell expansion. Fruit locules swell to full size through the growth of the transient perisperm, which will later be consumed by the endosperm as it fills the locules in future stages.[59,61] Endocarps, which will line the locules, begin to lignify. The size to which the locules swell depends greatly on the water status of the plants during this period; fruits that expand during wet weather become larger than fruits that expand in hot, dry weather.[42]

Figure 1.12 *C. arabica* L. fruit in the 'pinhead' stage.

After this rapid growth, the fruit enters the third stage, which is one of suspended and slow growth and lasts for only two weeks. In this stage, though the final fruit size is obtained, the amount of dry matter is still low.[45]

In the fourth stage, the endosperm fills in the locules, consuming all but a small amount of the perisperm that had previously occupied this space.[59] The remnants of the perisperm will become the silverskin that comes off as chaff when the coffee is eventually roasted. In arabica, this stage generally occurs between 17 and 28 weeks after flowering.[45]

The final stage of development is the ripe stage. Changes in this stage occur mostly in the pericarp, in particular an increase in the dry weight, the breakdown of the mesocarp leading to a softening of the fruit and the change in colour of the exocarp from green to red, yellow or in some cases pink or orange, depending on the flavonoid compounds associated with the genotype.

1.4.5.2 *Fruit Anatomy*

Knowledge of the anatomical aspects of the coffee fruit is relevant to determine how interactions between the anatomical components impact coffee quality, as well as to accurately study how quality can be maximized both during fruit development and in removing and drying the bean. The mature coffee fruit consists of a pericarp, comprising the outer layers of the coffee fruit (exocarp, mesocarp and endocarp) and the seed, comprising the embryo, endosperm and silverskin (Figure 1.13).[47,58]

Exocarp – The exocarp or epicarp, commonly called the skin or peel, is the outermost tissue of the coffee fruit. It is composed of a single layer of compact, polygonal parenchyma cells.[47,58] The exocarp is green for most of the fruit's development. Toward the end of maturation, chlorophyll pigments disappear, and after a transient yellow phase, the exocarp cells accumulate anthocyanin, bringing on a red coloration that can range from pink to

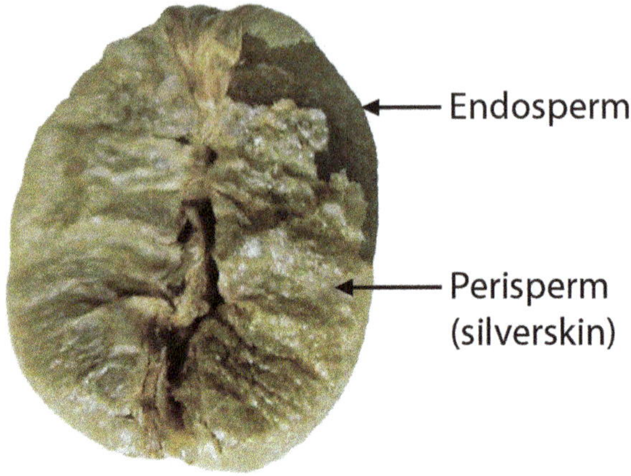

Figure 1.13 *C. arabica* L. seed showing the perisperm.[58]

burgundy. In the case of yellow fruit, leucoanthocyanin replaces anthocyanin, allowing exposure of the yellow pigment luteolin.[62]

Mesocarp – The mesocarp, also called the mucilage or 'pulp', is the fleshy part of the fruit between the parchment and the skin. In some literature, it is referred to as the 'true pulp',[59] and in other literature it is divided into an inner mesocarp, called mucilage, and an outer mesocarp, which is called the pulp *per se*.[63] However, popularly speaking, the part called pulp is the exocarp, the part of the mesocarp that is removed during the pulping process.

It is formed by parenchyma cells and vascular bundles and in general accounts for around 29% of the mass of the dry fruit.[64] Increases in altitude lead to higher concentrations of dry matter in the mucilage.[58] The mesocarp is hard in unripe coffee fruit. As the coffee matures, pectinolytic enzymes break down pectin chains, resulting in a hydrogel that is insoluble and rich in sugars and pectins. This difference is fundamental in the pulping process as it allows for the separation of unripe and ripe fruit.

Endocarp – The endocarp, more commonly called the parchment, is composed of sclerenchyma cells and completely envelops the seed. It is mostly composed of cellulosic material.[65] The endocarp is formed by 5–6 layers of intercrossing fibres, which give it extraordinary strength.[47] While it serves to protect the seed from mechanical damage, it is a barrier to both the transfer of chemical compounds from the pericarp to the endosperm, and the removal of water from the coffee seed during drying. It also acts as an impediment to germination, perhaps through mechanical resistance.[66] Nonetheless, the parchment is usually not removed since it is recommended to store coffee in parchment (or dried fruit pods), and the hulling process to remove the parchment can damage seeds, negatively impacting germination.[39]

Seed – Coffee seeds are generally elliptical and plane-convex in shape, with a longitudinal furrow on the plane surface. They comprise the silverskin, endosperm and embryo.

The silverskin, also called the perisperm or spermoderm, is the outermost layer of the seed and is composed of sclerenchyma cells. It is thought to serve in the accumulation and transport of biochemical compounds from the pericarp to the endosperm, although exactly which compounds are transferred and how this occurs is not well known.[59,61] As the fruit matures, the perisperm is consumed by the growing endosperm, and transforms into a thin pellicle that may become partially detached upon drying in *C. arabica*. This difference in adherence, as well as the colour of the silverskin after the coffee has dried, are used to determine the presence of immature coffee beans in several classification protocols.[67,68] In *C. canephora* the silverskin is adherent and brown.

The endosperm is the principal reserve tissue for initial plant growth after germination. It is a living tissue that is formed by the fusion of one spermatic nucleus and two polar nuclei, resulting in a triploid (3*n*) tissue.[47,65] Initially a liquid milky-coloured tissue with thin cell walls, as the coffee fruit develops, its cell walls thicken due to the deposition of complex polysaccharides. These thick and partially lignified cell walls do not present intercellular spaces, but are crossed by many plasmodesmata, which establish connections between these cells and play a key role in the transport of water and other substances.[69] The external part of the endosperm is composed of small polygonal cells that are rich in oils, and it is sometimes called the 'hard endosperm.' The internal part of the endosperm, sometimes referred to as the 'soft endosperm', is composed of larger rectangular cells with slightly thinner cell walls.[47,59]

The embryo is small (3–4 mm long in *C. arabica*), composed of a hypocotyl attached to two cotyledons, and localized close to the convex surface of the seed (Figure 1.14).[39,47,48] It contains few storage reserves and is therefore dependent upon the endosperm for nutrients during its initial growth.

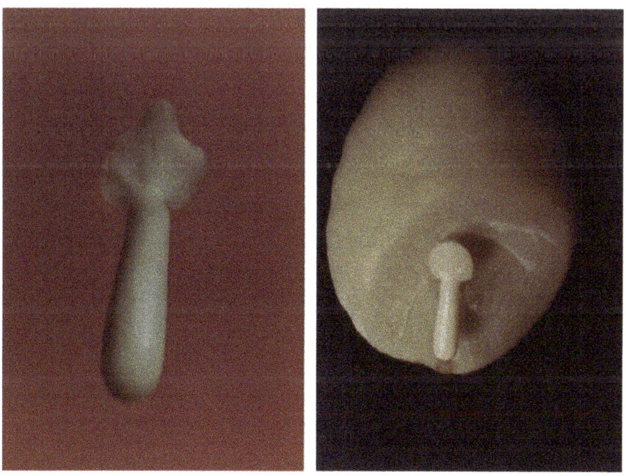

Figure 1.14 *C. arabica* L. embryo, (left) isolated and (right) with the outer surface of the endosperm cut away to expose the embryo.

Acknowledgements

The authors acknowledge the financial support and scholarship from the Coordination for the Improvement of Higher Education Personnel (CAPES); National Research Council (CNPq), Brazil; Carlos Chagas Filho Foundation for Research Support in the State of Rio de Janeiro (FAPERJ), Brazil; and Dr Aymbiré Fonseca, Pedro Malta Campos and Thompson Owen from Sweet Maria's Coffee for generously providing photos for this chapter.

References

1. A. P. Davis, R. Govaerts, M. D. Bridson and P. Stoffelen, *Bot. J. Linn. Soc.*, 2006, **152**, 65.
2. A. P. Davis, *Phytotaxa*, 2010, **10**, 41.
3. A. P. Davis, J. Tosh, N. Ruch and M. f. Fay, *Bot. J. Linn. Soc.*, 2011, **167**, 357.
4. A. Farah and T. S. Ferreira, in *Coffee in Health and Disease Prevention*, ed. V. R. Preedy, Academic Press is an imprint of Elsevier, USA, 2015, vol. 1, p. 5.
5. S. Krishnan, T. A. Ranker, A. P. Davis and J.-J. Rakotomalala, *Acta Hortic.*, 2015, **1101**, 15.
6. J. F. Leroy, *C. R. Hebd. Séances. Acad. Sci.*, 1967, **265**, 1043.
7. D. M. Bridson, *Kew. Bull.*, 1982, **36**, 817.
8. A. P. Davis, D. Brisdom and F. Rakotonasolo, in *A Festschrift for William G. D'Arcy: The Legacy of a Taxonomist*, ed. R. Keating, V. Hollowell and T. Croat, Missouri Botanical Garden, St. Louis, 2005, p. 399.
9. O. Maurin, A. P. Davis, M. Chester, E. F. Mvungi, Y. Jaufeerally-Fakim and M. F. Fay, *Ann. Bot.*, 2007, **100**, 1565.
10. A. P. Davis and F. Rakotonasolo, *Bot. J. Linn. Soc.*, 2008, **158**, 355.
11. M. D. Nowak, A. P. Davis and A. D. Yoder, *Syst. Bot.*, 2012, **37**(4), 995.
12. A. P. Davis, J. Tosh, N. Ruch and M. F. Fay, *Bot. J. Linn. Soc.*, 2011, **167**, 357.
13. P. S. Murthy and M. M. Naidu, *Resour., Conserv. Recycl.*, 2012, **66**, 45.
14. F. Anthony, E. L. C. Diniz, M. C. Combes and P. Lashermes, *Plant Syst. Evol.*, 2010, **285**, 51.
15. USDA. *United States Department of Agriculture. Natural Resources Conservation Service. Classification for Kingdom Plantae Down to Species* Coffea arabica *L.* http://plants.usda.gov/java/ClassificationServlet?source=display&classid=COAR2, last accessed May 2016.
16. Kew, *Royal Botanic Gardens, Word Checklist of Selected Plant Families,* C. arabica, 2015, http://apps.kew.org/wcsp/home.do, last accessed October 2015.
17. A. Charrier and J. Berthaud, in *Coffee Botany, Biochemistry and Production of Beans and Beverage*, ed. M. N. Clifford and K. C. Willson, Croom Helm, London, 1985, p. 13.
18. P. Lashermes, M. C. Combes, J. Robert, P. Trouslot, A. D'Hont, F. Anthony and A. Charrier, *Mol. Gen. Genet.*, 1999, **261**, 259.

19. F. Anthony, M. C. Combes, C. Astorga, B. Bertrand, G. Graziosi and P. Lashermes, *Theor. Appl. Genet.*, 2002, **104**, 894.
20. A. N'Diaye, V. Poncet, J. Louarn, S. Hamon and M. Noirot, *Plant Syst. Evol.*, 2005, **253**, 95.
21. F. Anthony, B. Bertrand, H. Etienne and P. Lashermes, in *Wild Crop Relatives: Genomic and Breeding Resources*, ed. C. Kole, Springer-Verlag, Berlin Heidelberg, 2011, p. 41.
22. M. Noirot, A. Charrier, P. Stoffelen and F. Anthony, *Tree*, 2016, **30**(3), 597.
23. F. Anthony, M. Noirot, E. Couturon and P. Stoffelen, *ASIC.*, 2006, Lausanne, 862.
24. F. Anthony, J. Berthaud, J. L. Guillaumet and M. Lourd, *Plant Genet. Resour. Newsl.*, 1987, **69**, 23.
25. P. Cubry, F. Bellis, K. Avia, S. Bouchet, D. Pot, M. Dufour, H. Legnate and T. Leroy, *BMC Genomics*, 2013, **14**, 13.
26. P. Cubry, F. Bellis, D. Pot, P. Musoli and T. Leroy, *Genet. Resour. Crop Evol.*, 2013, **60**, 483.
27. A. Lécolier, P. Besse, A. Charrier, T. N. Tchakaloff and M. Noirot, *Euphytica*, 2009, **168**, 1.
28. F. L. Wellman, *Coffee: Botany, Cultivation and Utilization*, Leonard Hill Books, London, 1961.
29. R. F. Smith, in *Coffee Botany, Biochemistry and Production of Beans and Beverage*, ed. M. N. Clifford and K. C. Willson, Croom Helm, London Sydney, 1985, p. 1.
30. F. Anzueto, T. W. Baumann, G. Graziosi, C. R. Piccin, M. R. Söndahl and H. A. M. van der Vossen, in *Espresso Coffee*, ed. A. Illy and R. Viani, Elsevier, London, 2005, p. 21.
31. H. A. M. Van der Vossen, B. Bertrand and A. Charrier, *Euphytica*, 2015, **204**, 243.
32. H. Willian and M. A. Ukers, in *All about Coffee*, The Tea and Coffee Trade Journal, New York, USA, 1922, vol. 15, p. 131.
33. A. Candolle, *Origin of Cultivated Plants*, ed. K. Paul, Trench, London, 1883.
34. B. Fausto, *A Concise History of Brazil*, Cambridge University Press, 1999.
35. M. Pendergrast, *Uncommon Grounds: The History of Coffee and How it Transformed Our World*, Basics Books, 2010.
36. ICO, International Coffee Organization, Total production by exporting countries, http://www.ico.org/prices/po-production.pdf, last accessed June 2016.
37. P. J. Kramer, *Water Relations of Plants*, Academic Press, New York, 1983.
38. A. Eshel, T. Beeckman, *Plant Roots the Hidden Half*, CRC Press, Boca Raton, 2013.
39. D. E. Livramento, in *Morfologia e fisiologia do cafeeiro*, ed. P. R. Reis and R. L. Cunha, U. R. EPAMIG SM, Lavras, 2010, p. 87.
40. K. L. Gallagher, in *Cellular Patterning of the Root Meristem: Genes and Signals*, ed. A. Eshel and T. Beeckman, CRC Press, Boca Raton, 2013.
41. M. G. R. Cannell, *Ann. Appl. Biol.*, 1971, **67**(1), 99.

42. M. G. R. Cannell, in *Physiology of the Coffee Crop*, ed. M. N. Clifford and K. C. Willson, AVI Publishing Co, Westport, 1985, p. 108.
43. M. Maestri and R. S. Barros, in *Coffee*, ed. P. T. Alvim and T. Kozlowski. Academic Press, Nova York, 1977, p. 249.
44. B. A. Rena, R. S. Barros, M. Maestri and M. R. Södahl, in *Coffee*, ed. B. Schaffer and P. C. Andersen, CRC press, Boca Rotan, 1994, p. 101.
45. F. M. DaMatta, C. P. Ronchi II, M. Maestri I and R. S. Barros, *Braz. J. Plant Physiol.*, 2007, **19**(4), 485.
46. P. Gregory, *Plant Roots: Growth, Activity and Interaction with Soils*, Blackwell Publishing, Oxford, 2006.
47. D. M. Dedecca, *Bragantia*, 1957, **16**(23), 315.
48. N. J. Wintegens, in *The Coffee Plant*, ed. J. N. Wintegens, Wiley-VCH, Weinheim, 2009, p. 3.
49. A. Carvalho and A. C. Krug, *Bragantia*, 1950, **10**(6), 151.
50. F. M. Damatta, *Field Crop. Res.*, 2004, **86**(2–3), 99.
51. C. H. C. Matos, A. Pallini, F. F. Chaves and C. Galbiati, *Neotrop. Entomol.*, 2004, **33**(1), 57.
52. C. H. C. Matos, A. Pallini, F. F. Chaves, J. H. Schoereder and A. Janssen, *Entomol. Exp. Appl.*, 2006, **118**, 185.
53. W. C. Dickison, *Integrative Plant Anatomy*, Academic Press, San Diego, 2000.
54. L. Taiz and E. Zeiger, *Plant Physiology*, Sinauer Associates, Sunderland, 2010.
55. M. P. Frischknecht, B. M. Eller and W. T. Baumann, *Planta*, 1982, **156**(4), 295.
56. S. S. Mosli Waldhauser, A. J. Kretschmar and W. T. Baumann, *Phytochemistry*, 1997, **44**(5), 854.
57. D. J. Alves, in *Cultivates de Café: Origem, Características e Recomendações*, ed. S. H. S. Siqueira, Embrapa café, Brasília, 2008, p. 35.
58. M. F. Borém, T. J Gracia Silva and A. E. Amaral da Silva, in *Handbook of Coffee Post-Harvest Technology*, ed. M. F. Borém, Gin Press, Norcross, 2014, p. 1.
59. D. R. De Castro and P. Marraccini, *Braz. J. Plant Physiol.*, 2006, **18**(1), 175.
60. P. H. Medina-Filho, in *Plant Breeding Reviews Volume 2*, ed. J. Janick, AVI Publishing Company, Westport, 1984, p. 157.
61. M. T. S. Eira, E. A. A. Silva, R. D. Castro, S. Dussert, C. Walters, J. D. Bewley and H. W. M. Hilhorst, *Braz. J. Plant Physiol.*, 2006, **18**(1), 149.
62. S. M. Marín-López, J. Arcila-Pulgarín, E. C. Montoya-Restrepo and C. E. Oliveros-Tascón, *Cenicafé*, 2003, **54**(3), 208.
63. S. Avallone, J. P. Guiraud, B. Guyot, E. Olguin and J. M. Brillouet, *J. Agric. Food Chem.*, 2001, **49**(11), 5559.
64. R. Bressani, in *Coffee Pulp : Composition, Technology, and Utilization*, ed. J. E. Braham and R. Bressani, IDRC, Ottawa, 1979, p. 5.
65. M. F. Borém, G. J. T. Salva and A. A. E. Silva, in *Pós-colheita do Café*, ed. M. F. Borém, Editora UFLA, 2008, p. 19.
66. M. F. I. Valio, *J. Seed Technol.*, 1980, **5**(1), 32.

67. ISO, International Organization for Standardization, International Standard 10470:2004, Green coffee - Defect Reference Chart ISO, 2004.
68. SCAA, Specialty Coffee Association of America, *Arabica Green Coffee Defect Handbook*, Specialty Coffee Association of America, Long Beach, 2013.
69. E. Dentan, in *Coffee: Botany, Biochemistry and Production of Beans and Beverage*, AVI Publishing Company, Westport, 1985, p. 284.

CHAPTER 2

Coffee Growing and Post-harvest Processing

RUBENS JOSÉ GUIMARÃES*[a], FLÁVIO MEIRA BORÉM[b], JOEL SHULER[b], ADRIANA FARAH[c] AND JOÃO CARLOS PERES ROMERO[d]

[a]Universidade Federal de Lavras/Departamento de Agricultura - Cx. Postal 3037 Lavras, MG 37200-000, Brazil; [b]Universidade Federal de Lavras, Departamento de Engenharia - Cx. Postal 3037 Lavras, MG 37200-000, Brazil; [c]Universidade Federal do Rio de Janeiro, Instituto de Nutrição, RJ 21941-902, Brazil; [d]Agronomy Consultant for Brazil and Latin America - Cx. Postal 2054 Ouro Fino, MG 37570-000, Brazil
*E-mail: rubensjoseguimaraes@gmail.com

2.1 Introduction

Coffee is not just a plant, fruit, or seed. It is also not just a drink option at a coffee shop, or restaurant. Coffee has played a role in the history of humanity ever since its discovery in Ethiopia, evolving with both new systems of cultivation and new forms of consumption, with contributions coming from nearly every continent. In the same way, coffee growing is not merely an agricultural activity. It gives meaning and passion to the lives of those who cultivate it, not to mention the pleasurable effects its consumption can render.

As with all perennial crops, care must be taken so that mistakes are not made from the seedling phase on. In order to ensure this, one should have

Coffee: Production, Quality and Chemistry
Edited by Adriana Farah

Published by the Royal Society of Chemistry, www.rsc.org

a basic understanding of both traditional and more recent coffee propagation technologies throughout the cultivation process. Preventative measures taken during planting can avoid future problems such as underdevelopment, phytosanitary problems, twisted tap roots, and even plant death.

This chapter covers the seedling and initial planting in the field including important aspects such as the understanding of soil conditions (texture, structure, depth, fertility), how to choose the ideal location for cultivation and the influence of climate on coffee growing. It also covers practical aspects related to irrigation, fertilization, crop interplanting, weed management, diseases and nematodes, companion planting, and pruning. Understanding factors that can predispose plants to attacks as well as crop management options are essential for more sustainable production. The chapter ends with a discussion on different aspects related to the harvest and post-harvest and the overall financial planning of a coffee farm from planting through the first harvests.

2.2 Adaptation and Improvements of the Main Commercial Species

As covered in Chapter 1, coffee's journey through history and throughout the world started with its discovery in Ethiopia, before making its way to the Arabian Peninsula, then east to India and Indonesia before heading to the Americas *via* Europe. Along this journey, diverse methods of cultivation have been used as the plant has been adapted to cultivation systems and cultivation systems adapted to the plant, leading to higher productivity and a greater facility of cultivation.

As coffee made its way to new origins, studies were conducted for adaptation and genetic improvements. In Brazil, between the years of 1727 and 1933, individual growers performed empirical selective breeding. After the creation of the Genetic Division of the Agronomic Institute of Campinas in 1933, selective breeding began to be performed using scientific methods, which increased coffee plant productivity by 396% compared with "Typica" variety that had been introduced in Brazil.[3] The Agronomic Institute of Campinas has contributed enormously to coffee growing worldwide and is historically considered to be one of coffee's most reputable institutions, especially in the area of genetic improvement. Other organizations have also contributed greatly to the improvement of the coffee plant, including CIRAD (Centre de Coopération Internationale en Recherche Agronomique pour le Développement); a French agricultural organization; Cenifcafé (Centro Nacional de Investigaciones de Café), the research arm of the Colombian Coffee Growers Federation (FNC); Anacafé (Asociación Nacional Del Café), and more recently World Coffee Research, an initiative to develop coffee varieties to alleviate constraints to the supply chain of high-quality coffee.

Today, with more modern techniques, improvements in genetics and seed quality, and advances in harvesting and post-harvest processing, productivity can reach over 60 bags of *C. arabica* and 120 bags of *C. canephora* per hectare with the use of irrigation;[4,5] however, average production levels worldwide remain far lower than this. For more information on genetic improvement, see Chapter 3.

2.3 The Basics of Coffee Plant Growth

A general description of the coffee plant and fruit anatomy is presented in Chapter 1. In this section we will build on those concepts by discussing the development of the coffee plant, and implications of this development on coffee production.

The coffee plant has peculiar characteristics that should be studied to gain a better understanding of its interaction with the environment and its reactions to biotic and abiotic stresses, especially in times of climatic changes such as poor distribution of rainfall throughout the year.[9]

For coffee seeds that will be taken to the nursery for seedling production, the fruit exocarp and mesocarp are removed (the endocarp remains intact) and the seeds are then dried (first in the sun, then in the shade) until moisture content levels reach around 14%, at which point they are ready for planting or for sale to other nurseries. In the nursery, the seeds are planted directly into their respective containers (usually polyethylene bags filled with a substrate composed of soil with organic matter and fertilizers).

Under normal conditions and with adequate irrigation, the seed radicle will begin to protrude 30 to 45 days after planting (Figure 2.1). The radicle grows downward, further into the soil (positive geotropism), providing support for the emergence of the seedling after hypocotyl torsion and subsequent elevation of the cotyledons above the ground, which occurs around 60 to 90 days after the seed is planted. In these phases of germination, the seedlings still have a root system that is not fully developed, and because of this

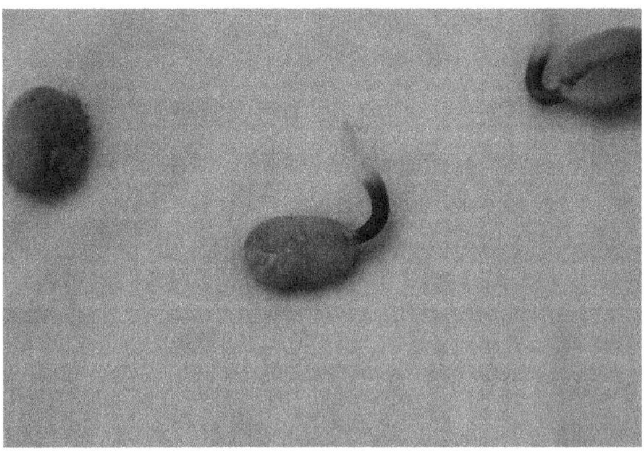

Figure 2.1 Germination of *C. arabica* L. seeds.

they are very sensitive to water deficits. Nursery managers must therefore pay careful attention to adequately water the seedlings.[2]

As the seedling begins to emerge from the soil, the hypocotyl hook pushes through. With proper development, this hook pulls the cotyledons (still contained in the remaining endosperm) out of the soil. In this phase of germination, it is important to have control over sun exposure inside the nursery to avoid rupture of the hypocotyl hook and subsequent plant death.

Upon emerging from the soil, seedlings still lack the ability to produce photoassimilates and are very vulnerable to soil fungi such as *Rhizoctonia solani*, which can cause damping off. Because of this, nursery workers must be aware when this germination phase has been reached, and apply appropriate chemical or natural controls if preventive measures (using fungus-free soil and irrigation control) are unsuccessful.[2]

After the cotyledonary leaves (Figure 2.2a) have appeared, "true" leaves (Figure 2.2b) emerge from the central (orthotropic) stem. Series buds form

(a)

(b)

Figure 2.2 (a) Cotyledonary leaves and (b) true leaves of *C. arabica* L.

on the leaf axils, and when induced these series buds develop into other orthotropic stems (suckers). Series buds, therefore, are important for renovation prunings, and the orthotropic stems that emerge from them may also be used as a source of seedlings for future plantings.[9]

The orthotropic stem is the main trunk of the plant. True leaves emerge from this stem and each subsequent set of leaves forms at an angle of 60° to the previous (lower set). Starting around the eighth to tenth node, "head-of-series" buds form above the series buds. These head-of-series buds form plagiotropic branches, which are the fruit-bearing branches of the coffee plant.[10] With seedlings, the emergence of the head-of-series buds is therefore an important sign for coffee growers that the plants are ready to begin their reproductive phase and can be taken to the field for transplanting.

As covered in Chapter 1, serial buds on orthotropic stems only form other orthotropic stems (Figure 2.3). It is common for orthotropic stems to form in excess if the plant suffers stresses from heat, mechanical harvesting damage, hail, *etc.* Coffee growers must remove these shoots one or more times a year (thinning). These orthotropic stems can also be used as cuttings for propagation both with *Coffea arabica* L. and *Coffea canephora* Pierre.[10]

As the plant grows and develops, new plagiotropic branches form. Serial buds form in the leaf axils of these branches and head-of-series buds form above each set of serial buds. The serial buds that form on the plagiotropic branches can form either more plagiotropic branches (secondary, tertiary, *etc.*) or inflorescences (which result in fruit). The head-of-series buds continue forming other plagiotropic branches. In other words, regardless

Figure 2.3 An orthotropic stem (sucker) forming from the serial bud of an orthotropic stem.

of whether they are on orthotropic or plagiotropic branches, head-of-series buds always form plagiotropic branches.[10]

It is important to observe that the inflorescences are formed by buds, and thus for continued production a plant must repetitively produce new buds. Since new buds are only formed on new branch growth, this branch growth is, therefore, necessary for the production of new fruit. With this knowledge, agronomists and coffee growers can establish strategies for managing coffee plants for optimal productivity, planning prunings, or calculating yields for the next year (which facilitates planning fertilizer needs and phytosanitary measures).

In years of high productivity, the plant's consumption of photoassimilates is also high and it is important to note that fruits are given priority over other parts of the plant, including roots and branches, which means that there is less branch growth. Since this branch growth is the basis for fruit formation, production will therefore be lower in the following year. Using the same logic, in a year of low productivity there will be enough photoassimilates to support vigorous branch growth until the following year, when this new growth will generate high productivities. This alternation between higher and lower productivity defines the biennial nature of the coffee plant as it repeats this cycle every two years, even with proper crop management.[9,10]

New branch growth is important not only for future productivity, but also for the formation of new leaves. Leaves are only produced by new branch growth. Old leaves that fall for any reason (age, pests, diseases, unfavorable climatic conditions) are not replaced by newly formed leaves on the branches from which they originated. It is only through the development of new leaves that the plant can return to its full photosynthetic capacity, and this is only achieved through the formation of new leaves on new branch growth formed by the apical meristem.

While fruits have high sink strength (ability to mobilize photoassimilates), which can compromise full plant growth, the root system has much lower sink strength.[10] This means that in years of high productivity the roots have access to a smaller quantity of photoassimilates for their growth or even maintenance. This is especially true in cases of improper crop management (poor nutrition, competition from weeds). Thus, given certain conditions and with inadequate crop management, years of high productivity can cause high rates of plant mortality.

2.4 Coffee Plant Propagation Techniques

This section describes traditional as well as more recent coffee plant propagation technologies, and discusses crop characteristics that allow the reader to grasp important coffee growing issues, from seedling production to field cultivation, such as underdevelopment, phytosanitary problems, twisted tap roots, and plant death.

Since coffee is a perennial plant, care should be taken to avoid errors during the planting out of the crop, as well as to ensure the development of healthy and vigorous seedlings, from nursery site selection to the transport of the seedlings to the field. Any errors made at the onset of cultivation, such as choosing a cultivar that is not well adapted to the region, or suboptimal plant spacing, can compromise production (lower productivity and consequently lower economic returns) throughout the life of the crop.[6,7]

One should also consider that the coffee plant begins its phase of significant production three years after the seed is planted in the nursery. Around six months are needed for the seedling to develop in the nursery, and, after transplanting to the field, around 30 months for the plants to yield their first commercial harvest. It is only after this period that growers will begin to see a return on their capital investment. For this reason, finances and the use of labor should be planned carefully and the crop must be well managed and well timed, with advice from an agronomist for the duration of the undertaking.[7,8]

In the past, coffee seeds were planted directly into prepared planting holes, and up to thirty seeds were used in order to guarantee a final count of six to eight seedlings per hole. Past accounts also indicate the common practice of transplanting seedlings sprouted from the fallen fruit of mature plants, called "natural nurseries", in other areas of the farm. Another common practice was to plant seeds under the crown of adult trees, which then protected the seedlings from excessive solar radiation.[1]

Over time, coffee growers felt the need to expand their plantations, and it became necessary to develop seedlings in nurseries. Initially, thin-walled wooden boxes were used, but these containers required chemical treatment (burnt oil or copper sulfate) and trussing with wire, making them difficult to handle. An alternative was to plant seeds in molded blocks of clay soil, a process that also proved difficult. Cow manure and clay were mixed in varying proportions, depending on the texture of the soil used, then put in metallic hexagonal forms to shape the blocks. These blocks were eventually replaced by polyethylene bags (Figure 2.4a), which have become perhaps the most widely used container for seedling production worldwide.[1]

Some disadvantages of the polyethylene bags are that their closed bottom can lead to root deformation and that they are not biodegradable, leading to environmental concerns after their use. An alternative container to the polyethylene bag that is currently used in the production of coffee seedlings is a rigid polyethylene cone (Figure 2.4b). Two key advantages of the cones over the bags are that their bottom is open, thus avoiding root deformation, and their rigidity facilitates handling in transport and planting, minimizing possible damage to the seedlings. A promising new option is biodegradable containers.[11]

Current recommended substrate mixtures vary by producing region, largely to incorporate materials that are readily available to the growers, though most recommendations include the elements of soil, manure, and oftentimes soil-correction elements depending on the soil used. Anacafé,

(a)

(b)

Figure 2.4 (a) Polyethylene bags and (b) rigid polyethylene cones used for seedling formation.

the national coffee association of Guatemala, recommends two-parts soil and one-part chicken manure for loamy soil, and two-parts soil, one-part sand, and one-part chicken manure for argillaceous soil, with all mixes passing through a ¼-inch sieve to avoid clods and foreign matter.[12] For the Southwest Monsoon areas in India, the Central Coffee Research Institute recommends a mixture of 10 kg well-dried manure or compost, 2 kg lime, and 0.5 kg rock phosphate.[13] In Brazil, the substrate used in seedling production has evolved considerably from the original mixture of 50% "forest soil" and 50% cow manure. Many studies have been conducted in search of the ideal substrate for coffee seedling production in polyethylene bags, then in 1999 the Commission on Soil Fertility for the State of Minas Gerais, Brazil (Comissão de Fertilidade do Solo do Estado de Minas Gerais),[14] recommended what is today considered to be the standard in Brazil: 700 liters of screened subsoil, 300 liters of composted and screened cow manure, 3 to 5 kg of single superphosphate, and 0.5 to 1 kg of potassium chloride.

Figure 2.5 Coffee fruit of various maturations, with nearly ripe and fully ripe fruit that is appropriate for seed production in the white box.

Coffea arabica L. seedlings are still produced principally through seeds, since arabica plants are autogamous (with around 10% allogamy), which facilitates the use of seeds in the production process.[15] These seeds should originate from seed production fields that are credentialed by the appropriate authority and under the supervision of the appropriate technician. Seedling production nurseries are regulated in most countries in order to guarantee that the seedlings are healthy, vigorous, and free of pathogens or physical damage that could compromise coffee production. In cases where this oversight is not available, extra care should be taken to ensure that seeds are taken from disease-free plants.

In the seed production process, fruits are harvested in the near ripe or ripe maturation state (Figure 2.5) when the seeds reach physiological maturity, which generally occurs around 220 days after flowering (DAF).[2]

Immediately after the harvest, fruits that will be used for seed production are pulped (the exocarp and part of the mesocarp are removed). They are then either put in water tanks for 12 to 48 hours to remove the remaining mesocarp (mucilage) through controlled fermentation or they are passed through mechanical demucilagers that remove the remaining mesocarp without the need for fermentation. When both the exocarp and mesocarp have been removed, the seeds, still enveloped by the endocarp, are then dried, commonly in the sun initially and then in the shade.[16]

It is important to remember that when stored in ambient conditions, the seeds of *Coffea arabica* L. quickly lose their ability to germinate after six months while those of *Coffea canephora* Pierre do so after only three months.[17] For this reason, seeds should be planted in the nursery as soon as they are ready. This guarantees vigor and germination while ensuring the availability of seedlings to transplant into the fields at the beginning of the rainy season. In general, six to seven months after seeds are planted in the nursery, the seedlings are ready to be transplanted into the fields.

Figure 2.6 Arabica coffee seedling produced using orthotropic cuttings.

Coffee can also be propagated vegetatively (a process commonly used with *Coffea canephora* Pierre) as well as through tissue cultures, or micropropagation. The practice of producing *Coffea canephora* Pierre seedlings through cuttings has a long history. If properly performed, nearly 100% of the cuttings render seedlings and the method also allows for the propagation of clones of genetically superior plants, since the species is allogamous. The seedling production process for *Coffea arabica* L. using orthotropic branch cuttings (Figure 2.6) is not as common, however, the technology is being developed at the Federal University of Lavras (Universidade Federal de Lavras) in Minas Gerais, Brazil.[2]

Plant grafting has been employed principally for growing *Coffea arabica* L. in areas where nematodes are a problem, typically using *Coffea canephora* Pierre as the root stock, since it is tolerant of some types of nematodes[2] (Figure 2.7).

With the use of biotechnology, it is also possible to produce seedlings through somatic embryogenesis, micro-cuttings (Figure 2.8), embryonic cultures, or anther cultures. While somatic embryogenesis has largely been confined to lab experiments, it is a technology that will likely be available to coffee growers in the near future.[2,18] In somatic embryogenesis, a callus is formed from fragments (*e.g.* pieces of leaves) of a mother plant. These are then put in a nutritive solution to promote differentiation and induce embryo formation. These embryos develop into plants that are then transplanted into nurseries where they complete the seedling development process.[19]

Since coffee is a perennial crop, the choice of which cultivars to plant in the field is one of the most important factors for the success of a coffee growing enterprise. Cultivars should have the potential for high productivity and, as much as possible, meet other needs such as: resistance to pests, diseases, and nematodes; adaptation to the region where they will be planted

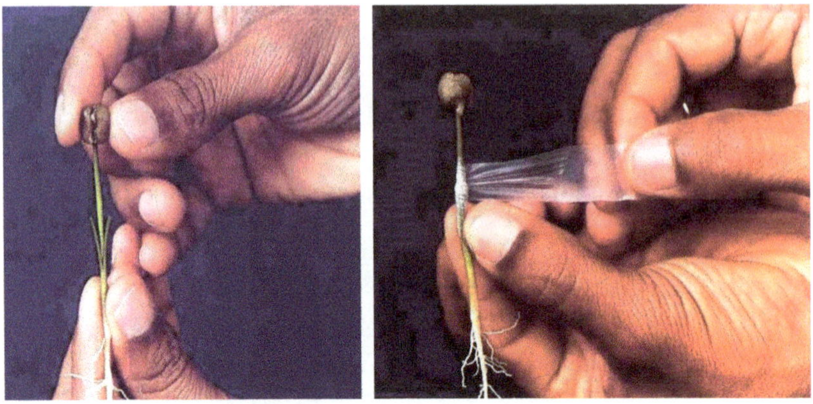

Figure 2.7 Grafting technique used for the production of seedlings resistant to certain types of nematodes.

Figure 2.8 Coffee propagation using micro-cuttings (tissue cultures).

(temperature, soil, and rainfall among others); adaptation to the planned crop spacing of the farm (which will vary with the declivity encountered and the level of mechanization); and more uniform fruit maturation that facilitates harvesting a higher quality final product.[20]

In some producing countries, coffee farms can potentially be established in areas where the plants propagate naturally (natural seed dispersion) or, more commonly, planted in understory areas with either direct seed planting or by using seedlings that grew naturally on the coffee farm (natural nursery). However, in commercial properties using more recent technologies, seedlings are produced in nurseries (Figure 2.9) and then taken to the fields for planting. Again, this facilitates the production of abundant seedlings that are healthy and vigorous.

Figure 2.9 Coffee seedling nursery at Finca La Merced, Guatemala.

The choice of location when building a coffee seedling nursery is important, as a good location can mitigate conditions that are favorable to pests, diseases, and abiotic stresses. The location should also feature easy access for receiving soil and chemical and organic fertilizers, and it should have a very subtle declivity to allow for some runoff while still ensuring that containers are in a vertical position and the formation of the seedlings' root systems is not compromised. The proximity of the nursery to water sources can minimize expenses associated with pumping water, such as equipment and electricity.[2]

Natural wind barriers are also desirable near the nursery, though nurseries are normally built with fences (bamboo or plastic covering), which also help control excess lateral solar radiation. Winds, especially cold winds, can cause foliar lesions that allow the entrance of pathogens, while excess solar radiation can cause scorching. However, the nursery should be in a sunny location. It should also be well drained so that the seedlings can fully develop without severe attacks from diseases that proliferate in humid environments, such as *Rhizoctonia solani* and bacteriosis.[2]

Nursery construction details that should be observed include:[2]

a) To avoid excessive humidity and limit pests and diseases, a five-meter strip that is free from any vegetation or debris should encircle the nursery.
b) The nursery should be protected against flooding by deep ditches or barriers around the entire area that also isolate it from possible contamination from nematodes or weeds.
c) The area where soil and manure are received should be downstream from the nursery.
d) Plant beds should be identified with numbering and by cultivar.
e) Nurseries should not be constructed below coffee fields or in lands infested with invasive plants that are difficult to control.

f) Maximum plant bed measurements should be 1.2 meters in width and 20 meters in length. Larger measurements can complicate maintenance such as hoeing, watering, spraying, and thinning.
g) Containers (usually bags) should be positioned vertically to allow perfect development of root systems with no contact between the roots and the sides of the bags. Damage to the tap roots can compromise plant growth and even cause plant death.
h) Plant beds should be delineated, oftentimes with broken or whole bamboo that are tied to buried stakes, leaving footpaths that should be 40 to 60 cm wide.
i) Lateral fences made of bamboo, plastic sheets, or other comparable material, should be constructed to inhibit the incursion of domesticated animals, which are commonly found in seedling production areas. These fences also protect the seedlings from cold winds and excessive sun exposure.
j) The cover of the nursery should be 1.8 to 2.2 meters high (to facilitate the work of employees) and provide 50% to 60% shade until the seedlings are hardened off (acclimated).[21] On smaller properties, local materials such as banana leaves, bamboo leaves, palm leaves and hay are commonly used as roof covering materials. However, many nurseries employ plastic shade cover to control sunlight intensity.

Nurseries produce two types of seedlings, depending on their duration in the nursery. Seedlings that are 6 to 7 months old are called "half-year" seedlings, and seedlings that are around one year of age are called "year seedlings". Half-year seedlings are more commonly used by growers since they are less costly and more easily transported. Year seedlings are used in special circumstances, such as replacing half-year seedlings that have died after several months in the field, since a year seedling allows for a more uniform crop than replanting another half-year seedling.[2]

The following numbers are provided to facilitate better comprehension of the seedling production process:[1,2]

- One cubic meter of substrate will fill 1200 to 1400 "half-year seedling" bags (which are 10 to 11 cm wide and 20 to 22 cm high) or 900 to 1100 "year seedling" bags (which are 15 cm wide and 25 cm high).
- In one day a worker can fill 600 to 800 "half-year seedling" or 400 to 600 "year seedling" bags with substrate. For every three workers filling bags, one works preparing and distributing the substrate.
- Using a watering can, one worker can water 100 000 seedlings a day.
- One kilogram of seeds comprises, on average, 4000 to 6000 seeds. Therefore, 1 kg is sufficient to plant 2000 to 3000 seedlings using two seeds per bag.
- One worker can plant seeds in up to 3000 bags in a day.

After the appearance of the second pair of true leaves, seedlings should be hardened off (acclimated) by gradually decreasing watering and increasing sun exposure to above 50% for around 30 days, until they are adapted and can withstand field conditions.

With conventional seedling production in areas of monomodal rainfall distribution, it is difficult to have seedlings ready by the beginning of the rainy season, the ideal time for planting. This is due to the slow germination of coffee seeds. However, seedlings from the prunings of healthy and productive plants provide key advantages: they have a higher guarantee of surviving once transplanted in the field, and they allow for earlier planting (during the rainy season when traditional seedlings are not yet ready).

Growers may use seedlings generated from prunings in an effort to provide a balance between the aerial and root systems of the plants.[22] Various studies have been conducted analyzing the use of pruning seedlings for planting coffee crops.[23–25] All of them have reported that seedlings derived from prunings showed development similar to or superior to traditional seedlings. Some of these studies examined pruning seedlings in the productive phase. By analyzing production data, the authors concluded that the production of pruning seedlings was the same or greater than plants originating from traditional seedlings.

Pruning seedlings developed in tubes showed superior growth to those directly seeded in a nursery.[24] The use of pruning seedlings in a nursery can allow for earlier planting, coinciding with the beginning of the rainy season in many growing countries. Pruning seedlings also produce less juvenility than traditional seedlings, which can lead to higher harvest yields in the first harvests.[25]

Most studies on the pruning height of orthotropic stems agree that the cutting should be made between the third and fourth pair of true leaves.[23] As with other seedling types, care should be taken to ensure that they are healthy and vigorous at the time of planting.

2.5 Planting the Coffee Crop

This section covers important aspects of choosing the ideal location for cultivating coffee, including soil analysis (texture, structure, depth, fertility), climate (locations with lower risks of frost and cold winds), and ease of access (transport of inputs and production).

The planting out of healthy and vigorous seedlings is no less important than their production. Since coffee plants are perennial, certain mistakes made during planting may lead to production losses throughout the life of the plant and can only be corrected with a new planting. For this reason, planning is essential for the success of the coffee farm, and this begins with choosing the right location for the coffee fields (Figure 2.10).

The first focus should be the region where the coffee will be planted. Average annual temperatures should be between 19 and 22 °C for cultivars of the species *Coffea arabica* L. and between 23 and 26 °C for cultivars of the species *Coffea canephora* Pierre. If *Coffea arabica* L. is planted in locations where average temperatures are above those recommended, or even if very hot periods occur, flowers may abort, decreasing productivity. If *Coffea canephora* Pierre

Figure 2.10 A coffee farm with field rows following the contours of the topography.

plants are planted in locations that are cooler than the recommended temperatures, this too can lead to decreased productivity.[26,27]

Even if occurring only a few times during the year, excessive heat or cold can compromise both productivity and final product quality. At temperatures below 16 °C, growth of the aerial part of the plant is affected by physiological disorders, with a drastic reduction in photoassimilate translocation, photosynthesis, and nitrogen assimilation by the leaves. At temperatures near 0 °C, the plant's cell walls freeze and rupture, causing plant death.[13]

Another factor to consider is that large bodies of water near coffee fields, such as reservoirs or other waterways, can influence crop management and even the quality of the final product. For example, higher relative humidity can increase the occurrence of diseases caused by fungi and bacteria, a risk that increases when the coffee is densely planted. The quality of the final product can also be compromised by undesirable fermentation of the fruit, even before the harvest. As a result, some coffee growers harvest their crop early to preserve the quality of the final product.

The altitude and topography of the area where the coffee will be planted is key to how that farm will be managed and it will impact, directly or indirectly, the quality of the coffee and how it will be classified.[8] While myriad factors are involved in coffee quality, higher altitudes are associated with cooler mean temperatures, and, in general, the production of coffees with higher acidity and better aroma characteristics.[28–30] In Guatemala, the altitude of the farm has a direct impact on coffee classification since coffees are classified by the altitude at which they were grown: (a) "Prima Lavado" (prime washed) is produced between 758 and 909 m; (b) "Extra Prima Lavado" (extra prime washed) is produced between 909 and 1060 m; (c) "Semiduro" (semi-hard bean) is produced between 1060 and 1212 m; (d) "Duro" (hard bean) is produced between 1.212 and 1.364 m; and (e) "Estritamente Duro" (strictly hard bean) is produced at over 1364 m.[12]

Another aspect of topography is its impact on mechanization. Declivities of over 30% limit options for mechanizing farm operations such as sprayings, fertilization applications, and harvesting. In regions with a more rolling topography, mechanization is often possible. On small farms, handheld machines and manual harvesting can be employed. On large-scale farms, more commonly found in Brazil, the mechanization of coffee production has provided an important counterbalance to the growing shortage of qualified rural labor without compromising job generation. It has facilitated both the work in the fields and in the harvest. The solutions and innovations involving mechanization are not limited to Brazil, and are being implemented in coffee growing regions throughout the world.[31]

In regards to rainfall and soil, preference should be given to areas with good rainfall distribution and soils with higher moisture retention capacity and a medium texture. In cases where these factors are not available growers should ensure that irrigation is available in periods of water deficits.[31]

The presence of pebbles and/or gravel in the top 30 to 40 cm of the field soil can limit the use of agricultural implements by increasing wear and tear. Pebbles also decrease the soil's water retention capacity. The effective soil depth, defined as the depth to which the plant's roots can easily and sufficiently penetrate in search of water and nutrients, should be at least 120 cm, provided that the soil texture is medium to clay-rich and contains no more than 15% rocks. In sand-rich soils or soils in dry regions, the effective soil depth should be deeper in order to avoid compromising development of the coffee plant.[32]

Medium texture soils (neither clay soil nor sandy soil) are preferable when choosing the cultivation area. Clay soils will require larger quantities of phosphate fertilizers and correction with higher quantities of lime. Sandy soils need a higher supply of micronutrients in addition to higher doses of fertilizer,[26] different from volcanic soils rich in organic matter, such as those typically found in Guatemala.

While soils with good natural fertility are desirable, they are not necessary given the availability of fertilizers (organic or chemical) and soil amendments to correct pH and aluminum. For example, in Brazil, soils with savanna vegetation were not used to plant coffee until the 1960s, since it was thought that coffee would only grow in fertile soils and under forest vegetation cover. Current technologies allow coffee growing in regions such as the Brazilian Cerrado, where the soil has low natural fertility, high acidity, low levels of organic material, low levels of phosphorus and calcium, low availability of micronutrients, and a low cation-exchange capacity.[26,32]

Other considerations when choosing a location for planting coffee include: (a) Avoid areas with compact soil, which limits root growth. Compact soil is common in areas of intense use of mechanized implements. (b) Avoid areas subject to constant winds and without natural wind barriers (*e.g.*, forest) or windbreaks. These windbreaks can be temporary (annual crops such as rice, soy, sorghum, or corn) or permanent (grevillea, bananas, and leucaena). (c) Use areas free of soil pests and nematodes.[26]

When determining the spacing between coffee plants, several factors should be taken into consideration, including the alignment of the plants in relation to the path of the sun, the cultivar that will be used (size and architecture), soil fertility, crop management techniques, pruning methods, and the declivity of the land, among others. The choice of spacing is definitive for the life of the crop and will influence the execution and cost of crop management, productivity, and the longevity of the coffee field that is being implemented. Proper plant arrangement, which is the combination of spacing between the plants in a line and the distance between the lines, can lead to higher yields and facilitate optimal crop management.

Currently plant spacing with higher density in the planting line results in around 6000 to 7000 plants per hectare. Mechanization of the crop management is possible at this density.[26] In some coffee growing regions, given the availability of sufficient manual labor, even more dense plantings may be used.

Field arrangement of plants for a desired density (plants per hectare) is accomplished through a combination of in-line plant spacing and spacing between the lines, while also taking into consideration: (a) the alignment of the coffee crop in relation to the sun's path (in-line plant spacing can be decreased if the plants receive more daytime sunlight); (b) the height and architecture of the chosen cultivar (taller cultivars require more spacing between rows than shorter cultivars, and failure to provide this spacing may affect maturation speed, uniformity, and productivity); (c) the altitude of the field; (d) crop management (spacing should facilitate the use of machines and/or equipment that will be implemented); (e) pruning (denser plantings will require more frequent pruning); (f) topography (if the ruggedness of the topography inhibits mechanization, closer spacing should be used in conjunction with shorter cultivars to facilitate harvesting); and (g) the total area of coffee production.[26]

Row spacing should be enough to allow for the passage of any machines that will be used when considering full plant growth, unless a pruning program will be implemented that allows for narrower rows. In general, ideal row spacing is the sum of the diameter of the fully grown crown of the chosen cultivar and the width of the machine/equipment that will be used.

Dense plant spacing allows for a larger number of plants per hectare and therefore, in general, higher yields. The densities used in coffee growing can be classified (in plants per hectare) as: (a) wide, or traditional (up to 3000); (b) semi-dense (3000 to 5000); (c) dense (5000 to 10 000); and (e) super dense (10 000 to 20 000).[26,33]

Rationales for choosing dense planting include: (a) higher productivity per area, principally with the first harvests, which allows for a faster return on capital investment; (b) better use of the planting area, which lowers investment costs in agricultural lands and frees up areas for other crops and/or livestock; (c) lower per-unit production cost, given the increased productivity; (d) more soil protection against erosion and improvement of the physical, chemical, and biological characteristics of the soil; (e) reduction of weed infestations due to increased shade; (f) less oscillation between high and low productivities (biennial production cycle) as a consequence of the lower production per plant (though the area has higher total production); (g) better fertilizer efficiencies; and (h) fewer attacks by coffee leaf miners (*Leucoptera coffeella*).[26,33]

However, in choosing dense spacing, certain factors should be considered: (a) higher initial investment in the planting as well as initial plant phases; (b) increased difficulties with or preclusion of mechanization in crop management; (c) increased plant maintenance workload (more frequent thinning and pruning); (d) increased difficulty in crop management, such as fertilization, spraying, and harvesting; (e) delayed and less uniform fruit maturation during harvest season; (f) higher risk of quality loss; (g) higher risk of attacks from coffee berry borer beetles in the fruit and rust on the leaves; (h) preclusion of interplanting with other crops.[26,33]

Given the pros and cons, the use of dense planting will depend on the conditions of a particular field, and should be considered on a case-by-case basis. In general, the practice is more common on small properties and in mountainous regions, where mechanization is impossible but labor is available. If dense planting is adopted, developing a regular pruning program is fundamental to maintaining productivity at an economical level.

Given the topography and conditions found in Guatemala, the ideal plant stand, or density, is generally at most 5000 per hectare, with spacing varying between 2.5 and 3.0 meters between rows and 0.7 to 0.8 meters between plants. A key factor to consider in planting on mountainous terrain, as is found in Guatemala, is the slope's aspect, or the direction it faces, as this can affect plant development (slope effect).[28] Another factor is altitude. With the lower average temperatures (below 19 °C) of areas above 1800 meters above sea level, while plants produce a lot of vegetation, they are not very productive.[12]

When planting a coffee field, a soil analysis is necessary to determine needed soil amendments (lime, gypsum, phosphorus) and to plan a balanced fertilization schedule. Soil should be sampled using rigorous criteria to guarantee a representative sample and reliable results that can then be used by an agronomist to make recommendations.[26]

The soil should be clean (free of wood debris and weeds) before proceeding with any conservation measures (terracing or contour plowing). Then plowing should be done at the end of the dry season, with the goal of incorporating vegetative remnants, lime and/or gypsum, and subsoil in cases of soil compacting.

When plowing, furrows (Figure 2.11) should be made along the contour. In areas of higher declivity, where mechanization is impossible, the grooves can be made using animal-driven plows and then completed with hand tools.[26]

In planning the initial layout of the fields, it is important to provide adequate space for completing field operations such as material transport and the comfortable transit of machines and vehicles. For this reason, larger rows should be implemented every 70 to 100 meters.

Figure 2.11 Fertilizing furrows before seedling planting.

Planting (Figure 2.12) should be done during a rainy period, when seedlings are fully formed. It is important that care is taken when planting to avoid future problems such as twisted tap roots, shallow planting (causing the plants to fall), or planting too deep (which can damage the base of the plants).[26]

Important planting recommendations include: (a) use seedlings with three to six pairs of leaves, except when replanting with seedlings specifically raised for replanting purposes (*e.g.*, the initial seedling did not survive); (b) harden off the seedlings by exposing them to more sunlight and reducing waterings until they are taken for transplant; (c) use care when transporting the seedlings to avoid damage; (d) plant seedlings at soil level so that the plants do not "fall" or "drown"; (e) align the plants in the planting furrow (or planting holes) to facilitate future operations such as mechanized crop management and possible mechanized harvesting; (f) cut 1 to 2 cm off the bottom of the seedling bags before transplanting to avoid any root system contortions; and (g) take care not to apply too much lateral pressure to the seedling root system when planting to avoid tap root twisting upon planting.[33]

There are options for semi-mechanized (planting platform attached to a tractor) or fully mechanized planting, but these are uncommon compared to manual planting.

Around 30 to 40 days after planting, the coffee plants should be evaluated to determine if replantings are needed and the initial surface fertilization (topdressing) should be done on the field. Replanting should continue until the crop is established (the first 2 years) in order to ensure a perfect crop stand.

Crop management during this formation period should be rigorous, with proper preventive measures taken against attacks from ants, coffee leaf miners, cutworms, cochineals, acari, and other pests, as well as diseases such as cercospora, bacteriosis, *etc.* Cover crops should be used to control erosion,

Figure 2.12 Alignment of planting holes.

maintain mild soil temperatures, and create an environment favorable to microbial growth in the soil; however, they should not compete with the coffee plants.[26]

2.6 Crop Management

This section addresses the general and practical aspects of crop management with the goal of providing the reader a broad vision of the activities necessary to cultivate coffee.

From the time of planting, care should be taken to ensure that the plants grow and develop properly and will therefore not only achieve quality production, but do so in a way that is economically, socially, and environmentally sustainable.

Since coffee is often planted during the rainy season, soil conservation measures should be considered. This includes ensuring that water retention terraces (and contour rows) are clean as well as the planting of cover crops to combat erosion, especially sheet erosion. A well-planned field will maximize the use of rain water, both allowing for soil infiltration and retaining the excess water in retention basins built in the fields.[26]

Water is important for the absorption and transport of nutrients and therefore should be well distributed over time and in sufficient quantity for plants to grow and develop.

Given the increasing water demand in urban and rural populations throughout the world, agricultural water usage should be done responsibly so that less is used while still maintaining high productivity. Furthermore, along with this increase in water demand, rainfall distribution throughout the year has not been satisfactory in many coffee growing regions worldwide. To mitigate the effects of the drought, agronomic techniques have been used such as mulching and drip irrigation (Figure 2.13), which consists of applying water directly to the root system of the plants in small quantities in order to optimize this important natural resource.[34] Irrigation can lead to gains of up to 120% depending on the region and the level of water stress.[35]

Fertilization that is balanced, rational, and therefore economical, should be a constant goal for coffee growers concerned with good crop management. Coffee plant nutrition is not only important for growth and development, but also indirectly impacts the plant's tolerance of pest and disease attacks, as well as prolonged dry periods.

Malavolta[36] summarizes "coffee field fertilization" with the following phrase: "Fertilization starts with **soil analysis**, continues with **acidity correction**, and ends with **fertilizer application**." In other words, a **soil analysis**

Figure 2.13 Drip irrigation in a coffee field.

where the coffee was or will be planted is fundamental so that an agronomist can see the current soil composition and make recommendations for soil corrections (lime, gypsum) and fertilizers (organic or chemical). **Soil corrections** (that alter pH and/or tie up elements that may be toxic to the plants) must be applied before the actual fertilizer since soil corrections allow for maximum absorption of macronutrients and micronutrients. Malavolta concludes that it is fundamental that qualified rural workers be used to **properly apply the fertilizer** around the roots of the plants. This will lead to maximum absorption of the fertilizers without excessive losses due to volatization or leaching.[36]

There are many types of fertilizer and many ways to apply them to the coffee plant: "green manure," organic fertilizers, chemical fertilizers, fertilization using irrigation water (fertigation), and fertilization by spraying the coffee plant leaves (foliar fertilization). Figure 2.14 shows an example of an animal-driven fertilizer spreader used to apply fertilizer onto the soil around the plants.

Green manure is the practice of planting other crops (especially legumes) between coffee rows and then incorporating them into the soil as a source of organic matter and nutrients (Figure 2.15). The green manure cover crop can be managed using brush cutters in order to decrease the competition of the

Figure 2.14 Animal-driven fertilizer spreader.

Figure 2.15 The use of cover crops in Guatemala for the cycling of nutrients and soil protection against erosion.

cover crop with the coffee while supplying organic matter that can be incorporated into the soil.[37,38]

It is also common in coffee crop management to use organic sources of nutrients: cattle manure, poultry manure, castor meal, and coffee hulls (residue from coffee milling comprising the exo-, meso-, and endocarps of the fruit).[37] However, the coffee plant's demand for organic fertilizers is usually greater than their availability, and the use of organic fertilizers is often limited to what can be produced on the farm or acquired from others as long as prices are competitive with chemical fertilizers.

Chemical fertilizers can be made with simple nutrients or in multinutrient formulations. They are applied to the plants (through the soil) during the rainy season for greater optimization and efficiency, or even applied *via* fertigation. The application of micronutrients can be done through the soil or through foliar applications *via* spraying.[14]

The management of the coffee fields should be integrated in a way that maximizes efficiency of inputs, environmental sustainability, and economic efficiency.

Proper weed control is essential, since weeds compete with coffee plants for available water, nutrients (fertilizations), and sunlight. However, weeds can potentially act in favor of both the coffee farm and the environment. If properly managed, they can help control erosion (especially during planting, when the soil is still stabilizing after plowing and disking); supply organic material to the soil, improving its physical structure and fertility; improve soil aeration and water retention; and recycle leached nutrients from the subsurface layers of the soil.[39–43]

Therefore, growers should incorporate an integrated weed control management program that includes preventive measures as well as mechanical and chemical controls. An integrated approach can be more effective and lead to both labor and capital savings.[44]

Preventive control measures entail avoiding the introduction of weeds that do not already exist in the area. These weeds or their seeds can enter by means of contaminated manure, with coffee seedlings, or they can even stick to machines and equipment used in infested areas.

Weeds should be controlled observing good practices for coffee management such as planting the coffee in the rainy season (which favors coffee over weeds in their competition for resources), adequate spacing, using the cultivar most adapted to the planting location, a balanced fertilization that is sufficient for good initial development of the plants, and irrigation when needed and available.[44]

Mechanical weed control occurs at the time of planting by means of soil preparation, but should also be done throughout the life of the crop. Weeds can be controlled manually (hoes) or mechanically (weed whackers, mowers, grates). Weeding machines can be hand-held or attached to tractors or even horses (Figure 2.16) to increase speed. Weeding should be done more frequently and between mowings during the hot and rainy part of the year, as vegetation grows more quickly.[44,45]

The advantage of using chemicals for weed control is that they are more efficient and, depending on the situation, can be cheaper than other methods. However, chemicals must be used responsibly. The chemical control of weeds should be done under the guidance of a qualified professional that can help the grower identify the main weeds, recommend the most effective and secure herbicide, and ensure that springs or waterways are protected. It is also important to properly qualify and train rural workers who will apply the chemical product, as this will be a key factor for the success of the operation and also ensure worker safety.[44,45]

Another way to manage the coffee crop between rows is by using intercropping, which is very common in most coffee producing countries, especially on small properties that employ little mechanization.

Coffee is a perennial crop with a high implementation cost. It renders significant economic returns only 2.5 years after planting. Upon planting, the

Figure 2.16 Animal-driven brush cutter.

coffee plants are still small and occupy little space, resulting in large gaps between the rows that will be subject to soil erosion and favor weed growth if not otherwise used.

Because of this, smallholder coffee growers often plant annual crops between the rows in their farms. Besides taking advantage of the space between the rows, it decreases relative labor costs per area and produces subsistence crops with the possibility of selling any excess to increase family income. In addition to these advantages, according to Guimarães, Mendes and Souza,[1] intercropping can decrease the initial cost of establishing the coffee farm, maintain soil covering, improve soil permeability and aeration, allow for the incorporation of crop remnants, serve as temporary windbreaks that protect the coffee, promote nutrient cycling, generate environmental benefits, and favor soil microbiology.

However, intercropping should only be used in situations where it does not promote competition with the coffee for water, nutrients, and light, which would decrease the productivity of one or both of the crops involved. Intercropping can also interfere with the application of insecticides, acaricides, or herbicides, in the case of non-organic crops, and with mechanical procedures such as fertilizations, spraying, or even the harvest.[46]

Crops used for intercropping include beans (Figure 2.17), corn, rice, soy, peanuts, cassava, sunflowers, cotton, tobacco, vegetables, and medicinal plants. In some regions intercropping can be done with other perennial crops such as rubber trees, papaya (or other fruit trees), or even lumber.

2.7 Coffee Cultivation in Agroforestry Systems

In its native habitat, *Coffea arabica* L. grows in the shade and these conditions were simulated as its cultivation began. As selections were made, it was often adapted to full sun conditions which rendered higher

Figure 2.17 Coffee interplanting with beans.

productivities. This full sun cultivation became more utilized around the world and cultivars were adapted to render higher production given full sun conditions.[47–49]

The return to shade-grown coffee in some parts of the world began in the 1990s when low coffee prices forced producing countries to develop strategies for economic recovery. One proposal was to incentivize the expansion of shade-grown coffee to reduce production costs per area.[50] Other advantages to doing this were the creation of environmental services and the increase in quality, thus allowing growers to improve their socioeconomic situation by producing specialty coffees, which fetch prices above the coffee commodity market.[51] Currently, the use of agricultural cultivation systems that preserve natural resources and favor crop diversification represent an alternative for growers who desire lower per area costs and a sustainable production model. Among the various production systems that aim to do this, agroforestry is one of the most popular.[52]

In some countries, coffee is grown under native forests, however, the productivity is often quite low.[53] On the other hand, shade-grown coffee in agroforestry systems that use silk oak (*Grevillea robusta* A. Cunn), rubber trees (*Hevea brasiliensis* Mull. Arg.), banana trees (*Musa*, sp.), and mahogany (*Swietenia macrophylla* King.), among others, have shown higher productivities.

2.8 *Coffea arabica* L. Prunings

Coffee is a perennial plant that can have economic viability for 20 to 30 years if well managed. Among management techniques, pruning is one of the most important, principally formative pruning to correct plant architecture or to eliminate parts affected by frost, hail, drought, or pests and diseases.

With dense planting, prunings should be planned in order to maintain a productive and well-aerated crop, given the diminished space between the

Figure 2.18 Loss of lower plagiotropic branches in a dense planting.

plants. Coffee growers or agronomists should be on the lookout for signs that indicate the need for pruning such as the loss of plagiotropic branches on the lower part of the plants (Figure 2.18).

Coffee plant prunings recuperate parts of the plant that have been damaged by either biotic (pest or disease attack) or abiotic factors (intense heat or cold, hail, excessive sunlight, water deficit, *etc.*). Prunings also lead to recuperation of high productivities (even with older plants); greater aeration and light interception (which can increase the quality of the fruit produced); control over plant height, which can decrease labor costs (principally during the harvest); increased efficiency of phytosanitary applications (more efficient applications of agrochemicals); and a better balance between the production and use of photoassimilates, thus attenuating the alterations in biennial production.[54]

As with any option for managing the coffee crop, the decision of whether or not to prune, or even to pull out the current crop entirely and replant, should be done carefully and be a shared decision between the agronomist and the coffee grower.

The first point to consider is if the coffee plants have the proper genetics for productivity. Pruning will only recuperate productivity up to the potential of the cultivar, but can't extend it. If it is determined that the cultivar has a low productive potential or that it is not well adapted to the environment, the best option is to pull up the crop and plant a different cultivar that is more productive and better adapted.

Similarly, in making the decision whether to prune or pull up the crop, one should pay attention to the original spacing and to the number of dead plants or empty spaces in the rows. It is not worthwhile to prune a field if there are insufficient plants to guarantee good future yields.[54]

After pruning, the plants need a suitable environment for their recuperation. Agronomists and coffee growers should ensure that the area is free from soil pests and heavy nematode pressure.

Ultimately the decision to prune or to pull up the crop and replant should be based on: (a) whether pruning is really necessary or if other crop management practices would be better (fertilization, for example) for recuperating high yields; (b) if the current market price is high enough to require quick recuperation (pruning), or if it is low enough to justify pulling up the crop; (c) if the planted cultivar does not meet the production potential of modern cultivar lines then the best option is to pull up the coffee plants; and (d) if the current plant stand (spacing and number of failed trees) will not meet the needs of the modern coffee grower then the crop should be pulled up. With this diagnostic, the technician or coffee grower should make the decision to take other measures, to prune, or to pull up the existing plants (for a new planting or not).[54]

Other factors that can justify pruning are: (a) the auto-shading of the plants in cases of closer spacing, with plagiotropic branch death in the lower third of the plant; (b) environmental damage (due to temperature, water, *etc.*); (c) plant decline through improper management (insufficient or unbalanced fertilization), competition from weeds, mechanical damage during the harvest, or even the advanced age of the plants; and (d) excessive plant height making crop management difficult (principally during the harvest).[54]

Prunings can be done on the orthotropic stems and/or on the plagiotropic branches, depending on the needs of the plants. Prunings of orthotropic stems can be classified in the following ways: (a) Clean stumping (Figure 2.19), also called stumping without a breather, or "without lungs". This pruning is done at a height of 40 to 50 centimeters on the plant, and the plagiotropic branches of the lower third are lost. (b) Partial stumping, also called stumping with a breather or "with lungs". This pruning is done at a height of 60 to 80 centimeters on the plant, and the plagiotropic branches on the lower third of the plant are maintained. (c) Low topping (or capping), a pruning similar to the high stumping, but done higher on the stem (usually in the middle third, depending on the necessity of the plant). (d) High topping (or

Figure 2.19 Clean stumping of a coffee field.

Figure 2.20 Equipment used to simultaneously perform capping and parrot pruning.

capping), done to limit the growth of tall plants, generally maintaining plant height at 1.8 to 2.2 meters (similar to a heading cut).[54]

Plagiotropic branches can be cut back in prunings such as "parrot perching". In this type of pruning, the plagiotropic branches that are attached to the orthotropic stem are cut at a distance of 20 to 40 centimeters from the main stem with the objective of renovating the branches of the coffee plant, permitting the return of high yields (with old coffee fields and in the "zero harvest system"). The "zero harvest system" is achieved by doing "parrot perchings" every two years or more, alternating years without production with years of high productivity with the goal of decreasing overall costs.[54] Figure 2.20 presents a machine that can simultaneously perform capping as well as pruning of the plagiotropic branches and is often used in performing the "zero harvest" prunings (Figure 2.20).

In years of high productivity, the plant's consumption of photoassimilates (compounds formed by assimilation using light-dependent reactions) is high and fruits are given priority over other parts of the plant, including roots and branches, which means that there is less branch growth. Since this branch growth is the basis for future fruit formation, production will therefore be lower in the following year. Using the same logic, in a year of low productivity there will be enough photoassimilates to support vigorous branch growth until the following year, when this new growth will generate high productivity. This alternation between higher and lower productivity defines the biennial nature of the coffee plant as it repeats this cycle every two years, even with proper crop management.[9,10]

The other type of plagiotropic branch pruning is of the branch's lead shoots. This is usually performed on newer coffee plants that have fewer branches. The productive branches are pruned in order to form more branches and thus generate higher productivity.

In the specific case of training plants in dense plantings, preventive pruning is recommended before a critical level of auto-shading is reached.[55] A

Figure 2.21 Plant-by-plant selective pruning, common to smallholder farms.

pruning program for dense plantings should consider: (a) elimination of alternating rows, allowing for mechanization; (b) stumping in alternating rows; (c) stumping in alternating double rows (and using super dense planting); (d) stumping of one-third of the rows; (e) alternating stumping and topping; (f) stumping of 20% to 25% of the lines per the Fukunaga model; (g) stumping of all rows in a particular field, and rotating the fields each year that they are stumped; (h) plant-by-plant selective prunings (Figure 2.21) to maximize the productive area of each plant (this is common only in smaller properties since fertilization, harvesting, and other crop management tasks are done plant-by-plant).

The most appropriate period for pruning coffee plants is shortly after the harvest since the plant will have more time to grow and recover for production in the following year. An exception should be made for depleted plants that first need to recover photoassimilate reserves before pruning.

As soon as the pruned plants put out shoots and these shoots reach a height of around 20 centimeters, they should be thinned. Thinning should be done gradually, in two to three rounds, in the year following the pruning, ending with one or two shoots per trunk[54] (Figure 2.22).

2.9 *Coffea canephora* Pierre Prunings

Plants of *Coffea canephora* Pierre have continual growth with exhaustion of fruit production on plagiotropic branches occurring after around three harvests, at which time they should be pruned. In these production prunings other branches are removed, such as broken branches, poorly located branches in the interior of the plants, or excess branches, all of which may negatively affect aeration and light penetration.[56]

Pruning is done at a height of around 20 to 30 centimeters from the soil, or directly above new shoots. As with *Coffee arabica* L., pruning should be

Figure 2.22 Selection of shoots from a coffee tree after stumping, leaving one shoot per plant.

recommended only after considering factors such as plant age, spacing, auto-shading, plant depletion, fruit load, and even the financial situation of the coffee grower. Also similar to *Coffee arabica* L., the best time to prune is shortly after the harvest so that the plants have more time to recover before the following harvest. Thinning is done so that only one shoot remains on each pruned branch. The recommended number of orthotropic stems per hectare is 12 000 for maximum productivity.[56]

2.10 Pests, Diseases, and Nematodes in Coffee Cultivation

In this section the interference of pests, diseases, and nematodes in coffee cultivation will be addressed, focusing on predisposing factors and management options in the quest for sustainable production.

2.10.1 Identification of Signs and Symptoms in Plants for Accurate Diagnosis

In plants, as with humans, the actions of pathogens can be masked by and sometimes confused with other concurrent conditions. In most cases, the problems that plants and people experience are caused by more than one factor, acting simultaneously and confounding accurate diagnosis. Disorders of plant nutrition, health, and physiology are conditions that can cause the coffee plant to present signs and symptoms that can confuse agronomists and coffee growers in the diagnosis of pest attacks and diseases. As such, this discussion will focus on the main pests, diseases, and nematodes of the coffee plant, concentrating on the factors that predispose plants to these conditions, and management techniques for minimizing their damage.[6]

2.10.2 Coffee Plant Pests

Among the pests that most interfere in the productivity and quality of coffee are the coffee leaf miner (*Leucoptera* spp.) (Lepidoptera: Lyonetiidae) and the coffee berry borer (*Hypothenemus hampei* Ferrari, 1867) (Coleoptera: Scolytidae).

2.10.2.1 *The Coffee Leaf Miner (*Leucoptera coffeella *Guérin-Mèneville & Perrottet, 1842;* Leucoptera caffeina *Washbourn, 1940;* Leucoptera meyricki *Ghesquière, 1940) (Lepidoptera: Lyonetiidae)*

The coffee leaf miner is a major coffee pest, particularly in the Neotropics, where *L. coffeella* predominates, and eastern Africa, where *L. caffeina* and *L. meyrocki* attack coffee and other members of the Rubiaceae family. Unlike *L. coffeella* and *L. meyrocki*, *L. caffeina* is associated with shade-grown coffee.[57,58]

In its adult form, the coffee leaf miner is a tiny moth whose larva feed on coffee leaf mesophyll tissue, where they also deposit their feces. The presence of the coffee leaf miner is signaled by necrotic areas (Figure 2.23) that diminish the foliar area for photosynthesis, resulting in loss of infected leaves, especially in the driest periods of the year. If the percentage of damaged leaves

Figure 2.23 Coffee leaf with necrotic area caused by coffee leaf miners (*Leucoptera coffeella*).

is greater than 30%, control measures should be taken so that productivity and economic sustainability are not compromised. However, if among the damaged leaves 40% of the leaves show signs of the presence of predatory wasps, chemical control is not needed as the natural control of the wasps will suffice.[59]

Control measures such as the construction of windbreaks, weed management, and irrigation are recommended. In drier regions or in fields with less shade, a recommended control method for *L. meyricki* and *L. coffeella* attacks is to use smaller spacings between the plants. This increases the relative humidity of the air within the crop, reducing insect infestations.

Given the widespread damage it causes, many efforts have been undertaken to develop cultivars that are resistant to the coffee leaf miner. One notable example of this effort is the cultivar Siriema 842, and collaborative works have been perfecting the process of its vegetative propagation for large-scale distribution.[60]

2.10.2.2 *Coffee Berry Borer (*Hypothenemus hampei *Ferrari, 1867) (Coleoptera: Scolytidae)*

The coffee berry borer (Figure 2.24) is a major pest that affects all species of *Coffea*, having spread from its initial habitat of Central Africa to nearly every producing origin, including Java (1909), Brazil (1913), Peru (1962), the Philippines (1963), Colombia (1989), Costa Rica (2000), and Hawaii (2010).[57,61,62] It is a small black beetle that lays its eggs in a small perforation in the region of the fruit crown. Subsequently the larvae begin to feed on the seeds, damaging both the seed productivity and quality of the final product. Control is called for when 3% to 5% of sampled fruit show infestation. However, there

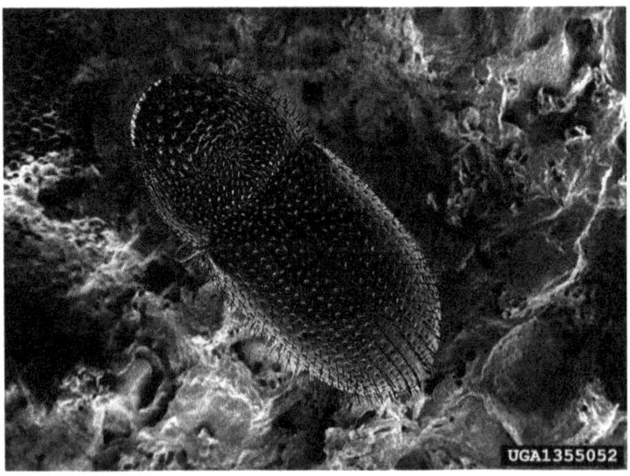

Figure 2.24 Coffee bean borer (*Hypothenemus hampei*). Photo: Eric Erbe, USDA Agricultural Research Service, https://www.bugwood.org, reproduced under the terms of the Creative Commons Attribution 3.0 License, http://www.forestryimages.org/browse/detail.cfm?imgnum=1355052.

have been cost and toxicity issues with the current chemical controls and efforts are underway to develop less toxic products that are more affordable and efficient.[59]

Unlike the environmental conditions that favor most species of coffee leaf miner, the coffee berry borer prefers conditions of high humidity, such as humid regions and, in particular, rainy summers. To minimize conditions favorable to the coffee berry borer, natural shelters of the pest should be eliminated and thus abandoned crops and remnant fruit should be removed from the coffee plant after the harvest to reduce the likelihood of the insect surviving until the next crop. Another good control measure is to begin the harvest in the most affected plots, thus diminishing the duration of the pest in the field and its dissemination to unaffected areas.[59]

Because humidity favors this pest, crops in humid regions should be planted with larger spacing, thus encouraging natural aeration.

2.10.2.3 *Cicadas (*Quesada *sp.,* Fidicina *sp.,* Carineta *sp.,* Dorisiana *sp.) (Homoptera: Cicadidae)*

Cicadas are found in some coffee regions. Sap suckers in their nymph phase, these insects suck on the roots of the coffee plant, causing deterioration of the plants, with symptoms of nutritional deficiency and early dropping of leaves on the apex of branches. Signs of the presence of cicadas in cultivated areas are: holes under the canopy of the coffee plant, molted exoskeletons attached to vertical branches, characteristic strident sounds emitted by male insects, and the presence of sap-sucking nymphs on the roots. When a sampling of the number of nymphs per plant reaches 35, controls in the form of systemic soil insecticides are recommended.[59]

2.10.2.4 *Mites –* Oligonychus ilicis *(Mcgregor, 1917) (Acari: Tetranychidae)* and Brevipalpus phoenicis *(Geijskes, 1939) (Acari: Tenuipalpidae)*

Mites occur on the coffee plant especially in periods of drought, with prolonged drought favoring their survival. These insects cause defoliation and damage to coffee plants, especially in fields in the planting phase. The mite *Brevipalpus phoenicis* (Geijskes, 1939) (Acari: Tenuipalpidae) causes ringspot brought by a virus that causes the coffee plant to present symptoms of severe defoliation within the canopy and loss of product quality.[59]

2.10.2.5 *Other Pests*

Other pests that occur on coffee farms and eventually require control include: (a) mealybugs, which can attack the coffee plant from the root to aerial parts; (b) coffee root fly (*Chiromyza vittata* Wiedmann, 1820 (Diptera: Stratiomyidae)); and (c) lizards that feed on coffee plant leaves.[59]

Ecological imbalances can lead to infestations by pests that normally do not cause problems for coffee cultivation. An example is rats, which feed on tender branches, flowers, and fruits, leaving signs such as bent and broken branches. No control is recommended other than seeking ways to re-establish ecological balance in the affected area.[63]

Slugs are another pest that can attack in certain situations. They appear in great numbers on coffee farms in conditions of high humidity, such as times of persistent rainfall. Aside from their physical presence, slugs leave behind stripped fruit on the ground near the coffee plant. These occasional attacks occur at night and should be of no concern to coffee growers. Because of the rarity and minimal damage of slug attacks, no controls are recommended.[59]

2.10.3 Coffee Plant Diseases

Certain diseases predominate in coffee plants. This section reviews both the environmental conditions favored by some more common diseases and management practices to deal with them.

2.10.3.1 *Damping Off (*Rhizoctonia solani*)*

A disease known as Rhizoctonia sheath blight, also called "damping off", is caused by the fungus *Rhizoctonia solani*, which survives for long periods in the soil and crop remnants. This disease occurs principally in seedling nurseries and is caused by the use of soil contaminated with the fungus. The characteristic symptom of this disease is the appearance of lesions that range in color from brown to black and encircle the stems of seedlings near the soil, resulting in the wilting and death of the plants.[64]

The following measures are recommended to prevent the disease or contain it to the nursery: (a) use uncontaminated soil in the substrate mixture to be used in crop formation; (b) use high-quality water (clear of pathogens); (c) do not reuse seedling containers or substrates; (d) periodically change the location of the nursery; (e) avoid excess humidity and shade in the nurseries; (f) remove areas of stunted growth and the surrounding plants; and (g) when moving seedlings from nurseries to be planted in the field, take care to rigorously avoid using any plants with signs of damping off.[64]

2.10.3.2 *Brown Eye (Cercospora) Spot (*Cercospora coffeicola *Berk et Cook.)*

While it may attack coffee trees of all ages, Cercospora is another disease that can severely attack seedlings in nurseries and young crops. Also known as brown eye spot, berry blotch, and Cercospora Blotch, it is caused by the fungus *Cercospora coffeicola* Berk et Cook.

The disease attacks both coffee leaves and fruit. On coffee leaves, Cercospora leaf spot symptoms appear as brown circular spots with grayish centers, usually surrounded by a yellow halo. On fruit the disease causes necrotic, depressed spots, brown to black in color, extending to the poles of the fruit. Damage caused by cercospora leaf spot includes: (a) leaf abscission and stunted growth in nursery seedlings, and defoliation and growth retardation of adult plants; and (b) fruit abscission and branch drying in new crops, and (c) in productive crops premature aging and premature fruit fall, resulting in quantitative and qualitative damage to the final product.[64]

Cercospora leaf spot can be prevented and controlled with proper plant management, both in the nursery and in the field. Nursery measures include: protecting seedlings from cold winds with side fences; using substrate of recommended composition to allow growth of properly nourished seedlings that will be less susceptible to the disease; and controlling irrigation and lighting inside nurseries. In the planting and production phase avoid planting in soils that are sandy, poor, compressed, or compacted; maintain sufficient and balanced plant nutrition using fertilizers; control the disease using fungicides, especially if the planting is done at the end of the rainy season; and manage the crop properly, avoiding damage or malformation of the root systems, which can indirectly affect plant nourishment and consequently favor development of Cercospora leaf spot. Chemical controls should be applied—to productive crops, nursery seedlings, and newly planted fields—when other preventive measures (fertilization, wind breaks, *etc.*) have proved insufficient to reduce the intensity of the disease.[64]

*2.10.3.3 Coffee Rust (*Hemileia vastatrix *Berk. et Br.)*

The disease known as coffee leaf rust (CLR), caused by the fungus *Hemileia vastatrix* Berk. et Br., is perhaps the most feared in coffee cultivation as it commonly occurs all around the world and can cause severe defoliation, which affects crop yields. Soon after first appearing in commercial coffee crops, coffee leaf rust devastated the coffee industry in Ceylon (now Sri Lanka) and in Central America coffee rust is currently considered the most important disease. Symptoms of the disease are orange-colored circular patches on the inferior (dorsal) surfaces of leaves, presenting as a powdery mass of uredospores (Figure 2.25). In advanced stages, some parts of the leaf tissue are destroyed and necrotic.[64]

In controlling coffee rust, it is important to note that the greater the plant vegetation, the higher the residual infestation of the disease. Also the higher the fruit load, the more intense the infestation. Moreover, crops planted with tighter spacing foster a microclimate of higher humidity that favors the spread of coffee rust. Thus, in addition to applying balanced fertilization, it is advisable to use recommended disease-tolerant cultivars, manage the plants for sufficient air flow, thin excess shoots to facilitate air movement within the crop, and apply fungicide treatments when needed.[64]

Figure 2.25 Symptoms of coffee rust in coffee leaves. *Hemilea vastatrix* Berk et Br.

2.10.3.4 *Bacterial Blight (*Pseudomonas syringae *Pv. Garcae)*

Bacterial blight, also called Elgon die-back since it was first observed on Mount Elgon in Western Kenya, can occur both in nursery seedlings and in adult plants.[65] In mature crops it occurs with highest intensity at high elevations and locations unprotected from winds that damage leaves and new branches, creating openings for the penetration of the bacteria. Hail and frost can also cause lesions in the plants that facilitate bacterial incursion.[64]

Symptoms of bacterial blight of coffee are brown spots surrounded by a yellow halo, with injured areas and leaf borders usually voided, creating a lace-like appearance. The bacteria are controlled with antibiotics mixed with copper fungicides (bacteriostatic), which increase the efficiency of the control.[64]

Suggested control measures include protecting nurseries from cold winds and protecting crops from cold and high winds (using windbreaks). Antibiotics should only be used when the presence of the bacteria is confirmed.

2.10.3.5 *Pseudomonas cichorii*

Leaves present irregular dark necrotic areas with surrounding tissue appearing moist. Normally, older leaves are more susceptible to these bacteria. High incidence of *Pseudomonas cichorii* is observed in hot and rainy periods in nursery seedlings. Excess irrigation water favors the disease, which usually penetrates through lesions caused by fungus (Cercospora and Phoma), coffee leaf miners, or other physical damage. Recommended controls include avoiding excess moisture and irrigation in nurseries, protecting seedlings from wind action that can cause leaf damage, and controlling brown eye spot, Phoma leaf spot, and coffee leaf miner attacks.[64]

2.10.3.6 *Phoma Leaf Spot (*Phoma tarda *Stewart, 1957;* Phoma costarricensis *Echandi, 1957)*

Phoma leaf spot, also known as leaf burn or blight, is a disease that causes defoliation, dropped fruit buds and flower petals, dry branches, and consequently diminished productivity. Leaf symptoms appear as dark circular spots that may feature concentric halos. While *Phoma tarda* is more common in Africa, Southeast Asia, and Brazil, *Phoma costarricensis* was mainly in Central America, but has been in India, Papua New Guinea, and higher altitudes in Brazil.[66,67] This disease has higher incidence in crops that are exposed to strong, cold winds, where the penetration of the fungus is facilitated by mechanical damage to plants (insects, leaf friction caused by wind, even harvesting operations).[64]

Recommended control measures include avoiding areas subject to cold winds when establishing crops, taking caution to install either temporary or permanent windbreaks; and balanced application of fertilizer and fungicides (before and after flowering) in periods favorable to disease.

2.10.3.7 *Coffee Berry Disease –* Colletotrichum kahawae *Waller et Bridge (Syn. Colletotrichum Coffeanum Noack Var. 'Virulans' Rayner (1952);* Colletotrichum coffeanum *Noack 'Sensu Hindorf' (Hindorf, 1970)*

Coffee berry disease (CBD) is a disease of arabica coffee and is caused by the fungus *Colletotrichum kahawae*. First discovered in Kenya in 1922, it has spread throughout most of the arabica coffee-producing regions of Africa. While currently limited to the African continent, its effects can be devastating, and the disease is a looming threat with the potential to spread to other coffee-growing zones.

There was a long period of confusion regarding its taxonomy, however, in 1993 Waller and Bridge described *C. kahawae* as the causal agent of CBD, distinguishing it from both *C. coffeanum* and other *Colletotrichum* isolates.[68]

While symptoms of the disease can be found on fruit of all maturations, in more mature fruit and as well as some underripe fruit, the disease forms a scab on the fruit exocarp, perhaps generated by resistance of the tissues to the disease.[69] Apart from the scabs, the characteristic symptoms are lesions on young, immature fruit that expand, causing the whole fruit to rot. Under humid conditions, pink spore masses can also be seen on the lesion's surface. While younger fruit may drop from the plant, older fruit may remain attached, becoming black and mummified.[66]

Cool and moist weather combined with warm temperatures and higher altitudes favor the development of the disease. As such, preventative measures for and treatment of CBD are similar to those for coffee leaf rust. Preventative measures include pruning to open up the canopy, as well as the planting of resistant cultivars, including those derived from Hibrido de Timor. One promising cultivar is Ruiru 11, a composite of hybrids including Rume Sudan, Hibrido de Timor, K7, SL 28, Sl34, and Bourbon that was developed at the Ruiru washing station in Kenya.[70] Fungicidal treatment, generally using copper-based fungicide, is a chemical measure that can be taken to suppress CBD outbreaks.

2.10.3.8 Other Diseases

Other diseases that occur in coffee cultivation include anthracnose and leaf spot (*Colletotrichum gloeosporioides*), rosellinia or "black root rot" (*Rosellinia* sp.) (Figure 2.26a and b), branch atrophy, and plant yellowing (*Xylella fastidiosa*).

2.10.4 Coffee Plant Nematodes

By damaging the roots of the coffee plant (Figure 2.27a), nematodes can cause symptoms in the aerial parts of plants (Figure 2.27b), including leaf fall, dry branches, and yellowing, which can lead to plant mortality. Symptoms at the

(a) (b)

Figure 2.26 Symptoms of *Rosellinia* sp. in coffee plants. (a) Initial symptoms of attack in plant beside healthy plant. (b) Lesions (black dots) in the plant stalk resulting from fungal attack.

Figure 2.27 Root and aerial detail of coffee plants infested with *Meloidogyne paranaensis* nematodes (Brazil).

root level can include the presence of galls, root thickening, and cracking and peeling, depending on the species and population density of the nematode and the susceptibility of the plant cultivar. When plant symptoms are observed and the species of nematode is identified, it is essential to prevent dissemination to other plots in order to avoid major losses.[71]

Certain preventive measures can be taken: avoid planting in infested areas (and avoid new plantings in areas formerly cultivated with coffee); use seedlings that are free of nematodes; divert water runoff; and sanitize machines and implements after use in infested areas. Ideally cultivation should be initiated in non-infested plots. Chemical control with nematicides is an alternative method for reducing nematode populations, but other techniques can be used, such as crop rotation, soil resting and tilling, destruction of compromised plants, soil supplementation with organic material, use of green fertilizers, as well as the use of resistant cultivars or seedlings grafted to resistant plants.[71]

2.11 Coffee Harvesting: Manual Selective, Manual Stripping, and Mechanical

The harvest is perhaps the most important time of the year, both for coffee growers who are harvesting the fruits of their labor executed throughout the year, as well as for the pickers, for whom the harvest often represents the most profitable time of the year. The cost of removing the fruit from the tree can reach 40% of total farming costs. Because of this, during this part of the year there is a heavy concentration of expenses for growers and consequently better income distribution for rural workers.[16]

In countries or regions where coffee fruit mature unevenly due to the occurrence of various flowerings throughout the year, the harvest must be done selectively, fruit by fruit. In countries or regions with more uniform fruit maturation, such as most coffee-growing areas of Brazil, it is possible to delay the harvest so that it is only performed once, removing all the fruits from the plagiotropic branches.[72]

2.11.1 Manual Selective Harvest

In manual selective harvesting, coffee growers pick only mature fruit each time, normally putting the harvested fruit in baskets that are tied around their waists (Figure 2.28a). Screens are sometimes used instead of baskets when hand-picking coffee, especially in Brazil (Figure 2.28b). This practice is common in countries near the equator, where coffee growers in general must perform eight to nine harvests per year, removing only ripe fruit each time, due to the lack of maturation uniformity. This considerably increases production costs since a large amount of labor is required to perform these harvests. However, in general post-harvest processes that separate unripe, ripe, and overripe fruits are not needed when manual selective harvest is used.[12]

2.11.2 Manual Strip Picking

When it is possible to easily perform just one round of picking because of more uniform maturation, as is the case with most farms in Brazil, growers have the option to strip pick the coffee either manually or mechanically. In this case, since the beginning of the harvest is delayed so that only one round of picking can be performed, some of the fruits from the first flowering will have fallen on the ground before the harvesting operation begins. These fruits as well as other fruits that fall on the ground during harvesting should be collected and processed separately from the other harvested fruits in order to avoid losses in final product quality.

With strip harvesting, growers will often base their decision to commence the harvest on the percentage of unripe fruit still on the tree. For specialty coffees, it is recommended that the percentage of unripe fruit be less than 5%.[73] However, in general, the standard threshold for commencing strip harvesting is less than 20% unripe fruit.

At the beginning of the harvest, coffee growers place a canvas (normally made of polyethylene) on the ground to avoid contact between the newly harvested fruits and both the ground and the fruits that were on the ground prior to the harvest (Figure 2.29). The canvas also serves to collect the fruit. The fruit harvested in this strip harvest can comprise fruits that have yet to reach physiological maturity, ripe fruits, and also fruits that have already dried on the tree. These maturation stages should be separated in post-harvest processing in order to preserve quality and meet the different demands of various consumer markets.

To finalize the harvest, the coffee on the ground must be collected. After this is complete, any material that was removed (leaves, grass clippings, *etc.*) should be returned to under the plant crown to maintain soil fertility.[73]

2.11.3 Mechanized Harvesting

With the development of harvesting machines, higher operational returns have been obtained without negatively impacting job growth. With this mechanization a degree of "selective harvesting" has been achieved that has

Figure 2.28 Selective harvesting of coffee fruit. (a) Coffee picker at Green Land Coffee Plantation, Myanmar. (b) Baskets full of selectively harvested ripe fruit, Shakiso, Ethiopia (photo courtesy of Thompson Owen, Sweet Maria's Coffee). (c) The sorting of the coffee after picking and before delivering to the mill to remove any unripe and overripe fruit as well as foreign material.

Figure 2.29 Preparation for manual strip harvesting by placing a canvas under the plants.

allowed large farms and farms in areas with high labor costs to become more competitive in meeting the various demands of the world coffee market.[74]

2.11.3.1 *Mechanical Harvesting with Portable Harvesters*

In areas where steep slopes inhibit the use of large mechanical harvesters or where adequate qualified labor is not available for manual harvesting, portable hand-held harvesters (Figure 2.30) may be used to facilitate the harvest.[75]

While this type of harvesting is still considered strip picking, the coffee fruit is actually removed from the trees using the hand-held harvesters.

Portable hand-held harvesters are machines with vibrating rods that knock the coffee off the branch. They can be powered pneumatically or motorized. The pneumatic models use an air compressor, powered *via* the tractor or through the harvester's own motor, which vibrates the rods causing the fruits to fall. Motorized harvesters operate on the same principle; however, they are powered directly by portable motors.[31]

2.11.3.2 *Large Tractor-pulled or Self-propelled Mechanical Harvesters*

Harvesting in extensive areas or even smaller individual plots that are worked cooperatively may, given the right conditions, be done with the use of large tractor-pulled or self-propelled mechanical harvesters. Various options are

Figure 2.30 Portable hand-held harvesters.

available, but the ultimate efficiency of any large harvester will depend on adapting the coffee field to it, for example, by limiting the number of turns the harvester must make and increasing the number and length of straight stretches (Figure 2.31a and b).[31]

Silva *et al.*[31] provide a review of harvesting machines that can be used for specific or combined-use activities. Specific-use machines can perform individual operations such as row cleaning, picking, and winnowing. Combined-use machines can pick, gather, and bag the harvested coffee, all in one operation. Examples of other machines used in coffee harvesting are: (a) row cleaners/blowers (equipment used to clean under the trees before the harvest); (b) conveyor sweepers (machines that remove fallen coffee from the ground); (c) winnowers or pneumatic separators (machines that remove impurities such as leaves, twigs, dirt clods, and stones from the recently harvested coffee using screens and air movement). There are several general advantages and disadvantages to mechanized harvesting. Advantages include: lower operational costs; shorter harvesting period, providing more time for the plants to recuperate before the next harvest; more steady flow of coffee from field to post-harvest operations; facilitation of night work; maximizing of the labor force already on the farm; and facilitation of overall harvest management. Disadvantages to mechanized harvesting include: requirement that both the land and the coffee plants meet certain specifications; follow-up manual labor still required for some operations; high initial investment costs, depending on the system used; increased machine maintenance costs; and a change in harvest management systems.

Depending on which of these methods is used, and when the coffee is harvested, the resulting coffee crop can consist of various combinations of ripe,

Figure 2.31 Mechanical harvesters. (a) Tractor-pulled harvester. (b) Self-propelled harvester.

semi-ripe, unripe, and overripe fruits, as well as other debris. For example, stripping, when done too early, produces coffees with a high percentage of unripe fruits, and when done too late results in a large quantity of overripe fruits; in these two cases the resulting coffee will tend to be of inferior quality (Figure 2.32).

There are stark differences in the anatomy, chemical composition, and moisture content of coffee fruits in different stages of maturation. The more homogeneous the harvested lot, the more efficient coffee processing becomes across all post-harvest procedures.

Figure 2.32 Harvested coffee fruit. (a) Strip-harvested coffee of various maturations and the presence of foreign debris. (b) Selectively harvested ripe coffee.

2.12 Coffee Post-harvest Processing

The choice of processing method directly affects the profitability of coffee production and depends on diverse factors such as regional climatic conditions; available capital, technology and equipment; consumer demand for specific quality characteristics; water usage rights; and the availability of technology for treating residual water. The three fundamental considerations in choosing a coffee-processing method are the cost/benefit analysis of the production method, the need to adhere to environmental legislation, and the desired quality standard of the coffee.[16]

Historically, the two methods employed to process coffee are dry processing and wet processing. In dry processing, coffee fruits are processed whole, producing dry fruit pods known as natural coffee. In wet processing, parchment coffee is produced. The dry process is the predominant process used for *C. arabica* seeds in Brazil, Ethiopia, and Yemen, as well as for most *C. canephora* crops worldwide.[76] The wet process is the predominant method for arabica coffees in Colombia, Costa Rica, Guatemala, Mexico, El Salvador, Kenya, and recently a small percentage of Canephora coffees, especially in India and Indonesia.[76–79] Figure 2.33 presents a simplified flowchart of coffee processing.

2.12.1 Winnowing and Coffee Separation

Manual or mechanical winnowing is performed to separate light impurities, such as leaves, sticks, and other debris, from the fruits. Manual winnowing is still employed by small producers, who conduct the winnowing in the fields

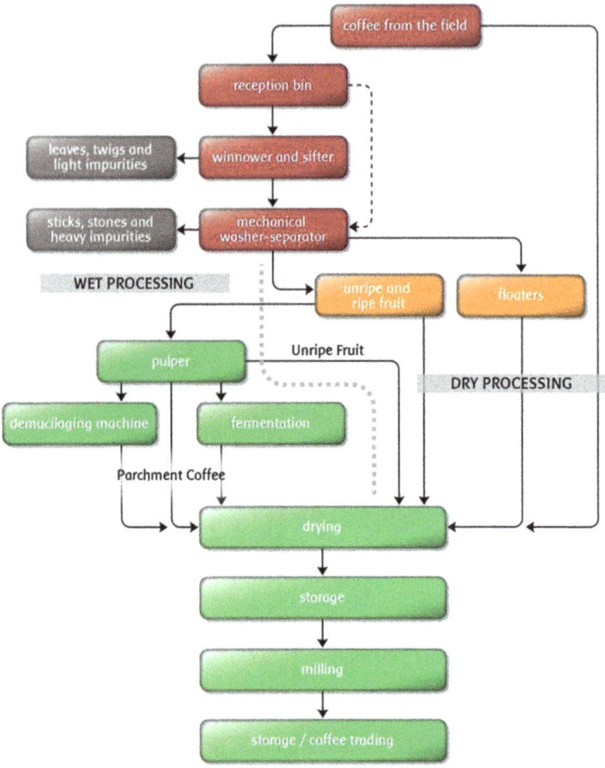

Figure 2.33 Coffee processing flowchart. Reproduced from ref. 16 with permission from Gin Press.

Figure 2.34 Manual winnowing of coffee fruit in the field – Fazenda Recanto, Brazil.

Figure 2.35 Density separation using water.

using screens (Figure 2.34). Mechanical winnowing is done by mobile or stationary machines that move air across the winnowing surface, usually a screen, through either suction or fan-blown air. After impurities are removed through winnowing, the coffee crop is ready for the hydraulic separator. Fruits that have fallen onto the ground both before and during harvesting should never be mixed with other lots since they are generally of inferior quality.

Density separation using water, sometimes called hydraulic separation or mechanical washing and separating, is one of the most important stages in coffee processing and employs flotation to separate the denser unripe and ripe fruits from the less dense fruits, known as floaters (Figure 2.35). It also removes material such as sticks and light impurities that were not removed in previous stages, as well as denser material such as soil and stones. These separators often consist of two water tanks connected at the bottom, a system to move the coffee and recirculate the water, two front exits, and one side exit.[76,80]

In addition to separation by density, floaters can be further sorted by size, using a cylindrical sieve with circular perforations that is placed just after the hydraulic separators. Growers now also have the option of electronic color separation to sort out unripe coffee fruits without using water or pulping equipment.

After sifting, hydraulic separation, and size separation, the coffee is then dried or pulped, depending on the processing method chosen by the producer.

2.12.2 The Dry Process Method – Natural Coffee

The production of natural coffee, traditionally known as the dry method or dry process but sometimes referred to as sun-dried or unwashed coffee, is the oldest and simplest coffee-processing method. It entails drying the entire coffee fruit intact (Figure 2.36) and is largely used in tropical regions where the dry season coincides with the harvest period.[81]

The history of the dry processing method can be divided into three stages. The first stage began with the initial establishment of coffee as a crop and

Figure 2.36 Dried fruit pods that are the result of the dry process, prior to dry milling to remove the seed, or bean.

lasted until the proliferation of the wet process in the 19th century. With the advent of the wet process, a second stage began that was characterized by the displacement of natural coffees by washed coffees. As the wet process became the norm in most producing countries, natural coffees were largely marginalized. The dry process was used mainly by growers who could not afford drying technologies or did not perform selective harvesting, as well as for fruits that were the by-product of the wet method (mostly unripe, overripe, or hollow fruits that could not be pulped). Because of this, natural coffees were largely viewed as an inferior product to washed coffees and the lower prices generally paid for natural coffees did little to incentivize quality production. With the growth of the specialty coffee industry in the late 20th and early 21st centuries, a third stage emerged defined by renewed interest in the dry processing method. The specialty coffee movement brought a demand for high-quality coffees with unique flavor profiles, as well as increased espresso consumption (the blends of which natural coffees often comprise a large part). It also brought the premiums paid for specialty coffees, allowing for higher quality control of natural coffees and an increased interest from traditional wet process growers looking to expand their offerings. Natural coffees are now produced not only in countries that traditionally produced them but also in countries throughout Central America, South America, and other regions that traditionally produced only washed coffees.[82]

Traditional literature defines the dry method as the drying of all coffee fruit immediately following the harvest[78] with no lot separation based on maturation or coffee quality. While this is the most common way to perform the dry process, it is just one of the many processing options available and is generally the option chosen by producers with inadequate coffee processing infrastructure. In fact, all coffee, whether composed of ripe, unripe, overripe, dried coffee, or any combination thereof, is considered to be natural coffee if it was dried with its pericarp intact.[16]

Figure 2.37 Dry process coffee being dried in a thin one-fruit-high layer at the onset of drying.

The final quality of a dry process coffee depends on many factors, including harvest method and care taken during processing and drying. When drying natural coffees on a drying patio, it is generally recommended (in the presence of abundant sunlight, low relative humidity, and good ventilation) to begin the drying by spreading the coffee out in a single fruit-high layer of around 14 L m^{-2} (Figure 2.37). This allows for a drastic reduction in moisture content, decreasing the risk of undesirable fermentation. Movement on and of the coffee should be avoided while the coffee is wet since even minimal pressure can rupture the exocarp, leading to non-uniform drying. After the fruit has withered, the coffee should be continually rotated (12 times a day) to guarantee uniform drying. Depending on the temperature and humidity, to ensure the coffee does not dry too quickly and/or to economize patio space, the thickness of the layer can gradually be increased up to a maximum of 5 cm for ripe coffee and 10 cm for floaters. Once the coffee has reached half dry (around a moisture content of 30% wet basis – wb), the coffee should be mounded or put into thick rows every afternoon and covered. This conserves and distributes the heat absorbed throughout the day, increasing uniformity and providing for better redistribution of moisture throughout the coffee mass. The next morning the coffee should be uncovered and the rotating recommenced. This process should be repeated until the coffee reaches 11% (wb) moisture content, the ideal level for coffee storage.

Mechanical dryers, generally rotary dryers, can be used to dry natural coffees. If the mechanical dryer will also be used as a pre-dryer (taking the coffee from its initial moisture content to half-dry) it is recommended that no heat be applied to the coffee mass for the first 30 minutes. After this time, heat can be applied, however before the coffee reaches half-dry, the coffee mass temperature should not exceed 30 °C. Once half-dry is reached, the coffee mass temperature may be increased to 45 °C for commercial grade naturals but no more than 40 °C for the production of higher quality specialty coffees.[83]

The lower quality often seen in natural coffees can be explained by two main factors: a lack of care during the harvest resulting in fruits of various qualities and maturations, and a higher risk of undesirable fermentation due to the elevated levels of sugar in the mucilage as well as the slower drying times.[84] When only ripe fruits are selectively harvested and then carefully dried, it is possible to produce high-quality dry process coffees. In general, quality natural coffees are considered to be sweeter and fuller-bodied coffees with flavors ranging from chocolatey and nutty to fruity and are greatly appreciated in espresso preparation.[81,82,84]

2.12.3 The Wet Processing Method

The first known use of the wet processing method was in 1730, in the islands currently known as Indonesia.[85] The wet processing method was developed in equatorial regions with continual precipitation during the harvest period, a condition not appropriate for dry processing. In these regions, the dry process would almost always result in coffee of inferior quality. The wet process will generally yield good quality coffee if only ripe fruit are harvested, if the skin and mucilage are properly removed, if fermentation is controlled, and if the coffee is carefully dried. Historically this method is associated with higher quality coffee and it is often used when the goal is the production of specialty coffees. This method is generally associated with selective harvesting for the production of arabica coffee, with the exception of Brazil, Hawaii, and Australia, as well as for robustas in several countries.[76] Today, the wet process is generally carried out in three distinct ways:

a) Fully washed coffees are wet processed coffees in which the pulp – the fruit skin (exocarp) and part of the mucilage (mesocarp) – is removed mechanically and then the remaining mucilage that adheres to the parchment (endocarp) and which is insoluble in water is removed through controlled fermentation and subsequent washing. This fermentation process can be completed by simply leaving the coffee in the fermentation tank by itself (dry fermentation), by soaking the coffee in water (wet fermentation) or through mixed fermentation. Schwan and Fleet[86] present a review of coffee fermentation. The duration of the fermentation process generally lasts between 12 and 36 hours, but will vary based on factors such as temperature, type of fermentation, maturation level of the coffee lot, height of the coffee layer, and coffee cultivar, among others. It is important to note that the wastewater from this process will have elevated levels of biochemical oxygen demand (BOD) and chemical oxygen demand (COD) and must be treated accordingly.[87] The remaining mucilage after the fermentation process can be removed by lightly scrubbing the coffee. Historically this was performed in channels that followed the fermentation tanks,

Figure 2.38 Parchment coffee that is the result of the wet process, prior to dry milling to remove the seed, or bean.

Figure 2.39 Pulped natural, honey process coffee (photo Courtesy of Thompson Owen, Sweet Maria's Coffee).

however in recent years it is becoming more common to pass the coffee through a demucilage machine. The resulting clean parchment coffee is then dried (Figure 2.38). Fully washed coffees are the most common wet processing method and their flavor is generally considered to be cleaner with a pleasant aroma, higher perceived acidity, and less body than dry process coffees.[78,88–90]

b) Pulped natural coffees, also referred to as semi-dry or honey(ed) coffees (Figure 2.39), are wet processed coffees in which, as with fully washed coffees, the fruit skin and part of the mucilage are removed mechanically. However, the remaining mucilage is not removed but rather is dried intact with the parchment coffee. Commonly used in Brazil since the 1990s, other countries have recently adopted this method. Pulped natural coffees can be produced in various ways, and many growers experiment with different variations in order to positively alter the sensorial qualities of the resulting beverage, in particular by varying the amount of mucilage left on the parchment, covering the parchment coffee, or putting it in closed bags during the drying process, and altering the thickness of the drying layer. Some of these treatments alter the color of the resulting parchment coffee, and are oftentimes referred to as yellow honey, red honey, and black honey coffees accordingly. While the pulped natural method is considered here as part of the wet process, it is sometimes considered to be a separate category, separate from both the wet and dry methods. Although the term semi-washed may sometimes be applied to pulped natural coffees, coffees processed in this way do not normally go through a "washing" process to remove the mucilage. Furthermore, there has been a more consistent use of the term "semi-dry" in the scientific community.[91–93] The flavor of pulped natural coffees is often considered to be an intermediate profile between fully washed and dry process coffees, having a "cleaner" flavor than standard naturals, but more body than most fully washed coffees.[76,89]

c) Semi-washed coffees are wet process coffees in which the skin and all of the mucilage are removed mechanically. They are sometimes referred to as demucilaged or mechanically demucilaged coffees.[68] Advantages to this process are the decreased amount of wastewater generated during processing and the ease of raking and rotating the parchment compared to the pulped natural method, in which the coffee clumps together. There is not a consensus as to the effects of this process on the flavor compared to traditional fermentation, and studies have shown mixed results.[91,94] However, with increasing limitations of water availability and usage as well as wastewater disposal, this method is becoming more common.

Parchment coffee resulting from the wet process should initially be spread out in one-bean-high layers of around 7 L m^{-2}, permitting rapid removal of the water from the parchment surface as well as dehydration of any remaining mucilage. The coffee should not be covered during the first night, rather the covering should start on the second night to avoid exposure to fog and dew. The parchment coffee should be kept in thin layers and covered at night until it reaches half-dry, which for parchment coffee is around 25% (wb) moisture content. At this point the coffee should then be piled and covered at night, while during the day the coffee should be kept in thicker layers, rotating it frequently. Once the endosperm detaches from the parchment, layer thickness should increase every day during drying. Tools used to rotate the coffee should be flat and light to avoid cracking the parchment as this causes evaporation rates to be higher and may also damage the endosperm, which in turn causes the bean to whiten. As with natural coffees, when drying

Figure 2.40 Wet-hulled coffee (photo courtesy of Thompson Owen, Sweet Maria's Coffee).

parchment coffee in mechanical dryers, the temperature of the coffee mass should not exceed 40 °C.[84]

2.12.4 The Wet-hulled Method

The wet-hulled method, called the *Giling Basah* method in Indonesia where it is almost exclusively employed, consists of hulling the coffee when the moisture content is still high, generally between 25 and 35% (wb). It is customary to complete this process in two steps. As with the wet method, the fruit is initially pulped after harvesting. However, after a short period of time (one to two days) in which the parchment coffee is either set to dry or soaked in buckets to remove some of the mucilage, the coffee is hulled and the naked beans are dried to completion. While there is little research and no consensus as to its impact on the flavor profile of the coffee, this process results in a deep bluish-green colored coffee (Figure 2.40).

Recently, research has been conducted at the Federal University of Lavras using a similar method of hulling the coffee with a high moisture content. Initial results are promising, showing decreased drying times without compromising the quality of the coffee.[95]

2.12.5 Animal Processing

Another type of processing that is not normally considered as such is bio-processed coffee, or coffee in which an animal consumes the coffee fruit, and the pulping of the fruit as well as the mucilage removal occurs *via* the animal's digestion process. The parchment coffee is expelled with the feces of the animal, where it is then collected and hulled. Various animal-processed coffees

Figure 2.41 Feces of the jacu bird, containing parchment coffee (photo courtesy of Paula Magalhães, Fazenda Recanto, Brazil).

are offered on the consumer market, most famously from the palm civet of Indonesia, also known as the kopi luwak, but also from the jacu bird of Brazil (Figure 2.41), Thai elephants, and Indian Rhesus monkeys (which, unlike the others mentioned here, consume the fruit but spit out the seeds).

While these are often mentioned as exotic or novelty coffees, fetching high prices, this process can also be viewed in the context of the natural evolution and dissemination of the coffee plant. There is significant documentation and wide acceptance of the role of frugivorous animals in the dissemination of plant species. As coffee fruit is indehiscent, its dispersion is not through air movement, but rather through the active transport of its seed by other organisms. Its development of a colorful, succulent fruit that is attractive to frugivorous animals likely facilitated its initial dispersion, long before human cultivation.

The limited research of these coffees has largely looked at ways to confirm their authenticity.[96–98] One study concluded that coffee processed by civet cats was different from control coffee in several ways, including: the presence of micropitting on the surface of the beans; the beans were harder and more brittle in nature, indicating that digestive juices were entering into the beans and modifying their micro-structural properties; proteolytic enzymes were penetrating into all the beans, causing substantial breakdown of storage protein. This study also concluded that coffees processed by civet cats in both Indonesia and Ethiopia were discernible from control coffees, both in using an electronic nose and a trained coffee cupper, with a noted decrease in both body and acidity in the coffees processed by the civet cats.[97]

It should be noted that while some of these coffees are noted for their high prices, recently the treatment of the animals in producing these coffees has become a point of contention in their production.

2.13 Dry Milling

With the exception of wet-hulled coffee, coffee from all of the processing methods mentioned above must go through at least one additional step before it can be commercialized: the remaining pericarp must be removed, leaving only the bean. In the case of dry processed coffees, the entire pericarp must be removed, while with wet processed coffees the endocarp must be removed (along with any mucilage that still adheres to it). Dry milling may be done at any time once the coffee is dry, however, it is recommended that the growers let the coffee rest, or cure, for at least one month before initiating the process.[16] The dry milling is a two-part process: first the coffee passes through a mill that initiates the hulling, then, using differences in size and density, the bean and pericarp remnants are further separated. Depending on the harvest method used (selective or strip) and the quality of coffee desired, the coffee may undergo further steps to remove defects and separate the coffee by size (see Chapter 6, Coffee Grading and Marketing).

2.14 Defects

The ultimate value of a coffee in relation to the market is based on many factors, but largely on its physical and sensorial evaluation. The presence of physical defects (Figure 2.42a–e) can negatively affect both of these evaluations. While proper care taken in the field, mill, and drying patio does not guarantee the absence of defects, by following the practices recommended above, it does diminish the likelihood of their occurrence, and thus increase the chance of producing a higher quality product.

Some common defects are:[99–101]

(a) Black Bean: a bean that has turned a dull black color. This defect can be caused by several factors, and can either originate in the field for various reasons including poor plant physiology (due to weather or poor crop management) and fermentation; or in the post-harvest through acute fermentation, poor drying, or re-wetting of the coffee.
(b) Sour Bean: a coffee bean that has experienced excess fermentation, turning it light to dark brown. This fermentation often occurs in beans if the fruits fall to the ground prior to harvest, but can also occur on the tree and/or in the post-harvest in the presence of excess heat and/or prolonged drying times.
(c) Unripe or Immature Bean: a bean from an unripe coffee fruit. These beans are characterized by a strong adherence of the silverskin to the endosperm, as well as its shiny green color.

(a)
(b)
(c)
(d)
(e)

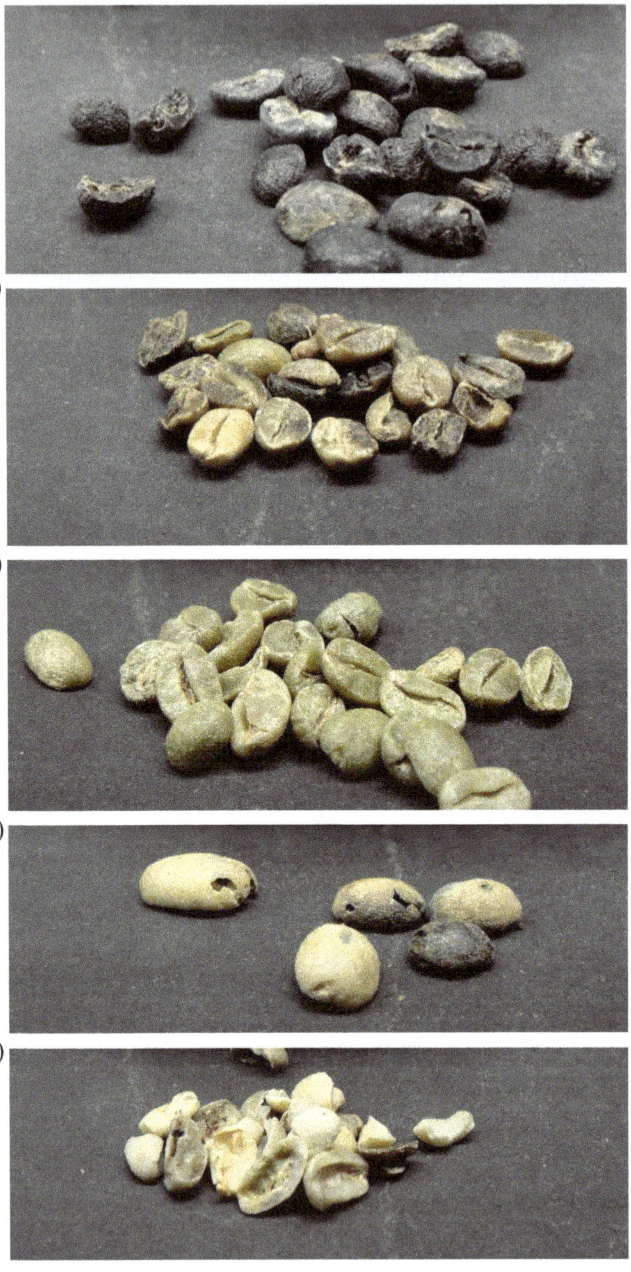

Figure 2.42 Common defects: (a) black, (b) sour, (c) immature, (d) insect-damaged and (e) broken/chipped.

(d) Insect-damaged Bean: a bean that was attacked by insects. Apart from the physical damage, the attack can facilitate further damage to the bean by opportunistic pathogens and fungi, as well as cause the fruit to fall on the ground, where it is subject to further deterioration.

(e) Broken or Chipped Bean: a bean that is the result of mechanical damage from milling or bean transport in the post-harvest. It should be noted that coffee dried to under 11% is more brittle and thus more susceptible to this damage.

(f) Non-coffee Defect: any non-coffee matter that was not removed in the post-harvest due to improper cleaning of the green coffee such as sorting, sieving, and densimetric separation.

References

1. R. J. Guimarães, A. N. G. Mendes and C. A. S. Souza, in *Cafeicultura*, ed. R. J. Guimarães, A. N. G. Mendes and C. A. S. Souza, UFLA/FAEPE, Lavras, 1st edn, 2002, pp. 139–159.
2. E. M. da Silva, J. C. de Rezende, Â. M. Nogueira and G. R. Carvalho, in *Café Arábica do Plantio à Colheita*, ed. P. R. Reis and R. L. Cunha, Epamig, Lavras, 1st edn, 2010, pp. 223–282.
3. A. Carvalho, *O Agronômico*, 1985, **37**, 7–11.
4. R. G. Ferrão, A. F. A. da Fonseca, M. A. G. Ferrão, S. M. Bragança, A. C. Verdin Filho and P. S. Volpi, in *Café Conilon*, ed. R. G. Ferrão, A. F. A. da Fonseca, S. M. Bragança, M. A. G. Ferrão and L. H. De Muner, Incaper, Vitoria, 1st edn, 2007, pp. 198–221.
5. A. A. Pereira, G. R. Carvalho, W. de M. Moura, C. E. Botelho, J. C. de Rezende, A. C. B. de Oliveira and F. L. da Silva, in *Café Arábica do Plantio à Colheita Volume 1*, ed. P. R. Reis and R. L. da Cunha, Epamig, Lavras, 1st edn, 2010, pp. 163–222.
6. R. J. Guimarães, A. N. G. Mendes and D. P. Baliza, *Semiologia do Cafeeiro: Sintomas de Desordens Nutricionais, Fitossanitárias e Fisiológicas*, Editora UFLA, Lavras, 1st edn, 2010.
7. *Café Arábica do Plantio à Colheita*, ed. P. R. Reis and R. L. da Cunha, Epamig, Lavras, 1st edn, 2010, vol. 1.
8. R. J. Guimarães, A. C. Fraga, A. N. G. Mendes, M. L. M. Carvalho, M. Pasqual and G. R. Carvalho, *Ciênc. Agrotecnol.*, 1998, **22**, 390–396.
9. D. E. Livramento, in *Café Arábica do Plantio à Colheita*, ed. P. R. Reis and R. L. Da Cunha, Epamig, Lavras, 1st edn, 2010, vol. 1, pp. 87–162.
10. A. B. Rena and M. R. Maestri, in *Cultura do Cafeeiro, Fatores Que Afetam a Produtividade*, ed. A. B. Rena, E. Malavolta, M. Rocha and T. Yamada, Associação Brasileira para a Pesquisa e Potassa Fosfato, Piracicaba, 1st edn, 1986, pp. 13–85.
11. J. K. Muriuki, A. W. Kuria, C. W. Muthuri, A. Mukuralinda, A. J. Simons and R. H. Jamnadass, *Small-Scale For.*, 2014, **13**, 127–142.

12. *Guia Técnica de Caficultura*, Anacafé, Guatemala City, 2014th edn, 2014.
13. *Coffee Cultivation Guide for South-West Monsoon Area Growers in India (Coffee Kaipidi)*, Central Coffee Research Institute, Karnataka, 1st edn, 2008.
14. P. T. G. Guimarães, A. W. R. Garcia, V. V. H. Alvarez, L. C. Prezotti, A. S. Viana, A. E. Miguel, E. Malavolta, J. B. Corrêa, A. S. Lopes, F. D. Nogueira and A. V. C. Monteiro, in *Recomendações para o Uso de Corretivos e Fertilizantes em Minas Gerais 5a Aproximação*, ed. A. C. Ribeiro, P. T. Guimarães and V. H. Alvarez, CFSEMG, Viçosa, 5th edn, 1999, pp. 289–302.
15. H. P. Medina Filho, R. Bordignon and C. H. S. de Carvalho, in *Cultivares de Café: Origem, Características e Recomendações*, ed. C. H. S. de Carvalho, Embrapa Café, Brasilia, 1st edn, 2008, pp. 79–103.
16. *Handbook of Coffee Post-Harvest Technology*, ed. F. M. Borém, Gin Press, Norcross, 1st edn, 2014.
17. A. F. A. Fonseca, R. G. Ferrao, M. A. G. Ferrao, A. C. Verdin Filho, P. Volpi and M. L. C. Bittencourt, in *Café Conilon*, ed. R. G. Ferrão, A. F. A. da Fonseca, S. M. Bragança, M. A. G. Ferrão and L. H. De Muner, Incaper, Vitoria, 1st edn, 2007, pp. 222–252.
18. H. Etienne, B. Bertrand, E. Dechamp, P. Maurel, F. Georget, R. Guyot and J.-C. Breitler, *Hum. Genet. Embryol.*, 2016, **6**, 136.
19. E. T. Caixeta, C. H. S. de Carvalho, E. M. Zambolim, L. F. P. Pereira and N. S. Sakiyama, in *Cultivares de Café: Origem, Características e Recomendações*, ed. H. S. de Carvalho, Embrapa Café, Brasilia, 1st edn, 2008, pp. 102–128.
20. *Cultivares de Café: Origem, Características e Recomendações*, ed. H. S. de Carvalho, Embrapa Café, Brasilia, 1st edn, 2008.
21. L. C. Paiva, R. J. Guimarães, C. Alberto and S. Souza, *Ciênc. Agrotecnol.*, 2003, **27**, 134–140.
22. A. B. Rena, A. P. Nacif, R. T. G. Guimarães and A. A. Pereira, *Inf. Agropecu.*, 1998, **19**, 71–80.
23. M. M. Carvalho and L. A. Caldani, *Ciênc. Prát.*, 1984, **8**, 25–31.
24. A. Mendonça De Carvalho, R. J. Guimarães, C. Augusto De Moura, A. Nazareno, G. Mendes and G. Rodrigues De Carvalho, *Coffee Sci.*, 2007, **2**, 79–86.
25. D. P. Baliza, A. L. de Oliveira, R. A. Almeida Dias, R. J. Guimarães and C. R. Barbosa, *Coffee Sci.*, 2013, **8**, 61–68.
26. C. E. Botelho, J. C. de Rezende, G. R. Carvalho, P. de T. G. Guimarães, A. de P. Alvarenga and M. de F. Ribeiro, in *Café Arábica do Plantio à Colheita*, ed. P. R. Reis and R. L. da Cunha, Epamig, Lavras, 1st edn, 2010, pp. 283–342.
27. R. C. Taques and G. G. Dadalto, in *Café Conilon*, ed. R. G. Ferrão, A. F. A. da Fonseca, S. M. Bragança, M. A. G. Ferrão and L. H. De Muner, Incaper, Vitoria, 1st edn, 2007, pp. 51–63.
28. J. Avelino, B. Barboza, J. C. Araya, C. Fonseca, F. Davrieux, B. Guyot and C. Cilas, *J. Sci. Food Agric.*, 2005, **85**, 1869–1876.
29. B. Bertrand, P. Vaast, E. Alpizar, H. Etienne, F. Davrieux and P. Charmetant, *Tree Physiol.*, 2006, **26**, 1239–1248.

30. B. Bertrand, R. Boulanger, S. Dussert, F. Ribeyre, L. Berthiot, F. Descroix and T. Joët, *Food Chem.*, 2012, **135**, 2575–2583.
31. F. M. Silva, G. R. Carvalho and N. Salvador, *Inf. Agropecu.*, 1997, **18**, 43–54.
32. R. J. Guimarães, A. N. G. Mendes and C. A. S. de Souza, in *Cafeicultura*, ed. R. J. Guimarães, A. N. G. Mendes and C. A. S. Souza, UFLA/FAEPE, Lavras, 1st edn, 2002, pp. 124–138.
33. A. N. G. Mendes, R. J. Guimarães and C. A. S. Souza, in *Cafeicultura*, ed. R. J. Guimarães, A. N. G. Mendes and C. A. S. Souza, UFLA/FAEPE, Lavras, 1st edn, 2002, pp. 160–181.
34. M. R. Vicente, E. C. Mantovani, A. L. T. Fernandes, G. H. S. Vieira, G. C. Sediyama and E. M. Figueredo, *Coffee Sci.*, 2011, **6**, 147–158.
35. E. Lopes Serra, M. S. Scalco, R. J. Guimarães, A. Colombo, A. Ramalho De Morais and C. H. Mesquita De Carvalho, *Coffee Sci.*, 2013, **8**, 157–165.
36. E. Malavolta, *Nutrição Mineral e Adubação do Cafeeiro*, Agronômica Ceres, São Paulo, 1st edn, 1993.
37. V. C. de A. Theodoro, R. J. Guimarães and A. N. G. Mendes, *Coffee Sci.*, 2014, **9**, 300–311.
38. W. P. de Carvalho, G. J. de Carvalho, D. de Oliveira Abbade Neto and L. G. V. Teixeira, *Pesqui. Agropecu. Bras.*, 2013, **48**, 157–166.
39. R. H. da S. Siqueira, M. M. Ferreira, E. N. de Alcântara, R. C. da S. Carvalho and R. C. Silva, *Rev. Bras. Ciênc. Solo*, 2014, **38**, 1128–1134.
40. R. H. da S. Siqueira, M. M. Ferreira, E. N. de Alcântara, B. M. Silva and R. C. Silva, *Ciênc. Agrotecnol.*, 2014, **38**, 471–479.
41. R. Melloni, G. Belleze, A. M. S. Pinto, L. B. de P. Dias, E. M. Silve, E. G. P. Melloni, M. I. N. Alvarenga and E. N. de Alcântara, *Rev. Bras. Ciênc. Solo*, 2013, **37**, 66–75.
42. C. F. Araujo-Junior, M. de S. Dias-Junior, P. T. G. Guimarães and E. N. Alcântara, *Rev. Bras. Ciênc. Solo*, 2011, **35**, 115–131.
43. C. F. Araujo-Junior, M. de S. Dias-Junior, P. T. G. Guimarães and E. N. Alcântara, *Planta Daninha*, 2011, **29**, 499–513.
44. R. J. Guimarães, A. N. G. Mendes and C. A. S. de Souza, in *Cafeicultura*, ed. R. J. Guimarães, A. N. G. Mendes and C. A. S. de Souza, UFLA/FAEPE, Lavras, 1st edn, 2002, pp. 235–246.
45. E. N. de Alcântara and R. A. Silva, in *Café Arábica do Plantio à Colheita*, ed. P. R. Reis and R. L. da Cunha, Epamig, Lavras, 1st edn, 2010, pp. 519–572.
46. R. J. Guimarães, A. N. G. Mendes and C. A. S. de Souza, in *Cafeicultura*, ed. R. J. Guimarães, A. N. G. Mendes and C. A. S. de Souza, UFLA/FAEPE, Lavras, 1st edn, 2002, pp. 247–257.
47. F. M. DaMatta, *Field Crops Res.*, 2004, **86**, 99–114.
48. I. A. C. Gomes, E. M. de Castro, A. M. Soares, J. D. Alves, M. I. N. Alvarenga, E. Alves, J. P. R. A. D. Barbosa and D. D. Fries, *Ciênc. Rural*, 2008, **38**, 109–115.
49. R. Van Kanten and P. Vaast, *Agroforestry Syst.*, 2006, **67**, 187–202.
50. A. E. Lyngbaek, R. G. Muschler and F. L. Sinclair, *Agroforestry Syst.*, 2001, **53**, 205–213.

51. C. Jaramillo-Botero, E. H. P. Martinez and R. H. S. Santos, *Coffee Sci.*, 2006, **1**, 94–102.
52. C. R. M. de Oliveira, A. M. Soares, L. E. M. de Oliveira, E. M. de Castro and J. P. R. A. D. Barbosa, *Ciênc. Agrotecnol.*, 2004, **28**, 350–357.
53. M. M. Campanha, R. H. S. Santos, G. B. De Freitas, H. E. P. Martinez, S. L. R. Garcia and F. L. Finger, *Agroforestry Syst.*, 2004, **63**, 75–82.
54. R. L. da Cunha, M. de F. Ribeiro, V. L. de Carvalho and D. E. Livramento, in *Café Arábica do Plantio à Colheita*, ed. P. R. Reis and R. L. de Cunha, Epamig, Lavras, 1st edn, 2010, pp. 415–446.
55. J. B. Matiello, *Sistemas de produção na cafeicultura moderna*, MAARA/ PROCAFE, Rio de Janeiro, 1st edn, 1995.
56. A. F. Fonsenca, R. G. Ferrao, J. A. Lani, M. A. G. Ferrao, P. S. Volpi, A. C. Verdin Filho, C. P. Ronchi and A. Guarçoni Martins, in *Café Conilon*, ed. R. G. Ferrao, A. F. da Fonseca, S. M. Bragança, M. A. G. Ferrão and L. De Muner, Incaper, Vitoria, 1st edn, 2007, pp. 253–276.
57. T. J. Crowe, in *Coffee: Growing, Processing, Sustainable Production*, ed. J. N. Wintgens, Wiley-Vch, Weinheim, 2nd edn, 2009, pp. 425–462.
58. T. J. Crowe and D. J. Granthead, *East Afr. Agric. For. J.*, 1970, **35**, 364–371.
59. R. A. Silva, J. C. de Souza, P. R. Reis and L. V. C. Santa-Cecília, in *Semiologia do Cafeeiro: Sintomas de Desordens Nutricionais, Fitossanitárias e Fisiológicas*, ed. R. J. Guimarães, A. N. G. Mendes and D. P. Baliza, Editora UFLA, Lavras, 1st edn, 2010, pp. 107–142.
60. C. H. S. de Carvalho, L. C. Fazuoli, G. R. Carvalho, O. Guerreiro-Filho, A. A. Pereira, S. R. de Almeida, J. B. Matiello, G. F. Bartholo, T. Sera, W. de M. Moura, A. N. G. Mendes, J. C. de Rezende, A. F. A. de Fonseca, M. A. G. Ferrão, R. G. Ferrão, A. de P. Nacif, M. B. Silvarolla and M. T. Braghini, in *Cultivares do Café: Origem, Características e Recomendações*, ed. C. H. S. de Carvalho, Embrapa Café, Brasilia, 1st edn, 2008, pp. 157–226.
61. *Compendium of Coffee Diseases and Pests*, ed. A. L. Gaitán, M. A. Cristancho, B. L. C. Caicedo, C. A. Rivillas and G. C. Gómez, APS Press, St. Paul, 1st edn, 2015.
62. E. Burbano, M. Wright, D. E. Bright and F. E. Vega, *J. Insect Sci.*, 2011, **11**, 1–3.
63. J. D. Alves and R. J. Guimarães, in *Semiologia do Cafeeiro: Sintomas de Desordens Nutricionais, Fitossanitárias e Fisiológicas*, ed. R. J. Guimarães, A. N. G. Mendes and D. P. Baliza, Editora UFLA, Lavras, 1st edn, 2010, pp. 169–215.
64. E. A. Pozza, V. L. de Carvalho and S. M. Chalfoun, in *Semiologia do Cafeeiro: Sintomas de Desordens Nutricionais, Fitossanitárias e Fisiológicas*, ed. R. J. Guimarães, A. N. G. Mendes and D. P. Baliza, Editora UFLA, Lavras, 1st edn, 2010, pp. 67–106.
65. D. M. Okioga, *East Afr. Agric. For. J.*, 1976, **42**, 191–197.
66. J. M. Waller, M. Bigger and R. J. Hillocks, *Coffee Pests, Disease, and Their Management*, CAB International, Oxfordshire, 1st edn, 2007.

67. *Phoma Identification Manual. Differentiation of Specific and Infra-specific Taxa in Culture*, ed. G. H. Boerema, J. de Gruyter, M. E. Noordeloos and M. E. C. Hamers, CAB International, Oxfordshire, 1st edn, 2004.
68. J. M. Waller, P. D. Bridge, R. Black and G. Hakiza, *Mycol. Res.*, 1993, **97**, 989–994.
69. R. A. Muller, D. Berry, J. Avelino and D. Bieysse, in *Coffee: Growing, Processing and Sustainable Production*, ed. J. N. Wintgens, Wiley-Vch GmbH & Co., Weinheim, 2nd edn, 2009, pp. 495–549.
70. C. O. Omondi, P. O. Ayiecho, A. W. Mwang'ombe and H. Hindorf, *Euphytica*, 2001, **121**, 19–24.
71. S. M. L. Salgado and V. P. Campos, in *Semiologia do Cafeeiro: Sintomas de Desordens Nutricionais, Fitossanitárias e Fisiológicas*, ed. R. J. Guimarães, A. N. G. Mendes and D. P. Baliza, Editora UFLA, Lavras, 1st edn, 2010, pp. 143–168.
72. F. C. da Silva, F. da S. Moreira, A. C. da Silva, M. M. de Barros and M. A. Z. Palma, *Coffee Sci.*, 2013, **8**, 53–60.
73. M. R. Malta and S. J. de R. Chagas, in *Café Arábica do Plantio à Colheita*, ed. P. R. Reis and R. L. da Cunha, Epamig, Lavras, 1st edn, 2010, vol. 1, pp. 805–860.
74. F. Santinato, R. P. da Silva, M. T. Cassia and R. Santinato, *Coffee Sci.*, 2014, **9**, 495–505.
75. G. A. e S. Ferraz, F. C. da Silva, R. A. Nunes and P. F. Ponciano, *Coffee Sci.*, 2012, **8**, 276–283.
76. C. H. J. Brando, in *Coffee: Growing, Processing, Sustainable Production*, ed. J. N. Wintgens, Wiley-VCH, Weinheim, 2nd edn, 2009, pp. 610–723.
77. G. I. Puerta-Quintero, *Cenicafé*, 1996, **47**, 85–90.
78. J. C. Vincent, in *Coffee: Volume 2: Technology*, ed. R. J. Clarke and R. Macrae, Springer Netherlands, Dordrecht, 1987, p. 1.
79. R. Wilbaux, *Coffee Processing*, FAO, Rome, 1963.
80. F. M. Borém, in *Pós-Colheita do Café*, ed. F. M. Borém, Editora UFLA, Lavras, 1st edn, 2008, pp. 127–158.
81. A. A. Teixiera, C. H. J. Brando, R. A. Thomaziello and R. Teixeira, in *Espresso Coffee: The Science of Quality*, ed. A. Illy and R. Viani, Elsevier Academic Press, San Diego, 2nd edn, 2005, pp. 91–96.
82. M. R. Fernandez Alduenda, Effect of Processing on the Flavour Character of Arabica Natural Coffee, PhD Thesis, University of Otago, 2015.
83. F. M. Borém, E. R. Marques and E. Alves, *Biosyst. Eng.*, 2008, 62–66.
84. F. M. Borém, C. H. R. Reinato and É. P. Isquierdo, in *Handbook of Coffee Post-Harvest Technology*, ed. F. M. Borém, Gin Press, Norcross, 1st edn, 2014, pp. 97–118.
85. A. Tosello, Colheita, preparo por via seca e armazenamento de café, in *Curso de Cafeicultura,* Instituto Agronômico, São Paulo, 1957, pp. 247–257.
86. *Cocoa and Coffee Fermentations*, ed. R. F. Schwan and G. H. Fleet, CRC Press, Boca Raton, 1st edn, 2014.

87. A. de Matos, R. Fia, F. A. R. Fia and P. A. V. Lo Manaco, in *Handbook of Coffee Post-Harvest Technology*, ed. F. M. Borém, Gin Press, Norcross, 1st edn, 2014, pp. 49–68.
88. D. Selmar, M. Kleinwächter and G. Bytof, in *Cocoa and Coffee Fermentations*, ed. R. F. Schwan and G. H. Fleet, CRC Press, Boca Raton, 1st edn, 2015, pp. 431–476.
89. A. A. Teixeira, C. H. J. Brando, R. A. Thomaziello and R. Teixeira, in *Espresso Coffee: The Science of Quality*, ed. A. Illy and R. Viani, Elsevier Academic Press, San Diego, 2nd edn, 2005, pp. 96–101.
90. R. Clarke, J. Jackson, J. Franck, D. Duris, J. Sherman, E. Vargas, G. Van der Stegen, G. P. Quintero and J. de Souza, *Good Hygiene Practices Along the Coffee Chain*, Rome, 2006.
91. C. H. J. Brando and M. F. P. Brando, in *Coffee and Coffee Fermentations*, ed. R. Schwan and G. H. Fleet, CRC Press, Boca Raton, 1st edn, 2015, pp. 367–396.
92. F. M. Borém, É. P. Isquierdo and J. H. da S. Taveira, in *Handbook of Coffee Post-Harvest Technology*, ed. F. M. Borém, Gin Press, Norcross, 1st edn, 2014, pp. 49–68.
93. G. S. Duarte, A. A. Pereira and A. Farah, *Food Chem.*, 2010, 851–855.
94. O. Gonzalez-Rios, M. L. Suarez-Quiroz, R. Boulanger, M. Barel, B. Guyot, J. P. Guiraud and S. Schorr-Galindo, *J. Food Compos. Anal.*, 2007, **20**, 289–296.
95. V. C. Siqueira, F. M. Borem, E. P. Isquierdo, G. E. Alves, D. E. Ribeiro, A. Celso, F. Pinto, J. H. Da and S. Taveira, *Afr. J. Agric. Res.*, 2016, **11**, 2903–2911.
96. U. Jumhawan, S. P. Putri, Y. Yusianto, E. Marwanni, T. Bamba and E. Fukusaki, *J. Agric. Food Chem.*, 2013, **61**, 7994–8001.
97. M. F. Marcone, *Food Res. Int.*, 2004, **37**, 901–912.
98. Ö. Özdestan, S. M. van Ruth, M. Alewijn, A. Koot, A. Romano, L. Cappellin and F. Biasioli, *Food Res. Int.*, 2013, **53**, 433–439.
99. *Instrução Normativa No 8*, Ministro de Estado da Agricultura, Pecuária e Abastecimento, 2003.
100. 10470, ISO, 2004, 20.
101. Specialty Coffee Association of America, *Arabica Green Coffee Defect Handbook*, Specialty Coffee Association of America, Long Beach, 2nd edn, 2013.

CHAPTER 3

Breeding Strategies

OLIVEIRO GUERREIRO-FILHO*[a] AND MIRIAN PEREZ MALUF[b]

[a]Agronomic Institute, Coffee Center, Campinas, São Paulo, 13020-971, Brazil; [b]EMBRAPA, Coffee Unit, Brasília, Distrito Federal, 70770-901, Brazil
*E-mail: oliveiro@iac.sp.gov.br

3.1 Introduction: *Coffea* Species

Coffee is probably the most famous beverage around the world, consumed mainly due to its stimulating and comforting properties. Even though its consumption is so extensive, cultivation of coffee plants is geographically restricted, as botanical aspects limit the cultivation to very specific environmental conditions. Depending on the plant species and variety, place of cultivation (and *terroir*), harvesting and post-harvesting processing methods, coffee final attributes may be surprisingly different. Among these aspects, botanical aspects and genetics play a very important role and will be discussed here.

Despite the increasing number of species identified as belonging to the genus *Coffea* (so far there are 125 species[1]), only two are actually responsible for the major production of commercially available coffee seeds: *Coffea arabica* and *Coffea canephora*. Both species are native of the central and equatorial regions of Africa, and thanks to intense breeding efforts they are currently cultivated in several other regions around the world, including Central and South America as well as Central and South Asia.

Both *C. arabica* L. and *C. canephora* Pierre (popularly known as robusta, the main cultivars of this species) provide most of the coffee consumed. Each

Coffee: Production, Quality and Chemistry
Edited by Adriana Farah

Published by the Royal Society of Chemistry, www.rsc.org

species has particular attributes to meet the needs of a different market niche. Arabica coffees are known for their fine cup quality, while robusta coffees are normally regarded as bitter, with poor cup quality. However, robusta coffees can also present excellent quality once they are produced with care. Additionally, *C. canephora* is highly productive, with fewer cultivation restrictions, and resistant to major pathogens often found in plantations, and therefore its cost of production is normally lower compared to arabica's, which is commonly susceptible to pathogens and is less productive. Thus, most commercial coffee blends normally contain some robusta coffees in order to keep a low retail price.

Other *Coffea* species are not economically cultivated but are largely used by breeding programs due to their variability for many characteristics especially resistance to diseases, insects and nematodes. Most wild *Coffea* species have reduced flower production, which accounts for the low fruit yield, and produce poor cup quality, limiting their use for breeding. Among those, the species *C. eugeniodes* has a promising cup quality.[2] The species *C. racemosa* exhibits a good efficiency on crosses with *C. arabica*, producing natural or artificial triploids, and is a source of resistance to leaf miner (*Leucoptera coffeella*) and to drought.[3] Another useful species is *C. dewevrei*, which besides the resistance to rust, nematodes and leaf miner exhibits a long fruit maturation period. Some of the species are caffeine-free, such as *C. pseudozangebariae* and *C. richardii*[4] or have a low level of caffeine, such as *C. dewevrei* (1.0%), *C. eugenioides* (0.4%), *C. salvatrix* (0.7%) and *C. racemosa* (0.8%).[5]

3.2 Biological Aspects of *Coffea arabica* and *Coffea canephora*

Coffea plants have a tree form, with plagiotropic branching, axillary-paired inflorescences, hermaphrodite flowers with usually white corollas and berry fruits containing two seeds, each seed having a deep groove on its flat side.[1,6] The mature seed has a true triploid endosperm,[7] with an embryo and two cotyledons, protected by a silverskin, the perisperm, and endocarp or parchment.[8] All species are diploid, with $2n = 22$ chromosomes, except *C. arabica* with $2n = 44$.

Coffea arabica is an allotetraploid species, resulting from a natural interspecific hybridization of *C. canephora* and *C. eugenioides*.[9] In addition, *C. arabica* is predominantly autogamous, with a percentage of natural cross-fertilization around 10%.[10] Fertilization in *C. arabica* occurs around 24 hours after pollination and the first cell division of the endosperm occurs 21 to 27 days after fertilization. The first zygote division occurs 60 to 70 days after pollination. The species *C. canephora* is self-incompatible and allogamous.[11–13]

The type of reproduction and compatibility for intra- and inter-specific crosses have a great impact on breeding strategies. The choice of parental species and genotypes, crossing type and direction and selection method, depends on those characteristics. For instance, *C. racemosa*, *C. stenophylla* and *C. salvatrix* are all natural sources of resistance to leaf miner, an insect that infests coffee leaves. However, as interspecific cross between *C. racemosa* and *C. arabica* is the most efficient, the first species is the donor of resistant genes in crosses aiming to develop leaf-miner-resistant cultivars.[2]

3.3 Genetics Aspects Associated with Fruit Development and Cup Quality

The cup quality, the ultimate coffee agronomic trait, is largely dependent on the overall conditions during fruit development. Biological aspects such as synchronization of maturation and genetic background affect final seed chemical composition. Environmental conditions, such as cultivation region, rainfall pattern and agronomic traits will influence both productivity and the seeds' chemical composition.[14–16] The interaction of those biological and environmental aspects will in turn outline beverage attributes. As these features will be discussed in detail in other chapters, here we would like to highlight relevant biological aspects.

The development of coffee fruits is a long and well-orchestrated process, with several defined steps, reviewed and illustrated by De Castro and Marraccini.[17] After the fertilization and embryo formation, the fruit is enlarged due to perisperm development and cellular division of endosperm cells, defining the growth phase. Maturation, the following step, represents the physiological maturity, and at this point the fruit continues its development even if harvested. Ripening comprises the stage where global characteristics related to fruit appearance and quality, such as chemical composition, color, texture, flavor and aroma, are determined. Senescence is the last stage and includes a series of physiological events that culminates with cellular death.[17] Studies on ethylene accumulation pattern along fruit development indicated that coffee fruits are climacteric.[18]

The genetic control of fruit development has been the focus of several studies aiming also to identify key genes associated with cup quality. Identification of which genes are responsible for sensory aspects of coffee quality is a long desired goal by breeding programs. Once known, those genes could be used as genomic markers by breeding programs aiming to select coffee cultivars with potentially defined cup attributes. Using high-throughput methodologies, such as microarrays, RNAseq and real-time RT-PCR, the common strategy of those studies is to compare the transcriptome of coffee fruits and seeds from different genetic backgrounds, cultivated in different geographic regions or submitted to variable post-harvesting treatments, in order to identify gene expression differences that may correlate with sensory cup quality. Following this approach, two studies designed putative metabolic maps associated with gene expression profiles during coffee fruit development.[19,20] The statistical analyses of gene expression profiles along fruit development allowed the clusterization of genes in functional groups associated with seed development events. Also, the authors suggest the occurrence of genetic mechanisms controlling the transcriptional transitions throughout fruit development, and identified several candidate-genes to regulate these events.[19] Association of these expression profiles with the synthesis patterns of important metabolites such as cell-wall polysaccharides, soluble sugars, chlorogenic acids and trigonellines allowed the design of metabolic maps of coffee fruits.[20] However, as those analyses were performed on fruits from the *C. arabica* Laurina variety, known for its compact plants, with slender branches, reduced caffeine levels and pointed seeds, the maps may not represent the regular development of a normal coffee fruit.

In another study, the goal was to identify genes associated with different stages of fruit development, and which may potentially be used as phenological markers.[21] The authors evaluated the expression profile of 28 candidate genes, in four different *C. arabica* cultivars, during three harvesting years. Upon those analyses, the authors selected four genes as potential markers: the α-galactosidase as a marker of green stage, caffeine synthase as a marker of transition to green and green stages and isocitrate lyase and ethylene-receptor 3 as markers of late maturation. These genes, in association with other phenological and agronomic attributes, represent a molecular parameter for a selective harvesting of fruits from any specific maturation phase. The selection of a precise fruit maturation phase may help the identification of ripening conditions resulting in coffee grains with improved cup quality.

3.4 The Importance of Germoplasm Collections

The successful cultivation of coffee in regions all over the world, distant from the African forests, the center of coffee origin, results from intense efforts by breeding programs conducted mainly in research institutions from coffee-producing countries. Thanks to those initiatives, coffee cultivars acquired novel plant architectures and physiological aspects, such as organized branches, shorter heights, controlled flowering and maturation, to mention a few, which favored the large-scale cultivation in non-origin regions. All those achievements were possible thanks to the establishment of *ex situ Coffea* germoplasm collections, which includes whenever possible most of the diversity for the main *Coffea* species. *Coffea* species are perennial plants, with long life-cycles, and their seeds have very low rates of germination upon long periods of storage. These aspects explains why the conservation of *Coffea* germoplasm has a high cost, demanding large areas, continuous management and characterization. The establishment of *ex situ* collections, out of Africa, located in Latin America (IAC, Campinas, Brazil, CATIE, Costa Rica, and Cenicafé, Colombia) and Asia (CCRI, India, and ICCRI, Indonesia) resulted from exchange of genetic resources and authorized collecting missions, organized by FAO in the late 1960s. Today, with the increasing degradation of African forests, especially in Ethiopia, the birth place of *C. arabica*, the existent BAG of *Coffea* are becoming the major repository of genetic diversity for those species.[22]

Although *C. arabica* and *C. canephora* are the more represented species in those germplasm collections, they also contain a large number of accessions from other *Coffea* species, which are the source of variability for interesting traits such as resistance to pathogens (*Coffea racemosa*, *C. congensis*, *C. dewrevrei*, *etc.*), productivity and fruit chemical composition.

3.4.1 Natural Genetic Variability of Coffee Fruits and Seeds

The collections also maintain coffee variants of *C. arabica* that represent most of the known genetics. For instance, the *ex situ* germplasm collections of *Coffea* maintained by the IAC include a large number of introductions from the center of origin and variability of the species,[23] exotic varieties,[24] botanical forms and mutants.[25] These accessions represent variations of morphological, anatomical and physiological features, among others, in specific organs or in various parts of the plants. Classical genetic analyses established the inheritance pattern of more than 40 mutants,[26] and about a quarter of these changes occur in fruits and/or in seeds. Several of these mutations, associated with fruit and cup quality, are listed in Table 3.1.

The mutants *xanthocarp*, *laurina*, *mokka* and *maragogipe* had a great importance for coffee breeding programs conducted in Brazil. Several cultivars that nowadays comprise the Brazilian coffee plantations have fruits with yellow pericarp, a monogenic characteristic simply selected, resulting from expression of the xanthocarp allele.[27] In addition, some authors claim that this mutation may influence positively the sensory cup quality.[28] The cultivars IAC 4761 Ibairi and Laurina IAC 870, resultant from selections made in the germplasm *mokka* and *laurina* respectively, despite their lower productivity, are known for the superior organoleptic quality of their coffee brew.[29] Maragogipe coffee, also less productive than the main Brazilian arabica cultivars,[30] presents higher average sieve seeds, which makes it very attractive in the current world market.[31]

3.4.2 Use of Natural Genetic Resources in Breeding for Quality

Whenever an agronomic problem occurs in the coffee field, or a special market niche arises, the solution probably involves the use of novel coffee cultivars. In order to fulfill those needs, breeding programs are continuously evaluating and selecting coffee plants bearing diverse agronomic traits and attributes. The genetic variability of any given species, or of species closely related, is the source of those diverse characteristics. There are species with such high genetic variability that there is no need to perform inter-specific crosses. This is not the case of *Coffea arabica*, a species with very restricted

Table 3.1 Genetic constitution, main features, origin and type of mutation of some *C. arabica* mutants.[a]

Germoplasm	Genotype	Origin	Inheritance	Characteristics
Standards				
Typica	*TT nana*			
Bourbon	*tt NaNa*			
Mutants				
Semperflorens	*sfsf*	Typica	Recessive	Continuous flowering throughout the year
Xanthocarp	*xcxc*	Typica	Recessive	Ripe yellow fruits
Guava	*Sdsd*	Typica	Recessive	Developed and persistent sepals in fruits
Macrodisc	*MdMd* ou *Mdmd*	Typica	Dominant	Large bracts in the inflorescence and very developed disc in fruits
Wax	*cecece*	Typica	Recessive	Dry seeds with yellow endosperm
Laurina	*Lrlr MoMo*	Bourbon	Recessive	Altered branching architecture; small fruits and seeds with narrowing base
Mokka	*LrLr momo* or *lrlr momo*	Bourbon	Recessive	Altered branching architecture; small and rounded fruits and small seeds
Purpurascens	*prpr*	Typica	Recessive	Purple organs in young plants and violet-green in adult plants
Maragogipe	*MgMg* or *Mgmg*	Typica	Dominant	Very large leaves, fruits and seeds
Fasciata	*FsFs*	nd	Partially dominant	Fasciation on branches and fruits with large number of seeds

[a]Data from Carvalho *et al.* (1991); nd = not determined.

genetic diversity, which normally requires the use of variability found in other *Coffea* species. In this way, resistance to pathogens, such as nematodes and leaf-rust, was transferred from the species *C. canephora*. However, although limited, there is genetic variability to explore within each species. Next, we describe examples of how to explore the natural variability of *C. arabica* to develop novel cultivars or by-products.

3.4.3 Naturally Caffeine-free Mutant – a Success Case of Wild-type Resource Use

During the past years, the pursuit of low-caffeine cultivars has been a common challenge for coffee breeders. Most strategies focused on transferring to *C. arabica* the low caffeine trait from other species. These strategies, however, have had no success so far, mainly due to limitations of inter-specific crosses.[2,5]

Besides these, breeding programs are extensively looking for phenotypic variability on accessions from *Coffea* germplasm collections, including natural variability for caffeine content.[2] In the collection of *Coffea* from the IAC, research identified three coffee plants out of more than 3000 accessions, namely AC1, AC2 and AC3, that were nearly caffeine-free (AC stands for absence of caffeine and Alcides Carvalho, the most important coffee geneticist, who greatly contributed to *Coffea* genetics and breeding).[32] Biochemical analysis showed that leaves of AC1 accumulated the caffeine immediate precursor theobromine, and that no caffeine synthase activity was present in both fruits and leaves. These results were very exciting, since they represent a possibility for development of naturally caffeine-free *C. arabica* cultivars. The strategies adopted by the IAC Breeding Program to develop caffeine-free cultivars included: *in vitro* cloning of wild-type accessions in order to develop a clonal cultivar,[33] transfer of the low-caffeine trait to already available coffee cultivars through traditional breeding methods and a search for molecular-markers associated with the desired trait for use on assisted-selection. All strategies were successful. A clonal variety with caffeine-free seeds has been licensed.[34] The commercial use is limited, though, since the cultivar is not highly productive. On the other hand, molecular analysis of caffeine biosynthetic genes allowed the identification of several polymorphisms, which occurred only on AC plants.[35] Further genotyping evaluation of these single nucleotide polymorphisms (SNP) on breeding populations confirmed their association with the caffeine-free trait.[36] Thus, the IAC Breeding Program is now using these SNPs as markers in assisted-selection for novel and more productive caffeine-free cultivars.

3.4.4 Selection of High-Oil Plants

For many decades, coffee breeding programs in Brazil prioritized the development of more efficient cultivars, with higher productivity and lower production costs. As a result, the current 129 arabica cultivars registered at the

Ministry of Agriculture, Livestock and Food Supply,[34] resulted from the sum of diverse traits, especially those related to plant architecture, yield, precocity of fruit maturation and resistance to major biotic agents.

Nowadays, to increase the competitiveness of Brazilian coffee in global markets, the focus of current breeding programs is to release cultivars with higher added value, either by improving the characteristics related to physical and sensory quality or by devising alternative uses or products for coffee.

One example of such novel use is the selection, by the IAC, of a new variety with higher oil content in seeds. The oil represents about 15% of the coffee endosperm and has many applications, especially in the food and pharmaceutical industries. The oil content in coffee endosperm results from a set of environmental components, and therefore may vary in the same cultivar. Those components include the crop year,[37] the stage of fruit development[38] or, moreover, the type of post-harvest treatment of the fruit.[39] The trait is also under a genetic control and varies in *Coffea* species: it is higher in *C. arabica* and lower in *C. canephora*, and in botanical varieties of these species.[40,41] On average, the total oil content varies between 8 and 14% in arabica coffees, but a study conducted by Wagemaker[42] showed that some coffee trees exhibit higher average values. In this study, the authors used nuclear magnetic resonance analysis, a non-destructive method, to determine the oil content of individual seeds and confirmed the occurrence of variability for this trait among seeds of the same plant (Figure 3.1). Thus, this method allows the screening and planting of selected seeds, assisting the development of high-oil coffee cultivars.

3.4.5 Genetic Diversity for Fat Components

Other related studies, performed on the *Coffea* collection from IAC, include the characterization of wax and unsaponifiable matter in coffee beans, as well as the sun protection factor given by the oil present in the seed.[40] The

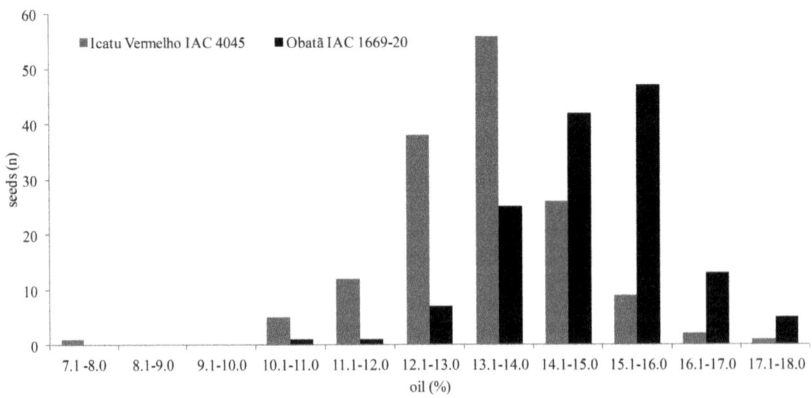

Figure 3.1 Oil content in coffee seeds determined by nuclear magnetic resonance in *C. arabica* cultivars IAC Icatu Vermelho 4045 and IAC Obatã 1669-20.

Table 3.2 Variability in wax content and unsaponifiable matter observed in *Coffea* species from the collection at Instituto Agronômico (IAC).[a]

Species	Wax content[b] (%)	Unsaponifiable matter content[c] (%)	Sun protection factor[d]
C. arabica	0.24	13.54	1.50
C. canephora	0.08	4.23	0.35
C. congensis	2.55	10.54	1.08
C. eugenioides	2.34	1.93	2.59
C. heterocalyx	1.24	3.90	2.37
C. kapakata	0.56	0.36	0.06
C. liberica var. *dewevrei*	0.91	0.28	0.88
C. liberica var. *liberica*	1.91	5.36	0.48
C. racemosa	1.00	2.19	1.59
C. salvatrix	1.58	10.71	2.54
C. stenophylla	1.68	4.36	2.45

[a]Data from Wagemaker *et al.* (2011).
[b]Dry green beans basis.
[c]Dry oil basis.
[d]Dry oil basis.

observed variability on levels of these compounds is considerable among the different *Coffea* species (Table 3.2).

Another study[43] characterized those accessions regarding the diversity of kahveol and cafestol, specific coffee diterpenes, with outstanding importance for plant vigor, since they act as anti-reactive oxygen species (anti-ROS) compounds during biotic and abiotic stress response in plants. Additionally, such compounds may act as an anti-carcinogenic in humans, although their consumption has been related to an increase in blood cholesterol levels.[44]

The diversity observed in *Coffea* species opens a possibility for breeding programs to use these accessions as donors of either low or high content of fat material, meeting the demands of the industrial sector.

We selected here examples to illustrate the importance of using natural genetic resources to improve coffee quality. However, the genetic diversity strategically preserved in germplasm collections likewise grants the development of novel and more efficient cultivars, adapted to the most diverse growing regions. These features, associated with advanced agricultural technologies, ensure sustainable large-scale coffee production with competitive capacity.

References

1. S. Krishnan, T. A. Ranker, A. P. Davis and J.-J. Rakotomalala, *Acta Hortic.*, 2015, **1101**, 15.
2. H. P. Medina-Filho, R. Bordignon, O. Guerreiro-Filho, M. P. Maluf and L. C. Fazuoli, *Acta Hortic.*, 2007, **745**, 393.
3. H. P. Medina-Filho, A. Carvalho and D. M. Medina-Filho, *Bragantia*, 1977, **36**, 43.

4. C. Campa, S. Doulbeau, S. Dussert, S. Hamon and M. Noirot, *Food Chem.*, 2005, **93**, 135.
5. P. Mazzafera and A. Carvalho, *Euphytica*, 1992, **59**, 55.
6. A. P. Davis, R. Govaerts, D. M. Bridson and P. Stoffelen, *Bot. J. Linn. Soc.*, 2006, **152**, 465.
7. A. Carvalho and C. A. Krug, *Bragantia*, 1949, **9**, 193.
8. D. M. Dedecca, *Bragantia*, 1957, **6**, 315.
9. P. Lashermes, M. C. Combes, J. Robert and P. Trouslot, *Mol. Gen. Genet.*, 1999, **261**, 259.
10. A. Carvalho and L. C. Monaco, Proc. Int. Hortic. Congr., 16th, Brussels, 1962, 447–449.
11. M. Devreux, G. Vallayes, P. Pochet and A. Gilles, *Bull. Inf. INEAC*, 1959, **78**, 1–44.
12. C. H. T. M. Conagin and A. J. T. Mendes, *Bragantia*, 1961, **20**, 787.
13. J. Berthaud, *Cafe, Cacao, The*, 1980, **24**, 267.
14. E. A. Da Silva, P. Mazzafera, O. Brunini, E. Sakai, F. B. Arruda, L. H. C. Matoso, C. R. L. Carvalho and R. C. M. Pires, *Braz. J. Plant Physiol.*, 2005, **17**, 229.
15. M. B. P. Camargo, *Bragantia*, 2010, **69**, 239.
16. L. P. Figueiredo, F. M. Borém, M. A. Cirillo, F. C. Ribeiro, G. S. Giomo and T. J. G. Salva, *J. Agric. Sci.*, 2013, **5**, 10.
17. R. De Castro and P. Marraccini, *Braz. J. Plant Physiol.*, 2006, **18**, 175.
18. L. F. P. Pereira, R. M. Galvão, A. K. Kobayashi, S. M. B. Cação and L. G. E. Vieira, *Braz. J. Plant Physiol.*, 2005, **17**, 283.
19. J. Salmona, S. Dussert, F. Descroix, A. Kochko, B. Bertrand and T. Joët, *Plant Mol. Biol.*, 2008, **66**, 105.
20. T. Joët, A. Laffargue, J. Salmona, S. Doulbeau, F. Descroix, B. Bertrand, A. Kochko and S. Dussert, *New Phytol.*, 2009, **182**, 146.
21. C. Gaspari-Pezzopane, N. Bounturi, O. Guerreiro-Filho, J. L. Favarin and M. P. Maluf, *Pesqui. Agropecu. Bras.*, 2012, **47**, 972.
22. T. W. Gole and D. Teketay, *Biol. Soc. Ethiop.*, 2001, 131.
23. A. Carvalho, *Bragantia*, 1959, **18**, 353.
24. A. J. Bettencourt and A. Carvalho, *Bragantia*, 1968, **27**, 35.
25. A. Carvalho and L. C. Fazuoli, *O melhoramento de plantas no Instituto Agronômico*, ed. A. M. C. Furlani and G. P. Viegas, Instituto Agronômico, Campinas, 1993, vol. 1, pp. 29–76.
26. A. Carvalho, H. P. Medina Filho, L. C. Fazuoli, O. Guerreiro-Filho and M. M. A. Lima, *Rev. Bras. Genet.*, 1991, **14**, 135.
27. C. A. Krug and A. Carvalho, *Bol. Tec. IAC., Campinas*, 1940, **82**, 1.
28. T. J. G. Salva, *Anais do Curso de Atualização em Café*, ed. R. A. Thomaziello, IAC, Campinas, 2005, pp. 1–16.
29. C. H. S. Carvalho, L. C. Fazuoli, G. R. Carvalho, O. Guerreiro-Filho, A. A. Pereira, S. R. Almeida, J. B. Matiello, G. F. Bartholo, T. Sera, W. M. Moura, A. N. G. Mendes, J. C. Rezende, A. F. A. Fonseca, M. A. G. Ferrão, R. G. Ferrão, A. P. Nacif, M. B. Silvarolla and M. T. Braghini, *Cultivares de café: origem, características e recomendações*, ed. C. H. S. Carvalho, Embrapa, Brasília, 2008, vol. 9, pp. 157–226.

30. L. C. Monaco, *Bragantia*, 1960, **19**, 458.
31. C. A. Krug and A. Carvalho, *Bragantia*, 1942, **2**, 231.
32. M. B. Silvarolla, P. Mazzafera and L. C. Fazuoli, *Nature*, 2004, **429**, 826.
33. J. A. A. Almeida and M. B. Silvarolla, *Int. J. Plant Dev. Biol.*, 2009, **3**, 5.
34. http://extranet.agricultura.gov.br/php/snpc/cultivarweb/detalhe_protecao.php?codsr=3642, last accessed November 2015.
35. M. P. Maluf, C. C. Silva, M. P. A. Oliveira, A. G. Tavares, M. B. Silvarolla and O. Guerreiro-Filho, *Genet. Mol. Biol.*, 2009, **32**, 802.
36. M. P. Maluf and M. B. Silvarolla, *Br. Pat.* BR1020130323179, 2013.
37. T. A. Wagemaker, *MSc. Thesis Dissertation*, Instituto Agronômico, IAC, Campinas, 2009.
38. H. Fonseca and L. E. Gutierrez, *Anais ESALQ*, 1971, **28**, 313.
39. J. S. Tango and A. Carvalho, *Bragantia*, 1963, **22**, 793.
40. M. N. Clifford, *Coffee: Botany, Biochemistry and Production of Beans and Beverage*. ed. M. N. Clifford and K. C. Willson, Avi Publishing, Westport, Connecticut, 1985, pp. 305–374.
41. P. Mazzafera, D. Soave, M. A. T. Zullo and O. Guerreiro-Filho, *Bragantia*, 1998, **57**, 45.
42. T. A. L. Wagemaker, C. R. L. Carvalho, N. B. Maia and O. Guerreiro-Filho, *Ind. Crops Prod.*, 2011, **33**, 469.
43. G. A. Ogasawara, *MSc. Thesis Dissertation*, Instituto Agronômico, IAC, Campinas, 2015.
44. A. A. A. Bak and D. E. Grobbee, *N. Engl. J. Med.*, 1989, **321**, 1432.

CHAPTER 4

Coffee Plant Biochemistry

HIROSHI ASHIHARA*[a], TATSUHITO FUJIMURA[b] AND ALAN CROZIER[c]

[a]Department of Biology, Ochanomizu University, Otsuka, Bunkyo-ku, Tokyo, 112-8610, Japan; [b]Faculty of Life and Environmental Sciences, University of Tsukuba, Tsukuba 305-8572, Japan; [c]Department of Nutrition, University of California, Davis, CA 95616-5270, USA
*E-mail: ashihara.hiroshi@ocha.ac.jp

4.1 Introduction

Coffee seeds contain various compounds including primary and secondary metabolites which are characteristic to some *Coffea* species. Figure 4.1 shows the typical composition of seeds of *C. arabica* and *C. canephora* expressed as a percentage of dry weight, according to the data of Farah.[1] Polysaccharides including mannan, galactan, glucan and araban comprise nearly half of the total weight. Lipids, proteins, sugars (mainly sucrose) and acids (mainly chlorogenic acids) comprise, respectively, 16, 11, 8 and 7% of dry weight in *C. arabica*, and 10, 11, 4 and 10% in *C. canephora*. Coffee seeds also contain caffeine and trigonelline, each in amounts corresponding to ~1–2% of dry weight.

In this chapter, carbohydrate and nitrogen metabolism are discussed briefly, and then the biosynthesis and metabolism of caffeine, trigonelline and chlorogenic acids in *Coffea* species are considered in detail. Physiological aspects of metabolism of these compounds in the developing and ripening fruits of *C. arabica* and *C. canephora* are described. According to

Coffee: Production, Quality and Chemistry
Edited by Adriana Farah

Published by the Royal Society of Chemistry, www.rsc.org

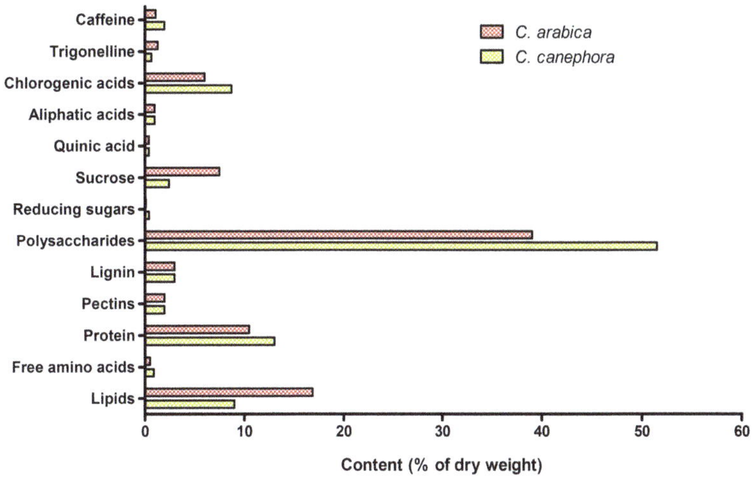

Figure 4.1 Chemical composition of seeds of *Coffea arabica* and *C. canephora*. Contents are expressed as % of dry weight.[1]

Cannell,[2] the stage of growth of coffee fruits can be classified as the pinhead stage, rapid expansion and pericarp growth stage, bean (endosperm) formation stage, bean dry matter accumulation stage and fruit ripened stage. The last stages are further divided into green-, pink- or yellow- and red-coloured stages which accompany ripening (Figure 4.2).[3] This classification is used throughout the text.

4.2 Carbohydrate Metabolism in Coffee

The literature on carbohydrate metabolism in coffee plants is limited and we first outline sugar metabolism which may occur in coffee and then molecular studies related to the biosynthesis of sucrose and cell wall storage polysaccharides in coffee fruits are discussed.

In coffee, as well as many other plant species, carbohydrates are produced from atmospheric CO_2 and water in photosynthetic tissues of leaves and fruits and are then transported to sink tissues/organs. Coffee is categorised as a C_3 photosynthetic species which directly fixes atmospheric CO_2 by ribulose-1,5-bisphosphate carboxylase/oxygenase.[4] The product, glycerate-3-P, is metabolised in chloroplasts by the Calvin–Benson cycle (Figure 4.3). Some intermediates of the cycle are transported to the cytosol of the mesophyll cells by a number of different transporters located in the chloroplast membranes. The best characterised transporter is the triose-P and inorganic phosphate (Pi) anti-porter protein that transports dihydroxyacetone phosphate (DHAP) out of the stroma in exchange for an influx of Pi ions. Sucrose biosynthesis from DHAP in cytosol is shown in Figure 4.3. Sucrose synthesised in leaves is transported into phloem by the H^+-sucrose

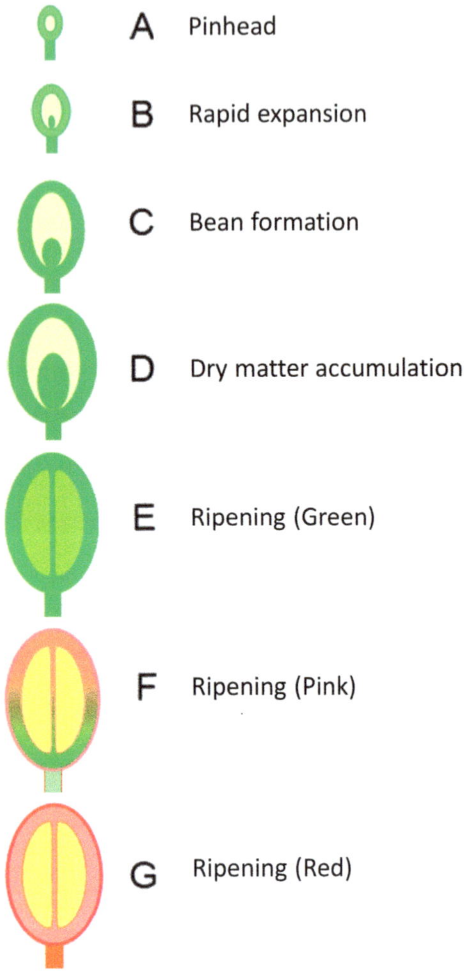

Figure 4.2 The growth stages of coffee fruits. In the present chapter, the growth stages are classified from stage A to G according to Koshiro *et al.*[3] These stages correspond approximately to the stages described by Cannell;[2] (A) the pinhead stage; (B) rapid expansion and pericarp growth stage; (C) bean (endosperm) formation stage; (D) bean dry matter accumulation stage; and (E–G) three fruit ripening stages.

co-transporter, and translocated to the flesh of fruit where it is unloaded into parenchyma tissue.[5]

Sucrose is not a reactive sugar, because both anomeric carbons are linked in the glycosidic band. In sink tissues, sucrose is converted to constituent sugars by invertase (EC 3.2.1.26, reaction 4.1) and/or sucrose synthase (EC 2.4.1.13, reaction 4.2).

$$\text{sucrose} \rightarrow \text{glucose} + \text{fructose} \tag{4.1}$$

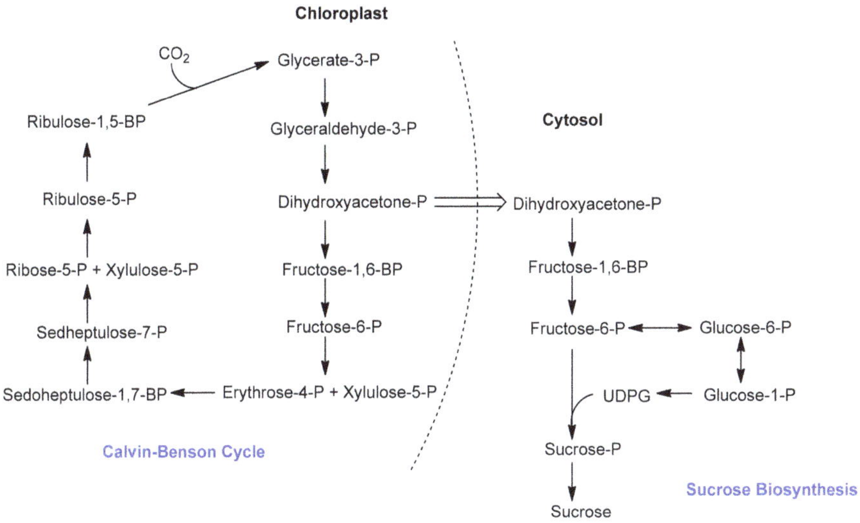

Figure 4.3 A possible route of CO_2 fixation and sucrose synthesis in *Coffea* leaves.[158]

$$\text{sucrose} + \text{UDP} \rightarrow \text{UDP-glucose} + \text{fructose} \tag{4.2}$$

Glucose and fructose are phosphorylated by various hexose kinases and sugar phosphates, glucose-6-P, glucose-1-P and fructose-6-P are produced, which act as precursors for the biosynthesis of carbohydrate and carbon skeletons of various compounds. Nucleotide sugars, such as UDP-glucose, formed from sugar phosphates also contribute to many glycosyltransferase reactions in biosynthetic pathways.

Compared with other crop plants, few studies have been carried out with coffee on photosynthetic CO_2 fixation and the subsequent reactions to form assorted carbohydrates. However, results obtained with ^{14}C-feeding experiments support the operation of the common pathway which occurs in many plants. Geromel *et al.*[6] established that $^{14}CO_2$ in coffee tree branches bearing young fruits was assimilated into both leaves and fruits. The radioactivity was found in sucrose and reducing sugars of the pulp, perisperms and endosperms of fruits even when $^{14}CO_2$ was fixed in leaves. The results indicate that CO_2 fixation occurred both in leaves and pulp of fruits, and assimilates were translocated to the endosperms of coffee fruits. Carbon partitioning in fruits was also studied by Geromel *et al.*[7] using pulse–chase experiments with ^{14}C-sugars which revealed high rates of sucrose turnover in perisperm and endosperm tissues. The feeding experiments with $^{14}CO_2$ showed that leaf photosynthesis contributed more to seed than to pericarp development in terms of photosynthate supply to the endosperm.

To investigate overall carbohydrate metabolism in coffee fruits, the metabolic fate of ^{14}C-glucose was examined in segments of pericarp and seed of two cultivars of *C. arabica* and *C. canephora* fruits at different growth and ripening stages.[8] The rates of uptake and metabolism of ^{14}C-glucose in developing fruit was higher than in the later ripening stages. Release of $^{14}CO_2$ from ^{14}C-glucose, which represents cellular respiration, was high in both pericarp and seeds during fruit development, but gradually declined and was lowest in the fully ripened fruits. This demonstrates that active carbohydrate metabolism occurs in developing fruits but it slows as fruits ripen. Radioactivity was also incorporated in various compounds including organic acids, amino acids and sugars and higher molecular weight components such as proteins and polysaccharides. In ripening fruits, a relatively higher rate of sucrose formation from glucose was detected in both pericarp and seeds. This may be reflected by the decrease of the activity of respiration and primary metabolism and, as a result, sucrose is accumulated as a storage compound.[8]

Rogers *et al.*[9] reported that major soluble sugars in developing coffee seeds are glucose and fructose, which are replaced by sucrose during ripening. Koshiro *et al.*[10] also investigated changes in the concentration of these sugars during the growth of coffee fruit. As shown in Figure 4.4, small amounts of glucose and fructose (stages B and C) but little or no sucrose were found in young fruits. In the dry matter accumulation stage (stage D) accumulation of sucrose began in seeds. Subsequently, the sugar content increased in pericarp and seeds (Figure 4.3). In *C. arabica* fruits, sucrose is the major free sugar in seeds, but similar amounts of sucrose, fructose and glucose are found in pericarp.

Geromel *et al.*[7] monitored activities of three enzymes related to sucrose metabolism, namely, acid invertase (EC 3.2.1.26), sucrose phosphate synthase (EC 2.4.1.14) and sucrose synthase (EC 2.4.1.13), in the fruit tissues (pericarp, perisperm and endosperm) of *C. arabica* during development. Among these enzymes, sucrose synthase showed the highest activities during the last stage of endosperm and pericarp development and this activity closely paralleled the accumulation of sucrose in these tissues. Therefore, the participating enzyme in sucrose synthesis in photosynthetic tissues (leaves and surface of fruits) is sucrose phosphate synthase while in seeds it is sucrose synthase.

Two genes *CaSUS1* and *CaSUS2* which encode sucrose synthase isoforms have been isolated and their expression profiles investigated.[7] The transcripts of *CaSUS1* accumulated mainly during the early development of perisperm and endosperm, as well as during pericarp growing phases, whereas those of *CaSUS2* paralleled sucrose synthase activity in the last stages of pericarp and endosperm development. These results indicate that *CaSUS2* plays an important role in the accumulation of sucrose in coffee fruit.

As noted above, glucose and fructose in coffee fruits are products of sucrose catabolism. This could occur either in the apoplasm, through the action of cell wall invertase, or intracellularly *via* invertase and/or sucrose synthase activities. Joët *et al.*[11] reported high levels of expression of cell wall and vacuolar invertase genes and low expression of sucrose synthase genes (*sus1* and *sus2*) in early developmental stages. This suggests that invertases, but not sucrose synthase, may have an important role in sucrose catabolism.

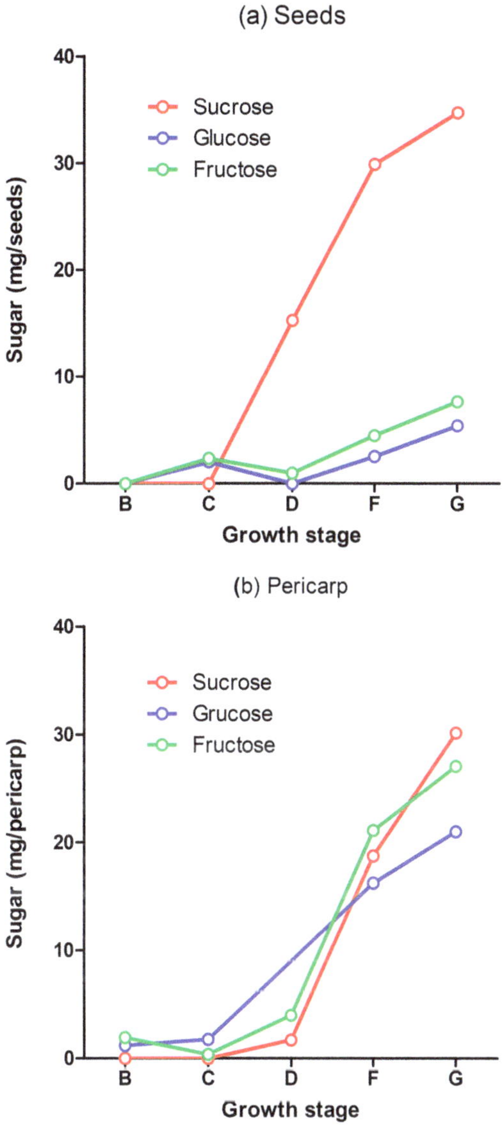

Figure 4.4 Changes in sugar contents in seeds (a) and pericarp (b) of *C. arabica* fruits during developing and ripening. The values are expressed in mg per fruit, *i.e.*, two seeds and pericarp. Based on data from Koshiro *et al.*[10]

Joët *et al.*[11] carried out a detailed investigation of the expression profiles of various coffee genes and compared them with metabolite contents. The changes of metabolite levels in coffee fruits during development are sometimes related to the gene expression. For example, the pattern of expression of a fructokinase gene matched that of fructose content and, to a lesser extent, the drop in glucose was in line with a progressive decline in hexokinase transcript abundance. Obviously, metabolism is controlled not only by the expression of genes encoding corresponding enzymes but also by post-transcriptional and other fine control mechanisms. Therefore, further detailed studies are required to elucidate the carbohydrate metabolism and accumulation of sugars in coffee seeds.

Coffee seeds contain large amounts of polysaccharides which make up ~50% of the dry weight and consist of three major types: mannans or galactomannans, arabinogalactan-proteins and cellulose. In addition, small amounts of pectic polysaccharides and xyloglucan also occur.[12] Recently, Joët *et al.*[13] reported the transcriptional regulation of galactomannan biosynthesis in *Coffea arabica* seeds. Coffee seeds accumulate large amounts of storage polysaccharides of the mannan family in the cell walls of the endosperm. The expression of five genes involved in galactomannan synthesis, namely genes coding mannan synthase, galactosyltransferase, α-galactosidase, mannose-1-phosphate guanylyltransferase and UDP-glucose 4′-epimerase, are closely related to the level of cell wall storage polysaccharides stored in the endosperm at the onset of their accumulation. This analysis also suggests a role for sorbitol and raffinose family oligosaccharides as transient auxiliary sources of building blocks for galactomannan synthesis. Based on these findings, the potential metabolic pathways of these polysaccharides are illustrated in Figure 4.5.

4.3 Nitrogen Metabolism

In addition to amino acids, nucleotides, proteins and nucleic acids, coffee plants produce some characteristic nitrogen-containing secondary metabolites, namely caffeine and trigonelline. Although it has long been known that coffee is a high nitrogen-demanding plant species, only a few reports

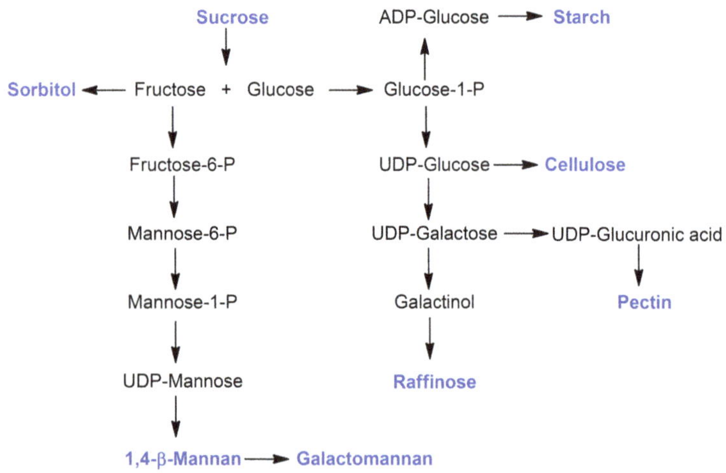

Figure 4.5 Possible biosynthetic routes of polysaccharides in *Coffea* plants.

concerning nitrogen metabolism in coffee have been published.[14] Here, we summarise what is known about the assimilation of nitrogen and amino acid biosynthesis in coffee plants.

Nitrate assimilation, conversion of nitrate (NO_3^-) to ammonium (NH_4^+), is performed by two enzymes: nitrate reductase (EC 1.7.1.1) and nitrite reductase (EC 1.7.7.1) (Figure 4.6). In many plants, nitrate reductase occurs in the cytosol and catalyses the reaction:

$$\text{nitrate} + \text{NADH} + \text{H}^+ \rightarrow \text{nitrite} + \text{NAD}^+ + \text{H}_2\text{O}$$

In contrast, nitrite reductase occurs in the chloroplast and other plastids. This reduction requires six electrons donated by reduced ferredoxin. The reaction catalysed is:

$$\text{nitrite} + 6\ \text{reduced ferredoxin} + 7\text{H}^+ \rightarrow \text{NH}_3 + 2\text{H}_2\text{O} + 6\ \text{oxidised ferredoxin}$$

NH_4^+ is assimilated by glutamine synthetase (GS, EC 6.3.1.2) and glutamate synthase (L-glutamine: 2-oxoglutarate aminotransferase, GOGAT, EC 1.4.1.13) and glutamic acid is formed (Figure 4.6). These two enzymes catalyse the following reactions:

$$\text{ATP} + \text{L-glutamate} + \text{NH}_3 \rightarrow \text{ADP} + \text{phosphate} + \text{L-glutamine}$$

$$\text{L-glutamine} + \text{2-oxoglutarate} + \text{NADPH} + \text{H}^+ \rightarrow 2\ \text{L-glutamate} + \text{NADP}^+$$

Coffee plants have a high potential for nitrate assimilation in leaves and roots. Some reports have shown higher NO_3^- reduction and NO_4^+ assimilation in leaves while others suggest higher activity occurs in roots. This may

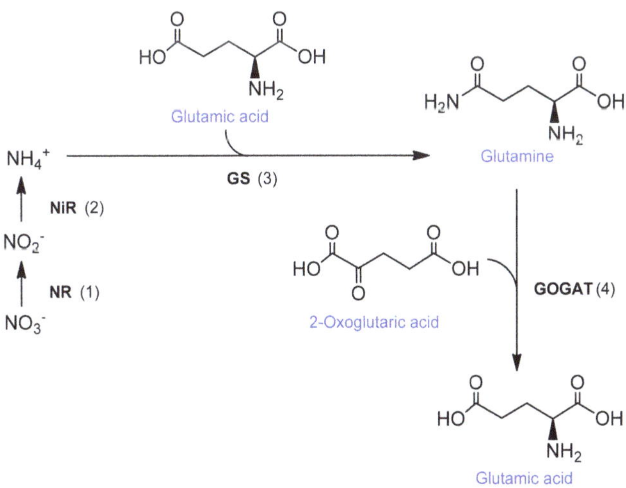

Figure 4.6 Nitrate reduction and assimilation of ammonia in plants. Enzymes shown are: (1) nitrate reductase (NR, EC 1.7.1.1); (2) nitrite reductase (NiR, 1.7.7.1); (3) glutamine synthetase (GS, 6.3.1.2); glutamate synthase (GOGAT, EC 1.4.1.13).

depend upon environmental conditions, such as light and/or age of tissues.[14] Nitrogen compounds in the xylem sap of coffee seedlings have been investigated by Mazzafera and Gonçalves.[15] The most abundant compounds were NO_3^- (52%) > asparagine (31%) > glutamine (7%). Sap did not contain NH_4^+. This is probably due to the fact that NH_4^+, produced by reduction of nitrate, is usually assimilated in the same cells in which it is generated. The data suggest that part of NO_3^- taken up by roots is converted to NH_4^+ and utilised for amino acid biosynthesis. Most amino acids are used for the protein synthesis and the biosynthesis of other compounds including phenolics in roots. Asparagine, glutamine and other storage amino acids and the remainder of NO_3^- are translocated to the leaves where they serve as sources for nitrogen compounds.

In plants, unlike animals, all protein constituent amino acids are synthesised from the intermediates of the glycolysis, pentose phosphate pathway and the TCA cycle (Figure 4.7).[16] As described in later sections, amino acids are the precursors of most secondary metabolites. For example, glutamine,

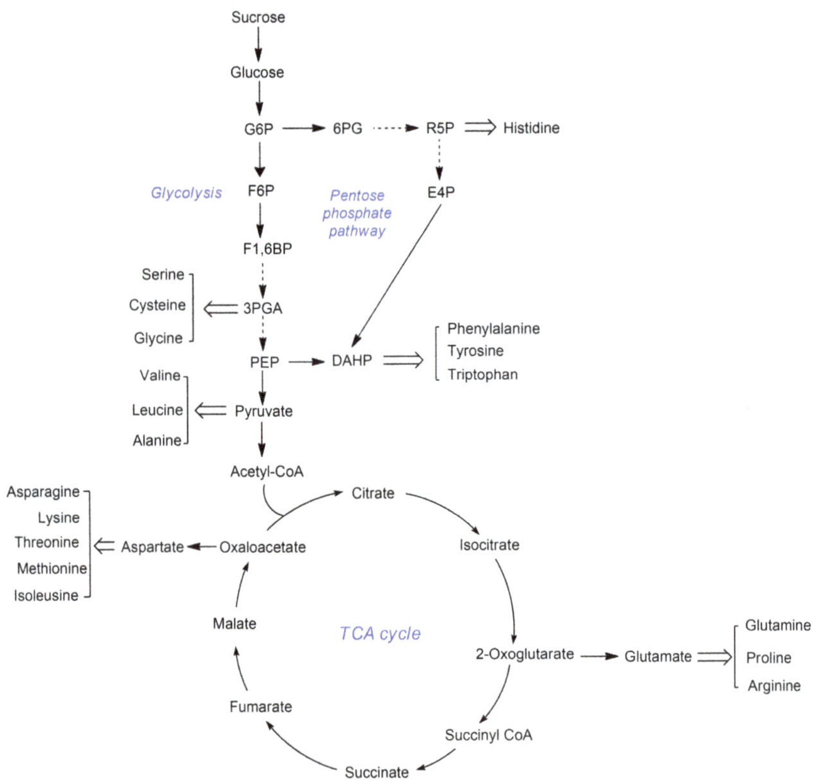

Figure 4.7 Outline of amino acids biosynthesis in plants. DAHP – 3-deoxy-D-arabino-heptulosonate 7-phosphate; E4P – erythrose-4-phosphate; F6P – fructose-6-phosphate; F1,6BP – fructose-1,6-bisphosphate; 6PG – 6-phosphogluconate; 3PGA – 3-phosphoglycerate; PEP – phosphoenolpyruvate.[158]

aspartate and glycine contribute to the formation of purine ring of caffeine. Aspartate and phenylalanine are precursors of trigonelline and chlorogenic acids, respectively. The genes and enzymes of nitrogen assimilation and amino acid biosynthesis have been well studied in plants.[17] Nevertheless, no definitive data are available with *Coffea* species.

4.4 Biosynthesis and Catabolism of Caffeine

The purine alkaloid, caffeine (1,3,7-trimethylxanthine), is one of the major constituents of coffee seeds which contain, as minor components, additional purine alkaloids, such as theobromine and theophylline. Other purine alkaloid-containing plants which also serve as the basis for the production of non-alcoholic beverages include tea (*Camellia sinensis*), cacao (*Theobroma cacao*) and maté (*Ilex paraguariensis*).[18]

The biosynthetic pathway of theobromine and caffeine has been the subject of much study over the years. Although early investigations up to the 1970s implied the involvement of nucleic acids as precursors in caffeine biosynthesis,[19,20] later investigations, mainly with tea leaves, indicated that caffeine is produced from xanthosine and that theobromine is its immediate precursor.[21–25]

In the 1990s, the biosynthesis of caffeine and theobromine became a topic of some controversy. Nazario and Lovatt[26] argued that theobromine was not the immediate precursor of caffeine in coffee leaves and that two separate *de novo* and salvage pools were involved in the biosynthesis of theobromine. However, subsequent detailed analysis of ^{14}C-metabolites,[27] characterisation of highly purified caffeine synthase[28] and cloning of the genes encoding biosynthesis enzymes[29] established the operation of a four-step xanthosine → 7-methylxanthosine → 7-methylxanthine → theobromine → caffeine pathway. This is the main caffeine biosynthesis pathway that was originally proposed by Suzuki and Takahashi in 1975 after experiments with a crude enzyme preparation that involved the use of ^{14}C-labelled tracers and analysis of radiolabelled products by paper chromatography.[30,31]

Compared to studies on biosynthesis, relatively little research has been carried out on the catabolism of caffeine. Although caffeine catabolism has been more thoroughly investigated in microorganisms,[32] this review will focus on findings that relate to coffee. More comprehensive reviews on caffeine biochemistry have been published elsewhere.[18,33]

4.4.1 The *De Novo* Biosynthetic Pathway of Caffeine

5-Phosphoribosyl-1-pyrophosphate (PRPP), which is produced from ribose-5-phosphate, an intermediate of the oxidative pentose phosphate pathway and the Calvin–Benson cycle, is the initial substrate for the biosynthesis of the purine ring of caffeine. The nitrogen and carbon atoms of the caffeine purine ring are supplied by glycine, glutamine and aspartate, 10-formyl tetrahydrofolate and CO_2 (Figure 4.8). In the 1960s it was reported that

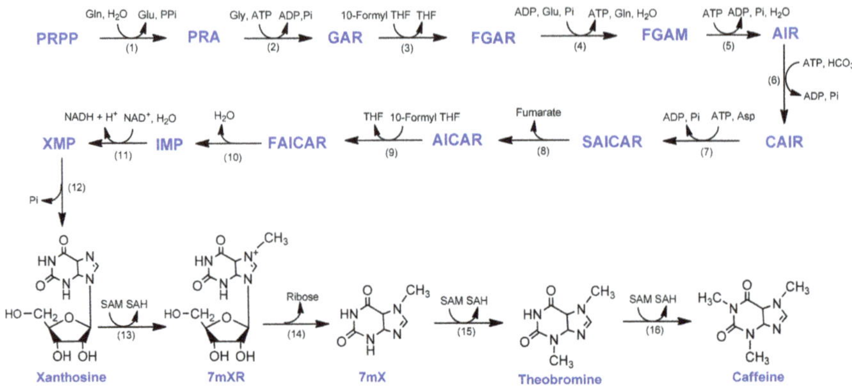

Figure 4.8 *De novo* biosynthetic pathway of caffeine in coffee plants. Enzymes (EC numbers) shown are: (1) PRPP amidotransferase (EC 2.4.2.14); (2) GAR synthetase (EC 6.3.4.13); (3) GAR formyl transferase (EC 2.1.2.2); (4) FGAM synthetase (EC 6.3.5.3); (5) AIR synthetase (EC 6.3.3.1); (6) AIR carboxylase (EC 4.1.1.21); (7) SAICAR synthetase (EC 6.3.2.6); (8) adenylosuccinate lyase (EC 4.3.2.2); (9) AICAR formyl transferase (EC 2.1.2.3); (10) IMP cyclohydrolase (EC 3.5.4.10); (11) IMP dehydrogenase (EC 1.1.1.205); (12) 5′-nucleotidase (EC 3.1.3.5); (13) 7-methylxanthosine synthase (EC 2.1.1.158); (14) *N*-methylnucleosidase (EC 3.2.2.25); (15) theobromine synthase (EC 2.1.1.159) and (16) caffeine synthase (EC 2.1.1.160). Steps 14 and 15 are also catalysed by 7-methylxanthosine synthase (EC 2.1.1.158) and caffeine synthase (EC 2.1.1.160), respectively (see text). Abbreviations: GAR – glycineamide ribonucleotide; FGAR – formylglycineamide ribonucleotide; FGRAM – formylglycine amidine ribonucleotide; AIR – 5-aminoimidazole ribonucleotide; CAIR – 5-aminoimidazole 4-carboxylate ribonucleotide; SCAIR – 5-aminoimidazole-4-*N*-succinocarboxyamide ribonucleotide; AICAR – 5-aminoimidazole-4-carboxyamide ribonucleotide; FAICAR – 5-formamidoimidazole-4-carboxyamide ribonucleotide; XMP – xanthosine-5′-monophosphate; 7mXR – 7-methylxanthosine; 7mX – 7-methylxanthine.

^{14}C-labelled serine, glycine, formaldehyde and formate were incorporated into caffeine in coffee and tea leaves.[34,35] The reactions of *de novo* caffeine biosynthesis up to XMP (steps 1–11 in Figure 4.8) are the same as the *de novo* biosynthetic pathway of guanine nucleotides, which also occurs in other organisms.[36]

The contribution of this pathway to caffeine biosynthesis was further demonstrated in the young tea leaf disks using ^{15}N-glycine, selected ^{14}C-labelled precursors and inhibitors of *de novo* purine biosynthesis.[37] Ribavirin, an inhibitor of IMP dehydrogenase (step 11 in Figure 4.8), reduced the rate of caffeine biosynthesis in leaf disks of tea and coffee.[38] These findings confirmed that the *de novo* pathway contributes to the caffeine biosynthesis *in planta*. However, it has not yet been established whether the *de novo* biosynthesis up to XMP formation (steps 1–11 in Figure 4.8) is specific for purine alkaloid formation or if the common *de novo* pathway of GMP synthesis is functional for both purine alkaloid and general purine nucleotide synthesis in plants. Nevertheless, operation of the *de novo* pathway is required for the net increase in purine compounds in caffeine-producing plants.

The later steps of the *de novo* caffeine biosynthesis (steps 12–16 in Figure 4.8) occur only in purine alkaloid-producing plants. In the *de novo* pathway, xanthosine is formed from XMP, but xanthosine is produced not only from XMP but also from guanosine (see Section 4.4.2). Hence, in the narrow sense, the caffeine biosynthetic pathway is defined in terms of the reactions from xanthosine to caffeine (steps 13–16 in Figure 4.8).

4.4.2 Caffeine Biosynthesis from Purine Nucleotides

Historically, caffeine biosynthesis has been investigated using radiolabelled purine bases and nucleosides. The radioactivity from ^{14}C-labelled adenine, adenosine, guanine and guanosine applied to leaf or fruit segments was efficiently incorporated into theobromine and caffeine. These purine nucleosides and bases are not the direct precursors of caffeine biosynthesis but are converted by so-called salvage enzymes to their respective nucleotides, AMP and GMP, which enter the purine alkaloid biosynthetic pathway.[18]

The synthesis of purine nucleotides from purine bases and nucleosides is referred to as "purine salvage", which functions as an efficient reutilisation of purines produced by the degradation of nucleotides and nucleic acids.[18] Plants inherently possess this characteristic and it may have important roles in certain fundamental physiological processes.[39] These tracer experiments indicated that caffeine biosynthetic pathways, initiated from AMP and GMP, are also operative.

In addition to *de novo* synthesis, a portion of the xanthosine used for caffeine biosynthesis is derived from cellular purine nucleotide pools. Based on current knowledge of plant nucleotide metabolism,[39] the following two pathways appear to be operative for the *in planta* biosynthesis of xanthosine:

$$\text{AMP} \rightarrow \text{IMP} \rightarrow \text{XMP} \rightarrow \text{xanthosine (AMP pathway)}$$

$$\text{GMP} \rightarrow \text{guanosine} \rightarrow \text{xanthosine (GMP pathway)}$$

AMP and GMP are produced by both *de novo* and salvage pathways of purine nucleotide biosynthesis.[39]

In addition, xanthosine is also derived from adenosine released from the *S*-adenosyl-L-methionine (SAM) cycle (Figure 4.9).[40] SAM is the methyl donor for the methylation reactions in the caffeine biosynthetic pathway (steps 11, 13 and 14 in Figure 4.9). In the process, SAM is converted to *S*-adenosyl-L-homocysteine (SAH) (step 2 in Figure 4.9), which is then hydrolysed to homocysteine and adenosine (step 3 in Figure 4.9). Homocysteine is recycled *via* the SAM cycle to replenish SAM levels (steps 4 and 1 in Figure 4.9), and adenosine released from the cycle (step 3 in Figure 4.9)

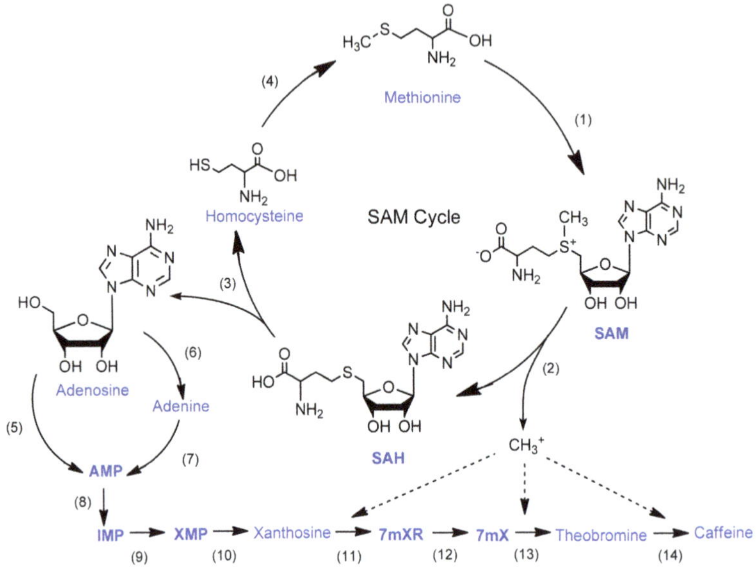

Figure 4.9 Possible caffeine biosynthetic pathway from adenosine released by *S*-adenosyl-L-methionine cycle. Enzymes (EC numbers) shown are: (1) SAM synthetase (EC 2.5.1.6); (2) SAM-dependent *N*-methyltransferases (EC 2.1.1.-); (3) SAH hydrolase (EC 3.3.1.1); (4) methionine synthase (EC 2.1.1.13); (5) adenosine kinase (EC 2.7.1.20); (6) AMP deaminase (EC 3.5.4.6); (7) IMP dehydrogenase (EC 1.1.1.205); (8) 5′-nucleotidase (EC 3.1.3.5); (9) 7-methylxanthosine synthase (EC 2.1.1.158); (10) *N*-methylnucleosidase (EC 3.2.2.25); (11) theobromine synthase (EC 2.1.1.159) and (12) caffeine synthase (EC 2.1.1.160). Abbreviations: SAM, *S*-adenosyl-L-methionine; SAH, *S*-adenosyl-L-homocysteine.

is then converted to AMP by the salvage enzymes (step 5 or steps 6 and 7 in Figure 4.9). AMP is further metabolised to xanthosine by the AMP pathway shown above (steps 8–10, Figure 4.9). Xanthosine is then utilised as the purine skeleton of caffeine. Since 3 moles of SAH are produced *via* the SAM cycle for each mole of caffeine that is synthesised, in theory this pathway has the capacity to be the sole source of both the purine skeleton and the methyl groups required for the three methylation steps in caffeine biosynthesis.[40]

The relative contributions of the *de novo* pathway and AMP, GMP and SAM pathways to caffeine biosynthesis may vary in different organs at different stages of development and the prevailing environmental conditions may also have an impact. A significant contribution of the *de novo* pathway to caffeine biosynthesis has been reported in young tissues.[37] It has also been reported caffeine is re-synthesised from theophylline, an intermediate of caffeine catabolism, *via* a theophylline → 3-methylxanthine → theobromine → caffeine pathway.[41]

4.4.3 *N*-Methyltransferases Involved in Caffeine Biosynthesis in Coffee Plants

The key enzymes of caffeine biosynthesis are *N*-methyltransferases which catalyse the sequential three-step methylation of xanthosine derivatives. There is no consistency in the nomenclature of *N*-methyltransferases in the literature. Ogawa *et al.*[42] used substrate names; xanthosine methyltransferase (XMT), monomethylxanthine methyltransferase (MXMT) and dimethylxanthine methyltransferase (DXMT). Kato *et al.*[29] used product names of each enzyme; 7-methylxanthosine synthase (XRS), theobromine synthase (TS) and caffeine synthase (CS). Since the latter system is registered with the IUBMB enzyme nomenclature, this review will use the terms 7-methylxanthosine synthase (EC 2.1.1.158), theobromine synthase (EC2.1.1.159) and caffeine synthase (EC 2.1.1.160).

4.4.3.1 *Genes of* N-*Methyltransferases*

Various genes encoding *N*-methyltransferases have been cloned from *C. arabica* and *C. canephora.*[42–45] Recently, Perrois *et al.*[46] isolated three different genes encoding *N*-methyltransferases in *C. canephora* and six genes in *C. arabica*. From the deduced protein sequences and previously published data, a phylogenic analysis was carried out. This revealed that the different *N*-methyltransferases involved in caffeine biosynthesis belong to three different clusters which align with the function of each enzyme. The clusters I, II and III correspond, respectively, to 7-methylxanthosine synthase, theobromine synthase and caffeine synthase (Figure 4.10).

The *N*-methyltransferases are involved in the newly characterised motif B′ methyltransferase family.[47] In contrast to the majority of plant SAM-dependent methyltransferases, which have three conserved motifs of the binding site of the methyl donor of SAM (motifs A, B and C),[48] the amino acid sequences of motif B′ methyltransferase family have motif A, motif B′, motif C and the YFFF. The motif B′ and YFFF region contains many specific hydrophobic amino acids. These types of amino acid sequences are also found in several methyltransferases which catalyse the formation of small, volatile methyl esters by using substrates with a carboxyl group as the methyl acceptor and SAM as the methyl donor.[49] Salicylic acid and benzoic acid *O*-methyltransferases and theobromine *N*-methyltransferase are included in this newly characterised methyltransferase family, which is referred to as the SABATH family.[49]

Recent genomic studies by Denoeud *et al.*[50] suggest that convergent evolution in caffeine biosynthesis occurred in different plant species, such as coffee, tea and cacao. These plants belong to several unrelated families, but they accumulate caffeine synthesised by a similar, if not identical, biosynthetic pathway. The genome sequence of caffeine biosynthesis indicates that the methyltransferase genes in some lineages have evolved

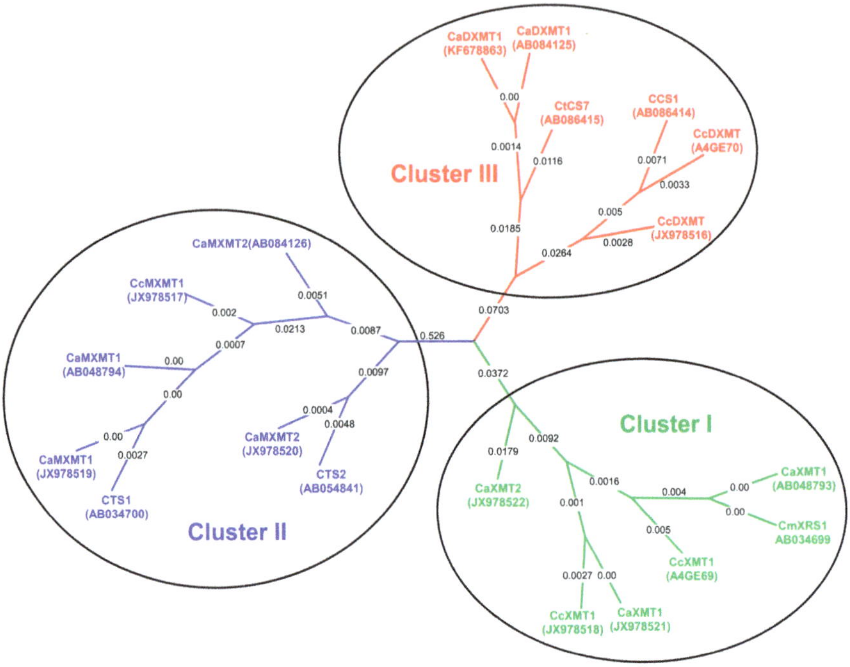

Figure 4.10 Unrooted maximum likelihood tree based on the alignment of *N*-methyltransferases involved in caffeine biosynthesis. Clusters I, II and III correspond to 7-methylxanthine synthase, theobromine synthase and caffeine synthase protein, respectively. Adapted from ref. 83 under the terms of the CC BY 4.0 licence, https://creativecommons.org/licenses/by/4.0/, © The authors.

independently from different branches of the SABATH methyltransferase gene family.[50,51]

Crystallographic data on salicylic acid carboxyl methyltransferase from *Clarkia breweri* suggest that members of this family of enzymes exist as dimers in solution.[52] Analysis of 7-methylxanthosine synthase and caffeine synthase from *C. canephora* have also revealed dimeric structures.[53,54] Despite the marked similarity in amino acid sequences of *N*-methyltransferases, each enzyme catalyses the methylation of specific substrate(s). Some reports suggest that a single amino acid residue of the *N*-methyltransferases decides the substrate specificity.[42,55]

4.4.3.2 Enzymatic Properties of Recombinant Enzymes

In contrast to tea caffeine synthase,[28] no highly purified, native *N*-methyltransferases of caffeine biosynthesis have been isolated from coffee. Therefore, recombinant enzyme proteins prepared with the coffee

N-methyltransferase gene sequences have been used to compare the properties of the three distinct enzymes. Here, we exclude the results obtained with partially purified native enzymes isolated from coffee plants, because the data were obtained with a mixture of plural enzymes, and as a consequence it is difficult to elucidate the properties of the individual *N*-methyltransferases. The properties of native enzymes from coffee plants were discussed in an earlier review.[18]

The first methylation step is catalysed by 7-methylxanthosine synthase (EC 2.1.1.158). The recombinant proteins (CmXRS1 and CaXMT1) involved in the Cluster I (Figure 4.10) have 7-methylxanthosine synthase activity. The methylation of the purine ring is initiated by the introduction of a methyl group at the 7 position of xanthosine (step 13 in Figure 4.8), after which ribose is released from 7-methylxanthosine and 7-methylxanthine is formed (step 14 in Figure 4.8). It has been suggested that this hydrolysis is carried out by a specific *N*-methylnucleosidase in tea leaves.[56]

Recombinant 7-methylxanthosine synthase proteins (CmXRS1 and CaXMT1) have been successfully prepared, independently, by two Japanese groups and some biochemical properties have been characterised.[43,44] The first isolation of the 7-methylxanthosine synthase gene from coffee leaves was claimed by Stiles and co-workers in the late 1990s.[57,58] However, Japanese groups have determined that the Stiles gene does not code 7-methylxanthosine synthase, but very closely resembles a lipase and the recombinant enzyme does not possess activity related to caffeine biosynthesis.[43,44]

The kinetics studies with recombinant 7-methylxanthosine synthase have indicated that the enzyme has relatively low Km values for xanthosine (~75 μM) and SAM (~10 μM) and that it is very specific for methylation of the 7*N* position (Table 4.1, Figure 4.11). Neither xanthine nor any methylxanthines are substrates of this enzyme.[43,44] On the basis of data obtained with a partially purified native enzyme preparation, Baumann and co-workers[59] argued

Table 4.1 The Km values in μM of recombinant *N*-methyltransferases from *C. arabica*.[a]

Enzymes			Substrates				
			XR	7mX	Px	Tb	SAM
I	7mXR synthase	CmXRS1	74				
		CaXMT1	78				13
II	Theobromine synthase	CTS1		873			
		CTS2		171			
		CaMXMT1		148	458		12
		CAMXMT2		251	738		14
III	Caffeine synthase	CCS1		126	31	157	
		CaDXMT1		916	973	1222	153

[a]XR, xanthosine; 7mX, 7-methylxanthine; Px, paraxanthine; Tb, theobromine; SAM, *S*-adenosyl-L-methionine.

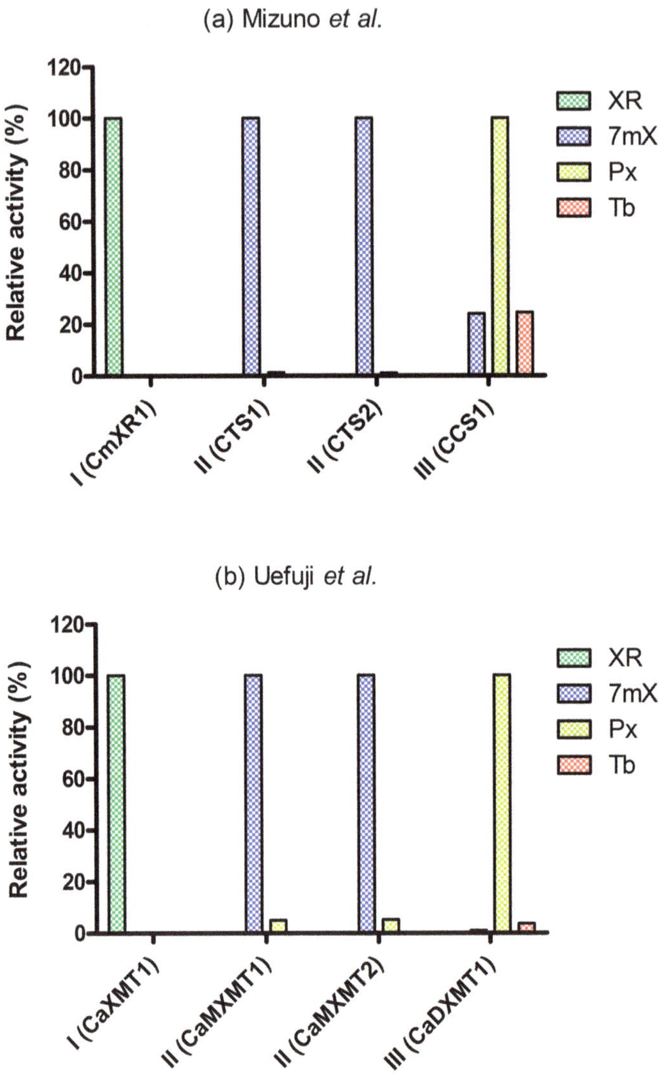

Figure 4.11 Substrate specificity of recombinant 7-methylxanthosine synthetase (I), theobromine synthase (II) and caffeine synthase (III). Based on data from Mizuno *et al.*[60] (a) and Uefuji *et al.*[43] (b). XR – xanthosine; 7mX – 7-methylxanthine; Px – paraxanthine; Tb – theobromine.

that caffeine biosynthesis in coffee begins with a 7*N*-methyltransferase converting XMP to 7-methyl-XMP, which is metabolised to 7-methylxanthine by a dephosphoribosylation reaction. However, XMP does not act as a substrate for the two recombinant 7-methylxanthosine synthases.[43,44] Although the possibility that unidentified coffee genes encode the XMP enzyme cannot be excluded, the Baumann results may be an artefact resulting from phosphatase contamination of the enzyme preparations.

The second *N*-methyltransferase involved in Cluster II (Figure 4.10) is theobromine synthase. Four recombinant enzymes (CTS1, CTS2, CaMXMT1 and CaMXTM2) have been characterised.[43,60] These enzyme preparations have similar kinetics. 7-Methylxanthine serves as the principal substrate. Paraxanthine (1,7-dimethylxanthine) is also a substrate, but its activity is only 1.1–5.3% of that of 7-methylxanthine (Figure 4.11). This enzyme, therefore, contributes mainly to 3*N*-methylation of 7-methylxanthine with theobromine (3,7-dimethylxanthine) being the resultant product. As a minor reaction, it also catalyses 3*N*-methylation of paraxanthine and, in this case, caffeine (1,3,7-trimethylxanthine) is formed. Since the Km value for paraxanthine is higher than that of 7-methylxanthine (Table 4.1), the affinity of theobromine synthase for paraxanthine is considered to be low, and as a consequence the enzyme appears to be used exclusively for theobromine synthesis.

The third *N*-methyltransferase is caffeine synthase. Two recombinant enzymes (CCS1 and CaDXMT1) belonging to Cluster III (Figure 4.10) have been produced and their properties were characterised by Mizuno *et al.*[45] and Uefuji *et al.*[43] In contrast to the first and second *N*-methyltransferases, the substrate specificity of this enzyme is broad. The best substrate for both enzymes is paraxanthine but the specificity of CCS1 is more diverse than that of CaDXMT1. To varying degrees, CCS1 can utilise several purine alkaloids as a substrate. The order of efficiency is: paraxanthine (100%) > theobromine (25%) > 7-methylxanthine (24%) > 3-methylxanthine (0.8%) >1-methylxanthine (0.5%). CCS1, therefore, catalyses the transfer of a methyl group to *N*3 and/or *N*1. *N*3 methyltransferase activity is higher than that of *N*1 activity. The Km values for three methylxanthines vary greatly between CCS1 and CaDXMT1 (Table 4.1). The Km values for 7-methylxanthine, paraxanthine and theobromine of purified native tea caffeine synthase are 186, 24 and 344 μM, respectively. The Km values of CCS1 are similar, but those of CaDXMT1 are 7–31 times higher than those of CCS1 (Table 4.1). This seems to be due, in part, to the varying degrees of purity of recombinant proteins.

Since caffeine synthetase has a dual function, possessing 1*N* and 3*N* methyltransferase activity, caffeine is synthesised from both theobromine and paraxanthine. However, *N*3 methylation activity is higher than *N*1 activity. Thus, formation of theobromine *via N*3 methylation of 7-methylxanthine predominates over paraxanthine production by methylation of 7-methylxanthine at *N*1 (III). Furthermore, theobromine synthase (II) preferentially produces theobromine. Therefore, a 7-methylxanthine → theobromine → caffeine pathway is catalysed by caffeine synthase. In addition, theophylline can be produced from 3-methylxanthine as a consequence of the broad substrate specificity of caffeine synthase.

The major and minor routes of the final stages of caffeine biosynthetic pathways are illustrated in Figure 4.12. The main route (steps 1–4 in Figure 4.12) is catalysed by three distinct *N*-methyltransferases (I, II and III). The nucleosidase reaction (step 2 in Figure 4.12) may be catalysed by a side

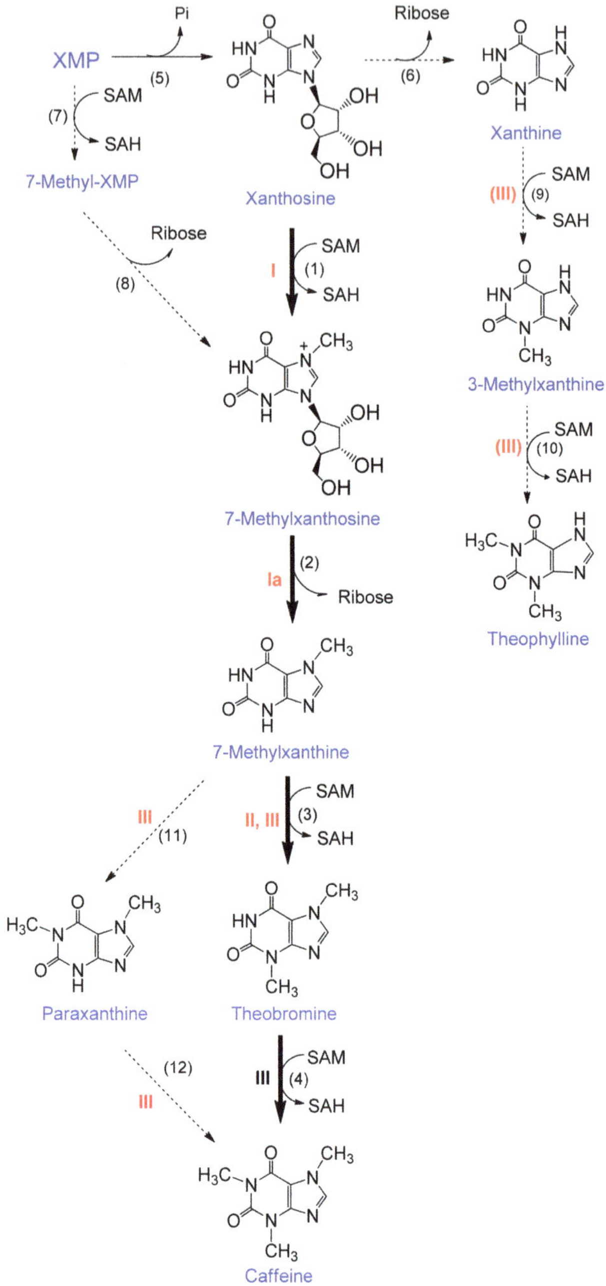

Figure 4.12 The biosynthetic pathways of caffeine from xanthosine. The major pathway consisting of four steps is shown in solid arrows (steps 1–4). Three types of *N*-methyltransferases, 7-methylxanthosine synthase, theobromine synthase and caffeine synthase, are shown as **I**, **II**, **III**. Conversion of 7-methylxanthosine to 7-methylxanthine (**Ia**) is catalysed by I or methylnucleosidase (see text). The EC numbers of enzymes involved are shown in the legend of Figure 4.1. Minor pathways, shown with dotted arrows, may occur because of the broad substrate specificities of the caffeine synthase (**III**). The route of 7-methylxanthosine formation from XMP *via* 7-methyl-XMP (steps 7–8) was not catalysed by any recombinant *N*-methyltransferases which involve the Clusters I to III.

reaction of the first *N*-methyltransferase (Ia). The reactions started from XMP (steps 7 and 8 in Figure 4.12) have not been demonstrated with recombinant enzymes. Minor routes of caffeine biosynthesis *via* paraxanthine (steps 11 and 12 in Figure 4.12) and conversion of 3-methylxanthine to theophylline (step 11 in Figure 4.12) may be due to the broad substrate specificity of caffeine synthase (III). Although conversion of xanthine to 3-methylxanthine was not detected in incubations with recombinant coffee *N*-methyltransfease,[43] this activity was detected in highly purified native enzyme from tea leaves.[28] The operation of this minor pathway in coffee plants needs to be confirmed.

4.4.4 Metabolism of Caffeine in *Coffea* Plants

In *C. arabica* and *C. canephora*, most of the synthesised caffeine accumulates as an end product. However, in some *Coffea* species, caffeine is catabolised initially by demethylation. Methyluric acids are also produced from caffeine in a small number of *Coffea* species.

4.4.4.1 Catabolic Pathways of Caffeine

Caffeine accumulates in leaves and seeds of coffee plants. Young leaves and fruits have a high capacity for caffeine biosynthesis but this declines markedly with age. Endogenous caffeine concentrations decrease as leaves and fruits mature, but this is due mainly to the increase in dry weight during the development of organs and substantial quantities of caffeine remain in mature leaves and fruits, even in aged tissues. Leaves and fruits of *C. arabica* and *C. canephora* have a very limited capacity for caffeine catabolism and, as a result, most caffeine that is produced accumulates and is not subjected to active turnover.[31,61]

Tracer experiments with ^{14}C-labelled caffeine, theophylline and theobromine have demonstrated that caffeine is degraded very slowly with cleavage of the three methyl groups resulting in the formation of xanthine.[62–67] Some low-caffeine containing *Coffea* species such as *C. eugenioides* possess high caffeine-degradation activity. To obtain further information on the detailed catabolic pathway of caffeine, pulse–chase experiments with [8-^{14}C]caffeine were carried out using disks of mature leaves of *C. eugenioides* (Figure 4.13).[67] Caffeine, theophylline and 3-methylxanthine were the most extensively labelled compounds after a 4 h pulse. The radioactivity associated with caffeine declined after the leaves were transferred to the non-radioactive medium. In contrast, ^{14}C-labelled theophylline, 3-methylxanthine, 1-methylxanthine, xanthine, allantoin, allantoic acid, urea and CO_2 increased after the 4 h chase, with >40% of the radioactivity taken up during the pulse being incorporated into 3-methylxanthine. After a further 20 h chase, radioactivity associated with theophylline, ureides and urea declined, whereas the ^{14}C incorporated into 3-methylxanthine, 1-methylxanthine and xanthine

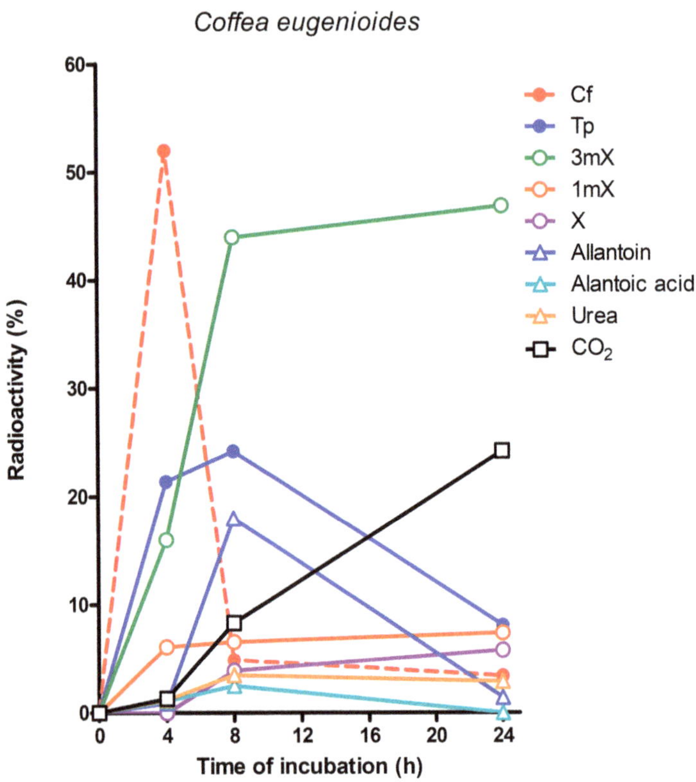

Figure 4.13 Metabolism of ^{14}C-labelled caffeine in a pulse–chase experiment with mature leaves of *C. eugenioides*. Leaf segments were incubated with [8-^{14}C]caffeine for 4 h (pulse), and then the incubation medium was replaced by fresh medium without tracer. The radioactivity was "chased" for a further 4 and 20 h. Incorporation of radioactivity into each compound is expressed as a percentage of the total radioactivity recovered. Cf – caffeine; Tp – theophylline; 3mX – 3-methylxanthine; 1mX – 1-methylxanthine; X – xanthine. Based on data from Ashihara and Crozier.[67]

changed little and $^{14}CO_2$ evolution increased from 8.3 to 24.2% of the recovered radioactivity. The results suggest that the major catabolic pathway is caffeine → theophylline → 3-methylxanthine → xanthine (steps 1–3 in Figure 4.14). In addition, a route *via* 1-methylxanthine (steps 1, 13 and 17 in Figure 4.14) is also functional. Xanthine is further degraded by the conventional purine catabolism pathway to CO_2 and NH_3 *via* uric acid, allantoin and allantoic acid (steps 4–7 in Figure 4.14).[65,67]

In contrast to caffeine, exogenous theophylline is readily degraded in *C. arabica*, demonstrating that the conversion of caffeine to theophylline is the major rate-limiting step of caffeine catabolism.[41,65] Theophylline is catabolised to xanthine mainly *via* 3-methylxanthine. In *C. arabica* leaves, small amounts of radioactivity from [8-^{14}C]theophylline and [2-^{14}C]xanthine were

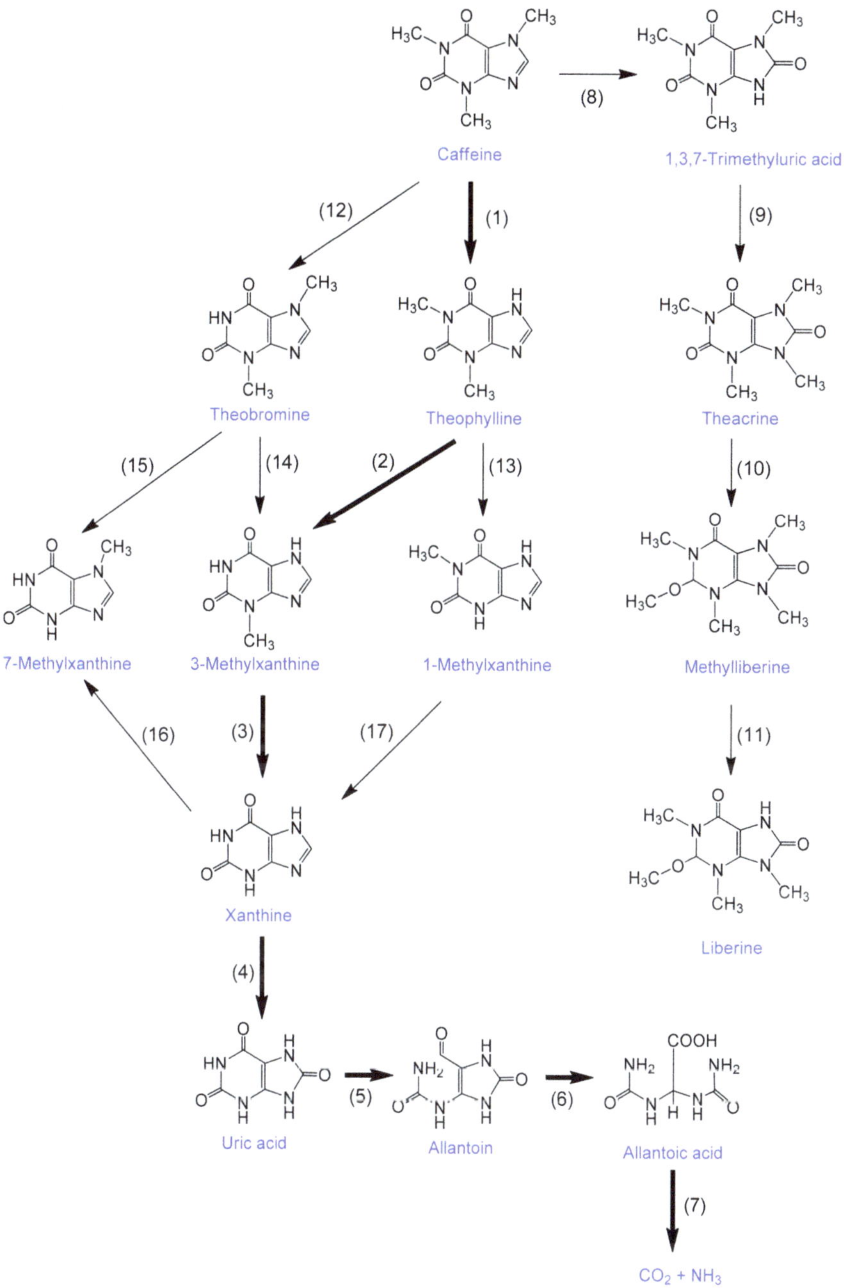

Figure 4.14 Possible routes for the catabolism of caffeine. After demethylation, xanthine enters the conventional oxidative purine catabolism pathway and is degraded to CO_2 and NH_3. The conversion of caffeine to theophylline is the rate-limiting step in *C. arabica* and *C. canephora*. Caffeine degrading pathway is operative in *C. eugenioides*. Solid arrows indicate major routes and thin arrow minor conversions. Some *Coffea* species, such as *C. abeokutae* and *C. dewevrei*, synthesise methyluric acids (steps 8–11). Minor routes observed in *Coffea* plants are also shown.

incorporated into 7-methylxanthine. This incorporation was enhanced by the treatment with allopurinol, an inhibitor of xanthine dehydrogenase, suggesting the operation of a unique theophylline → 3-methylxanthine → xanthine → 7-methylxanthine pathway (steps 2, 3 and 16 in Figure 4.14). This pathway has not been detected in species other than *C. arabica*. In tea and maté leaves, small amounts of theophylline are utilised for the caffeine synthesis *via* a theophylline → 3-methylxanthine → theobromine → caffeine pathway. In contrast to *C. arabica*, in other *Coffea* species, including *C. eugenioides*, *C. salvatrix*, *C. bengalensis* and *C. dewevrei*, [8-^{14}C]theophylline is neither converted to 7-methylxanthine nor utilised for the synthesis of caffeine.[67,68]

Although theobromine is an immediate precursor of caffeine, a small portion (<5% of total radioactivity taken up) of label from [8-^{14}C]theobromine applied to leaf disks of *C. arabica* was detected in CO_2.[65] Catabolism of theobromine to xanthine would be *via* either 3-methylxanthine or 7-methylxanthine (step 14 or 15 in Figure 4.14), although details of the pathway could not be determined because of the limited incorporation of radiolabel.[65] No enzyme activity related to the *N*-demethylation steps of caffeine catabolism has been isolated. Further studies on the genes encoding the caffeine demethylases are necessary to confirm the catabolic pathway(s).

4.4.4.2 *Biosynthesis of Methyluric Acid*

Mature leaves of *C. liberica*, *C. dewevrei* and *C. abeokutae* contain the methyluric acids theacrine (1,3,7,9-tetramethyluric acid), liberine (*O*[2], 1,9-trimethyluric acid) and methylliberine (*O*[2], 1,7,9-tetramethyluric acid).[69–72] Metabolism of purine alkaloids in the leaves of *C. dewevrei*, *C. liberica* and *C. abeokutae* was studied and the results indicate that the level of accumulated purine alkaloids changes with leaf age. Initially, caffeine accumulates in young leaves but it is gradually replaced by theacrine and liberine.[72] Data obtained in feeding experiments with leaf disks revealed that the radioactivity from [2-^{14}C]theobromine, [2-^{14}C]caffeine and [2-^{14}C]theacrine was incorporated into methylliberine and liberine. Although not detected because of a rapid rate of turnover, 1,3,7-trimethyluric acid may act as an intermediate with caffeine, formed from theobromine, being further metabolised by a caffeine → 1,3,7-trimethyluric acid → theacrine → methylliberine → liberine pathway (steps 8–11 in Figure 4.14).

4.4.5 Occurrence of Caffeine in *Coffea* Plants

The caffeine content of seeds of different *Coffea* species varies from 0 to 2.5% of dry weight (Figure 4.15).[73] The caffeine content of *C. canephora* seeds (1.7–2.4% of dry weight) is higher than that of *C. arabica* (Figure 4.15a). Most cultivars of *C. arabica* contain ~1% caffeine. However, Silvarolla *et al.*[74] discovered three naturally decaffeinated *C. arabica* plants, designated as AC1, AC2 and AC3, the seeds of which contained only 0.08% caffeine.

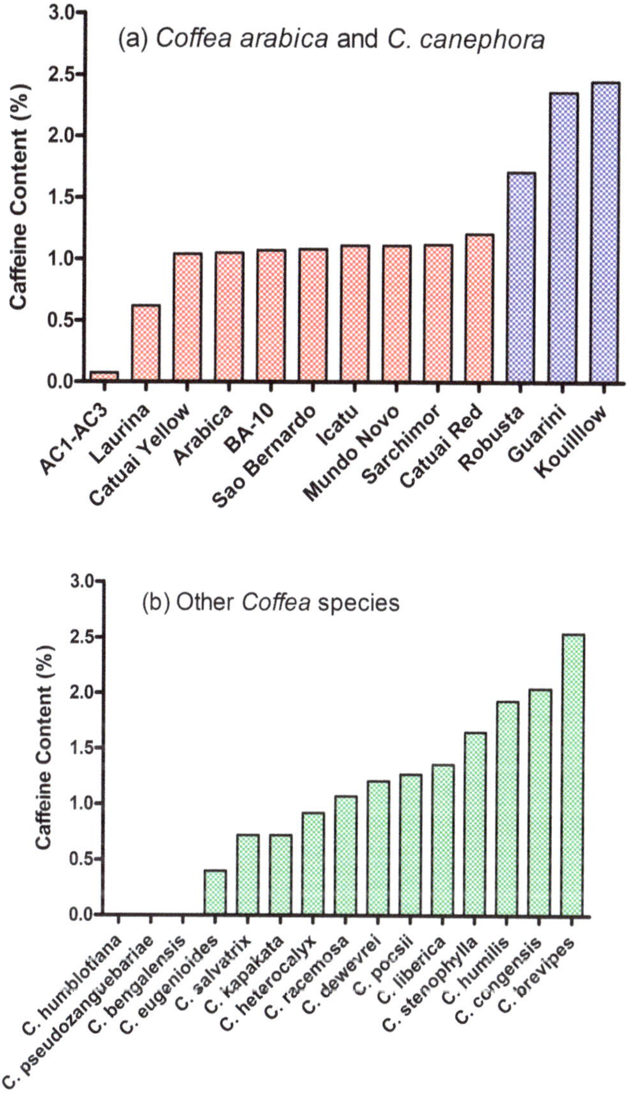

Figure 4.15 Concentration of caffeine in various cultivars of *C. arabica* (red bars) and *C. canephora* (blue bars) (a) and in different *Coffea* species (green bars) (b). Based on data from Mazzafera and Carvalho,[159] Mazzafera *et al.*[76] and Campa *et al.*[160]

In addition to *C. arabica* and *C. canephora*, which are used for commercial beverage production, the caffeine content of other *Coffea* species has been investigated (Figure 4.15b). No caffeine was detectable in *C. bengalensis* and a low caffeine content was found in *C. eugenioides* (0.4%), C. *salvatrix* (0.7%) and *C. racemosa* (0.8%). Higher concentrations of caffeine were detected in *C. congensis* (2.0%) and *C. stenophylla* (1.7%).[75]

Lower accumulation of caffeine is caused by reduced biosynthetic activity and/or enhanced degradation of caffeine. Mazzafera *et al.*[75] reported that caffeine was metabolised relatively slowly by immature endosperm of *C. arabica* and *C. canephora*. In contrast, the lower caffeine-accumulating species, *C. dewevrei*, *C. eugenioides*, *C. stenophylla*, *C. salvatrix* and *C. bengalensis*, all metabolised [8-^{3}H]caffeine much more rapidly, as the percentage recovery of the applied label was markedly lower and there was more extensive incorporation of radioactivity into theobromine, theophylline, 3-methylxanthine and two unidentified polar metabolites. The endogenous caffeine concentrations and the metabolism data indicate that there may be marked differences in the rate of turnover of caffeine in the various species of *Coffea*.

Mazzafera *et al.*[76] further investigated the [8-^{3}H]caffeine metabolism in leaves and immature fruits of *C. dewevrei* and *C. arabica* var. Mundo Novo and var. Laurina. Endosperms from *C. dewevrei*, containing 0.25% caffeine, had a much higher degradation rate than endosperm from *C. arabica* var. Mundo Novo, which had a 1.3% caffeine content. The caffeine contents in endosperms of mature fruits of these species were 1.2% and 1.1%, respectively. It was suggested, therefore, that the turnover of caffeine is high in endosperms of immature fruits of *C. dewevrei*, and decreases with maturation. The caffeine content of endosperms from immature and mature fruits of *C. arabica* var. Laurina were 0.8% and 0.6%, respectively, although they showed the same rate of degradation of [8-^{3}H]caffeine as *C. arabica* var. Mundo Novo. The difference in caffeine content between these two *C. arabica* varieties is therefore, arguably, due to a higher rate of biosynthesis in Mundo Novo than in Laurina.

Mazzafera *et al.*[76] also measured 1*N*- and 3*N*-methyltransferase activity in endosperms and leaves. These enzyme activities correspond to the dual functional caffeine synthase which catalyses both 1*N* and 3*N* methylation and theobromine synthase which catalyses 3*N* methylation. The enzyme activity in two *C. arabica* species was much higher than that in *C. dewevrei*. Therefore, biosynthetic activity also contributes to the accumulation of caffeine.

Ashihara and Crozier[67] studied the biosynthesis and catabolism of caffeine in leaf disks of *C. arabica* cv. Kent, and of three low-caffeine accumulating species, *C. salvatrix*, *C. eugenioides* and *C. bengalensis*, using various ^{14}C-precursors including [8-^{14}C]adenine and [8-^{14}C]caffeine. There was more extensive biosynthesis of caffeine from [8-^{14}C]adenine in young leaves of *C. arabica* than in *C. salvatrix*, *C. eugenioides* and *C. bengalensis*. Degradation of [8-^{14}C]caffeine, which is negligible in leaves of *C. arabica*, was also very slow in *C. salvatrix* and *C. bengalensis*. In contrast, [8-^{14}C]caffeine was catabolised rapidly by young and mature leaves of *C. eugenioides* (Figure 4.13). These results indicate that the low caffeine accumulation in *C. salvatrix*, *C. eugenioides* and *C. bengalensis* is a consequence of a slow rate of caffeine biosynthesis, whereas rapid degradation of caffeine also contributes to the low endogenous caffeine pool in *C. eugenioides*. The genes that regulate caffeine accumulation, therefore, appear to be those encoding 7-methylxanthosine synthase and caffeine demethylase activities.

Silvarolla *et al.*[74] reported that the low content of caffeine in *C. arabica* cv. AC was due not to enhanced degradation of caffeine but to a consequence of mutation of the gene encoding caffeine synthase which catalyses the last step of caffeine biosynthesis. Incorporation of [^{14}C]adenine into caffeine was not detected in the AC plants, instead, the ^{14}C label accumulated in theobromine.

These results suggest that in many cases, caffeine accumulation is dependent upon biosynthetic activity. Experiments using both [^{3}H]- and [^{14}C]caffeine established that the capacity to catabolise caffeine is limited to *Coffea* species such as *C. eugenioides*. Rapid conversion of caffeine to theophylline, theobromine and 3-methylxanthine in *C. salvatrix* and *C. bengalensis* was observed with [8-^{3}H]caffeine,[75] although catabolism of caffeine was not detected in the experiments with [8-^{14}C]caffeine.[67] The discrepancy might be due to substrates used with the much higher specific activity of [^{3}H]caffeine, enabling a lower dose to be administered than was the case with the [^{14}C]caffeine. In turn, this suggests a limited capacity to catabolise caffeine by these *Coffea* species.

Decaffeinated coffee plants have been produced by genetic engineering using RNAi or antisense technology to reduce the expression of caffeine biosynthetic enzymes.[77] An alternative approach to make transgenic caffeine-deficient *C. arabica* could be to enhance caffeine catabolism by the introduction of genes related to demethylation reactions. As noted above, the *N*-demethylase seems to be active in the leaves of *C. eugenioides*. Thus, the cloning of the demethylase from *C. eugenioides*, and its over-expression in transgenic *C. arabica*, is a further route to the production of caffeine-deficient coffee.

4.4.6 Physiological Aspects of Caffeine Metabolism in *Coffea* Plants

Accumulation of caffeine and caffeine biosynthetic activity fluctuates during growth and maturation of leaves and fruits of coffee plants. In general, caffeine biosynthesis occurs in young tissues, in particular leaves of flush shoots, young fruits and flower buds. Caffeine production is dependent mainly upon the expression of the genes encoding the *N*-methyltransferases involved in its biosynthesis pathway.

In six-month-old *C. arabica* seedlings, caffeine is found mainly in leaves and cotyledons at concentrations varying from 43–104 μmol g^{-1} dry weight. Essentially, caffeine is absent in roots or in older brown parts of the shoot. Caffeine biosynthetic activity, estimated from the incorporation of radioactivity from [8-^{14}C]adenosine into purine alkaloids, has shown that caffeine is synthesised *via* theobromine only in young leaves and young shoots including buds, but no such biosynthetic activity was found in roots or aged cotyledons.[78] A study with young and fully developed leaves from adult *C. arabica* trees showed they contain similar levels of caffeine but only the young leaves have the capacity to convert [8-^{14}C]adenosine to caffeine.[79]

Frischknecht *et al.*[80] proposed that the presence of caffeine in buds and leaves of *C. arabica* helps prevent predation by animals. Ashihara *et al.*[27] confirmed with *in situ* metabolism experiments that caffeine biosynthesis from xanthosine occurs only in young leaves of *C. arabica*, but the last step of the caffeine biosynthesis pathway, the conversion of theobromine to caffeine, occurs in both mature and aged coffee leaves. Caffeine synthase activity, therefore, appears to be present in coffee leaves even after maturation. This is different from tea leaves, in which caffeine synthase activity disappears once the leaves are fully developed.[81]

From a commercial perspective, studies on caffeine biosynthesis with coffee fruits are arguably more important than with leaves as it is the seeds which serve as the raw material for the production of the coffee beverage. Nonetheless, surprisingly few metabolic studies have been carried out using coffee fruits. On the basis of data obtained with the *in situ* metabolism of [methyl-^{14}C]methionine, Suzuki and Waller[64] suggested that caffeine biosynthesis occurred during the immature green stage of coffee fruit development.

More detailed studies on caffeine biosynthesis have been performed using fruits of *C. arabica* (cv. Mokka and Catimor) and *C. canephora*.[3] Fruits at seven different stages of development corresponding to (A) the pinhead and small size, (B) rapid expansion and pericarp growth, (C) endosperm formation, (D) early dry matter accumulation, (E) fully developed (green), (F) ripening (pink) and (G) fully ripened (red) were used as experimental materials (Figure 4.2).[2] Accumulation of caffeine and its biosynthetic activity are shown in Figure 4.16a–c. In all fruits, a marked increase in caffeine content was found between stages D and E accompanied by a parallel increase in dry weight. After stage E, 70–90% of caffeine was in seeds and the remainder was in pericarp. The concentration of caffeine in the ripened stage G seeds of the cultivars of *C. arabica* and *C. canephora* was 1.0% and 1.9% of dry weight, respectively. The biosynthetic activity of caffeine was estimated from the incorporation of radioactivity from [8-^{14}C]adenine into purine alkaloids. In stage A, no purine alkaloid synthesis was detected in *C. arabica* cv. Mokka, and only theobromine, the precursor of caffeine, was detected in *C. canephora*. Consequently, caffeine biosynthesis was low in the pinhead stage. Compared to *C. arabica* cv. Mokka, caffeine biosynthesis in cv. Catimor appeared to commence at an earlier stage of development as high biosynthetic activities of purine alkaloids were found in stages B and C which had almost disappeared at stages F and G.[3]

As shown in Section 4.4.3.1, three different types of genes encoding the *N*-methyltransferases for caffeine synthesis, namely *CmXRS*, *CTS* and *CCS* are present in *C. arabica*. Mizuno *et al.*[44] compared the expression of these three genes in developing endosperm, young leaves and flower buds using the RT-PCR, and found that the expression patterns of *CmXRS1* and *CCS1*,

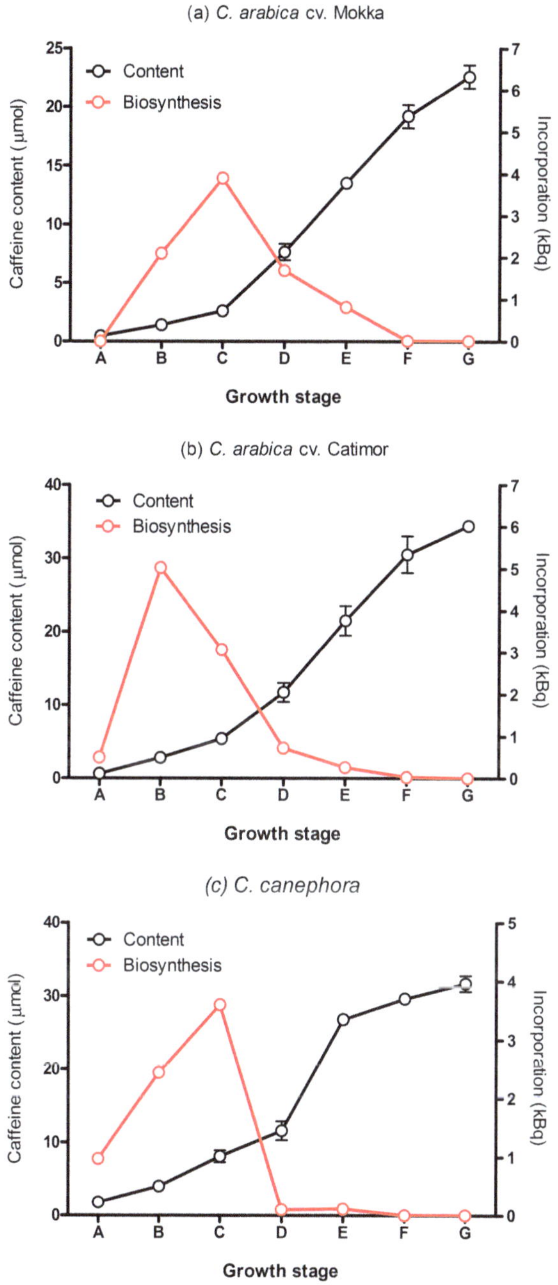

Figure 4.16 Changes in caffeine content and its biosynthetic activity in *C. arabica* cv. Mokka (a), *C. arabica* cv. Catimor (b) and *C. canephora* (c) during growth, development and ripening. The biosynthetic activity is estimated by the purine alkaloid synthesis from [8-^{14}C]adenine. Based on data from Koshiro *et al.*[3]

but not *CTS2*, were synchronised. Koshiro *et al.*[3] investigated the expression of genes encoding the three *N*-methyltransferases in *C. arabica* and *C. canephora*. In both species, the transcripts of *CmXRS1*, *CTS2* and *CCS1* were detected at every stage of growth, although the amounts decreased markedly at stage G. *CTS2* was the most weakly expressed gene. Since the second and third steps can be catalysed by the dual-functional CCS1,[45] the weak expression of *CTS2*, which encodes theobromine synthase, is unlikely to impact on caffeine synthesis in the fruits. Mizuno *et al.*[44] reported that *CTS2* was expressed strongly in flower buds of *C. arabica*.

To examine the site of caffeine biosynthesis in fruits, expression patterns of *CmXRS*, *CTS2* and *CCS1* in pericarp and in seeds from stages C–E fruits were investigated. Transcripts of the three genes encoding *N*-methyltransferases were higher in seeds than in pericarp (Figure 4.17a).[3] Higher expression of the genes in seeds than in pericarp correlated with higher activity of native caffeine synthase (*N*3-methyltransferase) in desalted protein extracts (Figure 4.17b) and higher *in situ* biosynthetic activity (Figure 4.17c). Expression of genes for *N*-methyltransferases for caffeine biosynthesis was generally related to the biosynthetic activity of caffeine. These findings demonstrate that caffeine accumulated in coffee seeds is synthesised mainly within seeds, although Baumann and Wanner[82] have reported the translocation of caffeine from the pericarp to seeds in *C. arabica* fruits.

Expression profiles of *N*-methyltransferase in coffee fruits have been investigated by two groups. Denoeud *et al.*[50] reported the expression of genes encoding 7-methylxanthosine synthase (*CcXMT*), theobromine synthase (*CcMXMT*) and caffeine synthase (*CcDXMT*) was higher in perisperm (the nutritive tissue surrounding the embryo) of *C. canephora* fruits (120–180 days after pollination) than in endosperm of mature fruits (180–320 days). Using *C. canephora*, Perrois *et al.*[83] carried out expression studies with four different stages of fruit development: small green, large green, yellow and red. Since there was no detailed description of the developing fruit, it is difficult to compare the findings with those of Koshiro *et al.*[84] However, large green, yellow and red stages may correspond to stages E–G of Koshiro *et al.*[84] The expression of the genes encoding 7-methylxanthine synthase (*CcXMT1*) and caffeine synthase (*CcDXMT*) was highest in younger fruits and decreased as the fruit ripened. The expression of theobromine synthase (*CcMXMT1*) was low. These results are essentially the same as those reported by Koshiro *et al.*[3] Compared with *C. canephora*, relative expression of genes from *C. arabica* is markedly low. Therefore, detailed comparison is difficult, other than the fact that the transcript levels of all genes decreased in the ripened red fruits.

In addition to the transcriptional control, a post-transcription process involving *N*-methyltransferases, or active supply of the substrate for caffeine biosynthesis from purine nucleotides, is important for the biosynthesis of caffeine in coffee fruits. These topics remain to be investigated.

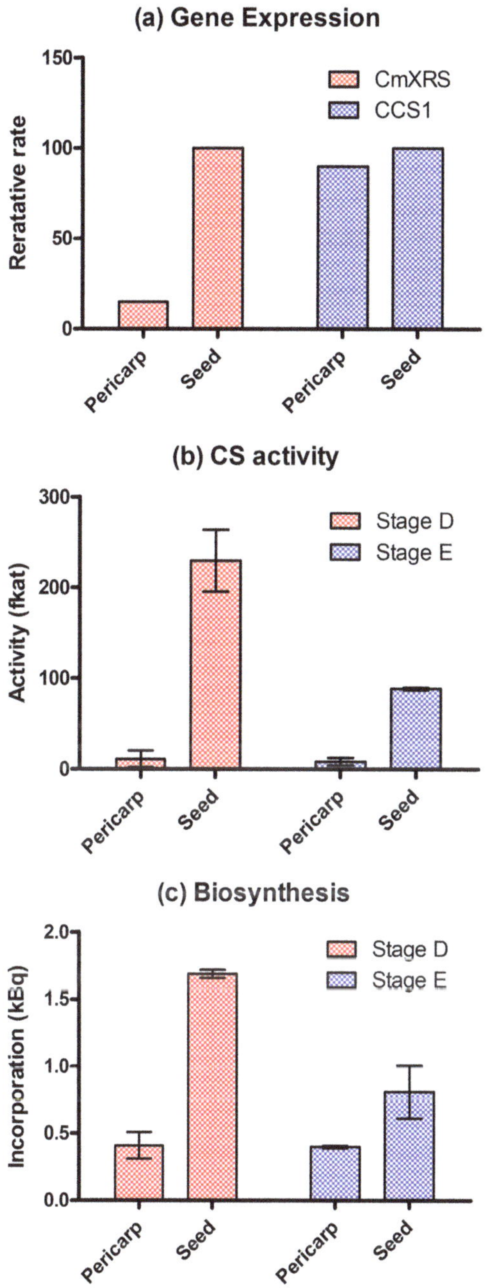

Figure 4.17 Comparison of the expression of genes encoding 7-methylxanthosine synthase (*CmXRS*) and caffeine synthase (*CCS1*) (a), caffeine synthase activity (b) and biosynthetic activity of caffeine (c) in pericarps and seeds of *C. arabica* cv. Mokka fruits. Data from stage D (a–c) and stage E (b and c) are shown. Based on data from Koshiro *et al.*[3]

4.5 Biosynthesis of Trigonelline

Trigonelline, *N*-methyl nicotinic acid, is one of the major nitrogen compounds in coffee seeds accumulating at almost the same concentration as caffeine at 1–3% of dry weight.[3,85–87] During roasting of seeds, trigonelline is partially degraded,[88,89] as a consequence, lower amounts (~0.5%) are found in roasted coffee beans and coffee beverage.[90] Large amounts of trigonelline also occur in leguminous species including fenugreek (*Trigonella foenum-graecum*),[91] alfalfa (*Medicago sativa*)[92] and soybean (*Glycine max*),[92,93] orange jessamine (*Murraya paniculata*)[94] and four o'clock flower (*Mirabilis jalapa*).[94] Unlike coffee, trigonelline does not accumulate in other beverage plants including tea,[95] cacao[96] and maté.[96,97]

The enzyme that catalyses the synthesis of trigonelline from nicotinic acid and *S*-adenosyl-L-methionine (SAM) was first discovered in the extracts of pea seeds.[98] Subsequently, trigonelline biosynthetic enzyme activity was detected in several plant species, including other legumes.[97,99] The nicotinic acid utilised for the trigonelline synthesis appears to be derived from nicotinamide, which is produced by the catabolism of NAD in *C. arabica*.[100,101] Recently, two genes encoding trigonelline synthase have been cloned from *C. arabica* by Mizuno *et al.*[102]

4.5.1 The *De Novo* Biosynthetic Pathway of Trigonelline

The biosynthesis of trigonelline in plants has been investigated in connection with nicotinic acid biosynthesis and NAD recycling, the so-called pyridine nucleotide cycle.[97] The *de novo* pathways of NAD biosynthesis operates in several plant species.[103,104] In bacteria and higher plants, quinolinic acid is synthesised from aspartate and glyceraldehyde-3-phosphate *via* the aspartate pathway (steps 1–5 in Figure 4.18).[97,104,105] In animals, however, quinolinic acid is formed by the tryptophan–kynurenine pathway.[106] Neither the enzymes nor the genome database of *de novo* pyridine nucleotide biosynthesis has been published for *Coffea* species, although, arguably, the pathway occurring in other plants may also function in coffee.[107,108]

Nicotinic acid mononucleotide (NaMN) is synthesised from quinolinic acid (step 3 in Figure 4.18). Theoretically, there are two possible pathways for synthesising nicotinic acid from NaMN. One is the pyridine nucleotide cycle, in which nicotinic acid is produced as a catabolite of NAD (steps 6–9 in Figure 4.18). The other is direct formation from NaMN produced by *de novo* pyridine nucleotide synthesis (steps 14–15 in Figure 4.18).

4.5.2 Pyridine Nucleotide Cycle for Nicotinic Acid Formation in *C. arabica*

Zheng *et al.*[100] studied the metabolism of [^{3}H]quinolinic acid, [^{14}C]nicotinamide and [^{14}C]nicotinic acid in fruits and leaf disks of *C. arabica*. All labelled precursors were efficiently converted to trigonelline and pyridine

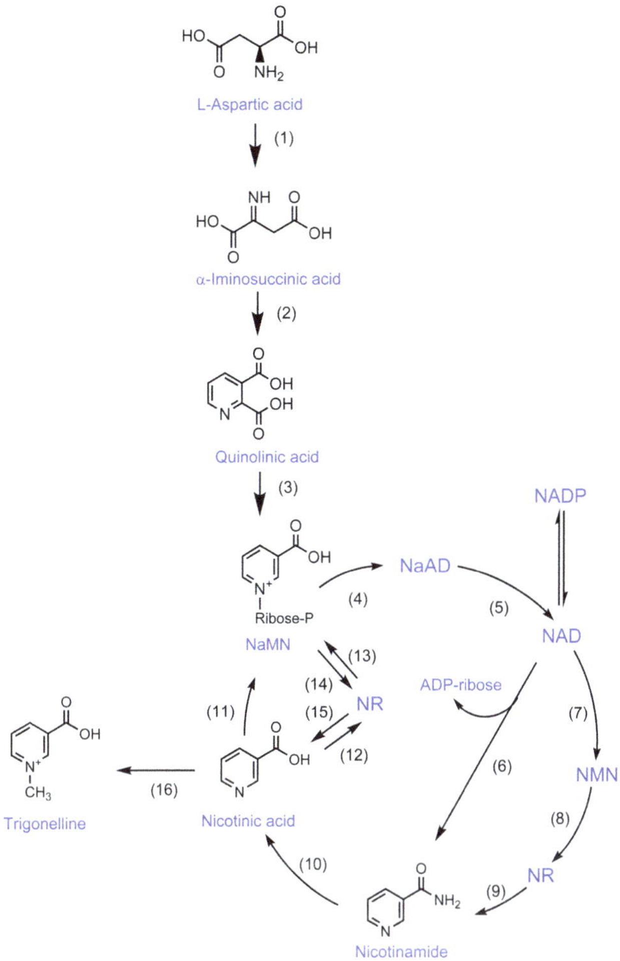

Figure 4.18 The *de novo* biosynthetic pathway of pyridine nucleotides, the pyridine nucleotide cycle and trigonelline synthesis in coffee plants. Enzymes (EC numbers) shown are: (1) L-aspartate oxidase (EC 1.4.3.16); (2) quinolinate synthase (EC 2.5.1.72); (3) quinolinate phosphoribosyltransferase (decarboxylating) (EC 2.4.2.19); (4) nicotinic acid mononucleotide adenylyltransferase (EC 2.7.7.18); (5) NAD synthase (EC 6.3.5.1); (6) poly(ADP-ribose) polymerases (EC 2.4.2.30), mono(ADP-ribosyl) transferase (EC 2.4.2.31) or NAD-dependent deacetylase (Sir2) (EC 3.5.1.-); (7) NAD pyrophosphatase (EC 3.6.1.22) (8) 5′-nucleotidase (EC 3.1.3.5); (9) nicotinamide riboside nucleosidase (EC 3.2.2.-); (10) nicotinamidase (3.5.1.19); (11) nicotinate phosphoribosyltransferase (2.4.2.11); (12) purine nucleoside phosphorylase (2.4.2.1); (13) NaR kinase (2.7.1.173) (14) 5′-nucleotidase (3.1.3.5); (15) nicotinic acid riboside nucleosidase (3.2.2.-); (16) trigonelline synthase (2.1.1.7). Substrate specificities of nucleotide and nucleoside hydrolysing enzymes are unknown; steps 7 and 13 may be catalysed by various phosphatases in addition to the 5′-nucleotidease, and steps 8 and 14 may be catalysed by tri-functional uridine nucleosidase/nicotinamide riboside hydrolase/nicotinic acid riboside hydrolase (3.2.2.3) shown in yeast.[161] Abbreviations: NaAD, nicotinic acid adenine dinucleotide; NaMN, nicotinic acid mononucleotide; NaR, nicotinic acid riboside; NMN, nicotinamide mononucleotide; NR, nicotinamide riboside.[97]

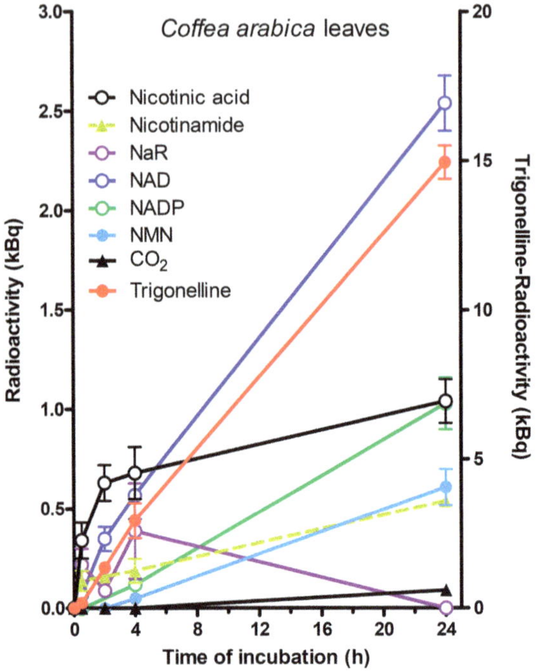

Figure 4.19 Metabolic fate of [carbonyl-^{14}C]nicotinamide in time-course experiments with almost fully developed young leaves of *Coffea arabica* cv. Yellow catuai. Incorporation of radioactivity into each compound is expressed as kBq per leaf disk (100 mg fresh weight). Scales for all metabolites except trigonelline are shown on left *y*-axis and trigonelline is on right *y*-axis. Based on data from Zheng *et al.*[100] Abbreviations: see legend for Figure 4.18.

nucleotides. The results suggest that in addition to the *de novo* pathway for NAD synthesis (steps 3–5 in Figure 4.18), the pyridine nucleotide cycle (steps 10–11and 4–5 in Figure 4.18) is also functional in coffee.

To study the pyridine nucleotide cycle, time-course experiments were carried out with [carbonyl-^{14}C]nicotinamide and disks of fully developed leaves of *C. arabica* (Figure 4.19). After 30 min. >85% of the [^{14}C]nicotinamide taken up by the leaf segments had been metabolised. The most heavily labelled metabolite was nicotinic acid (36%). Approximately 30% of the radioactivity was found in salvage products, mainly nicotinic acid riboside (NaR) and NAD, while 15% was incorporated into trigonelline.

When leaf disks were incubated with ^{14}C-nicotinamide for longer periods of time, most radioactivity was associated with trigonelline with ~60% and ~70% of the total radioactivity taken up at 4 h and 24 h. The remainder was distributed between NAD, NADP and nicotinamide mononucleotide (NMN). Incorporation of ^{14}C into these pyridine nucleotides increased with the duration of incubation. After 24 h only small amounts of radioactivity were evolved as CO_2. This points to nicotinamide being rapidly converted to nicotinic acid in a reaction catalysed by a plant specific nicotinamide deaminase (EC 3.5.1.19).[103,109,110] Part of the nicotinic acid pool is utilised for the salvage synthesis of NAD and NADP *via* nicotinic acid mononucleotide (NaMN).[103] In plants, formation of NaMN from nicotinic acid is catalysed by nicotinate phosphoribosyltransferase (step 11 in Figure 4.18).[103,109,111] However, the transient accumulation of radioactivity in NaR in this study suggests that the alternative nicotinic acid → NaR → NaMN (steps 12–13 in Figure 4.18) pathway operates in *C. arabica* leaves. Radiolabelled NMN found in the later stages of incubation may be a degradation product of NAD (step 7 in Figure 4.18).

A significant proportion of nicotinic acid is converted to trigonelline (step 16 in Figure 4.18) by trigonelline synthase (nicotinate *N*-methyltransferase, EC 2.1.1.7). These data indicate that the seven-component pyridine nucleotide cycle, NaMN → NaAD → NAD → NMN → nicotinamide riboside → nicotinamide → nicotinic acid → NaMN (steps 4–10 in Figure 4.18), operates in *C. arabica*. In addition to this cycle, an alternative route, catalysed by purine nucleoside phosphorylase and NaR kinase, appears to occur in *C. arabica*. NaR kinase (step 13 in Figure 4.18) was discovered recently in plants,[112] and exogenously supplied [carboxyl-^{14}C]NaR is metabolised to NAD *via* NaMN in *C. arabica* fruits.[113] Therefore, the eight-component pyridine nucleotide cycle which includes this alternative route (steps 12–13 in Figure 4.18) appears to operate in *C. arabica*. Although only a few studies have been performed with plants, direct release of nicotinamide from NAD by several adenosine diphosphate-ribosylation reactions (step 6 in Figure 4.18) may occur. In this case, a five component pyridine nucleotide cycle is functional. The relative importance of the different pyridine nucleotide cycles has not yet been determined in any plant species.

4.5.3 Direct Formation of Nicotinic Acid from NaMN

As shown above, nicotinic acid is derived from NAD. It is unclear, however, whether or not trigonelline is synthesised exclusively by this pathway in coffee. Wagner *et al.* proposed that the direct route from NaMN to nicotinic acid, catalysed by NaMN glycohydrolase, operated in nicotine-producing tobacco roots.[114] Another route for nicotinic acid formation from NaMN was proposed by Zheng *et al.*,[115] who used *in situ* tracer experiments with [^{14}C]NaR and enzyme extracts from mung bean seedlings. They showed the conversion of NaMN to trigonelline by three enzymatic systems (steps 14–16 in Figure 4.18): NaMN nucleotidase (or 5′-nucleotidase), NaR nucleosidase (or purine nucleosidase) and trigonelline synthase. NaMN produced by *de novo* biosynthesis (steps 1–3 in Figure 4.18) appears to be used preferentially for NAD synthesis, since NaMN adenylyltransferase activity (step 4 in Figure 4.18) is greater than that of NaMN nucleotidase (step 14 in Figure 4.18). Overflow of NaMN produced by the *de novo* pathway may be directed to trigonelline synthesis in the embryonic axis of mung bean seedlings.[115]

Ashihara *et al.*[101] recently investigated the metabolic fate of [carboxyl-^{14}C] NaR in developing fruit of *C. arabica*. [^{14}C]NaR was rapidly converted to NaMN, which was used for NAD synthesis. [^{14}C]NaR was also utilised for trigonelline synthesis, but this process took longer than NAD synthesis. These findings point to trigonelline being synthesised mainly from nicotinic acid produced by the degradation of NAD (steps 6–9 and 16 in Figure 4.18). The direct trigonelline formation from NaR (steps 14–16 in Figure 4.18) was not apparent in the developing coffee fruit.

4.5.4 Trigonelline Biosynthesis from Nicotinic Acid

The last step of trigonelline biosynthesis, nicotinic acid + SAM → trigonelline + SAH, is catalysed by trigonelline synthase (nicotinate 1*N*-methyltransferase, EC 2.1.1.7). This enzyme has been partially purified from cell cultures[116] and leaves[117] of soybean. The molecular weight of the native enzyme was estimated as ~85 000 by gel filtration and ~42 000 by sodium dodecyl sulfate–polyacrylamide gel electrophoresis,[116,117] suggesting that the enzyme is a dimer which consists of two identical subunits.

Mizuno *et al.*[102] reported that the *CTgS1* and *CTgS2* genes from *C. arabica* encoded trigonelline synthase. These enzymes belong to the SABATH (motif B′) gene family. *CTgS1* (AB054842) and *CTgS2* (AB054843) consist of 1373-bp and 1434-bp sequences, respectively, and encode 386 amino acid residues. Their amino acid sequences are very similar, exhibiting 95% homology. *CTgS1* shares a high degree of sequence identity (>80%) with 7-methylxanthosine synthase (*CmXRS1*), theobromine synthase (*CTS1*) and caffeine synthase (*CCS1*) of *C. arabica*. *CTgS2* had the same value of sequence identity as the caffeine biosynthesis enzymes. Thus, *CTgS1* and *CTgS2* are both highly homologous with the caffeine biosynthesis enzymes from coffee (Figure 4.20).

The molecular mass for both the recombinant *CTgS1* and *CTgS2* was 43.2 kDa. The recombinant enzymes are specific for nicotinic acid and no purine compounds function as substrates. The pH optimum for both *CTgS1* and *CTgS2* was 7.5. The Km values of *CTgS1* and *CTgS2* for nicotinic acid are 121 and 184 μM and for SAM are 68 and 120 μM, respectively.

4.5.5 Metabolism of Trigonelline in *Coffea* Plants

Although trigonelline is considered to be the end product of pyridine metabolism,[3,118] degradation of trigonelline has been postulated. Taguchi and Shimabayashi[119] first demonstrated trigonelline demethylating activity in several plants. Shimizu and Mazzafera[120] reported that trigonelline demethylase activity increases up to 4 weeks after germination of coffee seeds. The evidence, thus, suggests that at least some trigonelline is demethylated to nicotinic acid. Some of the nicotinic acid formed from trigonelline may be further degraded. However, trigonelline degradation activity is very low in *C. arabica* and *C. canephora* and rapid catabolism has yet to be detected in other *Coffea* species.

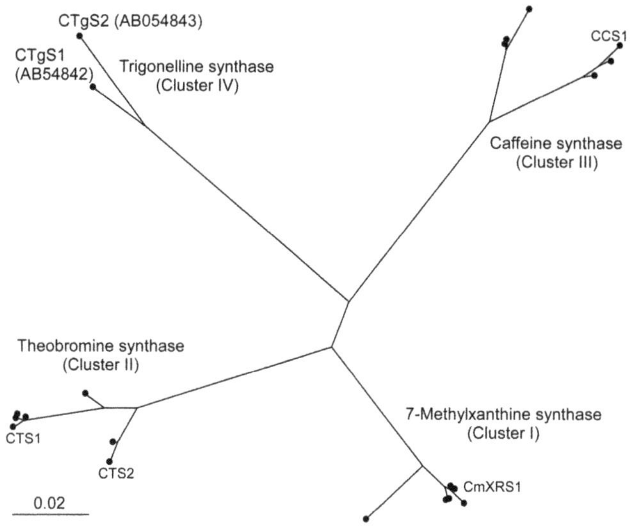

Figure 4.20 Phylogenetic analysis of the coffee *N*-methyltransferase family. *N*-methyltransferases involved in caffeine biosynthesis and those implicated in trigonelline biosynthesis are highly homologous and can be classified into four different clusters.[102]

4.5.6 Occurrence of Trigonelline in *Coffea* Plants

The trigonelline content of seeds of different *Coffea* species varies from 1.2% to 2.5% of dry weight, and most cultivars contain ~2% (Figure 4.21a).[3,86] The trigonelline content of *C. canephora* seeds is similar to or slightly lower than that of *C. arabica*,[3] but an exceptionally higher concentration has been reported to occur in *C. canephora* cv. Guariui (Figure 4.21b).[86] Ky *et al.* reported on the trigonelline content of green coffee beans in 76 genotypes of wild *C. arabica* and *C. canephora* accessions originated from African countries. The average and range of values were 1.19% (0.88–1.77%) and 1.01% (0.75–1.24%), respectively.[87]

In addition to *C. arabica* and *C. canephora*, the trigonelline content has been investigated in other *Coffea* species with low concentrations being found in three cultivars of *C. dewevrei* and *C. liberica* (Figure 4.21b).

4.5.7 Physiological Aspects of Trigonelline Metabolism in *Coffea* Plants

Accumulation of trigonelline and its biosynthetic activity fluctuate during the growth and maturation of leaves and fruits of coffee plants.[3,100] Zheng *et al.*[100] used leaves of *C. arabica* cv. Yellow catuai at six distinct stages of growth: (stage I) expanding buds; (stage II) small-sized young leaves; (stage III) almost fully developed young leaves; (stage IV) mature leaves from flash shoots; (stage V) aged leaves from 1 year-old shoots; and (stage VI)

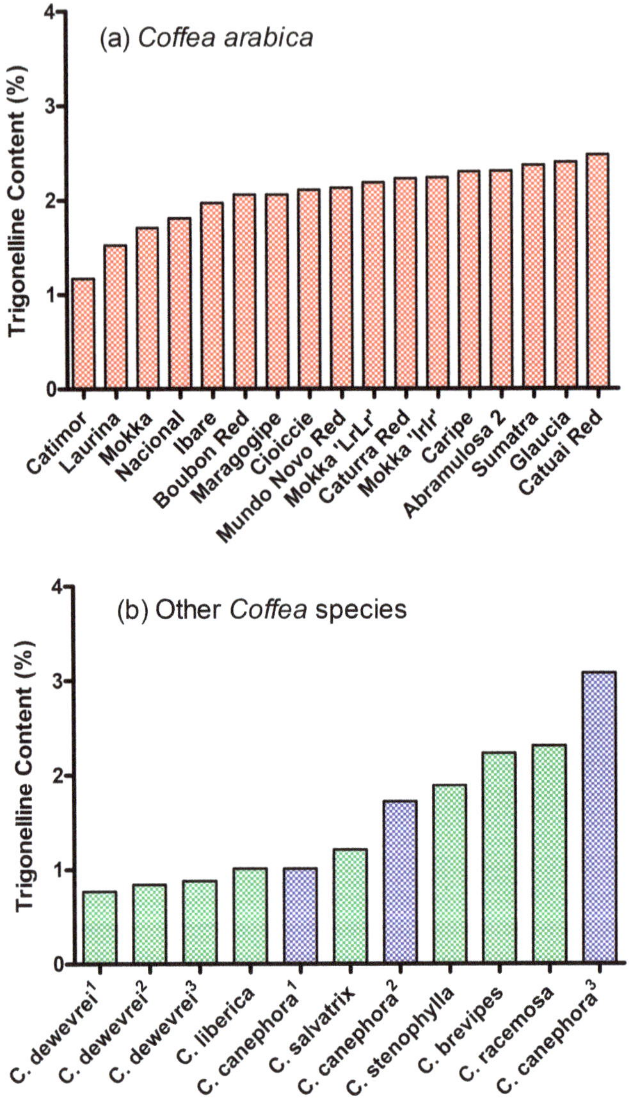

Figure 4.21 Concentration of trigonelline in various cultivars or varieties of (a) *C. arabica* and (b) *C. canephora* (blue bars) and wild *Coffea* species (green bars) (b). Based on data from Mazzafera,[86] Koshiro *et al.*[3] and Farah.[1]

fallen yellow leaves. Figure 4.22 shows the trigonelline content (μmole per leaf) and its concentration (μmole per g fresh weight) in tissues at these different stages of development. High concentrations were found in leaves at stages I and II, but the total content of trigonelline per leaf was highest in fully developed young leaves. Trigonelline accumulated during leaf development, but decreased markedly during leaf maturation and senescence.

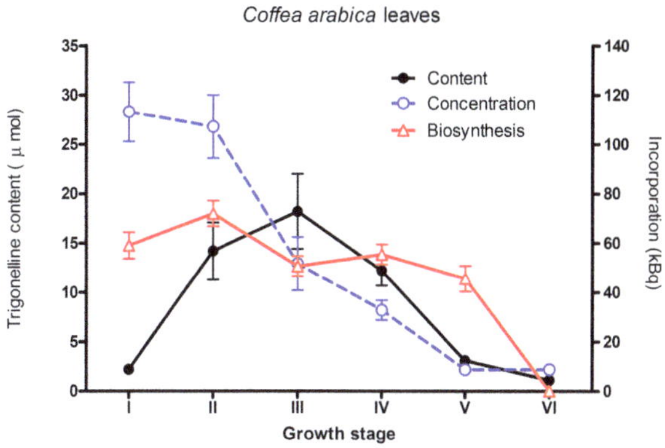

Figure 4.22 Changes in content and concentration of trigonelline and its biosynthetic activity in leaves of *C. arabica* cv. Yellow catuai during growth and senescence. The biosynthetic activity is estimated by trigonelline synthesis from [carbonyl-^{14}C]nicotinamide. Based on data from Zheng *et al.*[100]

Consistent with this finding, there is high trigonelline biosynthetic activity from [carbonyl-^{14}C]nicotinamide in developing leaves.[100] In young coffee leaves (stages I–III), *de novo* NAD biosynthesis and the pyridine nucleotide cycle are both very active.[100] Nicotinic acid, formed from NAD *via* nicotinamide, may be used preferentially for NAD formation and the remainder for the production of trigonelline. After the leaves mature there is no massive accumulation of trigonelline, and only a small amount remains in detached senescent leaves (stage VI). Trigonelline synthase activity, however, is high even in old leaves (stage V). The endogenous supply of nicotinamide in aged leaves *in planta* may be limited because of the decreased turnover of the pyridine nucleotide cycle.[100] As a result, production of trigonelline in aged leaves declines. The apparently high trigonelline synthetic activity in aged leaves, estimated with [^{14}C]nicotinic acid, may be related to the *in planta* conversion of nicotine to trigonelline possibly serving as a detoxification reaction to remove nicotinic acid, which is harmful in plant cells and tissues.[109,121] In aged leaves, sizable amounts of trigonelline may be transported to other organs, such as fruits, although it is difficult to preclude the possibility that trigonelline is demethylated to nicotinic acid and used for NAD recycling in aged leaves.

Trigonelline content and its biosynthetic activity in coffee seed and pericarp have been investigated using different varieties of coffee.[3,100] Koshiro *et al.*[3] used fruits of *C. arabica* cv. Mokka and cv. Catimor and *C. canephora*. Fruits were divided into seven stages according to growth and maturity as shown in Section 4.1. Accumulation of trigonelline and changes in trigonelline biosynthetic activity during the growth and ripening of the fruits of the two cultivars of *C. arabica* and *C. canephora* are shown in Figure 4.23a–c. There was an increase in trigonelline content after the bean dry matter accumulation stage (stage D), which continued to build up during ripening.

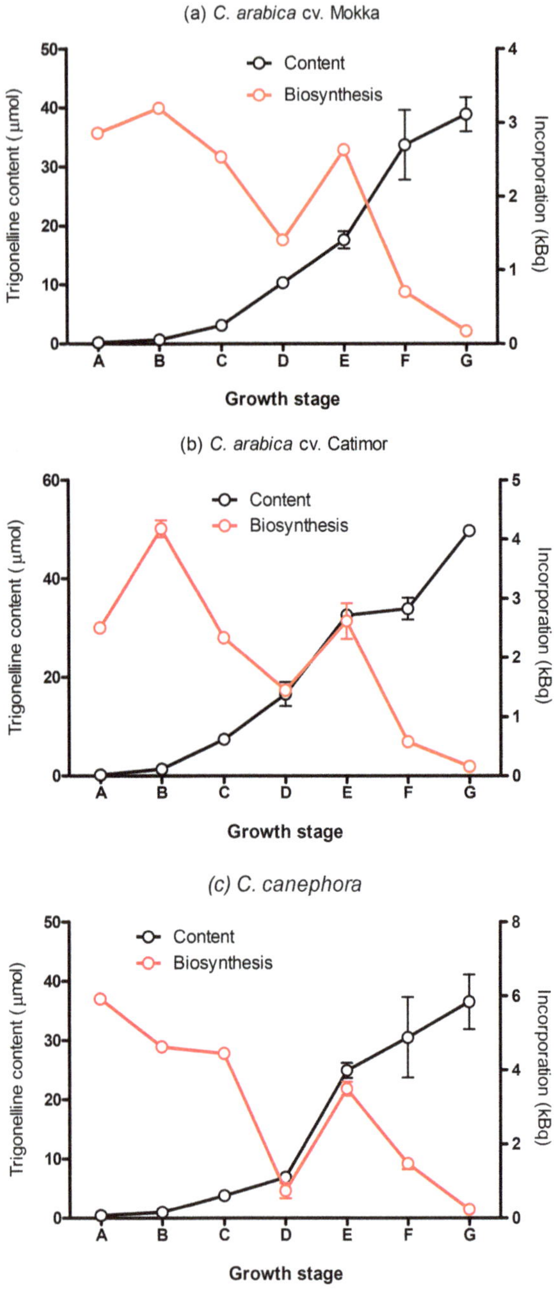

Figure 4.23 Changes in trigonelline content and its biosynthetic activity in fruits of (a) *C. arabica* cv. Mokka, (b) *C. arabica* cv. Catimor and (c) *C. canephora* during growth, development and ripening. The biosynthetic activity is estimated by trigonelline synthesis from [carbonyl-^{14}C]nicotinamide. Based on data from Koshiro *et al.*[3]

Changes in the biosynthetic activity of trigonelline in fruits during growth and ripening of the two cultivars of *C. arabica* and *C. canephora* are estimated from the incorporation of radioactivity from [carbonyl-^{14}C]nicotinamide into trigonelline and shown as "biosynthesis" in Figure 4.23. High *in situ* biosynthetic activity of trigonelline was found in fruits of (A) the pinhead, (B) rapid expansion and pericarp growth and (C) seed (endosperm) formation stages. The biosynthetic activity temporarily decreased at the seed dry matter accumulation stage (D) and increased again with the onset of the fruit ripening stage (E) before gradually declining with ripening of the fruits (stages F and G).

Distribution and biosynthetic activity of trigonelline in seeds and pericarp were investigated in the samples obtained from different stages of fruit ripening (Figure 4.24). After the early fruit ripening stage (stage E), 70–90% of trigonelline was found in seeds, and the remainder was recovered in pericarp (Figure 4.24a). Concentrations of trigonelline in seeds of ripe fruits (stage G) of *C. arabica* cv. Mokka, *C. arabica* cv. Catimor and *C. canephora* were *ca.* 1.3%, 1.0% and 1.4% of dry weight, respectively.

In all three *Coffea* species, biosynthetic activity of trigonelline was higher in pericarp than in seeds and the activity decreased with ripening (Figure 4.24b). This implies that net biosynthesis of trigonelline takes place in the pericarp, and that trigonelline is transported from the pericarp to the seeds during ripening.

4.5.8 *In Planta* Function of Trigonelline in *Coffea* Plants

Several *in planta* roles for trigonelline have been proposed in other plant species, namely: as a nutrient source, a compatible solute, a bioactive substance for nyctinasty, cell cycle regulation, signal transduction, detoxification of nicotinic acid and an eco-chemical function such as host selection by herbivores (see review by Ashihara *et al.*).[97] In addition to *Coffea* plants, large amounts of trigonelline accumulate in leaves and stems of *Murraya paniculata* where, it has been proposed, it acts as a chemical defence that protects soft tissues from herbivorous predators such as insect larvae.[122] However, the function of trigonelline in *Coffea* plants has yet to be elucidated.

4.6 Biosynthesis of Chlorogenic Acids

Chlorogenic acids are a family of compounds comprising esters of hydroxycinnamates and quinic acid, and together with caffeine and trigonelline are major secondary metabolites that accumulate in coffee seeds.[85,123] In addition to coffee, a number of other species including Yerba maté (*Ilex paraguariensis*), tomato (*Solanum lycopersicum*), eggplant (*Solanum melongena*), tobacco (*Nicotiana tabacum*), apple (*Malus domestica*), pear (*Pyrus communis*), plum (*Prunus domestica*) and artichoke (*Cynara cardunculus*) also produce chlorogenic acids.

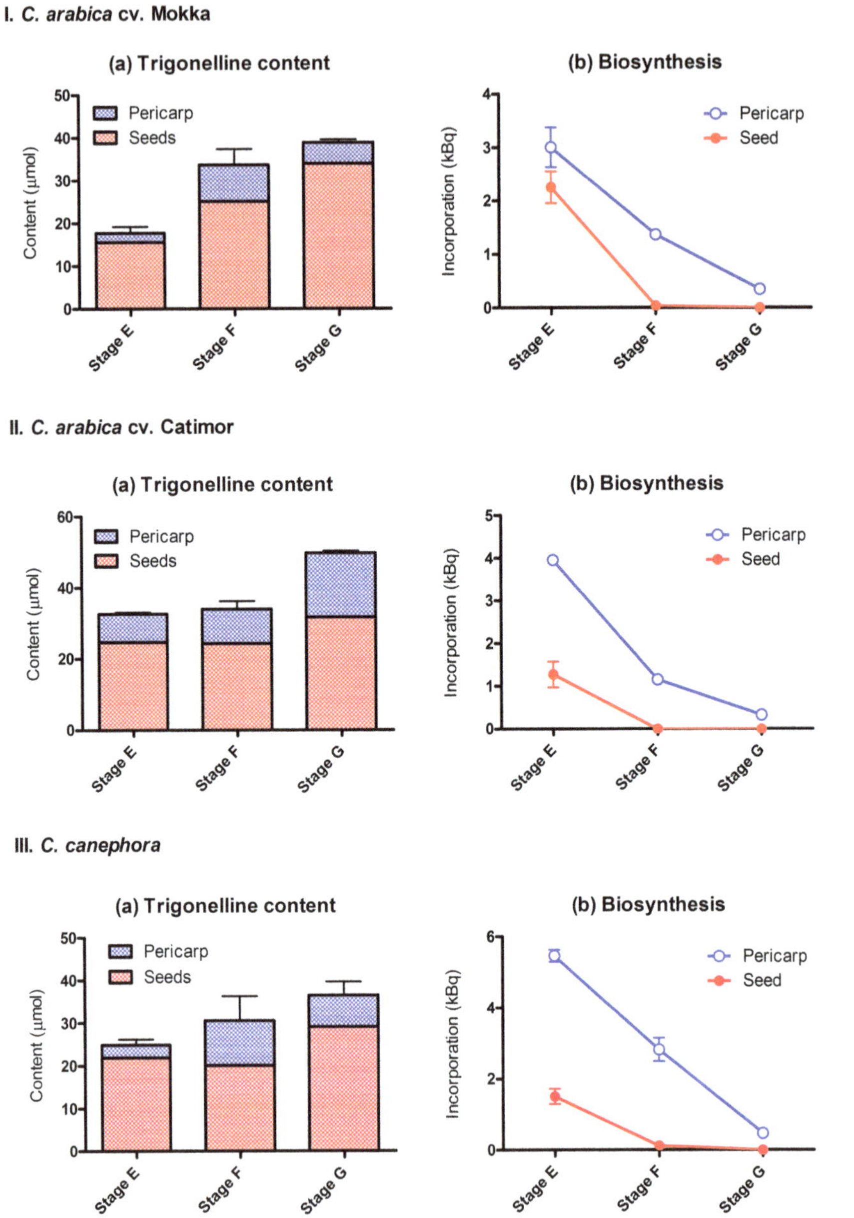

Figure 4.24 Comparison of the trigonelline content (a) and trigonelline biosynthetic activity (b) in pericarps and seeds of *C. arabica* cv. Mokka, *C. arabica* cv. Catimor and *C. canephora* during fruit ripening. Based on data from Koshiro *et al.*[3]

Caffeoylquinic acids (CQAs), which are depsides of caffeic acid and quinic acid, are major chlorogenic acids in seeds of coffee plants.[123,124] Three different groups of chlorogenic acids occur in *Coffea* plants, namely, CQAs, feruloylquinic acids (FQAs) and di-caffeoylquinic acids (diCQAs). The concentration of chlorogenic acids in seeds of *C. canephora* (7–14% of dry weight) is usually higher than that of *C. arabica* (3–8% of dry weight).[87,124]

4.6.1 Biosynthetic Pathways of Chlorogenic Acids

The biosynthetic pathway of chlorogenic acids starts with the core phenylpropanoid pathway, which converts phenylalanine to the simple phenylpropanoids, *trans*-cinnamic acid, *p*-coumaric acid and *p*-coumaroyl CoA (steps 1–3 in Figure 4.25).[125] The first step in the phenylpropanoid pathway is catalysed by phenylalanine ammonia-lyase (PAL, EC 4.3.1.24). The reaction is L-phenylalanine → *trans*-cinnamic acid + NH_3. The gene encoding PAL has been isolated from *C. canephora*.[126] The next step is catalysed by cinnamate 4-hydroxylase (C4H, EC 1.14.13.11). In this reaction, a hydroxyl group is introduced into the phenyl group of *trans*-cinnamic acid, and the first phenolic product, *p*-coumaric acid is formed. The total reaction is *trans*-cinnamate + NADPH + H^+ + O_2 → *p*-hydroxycinnamate + $NADP^+$ + H_2O. Although no report has been published on *Coffea*, this enzyme appears to be a cytochrome P450 monooxygenase. The final step in the core phenylpropanoid pathway is the formation of *p*-coumaroyl CoA from coumaric acid. This reaction is catalysed by 4-coumarate-CoA ligase (4CL, EC 6.2.1.12). The reaction is ATP + *p*-coumarate + CoA → AMP + PPi + *p*-coumaroyl-CoA.

Three routes have been proposed for the biosynthesis of chlorogenic acids from *p*-coumaroyl-CoA (Figure 4.25).[127] The route I is found in Solanaceous plants, including tomato, tobacco and potato. The pathway may be also operative in *Coffea*. The route is *p*-coumaroyl-CoA → 5-*O*-*p*-coumaroylshikimic acid → 5-*O*-caffeoylshikimic acid → caffeoyl-CoA → 5-*O*-caffeoylquinic acid (5-CQA) (steps 1–7 in Figure 4.25). The first step is catalysed by hydroxycinnamoyl CoA shikimate/quinate hydroxycinnamoyltransferase (HCT, EC 2.3.1.133). The reaction is *p*-coumaroyl-CoA + shikimate → CoA + 5-*O*-*p*-coumaroylshikimate. The next step is catalysed by *p*-coumaroylester 3′-hydroxylase (C3′H, *p*-coumaroylquinate/shikimate 3′-hydroxylase, EC 1.14.13.36). The last step involves the key enzyme of this pathway, a novel hydroxycinnamoyl-CoA quinate hydroxycinnamoyl transferase (HQT, EC 2.3.1), which catalyses the formation of 5-CQA from caffeoyl CoA and quinic acid. The reaction is caffeoyl-CoA + quinate → CoA + 5-CQA. This conversion resembles the reaction catalysed by hydroxycinnamoyl-CoA quinate transferase (HQT, EC 2.3.1.99), which uses feruloyl-CoA. Caffeoyl-CoA and *p*-coumaroyl-CoA can also act as donors, but the reaction is slower.[128]

Route II is found in sweet potato (*Ipomoea batatas*).[129] In this pathway, 1-*O*-caffeoylglucoside has been proposed as the activated intermediate (steps 10–12 in Figure 4.25). Route III operates in *Arabidopsis thaliana*, which produces, but does not accumulate, chlorogenic acids.[130,131] This route involves the synthesis of 5-*O*-*p*-coumaroylquinate by hydroxycinnamoyl CoA

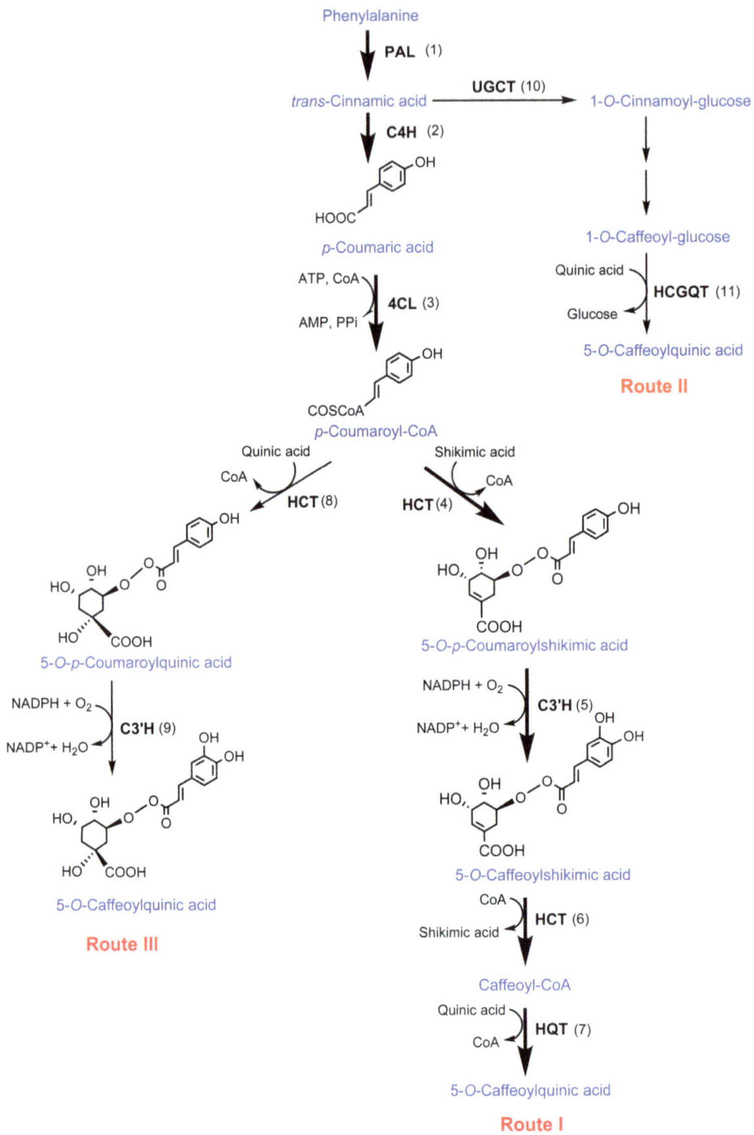

Figure 4.25 Possible biosynthetic pathways of chlorogenic acid (5-*O*-caffeoylquinic acid) in plants. The three different routes, I, II and III, are according to Niggeweg *et al.*[127] Routes I, II and III may be functional in *Coffea*, *Ipomoea batatas* and *Arabidopsis thaliana*, respectively. Abbreviations and EC numbers of enzymes: (1) phenylalanine ammonia lyase (PAL, EC 4.3.1.24); (2) cinnamate 4-hydroxylase (C4H, EC 1.14.13.11); (3) 4-hydroxycinnamoyl CoA ligase (4CL, EC 6.2.1.12); (4) hydroxycinnamoyl CoA shikimate/quinate hydroxycinnamoyltransferase (HCT, EC 2.3.1.113); (5) *p*-coumarate 3′-hydroxylase (C3′H, EC 1.14.13.36); (6) hydroxycinnamoyl CoA shikimate/quinate hydroxycinnamoyltransferase (HCT, EC 2.3.1.113); (7) hydroxycinnamoyl CoA quinate hydroxycinnamoyltransferase (HQT, EC 2.3.1.–); (8) hydroxycinnamoyl CoA shikimate/quinate hydroxycinnamoyltransferase (HCT, EC 2.3.1.113); (9) *p*-coumarate 3′-hydroxylase (C3′H, EC 1.14.13.36); (10) UDP-glucose:cinnamate glucosyl transferase (UGCT, EC 2.4.1.177); (11) hydroxycinnamoyl D-glucose: quinate hydroxycinnamoyl transferase (HCGQT, EC 2.3.1.–).

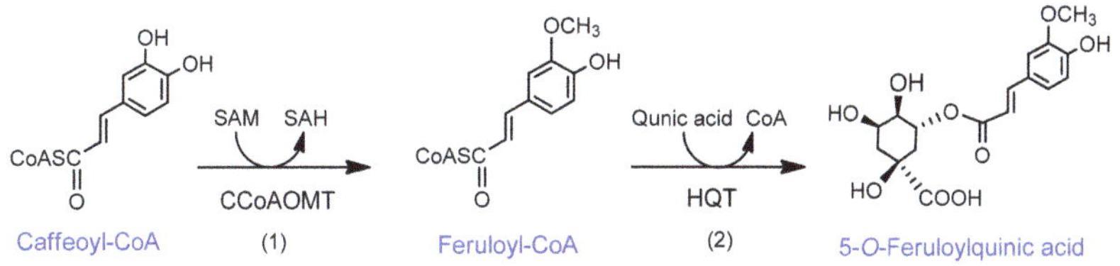

Figure 4.26 A possible biosynthetic pathway of 5-*O*-feruloylquinic acid. Abbreviations and EC numbers of enzymes: (1) Caffeoyl-CoA *O*-methyltransferase (CCoAOMT, EC 2.1.1.104); (2) hydroxycinnamoyl CoA quinate hydroxycinnamoyltransferase (HQT, EC 2.3.1.–).

shikimate/quinate hydroxycinnamoyltransferase (HCT) and subsequent hydroxylation by *p*-coumarate 3′-hydroxylase (C3′H) to form 5-CQA (steps 8–9 in Figure 4.25).

Biosynthesis of FQA starts with caffeoyl-CoA (Figure 4.26). Caffeoyl-CoA *O*-methyltransferase (CCoAOMT, EC 2.1.1.104) catalyses the reaction, SAM + caffeoyl-CoA → SAH + feruloyl-CoA (step 1 in Figure 4.26). 5-FQA is then formed in a reaction catalysed by HQT (step 2 in Figure 4.26). The gene encoding CCoAOMT from *C. canephora* and *C. arabica* has been sequenced although no enzymatic properties of the recombinant protein were reported.[132] Formation of diCQA is catalysed by HCT in *C. canephora*.[133]

4.6.2 Enzymes Involved in the Caffeoylquinic Acids Biosynthesis in *Coffea* Plants

Three key enzymes involved in the CQA biosynthesis in coffee plants have been characterised. The properties of these enzymes are as follows.

4.6.2.1 Hydroxycinnamoyl-CoA Shikimate/Quinate Hydroxycinnamoyltransferase (HCT)

In plants, there are two key acyl transferases which catalyse the formation and hydrolysis of 5-CQA; hydroxycinnamoyl-CoA shikimate/quinate hydroxycinnamoyltransferase (HCT, EC 2.3.1.133) and a novel hydroxycinnamoyl-CoA quinate hydroxycinnamoyltransferase (HQT, EC 2.3.1.). These enzymes share over 64% sequence identity and both can accept a range of acyl donors, including *p*-coumaroyl-CoA, caffeoyl-CoA and feruloyl-CoA, although they exhibit different acyl substrate preferences. [133] Sequence comparisons indicate that HCT and HQT both belong to the acyl-CoA dependent BAHD superfamily.[134] HCT has been purified and characterised from tobacco by Hoffmann *et al.*[130] The enzyme uses *p*-coumaroyl-CoA and caffeoyl-CoA as the preferred acyl group donors and the acyl-group is transferred more efficiently to shikimate than to quinate. The enzyme also catalysed the reverse reaction.

p-coumaroyl-CoA + shikimate → CoA + *p*-coumaroylshikimate
(step 4 in Figure 4.25)

CoA + *p*-caffeoylshikimate → caffeoyl-CoA + shikimate
(step 6 in Figure 4.25)

p-coumaroyl-CoA + quinate → CoA + coumaroyl quinate
(step 8 in Figure 4.25)

Lepelley *et al.*[135] cloned HCT from *C. canephora* and the recombinant protein was partially characterised, however, no enzymatic studies including the determination of Km values were reported.

4.6.2.2 *Hydroxycinnamoyl-CoA Quinate Hydroxycinnamoyltransferase (HQT)*

Niggeweg *et al.*[127] cloned cDNA encoding HQT from tobacco and tomato plants. Recombinant tobacco HQT was found to form coumaric or caffeic esters more efficiently with quinate than with shikimate, and was also capable of hydrolysing 5-CQA to form caffeoyl-CoA. Sonnante *et al.*[136] report on two genes encoding HQT from artichoke. The recombinant proteins (HQT1 and HQT2) showed much higher affinity for quinate over shikimate. For example, in the presence of saturated concentration of caffeoyl-CoA, the Km values of HQT1 for quinate and shikimate are 99 μM and 5000 μM, respectively. In the presence of *p*-coumaroyl-CoA, the Km of HQT1 for quinate is 216 μM, but no activity was found with shikimic acid. Therefore, this enzyme catalyses step 7 in the 5-CQA biosynthetic pathway (Figure 4.25). The CcHQT gene encoding HQT in *C. canephora* has been sequenced and cloned,[135] but there are no reports on the properties of the recombinant enzyme.

4.6.2.3 p-*Coumaroylester 3′-Hydroxylases*

In *Arabidopsis thaliana*, the 3′-hydroxylase reaction in the biosynthesis of 5-CQA can be catalysed by a cytochrome P450 enzyme CYP98A3, called *p*-coumaroylester 3′-hydroxylase.[131] This enzyme catalysed hydroxylation of the 3 position of coumarate, but only when it was esterified to either shikimate or quinate.[131,137] Two full-length cDNA clones (CYP98A35 and CYP98A36) encoding putative *p*-coumaroyl ester 3′-hydroxylases (C3′H) were isolated from *C. canephora* cDNA libraries.[138] Both recombinant enzymes catalysed the conversion of 5-*O*-*p*-coumaroylshikimate to 5-*O*-caffeoylshikimate (step 5 in Figure 4.25) at similar rates. The Km values of CYP98A35 and CYP98A36 for *p*-coumaroyl shikimate are 4.5 and 3.5 μM, respectively. However, only CYP98A35 can hydroxylate 5-*O*-*p*-coumaroyl quinate to 5-CQA, and the Km value is 10.3 μM (step 9 in Figure 4.25). These results suggest that two isozymes of C3′H occur in *C. canephora*; one participates in both routes I and III, but the other is specific for route III (Figure 4.25).

4.6.3 Shikimic Acid Pathway in Plants

Chlorogenic acids are derived from phenylalanine, shikimic acid and quinic acid. These compounds are produced by the shikimic acid pathway in plants (Figure 4.27).[139] To date, there are no reports on the shikimic acid pathway in coffee plants. Colonna[140] reported the conversion of $^{14}CO_2$ into ^{14}C-chlorogenic acids in coffee leaves but incorporation into the intermediates of the shikimic acid pathway was not shown. Therefore, an outline of this pathway will be explained briefly based on results obtained from other plants and their possible relationship to the CQA synthesis in coffee plants is discussed. The shikimic acid pathway occurs in higher plants, fungi and bacteria, but not in animals. The features of enzymes of this pathway, such as multi-enzyme complexes, vary in the different organisms.[139]

The initial substrates for the shikimic acid pathway are erythrose 4-phosphate (E4P) and phosphoenolpyruvate (PEP), which are intermediates of the pentose phosphate pathway and glycolysis, respectively. The pathway begins with a condensation reaction between E4P and PEP, catalysed by 3-deoxy-arabinoheptulosonate-7-phosphate (DAHP) synthase (EC 2.5.1.54, step 1 in Figure 4.27). The second step is catalysed by 3-dehydroquinate synthase (EC 4.2.3.4, step 2 in Figure 4.27). The reactions from 3-dehydroquinate to shikimic acid (steps 3–4 in Figure 4.27) are catalysed by a single dual-functional enzyme, 3-dehydroquinate dehydratase (EC 4.2.1.10)/shikimate dehydrogenase (EC 1.1.1.25). Shikimate is converted to shikimic acid 3-phosphate by shikimate kinase (EC 2.7.1.71, step 5 in Figure 4.3). The next step is EPSP formation using PEP by 5-enolpyruvylshikimic acid 3-phosphate (EPSP) synthase (EC 2.5.1.19, step 6 in Figure 4.27). EPSP is converted to chorismic acid by chorismate synthase (EC 4.2.3.5, step 7 in Figure 4.27). Chorismic acid is situated at a branch point in the pathway leading to the synthesis of aromatic amino acids, phenylalanine, tyrosine and tryptophan. The reaction that commits chorismic acid to the phenylalanine and tyrosine route is catalysed by chorismate mutase (EC 5.4.99.5, step 8 in Figure 4.27), which converts chorismic to prephenic acid. Prephenate aminotransferase (EC 2.6.7.78 or EC 2.6.7.79) transfers an amino group on to prephenic acid to form arogenic acid (step 9 in Figure 4.27). In the final step (step 10 in Figure 4.27), arogenic acid was dehydrated by arogenate dehydratase (EC 4.2.1.91) and L-phenylalanine, a substrate of the core phenylpropanoid pathway, is produced.

Shikimic acid, used in the biosynthesis of 5-CQA, is an intermediate in the shikimic acid pathway. Quinic acid seems to be formed from 3-hydroqunic acid, an intermediate of the pathway, in a reaction involving quinate dehydrogenase (EC 1.1.1.24, step 3a in Figure 4.27). The reaction is

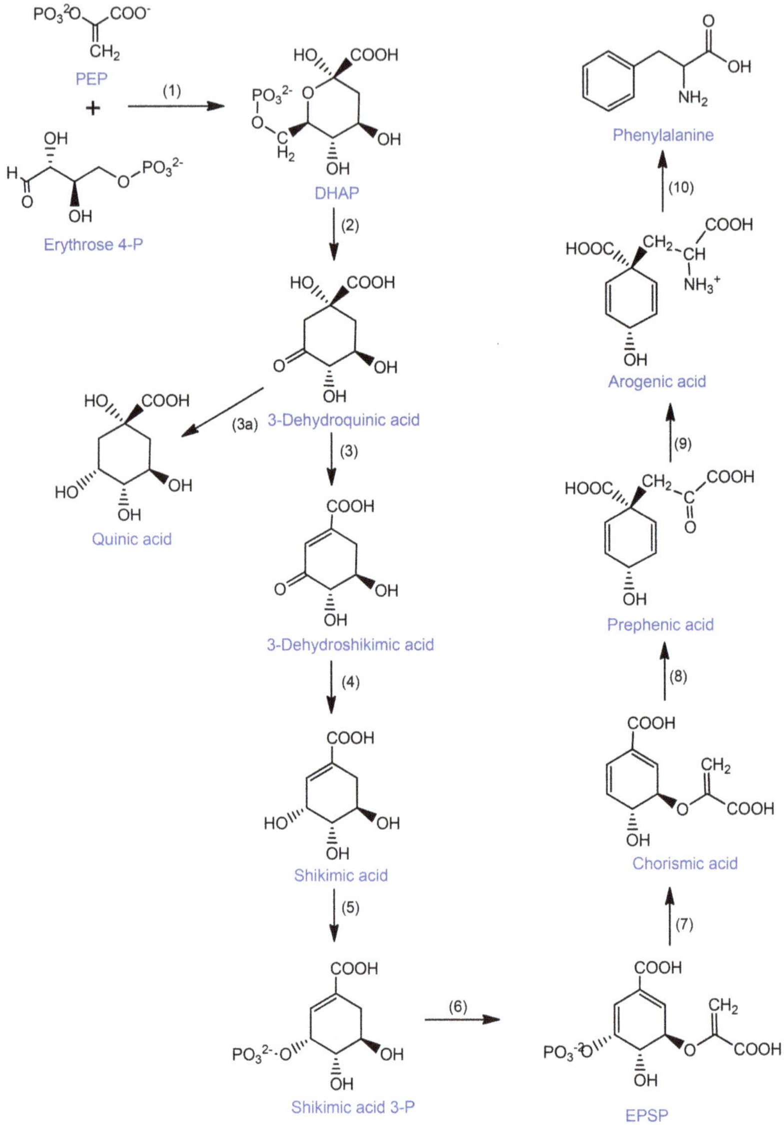

Figure 4.27 The shikimic acid pathway and the formation of quinic acid, shikimic acid and phenylalanine for the biosynthesis of chlorogenic acids. Enzymes and EC numbers: (1) 3-deoxyarabinoheptulosonate 7-phosphate (DAHP) synthase (EC 2.5.1.54); (2) 3-dehydroquinate synthase (EC 4.2.3.4); (3–4) 3-dehydroquinate dehydratase (EC 4.2.1.10)/shikimate dehydrogenase (EC 1.1.1.25) enzyme complex; (5) shikimate kinase (EC 2.7.1.71); (6) 5-enolpyruvylshikimic acid 3-phosphate (EPSP) synthase (EC 2.5.1.19); (7) chorismate synthase (EC 4.2.3.5); (8) chorismate mutase (EC 5.4.99.5); (9) prephenate aminotransferase (EC 2.6.7.78 or EC 2.6.7.79); (10) arogenate dehydratase (EC 4.2.1.91); (3a) quinate dehydrogenase (EC 1.1.1.24).

3-dehydroquinate + NADH + H^+ → L-quinate + NAD^+. Quinate dehydrogenase activity was found in enzyme extracts from cell cultures of mung bean (*Phaseolus aureus*).[141]

4.6.4 Metabolism of Chlorogenic Acids in *Coffea* Plants

Aerts and Baumann[142] found that sizable amounts of chlorogenic acids stored in coffee seeds decreased during germination and suggested that chlorogenic acids are utilised for the deposition of phenolic polymers, presumably lignin, in cotyledonary cell walls during the leaf expansion. Lignin consists of three monolignol monomers, *p*-coumaryl alcohol, coniferyl alcohol and sinapyl alcohol. These lignols are incorporated into lignin in the form of the phenylpropanoids, *p*-hydroxyphenyl (H), guaiacyl (G) and syringyl (S), respectively.[143] CQAs and FQAs are intermediate metabolites leading to G and S monolignols.

In *Coffea* seeds, a decrease in the concentration of chlorogenic acids occurs during seed ripening, but the total amounts of CGAs per seed remain stable or perhaps even increase.[11,144] Therefore, chlorogenic acids in matured seeds seem to be end products and are stored until germination. Although the metabolic activity of chlorogenic acids in *Coffea* seeds is very low, transcriptomic analysis suggest that the metabolic re-routing of these precursors towards lignin biosynthesis may be functional in the later stages of fruit development.[11] The genes potentially involved in lignin biosynthetic routes from chlorogenic acids, such as cinnamoyl-CoA reductase, ferulate 5-hydroxylase, caffeoyl-CoA 3-*O*-methyltransferase and cinnamyl alcohol dehydrogenase, exhibited maximal transcription levels at the late stages of fruit development. However, operation of the lignin pathway in *Coffea* seeds has not yet been clearly verified at the biochemical level.

4.6.5 Occurrence of Chlorogenic Acids in *Coffea* Plants

A number of chlorogenic acids occur in seeds of *Coffea* plants. Structures of main chlorogenic acids are illustrated in Figure 4.28. The most abundant chlorogenic acid in seeds is 5-*O*-caffeoylquinic acid (5-CQA), which normally comprises ~50% of the total content of chlorogenic acids. This is accompanied by significant amounts of 3-*O*- and 4-*O*-caffeoylquinic acid (3-CQA, 4-CQA), the three analogous FQAs and the 3,4-*O*-, 3,5-*O*- and 4,5-*O*-diCQA isomers, with smaller amounts of *p*-coumaroylquinic acids and lower quantities of the three isomeric monoacyl *p*-coumaroylquinic acids.[145] Recently numerous minor chlorogenic acids and cinnamoyl-amino acid conjugates have been characterised.[123]

Anthony *et al.*[146] investigated the diversity of chlorogenic acids in the genus *Coffea*. Examples of the occurrence of three major chlorogenic acids in different cultivars in *C. arabica* and other *Coffea* species are shown in Figure 4.29. In *C. arabica*, CQA, diCQA and FQA contents are, respectively, 4.7–6.1,

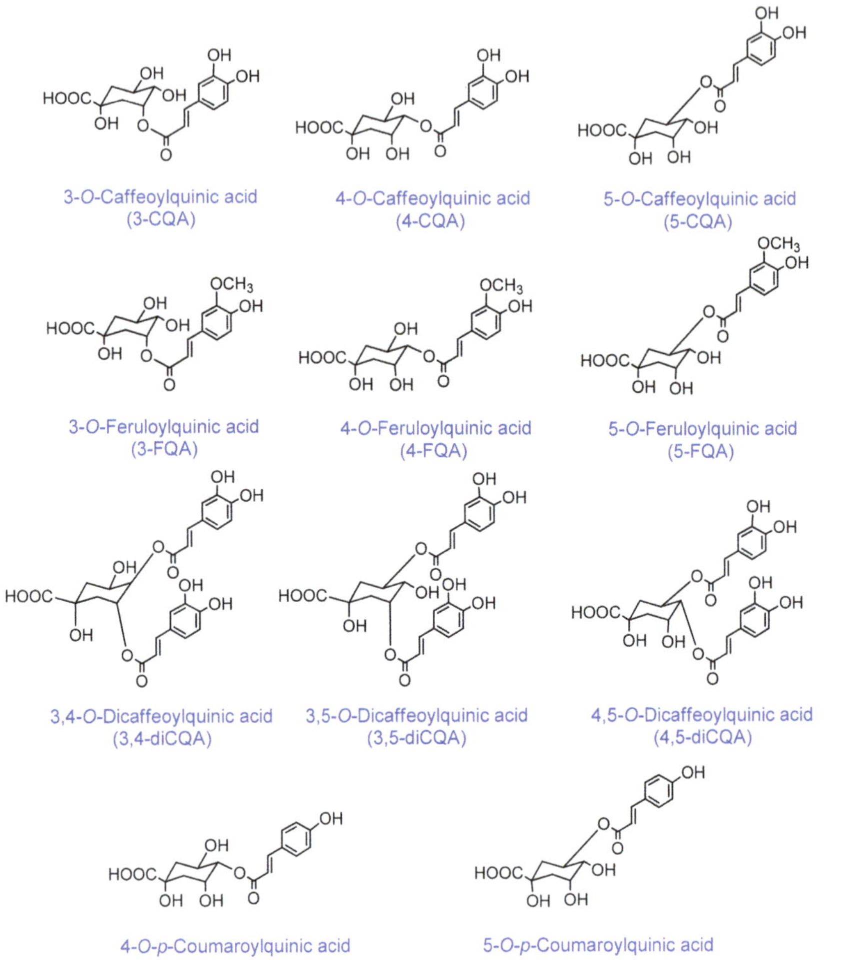

Figure 4.28 Chemical structures of major chlorogenic acids found in *Coffea* plants.[162]

0.5–1.2 and 0.2–0.5% of dry weight. In contrast, these values in *C. canephora* are 4.6–6.6, 0.8–1.6 and 0.9–1.0% of dry weight. Mean values of chlorogenic acids obtained from 21 different *C. arabica* samples (5.26, 0.64 and 0.31% for CQAs, diCQAs and FQAs, respectively) were lower than those from 6 samples from *C. canephora* (5.62, 1.29 and 0.89%, respectively).[146] Small quantities of chlorogenic acids occur in seeds of some *Coffea* species; *C. rhamnifolia* (0.1% of dry weight), *C. perrieri* (0.4%), *C. pseudozanguebariae* (0.9–1.8%) and *C. salvatrix* (1.7–2.3%).[146]

According to the review by Farah and Donangelo,[147] the order of concentration of main chlorogenic acids in coffee seeds is as follows: 5-CQA > 4-CQA and 3-CQA > 3,5-diCQA > 4,5-diCQA and 3,4-diCQA > 5-FQA > 4-FQA and 3-FQA. 5-CQA is responsible for about 56–62% of total chlorogenic acids.

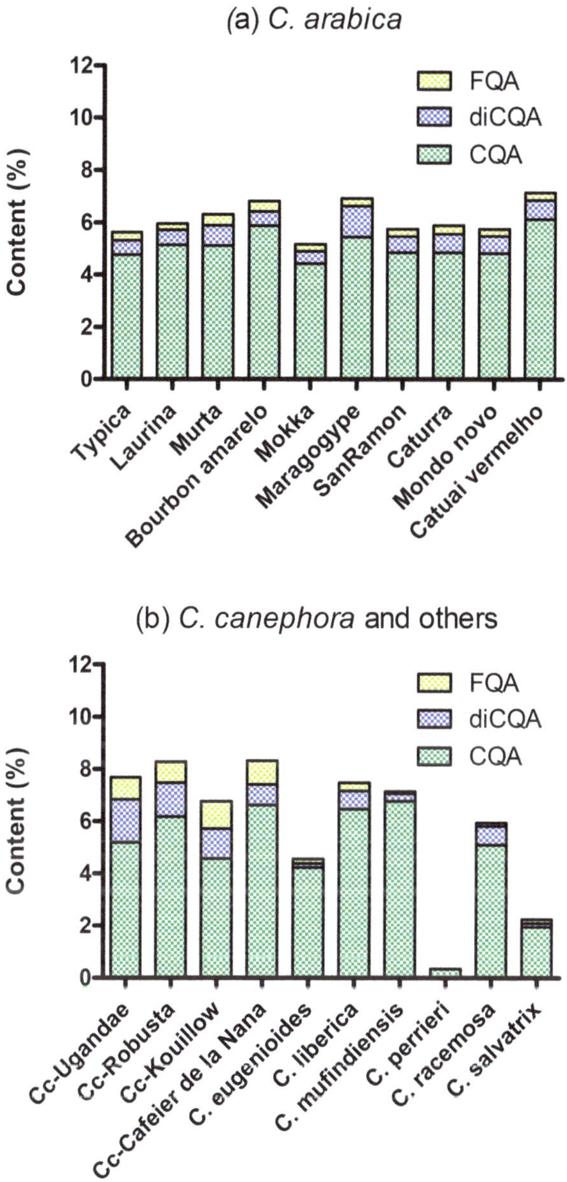

Figure 4.29 Concentration of CQAs, diCQAs and FQAs in (a) various cultivars of *C. arabica* and (b) other *Coffea* species. Based on data from Anthony *et al.*[146]

3-CQA and 4-CQA occur in broadly similar amounts, each comprising up to 10% of the total chlorogenic acid content while diCQA and FQA isomers account for 15–20% and 5–13%, respectively. Other minor chlorogenic acids are responsible for less than 1% of the total.

4.6.6 Physiological Aspects of Chlorogenic Acid Biosynthesis in *Coffea* Plants

Changes in contents and composition of chlorogenic acids and the expression of genes encoding enzymes involved in chlorogenic acid biosynthesis coffee seeds has been reported.[135,144,148–150] Koshiro *et al.*[144] reported the fluctuation on the levels of various chlorogenic acids and free quinic acids in seeds and pericarps of *C. arabica* and *C. canephora*. Coffee fruits at five different stages (C–G), as noted in Section 4.1 and Figure 4.2, were used. Changes in the total chlorogenic acid contents of fruits of *C. arabica* cv. Mokka, *C. arabica* cv. Catimor and *C. canephora* during fruit development and ripening are shown in Figure 4.30. The chlorogenic acid level in seeds increased gradually in *C. arabica* fruits, but a marked increase in content was found between stages D and E in *C. canephora*. In stage C, coffee fruits consist mainly of perisperm and pericarp, making it difficult to separate seeds and pericarp; but, after stage D, seeds were isolated more readily from coffee fruits. Most chlorogenic acids were located in seeds of both coffee species. In ripened fruits (stage G), 80–90% of total chlorogenic acid was located in seeds. Quinic acid, which is a polyol moiety of chlorogenic acid, was found in both pericarp and seed of both coffee species. The quinic acid content increased transiently at stage D and then decreased in *C. canephora* (Figure 4.30c).

The concentration of individual chlorogenic acids in *C. arabica* and *C. canephora* expressed as mg g^{-1} dry weight are illustrated in Figure 4.31. Three monocaffeoylquinic acids (3-CQA, 4-CQA and 5-CQA), three dicaffeoylquinic acids (3,4-diCQA, 3,5-diCQA and 4,5-diCQA) and one monoferuloylquinic acid (5-FQA) were detected in pericarp and seed of *C. arabica* and *C. canephora*. In stage C, chlorogenic acids were analysed in whole fruits, and the data are shown in the seed section, because seeds have not yet formed. The most abundant chlorogenic acid was 5-CQA, comprising 50–60% of the total chlorogenic acids in seeds of *C. arabica* and 45–50% in *C. canephora*. The content of the other CQAs, 3-CQA and 4-CQA, was low in pericarps and seed but the content increased with the growth of the seeds of both species. The content of diCQAs in *C. canephora* was much higher than in *C. arabica*, especially in stages D and E (Figure 4.31). 3,5-diCQA was the major di-CQA, especially in the early stages of fruit growth. A significant amount of 5-FQA was detected in seeds, although it was present in only trace amounts in whole fruits of stage C and in pericarps. A comparatively high 5-FQA content was found in *C. canephora* seeds after stage E. 5-FQA comprised 22% of the total chlorogenic acids in ripe *C. canephora* seeds. The proportion of 5-FQA in ripe seeds of *C. arabica* was lower, at <10% of total chlorogenic acids. The results show that the concentrations of chlorogenic acids in the stage G seeds of *C. arabica* are lower than in *C. canephora* (Figure 4.31). This is in keeping with previous observations.[87,147,148] A much higher content of 5-FQA in *C. canephora* seeds than in *C. arabica* seeds has also been observed (Figure 4.31).

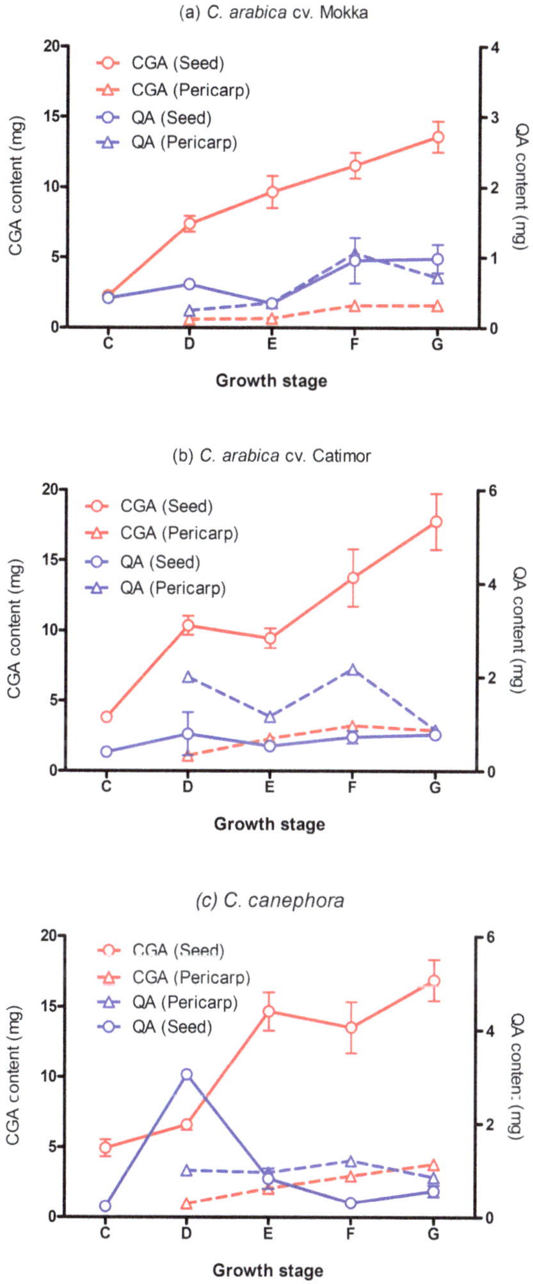

Figure 4.30 Changes in contents of chlorogenic acids (CGA) and quinic acid (QA) in seeds and pericarp of (a) *C. arabica* cv. Mokka, (b) *C. arabica* cv. Catimor and (c) *C. canephora* during growth, development and ripening. Contents are shown as mg per fruit (two seeds and pericarp). Whole fruits were used in stage C, and contents are shown on a per seed basis. Based on the data from Koshiro *et al.*[3,84]

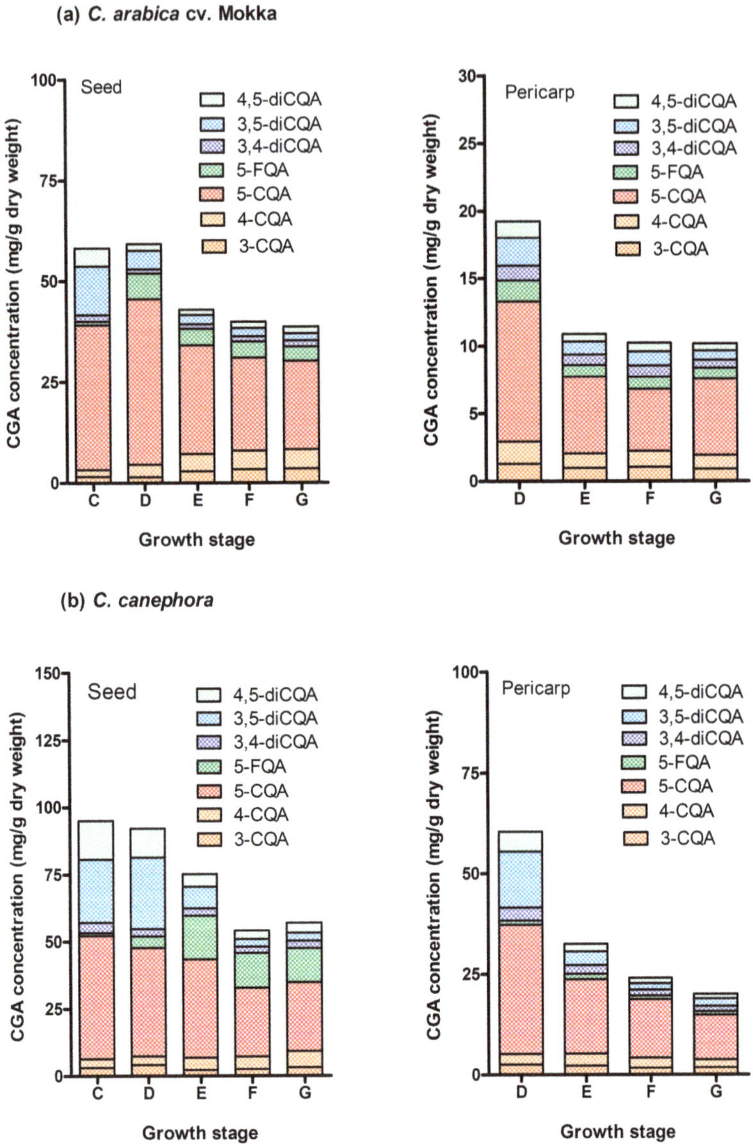

Figure 4.31 Concentration of major chlorogenic acids in pericarps and seeds of (a) *C. arabica* cv. Mokka and (b) *C. canephora* during fruit development and ripening. Based on the data from Koshiro *et al.*[3]

In order to estimate the biosynthetic activity of chlorogenic acid, the metabolic fate of [U-^{14}C]phenylalanine and [U-^{14}C]caffeic acid in the seed segments from various growth stages of fruits of *C. arabica* and *C. canephora* was investigated.[84,144] Distribution of the radioactivity into free amino acids, phenolics including 5-CQA and an insoluble fraction, which consisted of

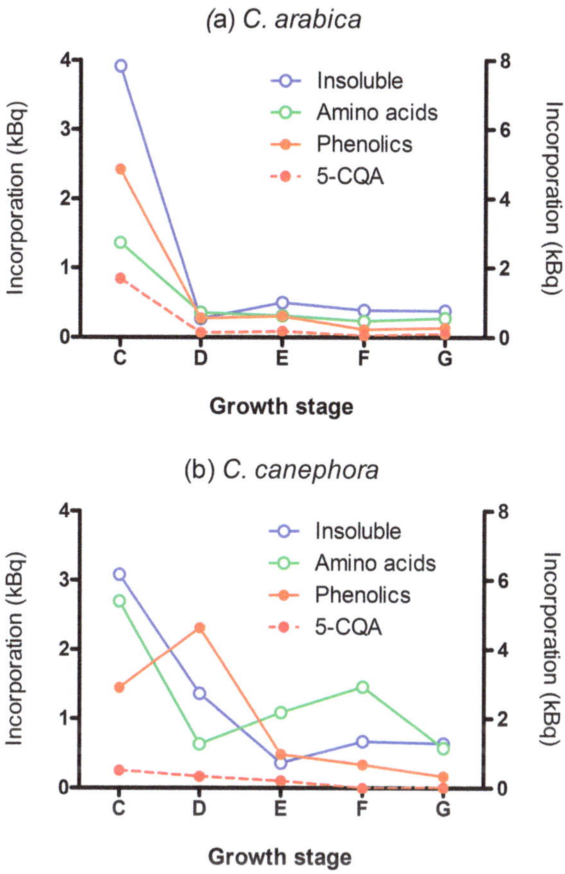

Figure 4.32 Changes in the metabolic fate of [U-^{14}C]phenylalanine in seeds of (a) *C. arabica* cv. Mokka and (b) *C. canephora* during fruit development and ripening. Incorporation of ^{14}C into the insoluble compounds (mainly protein and lignin) [right scale], free amino acids and total phenolic compounds [left scale] are shown. The incorporation of 5-*O*-caffeoylquinic acid (5-CQA) is also shown as a dotted line [left scale]. The incorporation of ^{14}C is expressed as kBq/100 mg fresh weight samples/18 h. Based on the data from Koshiro *et al.*[3]

proteins and lignin, was investigated over an 18 h period after administration of [^{14}C]phenylalanine (Figure 4.32). In both *C. arabica* and *C. canephora*, incorporation of ^{14}C into 5-CQA was limited but a higher incorporation of label was found in young fruits (stage C), which declined as the fruit developed. Significant amounts of radioactivity were found in the insoluble fraction at stage C. Since the incubation time is short (18 h), most radioactivity may be located in proteins, but not in lignin, at least in young fruits (stage C). Further detailed analysis of this fraction remains to elucidate. Overall, the tracer experiments suggest that chlorogenic acid synthesis occurs in the early stages of seed formation.

Using the same coffee materials, Koshiro *et al.*[144] investigated the expression of genes encoding *PAL1*, *C3'H* and *CCoAMT*. In both *Coffea* species, the transcripts of *PAL1*, *C3'H* and *CCoAMT* were detected in all stages of growth and ripening (stages D–G), although the amounts of the transcripts of *PAL1*, *C3'H* and *CCoAMT* all decreased markedly in stage G. Of these genes, *CCoAMT* was expressed more weakly in stage C in both species. The RT-PCR method was semi-quantitative, so it is difficult to find the absolute amounts of transcripts. Nevertheless, the expression pattern of a gene for FQA synthesis is not always the same as the genes for CQA synthesis: expression of C3'H was greater in the early stages, but expression of *CCoAMT* was more intense after stage D than in stage C. The expression patterns of *C3'H* and *CCoAMT* in pericarps and in seeds of *C. arabica* indicated that *C3'H* gene is similarly expressed, but *CCoAMT* is expressed only in seeds.

Recently, more detailed gene expression studies related to chlorogenic acid metabolism have been performed in *Coffea* plants.[11,135,150,151] Lepelley *et al.*[135] investigated the expression of *HCT*, *HQT*, *C3'H1* and *CCoAOMT1* in different growth stages of fruits obtained from three varieties of *C. canephora*. Their samples, small green, large green, yellow and red, seem to approximately correspond to stages C and E–G of the materials used by Koshiro *et al.*[144] Although expression profiles differed to some extent between the varieties, transcript levels of *HCT* and *C3H1* were much higher in the earlier stages (stages D and E) than in the later stages (stages F and G). The transcript level of *HQT* was high in the first (stage D) and the last stages (stage G). The level of *CCoAOMT1* was lower in small green seeds (stage D), increased in large green and yellow seeds (stages E and F) and then fell significantly in the red fruits (stage G). The expression profile suggested that a relatively higher rate of expression of *HQT*, *HCT* and *C3'H1* may support the high CQA biosynthesis in the early perisperm rich *C. canephora* seeds. The expression of *CCoAOMT1* may contribute to the rapid conversion of 5-CQA to 5-FQA in the later stage of fruit development.

From comprehensive transcriptomic analysis, Joët *et al.*[11] reported that genes encoding *PAL1*, *PAL2*, *C4H*, two *4CLs* and three *CCoAOMTs* showed high expression levels when the endosperm was undergoing rapid development (corresponding to stage C). Expression of these genes coincided with maximal chlorogenic acid biosynthetic activity suggesting they play a key role in their biosynthesis. Other genes encoding putative upstream enzymes such as *4CL* and *C3H* could also be related to CQA biosynthesis activity, as their expression at stage C of the fruit development was close to their maximal expression. Similarly, the bimodal expression pattern of *HCT1* suggests a dual role for this enzyme that could be directly involved in caffeoylquinate anabolism. In contrast, another key phenylpropanoid gene, *HQT*, did not show an expression profile that matches the chlorogenic acid accumulation pattern, suggesting that this transcript is restricted to the perisperm and not directly involved in CQA accumulation in the endosperm. These results point to the biosynthesis of CQAs being controlled mainly by the transcription of participating enzymes rather than the supply of precursors, such as phenylalanine and quinic acid.

4.6.7 *In Planta* Function of Chlorogenic Acids in *Coffea* Plants

It has been generally considered that chlorogenic acids are a storage form of cinnamic acid derivatives and precursors for the lignin biosynthetic pathway in plants[142] and that this may be their role during germination of coffee seeds. It has also been speculated that chlorogenic acids function as chemical defence compounds to protect against assorted pests and pathogens, since chlorogenic acids have antioxidant and antibiotic properties.[147] However, no rigorous evidence has been obtained on such a role in *Coffea* plants. Melo *et al.*[152] investigated the role of polyphenol oxidase related to the resistance against a leaf miner and coffee leaf rust disease in coffee trees. Although 5-CQA is the best substrate for coffee leaf polyphenol oxidase, the concentration of 5-CQA in leaves was not related to resistance. Ramiro *et al.*[153] also investigated the role of 5-CQA and the oxidase enzymes in the expression of resistance of coffee plants to an insect, *Leucoptera coffeella*. However, concentrations of 5-CQA and the activities of the oxidative enzymes were almost the same in leaves of *C. arabica*, *C. racemosa* and progenies of crosses between these species which have different levels of resistance to this insect. This suggests that the cellular concentration of 5-CQA is not related to the resistance to the coffee leaf miner.

Phenolic compounds have been identified as the most common allelochemicals produced by higher plants. Li *et al.*[154] reported that a relatively low concentration of 5-CQA (>100 μM) inhibited the growth of lettuce (*Lactuca cariola*) seedlings, but did not inhibit seed germination at up to 5 mM. In addition to 5-CQA, *trans*-cinnamic acid, *o*-, *m*-, *p*-coumaric acids and coumarin also have inhibitory effects on lettuce seed germination which are much stronger than that of 5-CQA. Therefore, it is currently not known whether or not 5-CQA in coffee seeds serves as an effective allelochemical in natural ecosystems.

Mondolot *et al.*[155] reported that concentration of chlorogenic acid in young leaves is much higher than in mature leaves of *C. canephora* and that CQAs and di-CQAs are closely associated with chloroplasts in very young leaves. The association with chloroplasts suggests that chlorogenic acids have a protective role against light damage. They may also be involved in the response to different abiotic stresses such as drought and temperature,[150,156] but the *in planta* function of chlorogenic acids remains to be elucidated.

4.7 Conclusions

This chapter describes the current status of our knowledge of coffee plant biochemistry especially with regard to secondary metabolism. Significant advances have been made in the last decade in studies on the biosynthesis of caffeine, trigonelline and chlorogenic acids. The genes of key enzymes of these pathways have been cloned, including those encoding *N*-methyltransferase involved in the biosynthesis of caffeine and trigonelline and acyl transferases associated with chlorogenic acid biosynthesis. In contrast to these developments, there have been relatively few studies on the biochemical and physiological events occurring during the development and maturation of coffee fruits. However, a series of biochemical studies using the fruits of *C. arabica* and *C. canephora* was discussed with the biosynthesis of caffeine, trigonelline and chlorogenic acids in the fruits of the same *Coffea* trees being compared. Caffeine biosynthesis is active in young fruits and the sites of biosynthesis may be perisperm and the developing endosperm. Although 70–90% of trigonelline is accumulated in seeds, active biosynthetic activity was found in pericarp. Rapid net increase of chlorogenic acids in seeds occurs in parallel with dry matter accumulation after active biosynthesis was observed at the bean formation stage. Information on transport mechanisms and location of these compounds within the coffee plant remains to be elucidated.

The contents of these three major secondary metabolites are closely related to the quality of coffee beverage and human health. Hence, an understanding of the mechanisms underlying the accumulation of these compounds in coffee beans is important not only for plant science but also for the biotechnology to produce metabolite-rich and metabolite-deficient beans. These topics are reviewed elsewhere.[33,77,97,127,157]

Acknowledgements

The authors thank Dr Kouichi Mizuno, Akita Prefectural University, and Professor Adriana Farah, Universidade Federal do Rio de Janeiro, for their valuable comments.

References

1. A. Farah, Coffee constituents, in *Coffee: Emerging Health Effects and Disease Prevention*, Wiley-Blackwell, 2012, pp. 21–58.
2. M. G. R. Cannell, in *Coffee: Botany, Biochemistry and Production of Beans and Beverage*, ed. M. N. Clifford and K. C. Willson, Croom Helm, London, 1985, pp. 108–134.
3. Y. Koshiro, X. Q. Zheng, M. Wang, C. Nagai and H. Ashihara, *Plant Sci.*, 2006, **171**, 242–250.
4. R. F. Sage and X.-G. Zhu, *J. Exp. Bot.*, 2011, **62**, 2989–3000.
5. S. Yamaki, *J. Jpn. Soc. Hortic. Sci.*, 2010, **79**, 1–15.
6. C. Geromel, L. P. Ferreira, A. Bottcher, D. Pot, L. F. P. Pereira, T. Leroy, L. G. E. Vieira, P. Mazzafera and P. Marraccini, *Ann. Appl. Biol.*, 2008, **152**, 179–187.
7. C. Geromel, L. P. Ferreira, S. M. C. Guerreiro, A. A. Cavalari, D. Pot, L. F. P. Pereira, T. Leroy, L. G. E. Vieira, P. Mazzafera and P. Marraccini, *J. Exp. Bot.*, 2006, **57**, 3243–3258.
8. Y. Koshiro, C. Nagai and H. Ashihara, *Eur. Chem. Bull.*, 2015, **4**, 296–302.

9. W. J. Rogers, S. Michaux, M. Bastin and P. Bucheli, *Plant Sci.*, 1999, **149**, 115–123.
10. Y. Koshiro, M. C. Jackson, C. Nagai and H. Ashihara, *Eur. Chem. Bull.*, 2015, **4**, 378–383.
11. T. Joët, A. Laffargue, J. Salmona, S. Doulbeau, F. Descroix, B. Bertrand, A. De Kochko and S. Dussert, *New Phytol.*, 2009, **182**, 146–162.
12. R. Redgwell and M. Fischer, *Braz. J. Plant Physiol.*, 2006, **18**, 165–174.
13. T. Joët, A. Laffargue, J. Salmona, S. Doulbeau, F. Descroix, B. Bertrand, P. Lashermes and S. Dussert, *J. Exp. Bot.*, 2014, **65**, 323–337.
14. M. L. C. Carelli, J. I. Fahl and J. D. C. Ramalho, *Braz. J. Plant Physiol.*, 2006, **18**, 9–21.
15. P. Mazzafera and K. V. Gonçalves, *Phytochemistry*, 1999, **50**, 383–386.
16. P. J. Lea and R. J. Ireland, in *Plant Amino Acids – Biochemistry and Biotechnology*, ed. B. K. Singh, Marcel Dekker, New York, 1999, pp. 1–47.
17. G. Coruzzi and R. Last, in *Biochemistry and Molecular Biology of Plants*, ed. B. B. Buchanan, W. Gruissem and R. L. Jones, American Society of Plant Physiologists, Rockville, 2000, pp. 358–410.
18. H. Ashihara and A. Crozier, *Adv. Bot. Res.*, 1999, **30**, 117–205.
19. T. Suzuki and E. Takahashi, *Biochem. J.*, 1976, **160**, 181–184.
20. D. B. A. Ogutuga and D. H. Northcote, *Biochem. J.*, 1970, **117**, 715–720.
21. T. Suzuki and E. Takahashi, *Drug Metab. Rev.*, 1977, **6**, 213–242.
22. T. Suzuki and E. Takahashi, *Phytochemistry*, 1976, **15**, 1235–1239.
23. H. Ashihara and H. Kubota, *Physiol. Plant.*, 1986, **68**, 275–281.
24. O. Negishi, T. Ozawa and H. Imagawa, *Biosci. Biotechnol. Biochem.*, 1992, **56**, 499–503.
25. O. Negishi, T. Ozawa and H. Imagawa, *Agric. Biol. Chem.*, 1985, **49**, 251–253.
26. G. M. Nazario and C. J. Lovatt, *Plant Physiol.*, 1993, **103**, 1203–1210.
27. H. Ashihara, A. M. Monteiro, F. M. Gillies and A. Crozier, *Plant Physiol.*, 1996, **111**, 747–753.
28. M. Kato, K. Mizuno, T. Fujimura, M. Iwama, M. Irie, A. Crozier and H. Ashihara, *Plant Physiol.*, 1999, **120**, 579–586.
29. M. Kato, K. Mizuno, A. Crozier, T. Fujimura and H. Ashihara, *Nature*, 2000, **406**, 956–957.
30. T. Suzuki and E. Takahashi, *Biochem. J.*, 1975, **146**, 87–96.
31. T. Suzuki and G. R. Waller, in *Analysis of Nonalcoholic Beverages*, ed. H. F. Linskens and J. F. Jackson, Modern Methods of Plant Analysis, New Series, Springer-Verlag, Berlin, 1988, vol. 8, pp. 184–220.
32. P. Mazzafera, *Front. Biosci.*, 2004, **9**, 1348–1359.
33. H. Ashihara, T. Yokota and A. Crozier, *Adv. Bot. Res.*, 2013, **68**, 111–138.
34. L. Anderson and M. Gibbs, *J. Biol. Chem.*, 1962, **237**, 1941–1944.
35. E. Proiser and G. P. Serenkov, *Biokhimiya (Moscow)*, 1963, **28**, 857–861.
36. J. F. Henderson and A. R. P. Paterson, *Nucleotide Metabolism – An Introduction*, Academic Press, New York, 1973.
37. E. Ito and H. Ashihara, *J. Plant Physiol.*, 1999, **254**, 145–151.

38. C. A. Keya, A. Crozier and H. Ashihara, *FEBS Lett.*, 2003, **554**, 473–477.
39. C. Stasolla, R. Katahira, T. A. Thorpe and H. Ashihara, *J. Plant Physiol.*, 2003, **160**, 1271–1295.
40. C. Koshiishi, A. Kato, S. Yama, A. Crozier and H. Ashihara, *FEBS Lett.*, 2001, **499**, 50–54.
41. E. Ito, A. Crozier and H. Ashihara, *Biochim. Biophys. Acta*, 1997, **1336**, 323–330.
42. M. Ogawa, Y. Herai, N. Koizumi, T. Kusano and H. Sano, *J. Biol. Chem.*, 2001, **276**, 8213–8218.
43. H. Uefuji, S. Ogita, Y. Yamaguchi, N. Koizumi and H. Sano, *Plant Physiol.*, 2003, **132**, 372–380.
44. K. Mizuno, M. Kato, F. Irino, N. Yoneyama, T. Fujimura and H. Ashihara, *FEBS Lett.*, 2003, **547**, 56–60.
45. K. Mizuno, A. Okuda, M. Kato, N. Yoneyama, H. Tanaka, H. Ashihara and T. Fujimura, *FEBS Lett.*, 2003, **534**, 75–81.
46. C. Perrois, S. Strickler, G. Mathieu, M. Lepelley, L. Bedon, S. Michaux, J. Husson, L. Mueller and I. Privat, *Planta*, 2014, 1–13.
47. M. Kato and K. Mizuno, *Front. Biosci.*, 2004, **9**, 1833–1842.
48. C. P. Joshi and V. L. Chiang, *Plant Mol. Biol.*, 1998, **37**, 663–674.
49. J. C. D'Auria, F. Chen and E. Pichersky, in *Recent Advances in Phytochemistry*, ed. T. R. John, Elsevier, 2003, vol. 37, pp. 253–283.
50. F. Denoeud, L. Carretero-Paulet, A. Dereeper, G. Droc, R. Guyot, M. Pietrella, C. Zheng, A. Alberti, F. Anthony, G. Aprea, J.-M. Aury, P. Bento, M. Bernard, S. Bocs, C. Campa, A. Cenci, M.-C. Combes, D. Crouzillat, C. Da Silva, L. Daddiego, F. De Bellis, S. Dussert, O. Garsmeur, T. Gayraud, V. Guignon, K. Jahn, V. Jamilloux, T. Joët, K. Labadie, T. Lan, J. Leclercq, M. Lepelley, T. Leroy, L.-T. Li, P. Librado, L. Lopez, A. Muñoz, B. Noel, A. Pallavicini, G. Perrotta, V. Poncet, D. Pot, Priyono, M. Rigoreau, M. Rouard, J. Rozas, C. Tranchant-Dubreuil, R. VanBuren, Q. Zhang, A. C. Andrade, X. Argout, B. Bertrand, A. de Kochko, G. Graziosi, R. J. Henry, Jayarama, R. Ming, C. Nagai, S. Rounsley, D. Sankoff, G. Giuliano, V. A. Albert, P. Wincker and P. Lashermes, *Science*, 2014, **345**, 1181–1184.
51. E. Pichersky and E. Lewinsohn, *Annu. Rev. Plant Biol.*, 2011, **62**, 549–566.
52. C. Zubieta, J. R. Ross, P. Koscheski, Y. Yang, E. Pichersky and J. P. Noel, *Plant Cell*, 2003, **15**, 1704–1716.
53. A. A. McCarthy, L. Biget, C. Lin, V. Petiard, S. D. Tanksley and J. G. McCarthy, *Acta Crystallogr., Sect. F*, 2007, **63**, 304–307.
54. A. A. McCarthy and J. G. McCarthy, *Plant Physiol.*, 2007, **144**, 879–889.
55. N. Yoneyama, H. Morimoto, C. X. Ye, H. Ashihara, K. Mizuno and M. Kato, *Mol. Genet. Genomics*, 2006, **275**, 125–135.
56. O. Negishi, T. Ozawa and H. Imagawa, *Agric. Biol. Chem.*, 1988, **52**, 169–175.
57. S. Moisyadi, K. R. Neupane and J. I. Stiles, *Acta Hortic.*, 1998, **461**, 367–377.
58. S. Moisyadi, K. R. Neupane and J. I. Stiles, *Proc. 18th ASIC Colloq.*, 1999, **18**, 327–331.

59. S. S. Mosli Waldhauser, F. M. Gillies, A. Crozier and T. W. Baumann, *Phytochemistry*, 1997, **45**, 1407–1414.
60. K. Mizuno, H. Tanaka, M. Kato, H. Ashihara and T. Fujimura, *Proc. 19th ASIC Colloq.*, 2001, **19**, 815–818.
61. T. Suzuki, H. Ashihara and G. R. Waller, *Phytochemistry*, 1992, **31**, 2575–2584.
62. P. Kalberer, *Nature*, 1965, **205**, 597–598.
63. T. Suzuki and G. R. Waller, *J. Sci. Food Agric.*, 1984, **35**, 66–70.
64. T. Suzuki and G. R. Waller, *J. Agric. Food Chem.*, 1984, **32**, 845–848.
65. H. Ashihara, A. M. Monteiro, T. Moritz, F. M. Gillies and A. Crozier, *Planta*, 1996, **198**, 334–339.
66. A. P. Vitoria and P. Mazzafera, *Pesqui. Agropecu. Bras.*, 1998, **33**, 1957–1961.
67. H. Ashihara and A. Crozier, *J. Agric. Food Chem.*, 1999, **47**, 3425–3431.
68. P. Mazzafera, *J. Agric. Food Chem.*, 1993, **41**, 1541–1543.
69. H. Wanner, M. Pešáková, T. W. Baumann, R. Charubala, A. Guggisberg, M. Hesse and H. Schmid, *Phytochemistry*, 1975, **14**, 747–750.
70. T. W. Baumann, M. Oechslin and H. Wanner, *Biochem. Physiol. Pflanz.*, 1976, **170**, 217–225.
71. J. Petermann, T. W. Baumann and H. Wanner, *Phytochemistry*, 1977, **16**, 620–621.
72. J. Petermann and T. W. Baumann, *Plant Physiol.*, 1983, **73**, 961–964.
73. P. Mazzafera and A. Carvalho, *Euphytica*, 1992, **59**, 55–60.
74. M. B. Silvarolla, P. Mazzafera and L. C. Fazuoli, *Nature*, 2004, **429**, 826.
75. P. Mazzafera, A. Crozier and A. C. Magalhães, *Phytochemistry*, 1991, **30**, 3913–3916.
76. P. Mazzafera, A. Crozier and G. Sandberg, *J. Agric. Food Chem.*, 1994, **42**, 1423–1427.
77. H. Ashihara, S. Ogita and A. Crozier, in *Plant Metabolism and Biotechnology*, ed. H. Ashihara, A. Crozier and A. Komamine, John Wiley & Sons, Ltd, Chichester, 2011, pp. 163–189.
78. X. Q. Zheng and H. Ashihara, *Plant Sci.*, 2004, **166**, 807–813.
79. N. Fujimori and H. Ashihara, *Phytochemistry*, 1994, **36**, 1359–1361.
80. P. M. Frischknecht, J. Ulmer-Dufek and T. W. Baumann, *Phytochemistry*, 1986, **25**, 613–616.
81. N. Fujimori, T. Suzuki and H. Ashihara, *Phytochemistry*, 1991, **30**, 2245–2248.
82. T. W. Baumann and H. Wanner, *Planta*, 1972, **108**, 11–19.
83. C. Perrois, S. Strickler, G. Mathieu, M. Lepelley, L. Bedon, S. Michaux, J. Husson, L. Mueller and I. Privat, *Planta*, 2015, **241**, 179–191.
84. Y. Koshiro, MSc thesis, Ochanomizu University, Tokyo, 2006.
85. M. N. Clifford, in *Coffee: Botany, Biochemistry and Production of Beans and Beverage*, ed. M. N. Clifford and K. C. Willson, Croom Helm, London, 1985, pp. 305–374.
86. P. Mazzafera, *Phytochemistry*, 1991, **30**, 2309–2310.
87. C. L. Ky, J. Louarn, S. Dussert, B. Guyot, S. Hamon and M. Noirot, *Food Chem.*, 2001, **75**, 223–230.

88. R. Viani and I. Horman, *J. Food Sci.*, 1974, **39**, 1216–1217.
89. R. H. Stadler, N. Varga, J. Hau, F. Arce Vera and D. H. Welti, *J. Agric. Food Chem.*, 2002, **50**, 1192–1199.
90. H. Taguchi, M. Sakaguchi and Y. Shimabayashi, *Agric. Biol. Chem.*, 1985, **49**, 3467–3471.
91. E. Johns, *Ber. Dtsch. Chem. Ges.*, 1885, **18**, 2518–2523.
92. A. Matsui, Y. Yin, K. Yamanaka, M. Iwasaki and H. Ashihara, *Physiol. Plant.*, 2007, **131**, 191–200.
93. Y. Cho, E. B. Turnipseed, D. A. Lightfoot and A. J. Wood, *Biol. Plant.*, 2008, **52**, 370–372.
94. H. Ashihara and S. Watanabe, *Nat. Prod. Commun.*, 2014, **9**, 795–798.
95. H. Ashihara and W.-W. Deng, *J. Plant Res.*, 2012, **125**, 781–791.
96. H. Ashihara, Y. Yin, R. Katahira, S. Watanabe, T. Mimura and H. Sasamoto, *Plant Physiol. Biochem.*, 2012, **60**, 190–195.
97. H. Ashihara, I. A. Ludwig, R. Katahira, T. Yokota, T. Fujimura and A. Crozier, *Phytochem. Rev.*, 2015, **14**, 765–798.
98. J. G. Joshi and P. Handler, *J. Biol. Chem.*, 1960, **235**, 2981–2983.
99. W. Barz, in *Primary and Secondary Metabolism of Plant Cell Cultures*, ed. K. H. Neumann, W. Barz and E. Reinhard, Springer-Verlag, Berlin, 1985, pp. 186–195.
100. X. Q. Zheng, C. Nagai and H. Ashihara, *Physiol. Plant.*, 2004, **122**, 404–411.
101. H. Ashihara, W. W. Deng and C. Nagai, *Phytochem. Lett.*, 2011, **4**, 235–239.
102. K. Mizuno, M. Matsuzaki, S. Kanazawa, T. Tokiwano, Y. Yoshizawa and M. Kato, *Biochem. Biophys. Res. Commun.*, 2014, **452**, 1060–1066.
103. H. Ashihara, C. Stasolla, Y. Yin, N. Loukanina and T. A. Thorpe, *Plant Sci.*, 2005, **169**, 107–114.
104. A. Katoh, K. Uenohara, M. Akita and T. Hashimoto, *Plant Physiol.*, 2006, **141**, 851–857.
105. R. Zrenner and H. Ashihara, in *Plant Metabolism and Biotechnology*, ed. H. Ashihara, A. Crozier and A. Komamine, John Wiley & Sons, Ltd, Chichester, 2011, pp. 135–162.
106. A. Katoh and T. Hashimoto, *Front. Biosci.*, 2004, **9**, 1577–1586.
107. G. Noctor, G. Queval and B. Gakiere, *J. Exp. Bot.*, 2006, **57**, 1603–1620.
108. G. Noctor, J. Hager and S. Li, *Adv. Bot. Res.*, 2011, **58**, 153–201.
109. X. Q. Zheng, E. Hayashibe and H. Ashihara, *J. Exp. Bot.*, 2005, **56**, 1615–1623.
110. G. Wang and E. Pichersky, *Plant J.*, 2007, **49**, 1020–1029.
111. R. Katahira and H. Ashihara, *Planta*, 2009, **231**, 35–42.
112. A. Matsui and H. Ashihara, *Plant Physiol. Biochem.*, 2008, **46**, 104–108.
113. H. Ashihara, W.-W. Deng and C. Nagai, *Phytochem. Lett.*, 2011, **4**, 235–239.
114. R. Wagner, F. Feth and K. G. Wagner, *Planta*, 1986, **167**, 226–232.
115. X. Q. Zheng, A. Matsui and H. Ashihara, *Phytochemistry*, 2008, **69**, 390–395.
116. B. Upmeier, W. Gross, S. Koster and W. Barz, *Arch. Biochem. Biophys.*, 1988, **262**, 445–454.

117. X. Chen and A. J. Wood, *Biol. Plant.*, 2004, **48**, 531–535.
118. H. Ashihara, in *Coffee in Health and Disease Prevention*, ed. V. R. Preedy, Academic Press, San Diego, 2015, pp. 19–28.
119. H. Taguchi and Y. Shimabayashi, *Biochem. Biophys. Res. Commun.*, 1983, **113**, 569–574.
120. M. M. Shimizu and P. Mazzafera, *Plant Biol.*, 2000, **2**, 605–611.
121. H. Sasamoto and H. Ashihara, *Phytochem. Lett.*, 2014, **7**, 38–41.
122. T. Nakayama and K. Honda, *Chemoecology*, 2004, **14**, 199–205.
123. A. Stalmach, M. N. Clifford, G. Williamson and A. Crozier, in *Teas, Cocoa and Coffee*, ed. A. Crozier, H. Ashihara and F. Tomas-Barberan, Wiley-Blackwell, Oxford, 2011, pp. 143–168.
124. M. N. Clifford, in *Coffee*, ed. R. J. Clarke and R. Macrae, Elsevier Applied Science, London, 1985, vol. 1, pp. 153–202.
125. C. Bowsher, M. Steer and A. Tobin, *Plant Biochemistry*, Garland Science, New York, 2012.
126. V. Mahesh, J. Rakotomalala, L. Ã. Le Gal, H. Ã. Vigne, A. De Kochko, S. Hamon, M. Noirot and C. Campa, *Plant Cell Rep.*, 2006, 1–7.
127. R. Niggeweg, C. Martin and A. J. Michael, *Nat. Biotechnol.*, 2004, **22**, 746–754.
128. D. Strack, H. Keller and G. Weissenbock, *J. Plant Physiol.*, 1987, **131**, 61–73.
129. R. J. A. Villegas and M. Kojima, *J. Biol. Chem.*, 1986, **261**, 8729–8733.
130. L. Hoffmann, S. Maury, F. Martz, P. Geoffroy and M. Legrand, *J. Biol. Chem.*, 2003, **278**, 95–103.
131. G. Schoch, S. Goepfert, M. Morant, A. Hehn, D. Meyer, P. Ullmann and D. Werck-Reichhart, *J. Biol. Chem.*, 2001, **276**, 36566–36574.
132. C. Campa, M. Noirot, M. Bourgeois, M. Pervent, C. L. Ky, H. Chrestin, S. Hamon and A. De Kochko, *Theor. Appl. Genet.*, 2003, **107**, 751–756.
133. L. A. Lallemand, C. Zubieta, S. G. Lee, Y. Wang, S. Acajjaoui, J. Timmins, S. McSweeney, J. M. Jez, J. G. McCarthy and A. A. McCarthy, *Plant Physiol.*, 2012, **160**, 249–260.
134. J. C. D'Auria, *Curr. Opin. Plant Biol.*, 2006, **9**, 331–340.
135. M. Lepelley, G. Cheminade, N. Tremillon, A. Simkin, V. Caillet and J. McCarthy, *Plant Sci.*, 2007, **172**, 978–996.
136. G. Sonnante, R. D'Amore, E. Blanco, C. L. Pierri, M. De Palma, J. Luo, M. Tucci and C. Martin, *Plant Physiol.*, 2010, **153**, 1224–1238.
137. N. Abdulrazzak, B. Pollet, J. Ehlting, K. Larsen, C. Asnaghi, S. Ronseau, C. Proux, M. Erhardt, V. Seltzer, J. P. Renou, P. Ullmann, M. Pauly, C. Lapierre and D. Werck-Reichhart, *Plant Physiol.*, 2006, **140**, 30–48.
138. V. Mahesh, R. Million-Rousseau, P. Ullmann, N. Chabrillange, J. Bustamante, L. Mondolot, M. Morant, M. Noirot, S. Hamon, A. de Kochko, D. Werck-Reichhart and C. Campa, *Plant Mol. Biol.*, 2007, **64**, 145–159.
139. K. M. Herrmann and L. M. Weaver, *Annu. Rev. Plant Physiol. Plant Mol. Biol.*, 1999, **50**, 473–503.
140. J.-P. Colonna, *Café, Cacao, Thé*, 1986, **30**, 83–92.
141. O. L. Gamborg, *Biochim. Biophys. Acta*, 1966, **128**, 483–491.
142. R. J. Aerts and T. W. Baumann, *J. Exp. Bot.*, 1994, **45**, 497–503.

143. W. Boerjan, J. Ralph and M. Baucher, *Annu. Rev. Plant Biol.*, 2003, **54**, 519–546.
144. Y. Koshiro, M. C. Jackson, R. Katahira, M. L. Wang, C. Nagai and H. Ashihara, *Z. Naturforsch.*, 2008, **62c**, 731–742.
145. M. N. Clifford, *J. Sci. Food Agric.*, 1999, **79**, 362–372.
146. F. Anthony, M. N. Clifford and M. Noirot, *Genet. Resour. Crop Evol.*, 1993, **40**, 61–70.
147. A. Farah and C. M. Donangelo, *Braz. J. Plant Physiol.*, 2006, **18**, 23–36.
148. M. N. Clifford and T. Kazi, *Food Chem.*, 1987, **26**, 59–69.
149. S. Smrke, I. Kroslakova, A. N. Gloess and C. Yeretzian, *Food Chem.*, 2015, **174**, 637–642.
150. T. JoËT, J. Salmona, A. Laffargue, F. Descroix and S. Dussert, *Plant, Cell Environ.*, 2010, **33**, 1220–1233.
151. J. Salmona, S. Dussert, F. Descroix, A. de Kochko, B. Bertrand and Joet, *Plant Mol. Biol.*, 2008, **66**, 105–124.
152. G. A. Melo, M. M. Shimizu and P. Mazzafera, *Phytochemistry*, 2006, **67**, 277–285.
153. D. Ramiro, O. Guerreiro-Filho and P. Mazzafera, *J. Chem. Ecol.*, 2006, **32**, 1977–1988.
154. H.-H. Li, M. Inoue, H. Nishimura, J. Mizutani and E. Tsuzuki, *J. Chem. Ecol.*, 1993, **19**, 1775–1787.
155. L. Mondolot, P. La Fisca, B. Buatois, E. Talansier, A. De Kochko and C. Campa, *Ann. Bot.*, 2006, **98**, 33–40.
156. S. C. Grace, B. A. Logan and W. W. Adams, *Plant, Cell Environ.*, 1998, **21**, 513–521.
157. H. Ashihara and A. Crozier, *Trends Plant Sci.*, 2001, **6**, 407–413.
158. H. Ashihara and M. Kato, *Thirty Lectures on Metabolism and Biosynthesis*, Asakura Shoten, Tokyo, 2011.
159. P. Mazzafera and A. Carvalho, *Euphytica*, 1991, **59**, 55–60.
160. C. Campa, S. Doulbeau, S. Dussert, S. Hamon and M. Noirot, *Food Chem.*, 2005, **91**, 633–637.
161. J.-E. Kurtz, F. Exinger, P. Erbs and R. Jund, *Curr. Genet.*, 2002, **41**, 132–141.
162. A. Crozier, H. Ashihara and F. Tomas-Barberan, *Teas, Cocoa and Coffee: Plant Secondary Metabolites and Health*, Wiley-Blackwell, Oxford, 2012.

CHAPTER 5

Mineral Nutrition and Fertilization

H. E. P. MARTINEZ*[a], J. C. L. NEVES[b], V. H. ALVAREZ V.[b] AND J. SHULER[c]

[a]Universidade Federal de Viçosa, Departamento de Fitotecnia, Av. P. H. Rolfs, s/n, Viçosa, 35570-000, Brazil; [b]Universidade Federal de Viçosa, Departamento de Solos, Av. P. H. Rolfs, s/n, Viçosa, 35570-000, Brazil; [c]Universidade Federal de Lavras, Departamento de Engenharia - Cx, Postal 3037 Lavras, MG, 37200-000, Brazil
*E-mail: herminia@ufv.br

5.1 Introduction

Cultivated plants require macro- and micronutrients in appropriate amounts and proportions to maximize their growth and production. Knowledge of the mineral nutrition requirements of a particular species of agronomic interest is the basis for fertilization recommendations and management. Studies of nutrient accumulation over time in each phenological stage of plant growth are of great importance, including the examination of how nutrients are partitioned within different plant organs, the plant's main sources and main sinks, nutrient internal cycling, and nutrient removal by the harvest.

The relationship between plant growth, crop production, and the nutrient content of the plant's tissues is well defined. This knowledge provides a basis for the use of tissue analysis in diagnosing nutritional status, a valuable tool

Coffee: Production, Quality and Chemistry
Edited by Adriana Farah

Published by the Royal Society of Chemistry, www.rsc.org

in evaluating and adjusting fertilization schedules. There is, however, often a lack of knowledge about how to conduct the sampling, when to sample, how to prepare and send the samples to specialized laboratories, and how to interpret analysis results.

Nutritional deficiencies and excesses can also translate into symptoms. Due to the metabolic functions of the nutrients, these symptoms often have similarities among different plant species, though they can also manifest in a peculiar way for a given species. Characterization of the symptoms of mineral deficiencies and excesses can be of great value at the field level. However, the presence of these symptoms represents the final stage of a metabolic disorder, meaning the plant is already suffering negative consequences for growth and production.

Soil is the natural environment for the growth of plants. Many complex reactions involving fertilizers can occur inside the heterogeneous soil matrix. Nutrients may be present in suitable amounts, but they may not be available for uptake by roots. To grow crops in an efficient and sustainable way, it is necessary to understand the dynamics of the nutrients in the soil and in the plant. Coffee is a high-nutrient-demanding crop. Significant amounts of nutrients, especially N and K, are removed by the fruits, which can lead to depletion of the soil nutrient reserves. Given coffee's increased needs, integrated knowledge about soils and mineral nutrition is even more important in providing fertilizer recommendations.

5.2 Nutrient Accumulation and Exportation

Supplying fertilizer to coffee plants to meet their nutritional needs is based mainly on the plant's requirements for maintaining vegetative growth and fruit development. Therefore, it is very important to know the amount of nutrients the plant accumulates throughout its lifecycle.

The coffee plant grows very slowly after transplantation to the field, and mineral accumulation and fertilizer needs are low in the first two years after planting. This is called the formation phase of the orchard. After the initial fertilization, which is done in the planting furrow or planting hole, fertilizations in this phase are done on a plant-by-plant basis. In general, given the spacing of the plants and their low nutritional demands in the initial stages, there is little competition between plants for nutrients. When the plants reach their first blooming, generally around 24 to 30 months after transplantation, the developing fruits become priority sinks of nutrients and photosynthates. The need for fertilization thus increases significantly—two to three times or more than the initial phase—to meet the demands of fruit loading as well as the vegetative growth of both orthotropic and plagiotropic branches. At this time, competition among plants should be considered, and fertilization doses are performed per-area as opposed to on a plant-by-plant basis.[1–3]

Since flower buds develop in the nodes of plagiotropic branches, the growth of these branches in any particular growing season is directly related to bean production in the subsequent one. Lack of or low doses of nutrients results in reduced formation of new nodes and consequently lower production in the following crop year. The lack of nutrients to fulfill the needs of vegetative and reproductive growth can also result in the death of fine roots. Without these fine roots, the root system cannot properly explore the surrounding soil, leading to negative feedback in branch growth and eventually in fruit growth. This feature is responsible for the biennial variation of coffee production, especially in plantations located in places like the southeastern region of Brazil, where vegetative and reproductive growth occurs in the same season. The sink-source competition is more severe when the coffee field is not shaded but rather receives full sun. Under full sun, the ratio of the phytohormones gibberellin and cytokinin favors intense flowering, generating more fruit (nutrient and photosynthate sinks) and thus exacerbating the biennial variation.[4,5]

Coffee plant nutrient accumulation, in decreasing order, is N > K > Ca > Mg > S > P > B > Zn > Cu. A coffee field containing plants that are 55 months old, with a mean of 5000 plants per hectare, takes from the soil 490 kg ha^{-1} of N, 330 kg ha^{-1} of K, 220 kg ha^{-1} of Ca, 66 kg ha^{-1} of Mg, 43 kg ha^{-1} of S, 30 kg ha^{-1} of P, 1600 g ha^{-1} of B, 770 g ha^{-1} of Zn, and 550 g ha^{-1} of Cu. These quantities may vary 25% for macronutrients and 30% for B, Zn, and Cu, depending on the cultivar planted (Figures 5.1 and 5.2).[6]

Figures 5.1 and 5.2 also present nutrient amounts exported when the coffee fruits are harvested. Of note are the amounts of N and K exported from the soil. The decreasing order of nutrients exported is K > N > Ca ≈ Mg > P > S > B ≈ Cu ≈ Zn. In an orchard with 5000 plants per hectare (rows 2 m apart and plants 1 m apart within each row), 25% of N, 37% of P, 46% of K, 5.5% of Ca, 18% of Mg, 19% of S, 7% of B, 20% of Cu, and 6% of Zn that had accumulated in the plant over the course of its 55 months are removed during the harvest. This translates to about 152 kg ha^{-1} of K, 125 kg ha^{-1} of N, 12 kg ha^{-1} of Ca, 12 kg ha^{-1} of Mg, 11 kg ha^{-1} of P, 8.5 kg ha^{-1} of S, 107 g ha^{-1} of Cu, 91 g ha^{-1} of B, and 44 g ha^{-1} of Zn that need to be returned by fertilization to maintain soil fertility. The husks that are separated from the beans in post-harvest processing contain a high percentage of the exported nutrients—over 50% of the K, Ca, and B. Returning the husks to the field can therefore significantly reduce the plants' subsequent fertilization needs.[1–6]

Figures 5.1 and 5.2 also show that the plants' nutritional needs, determined by examining the nutrient amounts that have accumulated in the plants, is much higher at 55 months than at 31 months after planting. In this example the mean coffee production 55 months after planting was 4650 kg ha^{-1}, whereas at 31 months after planting it was only 1248 kg ha^{-1}. This demonstrates that the beans are the main sinks for nutrients in the plant. Thus, when plants reach the production phase, fertilization programs must, in general, take into account the quantity of fruits in formation and the expected productivity to determine the accurate fertilization dose.[6]

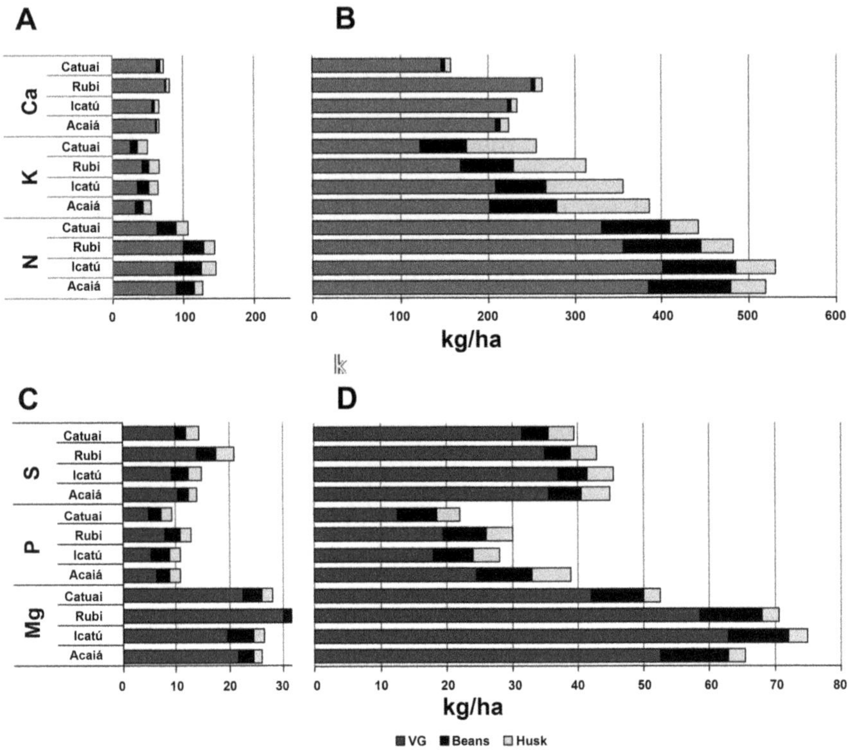

Figure 5.1 Amounts of N, K, Ca (A and B), Mg, P and S (C and D) accumulated (vegetative growth – VG) and exported (Beans and Husks) by shoots of coffee plants at 31 (A and C) and 55 (B and D) months after planting. Data from ref. 7.

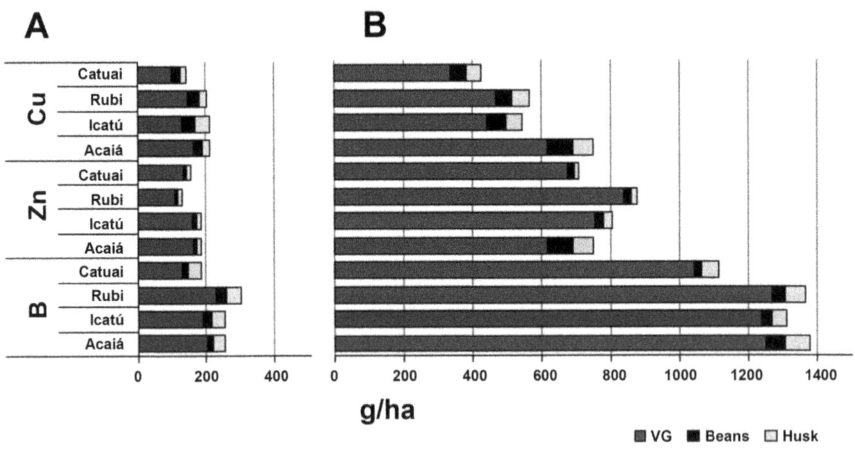

Figure 5.2 Amounts of B, Zn, and Cu accumulated (vegetative growth – VG) and exported (Beans and Husks) by shoots of coffee plants at 31 (A) and 55 (B) months after planting. Data from ref. 7.

5.3 Dynamic of Mineral Accumulation in Flowers and Fruits

In sub-tropical conditions, *Coffea arabica* L. takes two years to complete its reproductive cycle. In the first year, vegetative branches grow during the months with a longer photoperiod (day length). When the day length begins to diminish (January in the southern hemisphere), the shortened photoperiod induces the auxiliary buds into becoming reproductive buds. These flower buds grow and mature, then enter into a dormant state before blooming. The blooming will occur in the beginning of the second growth season (generally September/October in the southern hemisphere), after the first rains, or through the use of irrigation. The fruits present five development phases: first suspended growth, first rapid expansion, second suspended growth, filling, and maturation.[4,5]

Martinez *et al.* reported that floral bud growth from an initial 3 mm long until blooming (33 days) was accompanied by an increase in accumulated dry matter from 4.7 to 10.9 mg per bud. Until anthesis each bud accumulated a mean of 462, 38.5, 325, 35, 35, and 17.5 µg of N, P, K, Ca, Mg, and S, respectively. Considering an orchard with productivity of 5400 kg ha^{-1} (common in Brazil under high-technology conditions), such values represent 7.8, 0.69, 6.23, 0.56, 0.56, and 0.27 kg ha^{-1} of N, P, K, Ca, Mg, and S, respectively. Although these quantities are not unreasonably high, these nutrients are required even in such a period when the soil frequently has low water availability. This low availability can impair or limit the absorption of the demanded amounts. In some cases, internal redistribution of nutrients from roots, stems, and branches to the flowers may take place. Research performed in productive orchards does not show significant variation in the macronutrient content of leaves during flower development. Thus, plants with good nutritional status likely will not present problems in meeting the nutrient demand of the developing floral buds.[7–9]

The macronutrient accumulation in coffee beans, like dry matter accumulation, forms a double sigmoid shape curve (Figures 5.3 and 5.4). In one experiment, conducted in Brazil at 20° 45′ south, 42° 51′ west, and 640 m above sea

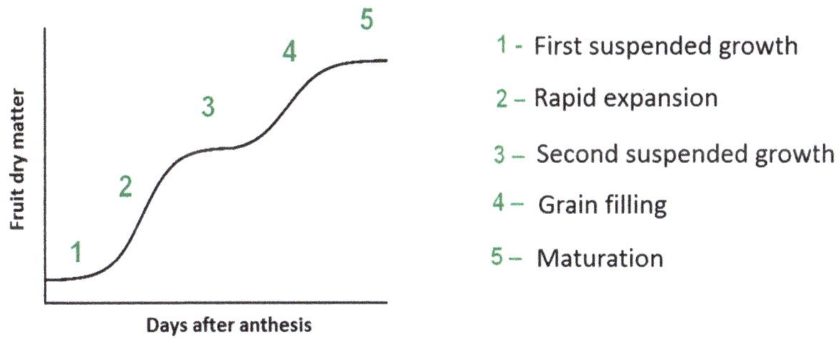

Figure 5.3 Phases of coffee-bean development. Data from ref. 7.

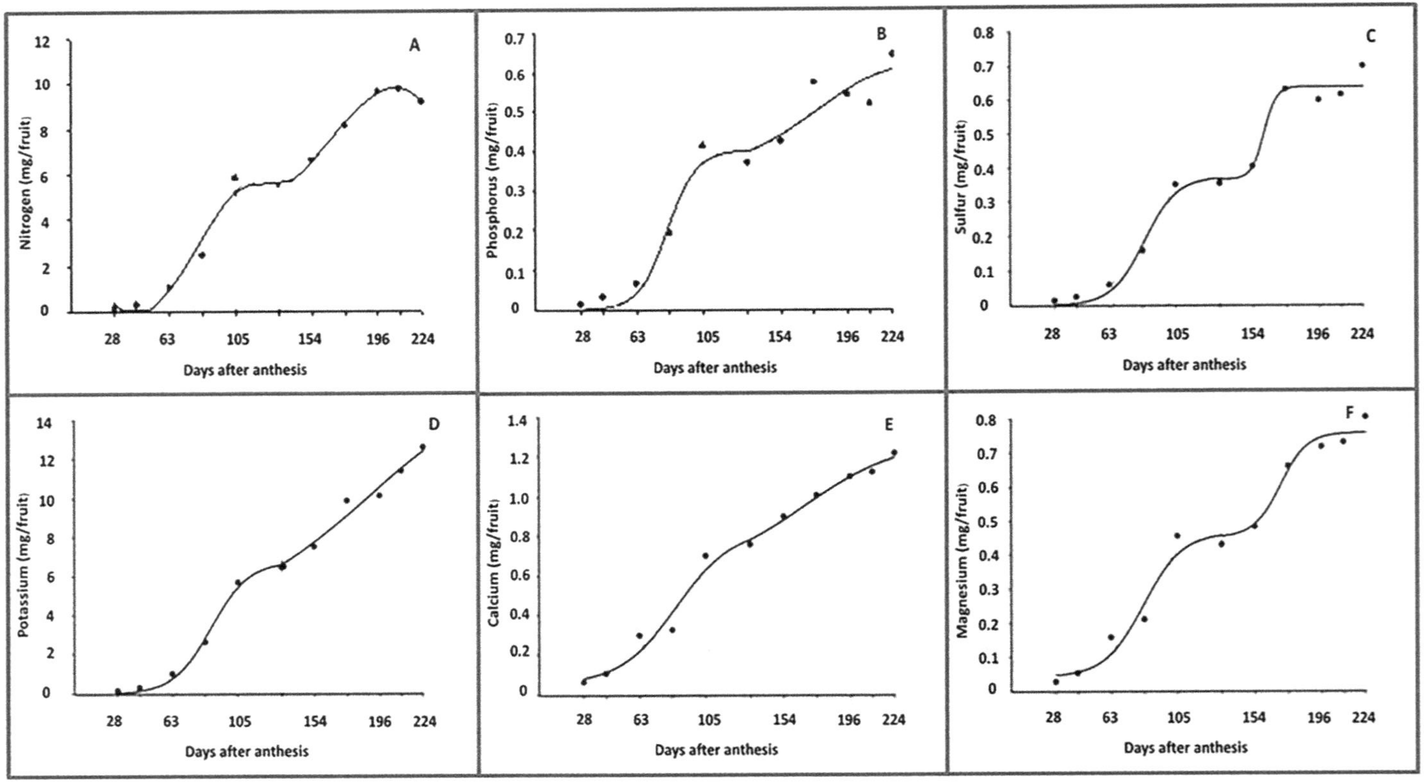

Figure 5.4 Coffee fruit N (A), P (B), S (C), K (D), Ca (E), and Mg (F) accumulation as a function of the number of days after the anthesis. Figure adapted from B. G. Laviola, H. E. P. Martinez, R. B. de Souza, L. C. C. Salomão, and C. D. Cruz, *J. Plant Nutr.*, 2009, **32**(6), 980–995.[12] Reprinted by permission of the publisher (Taylor & Francis Ltd, http://www.tandfonline.com).

level, the entire fructification cycle was 224 days. The first suspended growth phase after blooming was 42 days. In this period cellular division predominates over cellular expansion, so dry matter accumulation is inexpressive. During the first suspended growth phase, the fruit presents high respiratory rates, and most of the incoming photoassimilates are converted into energy for the formation of new cells, thus preventing the accumulation of reserves (Figure 5.4).[4–8]

In the same experiment, the first rapid expansion phase took place between 42 and 105 days after anthesis (DAA). In this phase, cellular expansion predominates over cellular division, leading to higher rates of water absorption as well as dry matter and nutrient accumulation. The increase of dry matter in the fruit during cellular expansion may be related to the increased polysaccharide synthesis that occurs for expanded cell wall formation. The cell wall polymers are continually synthesized during cell elongation, with concomitant expansion of the preexisting wall.[5–8]

The second suspended growth phase, in which dry matter accumulation temporarily ceases, occurred for 28 days, between 105 and 133 DAA. The lack of significant growth during this stage may be related to the recycling and synthesis of enzymes and intermediary compounds that were previously used in the synthesis of cell wall polymers. During the second suspended growth phase, they are recycled to be used as precursors in the synthesis of reserve compounds during the filling phase. The final two phases of the fruit lifecycle, the filling and maturation phases, are characterized by deposition of reserve substances, especially in the seeds (Figure 5.4). These phases began at 133 DAA and ended at 224 DAA. In these phases, dry matter and nutrient accumulation in the fruits were high, though in general the final amounts were attained in the filling phase, before the onset of the maturation phase.[5–8]

In this study, the maximum daily accumulation rates (MDAR) of dry matter and macronutrients were not observed during the filling phase (0.876 μg day^{-1}), as expected, but rather during the rapid expansion stage (6.72 μg day^{-1}), which occurred between 79 and 85 DAA. Therefore, it is during the rapid expansion phase, especially in highly productive orchards, that the coffee plant's nutrient demand will reach its highest levels. This can perhaps be explained by the high rates of water translocation into the fruits during this phase, when greater amounts of water are needed for cellular expansion. These high rates may also lead to the loading of mineral nutrients into the fruits. It follows that water deficiency during the expansion stage may hinder not only endocarp expansion, but also macronutrient accumulation (Figure 5.4).[5–8]

Since the rapid expansion phase is shorter (63 days) than the filling/maturation phase (91 days), a greater total accumulation of dry matter and nutrients occurs in the latter. Nevertheless, both phases are critical in supplying the coffee fruit with needed water and nutrients.[5–8]

Nutrient accumulation curves of the coffee plant's reproductive period are important tools in estimating nutritional requirements as well as identifying when to apply fertilizers. In Brazil, the coffee plant's fruit development normally occurs between September and the following June, though this will vary depending on factors such as climate and region. Fruit onset and development lengths of about 240 to 252 days were reported.

In general, at a given latitude, the higher the altitude above sea level, the longer the reproductive cycle and thus the slower the speed of accumulation of nutrients such as starch and total, reducing, and non-reducing sugars in fruits. For example, in Martins Soares, Minas Gerais, Brazil (20° 25′ south, 41° 85′ west), the reproductive cycle was 211 DAA at 720 m.a.s.l. and 266 DAA at 950 m.a.s.l. At the higher altitude, the filling/maturation phase was extended.[8–10] This means that in lower altitudes, the period of nutrient loading into the fruits is critical, since the metabolic processes will occur over a shorter time period.

The highest concentrations of N, K, Ca, and Mg in the fruit are attained in the first suspended growth phase. During the subsequent rapid expansion phase, the concentrations of nutrients previously accumulated are diluted as new tissues are formed. During the second suspended growth and filling/maturation phases (105 to 224 DAA), the concentration of macronutrients in fruit tissues tends to stabilize. N and K concentration fall again in the maturation phase.

Figure 5.4 shows that the demands for Ca and Mg are relatively greater than for other macronutrients in the first suspended growth phase. Thus, if necessary, limestone should be applied as soon as possible after the harvest, before the beginning of the new crop season. Given the high MDAR during the first expansion phase, the first fertilization should be completed before the first expansion phase begins to ensure availability of the necessary nutrients, and the fertilization dosage should account for the high MDAR.[8]

The accumulation of the micronutrients Cu, Fe, Mn, and Zn in coffee fruit fits to a single sigmoid model (Figure 5.5). Their accumulation occurs quickly, even at lower altitudes. Zn, in particular, accumulates quickly, with 60% of its accumulation occurring at the end of the first rapid expansion phase in an orchard located 720 m a.s.l.[9]

5.4 Macronutrients, Micronutrients, and Beneficial and Toxic Elements: Their Effect on Coffee Plant Growth, Production, and the Quality of its Beans

5.4.1 Nitrogen, Phosphorus, and Potassium

Nitrogen (N), the most required nutrient of the coffee plant, can be absorbed as NO_3^- or NH_4^+.[6] When properly nourished, coffee plant index leaves (the third and fourth pairs of leaves of productive branches, sampled before the first rapid expansion of the fruits) present average N concentrations between 26 and 30 g kg^{-1} of dry matter.[1–12] The main function of N is in the formation of amino acids that combine to form proteins. An adequate supply of N is

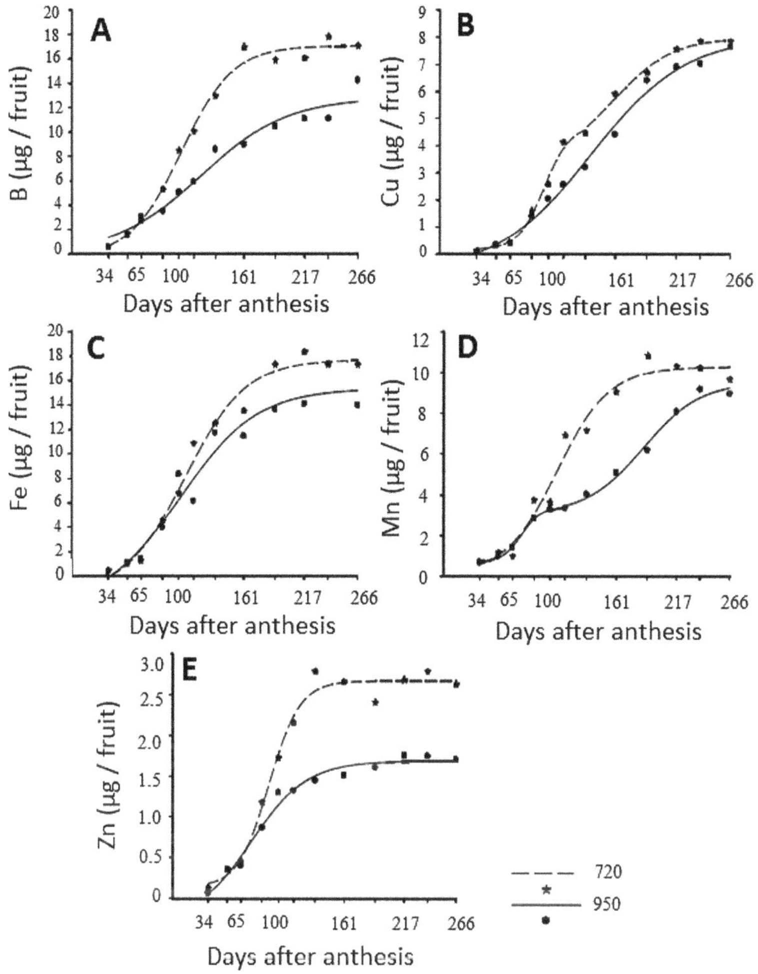

Figure 5.5 Coffee fruit B (A), Cu (B), Fe (C), Mn (D), and Zn (E) accumulation as a function of the number of days after the anthesis. Figure adapted from B. G. Laviola, H. E. P. Martinez, L. C. C. Salomão, C. D. Cruz, and S. M. Mendonça, *Rev. Bras. Ciênc. Solo*, 2007, **31**(6), 1451–1462, with permission from The Revista Brasileira de Ciência do Solo.[10]

important to ensure vegetative growth, flowering, and fruit filling. It therefore strongly influences productivity.[13,15] Plants deficient in N are small, grow slowly, and have widespread chlorosis (insufficient chlorophyll production in the leaves), which occurs initially in older leaves but progresses to younger leaves. When the deficiency turns severe, older leaves can fall.[1] Insufficient N is also associated with more severe attacks of Cercospora leaf spot[14] and leaf rust. On the other hand, excess N promotes abundant vegetative growth at the expense of reproductive growth (fruits). High doses of N can also favor the attack of pests such as the leaf miner[16] and green scale.[17]

Phosphorus (P) is mainly found in soils in its ionic form $H_2PO_4^-$ when soil pH is between 2.0 and 7.0. $H_2PO_4^-$ is also the main form of P that is absorbed by plants. Among its various functions, the participation of P in the ATP molecule is the most important. The light energy captured in photosynthesis is used in pyrophosphate synthesis, which can then be transferred to other compounds through the transfer of the phosphoryl group.[13] Like sulfur (S), P is a macronutrient that is found in low concentrations in the coffee plant.[1] The phosphorus index leaf concentration for optimal growth is in the range of 1.3 to 1.8 g kg^{-1} of dry matter.[12]

A shortage of P slows growth, and results in unsatisfactory development of the buds and opening of the flowers, resulting in the formation of fewer fruits and seeds. Older leaves may also become smaller and dark green. On the other hand, an excess of P reduces the plant absorption and translocation of Fe, Cu, and Zn.[13] P moves through the soil solution predominantly by diffusion, and its presence in adequate amounts is of great importance in the initial formation of a coffee field. In this initial phase, the root system is not well developed and thus cannot explore a soil volume large enough to meet the plant's P requirements.[18]

Among the major commercial species, *C. arabica* more efficiently absorbs and transports P than *C. canephora*, while *C. canephora* is more efficient in the actual utilization of P.[18] Cultivars also differ in both species. Experiments conducted with young plants showed that in cultivars of *C. arabica*, greater absorption efficiency was attained by plants presenting a greater number of root ramifications, greater root length, and root surface.[18]

After nitrogen, potassium (K) is the element most required by the coffee plant, a need which increases with age and productivity due to its increased accumulation in the fruit by means of translocation from the adjacent leaves. It is absorbed as K^+, and the index leaf concentration of K associated with optimal growth is between 21 and 29 g kg^{-1} of dry matter.[12] It functions in the plant as a free ion, participating as an enzyme activator in many reactions critical to growth, such as starch synthase. K is also primarily responsible for the change of turgor in guard cells, which regulate the opening and closing of stomata. Besides affecting the overall quantity of coffee bean production, adequate amounts of K positively affect the contents of caffeine, total phenols, and total and reducing sugars in the coffee bean. The presence of balanced amounts of K also may decrease electrical conductivity and K leaching, two factors that are associated with coffee bean degradation and that are commonly measured in evaluating coffee quality.

Under K-deficient conditions, soluble carbohydrates accumulate and starch content decreases. Since K participates in several steps of protein synthesis, K-deficient plants accumulate amino acids, amides, and nitrates.[13] Its deficiency may also result in chlorosis, stunted growth, low drought tolerance, reduced quality of the beans produced, the wilt of leaves and breakdown of plant stems, higher incidence of coffee leaf rust, and necrosis on the edges of older leaves. When necrosis occurs under K-deficient conditions, there is more protein degradation than synthesis, resulting in the accumulation of basic amines. High amounts of basic amines induce the activity of enzymes that regulate the synthesis of putrescine, which accumulates in the leaf edges.[1]

On the other hand, excessive amounts of K may restrict absorption of Ca and Mg, resulting in nutritional imbalance. For seedlings, restriction of Ca absorption caused by high doses of K can result in increased incidence of Cercospora leaf spot.[15]

Overall, the index leaf concentrations associated with the highest quality coffee beverages were 30 g kg^{-1} of N and 29.4 g kg^{-1} of K, with associated coffee bean concentrations of 22 g kg^{-1} of N and 18.2 g kg^{-1} of K.[19]

5.4.2 Calcium, Magnesium, and Sulfur

Calcium (Ca), which is absorbed as Ca^{2+}, acts as a structural element in the cell walls and is found in low concentrations in the cytoplasm. Its upward movement follows the transpiration stream through the xylem. In the phloem its mobility is considered null.[13] In roots, stems, and branches of the coffee plant, Ca content is on the same order as K, but in leaves and fruits it is much lower.[6] Index leaf concentrations of around 11 g kg^{-1} of dry matter are considered adequate.[12] When found in adequate concentrations in the plant tissues, Ca leads to good cell wall development. The cell wall can act as a physical barrier to the penetration of pathogens, and in experimental conditions, coffee seedlings submitted to increasing doses of Ca showed a linear decrease in damage caused by *Cercospora coffeicolla*.[15]

Deficiency symptoms of Ca appear first in young tissues and meristematic regions. They are characterized by the death of apical buds and root tips as well as deformed young leaves, which curl and present an off-white color in their margins which progresses to marginal and inter-vein blade chlorosis, while the veins themselves remain a darker green color.[1] When the concentration of Ca is low in the soil solution, cell division of root tips is impaired, and consequently the root system does not grow deep, and a limited soil volume is explored. However, excessive availability of Ca in the soil can induce deficiency of Fe and Zn.[13]

In Brazil, the world's largest coffee producer, symptoms of Ca deficiency are common, given the acidic soils (which are poor in bases and present high saturation of H^+ and Al^{3+}) of several of its main coffee regions and the heavy use of N fertilization, which promotes soil acidification.[20] In conditions of acidic soil, liming around the coffee plants is recommended to ensure adequate supply of Ca, though care should be taken as heavy doses of limestone may result in Fe deficiency. The limestone used should contain at least 12% MgO (dolomitic limestone) to avoid imbalance between cationic nutrients and inducement of magnesium (Mg) deficiency.[21]

Magnesium is absorbed as Mg^{2+}. In the leaves of well-nourished coffee plants, it appears in concentrations from 3.2 to 4.8 g kg^{-1}.[12] Mg occupies the center of the tetrapyrrole structure of the chlorophyll molecule, which is essential for photosynthesis. In the cytoplasm, it contributes in maintaining pH levels between 6.5 and 7.5 and has important functions in the activation of phosphatases, ATPases, and carboxylases. The synergistic effect of Mg on P uptake has also been reported.[13] Mg deficiency is characterized by interveinal blade chlorosis of fully expanded leaves. Tissue necrosis may follow the orange chlorosis.[1] Deficiency can occur when using calcitic limestone or with excessive doses of K.[20] As a mobile element in the phloem, the leaves adjacent to fruits under formation usually present the most severe symptoms.

Good sulfur (S) nutritional status is reported for plants whose foliar concentrations range between 1.4 and 2.0 g kg^{-1} of S in dry matter.[12] S is absorbed by the roots as SO_4^{2-} and is part of the composition of the amino acids cystine, cysteine, and methionine. In proteins where the C : S ratio is 34 : 1, S has important structural functions such as forming disulfide bridges, binding polypeptide chains, and maintaining the tertiary and quaternary structure of proteins.[13] S deficiency is characterized by generalized chlorosis, beginning in young leaves,[1] and may occur as a result of a lack of organic matter in the soil or when fertilization is performed using concentrated formulas that lack S in their composition.[20]

5.4.3 Micronutrients

Iron (Fe) is absorbed as Fe^{2+} and transported in the xylem as Fe^{3+}-citrate. The main functions of Fe are the formation of complexes and participation in redox systems, where $Fe^{3+} + e^- = Fe^{2+}$. It also operates in chlorophyll biosynthesis.[13] Foliar contents in the range of 68–121 mg kg^{-1} are considered sufficient,[12] although in plants grown in soils rich in iron oxides these concentrations may be higher, though the physiological demand for Fe is low. The physiologically active form is Fe^{2+}, so a correlation between foliar content and foliar deficiency symptoms does not always exist. Deficiencies may occur when employing high doses of lime and are characterized by a thin green network formed by leaf veins on a yellow-white background. The symptom can progress to complete whitening of young leaves followed by necrosis.[1–20] Fe deficiency can also occur when the soil presents high Mn availability. Under acidic conditions, Mn can present high availability even when liming neutralizes the toxic concentrations of Al^{3+}. When this occurs, Mn and Fe compete for the absorption sites, resulting in Fe deficiency.[13] The effect of excessive liming is difficult to correct. Deficient plants can be sprayed with iron sulfate. N top-dressing, which lowers soil pH, is also useful.[20]

After iron, manganese (Mn) is the most accumulated micronutrient.[1] It is absorbed as Mn^{2+} and serves several functions. It activates decarboxylases and dehydrogenases, participates in redox systems, plays a role in O_2 evolution during photosynthesis, and is also important in carbohydrate synthesis.[13]

The optimal concentration of Mn in index leaves is between 95 and 194 mg kg^{-1}.[12] When it occurs in excess, Mn is accumulated in old leaves, which in acidic soils often present high Mn contents. Although excesses may limit growth and production, toxicity symptoms are rare and are not manifested until the leaves present concentrations of about 1200 mg kg^{-1}.[22] Mn deficiency is characterized by interveinal paling of young leaves, which is easily confused with Fe deficiency.[1] While Mn deficiency is associated with high pH and high concentration of organic matter, Mn toxicity is associated with low pH and temporary flooding, which favors the reduction of Mn^{4+} to Mn^{2+}. Toxicity manifests as diffuse chlorotic spots with irregular borders on the leaf surface followed by drying and fall of the older leaves.[22]

Boron (B) occurs at concentrations between 36 and 57 mg kg^{-1} in the index leaves of coffee plants with good nutritional status.[12] It is likely absorbed as undissociated boric acid (H_3BO_3), and, like Ca, is considered immobile in phloem. Its functions in the apoplasm resemble those of Ca, acting in the regulation, synthesis, and stabilization of cell walls and plasma membrane. It participates in cell division and cell elongation, metabolism of nucleic acids, transport of sugars over short and long distances, tissue differentiation, auxin metabolism, and phenol metabolism.[13] Although not generally accumulated in very high amounts (Figure 5.2), B deficiency is common in coffee plantations and is characterized by the death of apical buds and root tips. The dead terminal sprout is replaced by another, originating from a bud in a lower position on the branch, which in turn also dies. This leads to the formations of branches with a rangy appearance. The leaves become small and twisted, with irregularly-shaped borders. Root meristems are also affected, with brown necrosis followed by death. The depth of the root system is thus limited as is the volume of soil explored by the roots.[1]

Deficiency can be quickly corrected through foliar applications. This procedure is especially recommended when soil water availability is limited. Nevertheless, its immobility in the phloem means that B must also be provided through soil fertilization. When correcting B deficiency, it is important to keep in mind that the boundary layer between deficiency and toxicity is narrow for this micronutrient.[6]

An excess of B causes premature leaf fall as well as a decrease in leaf area, and it can hinder the filling of the seeds, consequently affecting productivity. Marginal chlorosis and necrosis may appear in mature leaves, and mottled chlorosis in young ones.[1]

The appropriate content of copper (Cu) in index leaves is very small, between 17 and 37 mg kg^{-1} of dry matter.[12] Copper is absorbed as Cu^{2+}, and in xylem sap it appears mainly in the form of amino complexes. It participates in redox reactions and constitutes several copper proteins, many acting as enzymes (*e.g.* superoxide dismutases). It is also important for lignin synthesis and hence the integrity of support tissues.[13] For coffee plants, the role of Cu in controlling leaf rust, caused by *Hemileia vastatrix*, is well known.[15] Deficiency symptoms are characterized by irregular chlorotic spots in fully expanded leaves, which curl down and become easily detached.[1] When

Cu fungicides are used as part of a coffee leaf rust control program, toxicity of Cu is more likely to occur than deficiency. The toxic effects of Cu appear to be related to its ability to displace other metal ions, especially Fe, in physiologically important sites. Therefore, chlorosis resembling Fe deficiency is the most common symptom of Cu toxicity.[13]

Zinc (Zn) is absorbed as Zn^{2+}, and its functions in plants are largely related to its nature as a divalent cation with a strong tendency to form tetrahedral complexes. Zn acts as a metallic component of enzymes or as a cofactor in many of them. Alcohol dehydrogenase, superoxide dismutase, carbonic anhydrase, and RNA polymerase are examples of enzymes that contain Zn. As an enzymatic activator, Zn participates in the metabolism of carbohydrates, proteins, tryptophan, and indole acetic acid.[13] In Brazil, along with Cu and B, Zn is one of the micronutrients that most often limit coffee productivity.[6] Deficiency symptoms are characterized by the shortening of internodes, rosette formation at the apex of the branches, and small, narrow, and sometimes chlorotic leaves.[1] Low Zn availability also affects fruit production and bean size, perhaps due to the central role of Zn in the formation of the pollen tube.[23] When suitably supplied, it accumulates in stems or roots that can be considered reserve organs of this nutrient. When the supply is insufficient, the highest contents are observed in apical leaves.[24] Good nutrition with Zn contributes to coffee quality, reduces K leaching and coffee berry borer (*Hypothenemus hampei*) attacks.[25] Cultivars present differences in their demands for Zn. The cultivars Rubi and IPR 102 are less demanding in Zn, while the São Bernardo cultivar has low efficiency in the use of Zn.[24] A range between 9 and 19 mg kg^{-1} of dry matter is considered suitable content in index leaves.[12]

Molybdenum (Mo) appears in very low concentrations in coffee plant tissues, often less than 1 mg kg^{-1} of dry matter. Mo is an essential component of two important plant enzymes: nitrate reductase and nitrogenase;[13] however, its presence in the coffee plant has not been thoroughly studied. Coffee plants seem to have very low demand for Mo, and no reports of problems caused by Mo deficiencies are known. However, today's high productivities could result in future depletion of soil reserves, especially in regions such as the Brazilian Cerrado, where, as phosphate, molybdate becomes strongly adsorbed to clay minerals.[20]

Chlorine (Cl) is necessary for photosynthesis, as it plays a role in the water splitting that occurs in photosystem II. For the coffee plant, Cl may present a problem if it is supplied in excessive amounts as a companion ion of K^{+}. High doses of Cl^{-} are thought to reduce coffee productivity and quality. Concentration of Cl in leaves can reach 5000 mg kg^{-1} when employing high doses of KCl.[1]

The micronutrient nickel (Ni) was only considered essential to higher plants as late as 1987. It is absorbed as Ni^{2+} and is considered an element with high mobility within the plant, accumulating in leaves and seeds. Its functions are related to nitrogen metabolism, and its presence is of great importance in leguminous plants which provide symbiotic N fixation.[13]

Few studies have been conducted on the effects of Ni on the coffee plant. Deficiency symptoms have not been described, and the toxicity symptoms obtained by using a nutrient solution were characterized by chlorosis and many necrotic spots on young leaves and internodes, followed by premature leaf drop. Leaf contents in the range of 30 to 40 and 70 to 80 mg kg^{-1} were observed in plants showing toxicity symptoms. In field conditions, during the first expansion of the fruits, leaves with contents of 6 mg kg^{-1} and fruit with contents of 2.4 mg kg^{-1} were associated with the highest coffee production, high N and protein contents, nitrate reductase and glutamine synthetase activities.[26,27]

5.4.4 Silicon

Silicon (Si) is not considered an essential nutrient, but rather a beneficial element. It is absorbed as silicic acid (H_4SiO_4), and accumulates on the outside walls of the epidermal cells as amorphous silica or opal phytoliths, constituting a barrier, both to water loss through the cuticle and to fungal infection. Si also promotes better distribution of Mn in plant leaf tissues, increasing tolerance to excesses of this micronutrient.[13] In a nutrient solution, absorption of Si by coffee seedlings was very low. The largest portion of the absorbed Si accumulated in leaves, which showed concentrations of about 5.0 mg kg^{-1}.[28] There are reports of a positive effect of Si in reducing the incidence of *Cercospora coffeicolla* on coffee seedlings. The positive effect was attributed to an increase in the thickness of the cuticle and a well-developed epicuticular layer of wax.[15]

5.4.5 Aluminum

The species *Coffea arabica* seems to support relatively high aluminum (Al) content in the soil solution and, in fact, many coffee plantations are found in soils with medium or high acidity. (Acidic soils tend to have higher concentrations of Al, and Al is more soluble in acidic soils, increasing risk of toxicity.) Nonetheless, several experiments in nutritive solution showed the harmful effects of Al^{3+} on seedlings subjected to doses greater than 4 mg L^{-1}. This research also showed sensitivity differences among different cultivars. Cultivars IAC 91 Catuai Amarelo and IAC 4045 Icatu were classified as tolerant, while UFV 3880 Catimor was considered sensitive.[29] However, experiments in soil columns indicate that differences among cultivars are not as evident. When growing in columns in which the superficial layer of a latosol had a pH of 5.9, the base saturation (the fraction of exchangeable cations that are base cations) was 49%, and the Al saturation 0%, 6.5 month-old plants of the cultivar IAC 99 Catuai, previously classified as moderately tolerant, and IAC 4045 Icatu, previously classified as tolerant, presented normal growth and leaf mineral content, even when the pH, percentage of base saturation, and Al saturation in the subsurface (20–40 cm) were 3.9, 6.6%, and 93.3%, respectively. However, subsurface acidity did affect the distribution of the roots along the profile. The cultivar IAC 99 Catuai presented fewer roots and greater root thickening than the cultivar IAC 4045 Icatu, when growing in the soil layer that contained a high concentration of Al.[30]

Based on current research, correction and fertilization of the soil surface layer appear to be sufficient for proper development of the coffee plant if there is no water shortage. In cases of water stress, the presence of Al in the soil subsurface can be harmful, limiting the root depth, especially for the most sensitive varieties. Data from 152 coffee farms in the state of Minas Gerais, Brazil, showed that the presence of Al in low concentrations, both on the surface and in the subsurface, reduced the probability of obtaining high productivity.[31] Moreover, the growth of plants in soil columns uniformly corrected to pH 5.4 and 12.7% Al saturation was limited by restricted uptake of Cu and Zn.[32]

5.5 Diagnosis of Nutritional Status

Diagnosis of the nutritional status of plants is used to identify deficiencies, toxicity, or nutrient imbalances in the soil-plant system. The onset of deficiency may occur when the nutrient is present in insufficient amounts in the growth medium. Deficiency may also occur when the nutrient is present in adequate amounts, but it is not absorbed or metabolically incorporated into the plant due to unfavorable environmental conditions. Toxicity occurs because of high availability, imbalances, or unfavorable environmental conditions.

When demand for a particular nutrient is greater than its supply from the external environment, various metabolic adjustments are triggered by the plant to maintain biochemical and physiological homeostasis. These adjustments can be short-term or long-term responses designed to maintain a consistent concentration of the nutrient in the metabolic pool.

The adjustment mechanisms involve absorption, transport, and compartmentalization of ions in different organs and source/sink relationships. Similarly, the plant can display regulatory mechanisms to limit the absorption and/or excessive accumulation of nutrients and toxic elements in organs or parts of organs in which metabolism is intense. These adjustments generally involve energy costs and reductions in growth and production. If these adjustments fail, first the growth rate is reduced, then symptoms of deficiency or excess related to metabolic disorders appear.

For this reason, nutrient deficiency symptoms are quite similar in different species and can be used to diagnose the nutritional status of a culture. This technique is called visual diagnosis. However, as mentioned above, the appearance of the symptom is the final stage of a process in which growth and production can suffer irreversible losses.

Under conditions of intensive farming, the goal of diagnosing a plant's nutritional status is to identify deficiencies and/or toxicities before their visual symptoms appear, with the goal of correcting them before crop productivity is reduced. Such diagnosis is done by means of plant tissue analysis.

5.5.1 Visual Diagnosis

Visual diagnosis is rapid and inexpensive, but it is principally limited by the fact that by the time symptoms of deficiency or nutritional excess manifest visibly, a significant part of the crop production will have already been compromised. Another limitation is that under field conditions, more than one nutrient can be deficient or toxic at the same time, reflecting complex soil infertility or inadequate use of amendments and/or fertilizations. Furthermore, since water is the vehicle for absorption and transport of nutrients, the appearance of some deficiency symptoms in dry periods is common.[33]

The practice of visual diagnosis requires a careful analysis of biotic and/or abiotic conditions that may alter the nutritional status of the plant or induce patterns of damage similar to those developed in response to nutrient deficiency or toxicity. Among other factors, deficiency or excess of water; sudden changes in temperature, texture, and soil compaction; reactions among mixtures of pesticides; toxicity caused by herbicides; natural senescence of leaves; attack by pests and diseases; and poor farming practices can cause errors in the interpretation of foliar symptoms. Nutritional disorders are characterized by symmetry and present an intensity gradient from old to young leaves in the case of nutrients that are mobile in the phloem. The opposite occurs when the nutrient has low phloem mobility.[33]

Visual diagnosis, in most cases, is ineffective for determining appropriate corrective measures, but rather supports the chemical or biochemical analyses to better characterize the nutritional status of the crop.[33]

Nutrient concentrations associated with symptoms of deficiency or excess are useful as reference values for the interpretation of chemical analysis of tissues. Although the mineral nutrition of coffee has been widely studied, symptoms of nutritional deficiencies and excesses, as described in Section 5.4 and shown in Figures 5.6 and 5.7, are still easily found in the field.[33]

5.5.2 Diagnosis Based on Tissue Analysis

The theoretical curve relating plant growth or dry matter production to nutrient content in plant tissues shows well-defined regions, as illustrated in Figure 5.8. Regions I and II (regions of deficiency) are characterized by great increases in growth or dry matter production when nutrient content in tissues rises. Region III, called the region of adequate nutrition, is characterized by the slowdown in the increase in growth rate or dry matter production per unit of nutrient increase in the tissue, until it reaches a maximum point. After this maximum point, the luxury absorption region (IV) is attained. In this region, increases in nutrient content in plant tissues do not affect

Figure 5.6 Symptoms of deficiency of N (A), P (B), K (C), Ca (D), Mg (E), and S (F) in coffee-plants.

growth or production. If the availability of a nutrient continues to increase above the luxury region, it will become toxic (Region V). Region V is characterized by a decrease in the growth rate when nutrient content in the plant tissue increases. In regions I and II, the lower the concentration of nutrients in a tissue, the greater the intensity of visual symptoms. In the same way, in Region V, the higher the concentration of a nutrient in a tissue, the greater the intensity of toxicity symptoms.[33]

In Region I, the increase in the availability of nutrients in the external environment results in an increase in growth rate, resulting in a dry matter production per unit of nutrient that is greater than one. In other words, the plant grows faster than the nutrients are absorbed into the tissues, resulting in a decrease of the nutrient content in the tissue. In Region II, the increase in the absorption of a nutrient is linearly proportional to the increase in dry matter production. In Region III, the increase in the nutrient

Figure 5.7 Symptoms of deficiency of B (A, B, and C), Cu (D and E), Fe (F), Mn (G), and Zn (H). Symptoms of Mn toxicity (I).

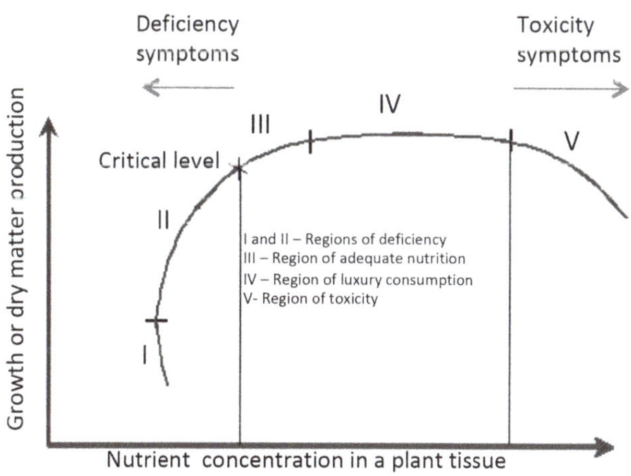

Figure 5.8 Relationship between growth or dry matter production and nutrient concentration in a plant tissue.

content in the tissue is proportionally greater than the dry matter increase. In this region we can see the critical value (CV), the tissue concentration corresponding to 90% of the maximum growth or production. In Region IV, the increase in absorption is not accompanied by an increase in growth or dry matter production, and if the nutrient availability, and consequently its concentration, continues to rise, it becomes toxic (Region V). In Region V, the reduction in growth or dry matter production can be caused directly by the effect of toxicity, or indirectly by interactions among the excess nutrient and other nutrients.[33]

From an economic perspective, the CV of a nutrient can be determined considering fertilizer prices and the economic return on production. Based on these values, the CV can be adjusted to a growth or productivity level that is higher than 90%; however, while this adjustment may increase the economic return, the efficiency of the use of a nutrient will certainly be lowered.[33]

The practical application of this knowledge is the establishment of ranges of nutrient concentrations measured in a particular part of the plant (usually the leaves). These ranges provide insight into whether the plant has adequate, insufficient, or toxic levels of a particular nutrient, and therefore they provide an assessment as to the potential for growth and production through tissue analysis.[33]

The interpretation of tissue analysis requires prior establishment of standards values—the nutrient contents in normal plants. Normal plants are defined as those that have in their tissues all the nutrients in proper amounts and proportions, and thus should be able to present high growth and yield; they should have a visual appearance similar to that of plants in highly productive crops. Normal plants can also be those grown under controlled nutrition conditions, receiving adequate amounts and proportions of essential nutrients.[1]

The mineral composition of plant tissues may, however, be influenced by a number of factors related to the plant itself as well as the surrounding environment: plant species, variety, or rootstock; age and growth phase; distribution, volume, and efficiency of root system; expected production; climate; water and nutrient availability in the soil; pest and disease attacks; type and management of the soil and interactions among nutrients.[33]

Thus, for diagnosis of nutritional status by means of tissue analysis, obtaining appropriate standards is of great importance. The standards refer to the sampling time, position on the plant, and number of leaves per plot.[33]

In general, newly mature leaves are considered the organ of the plant that best reflects their nutritional status. In addition to being a site of carbohydrate production by photosynthesis, these leaves play important roles in plant metabolism and are also the main site to which absorbed nutrients are carried.[33]

The analysis of flowers has also been successfully applied in the diagnosis of nutritional disorders. Early assessment of nutritional status through flower analysis is valuable because it enables producers to start adjusting the fertilization program at the beginning of the growing season, before the occurrence of irreversible losses in productivity and quality. Furthermore, since flowers are organs of short duration, in which there are no metabolic reactions as complex as in the leaves, they do not show marked differences between the total concentration of a nutrient and its physiologically active fraction, allowing for better diagnosis of the nutritional status of certain micronutrients, especially Fe and Zn.[34]

5.5.2.1 Sampling and Preparation Procedures

To evaluate the nutritional status of coffee orchards, leaf samples should be randomly picked at the stage between flowering and the first rapid expansion of fruit. Index leaves are the third and fourth leaf pairs, counting from the outermost leaves to the innermost leaves of productive branches, and are located in the median third of the plant (Figure 5.9). Leaves should be taken from parts of the plant facing all cardinal points. Between 40 and 50 pairs of leaves per homogeneous plot of up to a maximum area of 10 ha are sufficient.[1]

To diagnose by means of flower analysis, 100 to 200 complete flowers per plot should be taken from the first to the sixth rosette of branches located in the middle third of the tree and in all cardinal exposure faces.[34] Like the leaves, the flowers should be collected at random in homogenous plots not greater in area than 10 ha.

Packaging and shipment of leaf samples to the laboratory must be done carefully. It is important to stop or minimize tissue respiration, transpiration, and enzymatic activity as soon as possible. Ideally, samples are sent to the lab the day they are collected and are still green upon arrival. Samples should be wrapped in plastic bags and kept at a low temperature. If this is not possible, the samples should be packed in plastic bags and stored in a refrigerator at 5 °C.[33]

If sending fresh samples to the laboratory is not possible, samples should be washed with tap water, rinsed with filtered or distilled water, placed in

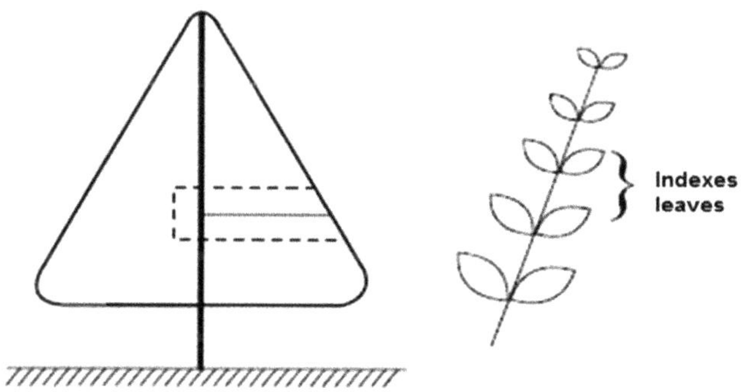

Figure 5.9 Schematic illustration of the desirable position of the index leaves in the plagiotropic branches, and of plagiotropic branches in the plant shoot. Data from ref. 7.

paper bags, then the material should be placed in the sun to dry. In all cases, samples should be identified with number, type, orchard location, collection date, nutrients to analyze, and return address.[33]

5.5.2.2 *Chemical Analysis of Tissue*

In the laboratory, after washing and drying, the plant material will be milled, submitted to acid extraction, and analyzed to determine macro- and micronutrient contents.

The sample should be washed by quickly immersing it in water, since K can be considerably leached into the washing solution (losses of around 40% of K have been reported).[14] Consistent washing is necessary to avoid errors derived from nutrients deposited on the leaf surface by dust or sprayings.

It is important to select a reliable laboratory that has rigorous systems of monitoring and quality assessment.

5.5.2.3 *Interpretation of Tissue Analysis Results*

Analytical results are interpreted by comparison with standards, which can be obtained from plant populations of the same species and cultivar presenting high productivity, or from plants cultivated under controlled conditions. In the absence of appropriate standards, standards may be created for a particular situation using plants that meet given soil, climate, and crop management criteria whose productivity is high.

Results can be interpreted by procedures that involve simple comparison of the concentration of a nutrient in a sample with the selected standard (such as the aforementioned critical levels and critical ranges) or by considering the relationships between two or more nutrients, as is done with DRIS (diagnosis and recommendation integrated system).

5.5.2.3.1 Critical Level and Sufficiency Ranges. The critical level of a nutrient in a given part of the plant is the level that is associated with 90% of the plant's maximum productivity or growth. If the tissue concentration falls below the critical value, the plant is considered to be deficient in that nutrient. The content of a nutrient in the leaves may change depending on a number of factors other than its availability in the soil, such as climate, genotype, availability of other nutrients, physical and chemical characteristics of the soil, and even sampling and handling techniques. Critical levels have the advantage of being easy, fast, and independent computational tools. On the other hand, the inability of relating the variation in the concentration of nutrients based on dry matter and the age of the plant is the major disadvantage of this method. The use of sufficiency ranges overcomes these and other limitations, improving the flexibility of the diagnosis.[35]

The results of the leaf sample laboratory analysis should be compared with the ranges stated as sufficient for coffee plants. Mineral nutrition is considered good when the nutrient content in the leaves is within the critical range. Values above the critical range indicate excesses and those below indicate deficiencies. It should be taken into account, however, that the results of any foliar analysis are influenced by a number of factors, and that lower leaf content of a nutrient does not necessarily indicate low availability in the soil. Water availability, interaction between nutrients, and root problems are some of the many factors that may be associated with this result. Because of this, leaf analysis does not replace soil analysis and *vice versa*. Both are necessary and complementary.[35]

In general, there is no great variation among critical ranges obtained in different regions or situations, but it is known that concentrations of nutrients in the soil solution affect critical levels or critical ranges.[14] This occurs because when the availability of a nutrient in the soil is high, absorption is higher than the metabolic demand. In this situation, storage of a nutrient in the vacuoles increases, leading to a high critical level compared to that obtained in situations where soil characteristics result in a low concentration of the nutrient in the soil solution. Additionally, universal critical ranges can be used (Table 5.1), and the regionalization of standards provides a reduction in the amplitude of sufficiency ranges allowing greater accuracy in diagnosis.[12]

Although not systematically used in coffee plant management, flower analysis allows early assessment of nutritional status and enables the start of fertilization program adjustment precisely in the beginning of the growing season, before irreversible losses in productivity and quality can occur. The reference values for the interpretation of results of coffee flower analysis are presented in Table 5.2.[34]

In Brazil, nutritional disorders in coffee plantations vary widely and are influenced by fertilizer prices and international coffee prices. As expected,

Table 5.1 Ranges of adequate contents of nutrient in index leaves of coffee plant.[a]

	Researcher					
Nut.	1	2	3	4	5	6
$g\ kg^{-1}$						
N	26.0–34.0	25.0–30.0	23.0–30.0	29.0–32.0	30.0–35.0	26.0–30.5
P	1.5–2.0	1.5–2.0	1.2–2.0	1.6–1.9	1.2–2.0	1.3–1.8
K	21.0–25.0	21.0–26.0	20.0–25.0	22.0–25.0	18.0–25.0	21.0–29.5
Ca	7.5–15.0	7.5–15.0	10.0–25.0	13.0–15.0	10.0–15.0	9.4–12.8
Mg	2.5–4.0	2.5–4.0	2.5–4.0	4.0–4.5	3.5–5.0	3.2–4.8
S	1.5–2.5	0.2–1.0	1.0–2.0	1.5–2.0	1.5–2.0	1.4–2.0
$mg\ kg^{-1}$						
Cu	7–20	16–20	10–25	11–14	10–50	17–37
Fe	70–200	70–200	70–125	100–130	100–200	68–121
Zn	15–30	15–30	12–30	15–20	10–20	9–19
Mn	50–100	50–100	50–200	80–100	50–100	95–194
B	40–90	40–100	40–75	50–60	40–80	36–57

[a]Data from 1. Willson (1985);[36] 2. Reuter and Robinson (1988);[37] 3. Mills and Jones Jr. (1996);[38] 4. Malavolta *et al.* (1997);[1] 5. Matiello (1997);[39] 6. Martinez *et al.* (2003).[12]

Table 5.2 Critical ranges of nutrient contents in coffee plant flowers.[a]

N	P	K	Ca	Mg	S
g kg^{-1}					
27.0–32.0	2.4–2.8	22.0–30.0	2.0–3.5	1.8–2.4	1.5–2.1

B	Cu	Fe	Mn	Zn
mg kg^{-1}				
30–50	12–30	80–140	90–150	9–18

[a]Adapted from Martinez *et al.* (2003)[34] and Zabini (2010).[7]

more productive crops have fewer nutritional problems. Macronutrient deficiencies are more common in middle and low productive crops, while problems with micronutrients, especially Cu, Zn, and B are widespread in a wide range of productivities. Diagnoses based on tissue analysis are especially useful for gauging or redirecting fertilization programs with micronutrients.[20]

Cu deficiencies may result from continued use of fungicides that do not contain copper to control leaf rust disease. In turn, excess may result from continued use of Cu-based fungicides without follow-up chemical analysis of leaf content. Low organic matter content associated with the poverty of soil parental material can also be associated with Cu deficiency.[20]

To avoid Zn deficiency, periodic spraying with products containing Zn is necessary, especially on high-clay soils, where the availability of the Zn applied to the soil is limited. For sandy soils, Zn can be supplied to the soil together with macronutrient fertilizers.[6]

For B, the threshold separating deficiency from excess is narrow, and the correction of a deficiency through continued spraying without proper monitoring by leaf analysis can result in toxicity. Although B should preferably be supplied by soil, due to its low mobility in the phloem, which limits transport to the active growth regions, severe deficiency can be corrected by concurrent foliar applications, allowing faster recovery of the plant.[6]

5.5.2.3.2 Diagnosis and Recommendation Integrated System (DRIS). Although very useful and relatively easy to apply, the interpretation of leaf analysis through critical levels, or critical ranges, involves the evaluation of the sufficiency of each nutrient without considering the balance among the nutrients. However, it is known that the nutrient content in leaves may change with age of plant, its growth stage, and certain soil and plant interactions that affect both nutrient absorption and translocation. Diagnosis using the DRIS is based on the calculation of an index for each nutrient considering its relationship with others. The ratios between each pair of nutrients in a tissue are compared with corresponding average ratios of standards, predetermined from a reference population. These ratios have historically presented less variation than the concentrations of nutrients in dry matter.[35]

Initially, the standards—*i.e.* the mean, standard deviation, and coefficient of variation of the direct and inverse relationship between all contents of nutrients, taken by pairs, of the reference population (high productivity)—are calculated.[40] The number of possible ratios (NR) is calculated by the following equation:

$$NR = n(n-1)$$

wherein:

n = number of nutrients studied.
If n = 11 (N, P, K, Ca, Mg, S, Cu, Fe, Zn, Mn, and B); NR = 110, with half being direct and half inverse relationships.

A direct relationship is that where the nutrient in question appears in the numerator (A/B) and an inverse relationship is that where the nutrient in question appears in the denominator (B/A).

After that, comparisons are made between the ratios of nutrient contents in the sample to be diagnosed with the ratios (standards) for the reference population, calculating the DRIS indexes according to the following formula (Alvarez V. and Leite, 1999):[41]

$$A\ \text{index} = \frac{+Z(A/B)+Z(A/C)+\ldots+Z(A/N)-Z(B/A)-Z(B/C)-\ldots-Z(B/N)}{2(n-1)}$$

wherein:
A index = DRIS index in the sample being diagnosed.
$Z(A/B) = [(A/B) - (a/b)].\ k/s$ (Jones, 1981):[40]

wherein:

$Z(A/B)$ = function of the ratio between contents of nutrients A and B in the sample being diagnosed;
A/B = numerical value of the ratio between the contents of nutrients A and B in the sample being diagnosed (direct relationship);
a/b = mean value obtained for the ratio A/B, derived from the population of plants with high productivity (standard);
n = number of nutrients involved in the nutritional diagnosis;
k = constant value (10), and
s = standard deviation of the values of A/B of the standard population.

The mean nutrient balance index (NBIm) is then calculated by dividing the sum of the absolute values of the DRIS indexes obtained for each nutrient according to the equation:

$$BNIm = [|A\ \text{index}|+|B\ \text{index}|+\ldots+|N\ \text{index}|]/n$$

DRIS indexes can be negative when there is a deficiency of a nutrient in relation to the others. On the other hand, positive values indicate excess, and the closer a nutrient's value is to zero, the better the balance of that nutrient in the plant. The method permits classifying a nutrient according to its degree of deficiency or toxicity. After a certain level, the lower the index the greater the deficiency; similarly, after a certain level, the higher the index the greater the toxicity.[35]

The sum of DRIS indexes, regardless of the positive or negative association, provides the "Nutritional Balance Index" (NBI), for comparing the nutritional balance of different orchards in a farm.[35]

However, the method does not allow for the calculation of the amount of a nutrient that should be applied to correct a deficiency. It only informs the restriction order, and this limitation can be due to lack or excess. The supply of the most limiting nutrient does not mean that the second element will become a major limitation because the relationships among nutrients can be changed.[35]

Tables 5.3 and 5.4 present DRIS standards obtained using 159 orchards producing more than 1800 kg ha^{-1}, and 55 orchards producing more than

Table 5.3 DRIS (Diagnosis Recommendation Integrated System) standards obtained using 159 orchards with average productivity greater than 1800 kg ha^{-1} [20]. [a]

Ratio	Mean	Standard deviation	CV (%)	Ratio	Mean	Standard deviation	CV (%)
N/P	18.792	4.466	23.77	S/Cu	0.009	0.005	51.39
N/K	1.159	0.262	22.61	S/Fe	0.002	0.001	45.39
N/Ca	2.638	0.537	20.37	S/Zn	0.017	0.011	60.73
N/Mg	7.435	2.192	29.48	S/Mn	0.002	0.001	59.17
N/S	16.681	4.587	27.50	S/B	0.004	0.002	52.19
N/Cu	0.147	0.065	44.51	Cu/N	9.659	9.091	94.12
N/Fe	0.036	0.014	39.94	Cu/P	190.158	217.208	114.22
N/Zn	0.266	0.132	49.48	Cu/K	10.635	8.739	82.17
N/Mn	0.027	0.014	53.42	Cu/Ca	24.717	22.483	90.96
N/B	0.061	0.026	42.48	Cu/Mg	68.569	58.213	84.90
P/N	0.057	0.015	26.99	Cu/S	165.072	175.511	106.32
P/K	0.064	0.018	27.35	Cu/Fe	0.308	0.252	81.79
P/Ca	0.148	0.042	28.45	Cu/Zn	2.079	1.188	57.16
P/Mg	0.418	0.155	37.14	Cu/Mn	0.262	0.321	122.38
P/S	0.917	0.252	27.51	Cu/B	0.564	0.509	90.24
P/Cu	0.008	0.005	54.69	Fe/N	33.722	18.679	55.39
P/Fe	0.002	0.001	48.74	Fe/P	651.581	430.057	66.00
P/Zn	0.015	0.008	51.92	Fe/K	38.249	22.110	57.81
P/Mn	0.002	0.001	59.35	Fe/Ca	86.779	49.277	56.78
P/B	0.003	0.002	45.57	Fe/Mg	240.818	134.271	55.76
K/N	0.903	0.190	20.99	Fe/S	579.748	390.225	67.31
K/P	16.724	4.697	28.08	Fe/Cu	4.598	3.042	66.16
K/Ca	2.353	0.573	24.34	Fe/Zn	8.230	5.031	61.13
K/Mg	6.639	2.149	32.37	Fe/Mn	0.903	0.782	86.61
K/S	15.021	5.086	33.86	Fe/B	2.035	1.334	65.56
K/Cu	0.129	0.059	45.66	Zn/N	5.114	3.876	75.79
K/Fe	0.032	0.012	38.85	Zn/P	99.152	91.166	91.95
K/Zn	0.231	0.104	44.98	Zn/K	5.682	4.192	73.77
K/Mn	0.025	0.015	60.35	Zn/Ca	13.187	9.846	74.67

Table 5.3 (*continued*)

K/B	0.054	0.024	44.06	Zn/Mg	37.470	32.192	85.91
Ca/N	0.395	0.079	20.13	Zn/S	89.203	85.839	96.23
Ca/P	7.362	2.237	30.39	Zn/Cu	0.644	0.420	65.31
Ca/K	0.452	0.123	27.20	Zn/Fe	0.165	0.106	63.87
Ca/Mg	2.877	0.850	29.56	Zn/Mn	0.137	0.117	85.69
Ca/S	6.526	2.032	31.14	Zn/B	0.285	0.189	66.28
Ca/Cu	0.057	0.026	46.01	Mn/N	50.503	34.220	67.76
Ca/Fe	0.014	0.006	40.32	Mn/P	947.657	678.795	71.63
Ca/Zn	0.103	0.054	52.15	Mn/K	59.944	50.219	83.78
Ca/Mn	0.011	0.006	55.96	Mn/Ca	136.124	109.241	80.25
Ca/B	0.023	0.010	41.32	Mn/Mg	380.870	293.235	76.99
Mg/N	0.146	0.044	30.41	Mn/S	840.847	602.404	71.64
Mg/P	2.738	1.047	38.25	Mn/Cu	7.371	6.227	84.48
Mg/K	0.168	0.062	36.74	Mn/Fe	1.793	1.510	84.23
Mg/Ca	0.378	0.112	29.61	Mn/Zn	14.408	16.235	112.68
Mg/S	2.441	1.042	42.70	Mn/B	3.290	3.440	104.57
Mg/Cu	0.021	0.011	53.16	B/N	19.319	7.795	40.35
Mg/Fe	0.005	0.002	40.32	B/P	350.657	134.334	38.31
Mg/Zn	0.038	0.022	58.14	B/K	21.924	8.988	40.99
Mg/Mn	0.004	0.002	62.52	B/Ca	49.367	18.442	37.36
Mg/B	0.009	0.005	50.55	B/Mg	145.701	84.218	57.80
S/N	0.064	0.016	24.74	B/S	320.469	150.444	46.95
S/P	1.174	0.322	27.41	B/Cu	2.823	1.759	62.32
S/K	0.074	0.025	34.07	B/Fe	0.696	0.407	58.48
S/Ca	0.168	0.050	29.52	B/Zn	4.811	2.526	52.50
S/Mg	0.474	0.167	35.29	B/Mn	0.538	0.376	69.95

[a]Data from Martinez *et al.* (2004).[20]

3000 kg ha^{-1} of coffee as a mean of two consecutive crop years.[20] The plant populations of these orchards, cultivated without irrigation, varied between 3000 and 5000 plants ha^{-1}. To calculate these standards, contents of macronutrients were expressed in dag kg^{-1} (%) and micronutrients in mg kg^{-1} (ppm).

5.5.2.3.3 Potential of Response to Fertilization. One of the difficulties of using the mean Nutritional Balance Index (IBNm) as a diagnostic tool is that the absolute values of the calculated indexes may vary with the calculation formula or the number of binary relations involved, preventing assessment in each case of the potential response to fertilization. To improve the interpretation of the results of DRIS indexes, we can use potential of response to fertilization (PRF).[42] This method defines five classes of probability of response to fertilization, comparing the index calculated for a given nutrient with the mean nutritional balance index (IBNm) as follows:

Class 1: positive response (P) is likely to occur when the DRIS index of a nutrient, being the lowest value, is simultaneously higher, in module, than the IBNm.

Class 2: null or positive response (PN) is likely to occur when the DRIS index of a nutrient is negative and although being higher, in module, than the IBNm is not the lowest index.

Table 5.4 DRIS (Diagnosis Recommendation Integrated System) standards obtained using 55 orchards with average productivity greater than 3000 kg ha^{-1} [20]. [a]

Ratio	Mean	Standard deviation	CV (%)	Ratio	Mean	Standard deviation	CV (%)
N/P	18.100	4.751	26.25	S/Cu	0.011	0.005	42.73
N/K	1.154	0.260	22.53	S/Fe	0.003	0.001	38.02
N/Ca	2.600	0.495	19.05	S/Zn	0.019	0.010	49.84
N/Mg	7.710	2.095	27.17	S/Mn	0.002	0.001	55.20
N/S	15.713	4.388	27.92	S/B	0.004	0.002	50.39
N/Cu	0.164	0.063	38.65	Cu/N	7.012	2.801	39.94
N/Fe	0.040	0.013	33.51	Cu/P	129.049	75.155	58.24
N/Zn	0.282	0.122	43.05	Cu/K	7.938	3.195	40.25
N/Mn	0.026	0.013	49.23	Cu/Ca	17.942	7.245	40.38
N/B	0.060	0.025	40.81	Cu/Mg	53.415	23.914	44.77
P/N	0.059	0.017	28.61	Cu/S	108.913	50.176	46.07
P/K	0.067	0.018	26.71	Cu/Fe	0.266	0.104	39.29
P/Ca	0.152	0.042	27.78	Cu/Zn	1.801	0.667	37.05
P/Mg	0.453	0.161	35.67	Cu/Mn	0.176	0.099	55.96
P/S	0.905	0.261	28.83	Cu/B	0.416	0.224	53.81
P/Cu	0.010	0.005	49.83	Fe/N	30.848	23.646	76.65
P/Fe	0.002	0.001	43.52	Fe/P	572.995	474.716	82.85
P/Zn	0.017	0.008	46.28	Fe/K	34.939	28.357	81.16
P/Mn	0.002	0.001	64.10	Fe/Ca	80.183	67.975	84.77
P/B	0.003	0.002	45.31	Fe/Mg	230.959	171.350	74.19
K/N	0.906	0.188	20.80	Fe/S	490.798	391.964	79.86
K/P	16.113	4.849	30.10	Fe/Cu	4.844	3.873	79.95
K/Ca	2.328	0.545	23.41	Fe/Zn	8.310	6.336	76.24
K/Mg	6.955	2.366	34.02	Fe/Mn	0.820	0.994	124.07
K/S	14.225	5.120	36.00	Fe/B	1.864	1.528	81.94
K/Cu	0.146	0.058	39.55	Zn/N	4.523	3.453	76.34
K/Fe	0.035	0.012	34.56	Zn/P	85.093	88.001	103.42
K/Zn	0.249	0.101	40.67	Zn/K	5.106	4.036	79.05
K/Mn	0.023	0.013	57.98	Zn/Ca	11.627	8.949	76.97
K/B	0.053	0.022	41.61	Zn/Mg	33.314	19.506	58.55
Ca/N	0.399	0.079	19.72	Zn/S	73.462	72.683	98.94
Ca/P	7.143	2.244	31.41	Zn/Cu	0.658	0.351	53.42
Ca/K	0.456	0.125	27.34	Zn/Fe	0.170	0.112	65.99
Ca/Mg	3.037	0.895	29.46	Zn/Mn	0.113	0.084	74.12
Ca/S	6.184	1.826	29.52	Zn/B	0.253	0.181	71.44
Ca/Cu	0.064	0.025	38.73	Mn/N	50.745	33.121	65.27
Ca/Fe	0.016	0.06	38.19	Mn/P	932.870	692.375	74.22
Ca/Zn	0.112	0.059	52.63	Mn/K	58.280	39.911	68.48
Ca/Mn	0.010	0.005	49.47	Mn/Ca	131.886	100.184	75.96
Ca/B	0.024	0.010	44.11	Mn/Mg	387.135	254.231	65.67
Mg/N	0.140	0.043	30.84	Mn/S	801.865	547.947	68.33
Mg/P	2.512	0.971	38.64	Mn/Cu	8.263	6.321	76.49
Mg/K	0.161	0.059	36.54	Mn/Fe	1.958	1.447	73.92
Mg/Ca	0.361	0.119	32.90	Mn/Zn	14.724	13.708	93.10
Mg/S	2.198	1.010	45.98	Mn/B	3.141	2.595	82.64
Mg/Cu	0.023	0.012	51.14	B/N	19.622	8.307	42.33
Mg/Fe	0.005	0.002	36.74	B/P	337.453	120.043	35.57

Table 5.4 *(continued)*

Mg/Zn	0.039	0.021	54.21	B/K	22.009	8.819	40.07
Mg/Mn	0.004	0.002	52.74	B/Ca	49.431	18.213	36.85
Mg/B	0.009	0.005	55.76	B/Mg	154.744	82.999	53.64
S/N	0.068	0.015	22.76	B/S	306.093	140.432	45.88
S/P	1.193	0.333	27.88	B/Cu	3.170	1.731	54.60
S/K	0.078	0.024	30.51	B/Fe	0.794	0.458	57.75
S/Ca	0.174	0.044	25.27	B/Zn	5.211	2.522	48.41
S/Mg	0.514	0.146	28.43	B/Mn	0.517	0.374	72.36

[a]Data from Martinez *et al.* (2004).[20]

Class 3: null response (N) is likely to occur when the DRIS index is, in module, lower than or equal to IBNm.

Class 4: negative or null response (NN) is likely to occur when the DRIS index is positive, and higher, in module, than the IBMm, but not the greatest DRIS index.

Class 5: negative response (N) is likely to occur when the DRIS index is, in module, higher than the IBNm and also the highest DRIS index.

5.6 Soil Requirements for Coffee Plant

5.6.1 Physical Characteristics

First, external physical conditions, *i.e.*, the topography, should be taken into account. These conditions are not limiting, but they determine how the orchard will be managed. In general, in flat areas row spacing is wider, with greater use of mechanization than in hilly areas.[43]

The best land for a coffee plantation is almost flat, with slopes from 5.5% to 12.0%. Strongly undulating areas (12% to 50% slopes) are also widely farmed, although mechanization is limited to between 15% and 20% slope. In slopes of 20% to 30%, only animal traction can be used. Above 30%, management must be manual, and it is particularly difficult in mountainous areas with slopes greater than 50%.[43]

Flat areas (from 0% to 2.5% slope) also present limitations for coffee cultivation. In these areas, aside from the possibility of cold air accumulation during the winter, soils are often heavy and poorly drained. Areas of plateaus, in turn, are subjected to winter cold winds.[43]

Areas with more than 15% stones and gravel in the surface layer of the soil are not recommended for coffee plantations, since they reduce the effective volume of soil that can be explored by the roots and hinder the movement of machines.[43]

Considering the internal physical conditions of soil, it is important that it be between 1.2 and 1.5 m deep, since the root system of the coffee plant reaches this depth and deeper. Effective soil depth may be limited by rock layers in the subsurface, compaction, and harmful chemical conditions. The occurrence of rock layers in the subsurface condemns the use of a specific area for coffee; however, soil compaction or harmful chemical conditions, *e.g.*, the presence of high amounts of aluminum and/or manganese, can be managed.[43]

Texture and structure determine the macro- and microporosity of the soil, and also interfere with its drainage. Clay contents in the soil ranging from 20% to 50% are suitable for the coffee plant. Soils with less than 20% clay drain excessively, and those with clay content above 50% may drain poorly. As the coffee plant does not tolerate waterlogged soils, the latter can be used only if well structured, with granular structure composed of large granules.[43]

For the coffee plant, the optimal volume of total pores (VTP) is around 50%, comprising ⅓ to ½ macropores and ½ to ⅔ micropores. Soils with porosity greater than 60% of VTP drain excessively, and soils with VTP less than 35% can be waterlogged.[43]

5.6.2 Chemical Characteristics

The chemical and physico-chemical characteristics of a soil determine its fertility, and chemical analysis is the main tool for its evaluation. Soil chemical analysis allows for the identification and quantification of adverse conditions for the development of crops, conditions such as acidity, salinity, and Al toxicity. It can thus predict the need for amendments and fertilizers and, together with other methods, infer the causes of nutritional disorders.[35]

However, we should keep in mind that although the soil is in most cases the natural medium for providing nutrients to the plant, soil analysis informs only the availability of nutrients contained therein, and does not evaluate whether these nutrients will be effectively acquired by the plant. Therefore, periodic analyses of soil and plant are necessary and complementary.[34]

For soil sampling, the area to be evaluated should be divided into homogeneous plots of no more than 10 hectares each, and 20 to 30 random sub-samples of soil should be taken. After being properly homogenized, these samples make up a representative sample of the soil of the field in evaluation. For fertilizer and lime recommendation purposes, analytical results of samples taken at a depth of 0–20 cm are employed. Samples at the 20–40 cm layer are used to evaluate subsurface acidic conditions.[35]

Soil chemical analysis is recommended before the orchard settlement and every crop year. In adult orchards, the samples should be taken in the crown projection area. To assess the soil chemical conditions of the sub-surface layer, samples from a depth of 20–40 cm are recommended every three or four years. This allows for identification of possible leaching and accumulation of K, Ca, and Mg, a condition promoted by the use of nitrate-based fertilizers, KCl, and/or inadequate doses of gypsum in the sub-surface.

Although reference ranges of chemical and physico-chemical soil characteristics are affected by the extraction method and may present large variation among different soil types, the values in Table 5.5 can be used as references to highly weathered oxisols. This table shows the critical concentrations of

Table 5.5 Fertility ranges for SOM, pH, Al^{3+}, K^{+}, Ca^{2+}, Mg^{2+}, and base saturation (BS) (0–20 cm) based in soil characteristics of 156 coffee orchards. Ranges obtained by the mathematical chance method described by Wadt *et al.* 1998[44].[a]

Fertility ranges	SOM[b] (g kg^{-1})	pH[c]	K^{+} [d] (mg dm^{-3})	Ca^{2+} [e] (mmol$_c$ dm^{-3})	Mg^{2+} [e] (mmol$_c$ dm^{-3})	BS %	Al^{3+} [e] Toxicity ranges	Al^{3+} [e] (mmol$_c$ dm^{-3})
0–20 cm								
Very low	<20	<4.1	<36	<10	<6	<13	—	—
Low	20–25	4.1–4.6	36–46	10–14	6–7	13–22	—	—
Median	25–32	4.6–5.0	46–80	14–19	7–9	22–34	Median	0.7–0.8
Adequate[f]	32–40	5.0–5.4	80–120	19–26	9–11	34–50	High	0.8–1.0
High	40–50	5.4–5.6	120–200	26–34	11–12	50–66	Very high	1.0–1.4
Very high (excess)	>50	>5.6	>200	>34	>12	>66	Toxic	>1.4
20–50 cm								
Very low	<7		<13	<3	<2	<7	—	—
Low	7–10		13–20	3–4	2–4	7–8	—	—
Median	10–16		20–33	4–5	4–6	8–11	Median	0.8–0.9
Adequate[f]	16–24		33–54	5–7	6–8	11–18	High	0.9–1.3
High	24–36		54–82	7–10	8–9	18–34	Very high	1.3–2.5
Very high (excess)	>36		>82	>10	>9	>34	Toxic	>2.5

[a] Data from Alves (2012).[31]
[b] Organic C determined by Walkey & Black method.
[c] Soil : water = 1 : 2.5.
[d] Mehlich I extractant.
[e] KCl 1 mol L^{-1} extractant.
[f] 90–100% of maximum productivity.

Table 5.6 Phosphorus soil fertility levels for coffee plant, according to clay content[1].[a]

Clay (%)	Critical value (mg dm^{-3})
<15	22.5
15–35	15.0
35–60	9.0

[a]Data from Mehlich I extractant. Guimarães *et al.* (1999).[3]

SOM, pH, Al^{3+}, K^+, Ca^{2+}, Mg^{2+}, and base sums obtained for the 0–20 cm and 20–50 cm layers of Brazilian oxisols cultivated with coffee plants. These critical concentrations were obtained by the method of mathematical chance,[44] sampling 156 orchards of Minas Gerais State, Brazil.[31] Critical concentrations of P are given in Table 5.6 and take into account soil texture.[3] It should be noted that, different from the other traits, the higher the content of aluminum (Al), the worse the performance of the culture.

5.7 Liming

In acid soils, liming aims to provide a defined amount of Ca and Mg, and raise pH to a range that allows good availability of all nutrients. In addition, liming reduces the toxicity of Al and Mn, raises the soil CEC, and favors the release of nutrients such as N, P, S, and B from decomposing organic matter. To fulfill these functions, dolomitic amendment must be used prior to the rainy and growth season.[3] The curves of nutrient accumulation show that in the first suspended growth phase, the coffee fruit accumulates a greater percentage of its total requirement of Ca and Mg than that of other nutrients in this same phase. This observation suggests that liming should be performed as early as possible to meet this demand, especially in upland orchards, where limestone reaction with the soil depends on rainfall.[8]

The lime dose (LD) can be calculated as:

LD = (BS2 − BS1/100) CEC where:
BS2 = Base saturation desired for such crop (50% – 0–20 soil layer, Table 5.5)
BS1 = Base saturation obtained in the 0–20 layer soil analysis
CEC = CEC at pH 7.0

The lime dose calculated thus considers the need to correct the volume of soil contained in a 0–20 cm layer of 1 ha. Since in perennial crops it is not possible to sufficiently deeply incorporate lime, and many times the correction does not reach the entire area, it is necessary to adjust the limestone dose considering the effective depth that will be reached by the correction and the effective area in which it will be applied. When deposited on the surface of the soil without incorporation, the effective depth achieved by liming is only 5–7 cm.[3]

Over-liming leads to negative results, promoting micronutrient deficiencies, particularly of Fe. Excess limestone leads to the disorder known as "lime-induced chlorosis", a condition that often results from the use of unadjusted doses of limestone.

5.8 Gypsum Use

Although the requirement of S by the coffee plant is low, similar to P, S deficiency may occur in coffee crops due to the use of N–P–K-concentrated formulas or soil poverty in organic matter, mainly in highly productive orchards. Deficiency can be corrected by the use of simple fertilizers containing S or by applying gypsum.[45]

Gypsum can provide S and Ca without significantly changing soil pH. It also neutralizes the effects of Al in deeper soil layers and carries bases to the soil sub-surface.[45]

Reaction of gypsum in the soil:

$$2CaSO_4 \cdot 2H_2O \xrightarrow{H_2O} Ca^{2+} + SO_4^{\ 2-} + CaSO_4^{\ o} + 2H_2O$$

In the soil solution, the Ca^{2+} ion may bind the soil exchange complex, displacing cations such as Al^{3+}, K^+, Mg^{2+}, and H^+, which in turn can react with the $SO_4^{\ 2-}$, forming $AlSO_4^{\ +}$ (which is a less harmful ion than other Al ionic forms) and neutral ionic pairs $K_2SO_4^{\ o}$, $CaSO_4^{\ o}$, $MgSO_4^{\ o}$. Because of their neutrality, these ionic pairs have great mobility along the soil profile, causing cation leaching down to deeper soil layers. As a consequence, neutralization of exchangeable Al^{3+} in deep layers of the soil can occur.[45]

After gypsum application, the following reactions are possible: Al precipitation in the form of $Al(OH)_3$ by release of OH^- in the solution due to the adsorption of sulfate; formation of the complex $AlSO_4^{\ +}$, which is less toxic to the plants; formation of the ionic pair $AlF_2^{\ +}$ due to the presence of F^- in the gypsum; precipitation of Al sulfate minerals such as alunite and basaluminite, for example, due to increased sulfate concentration in the soil solution.[45]

Gypsum recommendation is made based on results of analyses of a soil layer at 20–40 cm. For oxisols, the following conditions indicate advantages in gypsum use: Ca^{2+} <4 $mmol_c\ dm^{-3}$, Al^{3+} >5 $mmol_c\ dm^{-3}$, and aluminum saturation % >30. Optimal doses of gypsum are a matter of controversy as excessive amounts can lead to leaching of cations to depths beyond the reach of roots. Good results have been obtained using soil clay content to estimate the gypsum dose (Table 5.7). Gypsum should be applied after or together with limestone, as the soil should be enriched with Ca and Mg to avoid impoverishment caused by leaching of basic cations down in the soil profile.[45]

Table 5.7 Gypsum doses according to the soil clay content.[44 a]

Clay (%)	Gypsum dose (t ha^{-1})
0–15	0.0–0.4
15–35	0.4–0.8
35–60	0.8–1.2
60–100	1.2–1.6

[a]Data from Alvarez V. *et al.* (1999).[44]

5.9 Fertilization

5.9.1 Crop Settlement

Coffee seedlings should be planted considering spacing between rows, in furrows 30 cm deep or, alternatively, in holes of 40 × 40 × 40 cm. Liming and, if necessary, gypsum application to the total area, whether furrow or holes, should be made two to three months before planting to allow time for the reaction of limestone and gypsum with the soil. If lime is applied to the total area, an appropriate adjustment must be made if lime is also added directly to the planting hole.[3]

The holes or furrows should be filled with organic matter and mineral fertilizer mixed into the soil. Per-hole doses of 3–5 kg of cattle manure, 1–2 kg of chicken manure, 0.5–1.0 kg of castor bean cake, or 1–2 kg of coffee straw are recommended. About half of the recommended dose of P_2O_5 can be provided as natural reactive phosphate, mixed with the soil of the furrow or hole, with the other half placed at the bottom of the planting hole at the time of planting. In this phase, in which the root system explores a small volume of soil, critical levels of P in soil are higher than in subsequent growing phases. P_2O_5 recommended doses vary from 25 to 100 g per hole, according to the soil analysis and taking into account its clay content.[3]

N and K_2O should be provided after the establishment of the seedlings in three parcels during the rainy season. N doses vary from 9–15 g per hole and K_2O doses from 0–30 g per hole in the first growth season.[3] To obtain the fertilizer dose per linear meter of furrow, the quantities per hole must be multiplied by 2.5.

5.9.2 Crop Formation

Coffee crop formation is the period before which the coffee plant produces fruit. At this stage, nutritional requirements are lower than after the onset of fruit production, as outlined previously in this chapter; consequently, fertilizer requirements are also lower. For crops in formation, doses of 30–60 g of N per plant and doses of 0–60 g of K_2O per plant, split in three applications during the rainy season, are recommended.[3]

At crop formation, coffee plants do not compete with surrounding plants for nutrients, light, and water, so in this phase fertilizer doses are given per hole or plant. When fruit production begins, fertilizer doses are given per hectare.

5.9.3 Crop Production

Nutrient uptake data show great demand of the coffee plant for N and K. P is required in smaller quantities, and low levels of the element in the soil and plant can cause more severe damage in crops still in formation than in adult crops. In addition, nutrient export by the fruit is high, and fertilization needs to increase markedly in the production phase.[1] In this phase, fertilizer doses should be determined taking expected productivity into account. Doses of N and K_2O vary from 200–600 kg ha^{-1} per year and 50–450 kg ha^{-1} per year, respectively, according to soil analysis and expected productivity.[6] Doses should be portioned into three applications during the rainy season, from October to March in central-south of Brazil. First fertilization should be done before the first rapid expansion of the fruits. P_2O_5 doses vary from 10–110 kg ha^{-1} per year according to soil analysis, expected productivity, and clay content of the soil.[6] The total dose of P may be provided in the first fertilization, preferably localized in a shallow furrow lateral to the plant. In practice, the use of formulated fertilizer containing N, P_2O_5, and K_2O causes the P dose to be subdivided. Fertilizers must be applied in the crown projection area, which concentrates the roots of coffee plants.[3–6]

5.9.4 Fertilization with Micronutrients

Fertilization of coffee plants with micronutrients is quite dependent on information generated by studies of mineral nutrition. Tracking a fertilization program with tissue analysis is essential since critical levels of soil micronutrients are not reliable for several reasons: extraction methods during the analysis process are often problematic, there are still many questions to be answered in terms of the ideal levels of certain nutrients found in the soil, there is often not a consensus regarding the ideal extractant for micronutrients, and there is a lack of calibration curves for soil nutrient concentrations and plant productivities.

Manganese: In acid oxisoils, Mn can frequently be considered a problem due to the possibility of toxicity. However, excessive amounts of lime can lead to low availability of the nutrient, similar to what occurs with Fe. In case of deficiency, two to four foliar sprays with manganese sulfate (5 to 10 g L^{-1}) can correct the disorder. As soil fertilization, Mn doses from 0–15 kg ha^{-1}, based on soil analysis, are recommended.[3]

Copper: In the past, Cu deficiency has been a serious problem in Brazilian coffee plantations, but with the onset of leaf rust disease and the use of copper fungicides to control it, Cu toxicity has become a bigger problem than deficiency. Today, due to the increasing use of Cu-free fungicides, the shortage has come back in certain regions. Deficiency can be corrected with foliar sprays of 0.5% copper sulfate neutralized with lime-based solution. Correcting excess is always more difficult; procedures that raise the pH and decrease the availability of Cu can be used, but can result in deficiency of other micronutrients such as Fe, Zn, and Mn. Organic matter can also act as a complexing agent, mitigating some of the effects of excess Cu. Soil fertilization doses from 0–3 kg ha^{-1}, according to soil analysis, can be used.[1–3]

Boron: Deficiency is common in coffee plantations, especially those established in soils with low reserves of this element and poor in organic matter. Supply is often done by foliar spraying with 0.5% boric acid; however, as B is an element of lower mobility in the phloem, application is preferably done by soil fertilization, providing root absorption and continuous transport to active growing regions by transpiration flow. In cases of acute deficiency, foliar sprays, whose effect is faster and independent of soil conditions such as humidity, may be preferred.[1–3]

Zinc: Deficiency is quite common in coffee plantations. Typical deficiency symptoms include small and narrow leaves with short internodes, as well as growing branch tips that give the appearance of a rosette (Figure 5.7H). This characteristic symptom is derived from the role of Zn in tryptophan production, and thus with the production of IAA.[1]

Zinc is generally classified as partially mobile in the phloem. For the coffee plant, Zn mobility is minimal in both well-supplied and deficient plants. This leads to the conclusion that the appropriate way to supply Zn to coffee plant is by soil fertilization. However, in high-clay soils, Zn becomes unavailable, necessitating expeditious foliar applications.[24]

In sandy soils, applications of 2 to 6 kg ha^{-1} of zinc sulfate ($ZnSO_4$) can be used. For coffee plantations in high-clay soils during the growing season, 2–4 foliar sprays with 0.5% zinc sulfate mixed with other micronutrients are recommended.[3] It should be noted that the fruit completes the accumulation of Zn at the end of the first rapid expansion phase, so the first spraying must be done early in the growing season or just prior to first rains.[9] High concentrations in the spray mixture lead to excess, which can restrict photosynthesis and thus growth and production. When Zn is the only nutrient being sprayed, concentrations of about 0.2% zinc sulfate are sufficient. Adding 0.5% potassium chloride (KCl) enhances Zn absorption.[3] Zn chelate may also be used because it is less retained in the coffee plant leaf cuticles, promoting better absorption of the element compared to its supply as sulfate.[46]

It is common to use cocktails containing Zn, B, and Cu. Cu causes competitive inhibition of Zn absorption, and B causes non-competitive inhibition of Zn uptake. Competitive inhibition can be overcome by increasing the concentration of Zn in the spray, whereas non-competitive inhibition promoted by B is not overcome. In the latter case, the absorption sites are different, but the coupling of B on the absorption site determines changes in the sites of Zn absorption, making them inefficient to carry the element from the outside to the inner side of the plasma membrane.[1]

Fertilizer recommendations can also be obtained through modeling based on the difference between the crop requirement for a specific nutrient, and the supply of that nutrient in the soil. The crop requirement should take into consideration crop productivity, nutrient utilization efficiency, nutrient partition among plant organs, and the nutrient recovery rate. The soil requirement should consider the nutrient concentration in the soil, the volume of soil explored by the roots and the recovery rate of the nutrient by the chemical extractor.

References

1. H. E. P. Martinez, J. G. de Carvalho and R. B. de Souza, Diagnose Foliar. [A. do livro], in *Recomendações para o uso de corretivos e fertilizantes em Minas Gerais - 5a. aproximação*, ed. A. C. Ribeiro, P. T. G. Guimarães, V. H. Alvarez Comissão de Fertilidade do Slo do Estado de Minas Gerais, Viçosa, 1999, pp. 143–168.
2. E. Malavolta, *Nutrição Mineral e Adubação do Cafeeiro: Colheitas Econômicas Máximas*, Associação Brasileira para Pesquisa do Fosfato, São Paulo, 1997.
3. Y. P. Neves, *Productividad y acumulacion de materia seca*, 2002, **1**, 1.
4. C. Ribeiro A., *Recomendações para o uso de corretivos*, CFS do Estado de Minas Gerais, Viçosa, 1999, p. 3.
5. P. Camargo A, *Definiçao e esquematizaçao das fases*, 2001, vol. 60.
6. A. B. Rena, R. S. Barros and M. e Maestri, Desenvolvimento produtivo do cafeeiro, in *Tecnologias de produção de café com qualidade*, ed. L. Zambolim, UFV, Departamento de Fitopatologia, Viçosa, 2001, pp. 101–128.
7. H. E. P. Martinez, M. A. Tomaz. N. S. Kakiyama, *Guia de acompanhament das Aulas de Cafeicultura*, 2nd, Edotora UFV, Viçosa, 2007, p. 152.
8. A. Zabini, *Diagnóstico nutricional do cafeeiro por meio da análise de flores, folhas e extrato foliar*, Editora UFV, Viçosa, 2010, p. 78.
9. B. G. Laviola, Acúmulo de nutrientes em frutos de cafeeiro em quatro altutudes de cultivo, *Rev. Bras. Cienc. Solo*, 2007, **31**, 1451–1462.
10. B. G. Laviola and H. E. P. Martinez, Acúmulo de nutrientes em frutos, *Rev. Bras. Ciênc. Solo*, 2007, **31**, 1439–1449.
11. B. G. Laviola, H. E. P. Martinez, L. C. C. Salomão, C. D. Cruz, S. M. Mendonça and A. P. Neto, Alocação de fotoassimilados em folhas e frutos de cafeeiro cultivado em duas altitudes, *Pesqui. Agropecu. Bras.*, 2007, **42**(11), 1521.
12. B. G. Laviola, H. E. P. Martinez, R. B. Souza, L. C. Salomão and C. D. Cruz, Macronutrient accumulation in Coffee Fruits at Brasilian Zona da Mata Conditions, *J. Plant Nutr.*, 2009, **32**(6), 980.
13. H. E. P. Martinez and J. F. S. Menezes, Faixas críticas, *Pesqui. Agropecu. Bras.*, 2003, **38**, 703.

14. *Marschner's Mineral Nutrition*, ed. P. Marschner, Academic Press, 3rd edn., 2012, p. 651.
15. A. Pozza, Influencia da nutrição mineral na intensidade, *Pesqui. Agropecu. Bras.*, 2001, **36**, 53.
16. A. A. A. Pozza, P. T. G. Guimarães, V. L. de Carvalho, E. A. Pozza, J. G. de Carvalho, M. M. Romaniello, *Interação entre as doenças e o estado nutricional do cafeeiro*, EPAMIG: EPAMIG, Belo Horizonte, 2004, p. 84.
17. S. L. Caixeta and H. E. P. Martinez, Nutrição e vigor de mudas de cafeeiro infestação por bicho mieiro, *Ciênc. Rural*, 2004, **60**, 65.
18. F. L. Fernandes, M. C. Picanço, M. E. S. Fernandes, R. B. Queiroz, V. M. Xavier and H. E. P. Martinez, The effects of nutrients and secoundary compounds of Coffea arabica on the behavior and development of Coccus viridis, *Environ. Entomol.*, 2012, **41**(2), 333.
19. A. P. Neto, *Eficiencia do uso de P por cultivares de café*, ESALQ, Piracicaba, 2013, p. 97. Tese de doutorado.
20. R. A. Reis Jr. and H. E. P. Martinez, *Adição de Zn e asorção*, 2002, **59**(3), 537.
21. J. M. Clemente and H. E. P. Martinez, *Acta Sci., Agron.*, 2015, **37**(3), 297.
22. H. E. P. Martinez and R. B. Souza, *Nutrição Mineral, fertilidade do solo*, EPAMIG, Viçosa, 2004, p. 60, Bolletin=m Técnico n. 72.
23. A. V. Zabini and H. E. P. Martinez, Tolerancia de progenies de cafeeiros, *Coffee Science*, 2007, 87.
24. Y. Poltronieri and H. E. P. Martinez, Effect of Zinc and its form of, *J. Sci. Food Agric.*, 2011, 2431.
25. A. W. Pedrosa and H. E. P. Martinez, Charcterizing zinc use efficiency, *Acta Sci., Agron.*, 2013, 343.
26. Y. Poltronieri and H. E. P. Martinez, Zinc suplementation, production and quality of coffee beans, *Rev. Ceres*, 2013, **60**(2), 293.
27. T. Tezotto, J. L. Favarin, R. A. Azevedo, C. R. F. Alleoni and P. Mazzafera, Coffee is highly tolerant to cadmium, nickel and zinc : Plant and soil Nutritional status, metal distribution and bean yield, *Field Crop Res.*, 2012, **25**, 25.
28. A. C. M. C. M. da Cunha, M. L. de oliveira, E. C. Caballero, H. E. P. Martinez, P. C. R. Fontes and P. R. G. Perira, Growth and nutrient uptake of Coffee seedlings cultivated in nutrient solution with and without silicon addiction, *Rev. Ceres.*, 2012, **59**(3), 392.
29. M. C. L. Braccini and H. E. P. Martinez, Tolerância de genótipos de cafeeiro ao aluminio em solução nutritiva, *Rev. Bras. Cienc. Solo*, 1998, **22**, 435.
30. L. A. Rodrigues and H. E. P. Martinez, Growth response of coffee tree shoots and roots to subsurface liming, *Plant Soil*, 2001, **234**, 207.
31. L. C. Alves, *Faixas de suficiência*, UFV, Viçosa, 2012, p. 70. Tese Ms.
32. M. C. L. Braccini, Tese DS.
33. H. E. P. Martinez and R. B. Souza, Coffee-tree floral analysis as a mean, *J. Plant Nutr.*, 2003, **26**, 1467.
34. R. B. Cantarutti, N. F. de Barros, H. E. P. Martinez and R. F. Novais, Avaliação da Fertilidade do Solo e Recomendação de Fertilizantes, ed. V. H.

Alvarez V, N. F. Barros, R. L. F. Fontes, R. B. Cantarutti, J. C. L. Neves, R.F. Novais, *Fertilidade do Solo*, Sociedade Brsileira de Ciência do Solo, Viçosa, 2007, vol. XIII, pp. 769–850.

35. Proposed modifications, C. A. Jones, *Commun. Soil Sci. Plant Anal.*, 1981, **12**, 785.
36. Fundamentos estatísticos das fórmuloas, V. H. Alvarez V and R. A. Leite, *Boletim Informativo da Soc. Bras*, 1999, **24**, 20.
37. P. G. S. e Wadt and V. H. Alvarez V, Monitoramento nutricional. [A. do livro] Paulo Guilherme Salvador Wadt. (Org.), *Manejo do Solo e Recomendação de Adubação para o Estado do Acre*, Rio Branco : s.n., 2005, pp. 283–304.
38. K. C. Willson, Mineral nutrition and fertilizer needs, *Cofee Botany, Biochemistry and Production of Beans and Beverage*, ed. [A. do livro] K. C. Willson and N. N. Cliford, Croom Helm, London, 1985, p. 135.
39. D. J. Reuter and J. B. Robinson, *Plant analysis: An Interpretation Manual*, Inkata Press, Melbourne, 1988, p. 218.
40. H. A. Mills and J. B. Jones Jr, *Plant Analysis Handbook II: A Practical Sampling, Preparation Analysis and Interpretation Guide*, MicroMacro, Georgia, 1996, p. 422.
41. J. B. Matiello, *Gosto Do Meu Cafezal*, MAA/SDR/PROCAFE, Rio de Janeiro, 1997, p. 262.
42. A. Kupper, *Fatores Climaticos e edáficos na cultura Cafeeira. Nurtiçao e adubaçao do cafeeiro*, Instituto da Potassa & Fosfato, Piracicaba, 1981, pp. 28–54.
43. P. T. G. Guimarães, A. W. R. Garcia, V. H. Alvarez V, L. C. Prezotti, A. S. Viana, A. E. Miguel, E. Malavolta, J. B. Corrêa, A. S. Lopes, F. D. Nogueira, A. V. C. Monteiro and J. A. de Oliveira, *Cafeeiro. Recomendações para*, CFSEMG, Viçosa, 1999, pp. 289–302.
44. P. G. S. Wadt, V. H. Alvarez V, R. F. de Novais, R. F. de, S. Fonseca, S. e Barros and N. F. de, O Método da Chance Matemática Na Interpretação de Dados de Levantamento Nutricional de Eucalip, *Rev. Bras. Cienc. Solo*, 1998, 773–778.
45. V. H. Alvarez V., L. E. Dias, A. C. Ribeiro, R. B. de Souza, *Uso de gesso agrícola, Recomendações para o uso*, CFSMG, Viçosa, 1999, pp. 67–78.
46. I. A. L. Franco, H. E. P. Martiez and A. V. Zabini, *Cienc. Rural*, 2005, **35**(3), 491.

CHAPTER 6

Coffee Grading and Marketing

CARLOS HENRIQUE JORGE BRANDO*

P&A Ltda - Praça Rio Branco, 13 – E.S. do Pinhal – SP, CEP 13990-000, Brazil
*E-mail: peamarketing@peamarketing.com.br

6.1 Introduction

Although coffee grading may refer to different things and concepts, in this chapter it is understood as defect removal and bean sizing in order to refine the natural quality of the coffee lot and to bring it to standards demanded by different markets.

This type of grading is performed in three stages – cleaning, sizing, and sorting of defects, usually carried out in this order. Cleaning is undertaken by two machines – pre-cleaners and destoners – that are usually equipped with magnets, sizing by graders with screens of different sizes and shapes, and defect removal by densimetric separators and color sorters that take out beans with low density and unwanted colors, respectively.

Different markets require different bean size distributions and defect contents that are associated with appearance and, most importantly, quality. Bean size affects the aspect of coffee more than its cup quality that is related with the number and type of defects found in it as well as by processing, climate, variety, *etc.*

Some countries have established export types or qualities associated mostly with bean size and defect count, *e.g.* Colombia Supremo and Kenya AA. Other countries have this grading system also linked to cup quality and deliver mostly to clients' requirements, for example, Brazil. The usual

Coffee: Production, Quality and Chemistry
Edited by Adriana Farah

Published by the Royal Society of Chemistry, www.rsc.org

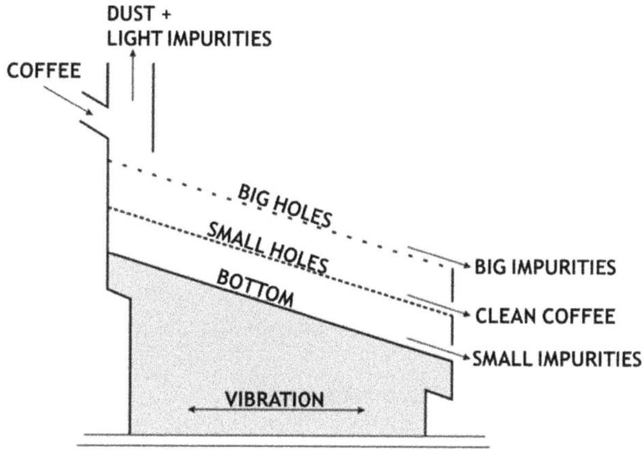

Figure 6.1 Pre-cleaner: operating scheme.

practice by roasters and soluble makers is to use coffees from different origins and qualities to make their own blends, with single-origin coffees still representing a very small part of the market in importing countries.

6.2 Cleaning

Impurities come from the coffee field (twigs, sand, stones, *etc.*), on-farm processing (husks, drying-ground fragments, *etc.*) and unfair trade practices (foreign matter added by sellers to increase weight and/or volume).

Pre-cleaners remove impurities that are larger and smaller than coffee with the help of screens whose hole sizes are selected according to the size of the impurities. Light impurities like dust and leaves are aspirated or blown with the help of airflows (Figure 6.1).

Impurities the same size as coffee, mostly stones, cannot be removed by sieving. Destoners separate stones whose density is larger than coffee that, unlike stones, float when air passes through it.

Finally, magnets are used to remove iron pieces the same size and density as coffee that are not removed by the other machines above. Magnets are usually installed before or after pre-cleaners and destoners (Figure 6.2).

Most coffee grades restrict the number and types of defects, *e.g.* small stone or husk piece, with quality and price inversely associated with the presence of such defects that are the easiest to identify by visual inspection. Even though a stone may not affect quality much, it may be specially damaging to the grinders used by roasters.

6.3 Separation by Size

There are several reasons to separate coffee lots according to the size of their beans: uniformity of roasting, appearance, separation of defects and quality.

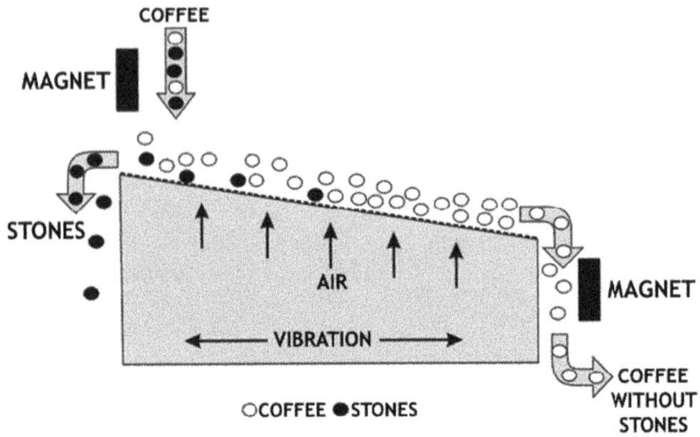

Figure 6.2 Stone remover: operating scheme.

The claim that lots with uniform bean size tend to roast in a more homogenous way is challenged by modern roasting technology. However, it may still hold for the majority of coffee roasting machines installed today.

Although most coffee sold today is in ground or soluble powder or agglomerate form, the tendency is for whole bean to gain space in coffee shops and outlets that supply consumers who have coffee grinders. That being the case, there is a growing demand for beans that are large and even those with specific shapes like the round peaberries.

The separation of coffee beans according to their size facilitates both density and color separation of defective beans that become more efficient when the lots to be processed are of uniform bean size. Even if these lots are to be later marketed with a combination of bean sizes, it may be justifiable to separate the beans according to size in order to process different sizes separately and then to blend them back together.

It is not clearly proven that coffee beans of large size have a better quality even though this seems to be behind a lot of marketing claims. However, it is a proven fact that large beans present fewer defects than small ones; a lot more processing work for the removal of defects is therefore required to bring a small-bean-size coffee lot to the same quality of a lot of bold, large beans.

Some countries export coffee with a range of sizes, usually above a given screen size, for example, Colombia and Central America. Other countries export mostly specific sizes, like Kenya and India. In any case, there is a tendency to export large bean sizes separately as a result of the growth of coffee shops, specialty coffee and, more recently, micro lots, which are small quantities of coffee of high quality and/or specific features.

Coffee beans are separated by size by sieving using screens of different sizes and shapes. Even though a few countries like India use the sizes of the screen holes in millimeters (mm), they are usually measured in 64ths of an inch, with the screen identified by the number of 64ths. For example, screen 17 has a width or diameter of 17/64″. The shape of the screen holes may be round, to separate the common "oval" beans, or elongated, to separate the round beans called peaberries, or simply PB. Peaberries, which come from coffee cherries with one single bean, tend to represent a small percentage of the whole coffee lot, usually under 10%, except for a few countries or when droughts occur during the cherry growth period.

Countries that export coffee by size usually offer peaberry grades. Countries that export mixed sizes usually have the peaberries included in these sizes. Some importing markets pay premium prices for peaberries and a few experts claim that they roast more uniformly because they are round.

6.4 Separation of Defects

Separation of defects comprises two different steps: mechanical separation by density and optical separation by color. Low-density coffee beans are associated with several types of defects: berry borer affected, malformed, hollow, fermented, and unhulled beans, shells, *etc.* Color defects are associated with immature beans/quakers, fermented, over-dried and black beans.

Separation by density can be performed by catadors and gravity tables that are also called densimetric separators. Catadors, which were much used in the past, are being progressively replaced by gravity tables that have a greater separation power and consume less electricity. The active separation principle is the passage of an air current through the product causing the less dense materials – mostly defective beans – to float and to be separated from the better quality denser material. The less dense materials to be separated are found in larger quantities mixed with small size beans that require a more intensive and efficient density sorting process (Figure 6.3).

Off-color beans are separated by an optical process that compares the color of the bean against reject patterns that once matched cause an electronic

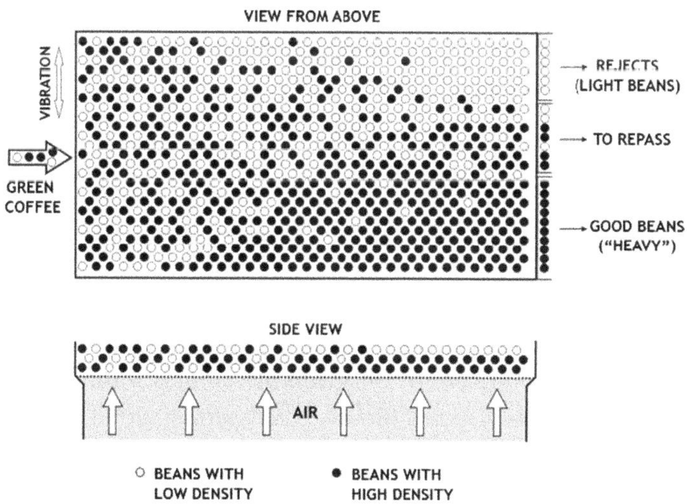

Figure 6.3 Separation by density: operating scheme.

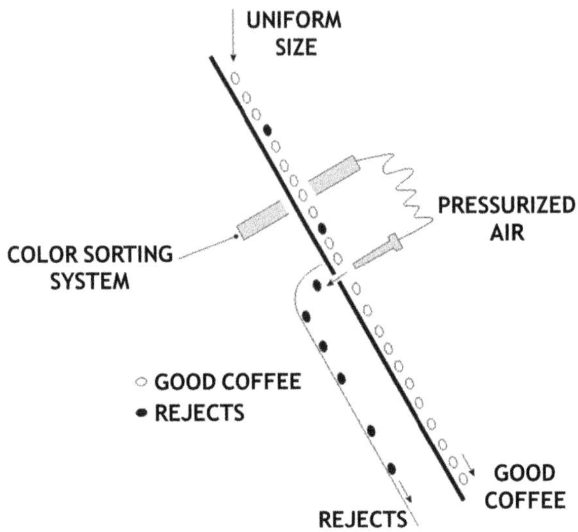

Figure 6.4 Separation by color: operating scheme.

signal to be emitted to activate a jet of air that separates these reject beans that either fall by gravity or travel on a conveyor belt along with the good quality coffee. When defects are in large numbers or there is little difference between accept and reject beans, the process must be repeated with different repassing sequences possible (Figure 6.4).

6.5 Examples of Grading Systems

Three examples of grading systems are presented below: the Brazil/New York Method, the Kenyan Grading and Classification and the Specialty Coffee Association of America Green Coffee Classification (extracted from *The Coffee Exporters' Guide*, International Trade Centre, 3rd edition, 2012, except the last two paragraphs).

These examples combine sizing and impurities and defect counts in different ways to achieve distinct results that appeal to markets that are not necessarily the same.

6.5.1 Brazil/New York Method

In the Brazilian method 300 grams of coffee is classified. The number of beans equivalent to one full defect is given in Tables 6.1 and 6.2. For example, every three shells counts as one full defect. On the other hand one large rock counts as five full defects. If a bean has more than one defect the highest defect is counted. For example, an insect damaged black bean counts as one full defect. After counting the number of defects use Table 6.3 to classify the type and its point rating.

Table 6.1 Brazil/New York method.

Intrinsic defect	Number	Full defects
Black bean	1	1
Sour (including stinker beans)	1	1
Shells	3	1
Green	5	1
Broken	5	1
Insect damage	5	1
Malformed	5	1

Table 6.2 Brazil/New York method.[a]

Foreign defect	Number	Full defects
Dried cherry	1	1
Floater	2	1
Large rock or stick	1	5
Medium rock or stick	1	2
Small rock or stick	1	1
Large skin or husk	1	1
Medium skin or husk	3	1
Small skin or husk	5	1

[a]Large rock or stick – screen size 18/19/20; medium rock or stick – screen size 15/16/17.

6.5.2 Kenyan Grading and Classification

At the mills the parchment skin surrounding each bean is removed followed by mechanical grading of the coffee into seven separate grades according to size, weight and shape of the bean (Table 6.4).

Mbuni is the coffee that has not gone through the wet process (unwashed or dry processed). It comprises about 10% of the total crop and is graded as either heavy mbuni (MH) or light mbuni (ML). This grade generally fetches lower prices and has a sour taste cup.

These grades are then classified based on a numerical reference system on a scale of 1 to 10. The quality of the raw, roast and liquor are analyzed and described based on this scale where one is the finest and best and ten is the least favored. The cup may be described as Fine, Fair to Good, Fair Average Quality (standard 4), Fair, Poor to Fair to Common Plain Liquors.

6.5.3 Specialty Coffee Association (SCA) Green Coffee Classification

The green coffee classification standard provided by the SCA accounts for the relationship between defect and cup quality. However, it leaves out a few of the important defects that can occur in coffee (see the Brazil/New York Method) (Tables 6.5 and 6.6).

Table 6.3 Brazil/New York method.

Defects	Type	Points	Defects	Type	Points
4	2	100	49	5–5	−55
4	2–5	95	53	5–10	−60
5	2–10	90	57	5–15	−65
6	2–15	85	61	5–2–	−70
7	2–20	80	64	5–25 5/6	−75
8	2–25 2/3	75	68	5–30	−80
9	2–30	70	71	5–35	−85
10	2–35	65	75	5–40	−90
11	2–40	60	79	5–45	−95
11	2–45	55	86	6	−100
12	3	50	93	6–5	−105
13	3–5	45	100	6–10	−110
15	3–10	40	108	6–15	−115
17	3–15	35	115	6–20	−120
18	3–20	30	123	6–25 6/7	−125
19	3–25 3/4	25	130	6–30	−130
20	3–30	20	138	6–35	−135
22	3–35	15	145	6–40	−140
23	3–40	10	153	6–45	−145
25	3–45	5	160	7	−150
26	4	0	180	7–5	−155
28	4–5	−5	200	7–10	−160
30	4–10	−10	220	7–15	−165
32	4–15	−15	240	7–20	−170
34	4–20	−20	260	7–25 7/8	−175
36	4–25 4/5	−25	280	7–30	−180
38	4–30	−30	300	7–35	−185
40	4–35	−35	320	7–40	−190
42	4–40	−40	340	7–45	−195
44	4–45	−45	360	8	−200
46	5	−50	>360	Above 8	

Table 6.4 Kenyan grading and classification.

PB	Round beans, usually one in a cherry
AA	Large beans (7.20 mm screen)
AB	This grade is a combination of A and B (6.80 mm screen)
C	Smaller bean than B
E	Elephants. The largest beans
TT	Any light coffee blown away from all grades including ears mostly from elephants
T	The smallest and thinnest beans mostly broken and faulty

To classify a coffee, 300 grams of properly hulled coffee is classified according to the standards given below. 100 grams of this coffee is sorted using screens 14, 15, 16, 17 and 18. The coffee remaining in each screen is weighed and the percentage is recorded. Since classifying 300 grams of coffee is very time consuming, 100 grams of coffee is typically used. It is recommend that

Table 6.5 Primary defects.

Primary defect	Number of occurrences equal to one full defect
Full black	1
Full sour	1
Pod/cherry	1
Large stones	2
Medium stones	5
Large sticks	2
Medium sticks	5

Table 6.6 Secondary defects.

Secondary defects	Number of occurrences equal to one full defect
Parchment	2–3
Hull/husk	2–3
Broken/chipped	5
Insect damage	2–5
Partial black	2–3
Partial sour	2–3
Floater	5
Shell	5
Small stones	1
Small sticks	1
Water damage	2–5

if the coffee is of high quality with few defects to use 300 grams. If the coffee is of a lower quality with many defects 100 grams will often suffice in a correct classification as either Below Standard Grade or Off Grade. The coffees then must be roasted and cupped to evaluate cup characteristics.

Specialty Grade (1): Not more than 5 full defects in 300 grams of coffee. No primary defects allowed. A maximum of 5% above or below screen size indicated is tolerated. Must possess at least one distinctive attribute in the body, flavor, aroma, or acidity. Must be free of faults and taints. No quakers are permitted. Moisture content is between 9 and 13%.

Premium Grade (2): No more than 8 full defects in 300 grams. Primary defects are permitted. A maximum of 5% above or below screen size indicated is tolerated. Must possess at least one distinctive attribute in the body, flavor, aroma, or acidity. Must be free of faults and may contain only 3 quakers. Moisture content is between 9 and 13%.

Exchange Grade (3): 9–23 full defects in 300 grams. Must have 50% by weight above screen size 15 with no more than 5% of screen size below 14. No cup faults are permitted and a maximum of 5 quakers are allowed. Moisture content is between 9 and 13%.

Below Standard Grade (4): 24–86 defects in 300 grams.

Off Grade (5): More than 86 defects in 300 grams.

The Brazil/New York method does not mention screen size or quality, which is the case of both the Kenyan and SCA systems. However, Brazil also adds both screen size(s), for example, Medium to Good Bean (beans retained by screens with holes of mid and large sizes), and a quality description – strictly soft, soft, hard, riado or rioy – to the type obtained using the defect count alone.

The inclusion of quality introduces a less objective criterion than those related to impurities, defects and bean size. In spite of recent advances in cupping procedures and criteria, sensory analysis is not an exact science as the wording in the SCA classification above demonstrates. Even though numbers are mentioned in the Kenyan system, the words used to describe cup quality do not say much either.

6.6 Grading and Quality

Figure 6.5 shows the phases of processing on one side and coffee features on the other side.

The processing stages mentioned early in this chapter – cleaning, separation by size, separation by density and color sorting – do not affect the quality of the beans that remain in the lot after processing but can enhance the quality of the lot itself by removing impurities and beans whose quality would jeopardize the quality of the full lot, *e.g.* insect-damaged, broken and off-color beans as well as those whose physical characteristics like density and color are associated with unwanted organoleptic features in the cup.

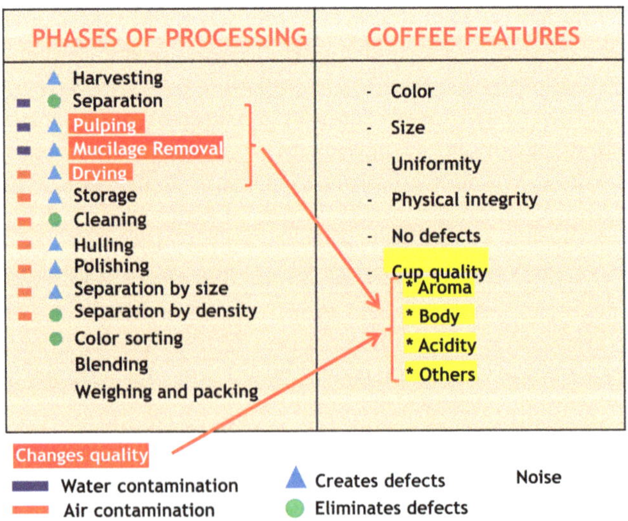

Figure 6.5 How does processing affect quality?

Only three processing phases – pulping, mucilage removal and drying – can modify or fine tune the intrinsic quality of the bean as imparted by terroir – climate, soil, altitude *etc.* – and variety. These post-harvest processing steps can emphasize characteristics like acidity or body without necessarily changing other organoleptic features of the beans. They can also be used to adapt the final product to the type of preparation, *e.g.* filter or espresso.

What happens to coffee immediately after it is harvested can have an important impact on cupping features. Coffee that is dried with the pulp and mucilage – *naturals* – will have body and sweetness emphasized. Coffee that is dried without the pulp and mucilage – *washed* – will have acidity and aroma emphasized. Coffee that is dried without the pulp but with all or some mucilage – *pulped natural*, *honey coffee* or *cereja descascado (CD)* – is closer to a natural but without the risk of having a harsh cup that could be caused by unripe cherries.

Post-harvest processing should also be a component of the grading system and in fact it is without being clearly mentioned because it tends to be associated with the country of origin of coffee. A Kenya AA and a Colombia Supremo are always washed because this is the post-harvest system prevailing in the country. Brazilian coffees are identified with naturals because this is the system most used in the country; pulped naturals and washed coffees are also offered by Brazil. Indian coffee is marketed with a clear identification of on-farm, post-harvest processing system because the country uses both the washed and natural systems for arabica and robusta. Robustas are mostly processed using the natural system, with few exceptions, primarily India itself.

6.7 Other Dimensions of Grading

Even though the core of the well-known and traditional grading systems is impurity and defect count and/or sizing, no coffee is traded today without two additional descriptors: post-harvest processing and cup quality. The first and the second descriptors have been addressed in this chapter but the third one – cup quality – is beyond its scope.

Specialty and high quality coffees are increasingly cupped using the Specialty Coffee Association (SCA) scores with cuppers trained by the Coffee Quality Institute (CQI). But the quality of commercial coffees is mostly described using country-specific scales and criteria that are difficult to summarize in a short chapter.

A fourth dimension of grading may be added: the name of the region where the coffee is grown, which conveys the expectation of some specific cup features. Geographic indication apart, it is usual to label a lot according to where it is grown within the country, *e.g.* Kiambu in Kenya, a Colombian Nariño, a Brazilian Cerrado or a Jamaican Blue Mountain. This last criterion in fact supersedes all the others in communication with the consumer because the region where coffee is grown is the descriptor that appears the most frequently in packages of high quality and specialty single-origin coffees.

Single-origin coffees still represent a very small part of the market but the consideration of the region where coffee is grown is also a major input for green coffee trading at all segments of the market irrespective of the fact that the leading brands, which account for the largest share of the market, do not mention the region(s) or even the country(ies) where the coffees they sell come from.

Reference

1. *The Coffee Exporters' Guide*, International Trade Centre, 3rd edn, 2012.

CHAPTER 7

Decaffeination and Irradiation Processes in Coffee Production

PEDRO F. LISBOA[b], CARLA RODRIGUES*[a], PEDRO C. SIMÕES[b] AND CLÁUDIA FIGUEIRA[a]

[a]Centro de Inovação Grupo Nabeiro, Alameda dos Oceanos, Condomínio Mar do Oriente, 65, 1.1, 1990-208 Lisboa, Portugal; [b]REQUIMTE, Departamento de Química, Faculdade de Ciências e Tecnologia, Universidade de Lisboa, Campus de Caparica, 2829-516 Caparica, Portugal
*E-mail: carla.rodrigues@grupo-nabeiro.pt

7.1 Introduction

Today, drinking coffee is a way of communicating ideas, emotions, memories, of sharing moments and opportunities, of creating new rituals. This important product is consumed all over the world and well-known chefs are learning how to create the best coffee-food pairing combinations. But how about the less beneficial effects of this product referred to by consumers such as, for example, the possibility of having mycotoxins present in your coffee or being sensitive to the caffeine present in the brew.

Epidemiological and clinical studies have attributed beneficial health effects to this beverage, mainly due to its high content of chlorogenic acids (CGA), which makes coffee one of the highest contributors to the intake of antioxidants in the western diet, in addition to other beneficial biological properties.[1] However, coffee may also contain undesirable contaminants produced prior to or during post-harvest or industrial processing. Coffee

Coffee: Production, Quality and Chemistry
Edited by Adriana Farah

Published by the Royal Society of Chemistry, www.rsc.org

seeds are amenable to mold infestation, which may lead to ochratoxin A (OTA)[1] contamination, due to improper and unhygienic drying or rehydration during storage and transportation.[3]

OTA's toxicity has been reviewed by the International Agency for Research on Cancer, which has classified it as a possible human carcinogen (group 2B).[1] The maximum limit permitted for OTA is 20, 10 and 5 ppb for green coffee beans, instant coffee and roasted coffee, respectively.[2]

Ferraz and coauthors[1] reported a study showing that roasting is an efficient procedure for OTA reduction in coffee, and its reduction depends directly on the degree of roasting. The proposed model can predict the thermal induced degradation of OTA and may become an important predicting tool in the coffee industry. However, it should be kept in mind that using contaminated beans will greatly affect the quality of the beverage and that roasting coffee beans severely to destroy OTA will also directly affect the levels of CGA. Moreover, this may compromise the development of coffee products with unpleasant sensory profiles. Therefore, even though roasting may destroy OTA, the coffee roasting industries are aware that the use of highly contaminated beans in coffee blends is not a recommended practice.[1]

In this sense, gamma radiation has been explored as a process to destroy *A. ochraceus* spores and preformed OTA in green coffee.[4] The minimum inhibitory doses (MID) of gamma radiation for inactivation of 10^4 and 10^8 spores of *A. ochraceous* strains in pre-sterilized coffee beans were found to be approximately 1 and 2.5 kGy, respectively.[4] Reduction of OTA through radiation inactivation varied significantly and was found to be commodity and condition dependent. Nonetheless, the sensory attributes of coffee prepared from treated (control, irradiated, and soaked-irradiated) samples were not found to vary significantly as compared to control, and did not indicate any off flavor due to the treatment.[4] OTA contamination in coffee is highly prevalent demanding good agriculture and post-harvest practices to prevent and reduce contamination. To achieve this, the application of irradiation techniques to green coffee may be a solution.

Another process that also impacts the functional and sensory properties of coffee products is decaffeination.

The properties of caffeine are numerous. In addition to its most popular role as a stimulant to the central nervous system, caffeine intake has been indicated as the main cause for the risk reduction of Parkinson and Alzheimer's diseases caused by coffee drinking.[3] Coffee is a major source of caffeine in the daily modern human diet.[3] Hamon and coauthors[3] refer to a caffeine content range from 0 to more than 3% dry matter basis (dmb) in beans of African *Coffea* species. However, the average amount of caffeine in *Coffea arabica* L. (hence called arabica) and *Coffea canephora* Pierre (hence called robusta), which are the most commercialized in the world, depend also on each coffee origin. In general, robusta seeds contain 40–50% more caffeine than arabica ones.[5]

Total coffee and caffeine consumption depends on many aspects such as age and cultural habits. There are only negligible losses of caffeine in the roasting process. Caffeine content may also vary considerably in a cup, depending on cup size and method of preparation,[6,7] an average content per 100 mL being 60 mg in caffeinated coffee and 3 mg[7] in decaffeinated coffee.

Quite a few decaffeination processes have been developed in the last decades to keep the desirable attributes of a coffee beverage when extracting caffeine from the seeds. To minimize flavor and aroma losses, the commercial decaffeination of coffee is at present carried out on the green coffee beans before roasting. These studies led to the improvement of coffee quality and therefore increase in coffee consumption. World consumption of decaffeinated coffee is difficult to gauge owing to the lack of separate data on this type of coffee in many importing countries. However, in the United States consumption of decaffeinated coffee has doubled in the last few years to 15.9% in 2009. Elsewhere, consumption of decaffeinated coffee has been static over the last decade, although this situation is not entirely clear-cut in that in many countries new low-caffeine coffee products have been introduced. More information on international coffee trade may be found in the International Trade Centre's *The Coffee Exporter's Guide – Third Edition* available for download (pdf version) at http://www.intracen.org/The-Coffee-Exporters-Guide---Third-Edition/. As per the report, decaffeinated coffee demand depends on the country. Brazil, Japan, Norway, and Sweden have low consumption of this type of coffee product whereas in countries such as Spain, The Netherlands, the United States, and the United Kingdom, the consumption of decaffeinated coffee as a percentage of total consumption in 2010 varied from 10 to 16%.

The aim of this chapter is to briefly discuss the main aspects concerning coffee decaffeination and irradiation and their influence in the functional and sensorial quality of the final product.

7.2 Decaffeination

The "perfect decaffeinated coffee" is expected to be one that is totally free of caffeine, and still able to reproduce the same organoleptic characteristics of a regular caffeinated coffee cup, without bringing harm to consumers' health. This concept fails right from the beginning, since the caffeine molecule contributes to the final cup sensory impact. Moreover, the total removal of caffeine from the green coffee beans using solvent-based methods would be unfeasible because of high operational costs to achieve such yield. Thus, decaffeinated coffee always contains small amounts of caffeine, its final cup concentration being strongly correlated with the decaffeination process applied to the green beans, the roasting degree, and the brewing methods.[5] In Europe, coffee manufacturers can only label their products as decaffeinated if the caffeine content (once coffee is roasted) does not exceed 0.1% of the coffee-based dry matter, while in the US more than 97% of the initial caffeine amount must be removed from the green coffee beans to be regarded as decaffeinated coffee.[7]

The history behind the removal of caffeine from green coffee beans spans more than two centuries, when in 1819 the German scientist Friedlieb Ferdinand Runge, while looking for coffee psychoactive components responsible for causing insomnia and restlessness, first isolated the caffeine molecule from coffee. Later, in the early 20th century, Ludwig Roselius found that by swelling the green coffee beans with steam and thereafter treating them with a water-immiscible and high caffeine-soluble organic solvent, it was possible to extract the caffeine.[8] Under the protection of his patent, Roselius founded the Kaffee HAG Company and started to commercialize decaffeinated coffee with the commercial name Sanka Coffee. Benzene was the first organic solvent used to extract the caffeine, but it was replaced by less toxic solvents such as trichloroethylene, methylene chloride, and ethyl acetate.[8]

In the 1970s the US National Cancer Institute (NCI) proved that trichloroethylene caused liver tumors in mice. Since that allegation, coffee manufactures stopped using trichloroethylene, and, notwithstanding the fact that more than 30 organic solvents have been tested in the literature for the green coffee bean decaffeination process, only the organic solvents methylene chloride and ethyl acetate are in use nowadays.[7] For many years, the choice of the solvent was strongly related with its price, caffeine solubility and selectivity, and solvent volatility. The public concerns related with the health effects of the use of chemicals in the food industry and their demand for more natural processes led coffee manufacturers to focus their attentions on alternatives such as water and carbon dioxide. Although methylene chloride is progressively being replaced by 100% natural solvents, it is still the most used solvent for decaffeination of coffee, and is also approved by FDA and other relevant food authorities. As per the European directive 2009/32/EC, a maximum of 2 mg kg^{-1} of methylene chloride residue is allowed in decaffeinated roasted coffee while the US FDA has established a maximum residue of 10 mg kg^{-1}. However, these limits are higher than those obtained in decaffeination plants that are fully committed to good manufacturing practices in which all or the major part of the solvent residues are removed from the food ingredient resulting in values below 0.3 mg kg^{-1}.[7]

Currently, the coffee decaffeination process can be divided into two main groups, the natural decaffeination process where water or carbon dioxide are employed, and the organic solvent decaffeination process where methylene chloride or ethyl acetate are the extraction solvents. Despite much progress, the original method suggested by Roselius is still very up to date: (i) decaffeination is performed using green coffee beans to avoid the loss of aroma; (ii) after the first pre-cleaning process wherein the silverskin is removed, the dry beans are wet steamed and soaked in water until they reach a final moisture content of 40–50% (w/w). The water promotes softening and opening of the bean pores, while it frees caffeine from chlorogenate potassium salt complex, which is insoluble in the non-polar solvents used.[9] This allows caffeine diffusional migration through the bean. After this initial step, the coffee beans undergo the caffeine extraction step, steam stripping of solvent residues, and reestablishment of their initial moisture content.

7.2.1 Decaffeination Process Using Organic Solvents

In the direct method in which methylene chloride is used,[10] the green coffee beans are placed in a battery of extractors and are steamed for between 15 and 60 minutes at a temperature of 105–110 °C to provide a bean moisture content of at least 16%. The steamed green coffee beans are then soaked in water to increase their moisture content up to a final value of 50% and subjected to extraction under direct contact with methylene chloride at temperatures between 60 and 100 °C and pressures of 0.35–0.55 MPa for 8 to 12 hours. The caffeine stream that exits from the extractor is downstream treated by distillation for caffeine removal and recirculation back to the oldest coffee extractor in line, free of caffeine. At the end of the extraction, the coffee beans are removed and steam treated from 1 to 4 hours to strip the solvent residues. On average, the amount of solvent required to decaffeinate 1 kg of green coffee beans is less than 8 kg.

In the indirect solvent method,[11] the green coffee beans are only in direct contact with an aqueous solution of green coffee bean extract. This process starts by soaking the green coffee beans with hot water until the solution gets saturated with all the coffee-soluble components, including caffeine. The solution is drained off and the caffeine is removed in a separate liquid–liquid extraction column using methylene chloride as a solvent. This caffeine-free coffee extract is recycled back to the other extractors filled with new raw green coffee beans. By using the coffee water extract as a solvent in this mode of extraction of water soluble components of the coffee bean, coffee aroma precursors are minimized.

Another way to extract caffeine is by using ethyl acetate in direct contact with the coffee beans. Ethyl acetate is accepted by the FDA, who did not impose any restrictions regarding the maximum residue allowed to be left in the coffee beans, and is also naturally abundant in many edible fruits, like banana, for example. The decaffeination process in which ethyl acetate is applied follows the same guideline procedures as those of the methylene chloride direct method with the advantage of not being considered toxic for human health. When compared with water, ethyl acetate has a lower boiling point, is more selective, and can dissolve slightly more caffeine. Nevertheless, natural ethyl acetate is very expensive and since it is flammable the extraction facilities have to be designed to be explosion proof. As a direct consequence, the investment in decaffeination plant construction is higher than for methylene chloride.[12]

7.2.2 Natural Processes: Water or Swiss Water Decaffeination

The concept behind the exclusive use of water in the decaffeination process is grounded in the idealization of a natural, chemically free process in which an odorless and tasteless solvent should be used. However, water is far from being a good solvent for decaffeination, since its solvation power and selectivity for caffeine are very low when compared with methylene chloride or ethyl acetate. More than 20% of green coffee compounds (dry weight) like proteins, amino-acids, chlorogenic acids and carboxylic acids, carbohydrates, and alkaloids are water-soluble.[12] Those compounds largely contribute to the final cup quality and should remain in the decaffeinated green coffee beans. Furthermore, water has a higher boiling point, which makes its regeneration more difficult and expensive. Water was first tested as a decaffeination solvent in 1933, but at the beginning of the 1980s the economically viable possibility of using water for decaffeination emerged.[12] In this application, the green coffee beans were soaked in an extraction vessel in hot water, 70–90 °C, to extract all water-soluble compounds including the caffeine. This green coffee extract was then forced to pass over a pre-loaded activated carbon bed where ideally only caffeine was adsorbed. After 6 to 12 h extraction, the green coffee beans were dried by hot air to 30% of the moisture content. The green coffee extract was then concentrated under vacuum distillation. The half-dried beans were mixed together with the concentrated coffee extract for 3 h at 70 °C to absorb completely all the solution. Finally, the green coffee beans were dried to 10% of the moisture content. Later, in 1988, the first water decaffeination was commissioned in Vancouver, British Columbia, in which a solution of aqueous green coffee extract was used to decaffeinate a battery of green coffee beans. The caffeine was separated from the flavor-charged water using battery-activated carbon filters. Many improvements have been made to this method throughout the years such as new adsorption filters for caffeine, improved hydrodynamics for less water usage, and better drying control of the green coffee beans. A simplified version of the Swiss water decaffeination process is shown in Figure 7.1.

7.2.3 Natural Process Using Supercritical CO_2

Caffeine extraction with supercritical carbon dioxide ($ScCO_2$) differs from the other methods in terms of the unique harsh process conditions, especially the high pressures involved. Nowadays, when referring to $ScCO_2$, one immediately associates it with the green coffee beans decaffeination process. This side by side coffee and $ScCO_2$ history has been ongoing since the 1960s when Kurt Zosel from the Max Planck Institute in Germany unintentionally, while studying $ScCO_2$ solvent power for mixtures separation, discovered that carbon dioxide at supercritical conditions could dissolve caffeine. This outcome has later resulted in the first patent in which carbon dioxide saturated with water above its critical point was used to remove caffeine from moist green coffee beans.[13] Luckily, this appeared at a perfect time, as coffee manufacturers were dealing at that time with a steeplechase towards the replacement of classic decaffeination organic solvents.

After the initial treatment of moistening the coffee beans to a water content between 40 and 50%, the swollen beans are then placed into a battery of high pressure vessels and fresh carbon dioxide saturated with water

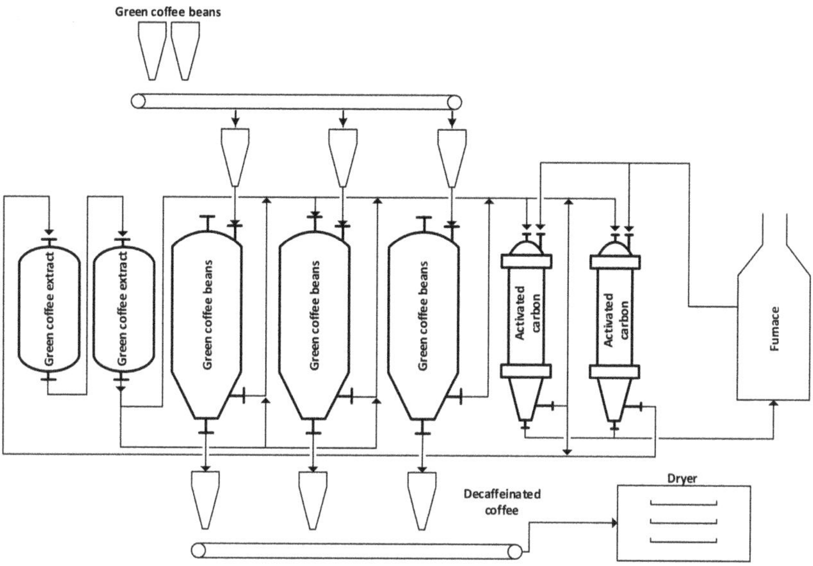

Figure 7.1 Simplified Swiss water process decaffeination flowsheet.

passes through the fixed bed as many times as needed to remove the caffeine present. According to the type of beans, at least 200–270 kg of carbon dioxide is needed to decaffeinate 1 kg of green coffee beans. This large excess of carbon dioxide is required because caffeine does not dissolve to its neat solubility,[13] and because the extraction is governed by the equilibrium that is present in the carbon dioxide-water-caffeine-coffee system.[14] Additionally, carbon dioxide should always be saturated with water as, if it is not, it will otherwise remove the water from the beans thus impairing the extraction from proceeding. In this situation, water acts as a polarity co-solvent modifier, while carbon dioxide acts as a transport vehicle to enhance the mass transfer. Knowing this, one should expect that other molecules similar to caffeine would be extracted as well, but theobromine, theophylline, and even the chlorogenic acids are more than 10-fold less soluble than caffeine.[13] Trigonelline, an important aroma precursor during roasting[15] and also a bioactive compound, displays a much lower affinity for $ScCO_2$ than caffeine.[16]

In the process where organic solvents or water are applied, the operating temperature is set close to the boiling point where the solubility of caffeine attains the maximum value. In the $ScCO_2$ decaffeination process the caffeine solubility and extraction rates are dependent on temperature and pressure.

Figure 7.2 shows a simplified process flow sheet for coffee decaffeination using carbon dioxide. The pre-wet coffee beans are loaded to the high-pressure extractor and carbon dioxide is continuously being fed to the extractor by a circulation pump. The caffeine-rich stream leaving the extractor enters

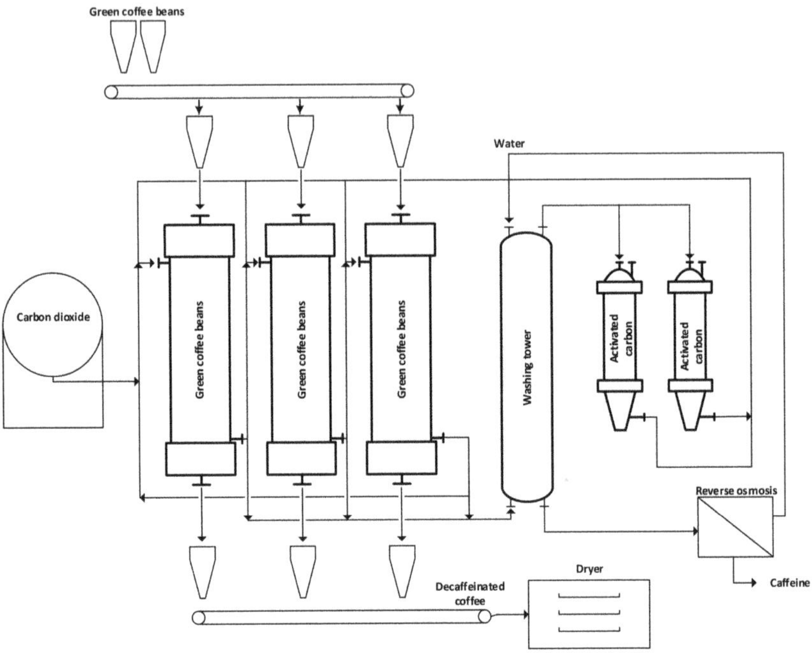

Figure 7.2 Simplified supercritical carbon dioxide decaffeination process.

the bottom of a vertical tower in which a shower of fine droplets of water are sprayed at the top. The aqueous droplets remove the caffeine from the $ScCO_2$ stream. To ensure that no trace of caffeine is present, when leaving the top of the column, the recycled $ScCO_2$ stream passes through activated carbon filters before being fed again to the coffee extractors. The caffeine can be concentrated by vacuum distillation or by reverse osmosis from the aqueous solution, and the evaporated or permeated water is sent back to the top of the column.

The quality of the decaffeinated coffee is claimed to be improved because of the chemical stability and inertness of carbon dioxide preventing any reaction with the coffee constituents. In addition, $ScCO_2$ has a very high selectivity for caffeine, avoiding losses of other non-caffeine solids and allowing different types of green coffee beans to maintain their identity. Product quality is claimed to be comparable to caffeinated coffee due to the avoidance of any aroma/flavor precursor loss during the decaffeination process.[17] Moreover, carbon dioxide is a gas at ambient pressure and temperature so no chemical residue is left in the beans after the decaffeination process.

7.2.4 Chemical Differences and Health Effects

As said before, the decaffeination process is always performed before the roasting process to minimize the loss of aroma and other soluble components. During the caffeine removal, the green coffee beans undergo many chemical changes which are highly dependent on the thermodynamic conditions of the method and of the solvent choice. However, there is scarce information in the literature related to the chemical composition changes related to the decaffeination process. The color change is the first immediate observable difference in the raw coffee beans which starts from a bluish-green color before decaffeination and ends with a dark brownish or dark greenish color after being processed. The color change occurs not only at the bean surface but throughout the whole bean structure. Because the raw beans are subjected to high temperatures at any given step of the decaffeination process, either by steam or by the hot solvent, the thermolabile compounds are affected and Maillard reactions are prone to occur. Analysis of the decaffeinated coffee showed that the dark color was strongly related with the formation of melanoidins,[17] and not, as first expected, from the oxidation reactions of coffee phenolics. Comparative studies between different decaffeination methods showed that changes in the chemical composition depend on the solvent. A study, in which decaffeinated coffee using dichloromethane was evaluated, revealed a significant difference in relative chemical composition especially in levels of sucrose and chlorogenic acids.[18] Due to the decaffeination process, losses of sucrose have reached 60% for arabica and 20% for robusta, while losses of chlorogenic acids have reached 16% for arabica and 11% for robusta coffees. Changes in the proteins and trigonelline were also observed with the decaffeination with methylene chloride. The changes in chlorogenic acids profiles and contents were also evaluated after water decaffeination.[19,20] Nevertheless, the gains or losses of the total amount of chlorogenic acids are rather difficult to measure with rigor, since many other compounds lixiviated by water may influence the results. In the two water decaffeination studies, one reported a total gain of 16% of chlorogenic acids,[20] while the second study with a different water decaffeination method reported an average loss of 20% of chlorogenic acids.[19] When supercritical carbon dioxide is used, the decaffeinated processed coffee beans appears not to compromise the original coffee composition, showing a relative gain of just 1.5% in chlorogenic acids.[19]

The world's specialty market is preoccupied with health issues surrounding coffee and the non-existence of more detailed information on the chemical composition of green coffee beans after decaffeination may create the misconception that active substances are extracted during the process and therefore the decaffeinated coffee intake may not provide health benefits. It is well known that caffeine is the most studied active substance present in coffee beans. Caffeine intake results in improved physical and cognitive performance during and after exercise.[20] In addition to its well-known physical and psychostimulant effects, it acts as a neuroprotective substance by reducing cognitive decline and dementia.[21] Moreover, it has been shown in human and animal studies that caffeine exerts protective effects against Alzheimer's and Parkinson's diseases by helping to keep the blood–brain barrier intact.[22] Caffeine also increases dopamine action in the brain by blocking the adenosine receptors, adenosine A2, a caffeine-like molecule,[23,24] and that may be one of the reasons why caffeine may palliate Parkinson's disease once it is installed. Moderate caffeine intake, below 300 mg per day, or 3 mg kg^{-1} in children, will not increase the risk for stroke, arrhythmia, hypertension, cardiovascular disease, cancer, infection, complications of pregnancy, calcium imbalance, bone disease, or kidney stones.[25] Still, caffeine may be associated with a variety of adverse but relatively inconsequential side effects such as wakefulness and sleep-disruption, heart palpitations, and urinary frequency. In a recent study, it was found that consumption of a caffeine dose equivalent to that in a double espresso three hours before habitual bedtime induced a ~40 min phase delay of the circadian melatonin rhythm in humans.[26] For humans who are more sensitive to the adverse effects of caffeine, decaffeinated coffee may be just as healthy as caffeinated. Besides caffeine, other agents in coffee may positively affect health, mainly chlorogenic acids and their derivatives; diterpenes (cafestol and kahweol), trigonelline, and niacin, among other compounds. Recent studies show that both caffeinated and decaffeinated coffee are associated with a reduced risk of diabetes type 2 [27,28] and both are associated with favorable liver health by exerting hepatoprotective effects.[29] Consumption of decaffeinated coffee may also have neuroprotective effects since chlorogenic acids, caffeic acid, and kahweol highly contribute to the neuropharmacological effects against Parkinson's[30,31] and Alzheimer's[32,33] diseases. For many years, the consumption of decaffeinated coffee has been associated with the increase of serum cholesterol levels[34] as well as the increase of gastrin under the allegation of being made of lower quality and more acidic robusta beans. Later studies have shown that an increase in serum levels and cholesterol were instead related to the consumption of unfiltered coffee.[35] since cafestol and, to a lesser extent, kahweol have been shown to raise total and LDL cholesterol in human serum.[36] Both caffeinated and decaffeinated coffees share strong gastrin-release, which may relate to health problems such as gastroesophageal reflux and ulcers.[37] Even though the consumption of decaffeinated coffee has resulted in total gastrin levels 1.7 times higher compared to control levels, caffeinated coffee has raised total gastrin levels 2.3 times.[38]

7.3 Irradiation

Although the concept of irradiating food to bring about beneficial outcomes has been considered for a century, it was not until the 1960s that commercially feasible sources of radiation became available. Initial interest was in using relatively high doses of irradiation as a replacement of canning for military rations, for space foods, and for hospital diets in immune-compromised patients. However, it soon became apparent that lower doses could be used more generally to improve food safety, increase food security (reduction of food losses and waste), and offer another option as a phytosanitary treatment of food moving across international or national borders.[39] The beneficial effects of food irradiation resulted from the ability of radiation to bring about the effects of inhibiting sprouting, to delay ripening, for pest control and parasite inactivation, to reduce spoilage organisms (extend shelf-life), to reduce non-sporulation pathogens, or to reduce pathogens to the point of sterility. Irradiation is one of the many physical processes applied to food, but it has several practical advantages that include versatility, high efficiency, the fact of being a cold process, and penetrating capacity, *i.e.*, if necessary, foods can be treated already in their final packaging, as target organisms are not protected by the packaging, its shape, or the food position inside the packaging. Moreover, the product distribution is relatively unimportant, and treating pallet loads is possible. Solid, raw foods can also be treated. Treatment does not involve chemicals or chemical residues. Food can be immediately distributed into the food supply chain after treatment. Despite these potential applications and advantages, irradiation has not become a major commercial food process.[39]

Several international agencies have reviewed safety issues on food irradiation such as the Joint Expert Committee on Food Irradiation (JECFI) 1981, the World Health Organization (WHO), and the European Food Safety Authority (EFSA) as well as other national food safety agencies in several countries. Because of the JECFI conclusions of 1981, Codex Alimentarius issued a General Standard for the Irradiation of Food, which was subsequently revised in 2003 (CAC, 2003). The Codex provisions (any food and any dose for a legitimate technical purpose) are rarely implemented in totality, but over 50 countries have approved the use of irradiation for at least one food or food class with a maximum dose dependent upon the purpose of treatment. Most irradiated food is consumed in the country of treatment. The only irradiated food that is traded internationally are fruits treated for quarantine purposes.[39]

Roberts[39] states that several strategies should be initiated if industry and retailers are to adopt a more open attitude towards irradiation. These include actions by irradiation proponents that are to stress the benefits to the food rather than the smartness of the technology (safer, reduced chemical residues, longer shelf-life), use labeling positively, always including the main benefit of the treatment on the label to offset any perception that the label is a warning, to discuss over-stringent labeling requirements with regulators, to recognize that food is a perishable commodity and that business models and attributes that are satisfactory for sterilization of medical products may not be adequate for food irradiation, and to build greater partnerships with the food industry so that some of the genuine practical barriers to food irradiation can be minimized.

In reference to coffee, insect infestation of coffee beans can cause significant losses. Chemical fumigants are not sufficiently effective against insect eggs and can leave residues that change taste and aroma, in addition to being detrimental to health. The irradiation of green coffee beans with doses as high as approximately 10 kGy does not cause detectable flavor change in the brew after conventional roasting. The freshly harvested coffee fruit or cherry is known to have very high moisture content (approximately 50%). In the dry post-harvest processing method, cherries are cleaned and dried in the open sun, which takes around four weeks to attain the optimum moisture level

(11%) and water activity (0.70). Overdried cherries become brittle and the respective beans are considered defective, whereas insufficiently dried ones may become moldy, which may lead to OTA formation during processing or further storage. Mold infestation can also take place during prolonged storage due to the hygroscopic nature of the product, which may, again, lead to toxin production.[40]

The ideal toxin inactivation method should be easy to use, economical, and non-toxigenic and should not affect the food quality.[41] The inclusion of gamma radiation treatment in the processing chain of green coffee following the drying step can reduce the toxicity of preformed OTA as well as eliminate toxigenic fungi.

It is essential to overcome microbial contamination of food products during all stages of processing, including the raw crop stage, harvesting, preservation, processing, packaging, distribution, and marketing. As per the United Nations Food and Agriculture Organization (FAO), microbial contamination of crops results in significant economic losses around the world every year. In developing countries, almost 75% of the food that is produced is lost on-farm and during transport and processing, because of spoilage caused by poor storage conditions or improper handling and processing of crops. Previously, chemical fumigation technology (using sulfur dioxide, or potassium nitrate) was used to preserve crops after harvest.[42] However, this technology has been banned by health authorities in many countries due to concerns regarding human health and environmental pollution. In this context, relevant government departments, and non-governmental research institutions of various countries, have been committed to developing more environmentally friendly and efficient food preservation processes and applications. To aid in this research, the Joint Expert Committee on Irradiation, comprised of the Food Research Organization (FAO), WHO, and the International Atomic Energy Agency (IAEA), proposed that "it is safe to appropriately use radiation for food decontamination and that food irradiated at a dose lower than 10 kGy has no toxic hazards and only a minor effect is posed on the nutrition". The irradiation technology has been suggested as an alternative to microbial decontamination in the food industry. On the other hand, according to a study by Amézqueta *et al.*,[42] an irradiation dose between 25 and 60 kGy does not cause any potential health risks or raise concerns regarding residual radiation, while retaining acceptable standards of nutritional value and sensory quality of food. Therefore, the irradiation technology cannot only improve food safety but also reduce crop-related economic losses.[43]

Early radiation treatment of food was mostly conducted using gamma rays (γ-rays). Despite its wide use in the preservation of various food products, consumers doubt the safety and efficacy of γ-ray-treatment of food products. The development of electron accelerators during the 1930s contributed to the research breakthrough by Cleland in the late 1950s.[43] Thereafter, accelerator technology matured, with lower production costs of equipment; therefore, this technology could be used for food irradiation. Electron accelerators appear to be more successful compared to γ-rays because of the following advantages: (1) source (equipment) that can suspend the irradiation at any time; (2) non-nuclear energy that can accelerate the generation of radiation when required; (3) little risk for occupational injuries; and (4) applicability in high-flow and high-dose irradiation.[44]

Electron-beam irradiation (EBI) is a novel food decontamination technology that uses low-dose ionizing radiation in the treatment of crops or other foods, to eliminate microbial contamination. Additionally, EBI inhibits the germination of crops and controls the ripening rate of vegetables and fruits, extending the shelf-life of these products. EBI is a low cost, environment friendly, and time effective alternative to the traditional thermal decontamination technology. EBI, which has been approved by the USFDA, can be applied as an alternative to chemical fumigation of food. EBI inhibits a variety of food-borne pathogens, and effectively maintains food quality, significantly extending the shelf-life. Better food preservation can be achieved by using EBI as a hurdle technology, in combination with other traditional or non-traditional food-processing technologies. EBI uses low-dose radiation for decontamination, which reduces the risk of microbial hazards in food. However, from the perspective of food safety, it must be proven that EBI exerts no adverse effect on the nutrition or residual radiation in the food, before it is applied in the food processing industry.[42] The main disadvantage of using EBI is the problematic low penetrability of the e-beam. The decontamination effect of EBI may be influenced by the size, thickness, direction (single- or double-side exposure), and packaging of the food. EBI treatment is especially effective in low-density and uniformly packaged food.[45] In order to ensure pathogen-free fresh and fresh-cut food, strict food safety measures must be observed throughout the production, processing, and marketing of food products. New processing technologies, such as the modified atmosphere package, and ozone, ultrasound, and ultraviolet treatments, have been used to improve the microbial safety of fresh fruit and vegetable products. However, most of these new technologies have limited applications, and are unavailable for commercial use. Previously, γ radiation has been used as an alternative to chemical preservatives for fruit and vegetable decontamination. Low-dose irradiation (<1.0 kGy) was recommended for disinfection and germination inhibition; intermediate-dose irradiation (1.0–3.0 kGy) is believed to be suitable for delaying the maturity or senescence of fresh fruits and vegetables and for eliminating microbial contamination; high-dose irradiation (>3.0 kGy) can be used in the extraction of bioactive compounds.[42]

The future trends of food processing cannot be considered independent of sustainability, eco-friendliness, innovation, and advanced technologies. EBI is significantly useful in decontamination, elimination of microbial contamination, and insect disinfestation, in a variety of food and agricultural products. EBI usage possesses many advantages, and can therefore be used as an effective alternative to chemical fumigation and γ-irradiation. The application of e-beam in food preservation has seen increasing popularity in recent times. The next stage of this technology requires a confirmation that the e-beam will not adversely affect the sensory flavor or nutritional quality of the food. EBI is regarded as an effective approach for food preservation. It can be combined with other common traditional techniques (*e.g.*, drying, canning, freezing, atmosphere packaging, and biological preservatives) to prevent the spoilage and deterioration of perishable food.[43] The main driver for novel and emerging technologies on food commercialization was better quality or added value on the products, the solution of safety issues and improvement to product shelf-life. Other drivers were an increase in product convenience, a decrease in price or other increase in competitiveness or cost saving in running costs, government or regulatory requirements, solving environmental or waste issues, global trade, availability of funding, results of basic research, and high quality of the equipment.[46]

7.4 Conclusions

Decaffeination and irradiation are processes that impact the characteristics of green coffee beans in what relates to quality and price. In both cases, the increase in information transparency in society, and through the publication of various articles and advocacy of seminars will promote a change in the attitude of consumers towards decaffeinated and irradiated coffees.

References

1. M. B. M. Ferraz, A. Farah, B. T. Iamanaka, D. Perrone, M. V. Copetti, V. X. Marques and M. H. Taniwaki, Kinetics of ochratoxin A destruction during coffee roasting, *Food Control*, 2010, **21**(6), 872–877.
2. T. N. Gopinandhan, K. Velmourougane, P. Panneerselvam, E. Keshamma and Y. Raghuramulu, Occurrence of ochratoxin-A (OT-A) in green and commercial coffee samples, *J. Food Sci. Technol.*, 2007, **44**(3), 247–249.
3. P. Hamon, J. Rakotomala, S. Akaffou, N. J. Razafinarivo, E. Couturon, R. Guyot, D. Crouzillat, S. Hamon and A. Kochko, Caffeine-free species in the Genus Coffea, in *Coffee in Health and Disease Prevention*, ed. V. R. Preedy, Elsevier, 2015.
4. S. Kumar, A. Kunwar, S. Gautam and A. Sharma, Inactivation of *A. ochraceus* spores and detoxification of Ochratoxin A in coffee beans by gamma irradiation, *J. Food Sci.*, 2012, **77**(2), 44–51.
5. F. Burdan, Caffeine in coffee, in *Coffee in Health and Disease Prevention*, ed. V. R. Preedy, Elsevier, 2015, Chapter 22, pp. 201–207.
6. J. J. Barone and H. R. Roberts, Caffeine consumption, *Food Chem. Toxicol.*, 1996, **34**(1), 119–129.
7. B. Stavric, R. Klassen, B. Watkinson, K. Karpinski, R. Stapley and P. Fried, Variability in caffeine consumption from coffee and tea: possible significance for epidemiological studies, *Food Chem. Toxicol.*, 1988, **26**(2), 111–118.

8. W. Heilmann, Technology II: decaffeination of coffee, in *Coffee: Recent Developments*, ed. R. J. Clarke and O. G. Vitzthum, Blackwell Science, 2001.
9. J. Johann Friedrich Meyer, L. Roselius and K. H. Wimmer, *Preparation of Coffee*, 1908.
10. S. N. Katz, Decaffeination of coffee, in *Coffee Volume 2: Technology*, ed. R. J. Clarke and R. Macrae, Elsevier Science Publishers Ltd, 1987.
11. J. M. Patel and A. B. Wolfson, *Semi-Continuous Countercurrent Decaffeination Process*, 1972.
12. N. E. Berry and R. H. Walters, *Process of Decaffeinating Coffee, US Pat.*, 2309092A, 1941.
13. K. Zosel, *Belgium Pat.* 646641, 1964.
14. M. Johannsen and G. Brunner, Solubilities of the xanthines caffeine, theophylline and theobromine in supercritical carbon dioxide, *Fluid Phase Equilib.*, 1994, **95**, 215–226.
15. M. A. McHugh and V. J. Krukonis, *Supercritical Fluid Extraction Principles and Practice*, Butterworth-Heinemann, 2nd edn, 1993.
16. H. Taguchi, M. Sakaguchi and Y. Shimabayashi, Trigonelline content in coffee beans and the thermal conversion of trigonelline into nicotinic acid during the roasting of coffee beans, *Agric. Biol. Chem.*, 1985, **49**(12), 3467–3471.
17. R. J. Clarke and R. Macrae, *Coffee Technology*, Elsevier Applied Science, London and New York, 1987.
18. A. Toci, A. Farah and C. Trugo, Efeito do processo de descafeinação com diclorometano sobre a composição química dos cafés arábica e robusta antes e após a torração, *Quim. Nova*, 2006, **29**(5), 965–971.
19. A. Farah, D. Perrone, J. Fernandes and J. Silanes, Chorogenic acids and lactones in coffees decaffeinated by water and supercritical CO_2 and roasted in a pilot plant scale fluidized bed roaster, in Proc. 23rd Int. Conf. Coffee Sci. 2010, Bali, Indonesia, 2010, pp. 367–372.
20. A. Farah, T. De Paulis, D. P. Moreira, L. C. Trugo and P. R. Martin, Chlorogenic acids and lactones in regular and water-decaffeinated arabica coffees, *J. Agric. Food Chem.*, 2006, **54**(2), 374–381.
21. E. Hogervorst, S. Bandelow, J. Schmitt, R. Jentjens, M. Oliveira, J. Allgrove and M. Gleeson, Caffeine improves physical and cognitive performance during exhaustive exercise, *Med. Sci. Sports Exercise*, 2008, **40**(10), 1841–1851.
22. K. Ritchie and I. Carrie, The neuroprotective effects of caffeine: a prospective population study (the Three City Study), *Neurology*, 2007, **69**(6), 536–545.
23. X. Chen, O. Ghribi and J. D. Geiger, Caffeine protects against disruptions of the blood–brain barrier in animal models of Alzheimer's and Parkinson's diseases, *J. Alzheimer's Dis.*, 2010, **9**(5), 636–650.
24. J. F. Chen, K. Xu, J. P. Petzer, R. Staal, Y. H. Xu, M. Beilstein and M. a. Schwarzschild, Neuroprotection by caffeine and A(2A) adenosine receptor inactivation in a model of Parkinson's disease, *J. Neurosci.*, 2001, **21**(10), RC143.

25. K. Sääksjärvi, P. Knekt, H. Rissanen, M. a. Laaksonen, A. Reunanen and S. Männistö, Prospective study of coffee consumption and risk of Parkinson's disease, *Eur. J. Clin. Nutr.*, 2008, **62**(7), 908–915.
26. L. W. Jones, Recent research on coffee and health, in *Coffee: A Comprehensive Guide to the Bean, the Beverage, and the Industry*, ed. R. W. Thurston, J. Morris and S. Steiman, Rowman & Littlefield, 2013.
27. T. M. Burke, R. R. Markwald, A. W. Mchill, E. D. Chinoy, J. A. Snider, S. C. Bessman Jr, C. M. Jung, J. S. O'Neill and P. W. Kenneth, Effects of caffeine on the human circadian clock in vivo and in vitro, *Sci. Transl. Med.*, 2015, 7(305), 305ra146.
28. M. Ding, S. N. Bhupathiraju, M. Chen, R. M. van Dam and F. B. Hu, Caffeinated and decaffeinated coffee consumption and risk of type 2 diabetes: a systematic review and a dose-response meta-analysis, *Diabetes Care*, 2014, **37**(2), 569–586.
29. K. Ohnaka, M. Ikeda, T. Maki, T. Okada, T. Shimazoe, M. Adachi, M. Nomura, R. Takayanagi and S. Kono, Effects of 16-week consumption of caffeinated and decaffeinated instant coffee on glucose metabolism in a randomized controlled trial, *J. Nutr. Metab.*, **2012**, 207426.
30. Q. Xiao, R. Sinha, B. I. Graubard and N. D. Freedman, Inverse associations of total and decaffeinated coffee with liver enzyme levels in National Health and Nutrition Examination Survey 1999–2010, *Hepatology*, 2014, **60**(6), 2091–2098.
31. Y. P. Hwang and H. G. Jeong, The coffee diterpene kahweol induces heme oxygenase-1 via the PI3K and p38/Nrf2 pathway to protect human dopaminergic neurons from 6-hydroxydopamine-derived oxidative stress, *FEBS Lett.*, 2008, **582**(17), 2655–2662.
32. K. Trinh, L. Andrews, J. Krause, T. Hanak, D. Lee, M. Gelb and L. Pallanck, Decaffeinated coffee and nicotine-free tobacco provide neuroprotection in Drosophila models of Parkinson's disease through an NRF2-dependent mechanism, *J. Neurosci.*, 2010, **30**(16), 5525–5532.
33. Y. J. Jang, J. Kim, J. Shim, C.-Y. Kim, J.-H. Jang, K. W. Lee and H. J. Lee, Decaffeinated coffee prevents scopolamine-induced memory impairment in rats, *Behav. Brain Res.*, 2013, **245**, 113–119.
34. S.-H. Kwon, H.-K. Lee, J.-A. Kim, S.-I. Hong, H.-C. Kim, T.-H. Jo, Y. I. Park, C. K. Lee, Y. B. Kim, S. Y. Lee and C.-G. Jang, Neuroprotective effects of chlorogenic acid on scopolamine-induced amnesia via anti-acetylcholinesterase and anti-oxidative activities in mice, *Eur. J. Pharmacol.*, 2010, **649**(1–3), 210–217.
35. H. R. Superko, W. Bortz, P. T. Williams, J. J. Albers and P. D. Wood, Caffeinated and decaffeinated coffee effects on plasma lipoprotein cholesterol, apolipoproteins, and lipase activity: a controlled, randomized trial, *Am. J. Clin. Nutr.*, 1991, **54**(3), 599–605.
36. S. H. Jee, J. He, L. J. Appel, P. K. Whelton, I. Suh and M. J. Klag, Coffee consumption and serum lipids: a meta-analysis of randomized, *Am. J. Epidemiol.*, 2001, **153**(4), 353–362.

37. R. Urgert and M. B. Katan, The cholesterol-raising factor from coffee beans, *Annu. Rev. Nutr.*, 1997, **17**, 305–324.
38. M. J. Rubach and V. Somoza, Impact of coffee on gastric acid secretion, in *Coffee Emerging Health Effects and Disease Prevention*, ed. Yi-F. Chu, IFT Press & Wiley-Blackwell, 2012.
39. P. B. Roberts, Food irradiation is safe: Half a century of studies, *Radiat. Phys. Chem.*, 2014, **105**, 78–82.
40. F. Acquaviva, A. DeFrancesco, A. Andriulli, P. Piantino, A. Arrigoni, P. Massarenti and F. Balzola, Effect of regular and decaffeinated coffee on serum gastrin levels, *J. Clin. Gastroenterol.*, 1986, **8**(2), 150–153.
41. M. L. Suárez-Quiroz, O. González-Rios, M. Barel, B. Guyot, S. Schorr-Galindo and J. P. Guiraud, Effect of chemical and environmental factors on *Aspergillus ochraceus* growth and toxigenesis in green coffee, *Food Microbiol.*, 2004, **21**(6), 629–634.
42. S. Amézqueta, E. González-Peñas, M. Murillo-Arbizu and A. López de Cerain, Ochratoxin A decontamination: a review, *Food Control*, 2009, **20**(4), 326–333.
43. S. D. Pillai and S. Shayanfar, Introduction to electron beam pasteurization in food processing, in *Electron Beam Pasteurization and Complementary Food Processing Technologies*, Woodhead Publishing Limited, 2015, vol. 2008, DOI: 10.1533/9781782421085.1.3.
44. H.-M. Lung, Y.-C. Cheng, Y.-H. Chang, H.-W. Huang, B. B. Yang and C.-Y. Wang, Microbial decontamination of food by electron beam irradiation, *Trends Food Sci. Technol.*, 2015, **44**(1), 66–78.
45. H. E. Clemmons, E. J. Clemmons and E. J. Brown, in *Electron Beam Processing Technology for Food Processing*, ed. S. Pillai & S. Shayanfar, Woodhead Publishing, 2015.
46. C. Jermann, T. Koutchma, E. Margas, C. Leadley and V. Ros-Polski, Mapping trends in novel and emerging food processing technologies around the world, *Innovative Food Sci. Emerging Technol.*, 2015, **31**, 14–27.

CHAPTER 8

Roasting

FERNANDO FERNANDES

Cia Lilla de Máquinas Indústria e Comércio, São Paulo, Brazil
*E-mail: f.fernandes@lilla.com.br

8.1 Introduction

A traditional legend says that coffee was discovered by some goats strayed from a herd. Apparently, they ate the fruits of a mid-sized bush, red in color, that made them more alert and stimulated. They could cover longer distances than others, showing fewer signs of tiredness, a gift for any goat shepherd in the 13th century AD somewhere in North Africa. Shepherds naturally paid closer attention to this fruit. It was a time when food processing was mostly done in monasteries, and a handful of these fruits were taken for some priests to examine. They thought they might be able to extract from those fruits something as useful as she-goats milk was to manufacture cheese. Besides their devotions, priests had food knowledge. Shepherds were goat behavior experts. Together, they tried to come up with something useful or tasteful out of this previously unnoticed bush. Their first attempt was to boil the fruits in order to extract a beverage, in the same way people have been doing for centuries with tea leaves and medicinal plants, but it was to no avail. Disgusted by the bad taste, they threw the fruits on the fire and left them behind as waste. Much to their surprise, their noses sensed a very distinctive smell. After the fruit pulp was burned, the bean itself was also burned. Then the aroma released into the monastery chambers captured their attention as well as their imagination. It was the astounding familiar effect that coffee has on people, the first ever in history.[1]

Coffee: Production, Quality and Chemistry
Edited by Adriana Farah

Published by the Royal Society of Chemistry, www.rsc.org

Nowadays, such an aroma experience is taken for granted in everyday life by millions of people around the globe. Even though that was an unsophisticated roasting method, with less-than-needed methodological skills, it was a new approach with a surprisingly tasteful outcome.

In *The Three Princes of Serendip*, a children's book mentioned in the 18th century by Horace Walpole in his letter to Horace Mann,[2] stories of fortunate ends with odd beginnings are told. The princes and their knights were on a quest for the Holy Grail. They bumped into good solutions for questions they were not asking. Since then, serendipity means to look for something of value and discover something of greater value, exactly what happened to our priests and shepherds.

Science has its own famous serendipity examples. One of the most famous is the discovery of penicillin that resulted from a careless experiment. Penicillin saved virtually millions of people. Alexander Fleming may have been careless, but he had an inquiring mind. What went wrong? Why? Although serendipity moments can represent some impressive improvements, it is not in itself a steady improvement. Much work is necessary from the failed experiment to obtain an injectable medicine that could be used to bring cure to various diseases. Serendipity may be a giant and impressive step, but it falls short in any process that requires sound methodology. There must be a systematic work after a smidgen of inspiration. We certainly drink better tasting coffee nowadays because we are better roasters than priests and goat shepherds from old North Africa.

That's why our questions are technical in nature. What thermal processes are implied in a successful roasting? What chemical features will be preserved or developed in the process? How much weight is lost, how much "drinking pleasure" will be gained? Priests and shepherds from long ago could afford to be careless, but the coffee industry that generates hundreds of millions of dollars annually certainly cannot.

There is a plethora of technical knowledge that must be taken into account if we are to take coffee roasting seriously. It may sound dry, but remember all along the reading that the intended roasting-to-perfection is a key component if we are to succeed. There is no real coffee expert if one does not master the basic principles of roasting as well as the available methods. Such technique development made possible the spread of the black gold all over the world. This chapter will approach the basic principles of roasting and, very briefly, the consequences of the process in chemical and sensory changes.

8.2 Chemical and Physical Transformations During Coffee Roasting

In order to understand the roasting principles and differences among the roasting techniques, it is important to go through an overview of the main chemical and physical changes that occur during roasting.

The roasting process can be divided into stages that follow a distinctive pattern. Such understanding is very helpful to roast masters because it enables them to control production and achieve a high quality standard in each batch. In this work, the division of the process is based on the intensity of the chemical and physical transformations as shown below.

8.2.1 Drying Process (up to 150 °C)

Drying is the process of evaporating the water present in green beans. Most water is eliminated before coffee beans reach 150 °C, although a small percentage is still retained inside the beans even after roasting is completed. It is counterintuitive that not all moisture is eliminated after water boiling temperature is reached (100 °C at sea level), but boiling temperature gets higher because of the beans' internal pressure. When the coffee moisture becomes vapor and other gases start to be produced inside the coffee beans, the beans' internal pressure becomes extremely high, so that the boiling temperature rises accordingly. After free water is dried up, coffee beans lose from 8% to 10% from their original weight, depending on the original moisture of the raw material.[2]

Besides some steps in Maillard reactions that occur between the amino groups of amino acids and the carbonyl groups of reducing sugars,[3] virtually no dramatic chemical reactions occur during this stage of the roasting process. Temperatures are still too low to cause pyrolysis. From the energy point of view, this is an endothermic process, *i.e.*, reactions and physical changes have to absorb energy in order to take place. Aroma starts developing at this phase, as well as a discrete color (Figure 8.1), through the formation of intermediate compounds from Maillard reaction and, later, melanoidins.[4] Acrylamide, which is an undesirable compound (see Chapter 30) is also formed at this stage, from the reaction of asparagine with reducing sugars.

Figure 8.1 Coffee color during drying phase.

Figure 8.2 Light roast.

Its concentration peaks at the beginning of roasting and decreases thereafter. Isomerization and lactonization of chlorogenic acids occur along with other reactions.[5,6] (These and other changes in coffee compounds will be approached in Chemistry section of this book.)

8.2.2 Roasting Initial Stage (150 °C–180 °C)

At this point Maillard reactions are still happening, beans get darker (Figure 8.2) and sugar caramelization starts in the coffee beans. Most types of sugar present in coffee, among which sucrose is the most important, will undergo thermal degradation, similar to preparing caramel heating table sugar in a skillet, at temperatures around 160 °C or higher. The exact temperature where caramelization begins depends on the bean temperature increasing rate. If temperature is increased by only a few degrees per minute, sucrose transformation will begin at 160 °C, but a faster temperature rising rate can change it. The faster the temperature rises the higher the caramelization temperature is, so that degradation of sucrose may start happening in temperatures above 180 °C.[9]

The beans' acidity increases during this phase because carbohydrates degrade into carboxylic acids. As acid disintegration is negligible in such temperatures, acid formation prevails and pH decreases.

Other thermolabile compounds like trigonelline and chlorogenic acids also start to degrade.[10]

8.2.3 Roasting – Stage 2 (180 °C–230 °C)

Coffee color continues getting darker as temperature increases (Figure 8.3). During the roasting process, there are two moments when coffee beans make a sound which resembles popcorn popping up, the coffee "cracks".

Figure 8.3 Moderately dark roast.

The first one happens in the beginning of this stage, usually close to 190 °C, but the crack temperature varies according to different coffee types. The crack is characterized by bean expansion and rupture of its internal structure because chemical reactions cause the release of gases like carbon dioxide, water, and organic volatile elements. As these gases force their way out, the coffee internal structure expands, along with its cell walls, rendering them more permeable, and making the soluble components extraction easier during coffee brewing.[11,12,15]

Pyrolysis takes place at this stage. Pyrolysis reactions are usually associated with organic substances heated in a low oxygen concentration environment.[7,8] This process is similar to controlled burning with intervening products being formed before organic compounds change into carbon dioxide, water and ashes. Such chemical reactions, of course, are interrupted before reaching this final stage, because the substances that produce a pleasant cup of coffee must be preserved. Pyrolysis degrades substances, so it will form smaller molecules when compared to those found in the initial green beans. It is worth noting that these are exothermal reactions, *i.e.*, coffee beans start producing heat while being roasted. This is an interesting phenomenon, because pyrolysis requires heat to happen, but at the same time it produces heat as result of its chemical transformations. Considering the whole roasting process, pyrolysis produces about 11% of the necessary heat for it to happen[13] and most of the energy released by beans becomes available from the first crack on. Varying the amount of heat supplied in each phase of the roasting process will produce different chemical and therefore sensorial results in the final product.

At this stage, caramelization progresses. Although this reaction is related to the perception of coffee's characteristic sweet and pleasant flavor in the cup, it may produce bitter results if this process is extended for too long. The darker the roasting degree, the more bitter is the taste.

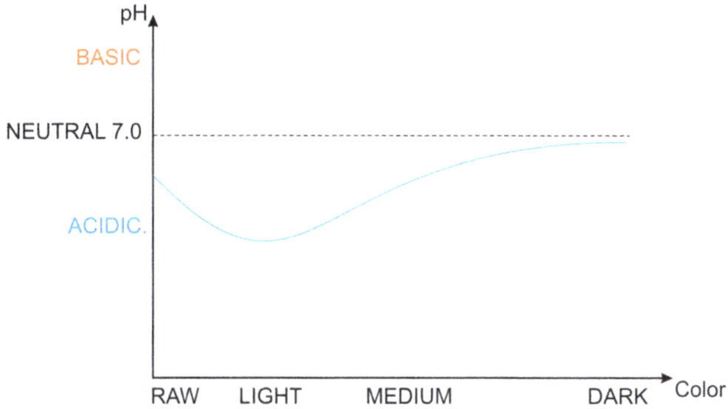

Figure 8.4 Coffee acidity as a function of roasting degree.

As happens to sweetness, acidity also drops as the roast gets darker because acids are degraded and volatized in this phase. While in the first stage of roasting acidity increases, from this moment on acidity decreases and pH increases, approaching 7.0 as roasting gets darker.[14] pH behavior along the roasting process is shown in Figure 8.4.

In addition to the weight loss due to the moisture evaporation, there is an extra weight loss (from 4% to 6% depending on final roasting color) due to carbon dioxide and water elimination, plus the elimination of other volatile and semi-volatile organic components from the beans. It is to be noted that most water eliminated in this stage is not free water, primarily present in the raw material. It is instead the result of chemical reactions that decompose large organic molecules into smaller molecules, carbon dioxide and water. This way the weight loss of the beans may correspond to 16% or a higher percentage in dark roasts or when the initial raw material water content is high. All the gas forcing its way out of the beans causes their volume to increase, so that the total swelling from the beginning to the end of the roast will vary between 40% and 60%. In cases of dark roasts the final roasting temperatures can be above 230 °C, and the volume can double.[11,15]

The roasting process may end in this phase. In industry, the final roasting degree is defined by color. The color quality control is made by colorimeters measuring the coffee color after grinding it. Online control is also necessary during the roasting process and it is not possible to be done by colorimeter because grinding the beans takes time and color changes quite fast during the process. As the monitoring of the beans' color is not accurate due to the color variation from bean to bean, roaster manufacturers have chosen the beans' temperature to monitor the roasting process. In a roasting plant, these two control systems work concomitantly. After roasting, the coffee color is checked in the laboratory and when deviation from the desired color occurs the set point of the final roasting temperature is readjusted in the roaster control system for the following roasting batches.

Table 8.1 Roast color classification system (Agtron).

Bulk roast classification	Agtron number	Color disk values
Very light	100/95	Tile # 95
Light	90/85	Tile # 85
Moderately light	80/75	Tile # 75
Light medium	70/65	Tile # 65
Medium	60/55	Tile # 55
Moderately dark	50/45	Tile # 45
Dark	40/35	Tile # 35
Very dark	30/25	Tile # 25

During the decades after industry began the use of colorimeters, there was no standardization for the color scale. Different manufacturers would produce their own colorimeters with different scales. Even today we find roasting industries working with diverse types of color gradations. However, in 1995, the Specialty Coffee Association of America, currently called the Specialty Coffee Association, created a new color classification method, the Agtron system, which is becoming more and more popular as a universal procedure for color detection. The lightest and darkest limits of the colorimeter scale were determined by cupping. The Agtron number reflects a narrowband infrared reflectance measurement, used with the purpose of detecting the grade of sucrose caramelization. At the same time these numbers are also translated to color disks with the main roasting degrees according to Table 8.1.[16]

Roasts ending during this phase are considered between light and medium dark roasts (Agtron 90 to 45). An example of it is one usually addressed as cinnamon roast (Agtron 90). It is light brown, with no oil on the beans' surface, its brew is typically acidic and aroma is not fully developed. Another roast with a little darker grade is named as American roast (Agtron 70–60) because it was extremely popular in the USA years ago. It has a medium brown color, still with no oil, and it gives a sweet and slightly acidic cup. The Full city roast ends in a temperature around 230 °C resulting in an Agtron color from 50 to 45.[17,18]

8.2.4 Roasting – Stage 3 (Above 230 °C)

The second crack occurs in the beginning of this stage. Both cracks happen due to high peaks of intensity in pyrolysis reaction. As mentioned above, these peaks are reflected in violent bursts of gas out of the beans, making a peculiar sound, the "crack".

Coffee beans are formed by cells, as are all living beings. In the case of plants, the cells' external walls have cellulose fibers which work as a filter controlling what comes in and out. The second crack causes severe damage to these fibers, stretching them and making them more and more permeable. Soluble components become easier to extract, increasing the effects initiated in the first crack. Because of the extra bean volume increase, the migration of coffee natural oils from bean center to its surface is inevitable. The greater beans expansion also leads to a greater release of aromatic and non-aromatic volatile compounds, but the perception of coffee's characteristic aroma changes with the degradation of volatile compounds, which happens in dark roasts.[11,19]

Some roasts will end during this phase, some during the second crack and, some after the second crack is completed. One of them is popularly named French roast (Agtron 45–40), which has less acidity then the American one and has the tang of the dark roasts. The darker Italian roast (Agtron 40–35)[18] is very oily, without acidity, and all the tang is obscured by carbonization.[17]

Pyrolysis exothermal effect drops at temperatures higher than 250 °C, indicating that a large part of organic compounds has already been carbonized at such a high temperature and coffee color is very dark, tending to black.[20] Carbonized beans result in undesired bitterness and are usually harmful to health, since they will probably contain carcinogenic polycyclic aromatic hydrocarbons.[21]

All roasting methods require heat to take place. The different roasting stages presented here have diverse chemical and physical transformations, some of them endothermic and some exothermic, each requiring specific levels of energy to take place. Fast cooling of beans must be performed in order to stop immediately the roasting reactions. For the majority of coffee roasted in the world this is done using water quenching, which is very efficient to extract heat due to the high latent heat of the water. Companies that produce specialty coffee tend solely to use cool air to cool down the coffee, because water can increase beans moisture and promote undesirable future reactions.

Over time it has been observed that controlling and varying the supplied heat transfer in the different roasting stages can modulate the final sensory results, taking full advantage of the beans' potential. Controlling the process also allows for good reproducibility.

8.3 Heat Transfer Systems and Types of Industrial Roasters

Heat is a form of energy that is always moving from hot to cold bodies, from higher to lower temperatures. There are different ways to transfer heat that depend on aspects such as distance between bodies as well as their physical state: solid, liquid, or gaseous. They are classified as radiation, conduction, and convection. Radiation occurs when heat is transferred at a distance by electromagnetic waves, like the sun's heat. Conduction is heat transfer between two solid bodies that are in contact with each other. An example is the heat transfer mode of cooking a food

Figure 8.5 Drum roaster.

on the grill or roasting a coffee using a casserole dish. In the case of convection, heat transfer happens by means of liquids or gases in contact with a solid body. Convection is what cooks potatoes immersed in hot water or the heat we feel in a steam room.[22]

Today, there is a huge variety of roaster models that can disorient even experts (Figures 8.5–8.7). They offer different roasting technologies using diverse forms of heat transfer. A glance at the history of roasting technological development will facilitate understanding of the evolution of heat transfer mechanisms used in coffee roasters.

8.3.1 A Brief History of Industrial Roasters Evolution

The last two centuries witnessed few improvements in roasting until the end of the 1960s. At that time, evolution of roasting design moved in slow, small steps. The first industrial roaster consisted of a metallic ball which could be manually rotated by handles (Figure 8.8). The sphere with coffee beans inside could be detached from the roaster structure to make coffee deployment on a plain surface for cooling easier. Fuel was either wood or coal. This was the principle of a coffee roaster in the middle of the 19th century. Besides the addition of an electric motor to rotate the ball (Figure 8.9), there were no major modifications until the beginning of the 20th century. This older roaster design had the air pass around the metal sphere. The hot steel transferred heat to the beans by conduction, *i.e.*, the hot metal was in direct contact with the beans. The drum was kept in constant rotation so that the beans would be evenly roasted, as much as possible, by a direct fire placed below the metallic ball.[23] The fire would provide thermal energy to the sphere by radiation and convection. Such design provided direct contact of

Figure 8.6 Convection drum roaster.

Figure 8.7 Fixed drum roaster.

the beans with the hot surface. Roasting would take 30 minutes or more to complete. Any attempt to roast faster using this technology would result in scorched coffee beans.

Another roaster design that came about in the beginning of the 20th century included the addition of forced ventilation around a roasting chamber shaped as a drum. This hot air flow was pumped by a blower. As seen in Figure 8.10, showing a typical roaster of the first half of the last century, these roasters had an external cooling tray for the roasted beans and this tray was equipped with moving paddles.

Figure 8.8 Manual ball roaster.

Figure 8.9 Ball roaster with electric motor drive.

Such a mechanical configuration represented progress, but on the other hand the heat transfer to the beans remained pure conduction. Although there was forced ventilation around the roasting chamber, inside nothing changed: a hot steel surface was in direct contact with the coffee beans. However, we must consider some drums fully made with perforated steel sheets that would allow some hot air to reach the beans supplying small amounts of heat by convection.[23,25]

A qualitative real improvement took place still in the first half of the 20th century, when the first roaster with air flow going through the roasting chamber was introduced in countries like the United States, Europe, and Brazil by different manufacturers. Now in addition to conduction, a small amount of

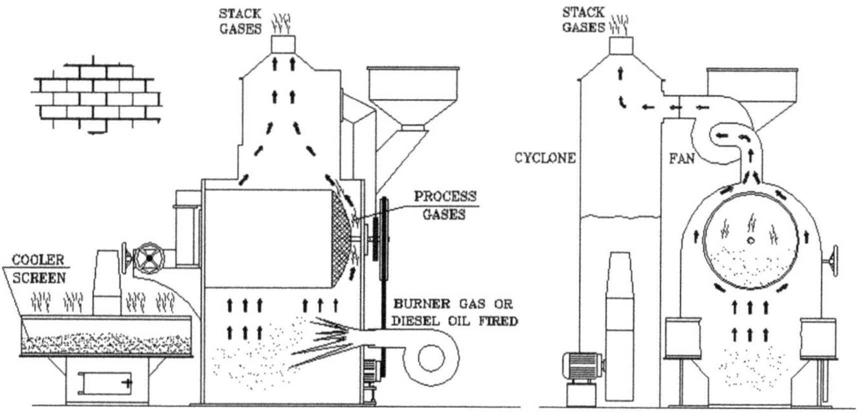

Figure 8.10 Direct fire drum roaster.

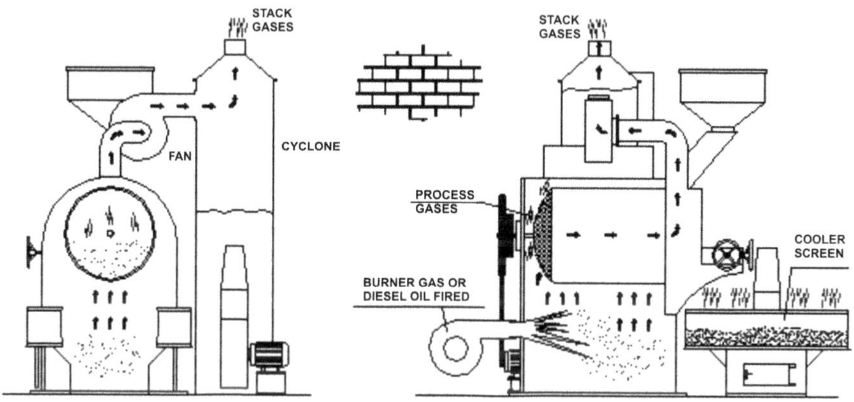

Figure 8.11 Direct fire drum roaster with conductive and convective heat transfer systems.

hot air could pass through the drum and be transferred to the beans by convection, a small but real improvement[23,25] (Figures 8.11–8.13).

The solution found for this design, to allow hot air to enter the drum and at the same time to prevent the beans coming out of it, was the use of a perforated metal plate in the back of the drum, as can be seen in the picture in Figure 8.14. The amount of air flow throughout the drum was small, because that design used a sieve covering all the drum back area, which was a barrier for the air passage which prevents higher air volumes passing through it. This system as a whole would always keep the percentage of convection secondary compared to conduction.

Amazingly, this type of technology is still currently in use for heat. Even though it does not represent the best design available to obtain the best roast quality, it has a considerable advantage: a very simple construction which makes it inexpensive. The technologies that followed, aiming for the complete

Figure 8.12 Picture of a drum roaster from the first half of the 20th century.

Figure 8.13 Picture of a drum roaster manufactured in this century.

elimination of conduction from the roasting process, resulted in much more complex equipment, increasing the manufacturing cost considerably.

The history of the era when drum roasters reigned almost absolutely wouldn't be complete without mentioning the continuous drum roasters. They came to reality during the 1940s in the United States. They consisted of long perforated drums where coffee was continuously fed in at one end and discharged at the other end. The coffee beans would be conveyed along the drum by paddles working like a screw conveyer. While moving, the coffee would be in contact with hot air and the coffee was discharged at the

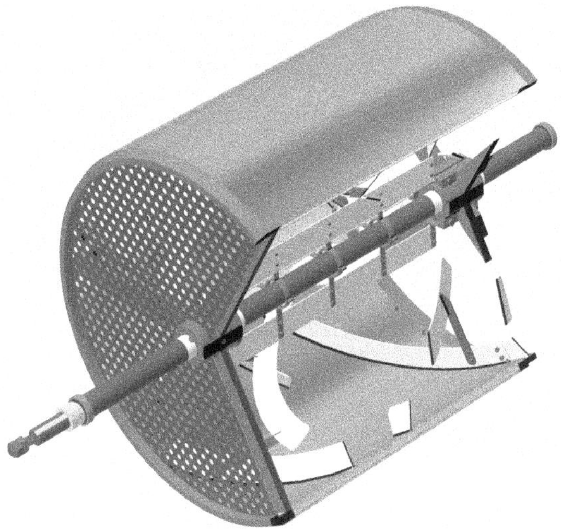

Figure 8.14 Typical drum design of a drum roaster with hybrid heat transfer system, using conduction and convection.

drum outlet when the final roasting color was achieved.[23,24] The beginning of the operation presented inaccuracy and fluctuations in the roasting degree, which would stabilize after some time of operation with adaptations. This same problem would appear in case of any change in the raw material or when starting the production of a new lot with different roasting degree. Therefore only companies with really high production volume found this kind of equipment suitable for their operation. It is probably because of these problems that this type of roaster design did not succeed in the long run.

The era initiated in the beginning of the 1970s witnessed a huge transformation in coffee roasters design. In order to eliminate or drastically reduce conduction, fixed drums, bowl, air jet, and other concepts were added to the new roasting equipment. The Got-Hot was the first roaster with fixed drum and rotating paddles. Currently, some manufacturers still produce machines using the same principle. This type of equipment uses convective roasting and a great amount of air, enough to help its paddles to move the beans while roasting as shown in Figure 8.15. As it contains pneumatic and mechanical elements to revolve coffee beans it is considered a semi-fluidized bed roaster and it provides shorter roasting times than traditional drum roasters.[23–25]

The bowl roaster is also a convective roaster which works with the semi-fluidized system. It is possible to see the roasting chamber in the shape of a bowl, which rotates in its vertical axel, in Figure 8.16. The hot air comes through perforations in the bowl, helping to move the beans and roasting them. It is also possible to see from the figure that the furnace is completely independent from the main roasting chamber.[24,25]

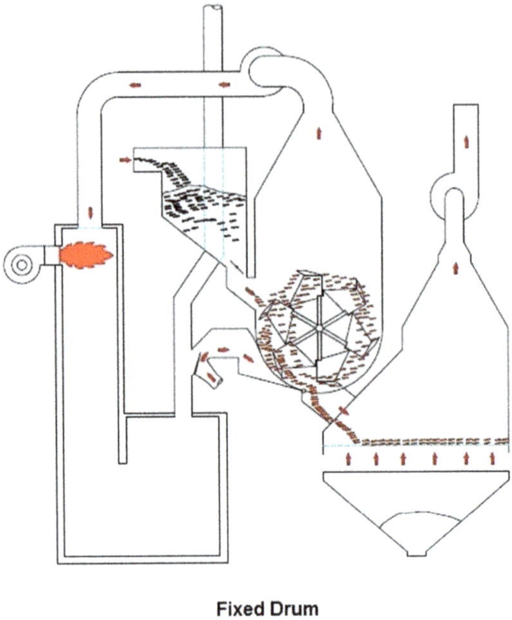

Figure 8.15 Fixed drum roaster.

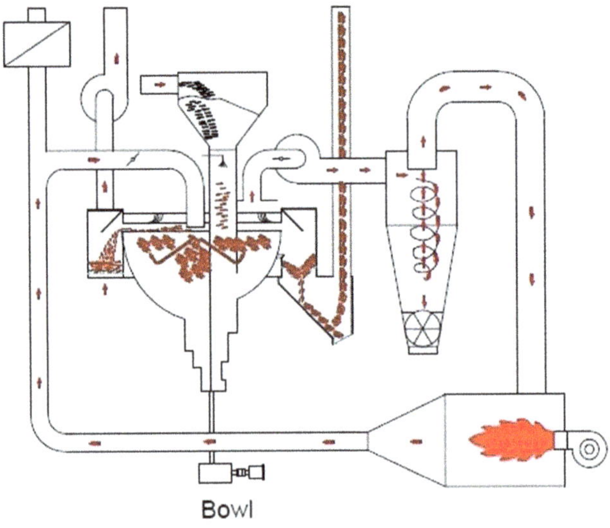

Figure 8.16 Bowl roaster.

The roasting chamber of an air jet roaster is shown in Figure 8.17. This type of roaster works with the fluidized bed system, because only air is used to revolve the beans during the process.[24,25] This requires higher air flow when compared to semi-fluidized roasters as its air speed must considerably

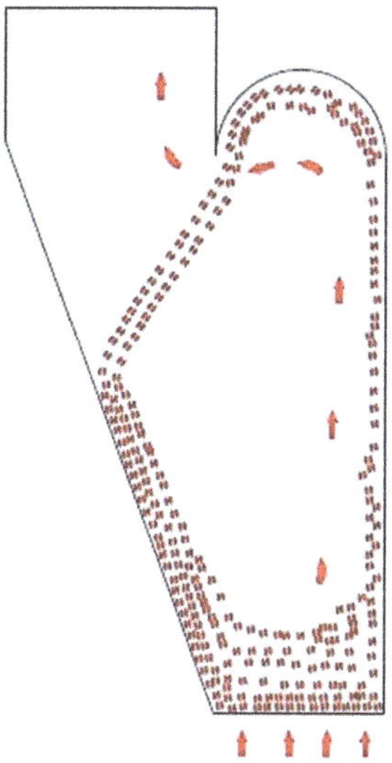

Figure 8.17 Air jet roaster.

exceed the necessary velocity to arrest the beans, moving them by the pneumatic effect of the air jet. This fact and the hot air temperature used by this type of roaster normally result in medium to high heat transfer rates allowing them to roast very fast (less than 3 minutes).

Although we name the air jet as a fluidized bed roaster, this name would be more appropriately applied to a roasting chamber shaped as a table as shown in Figure 8.18. This roasting system consists of a flat surface with perforations uniformly distributed along it. The air expelled through the holes forms an air layer between the metal plate and the coffee beans preventing them from touching the hot metal. The exclusive contact with the hot air implies a convective process. The name, fluidized bed, comes from the cinematic behavior of the particles suspended over the air layer which is similar to the cinematic behavior of fluids. This roasting principle also usually tends to work with high rates of heat transfer.

Still in the 1970s, the first drum roaster processing coffee exclusively by convection was invented. As can be seen in Figure 8.19, it was necessary to establish an external furnace so that nothing would heat up the drum under or around it. It was also necessary to eliminate the perforated plate at the drum air inlet, as shown in Figure 8.20, so that enough air flow would be allowed to enter the cylinder. This revolutionary design was necessary to increase the heat supplied by convection as conduction was eliminated. In this technology the hot air goes freely throughout the drum and only hot air transfers heat to the beans by convection. This kind

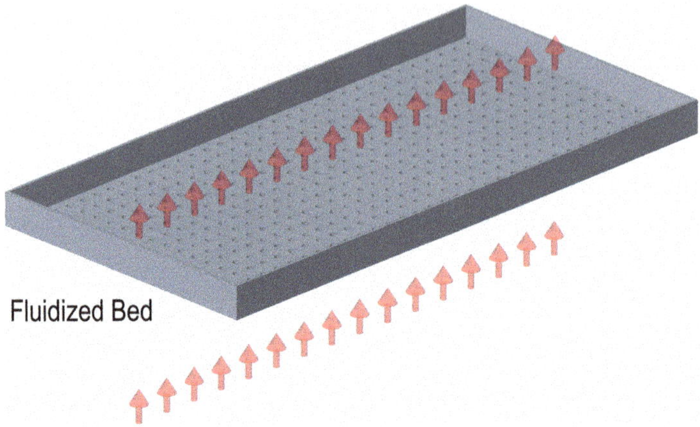

Figure 8.18 Fluidized bed roaster.

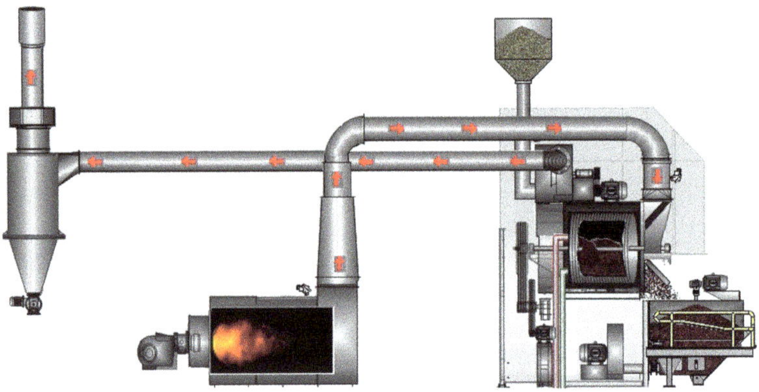

Figure 8.19 Drum roaster with open ventilation circuit.

of design allowed modern equipment to use high volumes of air flow, making the semi-fluidized bed system also possible for rotating drum roasting chambers.

It is worth noting that, in the last few decades, roasters allowed improved quality control because of increased hot air volume and air speed. These factors improved roasting flexibility because they allow roasters to work within a wider heat transfer range during the roasting process. This new technology can be divided into two categories: semi-fluidized and fully fluidized beds. In semi-fluidized roasters air flow is so intense that coffee beans movement is facilitated inside the roaster chamber, but mechanical action of some kind is still required to perfectly move the coffee beans. Fully fluidized bed roasters use solely air to move and mix the coffee while roasting. Semi-fluidized bed

Figure 8.20 Typical drum design of a drum roaster based on convection heat transfer system.

roasters have a wide variety of designs and even some rotating drum roasters can work applying this principle.[23,24]

8.3.2 Positive Aspects of Convection for the Coffee Roasting Process

As explained previously, conduction heat transfer depends on physical contact between two solid bodies. In the case of coffee, a drum hot steel surface is in contact with the beans surface. Picture a mildly curved hot surface and a small bean in contact with each other, and it becomes easy to realize that contact points will always be tiny, which results in an uneven roasting. Even having a high amount of kcal per square millimeter per second passing through these contact points tiny areas, the total heat being supplied to each coffee bean, counted in kcal per second, will be small, so causing the roasting time to be long. When trying to roast fast, plain conduction-based roasters produce burnt spots and scorched beans, and hence low quality roasting. Consequently, their roasting times must be 30 minutes or more.

Instead of using small spots to transfer heat, convection-based roasters use air flow that completely involves the beans, distributing heat evenly and efficiently. So convection is a better principle to build on a roaster, for it provides a higher roasting quality, opens the possibility of higher heat transfer rates and shorter roasting time for any given batch, preventing scorched beans.

Convection-based roasters have opened many opportunities to research for ideal heat transfer rates. This means not only the possibility for different roasting times, but also the option of modulating from low transfer rates to medium and high, all in the same roasting batch, providing the possibility of a customized roasting profile, depending on the raw material variety and on the desired characteristics for the final product. From years of hands-on experience, industry experts found that adjusting roasting profile to roasting times in the range from 8 to 12 minutes is ideal to produce both better and more complex taste and aroma for most types of coffees. However, longer and shorter roasting times are certainly useful for some cases.

8.4 In Roasting Profile, Control of Coffee Bean Temperature Is the Key

The next paragraphs in this section assume convection as the heat transfer method to be considered for all the arguments.

The coffee beans' temperature should be controlled during roasting in order to ensure high cup quality at the end of the process. This physical parameter has a primary relevance over hot air temperature, air flow, and even heat transfer rate. To evaluate this matter we first must consider what the roasting process is and why it happens. Then we will be able to identify key aspects that have a direct influence on roasting.

8.4.1 Hot Air Temperature, Hot Air Flow, Heat Transfer

Many make the wrong assumption that the temperature of hot air is what roasts the bean. It is true that, as we increase air temperature, roasting time gets shorter, but that is not a sound conclusion.

The flow of the hot air has a key importance in convection. In order to widen our understanding, we can perform a little experiment. First, we need a roaster that is able to keep the temperature of hot air constant throughout the roasting process. Secondly, we need this roaster to control the hot air flow passing through the roasting chamber. In the first batch run we will maintain a fixed temperature and a fixed air flow during roasting. In the second batch we will use the same hot air temperature, but now increasing sharply the air flow. The result will be a significantly shorter time to obtain the same roasting of the first run. So, it becomes clear that roasting depends directly on temperature and air flow.

Now the question is: are these two elements, temperature and air flow, the actual primary cause for convection roasting? Are they the final elements to be controlled in order to control roasting? It may be surprising that neither hot air temperature nor its speed are the direct agents causing chemical reactions to happen inside the coffee bean in the roasting process. There needs to be heat transfer from hot air to beans. Heat transfer relies on two variables: hot air temperature and flow. When one grasps

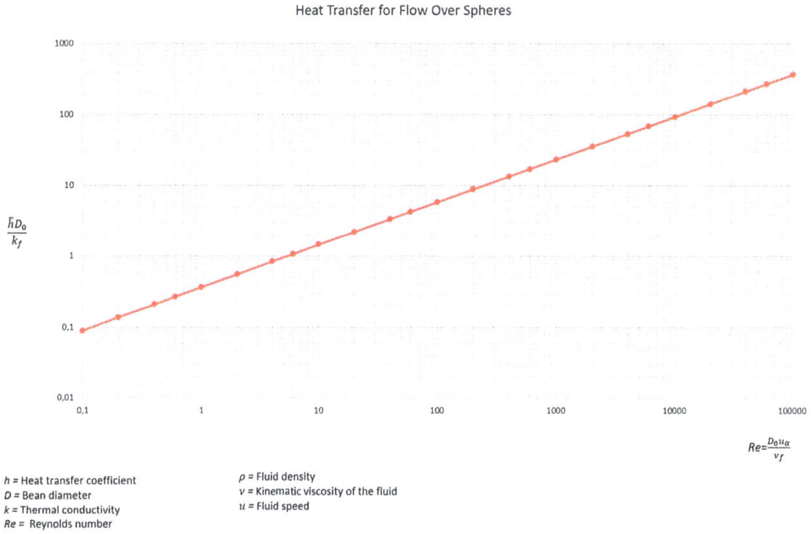

Figure 8.21 Heat transfer for flow over spheres as a function of Reynolds number.

the significance of such a concept, hot air temperature and air flow speed become important allies.

Heat transfer can be calculated when one knows the values of hot air temperature, the speed in which the air flows around the bean and the bean temperature, size, and shape. From the academic point of view, there are other thermodynamic variables involved in this calculation, but they are negligible. Building on the science of thermodynamics, heat transfer can be described by the following formula and in Figure 8.21.

$$Q = h \times A\ (T_{\mathrm{air}} - T_{\mathrm{bean}}),$$

where:[22]

h = heat transfer coefficient
A = coffee bean surface
T_{air} = hot air temperature
T_{bean} = bean temperature

The graph in Figure 8.21 was produced based on the following equation for heat transfer of gases flowing over spheres: $(h \times D)/k = [0.37 \times (u \times D)/\nu]^{0.6}$.[22]

According to the equations shown above, heat transfer is directly proportional to the difference of temperatures between hot air and coffee bean. The heat transfer coefficient does vary for different air temperatures, but this fact does not change the point we are making, that there is a direct relation between heat transfer and hot air temperature: when one goes up, the other does too.

If we examine the second equation closely, air speed causes an impact that is similar to the hot air temperature: the higher the speed is the larger will be the amount of heat transferred.

The influence of air speed on heat transfer can be felt when outdoor temperature is about 0 °C, but we experience it as if it were −10 °C, a lot colder. The outdoor temperature is different from thermal sensation, a phenomenon linked to wind. A stronger blowing wind will cause people to feel like it is colder than it really is, colder than the thermometer measurement would indicate. Our body temperature gets lower in a windy environment because it causes the body to transfer more heat to the cold wind.

If we put together all this information in a scientific point of view – the principles of thermodynamics – our conclusion will be the rebuttal of a great myth. Much has been said about the ideal hot air temperature to roast coffee properly, but like us coffee beans are not directly affected by it. What really matters is the bean temperature that is controlled by heat exchange rate with hot air flow. If one gets high hot air temperature and slow air flow, beans won't roast fast, as we do not experience being burned even in a hot sauna. The reason is simple: in both cases the heat exchange is minimal. On the other hand, a lower air temperature combined with too fast an air flow can even put to waste a bean batch for burning it badly. If we have a good grasp of the heat transfer notion, we will be able to define beforehand how much heat will be transferred to a batch in any given moment. Put your mind on setting the heat supply and do not get distracted by air temperature inside the drum.

As a consequence of what has been explained, it is possible to obtain the same heat exchange rate using different combinations of hot air temperature and air flow speed. The formula shown above enables us to calculate the amount of kcal that is transferred to the beans. A higher hot air temperature will result in a specific increase of heat transfer. Now, if we maintain this heightened air temperature, the formula can also show us the precise decrease in airflow in order to obtain the same initial heat transfer. In fact, there are countless combinations of hot air temperature and air flow speed that result in the same effect on the bean.

8.4.2 Bean Temperature Is What Roasting Is All About

Chemistry teaches us the concept of activation energy. It is the minimum energy amount required for any chemical reaction to happen. Each type of chemical change has an activation energy of its own.[26] It means that putting together the desired elements is not enough: we have to add the right amount of energy. Once the activation energy level is reached, it must be kept if reactions are to proceed. Thermodynamics teaches us that internal energy can be characterized by two properties,[27] temperature and pressure. A rough calculation, with very good precision, can be obtained for coffee beans, which contains solid and liquid elements, considering that the activation energy depends primarily on its temperature. For the gases that are

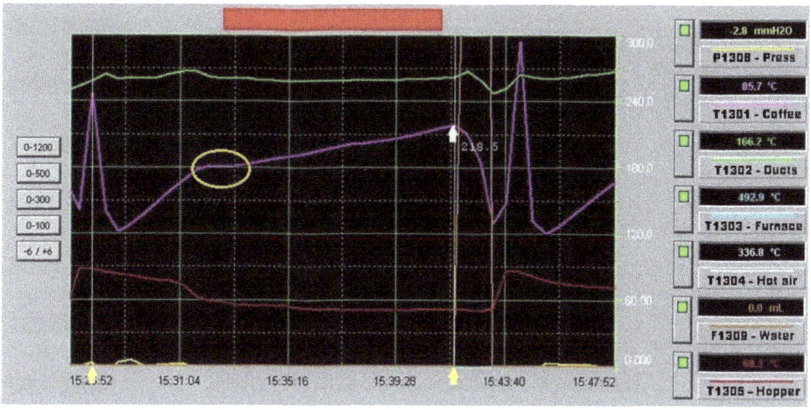

Figure 8.22 Example of coffee bean temperature evolution along the roasting process as a function of time for a roast of 16 minutes.

produced inside the beans, the internal pressure and temperature will define their energy which will activate the reactions. Nevertheless, this pressure is a function of the structure of the bean and the evolution of the temperature with time in a way that roasts faster will build up higher pressures than slow roasts. Therefore, roasting depends on coffee beans temperature profile with time[3] rather than on hot air temperature, air flow, or heat transfer. Examples of coffee temperature profiles are shown in Figures 8.22 and 8.23 where it is possible to observe the temperature of the coffee beans as a function of the roasting time.

Final roasting color is the main parameter to control cupping results for a defined raw material. However, the beans temperature raising rate is also an element of great importance to control coffee's final chemical composition which means it has influence over the cupping as well. Therefore it is possible to obtain diverse cuppings from the same raw material, roasted to the same final color, using different coffee beans temperature profile with time.

8.5 Environmental Aspects in Coffee Roasting

The coffee roasting process produces many volatile and semi-volatile components which are released into the roasting chamber during Maillard and pyrolytic reactions.[28] Most of these substances are not toxic, only a few of them present potential toxicity, like diacetyl and polycyclic aromatic hydrocarbon, and very few are actually toxic. Nevertheless, these gases cannot be freely discharged to the atmosphere and the emission of these roasting rejects is controlled by environmental agencies from most countries, like the United States[29] and European countries.[30,31]

One of the first and most traditional ways to eliminate the smoke derived from the coffee roasting process is by post-thermal oxidation.[28] This is

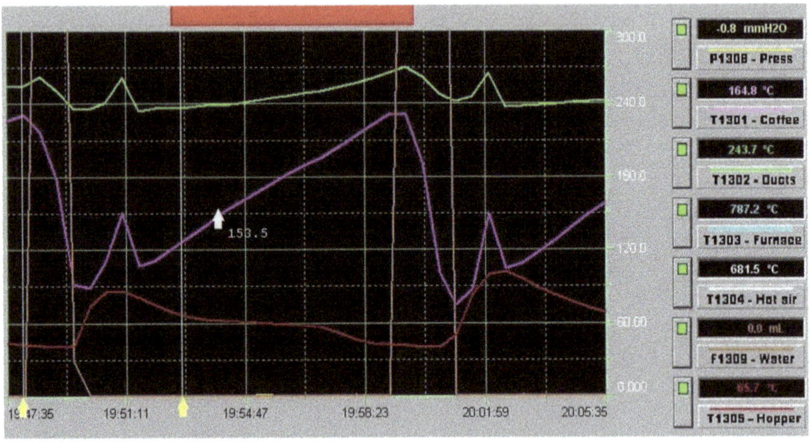

Figure 8.23 Example of coffee bean temperature evolution along the roasting process as a function of time for a roast of 8 minutes.

performed by a piece of equipment called an "after burner", which processes the roasting exhausting gases passing them throughout a combustion camera before venting them to the atmosphere. They are usually installed in roasters with an open ventilation circuit as shown in Figure 8.19 where they are regularly located in the chimney.[28]

Inside the after burner the effluent reaches temperatures high enough to promote a thermal oxidation of the organic substances, transforming the visible smoke into transparent and inodorous gases. The major volume of these gases is composed of carbon dioxide and water vapor, but some amount of carbon monoxide, nitrogen oxide, sulfur dioxide, and some other elements can also be found. There will also be traces of organic elements which are not completely oxidized.[27] The regulation for emission limits of each substance depends on the country's legislation and often they vary for different areas in the same country.[29,32]

A variation of the traditional after burner technology is the built-in after burner, which uses the same furnace that generates heat for the roasting process as the smoke oxidizing chamber. For this, the roaster must have a closed ventilation circuit where 100% of the gases coming from the roasting cycle are recirculated into the furnace as is shown in Figures 8.24 and 8.25. In this case, the exhaust gases leave the equipment through a chimney placed after the furnace.

The recirculation of all the air flow to the roaster furnace (see Figure 8.24) results in a more economic process from the energy consumption point of view. The open ventilation circuit (see Figure 8.19) vents all the gases from the roasting process to the atmosphere and a large amount of energy must be spent to heat them up to promote the smoke elimination. The closed ventilation system vents to the atmosphere only the mass equivalent to the system income flows like the fuel and air used for the combustion in the

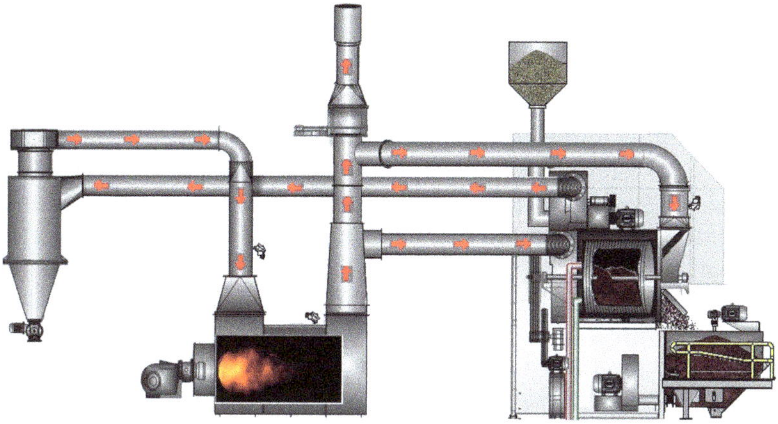

Figure 8.24 Drum roaster with closed ventilation circuit, recirculating 100% of gases back to the furnace.

Figure 8.25 Drum roaster with closed ventilation circuit.

furnace, the gases produced by coffee roasting processing, and any false air (undesired air flow caused by the collateral effect of equipment design) that might be introduced into the roaster. This amount of gases is just a small part of the total mass flow passing through the roasting chamber and, therefore, the air flow that is vented to the environment is much smaller. Consequently, the necessary energy to heat up these gases to oxidize the smoke is also small when compared to the open ventilation circuit.

A third well-known technology to oxidize smoke is the catalytic reactor.[28] Like the after burner this system is composed of a combustion chamber to raise the effluents' temperature in order to oxidize the organic matter.

However, the use of catalysts can reduce the necessary temperature to eliminate the smoke. This results in lower temperatures of the effluents and so leads to energy saving, but from an economic point of view the cost of the catalytic elements, which have a limited lifetime and must be replaced from time to time, must be also computed.

Besides the gaseous pollutants produced during roasting, there is also particulate matter carried by the air flow. Most of these particles come from the chaff, released along the roast, that originally is the silverskin which belongs to the green beans structure as its most external layer. As a result of thermal transformations, these skins detach from the beans and are easily conveyed by the roasting air flow due to their light density.

The most common used apparatus to clean the air, before venting it to the atmosphere, is the cyclonic separator. A cyclone can be observed in Figure 8.19 located just below the chimney. It forces the air flow to move in circles creating centrifugal forces that push the particles, heavier than the air, to the cyclone walls, as can be observed in Figure 8.26. Once touching the walls, an area where the air speed is lower, the particulate matter falls to the bottom of the cyclone by gravity. The collecting efficiency of a cyclone typically varies from 96% to 99% depending on the size of the particles and its design.[33]

The above described technologies are not exclusive, but they are the most used and they are effective to attend to emission regulations for coffee

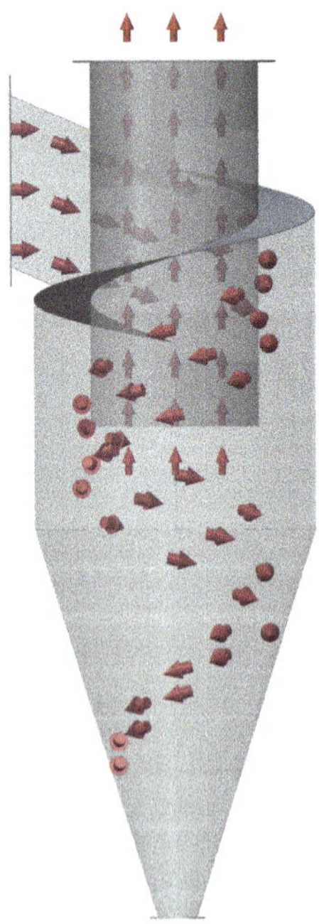

Figure 8.26 Cyclone used to collect particulate matter released by the coffee roasting process.

roasters found worldwide. Nevertheless the trend of effluent regulations is to force the reduction of emission limits aiming at the reduction of pollutants as the number of industries grows. Consequently, research for new and better technology to improve emission control is necessary.

References

1. G. Wrigley, *Coffee*, Longman Scientific & Technical, Longman Group UK Limited, 1988, pp. 10–11.
2. R. K. Merton and E. Barber, *The Travels and Adventures of Serendipity*, Princeton University Press, 2006.
3. S. I. F. S. Martins, W. M. F. Jongen and M. A. J. S. van Boekel, A review of Maillard reaction in food and implications to kinetic modelling, *Trends Food Sci. Technol.*, 2001, **11**, 364–373, Elsevier Science Ltd.
4. T. Hofmann, W. Bors and K. Stettmaier, Studies on radical intermediates in the early stage of the nonenzymatic browning reaction of carbohydrates and amino acids, *J. Agric. Food Chem.*, 1999, **47**, 379–390.
5. V. Gökmen, *Acrylamide in Food: Analysis, Content and Potential Health Effects*, Elsevier Science Ltd., 2016, pp. 1–16; 181–192; 433.
6. C. Yeretziana, E. C. Pascualb and B. A. Goodman, Effect of roasting conditions and grinding on free radical contents of coffee beans stored in air, *Food Chem.*, 2012, **131**, 811–816, Elsevier Science Ltd.
7. M. Jahirul, M. G. Rasul, A. Ahmed Chowdhury and N. Ashwath, *Biofuels Production Through Biomass Pyrolysis —A Technological Review in Energies*, 2012, ISSN 1996-1073, p. 4954, http://www.mdpi.com/journal/energies.
8. M. X. Fang, D. K. Shen, Y. X. Li, C. J. Yu, Z. Y. Luo and K. F. Cen, Kinetic study on pyrolysis and combustion of wood under different oxygen concentrations by using TG-FTIR analysis, *J. Anal. Appl. Pyrolysis*, 2006, 22–27.
9. J. Won Lee, Investigation of thermal decomposition as the cause of the loss of crystalline structure in sucrose, glucose and fructose, PhD Thesis, Graduate College of the University of Illinois at Urbana-Champaign, 2010.
10. A. Farah, T. de Paulis, L. C. Trugo and P. R. Martin, Effect of roasting on the formation of chlorogenic acid lactones in coffee, *J. Agric. Food Chem.*, 2005, 1505–1513.
11. M. S. Michael Sivetz and H. Elliott Foote Ph.D., *Coffee Processing Technology*, The Avi Publishing Company, Inc., 1963, vol. 1, pp. 229–239.
12. A. Illy and R. Viani, *Espresso Coffee: The Science of Quality*, Elsevier Academic Press, 2005, pp. 179–184.
13. R. J. Clarke and R. Macrae, *Coffee, Volume 2: Technology*, Elsevier Applied Science Publishers Ltd, London and New York, Reprint 2011 of First edition, 1987, p. 82.
14. M. S. Michael Sivetz and H. Elliott Foote Ph.D., *Coffee Processing Technology*, The Avi Publishing Company, Inc., 1963, vol. 1, p. 232.

15. R. J. Clarke and R. Macrae, *Coffee, Volume 2: Technology*, Elsevier Applied Science Publishers Ltd, London and New York, Reprint 2011 of First edition, 1985, pp. 84–86.
16. C. Staub, *Roast Color Classification System*, SCAA (Specialty Coffee Association of America), 1995.
17. G. Wrigley, *Coffee*, Longman Scientific & Technical, Longman Group UK Limited, 1988, pp. 502–504.
18. K. Davids, *Saying Coffee: The Naming Revolution*, Article in Roast Magazine Site, 2010, http://www.roastmagazine.com/resources/Articles/Roast_NovDec10_SayingCoffee.pdf.
19. A. Illy and R. Viani, *Espresso Coffee: The Science of Quality*, Elsevier Academic Press, 2005, pp. 191–194.
20. R. J. Clarke and R. Macrae, *Coffee, Volume 2: Technology*, Elsevier Applied Science Publishers Ltd, London and New York, Reprint 2011 of the First edition 1987, 1987, p. 79.
21. A. Farah, Coffee components, in *Coffee: Emerging Health Effects and Disease Prevention*, ed. Yi-F. Chu, John Wiley & Sons, Inc., Published 2012 by Blackwell Publishing Ltd. Coffee, 1st edn, 2012, ch. 2, pp. 21–58.
22. J. P. Holman, *Heat Transfer*, McGraw-Hill Book Company, 10th edn 2009, 2009, ch. 5 and 6.
23. M. S. Michael Sivetz and H. Elliott Foote Ph.D., *Coffee Processing Technology*, The Avi Publishing Company, Inc., 1963, vol. 1, pp. 203–214.
24. R. J. Clarke and R. Macrae, *Coffee, Volume 2: Technology*, Elsevier Applied Science Publishers Ltd, London and New York, Reprint 2011 of the First edition 1987, 1987, pp. 89–96.
25. A. Illy and R. Viani, *Espresso Coffee: The Science of Quality*, Elsevier Academic Press, 2005, p. 186.
26. D. A. McQuarrie, P. A. Rock and E. B. Gallogly, *General Chemistry*, University Science Books, 4th edn, June 1, 2011, p. 652.
27. G. J. Van Wylen and R. E. Sonntag, *Fundamentals of Classical Thermodynamics*, Wiley, 3rd edn, 1985.
28. M. S. Michael Sivetz and H. Elliott Foote Ph.D., *Coffee Processing Technology*, The Avi Publishing Company, Inc., 1963, vol. 1, pp. 222–226.
29. New Jersey State Department of Environmental Protection, New Jersey Administrative Code, Title 7, Chapter 27, PROHIBITION OF AIR POLLUTION.
30. Ordinance on Air Polution Control (OAPC) of 16 December 1985 (Status as 15 July 2010) The Swiss Federal Council, on the basis of articles 12, 13, 16 and 39 of the federal act of 7 October 1983 on the protection of the environment.
31. Federal Ministry for Environment, Nature Conservation and Nuclear Safety First General Administrative Regulation Pertaining the Federal Immission Control Act (Technical Instructions on Air Quality Control - *TA Luft*) of 24 July 2002 (GMBl. [Gemeinsames Ministerialblatt - Joint Ministerial Gazette] p. 511) (Technische Anleitung zur Reinhaltung der Luft - *TA Luft*).

32. Bay Area Air Quality Managment District, http://www.baaqmd.gov/rules-and-compliance/current-rules.
33. L. J. Wang, Theoretical Study of Cyclone Design, PhD Thesis, Office of Graduated Studies of Texas A&M University, 2004.

CHAPTER 9

Post-roasting Processing: Grinding, Packaging and Storage

CARLA RODRIGUES*, FILIPE CORREIA, TIAGO MENDES, JESUS MEDINA AND CLÁUDIA FIGUEIRA

Centro de Inovação Grupo Nabeiro, Alameda dos Oceanos, Condomínio Mar do Oriente, 65, 1.1, 1990-208 Lisboa, Portugal
*E-mail: carla.rodrigues@grupo-nabeiro.pt

9.1 Introduction

The conversion of green coffee beans into a beverage involves a series of main operations such as roasting, grinding, degassing, packaging and extraction. The word "coffee" is a general term that comprises roasted coffee (including decaffeinated coffee) and derived beverages, as well as a wide variety of convenience and semi-manufactured products such as instant coffee and coffee concentrates. Thus, this term is synonymous with coffee products and encompasses many technological processes responsible for the great compositional complexity of the derived products. Nonetheless, in terms of volumes sold, roasted whole and ground coffees remain the main coffee products present in the market.[1] These products come in attractive and convenient physical form and have a very long shelf-life in certain aspects, deriving from the research and innovation in new packaging technologies.

Coffee: Production, Quality and Chemistry
Edited by Adriana Farah

Published by the Royal Society of Chemistry, www.rsc.org

From the industry point of view, as Steiman[2] poses, "we accept that coffee is dynamic and that we are not going to have just one consistent, unchanging, and unwavering taste profile. So rather than try to fight a losing battle for consistency in taste, we try to create something that meets a certain level of expectation in terms of quality. We want layers of flavor and a substantial mouth-feel and a pleasurable texture on our tongue. If we can achieve that, we will accept a little bit of variability in the type of nuance and the type of flavor if it is adhering to these core criteria."

Preserving these desirable sensory attributes essentially depends on the storage conditions of the coffee, and a large part of the coffee that is produced passes through a storage period. Because storage is one of the steps that follows production but precedes the marketing of the coffee beans, storage is considered one of the most important steps in maintaining the quality of the final product, in addition to meeting demand between harvests and ensuring that the producer receives the best market price.[3]

Depending on the storage conditions, the initial characteristics of the coffee change due to physical, chemical and sensory transformations that intensify as the storage period increases and with varying environmental factors. Coffee is very complex, both from its chemical composition point of view as well as in what relates to the series of steps it must go through until it is available for consumption. Per literature, there are hundreds of volatile organic compounds present in roasted coffee[4] and many of them remain present in the brew solution. The volatile compounds' profiles of different coffee blends are influenced by several factors, such as grinding, degassing and subsequent packaging and storage. Considering all the combinations of steps and chemicals, coffee's journey to the cup has many pathways and destinations.[2] That's a lot of different things to understand, scientifically.

Extensive studies have been conducted since the beginning of the 20th century to discover the volatile compounds responsible for coffee aroma and flavor in roasted, ground and brewed coffees.[5,6] During roasting, once the coffee bean is heated, thermal decomposition and chemical changes occur. Carbon dioxide, aldehydes, ketones, ethers, acetic acid, methanol, oils and glycerol, among other compounds, are volatilized from the bean. Different volatile compounds break down at different temperatures, and the flavor and aroma of the coffee bean continues to develop and degrade as roasting progresses.[7] Coffee volatile chemicals vary from very low molecular weight compounds to relatively less volatile compounds. They include furans, pyrroles, pyrazines, pyridines, thiophenes, thiazoles, phenols and oxazoles, which, among others, contribute to the characteristic roasted coffee flavor.[8] Coffee volatile components are particularly important in coffee beverages as they are major constituents of the sensory experience of coffee drinkers, determining their perception of the product and purchasing choices.

While the importance of the human factor (psychology, history, culture, emotions) on the taste experience cannot be understated, ultimately what produces physical stimulation is the combination of these chemicals that constitute flavor, a major factor in purchase decision.[5] In this sense, packaging systems are designed to maintain the benefits of coffee processing after the process is complete, enabling it to travel safely for long distances from the point of origin and still be wholesome at the time of consumption. Packaging also plays an important role in creating a product brand and in communicating with the consumer.

A package design must be carried out considering the issues not only of cost, shelf-life, safety and practicality, but also of environmental sustainability.[9] The Life Cycle Assessment (LCA) methodology has been used in parallel with the design for finding and assessing technical solutions for reducing the impacts due to the different phases of production.[10,11] A sustainable production of goods involves the definition and the design of all their life cycle phases, as the technologies and the materials used for the production may adversely affect the environmental quality of the other phases, such as the use and the end of life.[11] Within the European Community, packaging is directly regulated by the Directive 94/62/EC amended.[12] The Directive aims at providing a high level of environmental protection and ensuring the functioning of the internal market by avoiding obstacles to trade and distortion and restriction of competition. Since 2004, the Directive suffered amendments and revisions to provide criteria clarifying the definition of the term "packaging" and increase the targets for recovery and recycling of packaging waste. Alongside a number of other waste stream Directives, the Packaging and Packaging Waste Directive was subject to review of waste policy and legislation in 2014, covering a review of key targets and related elements and an ex-post evaluation. It is possible to access the document at the European Organization for Packaging and the Environment (EUROPEN) website.[13] In short, quality, shelf-life and packaging sustainability are important aspects influencing coffee new packaging development.

9.2 Grinding

9.2.1 Particle Size

Grinding is the operation that converts whole roasted beans into smaller fragments to increase the specific extraction surface area and thus facilitate the transfer of soluble and emulsifiable substances from the coffee matrix to water during the brew extraction.[14] The control of the grinding conditions is critical as it influences the properties of the ground coffee and respective brew flavor. At the industrial level, after roasting, the whole coffee beans are conditioned to promote cooling and degassing. The grinding process will also promote the degassing of coffee. After grinding, the coffee remains in silos for further degassing prior to packaging (this process may take from 4 to 24 hours). Additionally, the particle size must be controlled to ensure that the desired flavor is achieved during brewing within a certain period. The exponential increase in surface area created by breaking the coffee into smaller particles speeds up the rate of dissolution. Larger coffee particles have less total surface area available for contact and subsequently have a faster and less efficient extraction process. Smaller particles have the opposite effect on the rate at which water flows through coffee grounds and allow a more efficient extraction. Traditionally, the particle size of coffee has been measured using a set of graded-mesh sieves with incrementally smaller-sized openings. Size distribution is reported as the mass of the material retained on a mesh of given size, but may also be reported as the cumulative mass retained on all sieves above a mesh size.[15a] According to the SCAA Brewing Handbook, the ratio of particle size increases over 10 000 times in an espresso grind when compared to the whole bean.[15b] However, this method does not provide an accurate assessment of the amount of particles produced during milling, a limitation in controlling the quality of the final product. Other sources of error with this method may also be related with the coffee powder characteristics which can cause particles to agglomerate. To overcome this difficulty, laser diffraction (dry method) has been used for the characterization of ground coffee (Figure 9.1). The particle size measurement with laser diffraction is faster and provides the full particle size distribution of the samples. Laser diffraction is a non-destructive particle sizing method based on the Mie theory, which describes the scattering of light by particles that constitute a region with refractive index differing from the refractive index of its surroundings. Coffee particles pass through

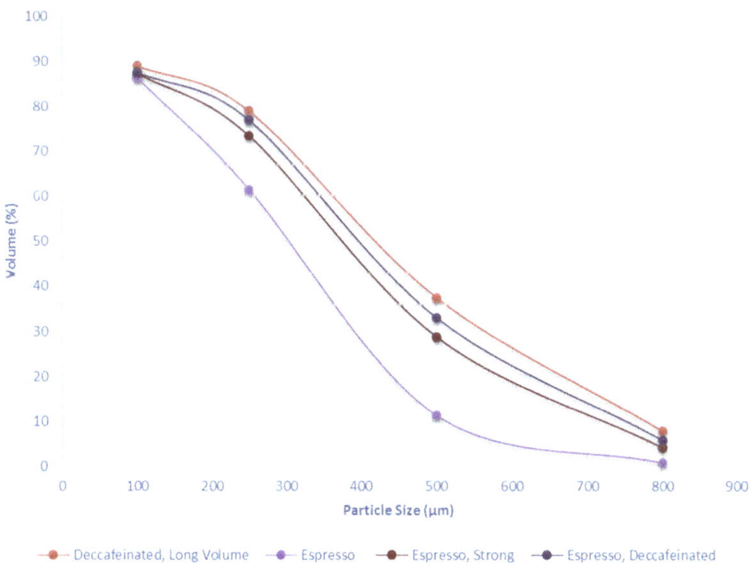

Figure 9.1 Examples of particle size distribution in random commercial blends determined by laser diffraction dry method (data obtained with a Malvern Mastersizer 3000 system, UK).

a beam producing a scattering, which disperses an amount of light. The larger particles scatter the light intensely at a narrow angle in contrast to the smaller particles that scatter more widely at lower intensities. Laser diffraction analyzers detect the scattered light pattern produced by a coffee sample and apply Mie theory to calculate the size distribution of the particle population. The limitation of laser diffraction remains the inability to distinguish between dispersed particles and agglomerates. The general principles of the laser diffraction method for particle size determination are described at NIST Practical Guide for Particle Size Characterization.[15] There is a link between the particle size distribution, the brewing time and the final taste of the coffee drink. Depending on extraction conditions, *e.g.* water temperature and pressure, the results will be different in terms of brew quality. For this reason, the grinding process has to be adjusted to the proper extraction technique for each coffee product. In the case of espresso, two contradictory needs must be satisfied: on one hand a short percolation time is required while, on the other hand, a high concentration of soluble solids (efficient extraction) must be reached. Both requirements can only be attained if a close contact between solid particles and extraction water can be achieved which demands a plurimodal particle size distribution, where finer particles enhance the exposed extraction surface for molecules extraction (chemical need) and the coarser ones allow the water flow (physical need).[14]

9.2.2 Grinding Equipment

In respect to the grinding equipment, different industrial grinders are available in a variety of sizes to be used with any type of roasted coffee, allowing for continuous adjustment depending on the ground coffee production rate, and also allowing a complete range of consistent grinds at capacities ranging from approximately 50 to 5000 kg per hour to be achieved. These grinders may utilize water-cooling technology to maintain a low operating temperature, preserving the sensorial properties of the coffee, although this may increase the moisture level of the coffee, decreasing its shelf-life. This also applies to commercial grinders, an important piece of equipment in an espresso bar. These usually work as gap grinders with the dropping of the beans through a gap between moving cutting tools that may be conical or flat cutters. Most commercial grinders are designed to pre-grind, with a dosing chamber kept full of grounds so that the barista simply needs to pull the lever to dose the required amount of coffee for brewing (Figure 9.2).

This type of system is very fast and convenient, but it has two significant flaws: first, the weight of each dose is affected by how much ground coffee is in the dosing chamber, and that amount constantly varies. Second, the ebb and flow of business causes the grounds to spend a variable amount of time degassing after grinding and before infusion.[16]

Nowadays, automatic grinders allow the amount of ground coffee available in a dosing chamber to be controlled. A quality grinder must produce the

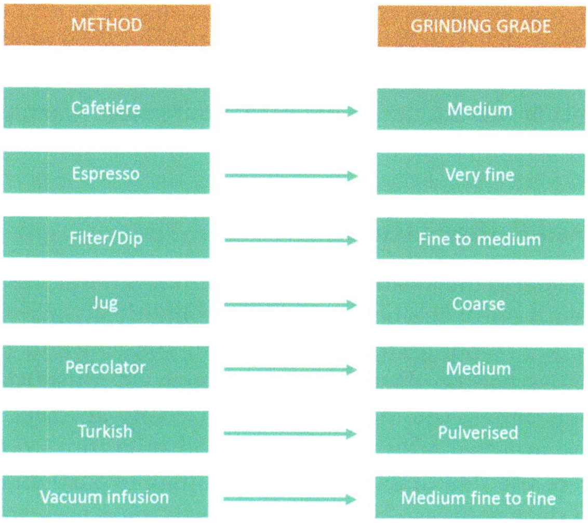

Figure 9.2 Most suitable grinder for several common methods of brewing coffee.

proper particle size to provide bimodal distribution of particle sizes, cause minimal heating of grounds during grinding and limit production of very fine particles, known as fines. The brewing water can transport and deposit fines lower in the coffee bed during percolation. When fines and large insoluble molecules are deposited at the bottom of the coffee bed they can form a compact layer that clogs holes at the bottom of the percolation filter resulting in obstruction of flow paths, uneven resistance to flow and channeling.[16] This will affect the brew quality.

9.2.3 Roasted and Ground Beans Degassing

Once coffee is ground, degassing dramatically accelerates. The formation of volatiles and carbon dioxide (CO_2) during roasting causes the expansion of the beans due to internal buildup of gases, which, along with the high temperatures, allows internal pores and pockets to be formed. The porous structure developed, which is dependent on the roasting temperature–time conditions applied, determines the residual CO_2 content after roasting, as well as the subsequent CO_2 mass transport during storage.[17,18] For this reason, a degassing step is carried out on whole and ground coffee before packaging to avoid the swelling of the packages during storage.[1] Wang *et al.*[19] showed that the amounts of CO_2 retained in roasted coffee beans, at any given roast degree, were independent of the roast temperature when 230 and 250 °C were compared. However, the CO_2 degassing rates for coffee beans roasted at higher temperature were significantly faster than those roasted at lower temperature. Also, as the roasted beans were ground from coarse to fine grinds, 26 to 59% of CO_2 was lost, respectively. As expected, the degassing rates of ground coffee were greater than in the whole beans due to increased particle surface area. The results from this study are useful for optimizing the degassing process and packaging of roasted coffee. Also, the degassing rate is inversely related to time from roasting. The massive degassing that takes place in the early hours after roasting slows down gradually.

9.2.4 Ground Coffee Oxidation

Ground coffee is particularly sensitive to oxidation, while at the same time the contact of oil with air in the bean's surface dramatically increases, promoting oxygen uptake.[20] Oil migration starts during roasting and goes on during degassing because carbon dioxide tends to push oil outwards through the cell pores. Nonetheless, the increase in oil viscosity at lower temperatures slows down the process. After the grinding, oil migration to the surface of the beans, where the risk of oxidation is maximal, is particularly important in fine ground dark espresso blends, since dark roasting leads both to fast degassing and to increased porosity, from disruption of cell walls. A further problem linked with oil exudation is the increase in stickiness of the particles, which tend to aggregate into lumps, making brewing irregular. Additionally, the aggregation of particles on storage is worsened by the absorption of moisture. Shelf-life studies allow to predict at what time of storage period these deterioration processes imply sensorial changes that are reflected in the brew (Section 9.4).

9.3 Packaging

9.3.1 Packaging Materials and Techniques

The packaging process refers to the selection of the packaging materials and techniques, the filling and sealing of the packages and the storage and transportation to the place of consumption. Each step of the process must be monitored so that the quality of the product meets with regulations and the consumers' expectations.

The oldest types of packaging used to store and sell roasted whole beans were simple cardboard bags.[21] With the development of large-scale manufacturing and the increased complexity of the distribution chain, a longer shelf-life for roasted coffee became a demand. The choice of the packaging materials and techniques is crucial for delivering the required shelf-life. The shelf-life of roasted coffee is the result of the interaction between the coffee matrix and the packaging, depending on the environmental conditions inside the package. Roasted coffee exposed to air will lose some flavor through oxidation, the staling effect. Coffee staling reflects the oxidation of many of the pleasant volatiles and the loss of others.[21] The most important physical and chemical events involved in roasted coffee staling during storage are volatiles and carbon dioxide release, surface oil migration, hydrolysis and oxidation reactions.[1,22] Thus, the packaging material must be greaseproof due to the presence of oil on the surface of the coffee. To meet total quality requirements, the package should also act as a barrier against water and moisture and against atmospheric oxygen. Moreover, the package must preserve the coffee aroma, but at the same time allow the carbon dioxide released during degassing to escape. Its material must be chemically inert and suitable for foodstuffs as well as environmentally and consumer friendly.[20] Most of these requirements are aimed at preventing coffee spoilage, others intended to add value to consumers. Furthermore, the headspace volume and the resistance to increases in internal pressure of the package can play a critical role in the selection of the packaging materials and procedures.[1] Commonly used materials for packaging of roasted coffee are tinplate, glass, aluminum and laminated materials such as flexible combined multiply polymers. Nowadays, the most commonly used materials are the inexpensive and easy-to-manage flexible polymer–aluminum multiply laminates, which permit both hard and soft packs. These materials ensure an efficient barrier due to the presence of a central layer of aluminum foil. The other layers are a waterproof film on the inner side and a rigid film that gives mechanical strength on the outer side.[1] A typical construction is polyethylene terephthalate (PET)/alufoil/low density polyethylene (LDPE). Other materials used with comparable performance are metalized PET laminated to LDPE or four-ply structures based on biaxially oriented polypropylene (BOPP) or biaxially oriented nylon (BON) in addition to biaxially oriented polyethylene terephthalate (OPET), alufoil and LDPE.[23] These packages can be fitted with a one-way valve that opens at a preset pressure to release gases but does not allow atmospheric air to penetrate the package. More recently, coffee capsules have been introduced into the market. Designed for specific espresso machines, the capsules can be made of aluminum coated on the inside with a protective film. Before sealing they are saturated with nitrogen (N_2) to improve shelf-life. Alternative capsule systems involve injected plastic, with a cellulose filter on the inside to prevent coffee fines migration to the brew and an aluminum film that increases the gas barrier. Today, many plastic suppliers are presenting innovative solutions to this type of packaging as consumers are demanding a package more ecological. Several coffee roaster companies already sell coffee in bio capsules of 100% renewable materials.

There are different techniques used for packaging roasted coffee such as air packaging, vacuum packaging, inert gas packaging and pressurization as well as the combined use of one of the previous techniques with active packaging. Air packaging consists of simply filling and hermetically sealing the package; coffee is protected against humidity, external off-flavors and light, but the presence of air inside the package means high oxygen levels and consequently shortened shelf-life. Air packaging using a one-way safety valve is an acceptable technique for air-cooled coffee beans since they contain large quantities of gas.[20] However, when using the one-way safety valve, loss of aroma volatile compounds occurs, since it only blocks the air from entering the package. Vacuum packaging allows for air extraction with lowering of oxygen level. The technique, which can also be used with rigid materials such as tinplate, is commonly applied to flexible materials to make the coffee "bricks" sold in supermarkets. In inert gas packaging the air inside the container is replaced by inert gas, either through the compensated vacuum technique or by flushing the inside of the package with inert gas. This process generally uses N_2 or CO_2, which, although not an inert gas, behaves as such in a moisture-free environment and, moreover, is naturally present in roast coffee.[20] The packaging technique must be chosen as a function of the desired shelf-life. O_2 is a prime determinant of shelf-life and there are three main ways of lowering its concentration inside a package. The first method is to apply a high vacuum immediately after filling into the package and then sealing. The second is to flush the roasted and ground coffee and package with an inert gas immediately prior to sealing. The third is to place an O_2-absorbing sachet inside the package, a form of active packaging.[22,24] The emergence of active packaging has required reappraisal of the normal requirement that the package should not interact with the packaged product. For example, the introduction of a new EU Regulation (1934/2004) repealing the earlier relevant EU Directives for food contact materials (89/109/EEC and 80/590/EEC) attempts to reconcile the EU's philosophy that food contact materials should not give rise to chemical reactions that alter the initial composition or organoleptic properties of the food, while recognizing the potential benefits of active packaging technologies to enhance the preservation of packaged food. Active packaging introduced so far represents substantial fine-tuning in the matching of packaging properties to the requirements of the product. Accordingly, it will be seen increasingly in niche markets and in wider applications in which specific problems are inhibiting the marketing of the product.[24]

When used by consumers, packages of coffee will be opened and closed frequently. In such situations, the rate of coffee degradation increases rapidly owing to modification of the conditions inside the package as a result of interaction with air and moisture. The length of time after opening of the package during which coffee maintains acceptable quality is referred to as secondary shelf-life.[1] Anese *et al.*[25] showed that the end of secondary shelf-life may be almost constant, at around 20 days, at water activity (a_w) values below 0.36. At higher a_w values, the secondary shelf-life greatly decreased to about 13 days at an a_w of 0.44. This points out the importance of careful selection of packaging materials and procedures as well as package design per final usage of the coffee product. Moreover, it is important to refer that the packaging's function to protect the contents and facilitate storage and transport is not sufficient from a customer perspective. The packaging also must be informative and easy to use and, hence, desirable for the customers. If packaging developers succeed in integrating customer demands in the development process of new packaging, it would be interesting to know more about what happens to the environmental impact from a life-cycle perspective.[12] In order to improve a quality attribute of the packaging, it is important to consider the technical changes and possible solutions for the mechanical protection of contents, for higher barriers to improve durability of contents, the necessary changes to achieve new textures improving printing ability, as well as changes of size and introduction of new opening and reclosing possibilities and improvement of recycling ability.[12]

9.4 Storage

For most foods and beverages in which quality decreases with time, it follows that there will be a finite length of time before the product becomes unacceptable. This time from production to unacceptability is referred to as *shelf-life*.[22] Roasted coffee is a shelf-stable product. Due to the high temperatures attained in the roasting process and to its low water activity (a_w), no enzymatic and microbial spoilage occurs.

However, during storage, coffee undergoes important chemical and physical changes, which greatly affect quality and acceptability of the brew.[20] The adoption of proper grinding and packaging conditions can greatly slow down the staling reactions that lead to the sweet but unpleasant flavor and aroma of roasted coffee, which reflects the oxidation of many of the pleasant volatiles and the loss of others.[21]

Volatile solubilization, adsorption and release, CO_2 release, oxidation and oil migration are the main physical and chemical changes occurring in roasted coffee during storage. Although many of these changes are considered unavoidable, the rate at which they occur mostly depends on some environmental and processing variables such as oxygen and moisture availability, temperature, exposed surface as well as packaging conditions, as previously stated. Since, during coffee roasting, hydrolytic enzymes are thermally inactivated, moisture and temperature are the main factors that will govern hydrolysis reactions in roasted coffee. When moisture in the storage system is low, entropy decreases in the system, which leads to a decrease in the kinetic energy of the molecules and thus in the rates of all types of reactions. However, when storage temperature is high, entropy increases, accompanied by a raise in the rate of degradation reactions. On the other hand, oxidation reactions are facilitated by the presence of oxygen.[26]

Toci and co-authors[27] confirmed the hypothesis of hydrolysis of triacylglycerols (TAG) and the oxidation of free fatty acids (FFA) during storage of roasted coffee. Both atmosphere and temperature influenced these changes when associated with storage time. The use of inert atmosphere and low temperature contributed to a slower loss of FFA. The authors referred that the changes observed in the ratio between unsaturated and saturated fatty acids from TAG and FFA fractions during coffee storage might potentially be used as a tool to establish the shelf-life for ground roasted coffee. However, sensorial implications of these changes should also be investigated before shelf-life reevaluation.

The shelf-life assessment of a coffee product requires the exact definition of criteria determining the end of the product life. To follow this approach, the questions to be answered, as proposed by Manzocco and Lagazio,[26] are:

(i) how do consumer acceptability and analytical indices evolve during coffee storage? (ii) Which are the analytical indices best correlating with consumer acceptability during coffee storage? (iii) Which is the value these analytical indices should reach for consumers to reject the product? And finally, (iv) which is the mathematical model of shelf-life accounting for consumer acceptability? This integrated approach to coffee shelf-life determination requires sensory evaluation of the product not only by a trained sensory panel but also by a consumer panel and at different times of the shelf-life study.

The results of the sensory evaluation may be complemented with the determination of a series of physico-chemical parameters of the product, *e.g.* relative humidity and volatile compounds' profile at the same control points used for sensory evaluation. It is also common to perform accelerated shelf-life studies. The reason behind the need for accelerated shelf-life studies is that coffee products typically have a shelf-life of at least one year. Evaluating the effect on shelf-life of a change in the coffee products, for example, O_2 level or relative humidity, the process or the packaging, would require shelf-life trials lasting at least the required shelf-life of the product. Companies cannot afford to wait for such long periods to know whether the new product, process or packaging will give an adequate shelf-life, and therefore accelerated shelf-life studies are used.[24] Additionally, each new packaging material has to be tested for (i) overall migration into aqueous food simulants by total immersion (British Standards Institution) (EN 1186-3:2002), (ii) according to the standard test method for oxygen gas transmission rate through plastic film and sheeting using a colorimetric sensor (American Society for Testing) (ASTM D 3985-05 (2010) e1) and (iii) the standard test method for water vapor transmission rate through plastic film and sheeting using a modulated infrared sensor (ASTM F 1249-06 (2011)), all using deionized water as foodstuff model. Other migration tests may be required such as the determination of global migration with aqueous simulants by total immersion, the determination of 1-hexen, irganox 1076 and vinyl acetate migration in aqueous simulants, the determination of total primary aromatic amines migration in aqueous simulants, the determination of metal migration (barium, zinc, copper, cobalt, manganese, iron and lithium) in aqueous simulants and the determination of heavy metal concentration (chromium, lead, cadmium and mercury) in the plastic material. The packaging material must show concentration values for the target compounds below the legal limits (EU regulation 10/2011 of 14.01.2011). The shelf-life studies may be performed at different new product development phases. For example, at the time of the product concept, to give general stability information on the packaging materials. Alternatively, they may be performed at the phase of prototype development before advancing to pilot line testing. However, during trials for line scale-up, accelerated shelf-life studies may be useful to have an earlier estimation of the real shelf-life of the coffee product. The results of the accelerated shelf-life studies are confirmed with the correspondent control at ambient temperature and relative humidity. During the shelf-life study, it is possible to monitor other criteria identified as critical to process, for example, volatile compounds' profile, to evaluate and identify ageing indexes for the roasted coffee on the package.

In short, the storage conditions of the coffee products must be controlled in order to decrease water vapor and O_2 transfer rates to the inside of the package, the effect of high temperatures and possible physical and chemical damages to the package. These are important aspects to consider in logistics with current larger-scale manufacturing and complex logistical networks due to the needs for faster delivery and seeking new markets.[28]

9.5 Conclusions

This chapter discusses general aspects concerning the main technological steps of grinding, packaging and storage of roasted coffee. Coffee products packaging demands a multidisciplinary approach addressing its main aspects of design, materials and techniques selection, coffee product characteristics, its final use and consumer needs. Stakeholder management, sustainability, supply-chain management and sensory evaluation are also important aspects to consider for coffee packaging.

Nonetheless, in the end, the goal is to offer the consumer the best coffee product that meets the expectations in terms of quality and fine taste in a package that communicates the company brand. Nowadays, packaging materials that are more ecological and more sustainable are being applied in different fields of the food industry. Several brands already use, for example, recyclable coffee capsules. We believe that in the future more sustainable packaging materials offering similar protection of the coffee products quality will be used, with less impact to the environment.

References

1. M. C. Nicoli, L. Manzocco and S. Calligaris, Packaging and the Shelf Life of Coffee, in *Food Packaging and Shelf Life, A Practical Guide*, ed. G. L. Robertson, CRC Press, Boca Raton, 2010.
2. S. Steiman, What is specialty coffee?, in *Coffee, A Comprehensive Guide to the Bean, the Beverage, and the Industry*, ed. R. W. Thurston, J. Morris and S. Steiman, Rowman & Littlefield, New York, 2013.
3. F. M. Borém, F. C. Ribeiro, L. P. Figueiredo, G. S. Giomo, V. A. Fortunato and E. P. Isquierdo, Evaluation of the sensory and color quality of coffee beans stored in the hermetic packaging, *J. Stored Prod. Res.*, 2013, **52**, 1–6.
4. C. Yeretzian, E. C. Pascual and B. A. Goodman, Effect of roasting conditions and grinding on free radical contents of coffee beans stored in air, *Food Chem.*, 2012, **131**, 811–816.

5. N. Bhumiratana, K. Adhikari and E. Chambers, Evolution of sensory aroma attributes from coffee beans to brewed coffee, *LWT--Food Sci. Technol.*, 2011, **44**, 2185–2192.
6. A. N. Gloess, A. Vietri, F. Wieland, S. Smrke, B. Schönbächler, J. A. S. López, S. Petrozzi, S. Bongers, T. Koziorowski and C. Yeretzian, Evidence of different flavor formation dynamics by roasting coffee from different origins: On-line analysis with PTR-ToF-MS, *Int. J. Mass Spectrom.*, 2014, **365–366**, 324–337.
7. Yeretzian, *et al.*, Progress on Coffee Roasting: A process control tool for a consistent roast degree-roast after roast, *New Food*, 2014, **15**(3), 22–26.
8. T. Shibamoto, Volatile Chemicals from Thermal Degradation of Less Volatile Coffee Components, in *Coffee in Health and Disease Prevention*, Elsevier, London, 2015.
9. L. Zampori and G. Dotelli, Design of a sustainable packaging in the food sector by applying LCA, *Int. J. Life Sci. Assess.*, 2014, **19**, 206–217.
10. M. De Monte, E. Padoano and D. Pozzetto, Alternative coffee packaging: an analysis from a cycle point of view, *J. Food Eng.*, 2005, **66**, 405–411.
11. V. Siracusa, C. Ingrao, A. Lo Giudice, C. Mbohwa and M. Dalla Rosa, Environmental assessment of a multilayer polymer bag for food packaging and preservation: Na LCA approach, *Food Res. Int.*, 2014, **62**, 151–161.
12. H. Williams, F. Wikström and M. Löfgren, A life cycle perspective on environmental effects of customer focused packaging development, *J. Cleaner Prod.*, 2008, **16**, 853–859.
13. EUROPEN, http://www.europen-packaging.eu/news/news/77-joint-statement-by-packaging-value-chain-industries-on-the-eu-waste-package.html, Retrived at 14.08.2015.
14. M. Petracco, *Grinding in Espresso Coffee, The Science of Quality*, ed. A. Illy, R. Viani, Elsevier Academic Press, Amsterdam, 2005.
15. (a) A. Jillavenkatesa, S. J. Dapkunas and L.-S. H. Lum, *Particle Size Characterization. Practice Guide*. NIST Special Publication 960-1, 2001; (b) *SCAA Brewing Handbook*, https://www.coffeechemistry.com/quality/grinding/grinding-fundamentals.
16. S. Rao, *The Professional Barista's Handbook. An Expert's Guide to Preparing Espresso, Coffee, and Tea*, 2008.
17. R. J. Clarke, O. G. Vitzthum, *Coffee, Recent Developments*, Blackwell Science, Oxford, 2001.
18. R. Geiger, R. Perren, R. Kuenzli and F. Escher, Carbon dioxide evolution and moisture evaporation during roasting of coffee beans, *Food Eng. Phys. Prop.*, 2005, **70**, 124–130.
19. X. Wang and L.-T. Lim, Effect of roasting conditions on carbon dioxide degassing behavior, *Food Res. Int.*, 2014, **61**, 144–151.
20. M. C. Nicoli and O. Savonitti, *Storage and Packaging in Espresso Coffee, The Science of Quality*, ed. A. Illy and R. Viani, Elsevier Academic Press, Amsterdam, 2005.
21. R. A. Buffo and C. Cardelli-Freire, Coffee Flavour: an overview, *Flavour Fragrance J.*, 2004, **19**, 99–104.

22. G. L. Robertson, *Packaging of Beverages in Food Packaging, Principles and Practice*, CRC Press, Boca Raton, 2013.
23. J. Kerry, *Aluminium Foil Packaging in Packaging Technology. Fundamentals, Materials and Processes*, ed. A. Emblem and H. Emblem, Woodland Publishing, Oxford, 2012.
24. A. Scully, Active Packaging, in *The Wiley Encyclopedia of Packaging Technology*, ed. K. L. Yam, Wiley, 2009.
25. M. Anese, L. Manzocco and M. C. Nicoli, Modeling the secondary shelf life of ground roasted coffee, *J. Agric. Food Chem.*, 2006, **54**, 5571–5576.
26. L. Manzocco and C. Lagazio, Coffee brew shelf life modelling by integration of acceptability and quality data, *Food Qual. Prefer.*, 2009, **20**, 24–29.
27. A. T. Toci, V. J. M. F. Neto, A. G. Torres and A. Farah, Changes in triacylglicerols and free fatty acids composition during storage of roasted coffee, *LWT--Food Sci. Technol.*, 2013, **50**, 581–590.
28. L. L. Massey, *Permeability Properties of Plastics and Elastomers: A Guide to Packaging and Barrier Materials. PDL Handbook Series*, Plastic Design Library/William Andrew Publishing, Norwich, New York, 2003.

CHAPTER 10

Beverage Preparation

M. P. DE PEÑA*[a], I. A. LUDWIG[b] AND C. CID[a]

[a]Department of Nutrition, Food Science and Physiology, School of Pharmacy, University of Navarra. IdiSNA, Navarra Institute for Health Research, E-31008, Pamplona, Spain; [b]Department of Food Technology, Universitat de Lleida, E-25198, Lleida, Spain
*E-mail: mpdepena@unav.es

10.1 Introduction

The origin of coffee is involved in several legends and it is still unclear when it became a beverage. However, its ability to enhance alertness, together with its pleasant and evocative aroma and taste, has increased its consumption from the very beginning. Today, a cup of coffee is synonymous with staying awake, but also with many social events. Coffee beverage evokes a relaxing, pleasant conversation after a meal, a family celebration, a nice meeting with friends, or colleagues, or in contrast, it takes part in the work of students in the previous days of the final exams, or in the expectation of a nervous father in the corridor of the delivery room. A cup of coffee can be a pleasant starting point for a business, but also for friendship or love. Independently of the presence of coffee or not, "coffee break" is the name used for breaks in conferences, courses or in any workplaces. Moreover, the increasing number of studies which substantiate the beneficial health effects of up to five cups of coffee per day contributes to promote its consumption. Today, coffee is one of the most consumed beverages in the world, and its consumption increases every year.[1]

Coffee: Production, Quality and Chemistry
Edited by Adriana Farah

Published by the Royal Society of Chemistry, www.rsc.org

This chapter describes the different brewing methods and the requirements on water pressure, grinding grade, extraction time and water quality. Further, physico-chemical characteristics and chemical composition of the most common brewing methods will be discussed in detail.

10.2 Coffee Brewing Methodology

The preparation of a cup of coffee involves a solid–liquid extraction where not only soluble but also some insoluble coffee compounds pass to the coffee brew. According to the methodology used for coffee brewing, the physico-chemical characteristics and chemical composition of coffee brews, and consequently their sensory properties, can substantially vary from one method to another.

According to the methodology used, coffee brewing techniques can be classified in many ways.[2–4] Five types of methods may be distinguished: (1) decoction methods where ground coffee is in direct contact with hot water for a considerable amount of time, such as in boiled Turkish and vacuum coffee; (2) infusion methods, which are very similar to decoction methods, because there is a direct contact between ground coffee and water, but in this case for a short time after which coffee grounds are separated from the brew by a metal strainer or filter such as in plunger coffee; (3) percolation where continuously recirculating boiling water extracts soluble material from ground coffee; (4) filtration methods, in which hot water flows through a bed of ground coffee in a filter paper or metal strainer, allowing to extract soluble coffee compounds, such as in filter or napoletana coffee; and (5) pressure methods where higher pressure than atmospheric is required to separate coffee beverage from grounds (Figure 10.1). Two brewing methods apply pressure during extraction. Mocha coffee is prepared by forcing water heated to boiling point through a bed of ground coffee by slight excess pressure (1 bar), whereas espresso coffee uses water at 92–95 °C, which is pumped at 8–12 bars through a compacted bed of finely ground coffee (cake) for a very short time of just 15–30 s.[2–4] The lines between the different classifications, however, are not clear cut. For example, in a broad sense, the term "filtration" may include not only those methods where ground coffee is in a filter paper or metal strainer, but also many others that apply filtration as the final step after percolation and other types of extraction. Similarly, "percolation" may be used in a general sense, not only when there is recirculation. Plunger coffee, for example, can also be listed under pressure methods since approximately 0.5 bar is applied when the metal strainer is pushed down to separate the coffee beverage from the brew.[2] Similarly, vacuum brewing technique may be also classified as a pressure method.

Although less common, coffee can also be prepared with cold water. The so called cold brew coffee is made by the infusion method where ground coffee is soaked in cold water for a prolonged time of 8–24 hours. The grounds are then removed by pouring the suspension over a paper filter or metal sieve. The use of cold water produces a brew with a different chemical profile from

Figure 10.1 Most common coffeemakers: (a) Plunger coffeemaker before and after the filter is pushed down; (b) filter coffeemaker; (c) mocha coffeemaker; (d) espresso coffeemaker.

conventional brewing methods, which shows a lower acidity and lower caffeine content.[2]

Each coffee brewing method has specific requirements in order to prepare a good cup quality coffee and the most common are described below. Table 10.1 summarizes some of these requirements. However, it should be noted that there are as many ways to prepare a cup of coffee as people who drink coffee.

10.2.1 Boiled Coffee

Boiled coffee is probably the most basic way of preparing a coffee brew. It consists of warming up to boiling point a pot of water with coarse ground coffee in it. The resulting beverage is poured over a strainer to remove floating

Table 10.1 Traditional coffee brewing methods.[2,3,9]

Method	Type	Coffee	Water	Other requirements	Region
Boiled	Decoction	Coarsely ground	Warmed up to boiling point	Pot in stove	Nordic countries (and many others)
Turkish	Decoction	Very fine	Warmed up to boiling point 3 times	IbrikAddition of sugar	Mediterranean countries from Slovenia to Morocco
Vacuum	Decoction/pressure			Cona	
Percolator	Percolation		In continuous reflux	Percolator	Hotels and catering
Filter	Infusion	Medium ground	Boiled water	Paper filter	Worldwide
Napoletana	Infusion	Medium to coarse ground	Boiled water	*Macchinetta napoletana* (flip drip pot)	Italy
Cold brewed coffee	Infusion		Cold or room temperature	Toddy	
Plunger (French press)	Pressure/infusion	Medium to coarse ground	Boiling water	Plunger2–4 min	Worldwide specialties
Mocha	Pressure	Medium to fine ground	Above 100 °C	Mocha coffeemaker	Italy (origin), Spain
Espresso	Pressure	Very fine	Boiling water at high pressure (up to 10 atm)	Espresso coffeemaker	Italy (origin), worldwide

grounds. Boiled coffee was very common in the Nordic countries but has lost popularity over the last decades, partly due to its high content of the diterpenes cafestol and kahweol, which have been implicated in the cholesterol-raising effect of coffee.[5]

10.2.2 Turkish Coffee

Similar to boiled coffee, grounds are heated directly with water in a pot. However, Turkish coffee has some peculiarities that distinguish it from boiled coffee. The first one is the use of very fine coffee grounds, which are unable to float due to their high density.[2] This allows both a strong extraction due to the increased surface and the settling of the grounds at the bottom of the vessel. The second peculiarity is the way of brewing. Turkish coffee is usually prepared in an ibrik, a conical and long-handled pot, traditionally made of copper. Coffee grounds are place in the pot with sugar (optional) and cold water and heated to boiling point. The pot is removed from the heat, and replaced on the heat source for a second and third boiling. This method produces an intense, dark and full-bodied coffee, which, when prepared with the addition of sugar, results in a certainly unique drink.[2,4]

10.2.3 Vacuum Coffee

The brewing of vacuum coffee is characterized by the consecutive use of vapor pressure and vacuum. The apparatus used to prepare vacuum coffee is often referred to by the brand name "Cona", the main producer of this brewing device. It consists of two chambers; the lower one contains the water and the upper one the ground coffee. When the device is heated to boiling point, vapor pressure forces hot water from the lower chamber up through a filter and into the upper chamber, where the extraction process starts. The device is removed from the heat source and extraction continues until steam condenses in the lower chamber creating a vacuum that pulls the beverage down.[2–4]

10.2.4 Plunger Coffee

The plunger coffeemaker, also called French press, consists of a glass cylinder equipped with a plunger that fits tightly in the cylinder and has a fine wire or nylon mesh filter. The ground coffee is placed in the cylinder and infused with boiling water. The coffee/water suspension should be allowed to stand for a few minutes before the plunger with the filter is slowly pushed down to separate the beverage from the grounds. While pushing down the plunger slight pressure of approximately 0.5 bar has to be applied, which depends, besides force exerted on the plunger, also on the grinding grade of the coffee, up to the extreme case of too finely ground coffee hindering the beverage passing through the filter no matter how much force is applied. Plunger coffee is characterized by fine suspended particles and oil droplets, giving a full bodied turbid beverage.[2]

10.2.5 Percolator Coffee

The percolator is a device in which a liquid recirculates through a bed of ground coffee. It consists of a vessel fitted with a vertical tube that leads to the upper part of the device. Just below the upper end of the tube is a perforated chamber where the ground coffee is placed while water is poured into the lower chamber. When heated, water is forced through the tube due to a tiny steam pressure gushing over the coffee grounds in the perforated upper chamber. After passing through the coffee grounds the liquid trickles down to the lower chamber and the process starts again. Due to recirculation almost all soluble material present in coffee is extracted into the beverage and repeated heating causes loss of volatile compounds. The resultant coffee brew can taste harsh and astringent with little aroma.[2,3]

10.2.6 Filter Coffee/Drip Coffee

Filter coffeemakers are simple devices consisting of a container that serves both as extraction chamber and as a means of separating the grounds from the resultant beverage, similar to the percolator but in this method water passes through the coffee grounds just once. Ground coffee is placed in a filter placed in a cone-shaped holder. Then hot water is poured over the coffee, seeps through the bed of grounds and drips from the brewing chamber into a pot placed below it. The brew volume dripping out from the extraction chamber depends on the water amount, and consequently on the water pressure in the extraction chamber of the coffeemaker according to Darcy's law.[6] Therefore, at the beginning of extraction, during wettability, low coffee brew volume is obtained. With time, water fills the extraction chamber inducing turbulence that prevents water from becoming saturated, and increasing the pressure, favoring that water passes through the coffee bed, yielding higher volumes in the middle brewing process.[7] At the end of the brewing procedure, pressure decreases when the water reservoir depletes, until the flow of beverage dripping out of the brewing chamber stops. The resulting coffee brew is a clean and transparent beverage.[2–4,6]

10.2.7 Napoletana Coffee

The macchinetta napoletana, also known as a “flip drip pot”, consists of three parts: the base section filled with water, the middle section perforated on both sides, which contains the coffee and serves as brewing chamber, and an upside-down pot at the top of the device. The macchinetta napoletana is heated until the water in the bottom section reaches boiling point and is then removed from the heat source. By turning the device upside down the hot water flows through the bed of ground coffee and the beverage drips into the pot that at the same time is used to serve the beverage through a spout. The napoletana method is similar to the filter method but the resulting coffee beverage is stronger with a bitter flavor, mostly because the ground coffee undergoes some steam heating during the time spent to heat the water up to boiling point.[2]

10.2.8 Mocha Coffee

The brewing technique to produce mocha coffee resembles that described above for napoletana coffee but with one decisive difference: the lower and upper parts of the brewing device are screwed and sealed by a rubber gasket. This, together with a funnel-shaped extraction chamber containing a bed of ground coffee and fitted between the lower and the upper parts, allows pressure to build up in the lower section. The pot is placed on a heat source and the water is brought to its boiling point. The steam created eventually reaches a high enough pressure to gradually force the surrounding boiling water of approximately 110 °C up the funnel and through the coffee grounds. Mocha coffee is a strong brew characterized by high extraction yield.

10.2.9 Espresso Coffee

Espresso is likely one of the most appreciated coffee beverages, probably because of its very distinct characteristics when compared with coffee brews obtained by other methods. The espresso method is based on pressure-induced extraction of a limited amount of hot water through a compact ground coffee cake in the brewing chamber, where the energy of water pressure is spent within the cake itself.[2] As the name says, espresso is to be freshly prepared and consumed immediately. Extraction times are very short and should not exceed 30 seconds. Typical water temperatures applied during extraction range between 88 °C and 93 °C and optimal pressure ranges between 9 and 10 atmospheres. The combination of heat and pressure extracts soluble flavoring material, emulsifies insoluble oils and suspends both ultra-fine bean fiber particles and gas bubbles.[3] The result is a polyphasic beverage consisting of a foam layer of small bubbles, also called crema, on the top of an emulsion of microscopic oil droplets in an aqueous solution of sugars, acids, protein-like material and caffeine, with dispersed gas bubbles and colloidal solids.[8] Altogether, these characteristics confer espresso coffee its particular sensorial properties which include a strong body, a full fine aroma, a balanced bitter/acid taste and a pleasant lingering aftertaste.

Factors influencing the quality of espresso coffee are much more complex than for other methods and have been published on several occasions[6,9] and will therefore not be discussed here.

During last decade, espresso coffeemakers adapted for capsules have been extended at a domestic level and also in workplaces, restaurants, *etc.* One of the main advantages of these machines is the standardization of the technological parameters, including coffee grinding and the pressure to obtain the coffee cake, which depends on the barista in professional espresso coffeemakers and consequently is one of the most variable factors to have a good quality coffee cup. Other advantages are the diversity of coffee blends adapted to the consumers' likes, and the easy, clean and fast ("espresso") preparation. However, the high price and the high amount of package waste generated can be considered as two disadvantages.

10.3 Coffee Brewing Extraction

Coffee brewing is a solid–liquid extraction process where at the beginning of the process, ground roasted coffee (the solid phase) is placed in contact with water (the solvent), which permeates the matrix and dissolves the soluble compounds. During wettability, as a general rule, 1 g of coffee will absorb 2 mL of water.[3] Once the water has completely surrounded a coffee particle, both inside and out, the coffee extractable material begins to move out of the bean's cellular structure and into the surrounding water inducing a mass transfer to reach an equilibrium state where concentrations in the interior and exterior of the solid would be identical.[4] Then, to obtain the coffee brew, the separation of the liquid from the residual solid is necessary, breaking this equilibrium.

The extraction process differs to some extent when pressure is applied during the coffee brewing. In an espresso coffeemaker water is forced to go through the coffee cake applying a constant pressure and flow, and inducing coffee brewing almost immediately after the starting point. The short extraction time of the espresso method does not allow equilibrium to be reached.[6] However, the high coffee/water ratio, the fine particle size of the ground coffee and the pressure applied during brewing result in a beverage with extremely high soluble concentration (strength), emulsified insoluble oils and suspended ultrafine bean fiber and gas bubbles.[3]

A common way to evaluate the coffee brew quality is the use of coffee brewing control charts. These charts represent a graphic display of the inter-relationship between strength (soluble concentration), extraction (yield) and brewing formula (water/coffee ratio). The aim is to evaluate if the optimum flavor, in which the soluble concentration is in balance with the soluble yield, has been reached in the coffee brew, which means that the most flavorful mixture is present at the most pleasing level of concentration.[3] Strength represents the amount of total dissolved solids in the final coffee brew and is usually expressed as g per 100 g of brew. Extraction is measured as the amount of organic and inorganic matter contained in roasted coffee that will dissolve in water during the brewing process and is usually expressed as g per 100 g of coffee grounds used during the extraction process. Optimal strength is considered to range between 1.15% and 1.35% of flavoring material, while optimum extraction ranges from 18 to 22%. Coffee brewing control charts are very useful to evaluate appropriate grinds for use with specific types of coffee brewing methods.[3] However, it should be noted that some of these types of charts can be expressed in different units such as ounces per pound for extraction or gallons per pound for brewing formula.

10.4 Coffee Brewing Quality

The quality of a cup of coffee depends further on many factors related to coffee, water and the coffeemaker. Coffee species (*Coffea arabica* L. or arabica coffee and *Coffea canephora* Pierre popularly called robusta coffee), variety and origin have a clear influence on chemical composition and quality of coffee brew, but many other factors from harvesting to roasting, grinding and brewing processes can contribute to maintain the high quality of the coffee beans or, on the contrary, to decrease or even ruin it. The optimal combination of grinding grade and brewing method allows exposure of the maximum surface area to the action of water for the obtainment of a high-quality coffee brew. For espresso coffee, for example, when ground coffee is too fine a low volume of a bitter, over-extracted coffee brew is obtained, due to agglomeration and insufficient wetting of particles. On the other hand, when ground coffee particles are too coarse extraction could also decrease, yielding under-extracted coffee due to the fact that the volume specific surface would be too small to retain water and allow coffee compounds solubilization and emulsification. The grinding grade must be adapted to the brewing technique applied. In this sense, medium-coarse grinds are required for boiled, filter and napoletana coffee brews, fine grinds are needed for espresso coffee and extremely fine grinds are required for Turkish coffee. Furthermore, to prepare espresso coffee, a bimodal or plurimodal particle size distribution is needed, with coarse particles fixing a structure that allows the correct flow through the cake and retains finer particles which facilitate the extraction of large amounts of emulsifiable soluble substances.[9,10]

10.5 Water Influence in Coffee Brewing

Water is the most abundant component of any coffee brew (>95%), but sometimes not much attention is paid to it. Water must be free from any unpleasant flavors due to both disinfecting treatments, such as chlorination, and further filtration, usually through activated carbon or resins which became saturated after intensive use. Despite sensory aspects, water hardness is crucial to maintain the proper heat transfer in the coffeemaker, because calcium and magnesium cations produce insoluble salts (mainly carbonates, but also sulfates and silicates) that tend to precipitate as compact plaques on heated surfaces affecting the heat exchange coefficient.[6] Weak acid solutions, like vinegar, can be used to remove these deposits in coffeemakers, including many home espresso machines, but after that they should be properly washed out to avoid off-flavor in the brews to be made. However, for professional espresso coffeemakers, softeners are employed to maintain constant water hardness, usually[9] French degrees, to guarantee a good percolation. Furthermore, water rich in bicarbonate ions used to prepare espresso coffee leads to the formation of a high volume of foam (usually called *crema*), which becomes evanescent due to the presence of undesirable large bubbles.[11] Despite the water filtration and the use of softener devices in professional espresso coffeemakers to maintain water quality, very few people control water or use a controlled mineral water to prepare a coffee brew. In fact, most consumers use tap water which might provide a cup of coffee with different properties from one town to another, even if the same coffee and brewing conditions are used.

Water pH might also affect coffee compounds extraction, including those involved in the formation of foam in espresso coffee, and obviously in coffee brew taste and flavor. Unfortunately, there are only a few studies on the influence of water composition, mainly electrolytes, and less about water pH in coffee brew quality.[12] Typically, during the brewing process water pH drops from around 7.0–7.5 to 4.85–5.15 in arabica coffee brew or 5.25–5.40 in robusta coffee brew.[13] Although depending on coffee origin, roasting, coffee/water ratio and many other factors, the optimum pH of a filter coffee brew is usually 4.9–5.2, becoming sour at pH < 4.9 and bitter and flat at pH > 5.2.[14] In good quality espresso coffee brews normal values of pH are higher and range from 5.2 to 5.9.[2,10,15–17]

10.6 Physico-chemical Characteristics of Coffee Beverages

Both coffee/water ratio and brewing procedure determine total solids and concentration yields in the coffee brew (Table 10.2). The higher the ground coffee dose is, the greater will be the total solids content in the brew,[17] even though this is not a linear correlation. As regards the brewing procedure, the pressure applied influences the extraction of total solids and concentration yields. Filter coffee is an infusion method and plunger and mocha coffeemakers apply pressure at 0.5–1 bar, respectively. Espresso professional machines work at a pressure of up to 15 bar, which allows to extract more total solids.[18] Percentages of extraction ranging from 18 to 22% have been proposed as the most acceptable, while the coffee brews below 16% are considered to be underdeveloped and those above 24% to be overextracted.[3] Nevertheless, extraction yields >24% in espresso coffees prepared with torrefacto blends, due to the solubilization of caramelized sugar and melanoidins, and in mocha coffee brews, did not result in bitter and astringent coffee brews. Therefore, for torrefacto roasted coffees, the range for an acceptable extraction yield could be extended to 25% in espresso coffee and for mocha coffee brews the range should be higher (28–30%).[10] In fact, coffee consumers who like strong coffee brews usually choose espresso or mocha ones.

The absorbance at 420 nm is a convenient index to measure the browned compounds formed during roasting due to caramelization and Maillard reactions, and extracted by coffee brewing. Torrefacto roasted coffee (*i.e.* with sugar addition during roasting) has higher values of this index than conventional coffee because the addition of sugar favors both the development of Maillard reaction products, such as melanoidins, and caramelization. Despite the type of coffee and roasting, the increases of the coffee/water ratio and the pressure employed by each coffeemaker are positively associated with the browned compounds index measured.[18]

The presence of foam is an essential characteristic in a good cup of espresso coffee, but is almost or totally absent in other coffee brews. A fine espresso coffee should have a great amount of persistent, consistent and hazelnut foam with "tiger-skin" effect.[9] Foam is responsible for the

Table 10.2 Physico-chemical characteristics of the most common coffee brews.[10,15,17,18,20]

	Filter	Plunger	Mocha	Espresso
pH	5.4–5.5	5.1–5.3	5.0–5.2	5.1–5.8
Total solids (mg mL^{-1})	16.2–17.7	16.9–20.0	22.9–37.3	27.1–43.3
Extraction (%)	21.7–23.6	21.7–24.3	27.6–30.7	15.5–25.0
Abs 420 nm	0.29–0.47	0.30–0.60	0.65–0.92	0.78–1.67
L	20.75–21.22	21.50–23.16	22.05–23.34	20.76–22.34
*a**	0.29–1.06	0.86–1.07	1.02–2.54	0.74–2.13
*b**	0.43–0.82	0.80–1.47	1.10–3.22	0.63–2.24

visual acceptance of the coffee brew and contributes to its intense aroma because it doses the emission to the atmosphere of the volatilized aromas that are trapped in it. An abundant foam is a freshness marker for ground coffee, which has not yet released all the carbon dioxide (CO_2) formed during roasting.[2] CO_2 and other dissolved gases tend to form a foam layer helped by tensioactive compounds present in the coffee brew. Foamability expressed as foam index (defined as the ratio, in percentage, of espresso coffee foam and liquid volumes measured immediately after the extraction) should be higher than 10% in a good espresso coffee.[9] Foam should remain at least two minutes before breaking and leaving a first uncovered black spot on the surface of the beverage.[2] Coffee foamability is mainly influenced by the melanoidin type subfraction and protein content, whereas foam stability is related to the amount of galactomannan and arabinogalactan present.[19] Furthermore, foam index has been correlated with pH and total solids.[16,19]

Coffee brew density and viscosity are physico-chemical parameters related with body and mouthfeel. The density is slightly higher than that of water (around 1.010), whereas the viscosity is considerably higher (1.14–1.34 mN m^{-2} s) due to both total solids and the presence of lipid droplets in the emulsion, especially in the case of espresso coffee. A higher viscosity in arabica espresso coffee than in robusta has also been reported, probably due to higher amounts of fat.[17] These physico-chemical characteristics increase with grinding degree, coffee/water ratio and water pressure,[10,17,20] but are less affected by temperature.[15]

10.7 Caffeine Extraction

Caffeine is a purine alkaloid (1,3,7-trimethylxanthine) easily extracted with hot water during coffee brewing. Despite the well-known fact that robusta coffee contains on average twice the amount of caffeine in arabica coffee,[14] coffee brewing methods and their corresponding technological parameters influence caffeine extraction, and consequently many variations in caffeine content among coffee cups have been found. The mocha coffeemaker extracts the majority of the caffeine present in coffee leaving undetectable caffeine amounts in the spent coffee grounds, whereas filter, espresso and plunger coffeemakers extract 66–76% of the total caffeine.[21] Other authors reported 75–85% of caffeine extraction for espresso coffeemaker,[2] and higher yields for others up to 100% for filter coffeemaker.[22]

In order to illustrate the wide variations in the caffeine content per serving, Table 10.3 summarizes the caffeine content in coffee cups according to the type of coffee, the brewing methodology and the technological conditions applied by our research group, as well as those found by other authors in coffee brews from coffee shops in recent years.

10.8 Phenolic Compounds and Non-phenolic Acids Extraction

Chlorogenic acids (CGAs) are the main phenolic components present in coffee. They are water soluble esters formed between quinic acid and one or two molecules of *trans*-cinnamic acids, such as caffeic, ferulic and coumaric acids. Caffeoylquinic acids (CQAs), 5-CQA, 4-CQA and 3-CQA, dominate along with lower amounts of feruloylquinic acids, *p*-coumaroylquinic acids, dicaffeoylquinic acids (diCQA) and caffeoyl-feruloylquinic acids. In addition, caffeoylquinic acid lactones (CQLs), also named caffeoylquinides, can also occur in coffee in significant amounts.[23] Many other minor chlorogenic acids have been detected in coffee brews.[23]

Coffee brew is probably the richest dietary source of CGAs.[5,24,25] However, the variety and origin of coffee beans and the degree of roasting, together with the extraction methodology applied and the significant differences in cup size, induce extremely wide variations in the CGAs content in coffee brews (Table 10.3). Moreover, most research studies only measured the amount of 5-CQA that is certainly the most abundant chlorogenic acid, but it can account for only 24% of the total CGAs.[23]

During roasting, CGAs suffer substantial losses (up to 95%) depending on the roasting degree.[23,26] It is generally assumed that robusta coffees have higher amounts of CQAs than arabica ones, but the several origins of coffee and the higher losses of chlorogenic acids in robusta coffee during the roasting process,[26–28] along with many other factors, might explain the higher amount of 5-CQA found in some arabica coffee brews.[7,29] Similar to caffeine, the mocha coffeemaker extracts practically the total of CQAs, the amount of these phenolic compounds in the spent coffee grounds being negligible. When coffee brew is prepared with filter, espresso or plunger coffeemakers, the CQAs are extracted to a great extent during brewing with extraction yields (68–78%) similar to those of caffeine, whereas the majority of diCQAs remain in spent coffee grounds.[21] This is due to the fact that diCQAs or their lactones (diCQLs) are extracted rather slowly from coffee in comparison to CQAs.[7,30] The esterification of an additional caffeic acid molecule in diCQA (and in diCQLs) increases the number of hydroxyl groups that can be bound with melanoidins and other polymeric compounds.[31–33] The high water pressure

Table 10.3 Caffeine and phenolic compounds (mg per serving) in coffee brews according to the type of coffee used and the technological parameters applied.[a]

Coffee brew	Type of coffee	Technological or other conditions	mL per serving	Caffeine content (mg per serving)	Phenolics content (mg per serving)	Reference
Espresso coffee	Arabica (Colombia) and blends	7.5 g 40 mL^{-1}. 9 atm. 96 °C	40^b	84–118	52–60 (5-CQA)	Maeztu *et al.*, 2001[16]
	Arabica (Colombia) and blends	Conventional and torrefacto roasting. Very fine to coarse grinding. 6.5–8.5 g 40 mL^{-1}. 7–11 atm. 88–98 °C	40^b	72–152	32–72 (5-CQA)	Andueza *et al.*, 2002, 2003a, b, 2007[10,15,17,20]
	Blends	Conventional and torrefacto roasting. 7 g 40 mL^{-1}. Domestic espresso coffeemaker (Saeco aroma, Italy)	40^b	25–150	12–27 (5-CQA)	Lopez-Galilea *et al.*, 2007[18]
	Unspecified	Coffee shops in Scotland	23–70	51–322	24–423 (CQAs)	Crozier *et al.*, 2012[24]
	Arabica (Guatemala) and robusta (Vietnam)	Medium roast. 7 g 45 mL^{-1}. Domestic espresso coffeemaker (Saeco aroma, Italy)	45^b	64–114	51–88 (CQAs) 56–94 (CQAs + diCQAs)	Ludwig *et al.*, 2012[7]
	Arabica (Guatemala) and robusta (Vietnam)	Medium roast. 7 g 40 mL^{-1}. Domestic espresso coffeemaker (Saeco aroma, Italy)	40^b	60–132	22–39 (CQAs) 24–42 (CQAs + diCQAs)	Bravo *et al.*, 2012[21]
	Arabica, robusta and blends	Coffee shops in Italy, Spain and Scotland	13–104	54–276	6–188 (CQAs)	Ludwig *et al.*, 2014b[23]
Filter coffee	Blends	Conventional and torrefacto roasting. 4 g 100 mL^{-1}. Domestic filter coffeemaker (KF147 aroma select, Braun, Spain)	100^b	22–110	22–42 (5-CQA)	Lopez-Galilea *et al.*, 2007[18]
	Arabica (Guatemala) and robusta (Vietnam)	6 g 100 mL^{-1}. Domestic filter coffeemaker (Avantis 70 aroma plus, Ufesa, Spain)	100^b	57–115	56–81 (CQAs) 61–90 (CQAs + diCQAs)	Ludwig *et al.*, 2012[7]
	Arabica (Guatemala) and robusta (Vietnam)	4 g 100 mL^{-1}. Domestic filter coffeemaker (Avantis 70 aroma plus, Ufesa, Spain)	100^b	60–132	57–100 (CQAs)	Bravo *et al.*, 2012[21]

Plunger coffee	Blends	Conventional and torrefacto roasting. 8 g 100 mL^{-1}. Plunger coffeemaker 1 L	100[b]	20–136	23–44 (5-CQA)	Lopez-Galilea *et al.*, 2007[18]
	Arabica (Guatemala) and robusta (Vietnam)	8 g 100 mL^{-1}. 98 °C. Plunger coffeemaker 1 L (Bodum, France)	100[b]	95–125	69–148 (CQAs)75–157 (CQAs + diCQAs)	Bravo *et al.*, 2012[21]
Mocha coffee	Blends	Conventional and torrefacto roasting. 8 g 100 mL^{-1}. Mocha coffeemaker (Valira, Spain)	40[b]	11–77	10–22 (5-CQA)	Lopez-Galilea *et al.*, 2007[18]
	Arabica (Guatemala) and robusta (Vietnam)	8 g 100 mL^{-1}. Mocha coffeemaker (bra, Spain)	40[b]	50–72	44–76 (CQAs) 49–82 (CQAs + diCQAs)	Bravo *et al.*, 2012[21]
Espresso and specialty coffees	Arabica and unspecified coffees	Coffee shops USA	30–473	58–259	—	McCusker *et al.*, 2003[62]
Instant coffees	Unspecified (commercial brands)	2 g 125 mL^{-1}.	125	48–88	35–152 (CQAs)	Ludwig *et al.*, 2014b[23]
	Unspecified	UK.		21–120		UK Food Standards Agency
Cappuccino and latte	Unspecified	Coffee shops Scotland	115–310	85–311	19–187 (CQAs)	Ludwig *et al.*, 2014b[23]

[a]CQAs is the sum of 5-CQA, 4-CQA and 3-CQA.
[b]Volume used to estimate the coffee compounds content per serving cup.

applied in espresso coffeemaker favors the extraction process. However, higher extraction values for phenolic compounds have been found in filter coffee brews compared to espresso.[7,34] These findings can be explained by the high coffee/water ratio, the shorter contact time between water and coffee grounds and the limited space in the coffee cake that do not allow equilibrium to be reached.[5] In contrast, longer time and turbulence in the extraction chamber of the filter coffeemaker allow the water in immediate contact with the coffee to extract additional compounds when it has not become so saturated with dissolved material. Both technological factors, contact time and turbulence, might favor the extraction of CQAs and diCQAs, free and bound with melanoidins, in filter coffeemaker.[7]

Despite the content of free CGAs in coffee brew, it should be taken into account that CGAs are also bound to melanoidins (20% in filter coffee brew) contributing to a higher total content than that usually quantitated.[35] This fact increases the value of the coffee brew as a source of these bioactive compounds, which are bioaccessible after their release from food matrices by gastrointestinal enzymatic action or further microbiota activity[36,37] and eventually contribute to health-related properties associated with the consumption of coffee.

Traditionally, both caffeine and chlorogenic acids have been proposed as the main compounds responsible for the bitterness of coffee brew. In fact, CQL and diCQL exhibit a coffee-typical bitter taste profile in coffee brews prepared with slight to medium roasted coffee, and they are degraded to generate harsh bitter-tasting 4-vinylcatechol oligomers when coffee is roasted stronger.[38,39] Moreover, as discussed above diCQLs are extracted rather more slowly than CQAs and CQLs but the 4-vinylcatechol oligomers are strongly retained by ground coffee.[30] This may explain, at least partly, why coffee brew bitterness increases with longer extractions, like in *lungo* espresso coffee or in espresso coffee prepared in northern European countries with higher volumes than in southern ones, and with strong roasting degrees.

Many other phenolic and non-phenolic acids are present in coffee brews in smaller amounts than CGAs. All of them contribute to the sensory acidity of coffee brew, but unfortunately neither pH nor titratable acidity correlate well with the sensory acidity.[13]

Other phenolic compounds have been found in coffee brews in much lower amounts, namely lignans and isoflavones.[40,41] In coffee brew, formononetin (7-hydroxy-4′-methoxyisoflavone) is present in higher amounts, followed by daidzein (4′,7-dihydroxyisoflavone) and genistein (4′,5,7-tridhydroxyisoflavone), in order. Similarly to the chlorogenic acids, robusta coffees have higher amounts of isoflavones than arabica ones, and during roasting substantially decrease. Moreover, the brewing process used to prepare a cup of coffee also induces isoflavones content variations, with the highest concentration in mocha and espresso coffee (*ca.* 550–600 μg 100 mL^{-1}), followed by plunger (*ca.* 220 μg 100 mL^{-1}) and filter (*ca.* 150 μg 100 mL^{-1}) ones.[42]

10.9 Carbohydrates and Melanoidins Extraction

Only soluble carbohydrates like free galactose and mannose that are released during brewing or the manufacture of soluble coffee powder by hydrolysis of polysaccharides are present in sizeable amounts in the coffee brews. Coffee brews also contain non-digestible polysaccharides, mainly galactomannans and type II arabinogalactans that are by definition part of the dietary fiber complex. The dietary fiber content of coffee brews range from 0.14 to 0.65 g per 100 mL depending on the type of coffee, the degree of roasting and grinding and the brewing procedure.[43] However, more recently, some studies concluded that coffee dietary fiber includes melanoidins and also that the content of coffee melanoidins includes a substantial part of dietary fiber.[44]

Melanoidins, which are generically defined as heterogeneous, brown-colored, nitrogen-containing, high molecular weight end products of the Maillard reaction and formed during coffee roasting, account for up to 25% of the total solids of the coffee brew.[45] Coffee brews are considered one of the main sources of melanoidins in the human diet with an intake of coffee melanoidins ranging between 0.5 and 2.0 g per day for moderate and heavy consumers, respectively.[46] Their exact composition is still unknown, but interactions by sugar and polysaccharides degradation products with amino acids, protein and CGAs are indicated, although it is still unclear how these different constituents (or their derivatives) are linked in the melanoidin structures.[32,47–50]

10.10 Lipids (Diterpenes) Extraction

Lipids are present in relevant amounts in roasted coffee, but only limited amounts are extracted during the brewing process, especially in the case of filtered coffees. Due to the use of hot water at high pressure, the espresso coffeemaker extracts lipids as an emulsion of microscopic oil droplets that contribute to the typical aroma, flavor and mouthfeel of espresso coffee. Depending on the type of coffee (arabica is richer than robusta), the grinding grade, coffee/water ratio and water temperature and pressure, the total content of lipids in espresso coffee as reported in most published studies ranges between 3.34 and 6.06 mg mL^{-1},[10,15–17,20] but lower amounts (1.59–2.95 mg mL^{-1}) have also been reported, with extraction yields around 7–9%.[51]

Coffee diterpenes, mainly cafestol and kahweol, are bioactive compounds that have been related with the increase in serum cholesterol[52] in those coffee consumers who usually drink boiled and unfiltered coffee (from 6 to 12 mg per cup). These lipid compounds are in very low amounts in filter coffee (0.6 mg per cup). Although espresso coffee is a filtered coffee, it has higher concentrations than filter coffee with total diterpenes accounting for 2.9–5.9 mg 100 mL^{-1}, and cafestol and kahweol for 1.9–2.4 mg 100 mL^{-1} and 1.7–3.5 mg 100 mL^{-1}, respectively,[51] but serving sizes are usually smaller.

Other minor lipid compounds found in coffee brews are α- and β-tocopherols, which account for a total of 3 μg 100 mL^{-1} for filtered brews, 23 μg 100 mL^{-1} for espresso brews and those prepared in plungers, 27 μg 100 mL^{-1} for boiled, 33 μg 100 mL^{-1} for mocha and 42 μg 100 mL^{-1} for Turkish coffees.[53]

10.11 Volatiles Extraction

One of the most contributory factors for the high acceptability of coffee by population is its aroma that involves more than 1000 volatile compounds. However, not all the volatiles in coffee are odorants, and their contribution to flavor is not usually directly related to their abundance. In coffee brew, around 100 volatile compounds have been identified, depending on the coffee and brewing process conditions, but also on the analytical methodology, with around 30 key odorants. Sulfur compounds are responsible for freshness aroma; aldehydes are related to fruity and malty flavors; some ketones like diones are associated with buttery flavor; and many pyrazines are related with roasty, but also with earthy/musty flavors.[54–58] The change in the aroma profile from the ground roasted coffee to the brew is mainly due to the different concentrations of the aroma compounds and not by the formation of new odorants. The aroma of coffee is unstable, with a rapid loss in the fresh notes. Methanothiol evaporates the fastest, followed by acetaldehyde.

Furan is a highly volatile compound, which has been classified as a possible carcinogenic to humans (group 2B) by the International Agency for Research on Cancer.[59] Coffee consumption is the major contributor to dietary furan exposure for adults. However, furan content in coffee brew ranging from 9 to 262 ng mL^{-1} [60,61] decreases up to 94% after stirring for 5 min[61] due to its high volatility.

Acknowledgements

The support from Spanish Ministry of Economy and Competitiveness (AGL2009-12052), Departamento de Educación, Cultura y Deporte of the Gobierno de Navarra, Association of Friends of the University of Navarra, Unión Tostadora S.A. is gratefully acknowledged. Iziar Ludwig is supported by a postdoctoral fellowship funded by the Spanish Ministry of Economy and Competitiveness (FJCI-2014-20689).

References

1. International Coffee Organization, http://www.ico.org, 2015.
2. M. Petracco, in *Coffee. Recent Developments*, ed. R. J. Clarke and O. G. Vitzhum, Blackwell Science, Oxford, UK, 2001, pp. 140–164.
3. T. R. Lingle, *The Coffee Brewing Handbook. A Systematic Guide to Coffee Preparation*, Specialty Coffee Association of America, Long Beach, CA, 1996.

4. G. Pictet, in *Coffee. Vol. 2. Technology*, ed. R. J. Clarke and R. Macrae, Elsevier, Essex, UK, 1987, pp. 221–256.
5. I. A. Ludwig, M. N. Clifford, M. E. J. Lean, H. Ashihara and A. Crozier, *Food Funct.*, 2014, **5**, 1695–1717.
6. M. Petracco, in *Espresso Coffee: The Science of Quality*, ed. A. Illy and R. Viani, Elsevier, Oxford, UK, 2005, pp. 259–289.
7. I. A. Ludwig, L. Sanchez, B. Caemmerer, L. W. Kroh, M. P. De Peña and C. Cid, *Food Res. Int.*, 2012, **48**, 57–64.
8. M. Petracco, in *Espresso Coffee: The Science of Quality*, ed. A. Illy and R. Viani, Elsevier, Oxford, UK, 2005, pp. 290–315.
9. A. Illy and R. Viani, *Espresso Coffee: The Chemistry of Quality*, Academic Press, London, 1995.
10. S. Andueza, M. P. De Peña and C. Cid, *J. Agric. Food Chem.*, 2003, **51**, 7034–7039.
11. L. Navarini and D. Rivetti, *Food Chem.*, 2010, **122**, 424–428.
12. R. M. Pangborn, *Lebensm.-Wiss. Technol.*, 1982, **15**, 161–168.
13. H. H. Balzer, in *Coffee. Recent Developments*, ed. R. J. Clarke and O. G. Vitzhum, Blackwell Science, Oxford, 2001, pp. 18–32.
14. H. D. Belitz, W. Grosch and P. Schieberle, *Food Chemistry*, Springer, Berlin, Germany, 2004.
15. S. Andueza, L. Maeztu, L. Pascual, C. Ibañez, M. P. de Peña and C. Cid, *J. Sci. Food Agric.*, 2003, **83**, 240–248.
16. L. Maeztu, S. Andueza, C. Ibañez, M. P. de Peña, J. Bello and C. Cid, *J. Agric. Food Chem.*, 2001, **49**, 4743–4747.
17. S. Andueza, M. A. Vila, M. P. de Peña and C. Cid, *J. Sci. Food Agric.*, 2007, **87**, 586–592.
18. I. Lopez-Galilea, M. P. de Peña and C. Cid, *J. Agric. Food Chem.*, 2007, **55**, 6110–6117.
19. F. Nunes, M. Coimbra, A. Duarte and I. Delgadillo, *J. Agric. Food Chem.*, 1997, **45**, 3238–3243.
20. S. Andueza, L. Maeztu, B. Dean, M. P. de Peña, J. Bello and C. Cid, *J. Agric. Food Chem.*, 2002, **50**, 7426–7431.
21. J. Bravo, I. Juaniz, C. Monente, B. Caemmerer, L. W. Kroh, M. P. De Peña and C. Cid, *J. Agric. Food Chem.*, 2012, **60**, 12565–12573.
22. A. Peters, in *Proceedings of the 14th ASIC Colloquium (San Francisco)*, ASIC, Paris, France, 1991, pp. 97–106.
23. I. A. Ludwig, M. P. de Peña, C. Cid and A. Crozier, *Biofactors*, 2013, **39**, 623–632.
24. I. A. Ludwig, P. Mena, L. Calani, C. Cid, D. Del Rio, M. E. J. Lean and A. Crozier, *Food Funct.*, 2014, **5**, 1718–1726.
25. T. W. M. Crozier, A. Stalmach, M. E. J. Lean and A. Crozier, *Food Funct.*, 2012, **3**, 30–33.
26. M. N. Clifford, *J. Sci. Food Agric.*, 1999, **79**, 362–372.
27. M. N. Clifford, *J. Sci. Food Agric.*, 2000, **80**, 1033–1043.
28. D. Perrone, R. Donangelo, C. M. Donangelo and A. Farah, *J. Agric. Food Chem.*, 2010, **58**, 12238–12243.

29. J. A. Vignoli, D. G. Bassoli and M. T. Benassi, *Food Chem.*, 2011, **124**, 863–868.
30. S. Blumberg, O. Frank and T. Hofmann, *J. Agric. Food Chem.*, 2010, **58**, 3720–3728.
31. E. K. Bekedam, H. A. Schols, M. A. J. S. Van Boekel and G. Smit, *J. Agric. Food Chem.*, 2008, **56**, 2055–2063.
32. F. Nunes and M. Coimbra, *Phytochem. Rev.*, 2010, **9**, 171–185.
33. N. G. Kroll, H. M. Rawel and S. Rohn, *Food Sci. Technol. Res.*, 2003, **9**, 205–218.
34. M. Perez-Martinez, B. Caemmerer, M. P. de Peña, C. Cid and L. Kroh, *J. Agric. Food Chem.*, 2010, **58**, 2958–2965.
35. C. Monente, I. A. Ludwig, A. Irigoyen, M. P. De Peña and C. Cid, *J. Agric. Food Chem.*, 2015, **63**, 4327–4334.
36. C. Manach, A. Scalbert, C. Morand, C. Remesy and L. Jimenez, *Am. J. Clin. Nutr.*, 2004, **79**, 727–747.
37. M. Andreasen, P. Kroon, G. Williamson and M. Garcia-Conesa, *Free Radical Biol. Med.*, 2001, **31**, 304–314.
38. O. Frank, S. Blumberg, C. Kunert, G. Zehentbauer and T. Hofmann, *J. Agric. Food Chem.*, 2007, **55**, 1945–1954.
39. O. Frank, G. Zehentbauer and T. Hofmann, *Eur. Food Res. Technol.*, 2006, **222**, 492–508.
40. L. U. Thompson, B. A. Boucher, Z. Liu, M. Cotterchio and N. Kreiger, *Nutr. Cancer*, 2006, **54**, 184–201.
41. W. Mazur, K. Wahala, S. Rasku, A. Salakka, T. Hase and H. Adlercreutz, *Br. J. Nutr.*, 1998, **79**, 37–45.
42. R. C. Alves, I. M. C. Almeida, S. Casal and M. B. P. P. Oliveira, *J. Agric. Food Chem.*, 2010, **58**, 3002–3007.
43. D. Gniechwitz, B. Brueckel, N. Reichardt, M. Blaut, H. Steinhart and M. Bunzel, *J. Agric. Food Chem.*, 2007, **55**, 11027–11034.
44. J. M. Silvan, F. J. Morales and F. Saura-Calixto, *J. Agric. Food Chem.*, 2010, **58**, 12244–12249.
45. R. C. Borrelli, A. Visconti, C. Mennella, M. Anese and V. Fogliano, *J. Agric. Food Chem.*, 2002, **50**, 6527–6533.
46. V. Fogliano and F. J. Morales, *Food Funct.*, 2011, **2**, 117–123.
47. D. Perrone, A. Farah and C. M. Donangelo, *J. Agric. Food Chem.*, 2012, **60**, 4265–4275.
48. M. Daglia, A. Papetti, C. Aceti, B. Sordelli, C. Gregotti and G. Gazzani, *J. Agric. Food Chem.*, 2008, **56**, 11653–11660.
49. A. S. P. Moreira, M. A. Coimbra, F. M. Nunes, C. P. Passos, S. A. O. Santos, A. J. D. Silvestre, A. M. N. Silva, M. Rangel and M. R. M. Domingues, *Food Chem.*, 2015, **185**, 135–144.
50. F. Nunes and M. Coimbra, *J. Agric. Food Chem.*, 2007, **55**, 3967–3977.
51. M. Moeenfard, J. A. Silva, N. Borges, A. Santos and A. Alves, *Eur. Food Res. Technol.*, 2015, **240**, 763–773.
52. R. Urgert, N. Essed, G. vanderWeg, T. KosmeijerSchuil and M. Katan, *Am. J. Clin. Nutr.*, 1997, **65**, 519–524.

53. R. C. Alves, S. Casal and M. B. P. P. Oliveira, *J. Food Compos. Anal.*, 2010, **23**, 802–808.
54. L. Maeztu, C. Sanz, S. Andueza, M. P. De Peña, J. Bello and C. Cid, *J. Agric. Food Chem.*, 2001, **49**, 5437–5444.
55. C. Sanz, M. Czerny, C. Cid and P. Schieberle, *Eur. Food Res. Technol.*, 2002, **214**, 299–302.
56. I. Lopez-Galilea, N. Fournier, C. Cid and E. Guichard, *J. Agric. Food Chem.*, 2006, **54**, 8560–8566.
57. P. Semmelroch and W. Grosch, *LWT--Food Sci. Technol.*, 1995, **28**, 310–313.
58. P. Semmelroch and W. Grosch, *J. Agric. Food Chem.*, 1996, **44**, 537–543.
59. IARC (International Agency for Research on Cancer), *IARC Monographs on the Evaluation of Carcinogenic Risks to Humans*, 1995, vol. 63, pp. 393–407.
60. T. Kuballa, S. Stier and N. Strichow, *Dtsch. Lebensm. -Rundsch.*, 2005, **101**, 229–235.
61. M. Mesias and F. J. Morales, *Food Res. Int.*, 2014, **61**, 257–263.
62. R. McCusker, B. Goldberger and E. Cone, *J. Anal. Toxicol.*, 2003, **27**, 520–522.

CHAPTER 11

Instant Coffee Production

DENISLEY G. BASSOLI

Starbucks Coffee Company, 2401 Utah Ave. South, Seattle, WA, USA
*E-mail: dgbassoli@gmail.com

11.1 Introduction

The oldest records on instant coffee (also known as soluble coffee) date from 1771, in the United Kingdom. However, it was only patented in 1890 in New Zealand, citing the patented "Dry Hot-Air" process, and sold locally named after its inventor, David Strang.[1]

By the end of the 19th century in England the use of liquid or concentrate extracts obtained by batch extraction of ground roasted coffee followed by addition of sugar and vacuum evaporation was common. The first North American similar product was developed around 1853, and L. D. Gale was granted a patent in 1865 to obtain an extract from ground roasted coffee, mixed with sugar. This invention was first shown publicly during the Pan American Exposition of 1901 although about 10 to 20 years before a small-scale trade of dried instant coffee already existed. Only in 1903 an American patent on the process of instant coffee powder production was filed by Satori Kato from Chicago. In 1906, George Constant Washington created "Red E Coffee", the first instant coffee produced on a large scale, introduced to the American market in 1909. During World War I (1914–1918), following the use of ground roasted coffee (adopted as a replacement for rum in wars after 1832) supplied to the troops during the American civil war (1861–1865), instant coffee was regarded as one of the most important articles of

Coffee: Production, Quality and Chemistry
Edited by Adriana Farah

Published by the Royal Society of Chemistry, www.rsc.org

subsistence, being the entire available production acquired by the army, and peaking at 20 t per day. However, the product was too dark and sticky.[2] In 1930, a Swiss company began to sell instant coffee produced in a battery of percolators and mixed with 50% of corn sugars before drying; they further invented the first freeze-dried instant coffee in 1938 while helping the Brazilian Government solve its problem of green coffee overproduction at the time. During World War II, American soldiers received instant coffee as part of their daily feed ration; the demand was very high (reaching over 13 000 t), promoting the appearance of newcomers to the market. By then, vacuum drying had been replaced by spray drying into towers and around 1950 a North American company began large-scale production with higher yields, obtaining a powdered product of good flow ability without addition of carbohydrates, setting the standard of producing instant coffee exclusively using coffee and water.

In 1966, freeze-dried instant coffee was introduced to the market and a couple of years later instant coffee granules appeared, basically improving the dissolution compared to the powder form and reducing the foam in the cup.[3]

In modern society, being fast and practical is often important; therefore instant coffee consumption has been continuously increasing along with quality improvement. Its production is composed of various unit operations and any thermal treatment affects significantly the product final quality. There are various technological approaches that can be adopted to improve its production and quality; the present chapter will address them.

11.2 Current Uses

Instant coffee is traded in bulk form either as a concentrated liquid extract packed into drums or in its dried forms – powder or granulated, herein packed into cardboard boxes with plastic liners or into super sacks.

It is found in the retail market most commonly in glass or polyethylene (PE)/polyethylene terephthalate (PET) plastic jars, aluminum foiled sachets or tins.

Some consumers drink both roasted and instant coffees, respectively black and under milk/cream, as this last one delivers a rather sweet caramel taste, much appreciated, for instance, in the southern region of Brazil and in the US.

It is used as an ingredient in the manufacturing of dairy beverages, bottled or freshly prepared – very common in Japan and South Korea – and with other dairy products such as ice-creams. It has been available since the 1970s in flavored varieties, with the most recent innovations including instant mixes for latte and mocha beverages and instant iced coffee products with vanilla, mocha and original coffee flavors, mixed with hot milk or boiling water. Some Asian countries commonly consume it as a pre-mix with non-dairy creamer and sugar single doses, known as "coffee mix" or "3 in 1".[4]

11.3 Definition

ISO/FDIS 3509 norm defines green coffee as "raw coffee, dry coffee plant seed", roasted coffee as "coffee obtained from roasting green coffee" and instant coffee as "dry product, water soluble, obtained exclusively from roasted coffee by physical methods using water as the only transport agent non derived from coffee".[5] Thus, instant coffee is the product of the aqueous extraction of roasted and ground coffee beans.

Along with its intrinsic definition, best manufacturing instant coffee practices may apply various international norms, such as acrylamide (ISO/CD 18862:2014), authenticity (ISO 24114:2011), caffeine (ISO 10095:1992), carbohydrates (ISO 11292:1995), density (ISO 8460:1987), moisture (ISO 3726:1983, ISO 20938:2008) and sampling (ISO 6670:2002).

Many of the presently available accreditations for the food industry together with some specific programs for coffee have been adopted by various instant coffee manufacturers such as ISO 9000, ISO 14000, ISO 18000, ISO 22000, BRC, 4Q, UTZ, organic, Fairtrade and Rainforest.

11.4 Production

Worldwide overall production of freeze dried instant coffee is currently estimated to be 200 000 t $year^{-1}$ while the production of powder or granules is in the range of 500 000 to 900 000 t $year^{-1}$.[6] Most industrial plants have the instant coffee production process as automated as possible, with data collection throughout the steps for quality and performance controlling.

Final product controlled attributes further comprise color, mycotoxins such as ochratoxin A, particle size distribution, sediments, carbonized or scorched particles, flow ability, microbiological and sensorial requirements, metal particles, pesticides, *etc.*

The flowchart presented in Figure 11.1 contains the different steps involved in instant coffee production.

Figure 11.1 also highlights possible unit operations throughout the processing where volatile components might be recovered and on the other hand processing points in which supposedly they might be added back. As can be observed, these might happen in parallel. For example, reincorporating aromas into the concentrated coffee extract would proportionally augment those volatiles retention during drying, therefore enhancing the volatiles content in the final product and leading to a superior flavor quality.

11.4.1 Green Coffee

The manufacturing of instant coffee begins with the selection of suitable raw materials, generally variable blends of *C. arabica* and *C. canephora* (robusta or other variety) beans from the various producing regions comprised of the so-called coffee belt to deliver the final desired taste

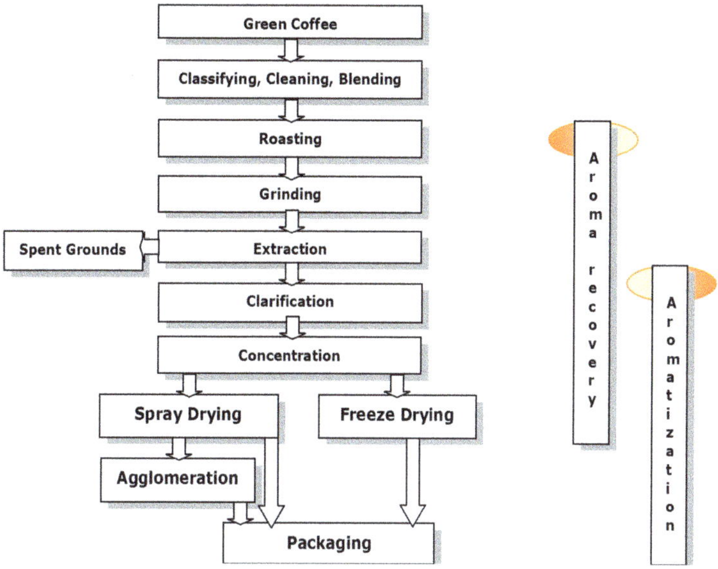

Figure 11.1 Simplified flowchart of the industrial processing of soluble coffee.

attributes. Predominantly, coffees from *C. canephora* species represent the most used beans since they deliver a higher overall yield, as detailed in Section 11.4.4.1.

Grading, storing and blending of the green coffees must observe the best manufacturing practices and controls; de-stoning is relevant to avoid any contamination and eventual broken or faulty equipment. On the sensory side, similarly to roasted coffee, black-green-stinker beans (known as coffee defects) are also crucial to the instant coffee final flavor and must be kept under strict control for better quality.

Typically, green or roasted beans are conveyed *via* bucket elevators, pneumatic transportation and or chain/belt conveyors inside the instant coffee manufacturing plants.

11.4.2 Roasting

Once selected, the beans are roasted, basically following the processes described in Chapter 8 mainly applying similar equipment to normal roasted coffee production. Here again the roasted beans undergo de-stoning and removal of eventual large metal particles (ferrous and non-ferrous), which is important for both product and equipment safety.

Once roasted, the beans might be degassed. Overall, standard roasting conditions lead to a high content of carbon dioxide (CO_2) produced during the roasting operation, mainly linked to the carbohydrates degradation, in addition to other compounds – approximately 10 liters of CO_2 per kilogram of roasted whole beans are produced.[7,8] Unlike in ground roasted coffee production where the CO_2 mainly affects packing, in the case of instant coffee, the CO_2 present will be directly linked to issues like forming thick and resilient foam on the extracts, eventually leading to final powder low density and light particle colors. The CO_2 release from the roasted beans follows a first-order degree reaction, and the right degassing period before extraction will help to avoid gas plugging on the pipelines and reduce foam, preventing relevant flavor losses or changes. The coffee aroma compounds are also released in accordance with a first-order degree reaction, but with a slower velocity when compared to the gases release mentioned above.[9]

Coffee aroma is a highly complex balance of diverse compounds impacted by the presence of water and temperature.[10]

11.4.3 Grinding

Once roasted the beans are ready to move forward into grinding, normally using high capacity roller mills or equivalent coupled or not with particles classifying devices. The desired particle size distribution, generally coarser than the one used to brew a French press coffee (typically coarse), will impact the behavior of the water flow through the coffee bed. For instance during extraction it may favor channeling formation and/or flow plugging as well as excessive pressure drops and presence of fines in the coffee extract, therefore directly linked to a smooth and steady process flow and extraction reaction yields.

11.4.4 Extraction

Industrial roasted coffee extraction is normally performed by percolating water through the ground coffee beans. The water must be potable or treated, in some cases softened – which means reducing the total content of mineral solids present, particularly calcium; eventually demineralized water can be applied.

The water is pumped sequentially through an array of multiple reactor vessels, "washing" the coffee beans, until reaching the total extraction of the soluble solids under the applied conditions; as temperatures at points would allow water ebullition, the reactors are pressurized in a way to always maintain liquid state throughout the process, favoring and optimizing the soluble solids extraction.

The reactors, referred to as extraction columns, normally are in a battery conformation set-up constituted of a series of 3 to 12 interconnected columns, filled up with ground coffee beans. In order to keep the process running continuously, a couple of cells already cleaned, cooled and loaded with fresh degassed and coarser ground coffee are kept waiting ready to undergo percolation through the first-in first-out process.[11]

The percolation battery normally characterizes a semi-continuous or batch counter-current process, where normally the coffee bed remains static

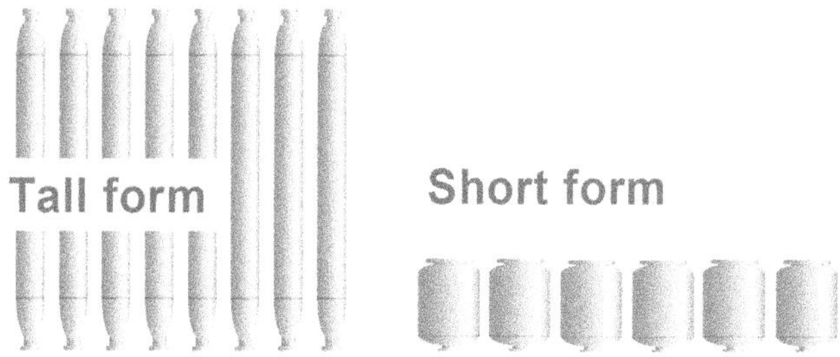

Figure 11.2 Extraction columns formats.

into each column and the water flows through the battery while changing the thermal load applied to each reactor accordingly.

The extraction columns may be arranged in a straight line or in a circular/semicircular way – for a more compact design; they might be more than 6 meters high and hold from a few hundred up to over a thousand kilograms of roasted and coarse ground roasted coffee. Their adopted geometry varies, such as "tall" form – ratio between the column length (L) and its inner diameter (D) up to 12 – or "short" form – L/D ratio might be as low as 1 (Figure 11.2).[11] The last ones will allow finer coffee bed grinds and faster extraction times and thus less heat damage, generally yielding better flavor – with a relatively lower yield as a setback.

On each extremity the columns are equipped with metallic filters or strainers to hold the ground coffee beans inside the column, yet allowing the aqueous coffee extract to flow through the openings.

The coarser ground roasted beans might get moistened before being transferred into the extraction columns, which will help the beans accommodation inside the column and settle down eventual fines and dust released during grinding. They are normally screw or chain conveyed fed into the extraction column from its top; in some installations a slight vacuum is exerted to the column bottom aiming for better coffee bed compaction.

There are various techniques described to extract coffee beans like "one feed water stream" and "two feed water streams", for example. In the "one feed water stream" technique there may be a conventional extraction (with a single draw-off stream) or a split extraction (the freshest initial portion of the extract is weighed separately, followed by a second part of more extracted coffee). The hottest water passes through several "hot" cells – from the hottest to the coldest with reaction temperatures typically between 180 and 140 °C.[12] The "hot" cells contain the most extracted/spent ground beans, which will be discarded sooner. The liquid flow then passes through two or more "cold" cells – inlet temperatures around or below 100 °C for extraction of the more flavorful compounds. The "cold" cells have the freshest or less extracted coffee beans.

In the "two feed water streams" technique, known as double extraction, the first extraction is performed in a set of columns with temperatures typically around 100 °C where the fresh coffee is, with a secondary extraction performed using water at temperatures up to 180 °C reaching the older columns with the coffee previously extracted in the first extraction. As a rule, the hot water enters the set at the most extracted cell and the extract leaves at the less extracted one where the fresh roasted beans are, in a way that improves the overall aroma and taste of the product.

The coffee extract is collected from the coolest end, immediately being thereafter cooled down below 15 °C, and weighed for process controlling purposes.

The extraction process comprises various critical variables, such as: extraction cycle time, number of on-line extraction columns, quantity of extract collected in every cycle named draw-off, temperature of each extraction column and the temperature profile throughout the entire battery. Typical conditions would be for extraction cycle times ranging from 15 to 50 minutes.

The higher the water to coffee ratio used the more acid the extract and the higher the expected yields. The use of various temperature ramps with intercooling or heating might be applied to the extraction set. The combination of the above variables will ultimately define the coffee extract flavor quality and the extraction yield – the content of soluble solids obtained from the roasted and ground coffee and the solids concentration reached in each batch. It is also possible to use continuous counter extraction processes, driven by screw conveyors that move the fresh coffee towards the direction of the hotter temperatures. These are more applicable when lower yields are envisaged.

Soon after extraction, the coffee extract is a dark brown color coffee-flavored liquid with a content of soluble solids ordinarily ranging from 10% up to 20%,[11] depending upon the processing conditions, being naturally slightly acid – typical pH range from 4.7 up to 5.4.[13] Therefore, from this unit operation onwards all the installation in contact with the coffee extract (pipelines, pumps, tanks, equipment, *etc.*) must be of sanitary grade stainless steel, allowing the necessary Clean-in-Place (CIP) systems efficiency as well as fulfilling Hazard Analysis and Critical Control Points (HACCP) requirements. Despite this, the coffee extract is not a good microbiological growth medium; on the contrary it tends to reduce contamination with time (due to its acidity, presence of some compounds with antimicrobial activity like caffeine and chlorogenic acids, for example).[13–15] However, biofilm formation and mold development occur and are the main points of attention.

When the oldest cell in the battery is considered wasted, it is isolated from the process and the coffee spent grounds are discharged, normally by releasing the pressurized column contents and collecting them into cyclones under lower pressure or atmospheric conditions. The emptied column is then ready to re-start into the percolation cycle.

The collected wet spent grounds (moisture around 80%) in general are mechanically expelled to reduce its moisture content to around 50% and then burned as a renewable energy source in order to generate steam, used as fertilizer upon further composting processes or as livestock feeding after composting.[16–18] Due to the high oil content (above 20%–25% in dry basis), the dry spent grounds have been recently considered for the production of biodiesel.[19,20]

11.4.5 Extract Clarification

The aqueous extract normally undergoes a filtering step that may be performed by using either centrifuges or filters, targeting the removal of insoluble pieces of the coffee beans that could have passed through the extraction columns filters and any other insoluble particles formed inside the percolators themselves. The obtained extract is often called clarified extract.

Additionally, a desludging decanter step might be adopted as well in order to recover more of the coffee solids from the discharged sediments obtained from normal centrifuges, maximizing yields by adding them back into the clean extract flow.

11.4.6 Extract Concentration

Considering either a spray or freeze drying process, in general the higher the infeed concentration the higher is the volatile compounds retention and there is less product exposition to thermal damage. Due to this and also considering cost constraints, the clarified coffee extract is treated in one or more of several ways to increase its concentration, targeting from 25% up to 60% of soluble solids concentration.[21] In most cases the extract is thermally evaporated normally under vacuum, using multiple stages falling film concentrators with thermal or mechanical vapor recompression, plate and frame evaporators, compact or centrifugal thermal evaporators.

In freeze concentration, the extract is slowly cooled down to freeze out pure ice crystals that are subsequently mechanically removed from the coffee concentrate. This leads to very high quality extracts, although reached concentrations are around 30% to 35% soluble solids, comparatively smaller than when applying thermal concentration.

Reverse osmosis or membrane filtration systems, crystallizers and infrared techniques might also be applied as alternative techniques to concentrate coffee extracts.[22]

Depending on the concentration and composition, the concentrated coffee extract (or thick extract) may present a high viscosity that eventually may hinder further processing and pumping, thus limiting the operational concentrations to equivalent viscosities around 0.3 Pa s.

Once the concentration is completed, the concentrated coffee extract might be commercialized and applied in vending machines or as an ingredient in beverages.

11.4.7 Aroma Recovery

Among various employed techniques, gas chromatography coupled to mass spectrometry and olfactometry (GC-MS/O) has been applied to roasted coffee with more than 1000 volatile compounds identified, positioning the coffee beverage as a high complexity drink.[23]

Generally speaking, several volatile components present in foodstuffs have little sensorial relevance and recent methodologies applied have made possible to researchers the identification and quantification of a comparatively reduced number of key aroma components, decisive of the odor finally sensed. In coffee, only a small percentage of volatile compounds is responsible for its typical aroma perception.[24,25]

At the end of the extraction and concentration processes, the total concentration of aroma impact compounds in instant coffee tends to be lower than in the original ground roasted coffee.[26] Summarizing, the main contributing factors for this would be: the non-volatile matrix present along with the volatile compounds in roasted and ground coffee, whose interactions would take a relevant role on the volatiles release; the amount of non-volatile soluble material that has been concentrated, whose concentration is around two and a half times higher for instant coffee than in ground roasted coffee; the losses of volatile compounds and further thermal treatments to which instant coffee is subjected during the several steps of the manufacturing process.[27,28]

Water is of utmost importance when extracting coffee, as it significantly influences extraction yield and flavor. Its mineral composition will impact the beverage brightness and, especially when consuming instant coffee with milk, it might significantly affect the color hue based on the ions present (for instance, particular concentrations of iron ions would turn the beverage color towards a different shade of red brown).

On the other hand, there are many opportunities for volatile compounds recovery along with the instant coffee manufacturing process itself. For instance, roasting or grinding gases can be trapped using cold temperatures or cryogenic techniques and be collected; ground, roasted coffee can be heated, steam stripped or solvent extracted or have part of the volatiles extracted using supercritical carbon dioxide; prior to extraction, aromatic coffee oil might be expelled or extracted with alcoholic solutions or liquid carbon dioxide; volatiles can be recovered from coffee extract during or after extraction, or collected from the water removed during concentration and then properly condensed.[29]

By selecting or combining the above-mentioned volatile fractions that eventually might be further processed for enrichment, these volatile compounds might be added back at a later step to try to maintain the aroma as close to the original as possible and produce an attractive and special instant coffee product – for example, these might be added back to the concentrate extract just before drying or perhaps onto the final product or added back directly in the final packaging. Special care might be observed for aromatized instant coffees such as protection from light and control of oxygen levels present in the inner atmosphere, preventing degradation and oxidation reactions.

11.4.8 Drying

Just before entering the so-called dry area, the concentrated coffee extract might be filtered, eliminating virtually all impurities present on the product while in the liquid state that may even come from the process itself, like metallic particles.

SPRAY or FREEZE drying are the two basic unit operations normally used for removing the final portion of water from the concentrated coffee extract until a powder around less than 5% moisture is achieved. Although contact times in spray drying are typically under one minute, it is done at higher temperatures which affect the taste of the final product, imparting caramel notes to it. It is less costly than freeze drying though, which uses high vacuum and longer residence times (up to 5 hours). Freeze drying would result in a higher quality product, under the same comparative basis.

In both processes metal detection and removal are important parts of the HACCP practices. Commercial instant coffee would have a compacted density from 220 to 240 kg m^{-3}, conforming to standard jars volumes worldwide.

Particularly applicable to freeze dried or agglomerated instant coffee, a common practice is to apply a fine atomized layer of roasted coffee oil onto the final dried product at a final content up to 0.3%; this will reduce the dust content allowing cleaner inner walls when the package is a transparent one with better product appearance.

The glass transition temperature, which can be defined as the temperature at which an amorphous system changes from the glassy to a rubbery state, is very important to be taken into account on both processes. Typically, dried instant coffee tends to start melting around 60 to 70 °C – also known as sticky temperature – and might adhere to surfaces and begin to create lumps.[30]

If instant coffee is exposed for longer periods to temperatures around those, it might begin such reactions – re-crystallization of sugars is relevant here as well. As the reactions are exothermal they will be self-sustaining an ongoing process for several days, even reaching product temperatures above 45 °C, forming big chunks of product on the already packed instant coffee.

Therefore, it is of utmost importance to cool down the instant coffee as soon as it leaves the drying chambers to product temperatures lower than 30 °C by adding dehumidified cold air.[11]

11.4.9 Spray Drying

Cooled, clarified liquid concentrate coffee extract is now ready for drying until reaching a moisture content under 5%. The liquid extract is atomized into small droplets to allow better heat and mass transfer, being sprayed through a nozzle or rotary atomization device at the top of a drying tower

Figure 11.3 Pressure atomization.

of various shapes and where inlet temperatures, extract concentration and extract composition allied to the type of nozzle and operating conditions will define the particle size distribution, density and color.

Various models of pressure nozzle atomization (Figure 11.3) can be used, each having its own advantages and disadvantages, as well as rotary atomizers (Figure 11.4) that would lead to higher processing volumes but requiring a wider chamber avoiding the heavier droplets to reach the drying chamber walls.

As can be seen in Figure 11.5, filtered heated inlet air up to 300 °C is typically blown downwards co-currently along with the mist of coffee extract atomized particles to evaporate the water, although counter current drying might also be practiced. The air is diverted out of the tower near the bottom at around 110 °C, together with the evaporated moisture, having its finer particles and dust removed into a cyclone or bag filter. The dry soluble matter that is collected at the bottom of the tower constitutes instant coffee powder – typically spherical brownish particles about 300 μm mean size.[11,31] The beads conformation will depend on the combination of variables like: inlet air temperature, turbulent or steady flow inside the drying chamber, total residence time inside drier, type of nozzle (orifice, chamber, design), nozzle positioning, number of nozzles used, atomization pressure, extract concentration and extract intrinsic composition (linked to surface tension and viscosity), extract inlet temperature, *etc.* This will affect the drying, yielding thicker or thinner crusts, porous or not, round or shriveled or exploded particles that ultimately will modify the powder density, color, particle size distribution and fines content as well powder flow ability.[32,33]

Spray drying may be followed by a further process step to build the powder into coarser particles that will dissolve more completely and readily in

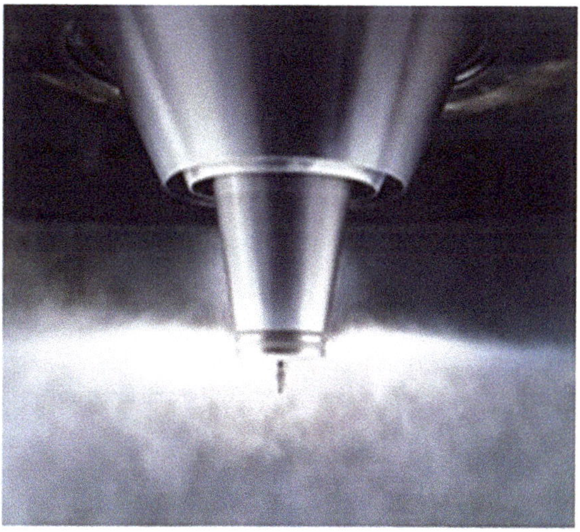

Figure 11.4 Rotary atomization.

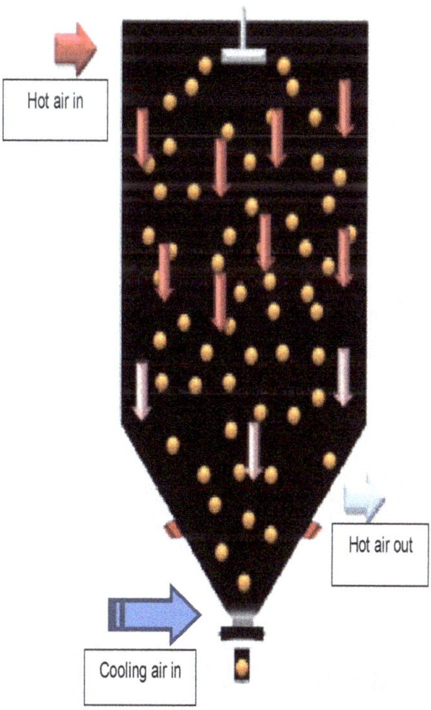

Figure 11.5 Co-current spray drying.

the consumer's cup. The agglomeration process basically takes further finer ground powder beads together with steam and heat-fuses them forming larger and more granular particles. This is accomplished by exposing the powder to steam or a fine mist, while tumbling it in the air, generally industrially performed either in a tower type equipment or horizontal tunnels called agglomeration units.

11.4.10 Freeze Drying

The freeze drying process involves four steps, beginning with chilling and "foaming" the concentrated coffee extract into a slushy, sorbet-like consistent fluid at about −6 °C, variable according to the extract composition and viscosity.

Then it is spread onto a steel belt or drum or placed into trays and further deep "freezing" until it reaches temperatures of −40 to −45 °C, now a solid block or slab. Quick initial cooling processes result in smaller, lighter colored products, while slower processes would generate larger, darker granules; this is directly linked to the ice crystals growth and melting equilibrium.

The thus formed ice chunks or slabs are hammered broken and "sieved" classified into particles of the proper size for the drying step – this is where the product final shape comes from, like brittle crushed caramelized sugar; excessive fine particles might be re-melted and reworked.

The frozen particles, often placed into special aluminum trays, are sent into a "drying" tunnel (Figure 11.6) – either continuous or batch type, where under high vacuum – typically 6600 Pa – they are submitted to sequential

Figure 11.6 Batch freeze drier.

heating zones defining a specific temperature profile for faster or more gentle drying. The water sublimates out, being condensed and removed, with the total drying process generally taking around five hours. After leaving the tunnel, the final product is then cooled down, sieved and finally packed. Control of metallic particles is carried out, since the product might get contaminated during the handling into trays for instance with aluminum, mostly used due to its heating conductivity properties.

Other instant coffee drying processes might be considered as well, such as vacuum drum drying, flash drying, rotary and fluid bed driers as well as infrared drying.

11.5 Packaging

Due to its intrinsic composition and the dried particles conformation (diameter, thickness, porosity, *etc.*), instant coffee might be very hygroscopic, quickly absorbing moisture from the air and eventually lumping and coalescing becoming liquefied. Consequently, it must be packaged under low humidity conditions in a controllable moisture container to keep the product dry until purchased and opened by the consumer. Also, to prevent loss of aroma and flavor and allow suitable product shelf-life, the inner atmosphere might be modified by either a reduced oxygen level usually achieved by flushing the headspace with carbon dioxide or nitrogen gaseous streams or applying partial vacuum.

11.6 Decaffeination

Commercial scale decaffeination of instant coffee almost always happens at the green coffee bean level, before the critical roasting process – which will determine the coffee's flavor and aroma – takes place, similarly to roasted and ground decaffeinated coffee production. Therefore, in this case the processing follows the normal conditions described in this book. Caffeine might also be removed directly from the coffee extract itself, widening the process flexibility.

Whichever is the decaffeination process followed, the overall consumption is not high and typically the same instant coffee processing facility would produce normal and decaffeinated instant coffees; thus, monitoring the caffeine content is very important to ensure in process segregation of the decaffeinated instant coffee production batches from the "non-decaffeinated" flows. Decaffeination processes are described in Chapter 7 of this book.

11.7 Trends

Instant coffee was brought about because of both the need of the coffee industry to expand the consumption itself and, on the other hand, the consumers' desire for such a hot beverage to be readily available. It is very convenient and practical to prepare, and can be considered as the opening door to new and emerging markets, such as China, where the traditional beverage is tea that presents a milder taste profile, more aligned with that of instant coffee.

In recent years, overall coffee consumption is growing in emerging countries as well as in coffee-producing countries and one in three cups of coffee worldwide is being made out of instant coffee. Considering instant coffee process yields, this is equivalent to approximately 20% or more of the world's green coffee beans production.[34]

The development of new technologies, such as aroma recovery, has facilitated the creation of new types of instant coffees with more coffee flavor, reaching in some cases quality levels comparable to brewed coffees; gourmet instant coffees such as Carte Noire® or VIA® product line – launched in 2009 – have revolutionized the instant coffee scenario with prospects of increased instant coffee consumption already evident in emerging countries and Asia. Overall coffee consumption is forecast by experts to increase up to 14 million bags of 60 kg in the next 10 years, and at least half of such an increase will come from instant coffee, with an expected 4% growth for instant coffee until 2017.[34,35]

References

1. K. Cheang, *Caffeine Time*, Published by Schiffer Publishing Ltd, 2014, p. 57.
2. K. Mason, The birth of the coffee nation: coffee in U.S. history since the revolutionary war, in *The Coffee Report: Can Fair Trade Solve the Coffee Crisis?* Carnegie Mellon University, 2005, p. 194.
3. M. Sivetz, *Coffee Technology*, AVI Pub. Co, 1979, p. 737.
4. P. Francis, *The Coffee Exporter's Guide*, International Trade Centre, Geneva, 3rd edn, 2011, p. 247.
5. International Organization for Standardization, *ISO/FDIS 3509. Coffee and its Products – Vocabulary*, Geneva, 2003, p. 21.
6. D. Bolton, *Global Solubles*, STiR Tea & Coffee International, 2015, vol. 4, pp. 34–36.
7. X. Wang, *Understanding the Formation of CO_2 and its Degassing Behaviours in Coffee*, PhD thesis in Food Science, University of Guelph, Ontario, 2014, p. 170.
8. P. Pollien, *et al.*, Liquid-air partitioning of volatile compounds in coffee: dynamic measurements using proton-transfer reaction mass spectrometry, *Int. J. Mass Spectrom.*, 2008, **228**, 69.
9. A. Brent, *et al.*, The diffusion kinetics of carbon dioxide in fresh roasted and ground coffee, *J. Food Eng.*, 2003, **59**, 71.
10. P. Semmelroch, *et al.*, Determination of potent odourant in roasted coffee by stable isotope dilution assays, *Flavour Fragrance J.*, 1995, **10**, 1.
11. R. Clarke and R. Macrae, *Coffee: Technology*, Elsevier Applied Science, NY, 2nd edn, 1989, vol. 2, p. 321.

12. R. Clarke and R. Macrae, *Coffee: Chemistry*, Elsevier Applied Science, NY, 2nd edn, 1989, vol. 1, p. 306.
13. A. Almeida, *et al.*, Influence of natural coffee compounds, coffee extracts and increased levels of caffeine on the inhibition of *Streptococcus mutans*, *Food Res. Int.*, 2012, **49**, 459.
14. A. Nonthakaew, *et al.*, Caffeine in foods and its antimicrobial activity, *Int. Food Res. J.*, 2015, **22**(1), 9.
15. G. Runti, *et al.*, Arabica coffee extract shows antibacterial activity against *Staphylococcus epidermidis* and *Enterococcus faecalis* and low toxicity towards a human cell line, *Food Sci. Technol.*, 2015, **62**, 108.
16. K. Liu and G. Price, Evaluation of three composting systems for the management of spent coffee grounds, *Bioresour. Technol.*, 2011, **102**, 7966.
17. M. Oliveira, *et al.*, Development of a green material for horticulture, *J. Polym. Eng.*, 2015, 7.
18. H. Didanna, A critical review on feed value of coffee waste for livestock feeding, *World J. Biol. Biol. Sci.*, 2014, **2**(5), 72.
19. R. Lago and S. Freitas, Extracao dos oleos de café verde e da borra com etanol comercial, *EMBRAPA - Comun. Tec.*, 2006, **92**, 2.
20. N. Kondamudi, S. Mohapatra and M. Misra, Spent coffee grounds as a versatile source of green energy, *J. Agric. Food Chem.*, 2008, **56**, 11757.
21. R. Clarke and O. Vitzhum, *Coffee – Recent Developments*, Blackwell Science, London, 2001, p. 257.
22. T. Mchugh and Z. Pan, Innovative infrared food processing, *Food Technol.*, 2015, **2**, 79.
23. I. Fisk, *et al.*, Discrimination of roast and ground coffee aroma, *Flavour*, 2012, **1**(14), 8.
24. J. Baggentoss, *et al.*, Advanced predictive analytical-sensory correlation: towards a better understanding of the perception of coffee flavor, 23rd International Conference on Coffee Science, Bali, 2010, pp. 108–115.
25. D. Bassoli and R. Silva, Key aroma compounds of soluble coffee, 21st International Conference on Coffee Science, Montpellier-France, 2006, pp. 340–348.
26. D. Bassoli, *Aromatic Impact of Soluble Coffee Volatile Components: An Analytical and Sensorial Approach*, DSc thesis in Food Science, Universidade Estadual de Londrina, Londrina, 2007, p. 198.
27. D. Bassoli, *et al.*, Instant coffee with natural aroma by spray-drying, 15th International Conference on Coffee Science, Montpellier, 1993, pp. 712–718.
28. N. Ohtani, *et al.*, Spray-drying instant coffee product at low temperature, 16th International Conference on Coffee Science, Kyoto, 1995, pp. 447–456.
29. A. Oliveira, *et al.*, Identification and recovery of volatiles organic compounds (VOCs) in the coffee-producing wastewater, *J. Water Resour. Prot.*, 2014, **6**, 375.
30. D. Heldmann, Encyclopedia of agricultural, *Food Biol. Eng.*, 2nd edn, October 2010, 388–389.

31. D. Huang, Modelling of Particle Formation During Spray Drying, European Drying Conference - Euro Drying 2011 Palma, Balearic Island, Spain, 26–28 October 2011, p. 3.
32. C. Anandharamakrishnan and I. Padma, *Spray Drying Techniques for Food Ingredient Encapsulation*, John Wiley & Sons, Ltd., 2015, p. 312.
33. D. Walton and C. Mumford, The morphology of spray-dried particles: the effect of process variables upon the morphology of spray-dried particles, *Chem. Eng. Res. Des.*, 1999, 21.
34. J. Ganes-Chase, *The Global Soluble Market through 2016*, J. Ganes Consulting LLC, March 2013, p. 87.
35. R. Colbert, *Coffee 2013: Ready for Take-Off. Overview of Coffee Trends in New Consumer Markets*, International Coffee Organization, 2013, p. 10.

CHAPTER 12

Coffee By-products

M. D. DEL CASTILLO*[a], B. FERNANDEZ-GOMEZ[a], N. MARTINEZ-SAEZ[a], A. IRIONDO-DEHOND[a] AND M. D. MESA[b]

[a]Institute of Food Science Research (UAM–CSIC), Nicolás Cabrera 9, 28049 Madrid, Spain; [b]Institute of Nutrition and Food Technology "José Mataix", University of Granada, Avd. del Conocimiento s/n, 18100 Granada, Spain
*E-mail: mdolores.delcastillo@csic.es

12.1 Introduction

Large amounts of coffee by-products are generated from the industrial processing of coffee cherries to obtain the coffee beverage.[1–4] Coffee is actually a cherry whose structure is shown in Figure 12.1. Coffee cherries are mainly used to prepare the beverage when they are processed. From farm to cup, coffee processing can be briefly summarized in ten key steps: planting, cherry harvesting, processing (wet and dry methods), drying the beans, milling, exporting, tasting, roasting, grinding and brewing (http://www.ncausa.org/). According to the method used to process the coffee beans (wet or dry method), different solid residues such as skin, pulp, husk, mucilage, parchment, silverskin and spent coffee grounds are obtained.

The steps from planting to exporting the beans are mainly carried out in coffee-producing countries like Brazil, Vietnam or Colombia. While most coffee-producing countries are developing countries, coffee-consuming countries are usually developed countries with local roasting industries based on green coffee imports. Therefore, two major classes of coffee by-products can be distinguished: those derived from green coffee production (skin, pulp, husks, mucilage and parchment) in producing countries, and those obtained

Coffee: Production, Quality and Chemistry
Edited by Adriana Farah

Published by the Royal Society of Chemistry, www.rsc.org

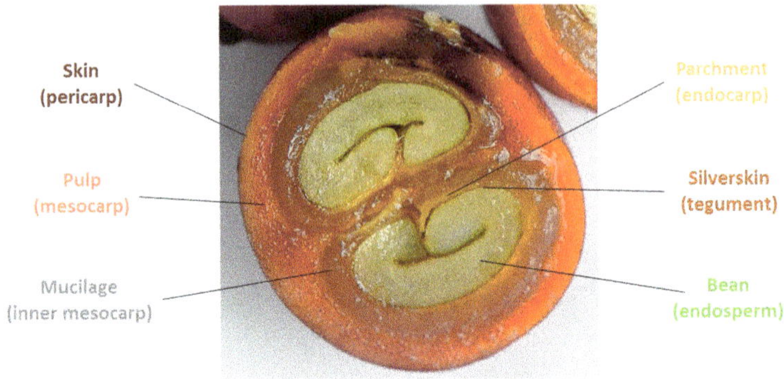

Figure 12.1 Transversal section of a ripe coffee cherry, showing its anatomic parts. Courtesy of Joey Gleason, Marigold Coffee, Portland, Oregon, USA.

after roasting (silverskin and spent coffee grounds) with a wider geographical distribution.

The sustainability of food production and consumption, defined as the exploration of innovative strategies to increase resource efficiency, providing consumers with healthier products of higher quality and safety while ensuring minimal waste in the food chain, is a research priority.[5] The agro-industrial and food sectors produce large quantities of liquid and solid waste. Since coffee is the second most valuable commodity exported by developing countries,[6] the coffee industry is responsible for the generation of large amounts of waste. Consequently, coffee by-products have attracted great attention because of their abundance and interesting chemical composition.

The study of coffee by-products generated during the different stages of processing is necessary to decrease the waste produced by this industry. The recovery of coffee by-products is mainly based on their use as a source of energy and biomass. Although these strategies are of interest, they do not consider valuable nutritional compounds that could improve consumers' health and increase the competitiveness and sustainability of coffee production.[7] Interest in the valorization of agronomical by-products into diverse and useful novel products to achieve a global sustainable world has been recently reported in the "*Food Waste Recovery*" book.[8]

The valorization of agricultural wastes, food processing by-products, wastes and effluents using the biorefinery approach represents the real contribution of many industries to sustainable and competitive development.[9] Biorefineries can be described as integrated biobased industries, which use a variety of technologies to make products such as chemicals, biofuels, food and feed ingredients, biomaterials, fibers, heat and power, aimed at maximizing the added value of the three pillars of sustainability (environment, economy and society).[10]

A brief description of these coffee by-products, their chemical composition and their applications is presented in this chapter.

12.2 Definition of Coffee By-products

The type of by-product generated depends on the process used to obtain the green coffee bean. In the case of wet processing, ripe cherries are depulped to eliminate the outer skin, eliminating most of the pulp fixed to the grains. Then, coffee beans undergo fermentation processes, are washed to remove the rest of the pulp, dried by sun exposure and peeled to remove the parchment. Here, skin and pulp are recovered in one fraction, and soluble sugars and mucilage are generated in another fraction. Finally, the parchment is obtained.[11] Dry processing involves sun drying the coffee cherries for two or three weeks, and green coffee beans are obtained by simply threshing the dried cherries. At this time, skin, pulp, mucilage and parchment are obtained in a single fraction, along with part of the silverskin.[12] The only by-product of coffee roasting is the silverskin.

12.2.1 Pulp

Coffee pulp is a by-product generated from wet coffee processing, and it represents 29% dry weight of the whole bean.[13] Coffee pulp consists of the outer skin or pericarp and most of the mesocarp (Figure 12.1), which is mechanically removed by pressing the coffee fruit in a depulper.[14] One ton of coffee pulp is obtained per two tons of coffee processed.[15]

12.2.2 Mucilage

The coffee mucilage fraction, also called the pectin layer (Figure 12.2), is located between the pulp and the parchment, and represents 5% dry weight of the berries.[16] It remains adhered to the coffee bean after depulping in wet

Figure 12.2 Ripped open coffee cherry, showing coffee pulp and mucilage. Courtesy of Andres Belalcazar, Pectcof B.V. Wageningen University, Netherlands (A). Coffee beans in parchment coated by mucilage. Courtesy of Sweet Maria's Coffee, Inc., West Oakland, California, USA (B).

Figure 12.3 Image of parchment-covered coffee beans, fragmented parchment and green coffee beans.

processing without enzymatic degradation. Since it is highly hydrated, it is an obstacle to further drying the beans. Thus, mucilage must be degraded to facilitate its elimination by washing, before the beans are dried and stored.[17] Wet processing allows the separation and concentration of this fraction.[14]

12.2.3 Parchment

This yellowish by-product is a strong fibrous endocarp (Figure 12.3) that covers both hemispheres of the coffee seed and separates them from each other. It represents 5.8% dry weight of the berries. In wet processing, the parchment is removed after drying and hulling in separate steps.[18] The latter process allows the parchment to be collected and used separately from other by-products.

12.2.4 Husks

Coffee husks are mainly obtained from the dry processing of coffee berries. This coffee by-product is composed of the outer skin, pulp and parchment of the coffee berry.[14] Coffee husks enclose the coffee beans and comprise nearly 45% of the berry.[13] About 0.18 ton of husk are produced from 1 ton of coffee fruits.[19]

Coffee husks are shown in Figure 12.4. Such by-products are generated in coffee-producing countries, which separate the coffee beans from the coffee cherry. Since most of these countries are developing, the diversification of agriculture and the coffee industry is particularly interesting from a socio-economic point of view.

12.2.5 Silverskin

Coffee silverskin (CS) is a thin tegument of the outer layer of the two beans forming the green coffee seed (Figure 12.5) obtained as a by-product of the roasting process.[2] It represents about 4.2% (w/w) of coffee beans. Coffee

Figure 12.4 Image of dried coffee fruits, skin and husks obtained from dry berries.

Figure 12.5 Coffee silverskin, the only by-product obtained during roasting.

silverskin is the only by product produced in the roasting process, and large amounts of CS are produced by large-scale coffee roasters in consuming countries.[20]

12.2.6 Spent Coffee Grounds

Spent coffee grounds (SCG) are the residual material obtained during the treatment of coffee powder with hot water to prepare coffee infusion or steam for instant coffee preparation (Figure 12.6). Almost 50% of worldwide coffee production is processed for soluble coffee preparation, generating around 6 million tons of SCG per year.[2] On average, 1 ton of green coffee generates about 650 kg of SCG, and about 2 kg of wet SCG are obtained for each kilogram of soluble coffee produced.[21]

Figure 12.6 Spent coffee grounds from the instant coffee brewing process.

12.3 Chemical Composition of Coffee By-products

Table 12.1 shows an overview of the work previously performed on the chemical characterization of food by-products. More details are provided in the present section of this chapter.

12.3.1 Pulp

Coffee pulp is mainly composed of carbohydrates (44–50%), proteins (10–12%) and fibers (18–21%), and it also contains appreciable amounts of polyphenols (1.48%) and caffeine (1.3%).[11,22–24]

Four major classes of polyphenols have been described in the fruit pulp of *C. arabica* L. (hence called arabica) beans: *viz.*, flavan-3-ols, hydroxycinnamic acids, flavonols and anthocyanidins.[25] The composition of phenolic compounds in fresh pulp has been analyzed by HPLC, and the obtained profile was chlorogenic acid (5-caffeoylquinic acid, according to IUPAC numbering) (42.2% of total identified phenolic compounds), epicatechin (21.6%), 3,4-dicaffeoylquinic acid (5.7%), 3,5-dicaffeoylquinic acid (19.3%), 4,5-dicaffeoylquinic acid (4.4%), catechin (2.2%), rutin (2.1%), protocatechuic acid (1.6%) and ferulic acid (1.0%).[26] Additionally, 5-feruloylquinic acid has been identified in coffee pulp.[27] The major anthocyanins present in the pulp derived from wet-processed fruits are cyanidin-3-rutinoside, cyanidin-3-glucoside and aglycone.[14] Several proanthocyanidins (condensed tannins) have also been isolated from coffee pulp. Tannins content has been found to increase throughout the drying process, and yellow coffee varieties are richer in condensed tannins than red varieties.[28] Interestingly, no hydrolyzable tannins were obtained in five samples of coffee pulp from different coffee beans.[29]

Table 12.1 Chemical composition and antioxidant capacity of coffee by-products (% w/w dry matter).[a]

Components	Pulp	Husks	Mucilage	Parchment	CS	SCG
Proteins	10–12[11,24]	5–11[23,32]	3.4–8.9[16,18]	–	16–19[4,34]	13–17[4,7,42]
Fats	2.5[11]	–	–	–	2.2–3.8[4,34]	1.6–2.3[4,7]
Carbohydrates	44–50[11,24]	35–85[23,32]	4.1–7.8[16,18]	–	62–65[34,35]	71–75[7,42]
Moisture	12.6[11]	–	84.2[18]	–	2.6–10.3[3,35]	–
Ash	8[11]	3–10[23,32]	0.7[18]	0.5–1[31]	5–7[4,34]	1.3–1.5[4,7,42]
TDF[b]	18–60[11,33]	43[33]	–	–	68–80[3,34]	54–60[4,7]
SDF[c]	18[33]	17–30.8[23,33]	0.91[18]	–	8–14[3,34]	6–16[4,7]
IDF[d]	10[33]	26[33]	–	–	46–80[3,34]	47–50[4,7]
Glucan	–	–	–	–	17.8[36]	–
Xylan	–	–	8.9–11.2[16]	–	4.7[36]	–
Arabian	–	–	14.7–52.5[16]	–	2[36]	1.7[42]
Galactan	–	–	–	–	3.8[36]	13.8[42]
Mannan	–	–	0.8–1.4[16]	–	2.6[36]	21.2[42]
Cellulose	17.7[11]	19–43[31,32]	8.9–9.1[16]	40–49[31]	23.8[4]	8.6[42]
Hemicellulose	2.3[11]	7–45[31,32]	–	25–32[31]	16.7[4]	36.7[42]
Lignin	17.5[11]	9–30[31,32]	–	33–35[31]	28.6–30.2[4,36]	24[4]
Caffeine	1.3[11,24]	–	–	–	0.8–1[3,38]	0.2–0.8[44,45]
Tannins	1.8–8.56[11]	5–9.3[23,32]	–	–	–	–
Flavonoids	0.6[25]	20[12]	–	–	–	–
CQAs	1.2–2.5[11,25,33]	2.5[33]	–	–	0.6–3[33,38]	0.3–1.4[7,44,45]
3-CQA	–	–	–	–	0.15[7,38]	0.063–0.14[7,45]
4-CQA	–	–	–	–	0.08–0.10[7,38]	0.097–0.25[7,45]
5-CQA	–	–	–	–	0.20–0.22[7,38]	0.12–0.34[7,45]
TPC[e]	1–3.79[25,33]	1.2[33]	–	–	0.7–1.7[3,7,38]	1–17[7,44,45]
ABTS[f]	–	–	–	–	19.2–598[7,34]	387[7]
FRAP[f] (μmol TEAC g^{-1})	–	–	–	–	654[7]	387[7]
ORAC[f]	–	–	–	–	–	1821–2594[44]
DPPH (%)	65–69[33]	50–65[33]		–	65–70[33]	61–89[33,44]
Melanoidins	–	–	–	–	–	15[7]
Ochratoxin A	–	–	–	–	<4 ppb[34]	–

[a]Superscript numbers correspond to the cited literature.
[b]TDF: total dietary fiber.
[c]SDF: soluble dietary fiber.
[d]IDF: insoluble dietary fiber.
[e]TPC: total phenolic compounds.
[f]ABTS, FRAP and ORAC units: μmol TEAC g^{-1}.

12.3.2 Mucilage

Mucilage is composed of water (84.2%), protein (8.9%), sugar (4.1%), pectic substances (0.91%) and ash (0.7%).[14] The polysaccharide composition of the alcohol insoluble fraction of mucilage from arabica coffee beans includes pectic substances (30%), cellulose (8%) and neutral non-cellulosic polysaccharides (18%). Crude pectins are formed by uronic acids (60%) with a high degree of methyl esterification (62%) and a moderate degree of acetylation (5%).[17,30]

12.3.3 Parchment

Coffee parchment is formed by (α-) cellulose (40–49%), hemicellulose (25–32%), lignin (33–35%) and ash (0.5–1%).[31]

12.3.4 Husks

As mentioned above, coffee husks, which are comprised of the outer skin, pulp and parchment, are the main residues obtained in the dry processing of the coffee berry. They have a high content of carbohydrates (35–85%), soluble fibers (30.8%), minerals (3–11%) and proteins (5–11%).[23,32] Coffee husks are also rich in insoluble dietary fiber, containing 24.5% cellulose, 29.7% hemicelluloses and 23.7% lignin.[31] They can also be a source of phytochemicals such as tannins (5–9%) and cyanidins (20%) for the food and pharmaceutical industries.[12,23,32] The amount of total polyphenols in coffee husks is 1.22%.[33]

12.3.5 Silverskin

The chemical composition of CS is currently being analyzed by different research groups. Coffee silverskin has a high dietary fiber content (68–80%), which includes about 85% insoluble dietary fiber and 15% soluble dietary fiber.[3,34] Polysaccharides are also abundant components (60–70%) in CS.[4,7,34,35] Sugars are polymerized into cellulose and hemicellulose structures which contain glucose, xylose, galactose, mannose and arabinose. The composition of sugars depends on the process used to extract carbohydrates. Some studies have found glucose to be the main monosaccharide in CS[4] while others reported fructose to be the main monosaccharide.[3] Total sugar content varies greatly in this by-product (1.6–12%).[3] Lignin is also a fraction present in a significant amount in CS (30%).[2,4,36]

Coffee silverskin contains protein, fat and ash, at 16.2–19.0%, 1.56–3.28% and 5–7%, respectively.[7,34,35,37] The ash in CS contains a variety of mineral elements including potassium, calcium, magnesium, sulfur, phosphorus, iron, manganese, boron and copper among others. Potassium is the most abundant mineral element followed by calcium and magnesium.[4] Caffeine content in CS is lower than in coffee beans. In general, the average caffeine content is 0.8–1.0%.[3,37]

CS is considered to be a good source of bioactive compounds, particularly chlorogenic acids.[33,38] The most relevant are 5-caffeoylquinic acids and 3-caffeoylquinic acids with amounts of 1.99 mg g^{-1} and 1.48 mg g^{-1}, respectively.[38]

The presence of melanoidins has also been reported in CS.[34] Melanoidins are the final product of the Maillard reaction, which occurs during the coffee bean roasting process. Coffee melanoidins are formed mainly by polysaccharides, proteins and chlorogenic acid. Conditions during coffee bean roasting give rise to the formation of different types and amounts of melanoidins.[39]

The extraction of bioactive compounds from natural products like CS is increasingly being used to prepare dietary supplements/nutraceuticals, food ingredients and some pharmaceutical products.[40] Due to its chemical composition, CS could be an important source of several bioactive compounds. The extraction yield of bioactive compounds from natural matrices greatly depends on variables such as type of solvent, solvent-to-solid ratio, time, temperature and pressure, which can be managed to optimize compound extraction.[20,41] Thus, the compound of interest would determine the conditions needed to achieve maximum extraction yield. The most used methods to obtain CS extracts are water extraction at temperature ranges between 25 and 100 °C and subcritical water extraction (SWE) with extraction temperatures ranging from 25 °C to 210 °C.[20,41] Table 12.2 summarizes the chemical composition of the CS extracts obtained using these environmentally friendly technologies.

12.3.6 Spent Coffee Grounds

The macronutrients, fiber and phenolic compounds of SCG are presented in Table 12.1. Polysaccharides are the main components of SCG derived from instant coffee production (75%). Hemicellulose (39%) and cellulose (12%) are the most abundant polysaccharides in SCG. The total sugars composition in SCG is 37% mannose, 32% galactose, 24% glucose and 7% arabinose.[4] Some authors have determined the dietary fiber content in SCG reporting 43–54% total dietary fiber, 47–50% insoluble dietary fiber and 6–16% soluble dietary fiber. These values differ from those of our experimental assay, which found higher amounts of total dietary fiber (82.8%) and insoluble dietary fiber (82.3%) and lower amounts of soluble dietary fiber (0.43%) (data not published).

Spent coffee grounds also contain protein, fat and ash (13.6–17.44%, 2.29% and 1.30–1.6%, respectively).[4,42] With regard to minerals, potassium is the major component, followed by magnesium and phosphorus.[4] Various caffeine concentrations (0.007–0.5%) have been reported depending on the caffeine extraction process and SCG variety.[43,44] The chemical composition was similar in SCG obtained from the preparation of coffee brews.[7,45] SCG from different coffeemakers (filtered, French press and espresso) contained between 0.2 and 0.8% caffeine with the exception of SCG from the regular coffeemaker, which did not present this compound.[45]

Table 12.2 Chemical composition and antioxidant capacity of coffee silverskin extracts (% w/w dry matter).[a]

	Water[65,66] (100 °C, 10 min)		SWE[c,20]	SWE[c,41]
Compounds	ACSE[b]	RCSE[b]	(210 °C, 1.5 MPa)	(200 °C, 10.3 MPa)
Proteins	5.36	0.99	53.5	–
Carbohydrates	5.44	13.43	22.8	–
Caffeine	3.02	3.39	1.4	0.38
Melanoidins	17.26	23.94	–	–
TDF[d]	28.69	36.21	–	–
SDF[e]	24.01	26.80	–	–
IDF[f]	4.67	9.41	–	–
CGAs	1.12	6.85	–	0.027
TPC[g]	3.10[h]	3.54[h]	12.4[i]	0.46
ORAC[k]	1194	1513	2321	109.9[j]
DPPH[k]	219.9	231.3	323	-
ABTS[k]	85.20	225.8	–	103.7[j]
FRAP[k]	829.8	640.1	–	–

[a]Superscript numbers correspond to the cited literature.
[b]ACSE: Arabica coffee silverskin extract; RCSE: Robusta coffee silverskin extract.
[c]SWE: subcritical water extraction.
[d]TDF: total dietary fiber.
[e]SDF: soluble dietary fiber.
[f]IDF: insoluble dietary fiber.
[g]TPC: total phenolic compounds.
[h]g CGA 100 g^{-1} CS extract.
[i]g GA 100 g^{-1} CS extract.
[j]μmol CGA g^{-1} CS extract.
[k]ORAC, DPPH, ABTS and FRAP units: μmol TEAC g^{-1} CS extract.

Different health-related chemicals bound to dietary fiber and proteins such as phenolic compounds, mainly chlorogenic acids, have been reported in SCG from different sources.[7,45,46] Monocaffeoylquinic acids (3-CQA, 4-CQA, 5-CQA) and dicaffeoylquinic acids (3,4-diCQA, 3,5-diCQA, 4,5-diCQA) have been identified and quantified in SGC obtained from different brewing processes and coffee species.[7,45] Total caffeoylquinic acids (CQA) contents ranging from 1.7 to 0.7% have been found in SGC,[7,45] and higher levels of total CQA have been found in arabica SGC than in *Coffea canephora* Pierre (hence called robusta) SGC.[45] Furthermore, the amount of total CQA is significantly lower in SCG from 100% torrefacto coffee. Polyphenols may interact with other coffee components in the torrefacto process[7] and CQAs may undergo chemical transformations during coffee processing, leading to reduced amounts of CQAs by transformation into quinolactones and melanoidins.[7]

The amount of coffee melanoidins that remained in SCG after coffee brewing ranged from 15 to 35%.[7] Coffee melanoidins fractions are diverse and possess different physico-chemical properties.[39] Likewise, coffee melanoidins have been hypothesized to become a part of the soluble dietary fiber fraction during the roasting process.[47]

12.4 Applications of Coffee By-products

Coffee by-products have attracted great attention in the last few years due to the large amounts generated. As these residues are derived from coffee beans, they are expected to have similar properties, which could be exploited for different industrial applications. In this sense, some alternatives have been proposed to reuse these coffee wastes.[4,13,22] Some of these value-added applications are only in their infancy, while others have already been patented and even used industrially. In this chapter, we will focus on applications related to food, health and derivatives from the application of the biorefinery concept in the coffee sector.

12.4.1 In Foods

The search for new technologies and ingredients with interesting characteristics and potential for incorporation into functional foods has emerged parallel to the environmental problem of water and land pollution caused by the unsafe disposal of coffee by-products in coffee-producing countries.[13] The rich chemical composition of coffee and its by-products remains underexploited at the different stages of the value chain.[14] Few applications of coffee by-products have been proposed in foods so far.

12.4.1.1 Coffee Pulp

Several applications have been proposed for coffee pulp. Medeiros *et al.*[48] reported a great potential use of coffee pulp as a substrate of *Ceratocystis fimbriata* 374.83 fungus to produce fruity aroma by solid state fermentation. A total of 13 compounds were produced from coffee pulp with and without previous thermal treatment. The major compounds obtained were ethyl acetate, ethanol and acetaldehyde, followed by ethyl propionate, propyl acetate, ethyl isobutyrate, butyl acetate and four compounds remained unidentified. More recently, Bonilla-Hermosa *et al.*[49] studied the production of aroma/flavor compounds that could be interesting for food by applying 8 yeast strains to coffee pulp. They detected 35 compounds corresponding to 6 groups of volatile compounds, higher alcohols, acetates, ethyl esters, aldehydes, terpenes and volatile acids contributing to floral and fruity notes with commercial interest.

Coffee pulp has also been assayed as a potential source of anthocyanins for application as a natural food colorant.[12,50] In the wet process, coffee pulp is removed prior to drying, and its color is rapidly degraded by the action of the enzymes liberated or by other oxidizing agents, such as oxygen. As a result, large amounts of natural colorants are wasted. Cyanidin-3-rutinoside was characterized as the dominant anthocyanin, responsible for the color red. Coffee anthocyanins have also been reported to have multiple biological effects.[50] Therefore, coffee pulp could potentially be used as a colorant and bioactive ingredient in formulated foods. A new interesting ingredient – coffee pulp flour – has been developed from coffee pulp.[51,52] Ramirez *et al.*[51] obtained coffee pulp flour with a high fiber and mineral content (18% and 8%, respectively) and low fat (1.6%). This recently developed CoffeeFlour® has been proposed for use in different food formulations such as breads, cookies, muffins, squares, brownies, pastas, sauces and beverages. It possesses five times more fiber than wholegrain wheat flour, 84% less fat and 42% more fiber than coconut flour and is gluten free. Regarding taste, this product does not have a coffee taste, but rather expresses more floral, citrus and roasted fruit-type notes.[52]

Furthermore, coffee pulp has been characterized as a source of soluble dietary fibers (SDF) for use in food and pharmaceutical applications.[53] Due to its properties as an emulsifier and stabilizer, this extracted coffee dietary fiber is a promising new ingredient for the food and beverage industry.

12.4.1.2 Coffee Mucilage

According to the International Coffee Organization (ICO), coffee mucilage could be used in foods as unrefined pectins, either thermo-reversible soluble gels or non-reversible cross-linked with different mouthfeel, antioxidants and flavonoid compounds such as chlorogenic acids and anthocyanin fruit color compounds and colorless pro-anthocyanins. All these compounds have become food additives of special interest for the food industry.[54]

Coffee mucilage has also been proposed for use as honey for human feed.[51] The chemical composition of this honey showed 30–40% moisture, 55 °Bx, 4% proteins, 2% fiber and a polyphenols content corresponding to 380 mg gallic acid equivalent 100 g^{-1}. This coffee honey with high sugar content was achieved by means of a vacuum dehydration step at a temperature below 65 °C, obtaining a product with minimum nutritional damage by heat, and high digestibility and palatability.

12.4.1.3 Coffee Husk

One of the first applications of coffee husks in the food industry was the production of value-added products such as citric acid.[55] A solid-state fermentation (SSF) system was carried out on coffee husks by employing *Aspergillus niger*. Data indicated that about 1.5 g citric acid was produced per 10 g dry coffee husk in the optimized medium with conversion reaching about 80% based on the sugar consumed. It was estimated that a commercial SSF plant could process 5 tons of coffee husk and produce 0.75 tons of citric acid per day, thereby making coffee husks an attractive substrate for the production of citric acid, an additive widely used in juices, candies and sauces.

Coffee husk has been also described as an important source of natural aroma compounds. Soares *et al.*[56,57] studied the production of fruity flavor by *Ceratocystis fimbriata* grown on steam-treated coffee husk supplemented with glucose, leucine, soybean oil and salt solution. Strong pineapple and banana aroma compounds were formed during fermentation, when different concentrations of glucose (20–46%) were used. Compounds such as acetaldehyde, ethanol, isopropanol, ethyl acetate, ethyl isobutyrate, isobutyl acetate, isoamyl acetate and ethyl-3-hexanoate were also identified. When leucine was added to the medium, total volatile production increased, resulting in a strong banana odor. In this sense, natural flavor from coffee by-products could also be used in the food industry.

12.4.1.4 Coffee Silverskin

Coffee silverskin has been proposed as a natural source of several ingredients such as prebiotic carbohydrates, dietary fiber and antioxidants.[4,13,14,32–34,37]

CS was used together with roasted coffee powder, cocoa powder and golden coffee to obtain innovative coffee blends.[58] The new blend was enriched in bioactive compounds such as chlorogenic acids, trigonelline, theobromine and caffeine. It also had a high antioxidant capacity and was favorably appreciated due to its sensory characteristics.

Coffee silverskin has been employed as dietary fiber for the formulation of innovative bakery products, specifically bread.[35] Results showed the feasibility of using the alkaline hydrogen peroxide CS as a food ingredient to reduce caloric density and increase the dietary fiber content of bread. CS has also been used in the formulation of novel biscuits.[59] Biscuit formulations were designed using stevia as a sweetener and CS as a natural colorant and source of dietary fiber. Coffee silverskin improved some of the quality attributes of the biscuits such as moisture, texture, thickness and color. Regarding the processing of chemical contaminants, hydroxymethylfurfural was greatly reduced and no bioaccessible acrylamide was detected in the digests of the new innovative biscuits. The nutritional value of the biscuits was also improved.

12.4.1.5 Spent Coffee Grounds

There is a rising search for new alternatives to add value to this by-product.

Sampaio *et al.*[60] successfully used SCG for the production of a distilled beverage with a coffee aroma. The process was based on the aqueous extraction of aromatic compounds from SCG, supplementation with sugar and the production of ethanol. The novel spirit then produced flavor and volatile compounds. Its organoleptic properties were acceptable and different from those of commercial spirits.

SCG has also had an antimicrobial effect on *S. aureus* and *E. coli*.[7] This antimicrobial activity might be related to the presence of coffee melanoidins in the SCG structure. In fact, the antimicrobial activity of coffee melanoidins extracted from SCG was 2–5 times higher than when assayed alone. Therefore, they could also be used as preservatives in foods at large concentrations. In this sense, other authors have obtained spent coffee extract powder with a high antioxidant capacity after defatting and extract lyophilization.[61] This powder could be used in the food industry as an ingredient or additive with potential preservation and functional properties.

Brazinha *et al.*[62] proposed a process for obtaining a natural bioactive extract from spent coffee grounds, enriched in caffeine, thereby obtaining a high-value product with known bioactive properties from an abundant material with low value. This extract has a "natural" label, which could be used in the market segment of energy drinks.

A granted patented application of SCG regards its use as an ingredient with a high level of dietary antioxidant fiber in healthy bakery products.[63] This ingredient could be directly applied in the manufacture of pastry and confectionery foods such as bread, biscuits and breakfast cereals, among others, making it a simple, low-cost method. The developed formulation employed SCG as source of antioxidant fiber in diverse combinations with other basic and/or innovative ingredients like stevia. The resulting formulas were rich in insoluble dietary fiber (3–7%) and had low acrylamide content. These products might be appropriate for special nutritional needs due to their low glycemic index and energetic value.

Common applications of several of the previously mentioned coffee by-products have been developed. Studies for the cultivation of edible mushrooms have been carried out using substrates of coffee cherry husk, coffee parchment, CS, SCG and dried leaves, with and without supplementation of agricultural wastes (wheat bran).[64] Coffee industry wastes were used either individually or in combination for mushroom cultivation. Individual use of the substrates led to low mushroom yields. The highest of these yields was obtained using coffee cherry waste, followed by SCG. Cultivation with a mixture of all coffee wastes yielded maximum mushroom production. Therefore, high mushroom productivity can be reached using coffee by-products.

Furthermore, coffee pulp, coffee husk, CS and SCG contain appreciable amounts of bioactive compounds, mainly chlorogenic acid and antioxidant dietary fiber.[33] Therefore, they represent an exciting opportunity to obtain new functional ingredients for use as natural antioxidant sources, nutraceuticals and preservatives in an enormous variety of food preparations with high nutritional value.

12.4.2 In Health

12.4.2.1 Coffee Pulp

Coffee pulp, enriched in anthocyanins, contains powerful inhibitors of glucosidase and amylase enzymes. As these enzymes play an important role in the management of glucose metabolism, the use of anthocyanin extracts from coffee pulp has been proposed to improve postprandial blood glucose metabolism.[50]

12.4.2.2 Coffee Silverskin

The highlighted chemical composition of CS suggests that it could be a good source of phenolic compounds, particularly chlorogenic acids.[34,65,66] Phenolic compounds have shown potential protective activity against several chronic diseases[67] associated with oxidative stress and inflammation.[68,69] Chlorogenic acids have shown relatively good bioavailability[70] and present several functions such as antifungal,[71,72] antibacterial,[69,70] anti-inflammatory,[73] antioxidant,[65] anti-glycative,[74] anti-carcinogenic[75] and neuroprotective[76] functions. CS extract is also rich in melanoidins, possessing antioxidant and other health-promoting properties.[7,20,34] Coffee silverskin antioxidants have exhibited a powerful antioxidant effect *in vitro*,[65,66] which may be bioavailable for reducing oxidative stress in humans, thereby decreasing the risk of chronic diseases such as cardiovascular diseases, cancer, type 2 diabetes, Alzheimer's and Parkinson's diseases.[67] Since AGEs (advanced glycation end products) and carbonyl stress are also associated with the pathogenesis of these diseases, the antiglycative properties of CS extract can also contribute to reducing the risk of such pathologies.[77]

Considering the high fraction of insoluble compounds, especially those enriched in high molecular weight polysaccharides and lignin, CS showed a relevant effect on gut microbiota. The prebiotic activity in CS increased the number of healthy bacteria such as *Lactobacillus* spp. and *Bifidobacterium* spp.[7] Hence, CS could be used as a source of a new functional ingredient with the capacity to promote human health through its potential effect on gut microbiota. Likewise, high molecular weight substances in CS, mainly acidic polysaccharides composed of uronic acid, exhibited a hyaluronidase-inhibiting effect. Hyaluronidase is a mucopolysaccharase related to inflammation by histamine released from mast cells. These results indicate that CS may also be beneficial in inflammatory conditions like allergies.[78]

Chlorogenic acids (CGA) and coffee melanoidins may also play a role in reducing and controlling body weight, and may therefore be of interest in treating and reducing the risk of obesity.[77,79] An antioxidant drink based on CS for body fat reduction and weight control has been developed. Health benefits were evaluated *in vitro* and *in vivo* using the animal model *Caenorhabditis elegans*. The resulting beverage contained physiological active concentrations of caffeine and chlorogenic acid to prevent the accumulation of body fat and possessed adequate sensorial properties.[66]

Very recently, CS extracts have also been associated with liporegulatory and glucoregulatory effects. Hence, CS has a great potential for treating metabolic syndrome and diabetes as well as its risk factors. Results were obtained *in vitro* and *in vivo*. The antidiabetic and antiobesity effect of CS extracts may be partially ascribed to the inhibition of α-glucosidase and lipase. Further studies are being carried out to gain knowledge on the mechanism of action of the CS extract in the pathogenesis of these chronic diseases and to determine the contribution of their bioactive components to health-promoting effects.[80]

12.4.2.3 *Spent Coffee Grounds*

The most important components of spent coffee grounds are polysaccharides, whose thermal hydrolysis may produce mannooligosaccharides (MOS). Previous experiments indicate that MOS obtained from SCG present prebiotics, since they resist digestion, are fermented to short chain fatty acids *in vivo* and promote bifidobacteria growth. Short chain fatty acids inhibited the growth of pathogenic bacteria in the colon by lowering pH, and bifidobacteria are thought to promote intestinal health.[81] Other health effects of MOS were a decrease in blood pressure, elevating the suppressing effect[82] and reducing body fat (especially abdominal fat).[81]

Studies have also found that spent coffee grounds are a good source of hydrophilic antioxidant compounds with antigenotoxic properties. Thus, these by-products could possibly be used to help protect against oxidative stress-related diseases, such as cancer.[61]

Spent coffee grounds can be a natural source of insoluble dietary antioxidant fiber.[63] This coffee by-product has nutritional characteristics, as insoluble dietary fiber material is indigestible by the enzymes of the gastrointestinal tract producing positive effects on health such as intestinal regulation, increase in the feeling of satiety and slimming. It also has a significant antioxidant capacity. The antioxidant properties of the dietary fiber of SCG have mainly been ascribed to the presence of phenolic compounds associated with coffee in their polymeric structure.

Furthermore, SCG exerts a positive effect on beneficial bacteria, increasing the numbers of lactobacilli and bifidobacteria. SCG could also be a good source of prebiotic compounds.[7]

12.4.3 Other Applications

Other feasible applications of coffee by-products are the production of biofuels, composts, animal feed and specific materials such as biosorbents, enzymes, chemicals and cosmetics among others (Figure 12.7; Table 12.3).

Coffee pulp,[49,51] coffee mucilage,[51,83,84] coffee parchment,[85] coffee husks,[54,86,87] and SCG[88–95] have been successfully used in the production of biofuels. The conversion of biomass into biofuels can reduce the strategic vulnerability of petroleum-based transportation systems. Bioethanol has received considerable attention over the last few years as a fuel extender and even as a neat liquid fuel.[96]

More attempts have been made to use coffee by-products as substrates in bioprocesses. Recent studies have shown the feasibility of using coffee pulp and husk in the production of enzymes and secondary metabolites by employing different microorganisms such as *Aspergillus oryzae* and *Penicillium* sp. One approach produced enzymes such as tannase from both pulp[97] and husk.[98] Pectinase,[99] α-amylase,[100] and xylanase[101] were obtained from coffee pulp and protease[102] and endoglucanase[103] from coffee husks. Moreover, Machado *et al.*,[104,105] used solid-state fermentation in coffee husk to produce gibberellic acid, a very potent hormone whose natural occurrence in plants controls their development. In addition, CS has also proved to be an excellent source of nutrients during fructooligosaccharides and β-fructofuranosidase production by *Aspergillus japonicus* under solid-state fermentation conditions. This process is a promising strategy to synthesize these two products at the industrial level.[106]

Recent studies on coffee by-products, specifically coffee pulp, coffee mucilage and CS, and their application in cosmetics have also been carried out.[41,51,107] Coffee pulp and mucilage have been described as raw materials for the production of coffee pulp flour and coffee honey, respectively, and may therefore be useful in cosmetics.[51] del Castillo *et al.*[41] prepared an emulsion by mixing light olive oil in 90% water and 0.4% powdered CS extract (granted patent). They obtained a solution which had a suitable pH for application to the skin (5.44), a phenolic compound content in the order of five times greater than that detected in the base emulsion (10.98 mg of Trolox 100 ml^{-1}) and a high antioxidant capacity (117.42 mg of Trolox 100 ml^{-1} of lotion). Rodrigues *et al.*[107] recently showed that CS is a safe source of natural antioxidants with antifungal and antibacterial activity and no cytotoxicity, thereby indicating another potential application in cosmetics.

Several authors have described the usefulness of coffee by-products as activated carbon and biosorbents. Irawaty and Hindarso[108] reported that pyrolysis of coffee pulp impregnated with phosphoric acid produced materials with high adsorption capacity. Coffee parchment proved to be an effective alternative material to remove acetic acid[109] and methylene blue dye in

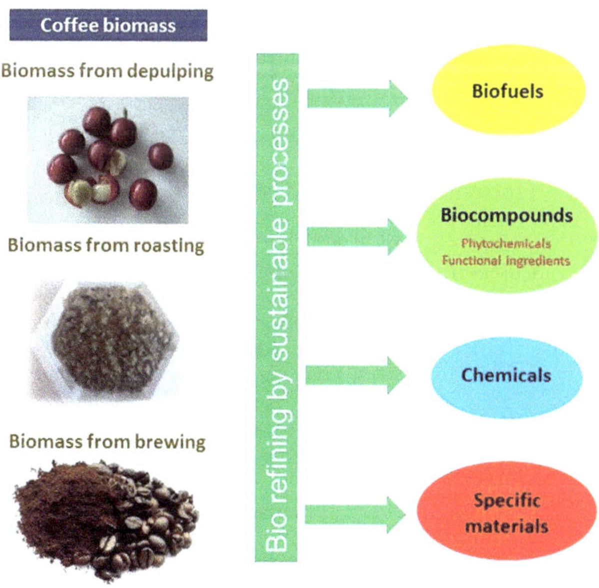

Figure 12.7 Application of the biorefinery concept in the coffee industry.

Table 12.3 Updated summary of proposed applications of different coffee by-products other than in food and health.[a]

Coffee by-product	Applications	Reference
Pulp		
	Biofuel	Bonilla-Hermosa *et al.*,[49] Ramirez *et al.*[51]
	Tannase	Bhoite *et al.*[97]
	α-Amylase	Murthy *et al.*[100]
	Pectinase	Murthy and Madhava Naidu[99]
	Xylanase	Murthy and Madhava Naidu[101]
	Cosmetics	Ramirez *et al.*[51]
	Biosorbents	Irawaty *et al.*[108]
	Compost	Nogueira *et al.*[117]
	Animal feed	Ramirez *et al.*,[51] Mazzafera,[120] Nurfeta,[121] Pedraza-Beltran *et al.*[122]
Mucilage		
	Biofuel	Ramirez *et al.*,[51] Pérez-Sariñana *et al.*[83,84]
	Cosmetics	Ramirez *et al.*[51]
	Livestock feed	Ramirez *et al.*[51]
	Thickening agent	Avallone *et al.*[16]
Parchment		
	Bioenergy	Luana Elis de Ramos *et al.*[85]
	Biosorbent	Hernández *et al.*,[109] Brum *et al.*,[110] de Matos *et al.*[111]
	Composite materials	Funabashi *et al.*[125]
Husk		
	Biofuels	Gouvea *et al.*,[32] Rathinavelu and Graziosi,[54] Saenger *et al.*,[86] Jayachandra *et al.*[87]
	Tannase	Battestin and Macedo[98]
	Protease	Murthy and Madhava Naidu[102]
	Endoglucanase	Navya *et al.*[103]
	Gibberellic acid	Machado *et al.*[104,105]
	Biosorbents	Oliveira *et al.*[112]
	Compost	Sathianarayana and Khan,[118] Adi and Noor[119]
	Particle board	Bekalo and Reinhardt[31]
Silverskin		
	Fructooligosaccharides and β-fructofuranosidase	Mussatto and Teixeira[106]
	Cosmetic ingredient	del Castillo *et al.*,[41] Rodrigues *et al.*[107]
Spent coffee grounds		
	Biofuels	Silva *et al.*,[88] Sendzikiene *et al.*,[89] Machado,[90] Sampaio,[91] Kondamudi *et al.*,[92] Rocha *et al.*,[93] Couto *et al.*,[94] Burton *et al.*[95]
	Biosorbents	Hirata *et al.*,[113] Franca *et al.*,[114] Nakamura *et al.*,[115] Namane *et al.*[116]
	Animal feed	Claude,[123] Givens and Barber[124]
	Ceramic manufacturing	Sena da Fonseca *et al.*[126]
	Composite materials	Funabashi *et al.*[125]

[a]Superscript numbers correspond to the cited literature.

aqueous medium.[110] Moreover, this by-product was proposed as a medium filter to remove suspended solids from the wastewater of coffee shrub cherry pulping.[111] An alternative use of coffee husk is as untreated sorbent for the removal of heavy metal ions from aqueous solutions, as reported by Oliveira *et al.*[112] SGC is also an inexpensive and easily available adsorbent for the removal of cationic dyes in wastewater treatments,[113,114] and a source of activated carbon.[115,116]

Regarding composting, coffee pulp solids are a good source of humus and organic carbon. Conversion of 350 000 tons of coffee pulp would yield approximately 87 000 tons of organic manure.[117] Moreover, coffee husks have proved to be very useful in vermicomposting. The high bacterial growth in earthworms' intestines improves soil fertility and stimulates plant growth making vermicasts good organic manure and potting media.[118,119]

Animal feed has become an important target of studies on the application of coffee by-products. Coffee pulp,[51,120–122] coffee mucilage[51] and coffee husk[120] can be used for feeding farm animals. Mazzafera[120] correlated the presence of tannins and caffeine in coffee pulp and husk to their lack of palatability and acceptability by animals. These coffee by-products were decaffeinated by microorganisms to use them in animal feed, replacing traditional components such as cereal grains. The possibility of using spent coffee grounds as animal feed for ruminants, pigs, chickens and rabbits has also been demonstrated by Claude[123] and Givens and Barber.[124]

Coffee husk has been used as special environment-friendly material in the production of particle boards[31] due to its high cellulose and hemicellulose content, which is almost comparable to that of wood. Furthermore, coffee parchment and SCG have been used as fillers of polyurethane composites,[125] and SCG has been proposed as an additive in the production of ceramic bricks, showing acceptable physical and mechanical performance and low thermal conductivity.[126]

From an agricultural point of view, the possibility of reutilizing spent coffee grounds as an easy and economically feasible soil amendment represents an exciting opportunity to obtain products of high nutritional value.[127] For instance, low amounts of composted SCG (up to 15% v/v) produced a relevant increase in essential macro-elements in lettuce, enhancing its quality features.

12.5 Safety Concerns in the Use of Coffee By-products as a Natural Source of Compounds

All foods might contain chemical and/or biological contaminants. Numerous processed foods such as French fries, chips, bread, cookies and coffee contain acrylamide, a chemical processing contaminant. The Maillard reaction has been shown to be the main pathway for acrylamide formation. This reaction takes place when precursors are present in raw materials, *e.g.* reducing sugars such as glucose and fructose, and asparagine, in combination with high temperature and cooking time.[128] Recent studies from García-Serna *et al.*[59] have shown that decreasing the quantity of sugar added to biscuits containing CS and stevia might be a good strategy for obtaining a safe, low-sugar product. To date, no other studies have focused on the impact of CS addition on acrylamide content. The addition of CS improved the physical properties and the nutritional value of stevia biscuits, and no bioaccessible acrylamide was detected in the digests of these new innovative biscuits. This indicates that CS could be used as a natural coloring and source of dietary fiber to achieve a healthier, nutritious and safe quality biscuit.[59]

With regard to the biological contaminants that could be present in coffee, ochratoxin A (OTA) is a mycotoxin produced by *Aspergillus ochraceus* and *Penicillium verrucosum* that tends to bioaccumulate along the food chain. Ochratoxin A can induce renal toxicity, nephropathy and immunosuppression, representing a risk for human safety. Therefore, its content in foods should be determined.[129] Coffee is contaminated with OTA when coffee fruits fall onto the soil or during storage. However, Ferraz *et al.*[129] demonstrated that OTA can be destroyed during roasting. Coffee is considered a secondary source of OTA in the human diet. Even when the coffee beverage is prepared from highly contaminated green beans, the coffee transforming process is able to reduce the amount of OTA that presents a risk for human health.[37] Research carried out by Toschi *et al.*[3] suggests that CS could be a safe source of bioactive compounds (such as fiber and polyphenols), which could be used as ingredients in the pharmaceutical/cosmetic industries or in the development of functional foods. However, as in other food ingredients, it is very important to establish rigorous quality controls and develop a suitable procedure to reduce OTA.

Heavy metals are widely dispersed in the environment and can be found in varying concentrations in human food. Food contamination is a serious problem, as heavy metals exert a harmful influence on many tissues. Metals disturb ionic balance and mineral regulation, induce oxidative damage to cell structures, produce injury to DNA and induce cancer transformations.[130] Nędzarek *et al.*[131] studied Mn, Co, Ni, Cr and Ag levels in coffee. Such levels were shown to be too low to influence human health. However, some coffees had high levels of Pb, which might be harmful if accumulated in the body. This indicates that such products need to be controlled for metal contamination.

The management of by-products is a key step to ensure the safety of these products. Proper management of collecting coffee by-products, cooling/freezing the material, drying or thermal stabilization and/or addition of chemical preservatives can provide solutions in the case of coffee by-products.

12.6 Conclusions

Coffee is not only for drinking, you can also use it for many beneficial purposes by applying the biorefinery concept (Figure 12.7). Coffee waste biorefinery is possible.

Acknowledgements

The SUSCOFFEE Project (AGL2014-57239-R), Sustainable coffee production and consumption: Valorization of coffee waste into food ingredients, funded this work.

References

1. J. M. V. Nabais, P. Nunes, P. J. M. Carrott, M. M. L. Ribeiro Carrott, A. M. García and M. A. Díaz-Díez, *Fuel Process. Technol.*, 2008, **89**, 262.
2. S. I. Mussatto, E. M. S. Machado, S. Martins and J. A. Teixeira, *Food Bioprocess Technol.*, 2011, **4**, 661.
3. T. G. Toschi, V. Cardenia, G. Bonaga, M. Mandrioli and M. T. Rodriguez-Estrada, *J. Agric. Food Chem.*, 2014, **62**, 10836.
4. L. F. Ballesteros, J. A. Teixeira and S. I. Mussatto, *Food Bioprocess Technol.*, 2014, **7**, 3493.
5. Confederation of the Food and Drink Industries of the EU, *Issues, Industry Action and Future Strategies*, Brussels, 2009.
6. J. M. Talbot, *Grounds for Agreement: The Political Economy of the Coffee Commodity Chain*, Rowman & Littlefield, Lanham, 2004.
7. A. Jiménez-Zamora, S. Pastoriza and J. A. Rufián-Henares, *LWT--Food Sci. Technol.*, 2015, **61**, 12.
8. C. M. Galanakis, N. Martinez-Saez, M. D. del Castillo, J. B. Francisco, S. V. Mitropoulou, in *Food Waste Recovery*, ed. C. M. Galanakis, Academic Press, London, San Diego, Waltham, Oxford, 2015, pp. 337–360.
9. F. Fava, G. Totaro, L. Diels, M. Reis, J. Duarte, O. B. Carioca, H. M. Poggi-Varaldo and B. S. Ferreira, *New Biotechnol.*, 2015, **32**, 100.
10. Biorefinery Euroview, Project Public Report, 2007.
11. R. Bressani, in *Coffee Pulp—Composition, Technology, and Utilization*, ed. J. E. Braham and R. Bressani, Ottawa, 1978, pp. 5–10.
12. E. R. B. A. Prata and L. S. Oliveira, *LWT--Food Sci. Technol.*, 2007, **40**, 1555.
13. P. S. Murthy and M. Madhava Naidu, *Resour., Conserv. Recycl.*, 2012, **66**, 45.
14. P. Esquivel and V. M. Jiménez, *Food Res. Int.*, 2012, **46**, 488.
15. S. Roussos, M. A. Aquiáhuatl, M. del Refugio Trejo-Hernández, I. Gaime Perraud, E. Favela, M. Ramakrishna, M. Raimbault and G. Viniegra-González, *Appl. Microbiol. Biotechnol.*, 1995, **42**, 756.
16. S. Avallone, J. P. Guiraud, B. Guyot, E. Olguin and J. M. Brillouet, *J. Food Sci.*, 2000, **65**, 1308.
17. S. Avallone, J. P. Guiraud, B. Guyot, E. Olguin and J. M. Brillouet, *J. Agric. Food Chem.*, 2001, **49**, 5556.
18. H. D. Belitz, W. Grosch and P. Schieberle, in *Food Chemistry*, ed. H. D. Belitz, W. Grosch and P. Schieberle, Springer, Leipzig, 4th edn, 2009, p. 938.
19. M. R. Adams and J. Dougan, *Trop. Sci.*, 1981, **23**, 177.
20. Y. Narita and K. Inouye, *Food Res. Int.*, 2014, **61**, 16.

21. R. A. Pfluger, in *Solid Wastes: Origin, Collection, Processing, and Disposal*, ed. C. L. Mantell, Wiley, New York, 1975.
22. A. Pandey, C. R. Soccol, P. Nigam, D. Brand, R. Mohan and S. Roussos, *Biochem. Eng. J.*, 2000, **6**, 153.
23. D. Brand, A. Pandey, J. A. Rodriguez-Leon, S. Roussos, I. Brand and C. R. Soccol, *Biotechnol. Prog.*, 2001, **17**, 1065.
24. C. Porres, D. Alvarez and J. Calzada, *Biotechnol. Adv.*, 1993, **11**, 519.
25. M. A. Ramirez-Coronel, N. Marnet, V. S. K. Kolli, S. Roussos, S. Guyot and C. Augur, *J. Agric. Food Chem.*, 2004, **52**, 1344.
26. J. R. Ramirez-Martinez, *J. Sci. Food Agric.*, 1988, **43**, 135.
27. M. N. Clifford and J. R. Ramirez-Martinez, *Food Chem.*, 1991, **40**, 35.
28. N. G. de Colmenares, J. R. Ramírez-Martínez, J. O. Aldana and M. N. Clifford, *J. Sci. Food Agric.*, 1994, **65**, 157.
29. D. Brand, A. Pandey, J. A. Rodriguez-Leon and C. R. Soccol, *Biotechnol. Prog.*, 2001, **17**, 1065.
30. S. Avallone, J. P. Guiraud, B. Guyot, E. Olguin and J. M. Brillouet, *J. Food Sci.*, 2000, **65**, 1308.
31. S. A. Bekalo and H. W. Reinhardt, *Mater. Struct.*, 2010, **43**, 1049.
32. B. M. Gouvea, C. Torres, A. S. Franca, L. S. Oliveira and E. S. Oliveira, *Biotechnol. Lett.*, 2009, **31**, 1315.
33. P. S. Murthy and M. M. Naidu, *Food Bioprocess Technol.*, 2012, **5**, 897.
34. R. C. Borrelli, F. Esposito, A. Napolitano, A. Ritieni and V. Fogliano, *J. Agric. Food Chem.*, 2004, **52**, 1338.
35. A. Pourfarzad, H. Mahdavian-Mehr and N. Sedaghat, *LWT--Food Sci. Technol.*, 2013, **50**, 599.
36. S. I. Mussatto, E. M. S. Machado, L. M. Carneiro and J. A. Teixeira, *Appl. Energy*, 2012, **92**, 763.
37. A. Napolitano, V. Fogliano, A. Tafuri and A. Ritieni, *J. Agric. Food Chem.*, 2007, **55**, 10499.
38. L. Bresciani, L. Calani, R. Bruni, F. Brighenti and D. del Rio, *Food Res. Int.*, 2014, **61**, 196.
39. A. S. P. Moreira, F. M. Nunes, M. R. Domingues and M. A. Coimbra, *Food Funct.*, 2012, **3**, 903.
40. J. Dai and R. J. Mumper, *Molecules*, 2010, **15**, 7313.
41. M. D. del Catillo, M. E. Ibañez, M. Amigo-Benavent, M. Herrero, M. Plaza and M. Ullate, *WO* 2013004873 A1, 2013.
42. S. I. Mussatto, L. M. Carneiro, J. P. A. Silva, I. C. Roberto and J. A. Teixeira, *Carbohydr. Polym.*, 2011, **83**, 368.
43. K. S. Andrade, R. T. Gonçalvez, M. Maraschin, R. M. Ribeiro-do-Valle, J. Martínez and S. R. S. Ferreira, *Talanta*, 2012, **88**, 544.
44. K. Ramalakshmi, L. J. M. Rao, Y. Takano-Ishikawa and M. Goto, *Food Chem.*, 2009, **115**, 79.
45. J. Bravo, I. Jua, C. Monente, B. Caemmerer and L. W. Kroh, *J. Agric. Food Chem.*, 2012, **60**, 12565.
46. J. Bravo, L. Arbillaga, M. P. De Peña and C. Cid, *Food Chem. Toxicol.*, 2013, **60**, 397.

47. J. M. Silván, F. J. Morales and F. Saura-Calixto, *J. Agric. Food Chem.*, 2010, **58**(23), 12244.
48. A. B. Pedroni Medeiros, P. Christen, S. Roussos, J. C. Gern and C. R. Soccol, *Braz. J. Microbiol.*, 2003, **34**, 245.
49. V. A. Bonilla-Hermosa, W. F. Duarte and R. F. Schwan, *Bioresour. Technol.*, 2014, **166**, 142.
50. P. S. Murthy, M. R. Manjunatha, G. Sulochannama and M. M. Naidu, *Eur. J. Biol. Sci.*, 2012, **4**, 13.
51. A. Ramirez Velez and J. C. Jaramillo Lopez, *US Pat.* 20150017270 A1, 2015.
52. www.coffeeflour.com.
53. www.pectcof.com.
54. R. Rathinavelu and G. Graziosi, *International Coffee Organization*, 2005, 1967/05.
55. V. S. Shankaranand and B. K. Lonsane, *World J. Microbiol. Biotechnol.*, 1994, **10**, 165.
56. M. Soares, P. Christen, A. Pandey and C. R. Soccol, *Process Biochem.*, 2000, **35**, 857.
57. M. Soares, P. Christen, A. Pandey, M. Raimbault and C. R. Soccol, *Bioprocess Eng.*, 2000, **23**, 695.
58. V. S. Ribeiro, A. E. Leitão, J. C. Ramalho and F. C. Lidon, *Food Res. Int.*, 2014, **61**, 39.
59. E. Garcia-Serna, N. Martinez-Saez, M. Mesias, F. J. Morales and M. D. Del Castillo, *Pol. J. Food Nutr. Sci.*, 2014, **64**, 243.
60. A. Sampaio, G. Dragone, M. Vilanova, J. M. Oliveira, J. A. Teixeira and S. I. Mussatto, *LWT--Food Sci. Technol.*, 2013, **54**, 557.
61. J. Bravo, C. Monente, I. Juániz, M. P. De Peña and C. Cid, *Food Res. Int.*, 2013, **50**, 610.
62. C. Brazinha, M. Cadima and J. G. Crespo, *J. Food Eng.*, 2015, **149**, 123.
63. M. D. del Castillo, N. Martinez-Saez and M. Ullate, *WO* 2014128320 A1, 2014.
64. S. M. Pushpa and H. K. Manonmani, *Res. J. Environ. Sci.*, 2008, **2**, 145.
65. M. Mesías, M. Navarro, N. Martínez-Saez, M. Ullate, M. D. del Castillo and F. J. Morales, *Food Res. Int.*, 2014, **62**, 1120.
66. N. Martinez-Saez, M. Ullate, M. A. Martin-Cabrejas, P. Martorell, S. Genovés, D. Ramon and M. D. Del Castillo, *Food Chem.*, 2014, **150**, 227.
67. D. Del Rio, A. Rodriguez-Mateos, J. P. E. Spencer, M. Tognolini, G. Borges and A. Crozier, *Antioxid. Redox Signaling*, 2013, **18**, 1818.
68. A. Scalbert, C. Manach, C. Morand, C. Rémésy and L. Jiménez, *Crit. Rev. Food Sci. Nutr.*, 2005, **45**, 287.
69. J. A. Vita, *Am. J. Clin. Nutr.*, 2005, **81**, 292S.
70. D. Del Rio, A. Stalmach, L. Calani and A. Crozier, *Nutrients*, 2010, **2**, 820.
71. C.-M. Ma, M. Kully, J. K. Khan, M. Hattori and M. Daneshtalab, *Bioorg. Med. Chem.*, 2007, **15**, 6830.
72. R. Puupponen-Pimiä, L. Nohynek, C. Meier, M. Kähkönen, M. Heinonen, A. Hopia and K. M. Oksman-Caldentey, *J. Appl. Microbiol.*, 2001, **90**, 494.

73. M. D. dos Santos, M. C. Almeida, N. P. Lopes and G. E. P. de Souza, *Biol. Pharm. Bull.*, 2006, **29**, 2236.
74. B. Fernandez-Gomez, M. Ullate, G. Picariello, P. Ferranti, M. D. Mesa and M. D. del Castillo, *Food Funct.*, 2015, **6**, 2081.
75. H. Shi, L. Dong, Y. Bai, J. Zhao, Y. Zhang and L. Zhang, *Eur. J. Pharmacol.*, 2009, **623**, 119.
76. J. Han, Y. Miyamae, H. Shigemori and H. Isoda, *Neuroscience*, 2010, **169**, 1039.
77. J. Uribarri, M. D. del Castillo, M. P. de la Maza, R. Filip, A. Gugliucci, C. Luevano-Contreras, M. H. Macías-Cervantes, D. H. Markowicz Bastos, A. Medrano, T. Menini, M. Portero-Otin, A. Rojas, G. R. Sampaio, K. Wrobel, K. Wrobel and M. E. Garay-Sevilla, *Adv. Nutr.*, 2015, **6**, 461–473.
78. M. Furusawa, Y. Narita, K. Iwai, T. Fukunaga and O. Nakagiri, *Biosci., Biotechnol., Biochem.*, 2011, **75**, 1205.
79. T. Murase, K. Misawa, Y. Minegishi, M. Aoki, H. Ominami, Y. Suzuki, Y. Shibuya and T. Hase, *Am. J. Physiol.: Endocrinol. Metab.*, 2011, **300**, E122.
80. M. D. Del Castillo, B. Fernandez-Gomez, M. Ullate and M. D. Mesa, *P201431848*, 2014.
81. I. Asano, S. Fujii, M. Katsuhito, T. Izumi, O. Kazuto, N. Kenichi and M. Toshiyuki, *Pat 036208*, 2006.
82. D. G. Stephen, T. Hatzold, I. Takao, I. Asano, S. Fujii, M. Kaneko and J. R. Nielson, *US Pat.* US8043645B2, 2009.
83. B. Perez-Sarinana, S. Saldana-Trinidad, F. Santis Espinosa, D. Pathiyamattom and J. S. Eapen, *Energy Procedia*, 2014, **57**, 950.
84. B. Perez-Sarinana, S. Saldana-Trinidad, C. Guerrero-Fajardo, L. Santis-Espinosa and J. S. Pathiyamattom, *Bioresources*, 2015, **10**, 2691.
85. L. E. de Ramos e Paula, P. F. Trugilho, A. Napoli and M. L. Bianchi, *Cerne*, 2011, **17**, 237.
86. M. Saenger, E. U. Hartge, J. Werther, T. Ogada and Z. Siagi, *Renewable Energy*, 2001, **23**(1), 103.
87. T. Jayachandra, C. Venugopal and K. A. Anu Appaiah, *Energy Sustainable Dev.*, 2011, **15**, 104.
88. M. A. Silva, S. A. Nebra, M. J. Machado Silva and C. G. Sanchez, *Biomass Bioenergy*, 1998, **14**, 457.
89. E. Sendzikiene, V. Makareviciene, P. Janulis and S. Kitrys, *Eur. J. Lipid Sci. Technol.*, 2004, **106**, 831.
90. E. Machado, "Reaproveitamento de resíduos da indústria do café como matériaprima para a producão de etanol," MSc thesis. Braga, Port., Dep. Biol. Eng., Univ. Minho, 2009.
91. A. Sampaio, Desenvolvimento de tecnologias para producão de etanol a partir do hidrolisado da borra de café, MSc thesis, University of Minho, Braga, Portugal, 2010.
92. N. Kondamudi, S. K. Mohapatra and M. Misra, *J. Agric. Food Chem.*, 2008, **56**, 11757.
93. M. V. P. Rocha, L. J. B. L. de Matos, L. P. De Lima, P. M. D. S. Figueiredo, I. L. Lucena, F. A. N. Fernandes and L. R. B. Gonçalves, *Bioresour. Technol.*, 2014, **167**, 343.

94. R. M. Couto, J. Fernandes, M. D. R. G. da Silva and P. C. Simões, *J. Supercrit. Fluids*, 2009, **51**, 159.
95. R. Burton, X. Fan and G. Austic, *Int. J. Green Energy*, 2010, **7**, 530.
96. V. M. Vučurović and R. N. Razmovski, *Ind. Crops Prod.*, 2012, **39**, 128.
97. R. N. Bhoite and P. S. Murthy, *Food Bioprod. Process.*, 2015, **94**, 727.
98. V. Battestin and G. A. Macedo, *Bioresour. Technol.*, 2007, **98**, 1832.
99. P. S. Murthy and M. M. Naidu, *Eur. J. Appl. Sci.*, 2011, **3**, 130.
100. P. S. Murthy, M. M. Naidu and P. Srinivas, *J. Chem. Technol. Biotechnol.*, 2009, **84**, 1246.
101. P. S. Murthy and M. M. Naidu, *Food Bioprocess Technol.*, 2012, **5**, 657.
102. P. S. Murthy and M. M. Naidu, *World Appl. Sci. J.*, 2010, **8**, 199.
103. P. N. Navya and S. M. Pushpa, *Bioprocess Biosyst. Eng.*, 2013, **36**, 1115.
104. C. M. M. Machado, B. H. Oliveira, A. Pandey and C. R. Soccol, "Coffee husk as substrate for the production of gibberellic acid by fermentation," in *Coffee Biotechnology and Quality*, ed. T. Sera, C. R. Soccol, A. Pandey and S. Roussos, Dordrecht, Netherlands: Springer, 2000, pp. 401–408.
105. C. M. M. Machado, C. R. Soccol, B. H. de Oliveira and A. Pandey, *Appl. Biochem. Biotechnol.*, 2002, **102–103**, 179.
106. S. I. Mussatto and J. a. Teixeira, *Biochem. Eng. J.*, 2010, **53**, 154.
107. F. Rodrigues, A. Palmeira-de-Oliveira, J. das Neves, B. Sarmento, M. H. Amaral and M. B. P. P. Oliveira, *Pharm. Biol.*, 2015, **53**, 386.
108. W. Irawaty and F. H. Hindarso, 10th APCCHE Congr., 2004, 17.
109. R. Hernández, P. Carrasco, R. Mujica and M. Espínola, *Rev. Fac. Ing., Univ. Cent. Venez.*, 2007, **22**, 31.
110. S. S. Brum, M. L. Bianchi, V. L. Da Silva, M. Gonçalves, M. C. Guerreiro and L. C. A. De Oliveira, *Quim. Nova*, 2008, **31**, 1048.
111. A. T. De Matos, M. A. Magalhães and D. C. Fukunaga, *Eng. Agríc.*, 2006, **26**, 610.
112. W. E. Oliveira, A. S. Franca, L. S. Oliveira and S. D. Rocha, *J. Hazard. Mater.*, 2008, **152**, 1073.
113. M. Hirata, N. Kawasaki, T. Nakamura, K. Matsumoto, M. Kabayama, T. Tamura and S. Tanada, *J. Colloid Interface Sci.*, 2002, **254**, 17.
114. A. S. Franca, L. S. Oliveira and M. E. Ferreira, *Desalination*, 2009, **249**, 267.
115. T. Nakamura, M. Hirata, N. Kawasaki, S. Tanada, T. Tamura and Y. Nakahori, *J. Environ. Sci. Health, Part A: Toxic/Hazard. Subst. Environ. Eng.*, 2003, **38**, 555.
116. A. Namane, A. Mekarzia, K. Benrachedi, N. Belhaneche-Bensemra and A. Hellal, *J. Hazard. Mater.*, 2005, **119**, 189.
117. W. Nogueira, *Water Sci. Technol.*, 1999, **40**, 113.
118. A. Sathianarayanan and A. B. Khan, *J. Environ. Prot. Sci.*, 2008, **2**, 36.
119. A. J. Adi and Z. M. Noor, *Bioresour. Technol.*, 2009, **100**, 1027.
120. P. Mazzafera, *Sci. Agric.*, 2002, **59**, 815.
121. A. Nurfeta, *Trop. Anim. Health Prod.*, 2010, **42**, 815.
122. P. Pedraza-Beltrán, J. G. Estrada-Flores, A. R. Martínez-Campos, I. Estrada-López, A. a. Rayas-Amor, G. Yong-Angel, M. Figueroa-Medina, F. A. Nova and O. a. Castelán-Ortega, *Trop. Anim. Health Prod.*, 2012, **44**, 329.

123. B. Claude, *Etude bibliographique: utilisation des sous-produits du café*, 1979.
124. D. I. Givens and W. P. Barber, *Agric. Wastes*, 1986, **18**, 69.
125. M. Funabashi, S. Hirose, T. Hatakeyama and H. Hatakeyama, *Macromol. Symp.*, 2003, **197**, 231.
126. B. Sena da Fonseca, A. Vilão, C. Galhano and J. a. R. Simão, *Adv. Appl. Ceram.*, 2014, **113**, 159.
127. R. Cruz, S. Morais, E. Mendes, J. A. Pereira, P. Baptista and S. Casal, *Food Chem.*, 2014, **148**, 294.
128. F. Pedreschi, M. S. Mariotti and K. Granby, *J. Sci. Food Agric.*, 2014, **94**, 9.
129. M. B. M. Ferraz, A. Farah, B. T. Iamanaka, D. Perrone, M. V. Copetti, V. X. Marques, A. a. Vitali and M. H. Taniwaki, *Food Control*, 2010, **21**, 872.
130. M. P. Waalkes, *Mutat. Res., Fundam. Mol. Mech. Mutagen.*, 2003, **533**, 107.
131. A. Nędzarek, A. Tórz, B. Karakiewicz, J. S. Clark, M. Laszczyńska, A. Kaleta and G. Adler, *Acta Biochim. Pol.*, 2013, **60**, 4.

Part II

Coffee Quality

CHAPTER 13

Coffee Cupping: Evaluation of Green Coffee Quality

ILDI REVI*

Ally Coffee and Purity Coffee, 1801 Rutherford Road, Greenville, South Carolina 29609, USA
*E-mail: ildi.revi@gmail.com

13.1 Introduction – Overview of Cupping

Tasting is natural for almost everyone. We do it every time we eat, and we have been deciding if we like or don't like a food for as long as most of us can remember. Those of us who have been drinking coffee for years have had great, mediocre and awful cups of coffee—and judged them as such. 'Cupping' coffee should certainly be what we've been doing for years anyway, shouldn't it? However, when many people observe a formal cupping for the first time, it seems anything but normal tasting. It appears to be an elaborate, numinous activity of slurping and spitting that culminates in a flurry of terms radiating poetically from the cupper's still-damp lips.

In reality, cupping is a specialization in the field of sensory analysis, and professional cuppers are trained panelists calibrated for the unique challenges of coffee. Evaluating food products professionally takes numerous hours of practice to achieve proficiency. The cupper is an instrument for sensory testing that can be trained, but may vary in capacity, depending on genetics, health, environment and life experiences that may have impacted the person's sensory systems both physically and psychologically. In other

Coffee: Production, Quality and Chemistry
Edited by Adriana Farah

Published by the Royal Society of Chemistry, www.rsc.org

words, some people have more capacity to cup than others. One of the important attributes of good cuppers is that they can focus intensely on their own senses, shutting out the rest of the world, quickly identifying and categorizing their experiences, then promptly articulating their discoveries in a discussion with others, who may or may not have experienced the same things, and then finally putting all the information together and agreeing on a numerical score (range) for the coffee.

Coffee has layers of complexity with aromas, tastes and textures that manifest from obvious to barely detectible, both pleasant and unpleasant. Preparation of coffee involves dozens of choices for the many variables that impact how the final brew tastes and smells. To approach the convolution of humans evaluating this complex food product objectively requires planning, tools, discipline, skills, knowledge and collaboration with others.

13.1.1 What is 'Coffee Cupping'?

'Coffee cupping' is the term used for the olfactory and gustatory evaluation of various qualities of a coffee, although professional cuppers may assess the coffee more broadly. In addition to taste and smell, some cuppers may also use their other senses before and during the coffee evaluation process, such as visual appraisals of the green (unroasted) beans, auditory cues made by the coffee in the sample roaster and tactile sensations of mouthfeel or 'body' of the coffee.

'Coffee tasting' and 'coffee cupping' are often used synonymously. The industry has begun to make a distinction between the two, although the definitions have not been adopted in any formal way.[1] This chapter will use the following distinctions:

- 'Coffee tasting' refers to the evaluation of coffee brewed by various methods for various purposes. A coffee taster may be called upon to judge espresso, batch-brewed coffee, manual pour-overs, coffee made in an ibrik, coffee brewed in home brewers, single-serve coffees and others. Often coffee tasters assess the brewing and offer solutions to problems or suggestions for optimizing the coffee flavor, and they have a base knowledge of cupping, roasting and brewing technology. Coffee tasters can be found in cupping rooms, quality control or product development labs, retail operations, training centers, competition spaces and other arenas where broad knowledge of coffee meets trained tasting ability. 'Coffee tasters contribute to an organization's buying program, establish quality control protocols, and define and protect a coffee program—building its reputation and maintaining its integrity.'[1]
- 'Coffee cupping' refers to the practice of evaluating coffee using specific protocols of water poured over ground coffee into a set number of cups per sample, eliminating as many variables (grind size, time, temperature or technique) as possible. Primarily, the practice of cupping

is used to determine the quality of a coffee sample being considered for sale or purchase. This practice is often referred to as 'cupping for green coffee quality', or 'green coffee cupping', which is a misnomer, because the coffee actually *is roasted* to specific standards—it is not cupped 'green'. In some coffee-producing countries, particularly African origins, 'cupping' is referred to as 'liquoring', which may be a more accurate term, although 'cupping' has supplanted it over the past few decades. The definition of 'to liquor' is 'to steep', and further, 'to steep' is 'to soak in liquid in order to extract a given property.'[2] In addition to cupping for green coffee quality, many companies practice production cupping, where the cuppers evaluate the roasted coffee that will be released for sale to consumers. Classifying and categorizing a coffee's tastes can help in decisions for blending or replacements, when a specific coffee is unavailable or too different from one crop year to the next for product consistency. Group work builds an evaluation team: 'Production cupping every roast creates a solid feedback loop that includes everyone in the process and builds excitement and shared responsibility.'[3]

- Coffee tasters may also be coffee cuppers, and *vice versa*, but 'cuppers' in this chapter will be used specifically to refer to those who are evaluating green coffee quality. That said, cupping methods may be used for evaluating roast level as well, but roasting outcomes are too numerous to address in this chapter.
- This chapter explains 'cupping' rather than 'coffee tasting', because cupping has a more mutually accepted methodology than coffee tasting, which is too broad. Cupping focuses squarely on the coffee itself, and from sample to sample the only variable that theoretically changes is the green coffee sample. Coffee tasting would involve discussing dozens of purposes, methods and processes and the rationale for each of them, and none are standardized.

Cupping for green coffee quality is intended to be repeatable and as objective as possible. Cuppers must train themselves to be impartial, use their anatomy and physiology of taste and smell as instruments, eliminate factors that produce bias, understand the limitations of their biology and psychology and employ the lexicon used by others in the coffee industry through collaborative cupping sessions. Striving for unbiased analysis is the best way in which trading partners can agree on selling price and acceptability, and self-monitoring partiality is an effective skill for cuppers to develop.

13.1.2 Why Does the Coffee Industry Cup?

Primarily, cuppers evaluate a coffee for its intrinsic quality, which will impact price discovery and purchase decisions. Cupping for green coffee quality must be performed consistently, not only for cuppers to build a repertoire of equivalent experiences from which they can make sound judgements, but also to create a platform for those in the chain of coffee custody to communicate reliably. In one of the classic historic texts on coffee, *Modern Coffee Production*, author A.E. Haarer wrote, 'Liquoring is a profession that safeguards the buyers and assists producers because a coffee which otherwise looks first class may be found to carry a taint which loses price and makes it of little use for blending with other high quality coffees.'[4]

The world benchmark for arabica coffee pricing is the 'Coffee C Contract' by the Intercontinental Exchange (ICE). The Coffee C Contract sets the price on physical delivery of exchange-grade green beans, from one of 20 countries of origin in a licensed warehouse to one of several ports in the US and Europe.[5] A *Notice of Certification* is issued based on testing the grade of the beans and by cup testing for flavor. The ICE uses certain coffees to establish the 'basis', or basic commodity grade and trading price for fair to average quality coffee. The trading price fluctuates according to market forces, and coffees judged better may receive a premium, while those judged inferior are at a discount.[5] The only way to determine quality is by physical analysis and cupping, and that concept is what motivates cuppers to hone their skills.

With such a vast amount of international business riding on the Coffee C pricing, it would seem that cupping should have a global standard that everyone uses. Similarly, it is not always apparent, when reading scientific analyses or research on coffee, what methods for roasting, brewing or cupping have been used. Roaster type, roast profile, water quality, temperature, steep time or grind size are not always divulged in the literature. Would it make sense for the coffee research community to use a standardized method?

Several coffee organizations in the world provide guidelines for cupping and some have their own standards and methods. The ICE's cupping method is not presented on its website, but the standards and cupping protocols of the Specialty Coffee Association (SCA) are published on their website. Many producers and exporters use the same cupping methods that have been handed down for decades by their predecessors with cuppers essentially learning as apprentices. Often importers and roasters develop their own cupping systems that meet their organizational needs. With coffee's hundreds of compounds that are impacted by numerous variables of roasting and brewing, how do buyers and sellers communicate effectively between each other from the opposite sides of the earth—or even within the same region—if they are using different procedures?

There is no universally employed standard for cupping coffee, but there is a generally used methodology. The International Organization for Standardization (ISO), a worldwide federation of national standards bodies, has a standard for sample preparation (ISO 6668) and a standard for coffee sensorial analysis vocabulary (ISO 18794:2018).[4a] The Specialty Coffee Association (SCA) has different (and more detailed) standards that are actively promoted through its programs and through the Coffee

Quality Institute (CQI). Many characteristics of most cupping methods are reflected in the system created by the SCA.[6] In 2016 the Specialty Coffee Association of America (SCAA) merged with the Specialty Coffee Association of Europe (SCAE) to form the Specialty Coffee Association (SCA). The standards of the former SCAA are referenced here as those of the 'SCA'. Originally the work of Ted Lingle, one of the founders of the SCAA, *The Coffee Cupper's Handbook* was the groundwork from which the SCA system began.[7] Over the past 30 years, facilitated by the SCA, dozens of coffee importers, roasters and quality control professionals developed and refined standards, protocols and etiquette of cupping through collaboration and consensus.

In 1996 the SCAA founded the Coffee Quality Institute (CQI), which created the Q Coffee System to introduce the concept of internationally accepted standards for quality, both cup and grade, particularly for the specialty coffee trade.[8] Their first focus was to help producers who had coffees that were on the borderline of 'specialty grade' to understand how to improve their coffees, and thus receive a higher price, and increase the supply of specialty grade coffee. They had to be able to explain, define, train and calibrate cuppers to 'specialty grade', so the method and protocol necessarily was refined and communicated over the years. CQI is now its own entity and has been instrumental in training and testing cuppers throughout the world on the SCA technical standards. Those who pass the rigorous exam, which consists of 19 tests, are called 'Q Graders'. They have been promulgating their method throughout the world since the 1990s.[8] As of this writing there are 19 countries as partners to the system with more than 4000 certified Q Graders from over 70 countries.

For ease of illustration and because it is a functional basis for arabica coffee sample evaluation, this chapter presents and explains cupping through the lens of the SCA/CQI system. It is a starting point and used here because it has the most public presence at the time of the publication of this book. However, this book does not claim to speak for or represent the SCA or CQI, and anyone looking for official information should go directly to the source for the most current standards, which will be updated regularly as the SCA continues its collaborative work with industry professionals around the world.[6] Also, please note that the SCA/CQI arabica coffee evaluation is different from that used for commercial-grade (the ICE) and for robusta coffee, which has different standards, ranking and attributes.

13.2 How to Cup Coffee

Cupping is a skill that develops over time if practiced often and regularly with other experienced, trained cuppers. One of the most important aspects of cupping is consistency, and the goal is that the only variable that should change from one cupping to the next is the coffee. By striving to achieve this consistency, the coffee is the highlight and focus of the cupping.

Professional cuppers seek to eliminate subjectivity and maximize objectivity. They develop a routine and try to stick to it:

- They use coffees sample-roasted to the same roast level for as close to the same number of minutes for every roast;
- They 'rest' the coffee for the same amount of time after roasting;
- They cup around the same time every day;
- They use the same number of cups with the same coffee weight, grind the coffee to a set screen size, pour the water at the same degree of temperature, steep the coffee for the same amount of time, break the crust with the same stroke pattern and follow the same order during the evaluation;
- They document their evaluations completely and file them methodically.

By systematizing their cupping routine, professional cuppers maximize their ability to achieve fluency in the coffee lexicon and build an internal catalogue of sensory experiences from which they can base descriptions and judgments.[9] When a routine is consistently performed over time, if a slight difference appears, it is more pronounced and identifiable. This is a valuable aptitude, especially if the cupper's assessment influences the purchasing decisions for thousands of pounds of coffee.

13.2.1 Basic Cupping

The core of the actual cupping process begins with the sensory evaluation of the dry coffee in its cups and ends after the coffee has been prepared, tasted and cooled, and the senses have done their jobs. However, unless there are several people involved, the cupping process starts much sooner than that and ends after the cupping bowls are back in the cupboard. There are dozens of choices to make and tasks to perform before the cups land on the table ready for assessment, and these choices impact what the coffee cupper tastes and smells.

Here are the stages of cupping (explained in detail in the following pages):

1. *Roast Samples*: Roast the samples, but only cup them from 8 to 24 hours after roasting.
2. *Prepare the Cupping Table*: Organize the environment to be ready to cup uninterrupted, arrange the cupping supplies, weigh the coffee as whole bean into cups and check the water supply.
3. *Cup the Coffees*: Perform the cupping protocol with good etiquette. Evaluate each cup of the sample in the same order, and consider the composite of all cups in the evaluation as a single coffee. Taste each sample at least three times at different temperatures to get a well-rounded understanding of the coffee's many qualities over time as it cools. Discuss evaluations with other cuppers to compare or align scoring and judgments.

4. *Clean up*: Dispose of the coffee grounds conscientiously, wash the cupping supplies and table thoroughly and store the equipment and supplies properly for the next cupping session.
5. *Document*: Record cupping results into the database or other system and report results to relevant farmers, importers, exporters, customers or suppliers.

Each of the stages of cupping above involves the following:

- *Materials*: Environment, equipment and supplies;
- *Skill*: Performing the protocols and etiquette;
- *Knowledge*: Cupping form terminology, scoring and lexicon;
- *Organization*: Record keeping.

The details of these categories are described below.

13.2.2 Materials: Environment, Equipment and Supplies

13.2.2.1 Environment

Ideally, cupping is done in a designated area that allows the cupper to concentrate entirely on the coffee and sensory information. The cupping room should be quiet, clean, well lit, a comfortable temperature and with limited distractions. There should be no interfering aromas in the space, and any roasting or other activities should be done in a separate room, not while a cupping is taking place. Cuppers should be encouraged to focus their senses and minds only on the coffee and their assessments. Silence must be observed always in the cupping room to allow cuppers to internalize information, reflect and remain free from bias.

The space should be big enough for cuppers to move around the cupping table with about a meter of space, to avoid interfering with each other or accidentally banging against the cupping tables. When deciding on cupping table designs, common sense, cuppers' specific physical characteristics, personal preference or organizational requirements should be considered. By SCA standard, cupping tables for six people must have a surface area of at least 10 square feet (0.93 m^2).

13.2.2.2 Equipment and Supplies

The following equipment and supplies support consistent cupping based on the recommendations and standards of the SCA. The SCA has a certification program for cupping labs, and much of the below reflects the requirements of the certification, although for exact, current details, please visit the SCA website directly. Again, the purpose of introducing the SCA standards here is to provide a starting point and answer common questions that refer to variables and choices.

- *Sample Roaster*:

The coffee sample roaster should be one manufactured specifically to roast and prepare small weight samples for evaluation and be able to manually adjust temperature to slow down or speed up the roast to meet target roast levels within 8 to 12 minutes (SCA standard). There should also be a way to visually observe the roast development. Of course, proper ventilation and a means of collecting chaff are also important, and the sample roaster must be cleaned and well maintained, so that any observed changes during roasting are due to the bean and not the functionality of the equipment (among other reasons, not the least of which is safety from fires).

- *Roast color measuring device*:

The roast level for cupping is measured between 30 minutes and 4 hours after roasting and ground for color measurement. The ground sample standard should meet a specific color, which has different scales, depending on the instrument doing the measuring (examples: Agtron Gourmet – 63.0; Colortrack – 62.0; ProbatColorette 3b – 96.0), with a tolerance of ±1.0 unit.[6] Again, consistency is emphasized. The best way to train and reinforce learning is by achieving as close to the same roast level in as close to the same amount of time as possible for every coffee to be sampled over time. The standard is an attempt to quantify a point in roast development that maximizes the inherent aromatic compounds of the coffee, while completely avoiding compounds and aromatics from charring or over-roasting. Most people in the coffee industry understand that roast color can be deceptive when predicting flavor and aroma, but because we have so few widely accessible objective or mechanical methods for measuring roasted coffee, this color standard will likely remain until a better indicator is developed.

- *Coffee Samples*:

Coffee samples offered by importers may generally be between 150 and 500 g. Only about 60–100 g of roasted coffee is needed for a cupping evaluation, which means 80–150 g of coffee should be charged into the sample roaster to adjust for weight loss from the roasting process. The standard coffee sample for green grading using the SCA method is 350 g. Green coffee grading includes analyzing the coffee for several features (including defects), which can help a cupper learn more about a coffee or make a connection about what they might be tasting at the table. After the green coffee is graded, it is important to keep, store, record and file some of the green coffee for reference later. This is particularly important when evaluating a pre-shipment sample of coffee, because upon delivery of the coffee, having a well-kept offer or pre-shipment sample with which to compare the arrival sample will reveal the impact of storage and transport on the coffee. This is not only good practice to build the sensory repertoire for cupping, but also good business.

- *Grinder*:

A well-maintained grinder with sharp burrs is recommended and should output within a 10% margin of difference from grind to grind using the same coffee. The standard reads, 'The coffee used for cupping shall be ground so that 70–75 percent of the grinds pass through the 20 mesh sieve.'[6] The sieves are the USA Standard or American Society for Testing Materials (ASTM) screen sizes, and '20' represents 841 microns.[11] The grinder must be able to achieve this.

- *Balance (Scale)*:

The scale should be able to measure from one coffee bean up to a ceramic cup tared, so the standard reads, 'A scale capable of precision of 0.01 grams and a capacity of 100–300 grams and that is ANSI certified to standard 169.'[10]

- *Cupping vessels with lids*:

It is important that all cups are identical. Sensory analysts are aware of the importance of uniformity to avoid bias. The cups should also have lids to keep the fragrance of the grinds within the cup before evaluation. Cups should be tempered glass or ceramic and have volume of 207 ml–266 ml (7 to 9 fluid ounces). The recommended diameter is between approximately 76 mm and 89 mm (3 in. and 3.5 in.). When water is poured on the grounds, they rise and come together to form a crust on top of the coffee, which contains hundreds of bubbles full of powerful aromatics. This crust should not auto-break, as it does when the cup is too wide, or the valuable aromatics will be lost to the cupper, who wants to break the crust purposefully.

One factor that differentiates one cupping method from another is the number of cups used per sample. Some companies use 2 or 3 cups, and some use 10. The SCA method uses 5 sample cups for statistical purposes, which we will not discuss in this chapter, but ultimately it helps evaluate sample uniformity and understand the possible extent of defects in a specific lot of coffee. Thus, using the SCA method, for each sample it is useful to have 6–7 cups: 5 cups for the coffee samples that will be cupped, and 1 cup each for the 'purge' coffee and dip cup (explained later), although these last two cups can be different from the 5 identical evaluation cups.

- *Cupping spoons and spittoons*:

The spoons should be of non-reactive metal, so as not to add any off flavors to the coffee and must be able to take 4 to 5 ml of coffee sample. Spittoons can be of any material, but the best spittoons allow for discreet spitting and limit the possibility of spillage. Recently, chewing tobacco spittoons have become popular in cupping labs, and often importers have 'signature' spittoons and spoons.

- *Water supply and hot water equipment*:

The water used for cupping is critical and can impact the flavors of the coffee in significant ways. Water used for cupping must be clean and odor free (no chlorine especially), but not distilled or softened. Usually, commercial spring water is acceptable, because ideal total dissolved solids (TDS) are between 125 and 175 ppm. If TDS is less than 100 ppm or more than 250 ppm, the flavor of the coffee can be impacted. More detailed information on the water standard can be found at the SCA website.[12]

The kettles or hot water dispensers should be able to heat an adequate amount of water for the number of cups being evaluated at the same time. If water runs out while it is being poured into the 5 cups, the cupping is spoiled since consistent evaluation is no longer possible.

The water should be freshly drawn and approximately 93 °C (200 °F) at the time it is poured onto the ground coffee. It is important if cupping multiple samples on the same table that the cupper pours the water quickly and carefully, so that each cup of the sample is as close to the same temperature as possible. More information on pouring is presented in the protocol later.

- *A stopwatch, pencils, clipboards and cupping forms (or a mobile app)*

Managing and monitoring time is an important aspect of cupping. Any type of stopwatch works and should be in plain view for the cuppers. Each cupper should have a couple of sharpened pencils with erasers, a clipboard and enough cupping forms for the samples on the table. Alternatively, there are several apps and electronic platforms available for recording cupping evaluations.[13] Although, having a smartphone or tablet at the cupping table may be awkward or impractical for some people and situations.

Once the materials and supplies are in order, a cupper has the tools needed to perform the cupping protocols fluidly and with good etiquette.

13.2.3 Skill: Performing the Protocols and Etiquette

Etiquette at the cupping table, like etiquette in any social system, has a collection of rules that describe and direct behavior. Etiquette at the cupping table must provide a safe, comfortable and stress-free place for cuppers to evaluate the coffees accurately. Etiquette is not a set of strict rules, but rather guidelines that are common sense and promote politeness, fairness and consideration for others.

Etiquette Prior to Cupping: Personal Habits

Cuppers should smell the coffee, not the people next to them. People in the cupping room should not introduce extra aromas, fragrances or odors. Primarily, a cupper should wear clean clothes and be free of body odors, but also should not have used aromatic soaps and shampoos immediately prior to cupping. No perfume, aftershave or scented deodorants should be worn in the cupping room, and there should be no smoking prior to cupping: not only might the cupper's tasting ability be impacted, but also smoke is often easily detected on a cupper's clothing and breath and is off-putting to other cuppers. Similarly, breath mints, chewing gum, toothpaste and mouthwash will impact the cupper's taste and can be detected by people near them.

Although it is important to have a little food in the stomach before cupping, it is also important to be mindful of odorous foods and kitchen smells that both the cupper and others can detect. Also, avoid foods that linger on the tongue, like garlic, onions, peppers or curry. Finally, cuppers who are not well or feel an illness coming on should not cup with others and are expected to excuse themselves.

Etiquette During Cupping: Behaviors

Many cuppers find it is best to have had a good night's sleep and to cup within 2 to 3 hours of waking up. Cuppers should prepare all materials and supplies before the evaluation is ready to begin (Figure 13.1), because once the cupping starts, there should be no interruptions to the routine. Silence should be observed from the time the first coffee is ground until the end, and beyond silence, facial expressions and gestures should be avoided. Cuppers should actively focus their senses, take good notes and be consistent from coffee to coffee. Here are other common practices of good etiquette:

- Use the dip cup after dipping the spoon in any cup of coffee every time.
- Do not interrupt other cuppers when they are evaluating a set of cups—stand back and allow them to evaluate on their own. Conversely, as soon as a cupper completes an evaluation of a set of cups, he or she should step away from the set to allow others to evaluate.
- DO NOT put a spoon in cups that are under evaluation by another cupper, even for just a small taste.
- Most cupping seems to progress from left to right, which means cuppers tend to move counter-clockwise around a table, although this is

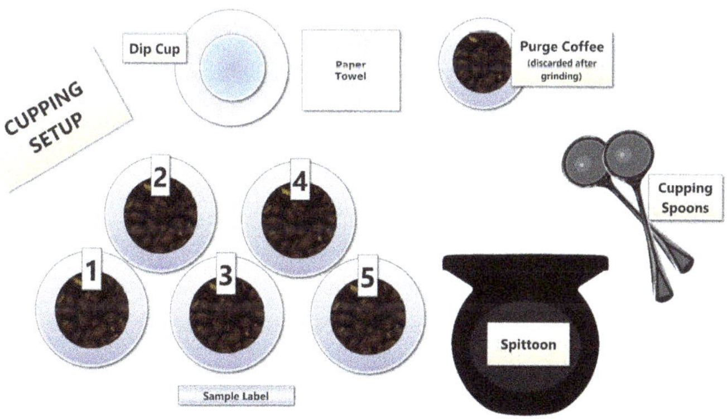

Figure 13.1 Cup numbering and setup.

not a formal rule. As noted in the SCA lab standard, the meter of space between a cupping table and the wall and a 2 meter space between cupping tables will help cuppers avoid bumping each other as they step back from and rotate around the table(s).
- Evaluate each cup of the 5-cup sample in the same order, but also imagine the composite of all 5 in the evaluation as a single cup coffee. The 5-cup setup has numbers for the cups from left to right. These numbers are for identifying defects.
- Once the cupping is complete, openly share findings and avoid reacting negatively or argumentatively to other cuppers' observations or scoring. Be mindful not to dominate the discussion, try not to be the first to talk always and do not force opinions on others. Balance this self-monitoring with neutral feedback and instruction for less experienced cuppers who seek to align scoring with those in seniority.

The cupping room should be a safe zone for both the physical and psychological aspects of cupping, because sensory analysis involves both.

13.2.3.1 Cupping Protocol

Once all the equipment and supplies are in place, it is important to focus only on the coffee by carefully and consistently following the protocol with good etiquette. The cupping protocol is almost a choreography for allowing the cupper's perceptions of attributes to flow freely and be retained. The cupping routine should become naturalized so that all attention is on the senses and what they are telling the cupper about the coffee, and then recording observations. Without a record, it is unlikely the cupper will remember details of the coffee even the following day.

The cupper analyses the quality of specific flavor attributes and rates them on a numeric scale. Coffees that receive higher scores should be noticeably better than coffees that receive lower scores, and cuppers should add detailed notes that bolster the rationales for the scoring. The scores between samples can be compared, and once all the activity is complete, discussing the coffees is a valuable aspect of the cupping process.

The general steps of cupping are listed below and detailed in the following pages:

1. Roast the sample and let rest 8–24 hours.
2. Weigh coffee in individual cups as whole beans.
3. Grind each cup separately to accurate grind size, and set cups uniformly in place.
4. Evaluate the fragrance of the coffee (the smell of the dry grounds) within 15 minutes of grinding.
5. Pour water at 92.2–94.4 °C (200 °F ± 2 °F) over the grounds in all 5 cups in order.
6. Pour water into dip cups.

7. Steep 3–5 minutes without disturbing cups: determine your target time limit for steeping and be consistent.
8. Break the crust of wet grounds and evaluate aroma (the smell of the wet grounds).
9. Skim the remaining surface bubbles and grounds.
10. Steep to a palatable temperature (±71 °C/160 °F).
11. Aspirate and taste; rinse, spit, document and repeat at least 'three passes' up until cooled to about room temperature ±21 °C (70 °F).
12. Complete documentation.
13. Discuss observations.
14. Clean and maintain equipment and supplies.

13.2.3.1.1 Sample Roasting. Roast the sample within 24 hours of cupping as covered in Section 2.2.2. (Agtron 63, measured ground, *etc.*).[6] Consistency in roasting is one of the hardest tasks to achieve, but the ideal outcome is that all the samples on a cupping table should be roasted to the same degree in the same way for green coffee evaluation. It is a useful way to compare one coffee to the next, or to evaluate a specific coffee over time to observe the sensory impact of its aging.

The sample-roasting standard states the roast should be completed in no less than 8 minutes and no more than 12 minutes. The sample should not be scorched or tipped, because that will impart flavors related to roasting, rather than flavors inherent in the beans. The sample should be immediately air-cooled (no water quenching) after roasting. When the sample reaches room temperature, it should be stored in an airtight container until cupping to minimize exposure to air and prevent contamination. Sample roasting should be done systematically, coding or labeling samples accurately and ensuring there is no mix-up or confusion.

13.2.3.1.2 Weighing. Weigh the coffee as whole beans into each cup plus an extra amount in a 'purge' cup. If a defect is present, it is important to isolate it into a single cup, and then count the number of cups in the sample with that defect. If coffee was ground en masse and dished out into cups already ground, the defect would likely be spread throughout the sample and be less detectible, if at all. Coffee should be in the correct ratio to cup volume and consistent for every cup and sample. The standard states: 'When cupping, the ratio of 8.25 grams (whole bean) coffee (±0.25 grams), to 150 ml (5.07 fluid ounces) water shall be used. When adjusting due to vessel size, a ratio of 1.63 grams (whole bean) coffee per 1 fluid ounce of water (or 0.055 g coffee per 1 ml water) shall be used.'[6] Whatever weights and ratios are chosen should be consistent among all cups and samples.

13.2.3.1.3 Grinding. Grind no more than 15 minutes prior to water contact (30 minutes with lids). When the time comes to grind the coffee, first

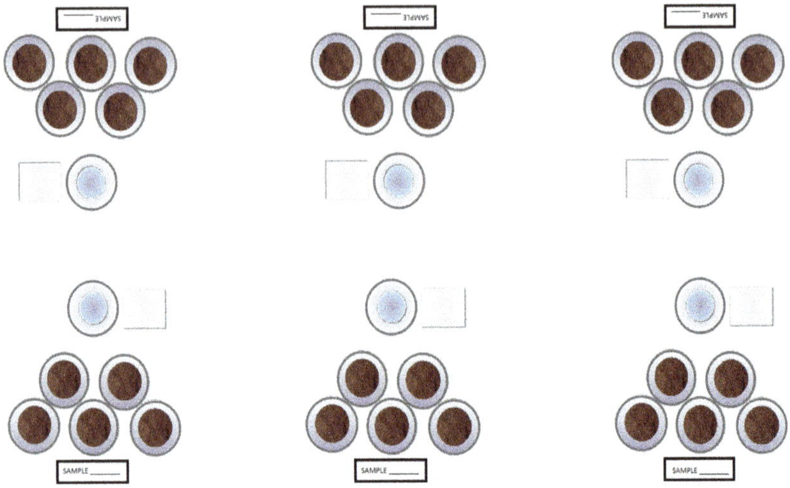

Figure 13.2 Cupping table setup diagram – example.

'season' the grinder by sending the beans from the purge cup through the grinder and tapping out anything that got stuck in the chute. After purging, grind each of the 5 sample cups separately, tapping the grinder chute gently between cups to ensure all the original quantity of coffee gets into each cup. Place a cover on the cups immediately after grinding, since some coffees give off stronger fragrance (the dry aroma given off after grinding) than others do when ground. Fragrance is ephemeral. If a sample has sat for too long a time, the intensity and qualities of the fragrance no longer can be evaluated accurately. Once one sample is done, if preparing another one, repeat the purge step to flush the grinder with the new sample before grinding the new sample cups of the batch.

When all the samples are ground, they should be placed uniformly on the table without any out of place, clearly demarcated with a dip cup nearby (Figure 13.2). Individuals should be assigned their spittoons and spoons, which are best to mark with names or other identifiers. Cuppers should keep track of their own materials and keep them close at hand, not placing them on the actual cupping table. The goal is that the space is uncluttered, and samples are presented identically.

Some companies, exporters and importers like their cupping tables to have a small tray of the green coffee sample and/or the roasted whole bean sample next to the cups. This helps cuppers make connections between what they are experiencing and different states of the coffee.

13.2.3.1.4 Evaluating Fragrance. Evaluate the fragrance of the dry coffee from 0–15 minutes after grinding, sniffing the dry grounds lightly. Using good etiquette, cuppers do not pick up, shake and slap the cups to release fragrance, because that renders aromatics lost to other cuppers. Ask cuppers for verbal confirmation that they have completed their fragrance evaluations before pouring the water.

13.2.3.1.5 Pouring. Pour SCA Standard (125–175 ppm TDS) water at 93.33 °C (200 °F) directly onto the grounds to the rim of the cup (Figure 13.3 and 13.4), making sure to wet all the grounds. Pour all the cups of the same sample quickly, carefully, uniformly and to the same volume without spilling or running out of water mid-cup or mid-sample.

In other words, all 5 cups must be filled completely at once with as little lag time between cups as practically possible. Because fresh roasted coffee 'blooms' with gases, when water is poured on the grounds, be mindful not to fill the cup to the brim, because the blooming grounds can spill out, yet it is

Figure 13.3 Pouring at the same time as much as possible. Photo courtesy Ally Coffee, South Carolina, by Alice Boswell Brown.

Figure 13.4 Pouring to the brim. Photo courtesy of Ally Coffee, South Carolina, by Alice Boswell Brown.

Figure 13.5 Crust with bubbles containing aroma compounds before 'breaking'. Photo courtesy of Ally Coffee, South Carolina, by Alice Boswell Brown.

useful to fill to as close to the brim as possible. Practice and experience are valuable in this action.

13.2.3.1.6 Filling Dip Cups. Pour hot water into dip cups after all sample cups have been filled. During the cupping, if the dip cups become too concentrated with coffee or begin to appear unappetizing, they should be replaced. It is common to place a paper towel next to the dip cup to blot excess water after dipping, thus avoiding adulterating cups with dip cup water.

13.2.3.1.7 Steeping. Determine your target time and be consistent. Let the coffee steep unbroken for 3–5 minutes before evaluation (Figure 13.5). This is the standard range of time, but consistency dictates picking a specific steep time and repeating it. It is an ideal, because to coordinate breaking five crusts per sample at the same steep time and evaluating each aromatic release is not practically possible. Cuppers may go around the table and evaluate what they can of the aroma at this point, being careful not to disturb the cups.

13.2.3.1.8 Breaking the Crust. Break the crust of grounds with three distinct rotating or swiping motions without stirring too much, and sniff the foam that runs down the back of the spoon after stirring. Evaluate aroma by getting as close to the crust as possible and sniffing while breaking without putting your nose into the grounds (Figure 13.6).

If more than one cupper is present, it is useful to agree on a specific cup number to break the crust. If more than one sample is on the table, each cupper can break the same cup number for each sample. Because there are five cups to each sample, up to five individuals get the opportunity to evaluate the aroma from the break, which contains the most valuable aromatics of the wet grounds. If more than five people are at the table, as they rotate from one sample to the next, they can break the crust of the next cup in order, but, of course, they will miss out on some.

Figure 13.6 Get as close to the cup as possible without wetting your nose. Photo courtesy of Ally Coffee, South Carolina, by Alice Boswell Brown.

Figure 13.7 Skimming residue from cups. Photo courtesy of Ally Coffee, South Carolina, by Alice Boswell Brown.

13.2.3.1.9 Skimming. Skim the surface bubbles and surface grounds without agitating the cup or unsettling the grounds (Figure 13.7), dipping the spoon(s) in water between each cup. Many cuppers use two spoons to clean the foamy remnants of the broken crust that may still be present. The skimmed residue is usually discarded into a spittoon.

13.2.3.1.10 Steeping (Continued). Before the first taste, the protocol recommends the cupper allows the coffee to cool to ±71 °C (160 °F). This recommendation is for the cupper to avoid burning the tongue, mouth or lips on hot coffee. Once the tongue is burnt, the cupping day has ended, since blistered parts of the tongue will prevent accurate tasting ability.

13.2.3.1.11 Tasting. Aspirate and taste; rinse, spit, document and repeat until cooled to 21 °C (70 °F). Cuppers often mention 'three passes', which refer to sampling the coffee three times at different temperatures.

Characteristics of the coffee present themselves in different ways as the coffee cools. When at the hottest point (71 °C), flavor and aftertaste present maximum intensity, then change as the coffee cools. Acidity, body and balance also change over time and are considered evaluated optimally between 60 and 71 °C (140–160 °F). Aspirating (slurping) is important to break up the mass of the liquid into tiny droplets to be more accessible to the olfactory receptors in the nasal cavity and disperse the liquid throughout the mouth. Because in some cultures slurping is considered rude, aspirating coffee can be an awkward skill to master both psychologically and sometimes even physically for some people. Similarly, spitting after each slurp can be awkward, although it is an important part of the process for both sensory and physical reasons.

The senses are interacting at this point, and the cupper's task is to deconstruct or analyze the experience as much as possible, then articulate it to share with others. This is the crucial point in the cupping where silence must be observed and there should be no distractions, because the slightest disturbance can cause a cupper on the cusp of identifying something to lose it entirely.

13.2.3.1.12 Documenting. Record as much as possible on the cupping forms. This practice will provide numerous benefits, both personal and professional. Try to write as much as possible about a coffee and anything of note during the cupping.

13.2.3.1.13 Cleaning up. Clean and maintain equipment and supplies to ensure that every cupping is uniform. Cuppers should not leave their spittoons for others to manage and should dispose of their own spittoon waste. Washing tabletops free of dry spilled coffee is as important as washing the cups themselves, and any soap used on the tables, cups and spoons should be fragrance-free and thoroughly rinsed off. After a sanitizing wash, cups, spoons and spittoons should be air dried before being put away.

13.2.4 Knowledge: Cupping Form Terminology, Scoring and Lexicon

Cupping is combining sensory analysis with judgement and recording information. A cupping form creates a common format or structure for analyzing coffee, which is important for communication. Any form that is used will require instruction on how to complete it. Numerical scales need to be clearly explained with experiential examples and a calibrated cupper to explain the scoring rationale. The goal is to align their scores, so that the communication is meaningful. Calibrated cuppers' analysis of a coffee should be within a narrow degree of discrepancy from other trained cuppers or a designated lead cupper.

There are several choices for standardized forms. The SCA has a standard paper-based form available from their website. The Alliance for Coffee

Excellence, which runs the Cup of Excellence competition in producing countries, also has a cupping form downloadable from its website.[14] In addition, many companies develop forms that meet their specific needs. For consistency of example, this chapter will use the SCA form. The job of a cupper is to identify and record clear, comprehensive descriptors on the form as they analyze each category. In addition, they must detect, identify and score defects accurately.

13.2.4.1 Cupping Form Terminology

The language of sensory analysis is difficult to explain by book alone. The best way to develop connections between the language of flavor and its meaning is to cup with experienced, professional cuppers.

The following criteria are commonly used when evaluating coffee and thus on the SCA form:[7]

Fragrance and aroma: Fragrance is the smell of the dry ground coffee, and aroma is the smell when the grounds are wet. A cupper identifies fragrances and aromas and determines the intensity of them as well.

Flavor: The combination of impressions of taste and smell makes up flavor. The five basic tastes generally used to evaluate specialty coffee are sweet, sour, salt, bitter and umami. There are a multitude of smells. Flavor represents main attributes and complexity of a coffee's character, when it is slurped strongly.

Aftertaste: After all the coffee is out of the mouth, the 'finish' that remains on the tongue and sometimes even in the nasal cavity is referred to as aftertaste.

Acidity: The sensation and taste of a lemon or vinegar is an example of acidity, which also has intensity. It is referred to as 'bright' or 'lively' when it is pleasant and considered 'sour' when overpowering or unpleasant.

Body: The thickness and feel of the coffee in the mouth is the body, or 'mouthfeel'. A common example of body is non-fat milk, *versus* partial fat milk, *versus* full cream milk. The intensity of body is also evaluated.

Balance: The way all aspects of the coffee interact with each other makes up the balance. For example: if a coffee does not have enough acidity or has an overpowering flavor, it would receive a lower score on balance.

Sweetness: The detection of sugars in the coffee is sweetness.

The SCA form measures 'clean cup' and 'uniformity' as well and allows the cupper an 'overall' judgment. These are explained further in the next section.

13.2.4.2 Scoring

The subject of assessment, judgement and scoring for sensory analysis is too vast to discuss in this chapter. Cuppers debate questions about subjectivity *versus* objectivity regularly. Ultimately, it is up to the cupper to use

Table 13.1 Quality scale.

Quality scale			
6.00 – Good	7.00 – Very good	8.00 – Excellent	9.00 – Outstanding
6.25	7.25	8.25	9.25
6.5	7.5	8.5	9.5
6.75	7.75	8.75	9.75

descriptive language from a set lexicon and be as non-judgmental in scoring as is humanly possible (which is the heart of the debate). We score coffees because assigning a numeric value to sensory attributes of the coffee allows us to compare coffees, communicate evaluations to others and remind us of our assessments at later dates outside the cupping room.

Many cupping forms use a scale of 1–10, and, in theory, the SCA scale ranges from a minimum value of 0 to a maximum value of 10 points. The lower end of the scale is below specialty grade. On the SCA form 'good' starts at 6.0, 7.0 is 'very good', 8.0 is 'excellent' and 9.0 is 'outstanding'. Only quarter points are used as fractions (Table 13.1).

When all the category scores are added together, and any defects are deducted, coffees at 80 points and above are classified as 'specialty'. Historically, the Exchange uses certain coffees to establish the 'basis', or a reference point for scoring, although it is difficult to nail down exactly what that number is on the SCA form. One way to demystify the SCA cupping form is to calibrate with Certified Q Arabica Graders, who are trained with tightly calibrated instructors and are tested on scoring. Another way to learn a scale is to cup with experienced cuppers who regularly evaluate both commodity and specialty grades, and afterward discuss notes and scores. Prior to the first time cupping together, cuppers might do an alignment round of one or more samples, then try to come to consensus on the scores for each of the categories. Once they begin the full cupping with several samples, they will refer to their scores on the alignment round to guide their decisions. This is referred to as 'peer calibration' or 'peer alignment'.

13.2.4.2.1 Evaluation Procedure. Below are general guidelines for scoring. The section that follows contains information on a lexicon for coffee that was released in 2016. On the SCA cupping form (Figure 13.8), intensity is marked on the small vertical scales within each category's box, and quality is numerically scored on the horizontal scales.

Roast color evaluation: Mark the cupping form with the roast level. If a color meter reading is available, note that, too.

Fragrance evaluation: Note the intensity of the dry coffee fragrance (low to high). Identify smells in the coffee and write them in the corresponding space. Score according to calibration or alignment; otherwise use the quality scale at the top of the form.

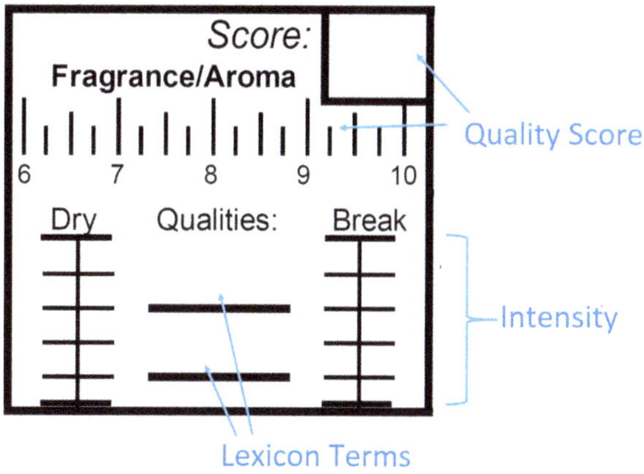

Figure 13.8 Cupping form category example from SCA official cupping form.[1]

Aroma evaluation: When breaking the crust, get as close to the grounds as possible to experience the most aromatics of the coffee available as the bubbles burst. Note the intensity and mark the score. Because fragrance and aroma of the crust-breaking are gone at the beginning, it is appropriate to mark their combined score at the time of evaluation, even though aroma continues to present itself as the coffee continues to steep (Figure 13.8). Allow the coffee to cool before proceeding to the next evaluation stage. Although mentioned before, it bears repeating: avoid burning your tongue.

Flavor, Aftertaste, Acidity, Body and Balance: When the sample has cooled to around 71 °C (160 °F), take a small amount of coffee to begin with and firmly aspirate (slurp) the liquor into the mouth, covering the tongue and upper palate. Make notes on flavor and aftertaste in corresponding spaces on the cupping form. The flavor score combines intensity, quality and complexity. For aftertaste, a long, positive flavor scores higher than a short or unpleasant experience. The evaluation should revisit each coffee through at least three passes as the cups cool, noting observations on the form. Acidity changes over time and should be scored with consideration of its intensity and characteristics in relation to the coffee's overall flavor profile. The score for body is usually higher for coffees that feel heavier in the mouth, although the score may be fairer using the intensity scale in conjunction with the quality rating in relation to the entire experience. The balance score should be based on how all the component categories of the coffee are blended and cooperate with each other. If some attributes overpower others, the cupper should score the balance lower.

Sweetness, Uniformity and Clean Cup: Once the brew cools to slightly above room temperature, check for the attributes of sweetness, uniformity and clean cup, marking 2 points per cup per category (10 points maximum score for each). That said, if cuppers detect a defect at any time during the session, they should identify and mark the box of the corresponding defective cup on the form. Defects are classified as taints or faults, are unpleasant and lessen the likeability of the coffee. A taint is considered detectable and generally manifests in flavor and/or aftertaste—2 points are deducted per taint. A fault is objectionable and is obvious in aroma, flavor and/or aftertaste—4 points are deducted per fault. When calculating defects, multiply the number of cups by the number and type of defect, and subtract this from the total score. Boxes to specify number of cups and intensity (taint or fault) are on the form.

Overall Score: The cupping ends when the coffee has cooled to equilibrium with the room temperature. At this point the cupper reviews the entire process again and awards a judgment of 'Overall Score'. Perhaps when they decided to put this category on the form they might have been thinking that 'the whole is greater than the sum of its parts.' This score usually is close to the mean of scores for other categories. However, when awarding these points, the cupper generally looks at the coffee in its entirety and judges whether the composite score up to that point is truly reflective of the overall impression they have of the coffee. They may score slightly higher or lower than the mean to bring the total score of the coffee more into line with their perception of the coffee.

Once this score is in place, the cupper adds all the scores together and fills in the Total Score. After filling in and calculating deductions for any defects, the cupper writes the Final Score in the lower right-hand box. This score determines the grade of the coffee, such as the 80-point minimum for 'specialty grade'.

13.2.4.3 Lexicon

In 2016 World Coffee Research (WCR) published a new flavor lexicon for coffee, which was developed at the Sensory Analysis Center at Kansas State University.[9] They embarked on the work to recognize the primary sensory qualities of coffee and to develop a reproducible method of measuring those qualities. 'Flavor lexicons are a widely used tool for documenting and describing sensory perception of a selected food.'[15] A lexicon is descriptive, quantifiable and replicable, as developed by a trained panel of sensory analysts.[15] Each attribute in a lexicon is defined with words, referenced specifically and given an intensity score.

The panelists of the WCR Sensory Lexicon spent more than 100 hours evaluating 105 coffee samples from 13 countries to end up with 110 attributes.[17] As coffee cuppers around the world begin to use this lexicon, communication and practice should advance the industry. The WCR website has a link to download the complete lexicon, which includes products, formulae and recipes to reproduce the representative aroma and/or flavor term and its intensity. In conjunction with their research, the SCA facilitated development of a new Coffee Taster's Flavor Wheel as a device to reference the lexicon. The wheel is a useful tool to have next to the cupping table to nudge the mind when it gets stuck trying to find the word for what the senses are detecting.[16]

13.2.5 Organization: Record-keeping

A cupper or evaluation team usually maintains a database of outcomes of cupping sessions for several reasons. Primarily, record-keeping enables verification and determination of quality with notes, so that as time passes and many additional cupping sessions take place, the results of a given coffee are accessible for reference. Documentation helps a cupper make connections between what they are tasting and what might have impacted coffee flavor. There are several online and mobile tools available for cuppers, in addition to spreadsheets and manual methods for recording.

Some record-keeping systems also document as much information about the coffee as possible, such as origin, specifics about the farm and the coffee's growing conditions, processing and drying method, exporter, bag numbers, importer, dates (harvest, shipping and arrival) and other details. Additionally, observations during sample roasting, and, of course, the cupping evaluation itself, are attached to a coffee. After evaluation, professional cuppers give feedback to the person who supplied the coffee sample, generally within a few days.

13.3 Conclusion

Coffee cupping is sensory analysis of a complicated food product. The goal of cupping is to evaluate a coffee as impartially as possible and make decisions based on the results of the assessment. By consistently practicing a method of coffee evaluation, cuppers hone their skills. With new tools available to the industry, such as the WCR Lexicon and the corresponding Coffee Taster's Flavor Wheel, being able to converse globally about cupping results may become more fluid. Many professional cuppers, whether they think about it daily or not, are gatekeepers who make judgments on behalf of consumers. Their evaluations may determine whether their company purchases a green coffee or not, which ultimately determines whether it lands in your morning cup or not.

References

1. Specialty Coffee Association of America, Specialty Coffee Association (SCA), SCA Education, http://SCAAeducation.org/pathways/coffee-taster/, last accessed February 2018.
2. Houghton Mifflin Harcourt Publishing Company American Heritage® Dictionary of the English Language, 5th edn,2011.
3. B. A. Caspersen, *Roast Magazine. A Well-rounded Palate: A Guide to the Coffee Taster's Flavour Wheel - Vol. May-June*, 2012, vol. 43, p. 46.

4. A. E. Haarer, *Modern Coffee Production*, 1956, **265–6**, 467(a) International Organization for Standardization, ISO 6668: Green Coffee—Preparation of Samples for Use in Sensory Analysis (Food Products, CH-1211). Geneva, Switzerland: ISO; and International Organization for Standardization. (2016). International Standard ISO 18794:2018—coffee—sensorial Analysis—Vocabulary, ISO, Geneva, Switzerland, 2008.
5. Intercontinental Exchange, https://www.theice.com/products/15/Coffee-C-Futures, last accessed February 2018.
6. Specialty Coffee Association of America (SCAA) Cupping Standards, https://sca.coffee/research/coffee-standards, last accessed March 2018.
7. T. R. Lingle, *The Coffee Cupper's Handbook*, 1985.
8. Coffee Quality Institute- Q Arabica Graders, http://database.coffeeinstitute.org/users/graders/Arabica, last accessed February 2018.
9. World Coffee Research, *Sensory Lexicon*, https://worldcoffeeresearch.org/media/documents/20170622_WCR_Sensory_Lexicon_2-0.pdf,lastaccessed February 2018.
10. Specialty Coffee Association of America (SCAA) How to Become a SCAA Campus, http://scaaeducation.org/program-updates/campus-program-update/, last accessed February 2018.
11. https://www.sigmaaldrich.com/chemistry/stockroom-reagents/learning-center/technical-library/particle-size-conversion.html, last accessed February 2018.
12. Specialty Coffee Association of America (SCAA) - Water Standard for Brewing Specialty Coffee, http://www.scaa.org/?page=resources&d=water-standards, last accessed February 2018.
13. K. Stewart, *Roast Magazine*, Sept–Oct 2013. http://www.roastmagazine.com/resources/Articles/Roast_SeptOct13_PointsOfTaste.pdf, last accessed September 2017.
14. Alliance for Coffee Excellence, http://www.allianceforcoffeeexcellence.org, last accessed September 2017.
15. M. A. Drake and G. V. Civille, Comprehensive Reviews in Food Science and Food Safety, Institute of Food Technologists, *Flavour Lexicons*, 2002, **33**, 40.
16. http://www.scaa.org/?page=resources&d=scaa-flavor-wheel, last accessed February 2018.
17. M. Spencer, E. Sage, M. Velez and J.-X. Guinard, Using Single Free Sorting and Multivariate Exploratory Methods to Design a New Coffee Taster's Flavor Wheel, *J. Food Sci.*, 2016, **81**, S2997–S3005.

CHAPTER 14

Coffee – Sensory Aspects and Consumer Perception

ROSIRES DELIZA

Embrapa Agroindústria de Alimentos, Av. das Américas 29501, CEP 23020-470, Rio de Janeiro - RJ, Brazil
*E-mail: rosires.deliza@embrapa.br

14.1 Introduction

The nice aroma of hot and fresh coffee is able to attract a huge number of people all around the world reaching an estimated world consumption of 152.1 million bags in 2015.[1] Despite the increasing consumption and production, it is observed that consumers are becoming more interested in quality. With regard to coffee, the term quality can be defined as a set of physical, chemical and sensory attributes that meet consumer needs.[2] The sensory quality of a product is related to the adequate levels of sensory attributes in all aspects of the appearance, aroma, flavour, texture and aftertaste[2] and it has to be taken into account under several perspectives, the consumer point of view being a very important one. It is known that consumers create subjective impressions on the quality of a product based on psychological processes.[3] These processes are influenced by many factors such as the level of previous knowledge and cognitive competencies of each individual consumer.[4] Thus, from a consumer perspective, quality refers to the perceived quality and not to quality in an objective way.

Coffee: Production, Quality and Chemistry
Edited by Adriana Farah

Published by the Royal Society of Chemistry, www.rsc.org

Several approaches have been used to study the perception process of food quality by consumers. Within the widely accepted multi-attribute approach, quality is a multi-dimensional phenomenon, described by a set of characteristics (attributes) that are subjectively perceived by consumers.[5,6] Perceived quality is generally considered as an overall concept and can be defined and analysed, according to Oude Ophuis and Van Trijp,[7] as the four '**Ps**': Perception, Product, Person and Place.

The first 'P' in the perceived quality refers to the *Perception* process, which is related to the overall judgment formed on the basis of visible and invisible product characteristics that may have been experienced, or are believed to be associated with the evaluated product. Perceived quality may differ depending on the *Product* or product category under investigation (second 'P'). This leads to the consideration that perceived quality is based on consumers' judgments, the *Person* factor (third 'P'), suggesting that perceived quality will vary among people, as they differ in their perceptual abilities, personal preferences, lifestyles and experiences.[8] The fourth 'P' is related to the context, and referred to as *Place*. There are many circumstances that affect perceived quality, such as availability, price, social facilitation, appropriateness of the consumption environment (*e.g.*, a specific coffee preparation may be judged as an excellent beverage to be consumed on a sunny and warm day, but inadequate after a meal).[9]

In this chapter, the perceived quality of coffee will be considered, taking into account intrinsic and extrinsic quality cues, and their effects on consumer judgment of the products and consumer food choice. The difference between intrinsic and extrinsic quality cues are presented and discussed in a broader scenario.

14.2 Extrinsic Factors Affecting Coffee Quality Perception

Extrinsic quality cues refer to product characteristics that are used to evaluate a product but are not physically part of it, such as price, brand, production and nutritional information, packaging design, country of origin, store and convenience. Extrinsic cues will become more important when products are very similar in appearance.[7] The intrinsic and extrinsic cues are categorised and integrated by consumers to establish the quality attributes of coffee beverage.

According to Steenkamp,[10] quality attributes can also be *experience* or *credence cues*. The experience originates from the actual experience with or consumption of the product (*e.g.*, aroma and taste, body, mouthfeel, flavour, aftertaste, *etc.*), whereas credence cannot be ascertained even after normal use (buying and consuming of the product). Examples of credence cues are fair trade label, Organic product, Rainforest Alliance. These cues are gaining importance due to the increased concerns of consumers on wellbeing, and ethical factors.

A relevant issue to consumers nowadays refers to the sustainability characteristics of food and beverages, which are credence attributes. Consumers are giving more importance to public values when evaluating food. Sustainability labels are growing all around the world and it is a sign of their increasing popularity.[11] Particularly, the coffee industry is considered a pioneer for sustainability certification, being a model among the commodities. Several initiatives in the sustainable certified coffee market are recognised such as fair trade and organic coffee. Although several studies have investigated the consumer liking and willingness to pay towards different sustainability labels,[12–15] there is still a lack of research on coffee.[11] The investigation on how consumers process sustainability information on coffee packages using eye-tracking technology to measure visual attention was carried out by Van Loo *et al.*[11] Results have shown that consumers are attracted to product attributes such as sustainability labels.

Lange *et al.*[16] reported that particularly the ethical aspects of production and trade are more valued by consumers. The authors investigated the impact of fair trade labelled coffee on the French consumer willingness to pay and liking for the product. Results revealed that the presence of the fair trade label increased the hedonic ratings and the willingness to pay when participants were exposed to ethical information. Products bearing the "fair trade" label are increasingly gaining importance in the market, and coffee is a well-established commodity on such a market.

Another extrinsic factor related to coffee consumption, which was recently under investigation, is the personal motivation to consume a food or beverage. Coffee drinking can be motivated by sensory enjoyment (hedonic motivation) provided by coffee flavour and the psychophysical stimulation, which comes from the caffeine.[17] Labbe *et al.*[18] investigated the impact of both motivations on consumer responses (*i.e.* pleasantness, emotions and importance and satisfaction for each of the five senses) during the entire experience of a coffee beverage. Results suggest that consumption motivation should be considered and controlled when performing consumer research on food or beverage because of its impact on consumer response.

14.2.1 Product Packaging and Label

The product sensory properties are of greatest importance for product acceptance; however, several other extrinsic factors affect consumer product perception.[9,19] The packaging and labelling of a food or beverage play an important role in its selection because it is the major source of information for consumers,[3,20] permitting them to make better choice decisions in the marketplace.[21] Many studies dealing with different label aspects have been carried out. They include food label legislation and government regulations to protect consumers from fraud and from products that could cause physical harm, brand name and its importance on food consumption;[22,23]

nutritional information on consumer attitude towards the product[24,25] and label format in order to provide more complete and useful information.[26] Given the large literature that has been devoted to consumer perception of labels, it is therefore expected that the role of the package attributes in determining sensory expectations is relevant to the subsequent experience with the product.

14.2.1.1 Consumer Expectations

Expectation is defined as the 'action of mentally looking for something' in the food product to come. It plays an important role because it may improve or degrade the perception of a product, even before it is tasted. The higher the expectation levels about the product, the greater chances it has of being purchased. On the other hand, low expectation can cause the product to be ignored.[20,27–29] After a choice, the consumer will taste the product, appreciating its sensory properties and other features that created the expectation. Once confirmed, the consumer is pleased and possibly purchases the product again in a future experience. It means that the effects of expectation are likely to be important variables in determining satisfaction with the product.[30] A common hypothesis is that satisfaction is achieved when the product matches the consumer's expectation.[31,32] The ability to determine such expectation about a particular product becomes a vital strategy in promoting consumers' sensory satisfaction.

A model summarising the role of expectations in product selection and evaluation is shown in Figure 14.1. The process begins with the previous information and experience consumers have, leading to prior expectations. Those prior expectations together with the product itself with its label, advertising and price will generate expectations, which can be high or low. Low expectation leads to product rejection; on the other hand high expectation contributes to product choice.[3]

As can be seen, if disconfirmation of the expectation occurs, a mismatch between the expected and the actual product evaluation will occur. If the expectation is low and the sensory quality of the product is high, there will be a positive disconfirmation and, hence, consumer satisfaction is achieved. Conversely, high expectation on a poor sensory quality product will lead to a negative disconfirmation. In both cases, depending on the individual's behaviour, the evaluation can be driven either towards or opposite to the initial expectation level.[3,27]

There are two main models to explain the influence of information on food acceptance – assimilation and contrast.[27] Assimilation is the most described model and seems to come up mostly in cases of higher expectation and lower sensory quality. The individual tastes the product and receives information about it. Both dimensions – sensory and information – are integrated in his or her mind. If the product is perceived as worse than expected, cognitive dissonance takes place and, to overcome it, the judgment moves towards

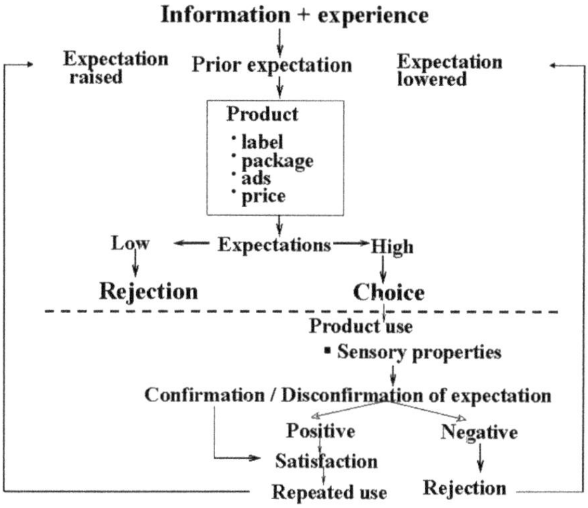

Figure 14.1 The illustrative representation of the effects of expectation on product selection and evaluation. Reproduced from R. Deliza and H. J. H. Macfie, The generation of sensory expectation by external cues and its effect on sensory perception and hedonic ratings: A review, *J. Sens. Stud.*, 1996, **11**(2), 103–128. With permission from John Wiley and Sons, © 1996 WILEY-VCH Verlag GmbH & Co. KGaA, Weinheim.

expectation.[3,29,33] The second model – contrast – can be observed when sensory quality plays the most important role. The individual has high expectation, evaluates the product and dislikes it. Therefore, the product will be perceived as worse than expected. This is called a negative contrast. On the other hand, a product perceived as good but with a low expectation level tends to be well accepted – a positive contrast.[28]

Two other models are cited in the literature: generalised negativity and assimilation–contrast. In the first case, when the consumer has a different perception from the expected, it will be rated worse than if there had been no expectation. Assimilation–contrast, however, is related to the size of the discrepancy between expected and actual product performance. When the perceived discrepancy is small, the person tends to ignore it and there will be assimilation. Large discrepancies are not acceptable, and contrast effect occurs.[34]

The importance of several features of the labels such as brand name, price and information on the expected product attributes is recognised in the literature and studies refer to this topic.[20,35] Previous research indicated that the majority of consumers are able to articulate clear sensory expectations about an unfamiliar product,[28] and therefore about a familiar product as well. However, familiarity with a product may create a less ambiguous experience and consumers' product perception from the label may follow a different pattern.

14.3 Sensory Evaluation and Consumer Studies. Methods Used in Sensory Evaluation – a Coffee Industry Perspective

Food products are developed, processed and marketed to appeal to the consumer, who is becoming more and more demanding about quality. However, the perception of food quality, particularly regarding coffee beverage, is a complex issue. Good quality coffee flavour has been described as a pleasant sensation, a balanced combination of flavour, body and aroma in the absence of faults.[2] Therefore, flavour is still the most important parameter when a consumer evaluates coffee.[36] In other words, the success of a product depends on its acceptance by consumers, because they are the users and, thus, the ones who will be willing to purchase the product. Therefore, professionals in the industries are very interested in understanding consumer perceptions and attitudes towards their products, a formulation change or a new process.

As mentioned before, the factors that affect food choice and intake are multidimensional and complex, but there is no doubt that the physical and sensory properties of a food must be among the main determinants. Therefore, it is very important that the methods used to measure the properties are accurate, not biased and take into account the special requirements involved when humans are used as scientific measuring instruments.[37] During sensory evaluation, the sensory properties of the food are measured and evaluated. The responses by the sensory professional (individual providing the connection between the internal world of technology and product development and the external world of marketplace) are also analysed and interpreted. This is done within the constraints of product marketing brief; in a way that specialists can anticipate the impact of product changes in the marketplace.[38]

The importance of sensory evaluation in the food area and, consequently, in the coffee sector is very well recognised, and considered a cost-effective tool. It has its own challenges and should be viewed in broad terms. Its contribution far exceeds questions such as which flavour is best, or whether process condition X can be replaced with process condition Z. This concept is especially important when looking at the impact of consumer response behaviour as developed by marketing research.

Several studies on the physiological and psychological approaches used in the measurement of consumer behaviour have been presented in the literature. Although research on sensory evaluation has only improved in recent years, much information on the physiology of the senses and the behavioural aspects of the perceptual process has been available for a while.[38] Comprehension of how sensory information is processed and integrated is important in understanding the evaluation process.[39]

Sensory evaluation can be used for several purposes in a company. In Table 14.1 there is a list of activities to which sensory evaluation can contribute, directly or indirectly.[38] Every activity is important to a company, and their

Table 14.1 Sensory evaluation activities within a food industry.[a]

Activities related to sensory evaluation
Product development
Product reformulation and cost reduction
Monitoring competition
Quality control
Quality assurance
Product sensory specification
Raw material specification
Storage stability – shelf-life studies
Process/ingredient/analytical/sensory relationship
Advertising claims

[a]Source: Stone, H. and Sidel, J.L., *Sensory Evaluation Practices*. Academic Press, San Diego, 2004.

use will depend on the kind of company, as well as on the purpose of the study. For some companies the emphasis may be on the marketplace, new products, cost reduction and reformulation, line extension, *etc.* For others, quality control is a primary focus.

In most companies sensory resources are located within the research and development area; however, it is sometimes part of the marketing research group. It will depend on the company and the relationship with the involved areas. There is no rule stipulating where sensory evaluation has to be placed, but since most sensory professionals have some technical training and most often are providing services to technical groups, sensory resources are frequently placed within the technical division. Regardless of the company goals and the sensory resources location, it is important to consider where the test will take place, who will serve as subjects and which methods will be used. The right establishment of all mentioned factors will be crucial to the achievement of the results.[38]

Adequate planning involving the sensory analyst and the requester/client is essential. The aim of the study has to be clear to the sensory analyst, *i.e.*, what is the purpose of the test; how many products will be evaluated; how the products will be prepared; and when the results are needed. In the next step, the test has to be organised in order to guarantee that everything that is needed will be available at the right time. Finally, the test has to be implemented and the data collected, analysed and a written report prepared, which will be given to the requester leading to actionable recommendations.

Before presenting the methods available in sensory evaluation, it is important to make some comments about the place where the tests are carried out – the Sensory Laboratory – including the sensory booth, and other facilities. The test area has to be easily accessed in order to make it easier to get people involved. A place in which most panel members pass on their way to lunch or to a break is a good location. If the panel members are coming from outside, the laboratory should be near the building entrance. Test rooms shouldn't have noise and sources of odour, as they can affect the senses of the panel.

The size of the booth, as well as the illumination, ventilation system, sinks and temperature, have to be adequate. The number of booths will depend on the number and type of tests, which can be estimated before the planning of the laboratory. The booths should be individual and adjacent to the product preparation area so that samples can be served efficiently. A preparation area has to contain at least refrigerated and frozen storage for ingredients and samples; cooking facility (electric or gas burners; conventional, convection or microwave ovens); hoods with charcoal filters or venting to the outside; dishwasher; garbage disposers; waste basket; sink and water sources. Other devices are specially used considering meat and meat products such as heating trays, meat cutters, scales, mixers. Another important area in the laboratory is the training area, designed to carry out training sessions with the panel. This area has to be close to the preparation area, and needs a round or rectangular table (big enough for 6 to 12 panellists), controlled illumination and ventilation (similar to the booth area), furnishings with neutral colour, non-odorous and easy to clean. More details about the facilities can be found in ISO[40] and Stone and Sidel.[38]

The preparation and presentation of the samples also have to be taken into account. Preliminary work is necessary to determine the method of sample preparation and preparation times, equipment and utensils, prior to testing. Every sample has to be tested at a given time and series. The experiments should be planned and served using exactly the same procedures, except for the factor under study.

The order of presentation of samples has to follow a design,[41] and a warm-up sample should be used when required.[38] The serving temperature can be a source of many problems. Therefore, it has to be well controlled throughout the entire test. Generally, it is best to serve at the temperature in which it is usually consumed. It is recommended to provide a liquid (mineral water at room temperature) to rinse the mouth between samples. Crackers are used to remove residual flavours. The time of the test is also relevant. If it is too early, it is difficult to evaluate hot spicy foods; however, if it is too late there is lack of motivation.

Motivation is an issue of concern to all sensory professionals, who need to develop a variety of practices to maintain the panel interested in every study. Several recommendations were given by Stone and Sidel[38] as useful guidelines on motivation, among them, subjects should be rewarded for participating, not for making correct scores; subjects' participation should be acknowledged on a regular basis, directly and indirectly; and management should visibly recognise sensory evaluation as a contributor to the company's growth and an indirect source of motivation.

14.3.1 Sensory Panel – Individuals Who Perform a Sensory Test

Physical and chemical properties are measured by instruments while sensory properties are measured by a sensory panel. The panel consists of individuals selected according to sensory acuity and ability to articulate the sensations experienced while viewing, smelling and eating or drinking the food and/or beverage. Individuals are trained to describe products in terms of sensory characteristics and perceived product attributes' intensity. A trained panel is normally used for discriminative and descriptive tests. However, when the aim of the study is to investigate whether the consumer perceived difference among samples or specific attribute of the sample, individuals who like and consume the product can be recruited. In addition, it is observed worldwide that more recently sensory description is performed by non-trained assessors, *i.e.* by consumers. The hypothesis is that the consumer is able to do the sensory characterisation of samples.[42] This recent approach is used as never before in the food industry, and is detailed in item 14.3.2.2. Preference and acceptability tests are carried out by consumers, who are the users or potential users of the product. Comments on the three main types of sensory tests are presented below.

14.3.1.1 Discriminative Tests

Focusing first on discriminative methods (a class of tests used to investigate whether there is a sensory difference between samples), a variety of specific methods can be found. The most common are the triangle test, duo-trio test and paired-comparison test. Triangle and duo-trio tests are used when the objective is to determine whether a sensory difference exists between two products. Paired-comparison is used to determine in which way a particular sensory characteristic differs between two samples (*e.g.* more or less bitter).

The importance of individuals' selection for a discriminative test has long been known. Using unqualified people results in substantial variability that will mask a difference, leading to a wrong decision.[38] However, when the goal of the study involves the perception of the consumer, she/he has to be recruited to take part in the test. It can be exemplified by mentioning the study that aimed at investigating the rejection threshold for defected coffee beans added to good quality ones.[43] The defected beans are known as black, green and sour beans and affect the quality of the coffee beverage. Results revealed that the addition of up to 16% of defected beans was not perceived by participants.[43] This finding came from a consumer panel where 79% declared to drink from one to three cups of black coffee a day. One can expect that the rejection threshold would be lower if a trained panel had been used.

14.3.1.2 Descriptive Tests

Descriptive sensory tests are among the most sophisticated tools available to the sensory professional and involve the detection and description of both qualitative and quantitative sensory components of a product by trained panels. A relevant point of the descriptive tests is its ability to allow the establishment of relationships between descriptive sensory and instrumental or consumer preference measurements. They can also be used to investigate product changes during its shelf-life, and to evaluate the effects of ingredients or processing variables on the final sensory properties of a product.[38]

There are several different methods of descriptive analysis, such as Flavor Profile®, Texture Profile®, Quantitative Descriptive Analysis (QDA™), the Spectrum™ and Free-choice Profiling. All of them require a panel with some degree of training and, in most of them (except for Free-choice Profiling), panellists are also required to have a reasonable level of sensory acuity. The selection and training of the panel and the monitoring of the performance of individuals are discussed in several publications.[38,44] The tests begin with the development of the descriptive terms, which describe the attributes of the product in a comprehensive and accurate way. It is achieved by exposing individuals to a wide range of products in the category under test. A consensus list of attributes is generated by panellists (except for Free-choice Profiling, in which each panellist has his/her own list of attributes), and will be used to evaluate samples. More details on descriptive tests can be gathered in Heymann *et al.*[45]

Descriptive analysis has been extensively used in coffee studies and many publications may be found.[9,46,47] They evaluated different types of coffee samples such as ground and roast, instant, decaffeinated. The lexicon used to describe the sensory characteristics of coffee beverages varied among the mentioned studies. Table 14.2 shows examples of studies focusing on coffee beverages using descriptive analysis.

The importance of the trained sensory panel was confirmed[58] in a study comparing two different sensory panels: a highly trained descriptive sensory panel and a group of Q-certified coffee cuppers. Their study demonstrated that "expert" coffee cupper data and trained sensory panel data cannot be used interchangeably. Thus, for research purposes sensory panel information is necessary for tracking changes in sensory properties.

14.3.2 Consumer Panel

As was mentioned before, the participation of consumers in descriptive tests is a trend in the sensory evaluation area. This approach provides quick answers to the industry, which has demonstrated an increased interest in consumer product profiling methods. One motivation for the application of the descriptive methods using consumers has come from the high cost and time-consuming aspect related to the trained panel, which is a big issue for both the industry and the academic.[59] There are several methods that were developed and adapted to be used with consumers. The pioneers were Free-choice Profiling – FCP[46] and Repertory Grid,[60] which first used non-trained assessors (consumers) for sensory description. It is worth mentioning that coffee beverages were the investigated products using FCP. The development of the emerging descriptive tools brought to the stage several methods such as flash profiling,[61] sorting,[62] projective mapping or Napping®,[63] the polarised sensory position (PSP)[64] and the check-all-that-apply (CATA) questions,[65] which is described in more detail in Section 14.3.2.2 due to its ease of being performed by non-trained assessors.

In addition, the tests focusing on liking, intention to purchase and perceived quality are also relevant to the field. These two types of tests that can be performed by consumers are presented below.

14.3.2.1 Hedonic Tests

The last type of sensory test described is the affective one, which has the primary goal of assessing the personal response (preference and/or acceptance) by users or potential users of a product or product idea. It is a valuable and important component of every sensory program and it is referred to as acceptance, preference or consumer testing. Subjects taking part in a sensory acceptance test have to be recruited based on demographic and usage criteria, and the number of participants in the test is an important issue in the design of the study. A large number of individuals is required; however, the number recommended in the literature varies, and about 100 consumers are usually considered adequate. Hough *et al.*[66] indicated 112 consumers to perform a test with confidence.

Acceptance testing means the measurement of liking or preference for a product. Preference can be measured directly by comparing two or more products with each other, *i.e.* which one of the two, three or more products is preferred. It can also be measured indirectly, by determining which product is scored higher (more liked) than another one in a multi-product test. By using appropriate scales it is possible to directly measure the degree of liking and to compare preference from these data. The two most used methods to measure preference and acceptance are the paired-comparison and the nine-point hedonic scale tests. The paired-comparison test is similar to the one described in the discrimination test but, in the present situation, the consumer will indicate which one of the two coded samples is preferred. The nine-point hedonic scale is probably the most useful and used worldwide in spite of some criticism about such a scale.[38]

Similar to what was mentioned for descriptive and discriminative tests, attention has to be paid concerning sample preparation and sample presentation order when an acceptance test is carried out. More details about affective testing are provided by Stone and Sidel.[38] After collecting the data, it is necessary to analyse them to get the results, and enable the researcher to prepare a report, with recommendations and further actions. A classical statistical method to analyse preference data is analysis of variance. However, a mean of sample preference is obtained, which does not take into account the individual's preference. Advanced statistical methods have been used to sort this problem out. Some tools such as preference mapping[67] and cluster analysis have been successfully used. Preference mapping has the advantage of identifying the sensory attributes that drive consumer preference, and cluster analysis identifies segments of

Table 14.2 Sensory attributes of coffee beverage using descriptive analysis.

Author	Product	Sensory attributes
Della Modesta *et al.*[48]	Ground and roast coffee	*Aroma and flavour*: peanut, animal, sour, burnt rubber, characteristic, caramel, cereal, chocolate, ash, citric, floral, wood, metallic, burnt, chemical, rancidity, tobacco, toast and green *Taste*: acid, bitter, sweet and sour *Mouthfeel*: astringent and bodied
Monteiro[49]	Ground and roast Brazilian coffee (Soft, Hard and Rio) by varying the roasting (light, dark and espresso).	*Appearance*: colour, oily and turbid *Aroma*: characteristic, green bean, sweet, caramelised, almond, fermented and burnt *Flavour*: characteristic, fermented and burnt *Taste*: residual bitter, sweet and acid *Mouthfeel*: astringent
Silva[50]	Ground and roast organic coffee	*Appearance*: colour and turbidity *Aroma*: caramelised, almond, fermented, green bean and burnt *Flavour*: burnt *Taste*: bitter, residual bitter and acid *Mouthfeel*: astringent
Geel *et al.*[51]	Instant coffee	*Appearance of the powder*: colour, granulometry, symmetry and density *Powder aroma*: fish *Beverage appearance*: solubility and turbidity *Beverage aroma*: cow leather/animal, cocoa, malt, cereal toast, nuts, earth, spices, acid, sweet, mushroom and root *Beverage flavour*: cocoa, malt, nuts, toast, acid, bitter and sweet *Mouthfeel*: body and astringency
Mendes[52]	Espresso coffee	*Appearance*: colour of the foam, creaminess of the foam, brown colour, powder on the surface, residual powder *Aroma*: coffee aroma intensity, toast, caramel, cereal, acid and burnt *Flavour*: coffee flavour, toast, cereal and burnt *Taste*: sweet, acid and bitter *Mouthfeel*: body and astringency

Moura *et al.*[53] used the attributes described by Howell[54]	Ground and roast arabica and conilon coffee and blends	*Aroma*: characteristic and fragrance of the powder *Flavour*: characteristic, caramel, chocolate, toast, bread, citrus fruits and residual flavour *Taste*: sweet, bitter and acid *Mouthfeel*: body
Perazzo *et al.*[55]	Decaffeinated instant coffee	*Appearance*: brown colour, presence of foam, gloss and turbidity *Aroma*: smoke, burnt, toast, typical of coffee, sweet, green and fruity *Flavour*: burnt, toast and typical *Taste*: acid, bitter and sweet *Texture*: astringency, body and creamy
ICO[56]	Ground and roast coffee	*Aroma*: animal, ash, burnt, smoke, chemical, medicinal, chocolate, caramel, cereal, malt, roast, earth, floral, fruit, citric, green grass, green, medicinal plants, nuts, rancid, rubber, spice, tobacco, wine and wood *Taste*: acid, bitter, sweet, salt and sour *Mouthfeel*: body and astringency
Santos *et al.*[57]	Arabica and conilon coffee blends	*Aroma*: Chocolate, characteristic, sweet, stale, cereal *Flavour*: Characteristic flavour, sweet taste, stale flavour, acid taste, bitter taste *Taste*: acid, bitter, sweet, acid *Mouthfeel*: astringency, body
Di Donfrancesco *et al.*[58]	Ground and roast Colombian coffee	*Aroma and flavour*: roasted, burnt, acid, smoky, ashy, woody, grain, malt, brown, spice brown, pepper, sweet aromatics, vanilla, honey, molasses, nutty, cocoa, chocolate, floral, fruity, overripe/near fermented, green, beany, tobacco, fermented, musty/dusty, stale, cardboard, caramelised *Aroma*: pungent, medicinal, sour aromatic, rubber-like, medicinal *Flavour*: syrup (maple), raw, astringent, metallic, bitter, sour, sweet *Aftertaste*: bitter, astringent, sour

consumers with similar preference, allowing the elaboration of marketing strategies. The study developed by Santos *et al.*[57] is an example of the usefulness of preference mapping applied to coffee. The authors evaluate the sensory characteristics and acceptability of coffee beverages prepared from different blends of arabica beans and conilon coffee beans. Beverages with different proportions of conilon, *i.e.* containing zero (100% arabica), 10, 20, 40, 60, 80 and 100% of conilon were evaluated. Segments of consumers with different liking were identified and a satisfactory acceptance for the beverages up to 40% of conilon beans was observed.

14.3.2.2 Descriptive Test with Consumers: the Use of Check-All-That-Apply

The main applications of consumer-descriptive studies are to achieve more thorough interpretation and understanding of consumer responses, to provide more specific product guidance. A recent methodology that enables to achieve this goal is CATA.

CATA questions are becoming a popular, quick and easy tool to acquire consumer-based sensory product characterisations. A list of attributes is presented to participants and they are asked to mark which words, phrases or terms are appropriate to describe their opinion about the sample under evaluation. One advantage of such a method is that it allows including hedonic and emotional terms, purchase intention, potential application, product positioning or any other term that the consumer might make any association with the sample, in addition to the sensory attributes.[68]

The consumers included in the study must be selected according to the purpose of the research because there are considerable differences for example between young and elderly individuals on sensory perception and pleasantness of food flavours, as well as related to cultural background and other individual characteristics.

The procedures related to the samples' preparation to apply the CATA questions have to be similar to any other sensory test, *i.e.*, when coffee samples are evaluated using such a method, it has to balance the order of samples' presentation,[41] to serve all samples at the same temperature, and monadically, as well as to code them using three-digit numbers. The attributes can be elicited using terms from the literature, and also asking consumers to identify them by presenting samples with different characteristics to allow the elicitation of distinct sensory attributes. It is carried out in preliminary sessions. According to Ares and Jaeger[69] it is recommended to balance the order of attributes. Furthermore, it is important that all terms are well understood by consumers and related to the vocabulary they normally use. The number of samples to be analysed using CATA questions can vary up to 12; however, to generate the sensory spaces at least five samples have to be taken into account in the study.[65]

It is possible to relate CATA results to consumer acceptance and, for that, CATA studies are often supplemented with liking questions and/or

Please, check all the terms or phrases that best describe the coffee you have just tried:

[ ] Burnt aroma	[ ] Caramel
[ ] Smoke aroma	[ ] Cardboard
[ ] Rubbery-like aroma	[ ] Astringent
[ ] Fermented aroma	[ ] Body
[ ] Fresh aroma	[ ] Weak
[ ] Medicinal aroma	[ ] Watery
[ ] Typical aroma	[ ] Stale
[ ] Sweet	[ ] Good flavour
[ ] Acid	[ ] Delicious
[ ] Bitter	[ ] Off flavour
[ ] Fermented	[ ] Perfect
[ ] Chocolate	[ ] Strong residual

Figure 14.2 Example of the evaluation sheet for CATA questions of coffee beverages.

might include the evaluation of an ideal (hypothetical) product, as well as to collect demographic data. As a consequence, the number of consumers usually used in CATA questions is the usual considered in hedonic tests (100–120). Figure 14.2 shows an example of the CATA questions for evaluating coffee beverages.

Correspondence analysis (CA) can be used to analyse the data from the CATA questions when five or more samples were considered in the design. It allows obtaining a sensory map of the samples and the CATA terms. This map enables the observation of the similarities and differences between samples, as well as their main sensory characteristics. When fewer than five samples were analysed, Cochran's Q test was used to investigate which attributes (terms) differed among samples.

CATA questions are becoming more and more popular in sensory science mainly because this methodology is useful when there is not enough time to train a sensory panel. Although they provide a sensory description of the samples, it should be mentioned that CATA questions do not give the intensity of the sensory attributes. (For more about CATA related to coffee, see Chapter 15.)

14.4 Concluding Remarks

Sensory properties (appearance, aroma, flavour and texture) are the most important quality attributes driving consumers' choices for coffee beverages. However, cultural, religious and food-related lifestyle can affect consumers' acceptance and expectation. Furthermore, there are several other factors like brand, price, nutritional composition and information, production information, advertising, convenience, *etc.*, which can create and affect the expectations and decisions towards coffee purchase.

Sensory evaluation plays an essential role on the assessment of coffee beverage quality, and several methods can be used depending on the goal of the project.

References

1. http://icocoffeeorg.tumblr.com/post/147286923545/coffee-consumption-increases-as-market-hits, last accessed July 2016.
2. E. E. M. Mori, N. Bragagnolo, M. A. Morgano, V. D. A. Anjos, K. Yotsuyanagi, E. V. Faria, *et al.*, *Foods Food Ingredients J. Jpn.*, 2003, **208**, 416.
3. R. Deliza and H. J. H. MacFie, *J. Sens. Stud.*, 1996, **11**(2), 103.
4. L. Bredahl, *Food Qual. Prefer.*, 2003, **15**(65), 2003.
5. A. Bernués, A. Olaizola and K. Corcoran, *Food Qual. Prefer.*, 2003, **14**, 265.
6. K. Hopper, R. Mai, S. Zahn, S. Hoffman and H. Rohm, *Appetite*, 2012, **59**, 949–955.
7. P. A. M. Oude Ophuis and H. C. M. Van Trijp, *Food Qual. Prefer.*, 1995, **6**(177), 1995.
8. K. G. Grunert, *Meat Sci.*, 2006, **74**, 149.
9. S.-E. Kim, S. M. Lee and K.-O. Kim, *Food Qual. Prefer.*, 2016, **52**, 124–132.
10. J.-B. E. M. Steenkamp, *J. Bus. Res.*, 1990, **21**(4), 311.
11. E. J. Van Loo, V. Caputo, R. M. Nayga Jr and H.-S. Seo, *Ecol. Econ.*, 2015, **118**, 215.
12. V. Caputo, R. M. Nayga and R. Scarpa, *Aust. J. Agric. Resour. Econ.*, 2013, **57**, 465.
13. L. Sirieix, M. Delanchy, H. Remaud, L. Zepeda and P. Gurviez, *Int. J. Consum. Stud.*, 2013, **37**, 143.
14. K. G. Grunert, S. Hieke and J. Wills, *Food Policy*, 2014, **44**, 177.
15. R. Vecchio and A. Annunziata, *J. Clean. Prod.*, 2015, **86**, 335.
16. C. Lange, P. Combris, S. Issanchou and P. Schlich, *Food Res. Int.*, 2015, **76**, 317–324.
17. J. L. Hsu and W.-C. Hung, *Food Qual. Prefer.*, 2005, **16**, 361.
18. D. Labbe, A. Ferrage, A. Rytz, J. Pace and N. Martin, *Food Qual. Prefer.*, 2015, **44**, 56.
19. M. Masson, J. Delarue, S. Bouillot and J.-M. Siefferman, *Food Qual. Prefer.*, 2016, **47**, 156.
20. P. Varela, G. Ares, A. Giménez and A. Gámbaro, *Food Qual. Prefer.*, 2010, **21**, 873.
21. H. R. Moskowitz, M. Reisner, J. B. Lawlor and R. Deliza, *Packaging Research in Food Product Design and Development*, 2009, Wiley-Blackwell, Hoboken.
22. S. Spinelli, C. Masi, G. P. Zoboli, J. Prescott and E. Monteleone, *Food Qual. Prefer.*, 2015, **42**, 1.

23. A. M. King and J. Cook, *Manage. Accounting*, 1990, **72**(5), 41–45.
24. G. A. Zarkin and D. W. Anderson, *Am. J. Agr. Econ.*, 1992, **74**, 1201.
25. B. Lundgren, in *Adv. in Catering Technol.*, ed. G. Glew, Applied Science Publishers Ltd, London, 1979, p. 453.
26. J. McCullough and R. Best, *J. Cons. Affairs*, 1980, **14**(1), 180.
27. A. V. Cardello and F. M. Sawyer, *J. Sens. Stud.*, 1992, **7**, 253.
28. R. Deliza, H. MacFie and D. Hedderley, *Braz. J. Food Technol.*, 1999, **2**(1–2), 63.
29. J. H. Behrens, N. D. M. Villanueva and M. A. A. P. Silva, *Int. J. Food Sci. Technol.*, 2007, **42**, 50.
30. R. Deliza and H. J. H. MacFie, in *Food, People and Society. A European Perspective of Consumers' Food Choices*, ed. L. J. Frewer, E. Risvik and H. Schifferstein, Springer-Verlag, Berlin, 2001, p. 55.
31. D. Martin, *Food Sci. Technol. Today: Proceed.*, 1990, **4**(1), 44.
32. P. K. Kopalle and D. R. Lehmann, *J. Marketing Res.*, 1995, **32**, 280.
33. S. Issanchou, *Meat Sci.*, 1996, **43S**, 5.
34. H. N. J. Schifferstein, in *Food, People and Society, a European Perspective of Consumers' Food Choices*, ed. L. Frewer, E. Risvik and H. Schifferstein, Springer Verlag, London, 2001.
35. R. Deliza, H. MacFie, A. Feria-Morales and D. Hedderley, *Braz. J. Food Technol.*, 2000, **3**, 97.
36. M. Charles, A. Romano, S. Yener, M. Barnabà, L. Navarini, T. D. Mark, F. Biasoli and F. Gasperi, *Food Res. Int.*, 2015, **69**, 9.
37. H. J. H. MacFie, *Nutr. Rev.*, 1990, **48**(2), 87.
38. H. Stone and J. L. Sidel, *Sensory Evaluation Practices*, Elsevier Academic Press, San Diego, USA, 2004, p. 377.
39. R. L. McBride and H. J. H. MacFie, *Psychological Basis of Sensory Evaluation*, Elsevier Science Publishing, New York, NY, 1990.
40. ISO, 8589, *Sensory Analysis – General Guidance for the Design of Test Rooms*, 2011, p. 16.
41. H. J. H. MacFie, N. Bratchell, K. Greenhoff and L. V. Vallis, *J. Sens. Stud.*, 1989, **4**, 129.
42. P. Varela and G. Ares, in *Novel Techniques in Sensory Characterization and Consumer Profiling*, ed. P. Varela and G. Ares, CRC Press, Taylor & Francis Group, Boca Raton, pp. 1–8.
43. A. M. O. Gonçalves, *Influence of the Defected Coffee Beans on the Beverage Sensory Characteristics Perception by Consumer*, Federal Rural University of Rio de Janeiro, Master Thesis, 2006, p. 67.
44. H. T. Lawless and H. Heymann, *Sensory Evaluation of Food. Principles and Practices*, New York, Springer, 2010.
45. H. Heymann, E. S. King and H. Hopfer, in *Novel Techniques in Sensory Characterization and Consumer Profiling*, ed. P. Varela and G. Ares, CRC Press, Taylor & Francis Group, Boca Raton, 2014, p. 9.
46. A. A. Williams and G. M. Arnold, *J. Sci. Food Agric.*, 1985, **36**, 204.
47. C. Narain, A. Paterson and E. Reid, Free choice and conventional profiling of commercial black filter coffees to explore consumer perceptions of character, *Food Qual. Prefer.*, 2003, **15**, 31.

48. R. C. Della Modesta, E. B. Gonçalves, P. B. Mattos and J. C. S. Ferreira, *Boletim de Pesquisa* EMBRAPA Agroindústria de Alimentos, Rio de Janeiro, 1999, vol. 22, p. 33.
49. M. A. M. Monteiro, PhD Thesis, Universidade Federal de Viçosa, Minas Gerais, Brazil, 2002.
50. A. F. Silva, Master Dissertation, Universidade Federal de Viçosa, Viçosa, Minas Gerais, Brazil, 2003.
51. L. Geel, M. Kinnear and H. L. De Kock, Relating consumer preferences to sensory attributes of instant coffee, *Food Qual. Prefer.*, 2005, **16**, 237.
52. L. C. Mendes, PhD Thesis, Universidade Estadual de Campinas, Campinas, São Paulo, Brazil, 2005.
53. S. C. S. R. Moura, S. P. M. Germer, V. D. A. Anjos, E. E. M. Mori, L. H. C. Mattoso, A. Firmino and C. J. F. Nascimento, *Braz. J. Food Technol.*, 2007, **10**(1), 17–25.
54. G. Howell, *10th Annual Conference & Exhibition "Peak of Perfection" – Presentation Handouts*, 1998, p. 17.
55. K. K. N. Perazzo, L. D. Carvalho, H. Oliveira, J. T. Tavares, F. L. Góis and M. E. O. Mamede, in *XVI Encontro nacional e II Congresso Latino-Americano de Analistas de Alimentos*, 2009.
56. International Coffee Organisation (ICO), *Vocabulary to describe the coffee beverage flavours: definitions*, Available at: http://www.ico.org/pt/vocab_p.asp, last accessed June 2016.
57. E. S. M. dos Santos, R. Deliza, D. D. G. C. Freitas and F. M. Corrêa, *Semina: Cienc. Agrar.*, 2013, **34**(5), 2297.
58. B. Di Donfrancesco, G. N. Guzman and E. I. V. Chambers, *J. Sens. Stud.*, 2014, **29**, 301.
59. P. Varela and G. Ares, *Food Res. Int.*, 2012, **48**, 893.
60. D. Thomson and J. McEwan, *Appetite*, 1988, **10**, 181.
61. V. Dairou and J.-M. Sieffermann, *J. Food Sci.*, 2002, **67**(2), 826.
62. S. Chollet, D. Valentin and H. Abdi, in *Novel Techniques in Sensory Characterization and Consumer Profiling*, ed. P. Varela and G. Ares, CRC Press, Taylor & Francis Group, Boca Raton, 2014, p. 207.
63. C. Dehlholm, in *Novel Techniques in Sensory Characterization and Consumer Profiling*, ed. P. Varela & G. Ares, CRC Press, Taylor & Francis Group, Boca Raton, 2014, p. 229.
64. E. Teillet, P. Schlich, C. Urbano, S. Cordelle and E. Guichard, *Food Qual. Prefer.*, 2010, **21**, 463.
65. G. Ares and S. R. Jaeger, in *Rapid Sensory Profiling Techniques and Related Methods. Applications in New Product Development and Consumer Research*, ed. J. Delarue, J. B. Lawlor and M. Rogeaux, Elsevier: Oxford, 2015, pp. 227–245.
66. G. Hough, I. Wakeling, A. Mucci, E. Chambers, I. Méndez, L. R. Alves, *Food Qual. Prefer.*, 2006, **17**(6), 522.
67. K. Greenhoff and H. J. H. MacFie, in *Measurement of Food Preferences*, ed. H. MacFie and D. M. H. Thomson, Blackie Academic & Professional, London, 1994.

68. M. Meyners and J. C. Castura, in *Novel Techniques in Sensory Characterization and Consumer Profiling*, ed. P. Varela and G. Ares, CRC Press, Taylor & Francis Group, Boca Raton, 2014, p. 272.
69. G. Ares and S. R. Jaeger, *Food Qual. Prefer.*, 2013, **28**, 141.

CHAPTER 15

An Emotion Lexicon for the Coffee Drinking Experience

K. ADHIKARI*[a], E. KENNEY[a], N. BHUMIRATANA[b] AND E. CHAMBERS IV[c]

[a]University of Georgia, Department of Food Science & Technology, 1109 Experiment Street, Griffin, GA 30223, USA; [b]PepsiCo AsiaPacific, Research & Development, 17th Floor Sukhamvit Road, KlongTon, KlongToey, Bangkok 10110, Thailand; [c]Kansas State University, Sensory Analysis Center, 1310 Research Park Drive, Ice Hall, Manhattan, KS 66502, USA
*E-mail: koushik7@uga.edu

15.1 Introduction

Coffee is the most consumed beverage after water with around two billion cups of coffee being consumed every day around the world. Coffee consumption has risen by ~43% in the last 15 years and represents a $100 billion industry.[1,2] Given its popularity in today's world, understanding the relationship between coffee preferences and the emotional experiences among consumers could be beneficial for researchers and product marketers across the globe. Coffee drinking is rooted in tradition with millions of devoted followers. For example, in the Americas, the Middle East and Europe, coffee is commonly consumed in a café and is a drink for social interaction.[3]

While there has been copious research about the sensory and chemical aspects of coffee and coffee beans, we know surprisingly little about consumer psychology during coffee consumption. What are people really experiencing

Coffee: Production, Quality and Chemistry
Edited by Adriana Farah

Published by the Royal Society of Chemistry, www.rsc.org

when they drink a cup of coffee? Emotions are a big part of the consumer experience, particularly with something as culturally entrenched as coffee. Developing an understanding of emotions will add a layer of explanation to how people truly perceive their coffee and allow us to better conceptualize coffee consumption.

15.2 Why Study Food-evoked Emotions?

Any influence that food has on emotions can have broad significance to consumer behavior. Measuring emotions is a relatively new field in sensory and consumer science, but one that has shown great promise by providing information about products that goes beyond traditional sensory and acceptability information. Foods are designed and marketed on the basis of potential emotional impact on consumers. Emotion profiling can help companies tailor products to different consumer segments or market specific categories or brands.[4] Additionally, knowledge of consumer emotions in response to specific foods can help to map a product category and relate a product to the brand essence.[5]

Measuring emotions can also reveal previously unknown aspects of sensory profiles and product attributes. Liking is one way to discriminate between products, but industry produces a lot of products with similar liking. Information about consumer emotions can discriminate between products with equivalent liking, thus providing an important source of information for product development and marketing.[6] Emotions are also critical in understanding consumer choice. Measuring emotions as compared to only using liking information adds predictive power and strength to a model predicting for consumer choice.[7]

15.2.1 Emotions and Their Origin

An emotion is a brief, intense physiological and mental reaction focused on a referent.[5] Appraisal theorists in the field of psychology explain emotions as positive or negative feelings based on a person's appraisal or evaluative judgment of a stimulus. These evaluations of external and internal input can be either conscious and deliberate or unconscious and automatic.[8] Emotions overall are subjective in nature, often accompanied by physiological responses, and may result in specific actions to address them. Moods, while they may still be categorized as 'feelings', are generally longer-lasting and less intense than emotions.[9]

There are five typical referents for food emotion: sensory attributes, experienced consequences, anticipated consequences, personal or cultural meaning and actions of associated agents that include people involved in the experience. When consumers were asked to indicate the product aspect that correlated with the emotion they were experiencing 49% cited smell and taste as eliciting emotion, while 23% credited the food quality and 14.6%

said it was the anticipated consequence and subsequent consumption of eating food. Important contextual factors included social events and consumption moments.[10]

There have been many efforts to organize emotion words into a hierarchy or model to facilitate conceptualization. The most common model used for food-evoked emotions, and the one found appropriate for this research, is the circumplex model of affect – a bipolar model that can be arranged on two intersecting axes of valence (positive to negative) and arousal (strong to weak). Most emotions are able to be mapped onto one of the four quadrants created by this model.[11,12]

15.2.2 Measuring Emotions

There is a wide array of instruments available to measure consumer emotions, including physiognomy or face-reading, visual surveys and verbal surveys. The most commonly utilized method for emotion research is a verbal survey. Verbal surveys or questionnaires ask consumers to self-report the emotions they are experiencing by checking them off of a list or scaling them from low to high intensity. Providing consumers with pre-existing lists of emotions results in a much higher quantity of data that can be subsequently analyzed (27 words on the average) whereas asking consumers to freely list the emotions they are experiencing causes them to struggle for words (averaging less than 4 words per person).[13]

According to the Oxford dictionaries, a lexicon is "the vocabulary of a person, language, or branch of knowledge." In emotions research, it refers to the list of emotions that have been found to be relevant to that particular product. Currently, the most widely used emotion lexicon is the EsSense Profile®, a list of 39 emotions that can be rated on a 5-point scale from 0 (not at all) to 4 (extremely). It is a general 'eating experience' scale that can be broadly applied to all foods. However, different foods have unique sensory characteristics and functional purposes that could induce or elicit distinctive sets of emotions that could have a widely different emotional impact on consumers. If the resources are available, it is recommended that a lexicon be developed for specific product categories and applications.[5]

15.3 An Emotion Lexicon for the Coffee Drinking Experience (CDE)

15.3.1 Developing the Initial Lexicon

A consumer-generated emotion lexicon was developed in order to fully describe the coffee drinking experience (CDE), and show that general food emotion lexicons such as the EsSense Profile® might not completely elicit the range of emotions,. The EsSense Profile® was used as a foundation for the CDE lexicon development throughout the study because it has been validated and shown to discriminate successfully among different categories and within the same food types.[5,14,15]

A series of five mini-focus groups, followed by two larger focus groups of heavy coffee users, were used to generate emotion terms related to the CDE. The participants were asked to share their coffee drinking habits and their feelings during coffee drinking. They gave definitions for a 'good' cup and a 'bad' cup of coffee, and discussed emotions they experienced during coffee consumption at different locations – such as a coffee shop, restaurant, home, office or work and on-the-go – in order to generate as many emotion terms as possible. These focus groups generated 79 new emotion adjectives related to the CDE, and when the 39 terms from the EsSense Profile® were included, the total was 118 terms.[14,15]

These 118 terms were further fine-tuned by performing a Check-All-That-Apply (CATA) test with 48 coffee consumers using two popular coffee samples (a Breakfast Blend and a Dark Roast). Emotion terms that were checked less than 10 times were eliminated from the lexicon, which resulted in a final list of 86 terms, including 39 EsSense Profile® terms. This list of generated terms shows an overwhelming trend towards positive emotions, known as hedonic asymmetry; and therefore upset, frustrated, sad, disappointed, annoyed and grouchy were included in the initial lexicon of 86 terms. This is commonly seen in food-evoked emotion measurement, as people respond to commercial products with primarily positive emotions.[10,14,15]

15.3.2 Refining the Initial Lexicon to Create the Final Lexicon

An emotion lexicon should have a sufficient number of terms to reveal significant differences in the testing conditions or products, but care should be taken not to overwhelm the participants. Emotion questionnaires usually list around 30–40 terms in order to avoid missing out on information and ensure consistent results.[16]

In order to further refine the list of terms and create a final lexicon, 94 coffee drinkers (3 times/week or higher) were asked twice to score the 86 emotion terms on a 5-point scale (1 = not at all, 5 = extremely) before the coffee was served and while drinking a cup of coffee. They were also asked to indicate overall acceptance while drinking the coffee. The difference in ratings before and during coffee drinking was calculated to determine the influence of coffee consumption on emotion. Six coffee samples, with roast levels ranging from light to dark, were used (Table 15.1), and consumers made six visits in total to evaluate the six coffee samples (one coffee sample per visit). A student lounge was used as a testing location to mimic the casual, social atmosphere of a coffee shop. Keurig® K-Cups were used in this study to enable a randomized design of products among consumers and ensure similar serving temperature.[14,15]

The first step in the data analysis involved clustering the consumers into six clusters based on their overall acceptability rating for the six coffee samples.

Table 15.1 Description of the coffee samples used to refine the initial lexicon.

Type/blend	Roast level
Breakfast	Light
Nantucket	Medium
Sumatra reserved	Dark
Kona	Medium
Italian roast	Dark
Newman – special blend	Medium/dark mix

Table 15.2 The final lexicon of 44 terms describing the coffee drinking experience.

Active[a]	Disgusted	Jolted	Relaxed
Annoyed	Educated	*Joyful*	Rested
Awake	Empowering	Jumpstart	Rewarded
Balanced	*Energetic*	*Merry*	*Satisfied*
Boosted	*Free*	Motivated	Social
Bored	Fulfilling	Nervous	Soothing
Clear-minded	Fun	Off-balance	Special
Comfortable	*Good*	*Peaceful*	*Understanding*
Content	Grouchy	*Pleasant*	*Warm*
Curious	*Guilty*	*Pleased*	*Wild*
Disappointed	In-control	Productive	*Worried*

[a]The emotion terms in *italics* were retained from the EsSense Profile®.

Then, stepwise regression analysis with forward selection was performed on the entire emotion data set by each consumer cluster and by each coffee sample, using overall acceptability scores as the dependent/predicted variable in the model. The significant terms at α 0.20 from the entire data set, and the terms that appeared at least three times in the clusters or in the coffee samples were retained and this list became the final lexicon (Table 15.2). This list of 44 emotion terms had 27 new terms and 17 terms that were retained from the EsSense Profile®.[14,15]

15.3.3 A Further Insight into the Final Lexicon

As mentioned above, consumers were grouped into six clusters based on their overall acceptability scores, depending on which coffees they liked or disliked. Emotion profiles for each coffee sample within the consumer cluster were distinct and the product-emotion Principal Components Analysis (PCA) bi-plots demonstrated that each consumer cluster responded differently to the coffees they rated the highest. This indicates that each group of coffee drinkers sought different affective feelings from the drinking experience. Results indicated that coffee drinkers consume coffee for three distinct reasons—high energy emotions, low-energy emotions or a focused mental state.[15]

Figure 15.1 Word Cloud or Wordle™ showing the frequency of usage of the 44 emotion terms in the final lexicon.

Figure 15.2 Word Cloud or Wordle™ showing the frequency of usage of the 12 most used terms out of the 44 emotion terms in the final lexicon.

To gain further insight into the final lexicon, the frequency of usage of all 44 terms was tabulated for the entire data set. This was then presented as a Wordle™ or a word cloud (Figure 15.1). Further, the 12 most used terms (Awake, Boosted, Energetic, Jolted, Jumpstart, Pleased, Productive, Relaxed, Rewarded, Satisfied, Soothing and Warm) were selected and also presented as a Wordle™ (Figure 15.2). Using the word clouds, this analysis showed the comparative importance of the emotion terms in a semi-quantitative visual approach, which is very easy to interpret.

Out of these 12 terms, the five most used words – 'boosted', 'productive', 'jump-start', 'jolted' and 'energetic' – were all positive and high energy terms that might be associated not only with the psychological impact of drinking coffee, but also with the physiological impact of a caffeinated beverage. All the 12 terms were positive out of which eight were developed in the study and only 'energetic', 'pleased', 'satisfied' and 'warm' were retained from the EsSense Profile®.

15.4 Conclusion

In conclusion, coffee drinking could be considered an emotional experience for coffee consumers. Coffee drinkers have varying preferences for coffee and appear to seek different emotional experiences from the beverage. Understanding emotions that occur during the coffee drinking experience, and how they relate to acceptance and consumption behavior could help the coffee industry gain better insight into the coffee consumer's psyche. The 44-term consumer-generated emotion lexicon developed in this study could be used to gain a more thorough understanding of the coffee drinking experience. Future research may focus on understanding differences in the emotional experience of drinking coffee among various demographic and cultural profiles around the world and identify the influence of these factors on coffee consumption.

References

1. http://www.businessinsider.com/facts-about-the-coffee-industry-2011-11#after-crude-oil-coffee-is-the-most-sought-commodity-in-the-world-1, last accessed June 2015.
2. http://www.bbc.com/news/magazine-32736366, last accessed June 2015.
3. http://ezinearticles.com/?Coffee-Culture-Around-the-Wold&id=4282795, last accessed June 2015.
4. S. Gutjar, C. de Graaf, V. Kooijman, R. A. de Wijk, A. Nys, G. J. ter Horst and G. Jager, *Food Res. Int.*, 2015, **76**(2), 216.
5. S. C. King and H. L. Meiselman, *Food Qual. Prefer.*, 2010, **21**(2), 168.
6. C. Chaya, C. Eaton, L. Hewson, R. F. Vázquez, V. Fernández-Ruiz, K. A. Smart and J. Hort, *Food Qual. Prefer.*, 2015, **45**, 100.
7. J. R. Dalenberg, S. Gutjar, G. J. ter Horst, K. de Graaf, R. J. Renken and G. Jager, *PLoS One*, 2014, **9**(12), e115388.
8. K. Mulligan and K. R. Scherer, *Emotion Rev.*, 2012, **4**(4), 345.
9. R. P. Bagozzi, M. Gopinath and P. U. Nyer, *J. Acad. Mark. Sci.*, 1999, **27**(2), 184.
10. P. M. A. Desmet and H. N. J. Schifferstein, *Appetite*, 2008, **50**(2–3), 290.
11. J. Russel, *J. Pers. Soc. Psychol.*, 1980, **39**(6), 1161.
12. M. Ng, C. Chaya and J. Hort, *Food Qual. Prefer.*, 2013, **28**(1), 193.
13. S. R. Jaeger, A. V. Cardello and H. G. Schutz, *Food Qual. Prefer.*, 2013, **30**(2), 229.
14. N. Bhumiratana, *PhD Dissertation*, Kansas State University, 2010.
15. N. Bhumiratana, K. Adhikari and E. Chambers IV, *Food Res. Int.*, 2014, **61**, 83.
16. S. C. King, H. L. Meiselman and B. T. Carr, *Food Qual. Prefer.*, **28**(1), 8.

CHAPTER 16

Influence of Genetics, Environmental Aspects and Post-harvesting Processing on Coffee Cup Quality

FLÁVIO MEIRA BORÉM*[a], HELENA MARIA RAMOS ALVES[b], DIEGO EGÍDIO RIBEIRO[a], GERSON SILVA GIOMO[c], MARGARETE MARIN LORDELO VOLPATO[d], ROSÂNGELA ALVES TRISTÃO BORÉM[e] AND JOSÉ HENRIQUE DA SILVA TAVEIRA[f]

[a]Departamento de Engenharia Agrícola da Universidade Federal de Lavras, Lavras, MG, Brazil; [b]Empresa Brasileira de Pesquisa Agropecuária, Embrapa Café, Brasília, DF, Brazil; [c]Instituto Agronomico de Campinas, Campinas, SP, Brazil; [d]Empresa de Pesquisa Agropecuária de Minas Gerais – EPAMIG, Lavras, MG, Brazil; [e]Departamento de Biologia da Universidade Federal de Lavras, Lavras, MG, Brazil; [f]Departamento de Engenharia Agrícola da Universidade Estadual de Goiás, Santa Helena, GO, Brazil
*E-mail: flavioborem@deg.ufla.br

16.1 Introduction

The quality of the coffee beverage is manifested differently depending on its geographic origin, and there are notable variations in sensory profile according to country of origin, microregions, and even different planting locations

Coffee: Production, Quality and Chemistry
Edited by Adriana Farah

Published by the Royal Society of Chemistry, www.rsc.org

in the same farm. Essentially, coffee is a product of its *terroir*. Coffee quality is directly influenced by environmental aspects, both natural and human.

The genetic traits of different varieties and the particular climate, soil, and relief conditions associated with local "know-how", represented by different crop management methods and harvest, processing, and drying techniques, create the identity of the beverage.

In this chapter, the contributions of climate, genotype, and processing on coffee beverage quality will be addressed. However, even more than these factors in themselves, the interaction of these factors is what really counts in expression of coffee quality.

Beverage quality and complexity are the main differentials of a specialty coffee; the more uncommon and exotic the pleasurable sensation and sensory perception the beverage provides to the consumer, the more valued it will be. This is mainly determined by the flavor and aroma formed during roasting from precursors present in the raw coffee bean.[1–6] However, beverage quality may be affected by fermentation, processing conditions, drying, milling, and storage. Depending on the climate and processing conditions, these changes often result in characteristics considered undesirable. In this case, lack of knowledge and low technological levels result in defective beverages. In this chapter, we have chosen not to discuss the factors that result in rotting, the occurrence of fungi, or undesirable fermentation, but rather focus on studies related to quality in connection with the pleasure the beverage may provide. Some insights will be presented in response to one of the most intriguing questions at this time: What are the reasons for the quality of specialty coffees?

Among various studies developed around the world, a case study developed in Brazil will be presented in greater detail so as to provide further understanding in regard to this issue.

16.2 Environment and Coffee Quality

16.2.1 Climatic Suitability and Coffee Quality

Agroclimatic zoning delimits areas with suitable climate and soil potential for crops, and this is fundamental for setting up any agricultural activity. To achieve economic productivity, each crop requires favorable conditions during its vegetative cycle, *i.e.*, it requires certain temperature ranges, a minimum amount of water, and a dry period in the maturation and harvest phases.[7–10]

Fundamental factors for definition of climatic suitability are temperature and moisture, represented by the mean annual temperature and by water deficit in critical crop periods. Meeting these demands is what will allow a determined region to be considered suitable for a given crop.[7–11]

The ideal environment for a species is the one nearest its region of origin. Therefore, knowing the environmental conditions of its natural habitat makes it easier to understand its climatic requirements.[12] Coffee belongs to the Rubiaceae family and, as the family name indicates, one of its characteristics is that its fruits, which generally turn red when mature, resemble rubies. The *Coffea* genus has over 100 species, all originating from the equatorial region of Africa.[13] Of these, two are the most widely produced worldwide for commercial purposes, *Coffea arabica* L. and *Coffea canephora* Pierre.

Among the species of the *Coffea* genus, *C. arabica* L. is the one that exhibits the most widely appreciated sensory profile internationally. The beverage obtained from this species is generally described as aromatic, flavorful, and pleasant to the taste. Nevertheless, some genotypes of *C. canephora* Pierre have exhibited beverages with sensory scores appreciated in various markets and classified as having good quality. This situation constitutes a motivational element for selection of new cultivars of this species. Surprising results have also been obtained from the coffee hybrid arabusta (interspecific hybrid between *C. arabica* L. and *C. canephora* Pierre). Qualitative analyses indicate that specific genotypes had sensory quality compatible with that of arabica coffee, though with slightly lower sensory scores.[14]

In general, arabica coffee is considered to be genetically conditioned to produce a naturally sweet, flavorful, and balanced beverage, without any type of defect. Lower quality beverages, with high astringency and even the presence of flavors considered as defective, such as phenol, common in some coffees from Brazil, potato flavor, common in some coffees from some African countries, moldy and fermented coffee are considered abnormalities for the species *C. arabica* L.[15] These alterations are associated with some flaws in the production, harvest, or post-harvest of the coffee. In addition, certain environmental adversities may also hold back full expression of the genetic potential of the cultivars.

C. arabica L. is understory vegetation from deciduous forests of the southwest of Ethiopia. These are high altitude regions, from 1300 m to 2800 m, with mean annual temperatures from 18 °C to 21 °C and total annual rainfall from 1300 to 1800 mm. The four driest months are November to February, with approximately 10% of total annual rainfall. This is the vegetative resting period for the coffee tree. Flowering is stimulated at the beginning of the rainy period, in March, associated with photoperiods greater than 12 hours daily. *C. arabica* L. is commercially produced in equatorial high plateau regions of Africa, Central America, South America, Mexico, India, and Indonesia, mainly under tree cover. In Kenya, Brazil, and Colombia, arabica coffee is mainly grown in full sun.[16,17]

C. canephora Pierre is vegetation from rainforests of the Congo River basin, near Lake Victoria, regions with altitudes of up to 1000 m, mean annual temperatures from 22 °C to 26 °C, and total annual rainfall greater than 2000 mm, *i.e.*, without occurrence of water deficit. It is produced commercially in equatorial regions of Africa, India, and Indonesia under tree cover. In Brazil, it is likewise mainly grown in full sun.[12,18] It is common to use the term robusta to designate the species *C. canephora* Pierre. However, robusta is only one of the varieties of this species, although, on the one hand, robusta is the variety most grown in the world. In Brazil, practically the entire production of *C. canephora* is derived from the Konillou variety, popularly known as conilon.[19]

Currently, as most commercial coffee production is situated in areas of latitudes greater than 4°, coffee is phenologically in tropical, not equatorial, regions. Under these conditions, the phenological cycle is well defined, *i.e.*, flowering in spring, fruit development in summer, maturation in fall, and harvest in winter.[20] Air temperature is the most important variable for defining climatic suitability for the coffee tree in commercial growing. In plants, temperature increase is directly proportional to photosynthetic activity and some biochemical reactions may be catalyzed enzymatically. Depending on the temperature level, the enzyme activity may be lost, a factor which is associated with plant heat intolerance.[21]

Mean annual air temperatures most favorable to production of arabica coffee are those from 18 °C to 22 °C. In areas with very low mean annual temperature, less than 17 °C, flower buds generally have a delayed dormancy period and slower fruit development. In this case, maturation may overlap the following flowering period, hindering coffee tree vegetation and production.[19,22] Minimum temperatures below 1 °C may lead to leaf and fruit cell death. Yet in regions with mean annual temperatures greater than 23 °C, fruit formation problems occur through flower abortion.[20,23] Frequent occurrence of maximum temperatures greater than 34 °C also causes flower abortion and, consequently, yield loss, especially in years in which the dry season extends for a longer time or is delayed.[20,24,25] Temperatures from 28 °C to 33 °C lead to reduced leaf production and photosynthetic activity of the coffee tree.[26] For *C. canephora* Pierre, mean annual temperatures below 20 °C and above 27 °C are unsuitable for production. Minimum daily temperatures less than 5 °C may lead to leaf and fruit cell death and reduce production.[12,18,20]

For both *C. arabica* and *C. canephora* species, the phases of greatest demand for water are those of leaf and flower bud formation and fruit filling. When water deficit becomes much greater, the coffee tree begins to show the following symptoms: wilting, leaf shedding, shriveling of branches, root death, and the appearance of nutrient-induced deficiencies. The consequence of these symptoms is a decline in yield as the plant normally needs to find moisture in the soil throughout the entire vegetation and fruit formation period. Soil characteristics are fundamental determinants of water availability for the coffee tree. For example, effective soil depth in Brazil should be from 1 to 1.5 m, even though most of the fine roots, which are most active, are concentrated in the first 30 cm of the soil profile.[27]

Rainfall is one of the factors of greatest influence on the phenological phases of the coffee tree and it is used to geographically determine suitable, restricted, or unsuitable areas for coffee growing. Currently, regions restricted by water availability but which adopt supplemental irrigation practices have obtained greater yields than the mean yield through dryland farming. Optimum annual rainfall for the coffee tree is from 1200 to

1800 mm; however, the plant grows and produces well from 800 mm to 2000 mm.[28,29] In the vegetation and fruit formation periods, the coffee tree requires greater soil moisture. In the harvest and resting phases, the need is small and the soil may remain at lower moisture without great harm to the plant.[27,30,31]

Agricultural zoning of the coffee crop must consider both macroclimatic aspects and thermal and moisture factors. Macroclimatic studies may result in areas mapped as suitable but that, nevertheless, have unfavorable local topographic and climatic features, with soil-related restrictions that prove to be ecologically unsuitable for coffee growing.[32] The incorporation of longer climatological series and the use of geotechnology and computer systems allow development of more refined zoning. Currently, zoning considers studies of climatic risks, focusing mainly on elements of water balance, variations in air temperature, and the occurrence of adverse events, using geotechnologies for spatial determination of the results.

The use of more complete meteorological data and a Geographical Information System allow determination of the climatic elements most adequate for zoning. For example, for zoning carried out in Brazil,[24] the elements used to determine suitability were the mean annual temperature range from 18 °C to 23 °C, minimum critical temperatures less than or equal to 2 °C in an instrument shelter, at the level of 30% probability, for the months from May to July, and annual water deficit for soil available water capacity of 125 mm.

However, zoning is a continual process that must always be updated whenever new data or methods that allow greater detailing and proximity to reality become available. Updating is carried out from supplying information from the meteorological data network available in the nationwide Agrometeorological Monitoring System. Programs of this nature are fundamental for strategic actions because they establish homogeneous zones with planting times of lower risk for the coffee crop.

In addition, climatic zoning is affected by global climate changes. The study of climate change on cultivation of arabica coffee (*Coffea arabica* L.) in 19 countries of four continents (America, Africa, Asia, and Oceania) foresees severe losses in production as imminent if coffee growing does not move in the direction of higher altitudes in many countries.[33] Such a move would harm many agricultural regions of the world, due to the limited availability of land, especially in parts of Central America and Asia. On the other hand, the increase in temperature and rain may make specific areas of land adequate for coffee growing, particularly those near the equator in South America. However, this turns into a paradox since migration of coffee growing to areas of greater altitude would lead to the cutting of native forest, which would increase even more the greenhouse effect (the main agent of climate change). Other alternatives need to be contemplated.

The effect of temperature increase on zoning of climate risks for coffee growing in the next 100 years has been studied for some coffee growing regions of Brazil.[34] Based on indications from the most recent report of the IPCC (International Panel on Climate Change), various simulations were carried out, and the impacts that a mean temperature increase of 1 °C, 3 °C, and 5.8 °C in air temperature and an increase of 15% in rainfall would have on Brazilian coffee growing were evaluated. The results indicate reduction from 75% to 95% of the area suitable for growing arabica coffee in the case of a temperature increase of 5.8 °C. These results are valid if the current genetic and physiological characteristics of the arabica coffee cultivars used in Brazil are maintained. These cultivars have a mean annual temperature range from 18 °C to 23 °C as a tolerance limit.

The suitability ranges for coffee production, however, are normally more extensive than those regarding beverage quality. The macroclimatic parameters considered as favorable for obtaining a high quality beverage may be strongly affected by oroclimatic and topoclimatic effects, which may increase ambient moisture and affect, among other qualities, the chemical composition of coffee, the type of microbial activity, and the intensity of the fermentation process. Thus, even regions appropriate for coffee production have climatic diversity that causes variations in beverage characteristics.

Temperature is a climatic factor important in pre-harvest, harvest, and post-harvest. In pre-harvest, high temperatures affect leaf shedding, especially during the fruit formation stage.[35] Leaf shedding during fruit formation may hurt the plant, which needs leaves for fruit filling. In harvest and post-harvest, the influence of temperature may be greater. In addition to the classic explanations of climate negatively modifying the beverage as a result of proliferation of microorganisms, new evidence indicates that determined genotypes synthesize and accumulate compounds under stress conditions, compounds that may be associated with lower sweetness, lower acidity, and greater astringency, which result in lower quality coffees, although the coffees might not be classified as defective beverages.[36]

Classical studies on the effects of climate and environment on coffee beverage quality have reported that it is strongly influenced by both temperature and rainfall. Low temperatures are responsible for delay in the maturation process which, in turn, leads to greater accumulation of chemical and biochemical compounds associated with improvement in coffee aroma.[37] Although this is a legitimate approach, it does not exhaust the phenomena that occur in the coffee plant and fruits. Physiological, metabolomic, and gene expression approaches may generate more evidence for understanding differences in coffee quality in different environments.

An increase in altitude is related to an increase in beverage quality as a result of reduced temperature. This phenomenon is recognized worldwide and has been described in different countries. In Colombia, studies described the relationship between altitude (1450 to 1650 m) and coffee quality and reported a significant improvement in the beverage from coffee grown at higher altitudes.[38] In Honduras, superior quality coffees were obtained in locations with altitudes higher than 1000 m and mean annual rainfall below 1500 mm. However, the quality of coffees from other regions studied in Honduras was not determined.[39] Positive correlations between quality and altitude are also described for other geographic regions.[7,40,41]

Sensory differences for the coffee beverage are associated with the difference between higher and lower altitude, described as "amplitude of altitude". However, specific values of altitude above which coffee quality significantly improves have not been determined.

Topographical relief of the field is another highly relevant environmental factor for coffee growing. In a single region, microclimates may be differentiated according to higher altitude ranges. In addition, there are hypotheses regarding the contribution of slope exposure and levels of solar radiation to final quality. However, few studies evaluate this influence on coffee quality. Studies on the effects of altitude and the geographic positioning of slopes for two different *terroirs* (Orosi and Santa Maria de Dota) were conducted in Costa Rica.[7] It was possible to observe differences in the coffee beverage produced from coffee beans from these different areas, confirming the evidence that the greater the altitude, the better the quality. Nevertheless, significant evidence was not obtained for an effect from different sunlight exposure profiles of the field.

The effect of altitude is consistent among all the studies conducted in different countries of the world, with results that report both sensory variations and variations in the allocation of photoassimilates in coffee leaves and fruit. Altitude and climate play an important role through temperature and availability of light and water, affecting the photosynthetic rate during the coffee maturation period.[40,41]

In addition to the effect of altitude on reduction of the mean temperature of a *terroir*, latitude has a marked effect. This may be observed throughout the world. Near the equator, all studies indicate altitudes greater than 1600 m for coffee growing. However, at increasing distance from the equator, the requirement for higher altitudes is lower.

In Brazil, this effect is easily observed in light of the fact that coffee growing is distributed across an extensive range of latitudes. In the state of Minas Gerais, the largest Brazilian coffee producer, coffee growing is distributed in four different environments. These regions have different characteristics, in relation to both the physical milieu and socio-economic conditions. Because of its vast territorial extension, this region includes several latitude ranges, from the extreme south, with milder temperatures, to the extreme north, with warmer temperatures. Also in reference to climate, high altitudes are synonymous with low temperatures. Considering that air temperature decreases at a mean rate of 0.6 °C for each 100 m of altitude,[43] diverse coffee-producing regions of the state are influenced by the interactions between latitude and altitude.

To demonstrate the effect of the interaction between altitude and latitude, coffee samples were collected from among the participants of coffee quality competitions carried out in the state of Minas Gerais.[44] The 1161 samples were evaluated spatially based on geographic location (latitude and longitude). The temperature, rainfall, and humidity index data were generated by the ZEE (Economic Ecological Zoning of Minas Gerais)[45] in the period of 1961–1990. Discrimination of high and low scores was observed as an outcome of environmental variables, showing the strong influence of temperature, rainfall, altitude, and latitude on the quality of the coffees studied (Figure 16.1). The results confirm that the greatest sensory scores are associated with greater altitudes. However, the greater the latitude, the lower the requirement for high altitudes to achieve the best scores.[44]

Although growing arabica coffee at higher altitudes is known to have a favorable effect on final beverage quality, quantitative and consistent data that describe the influence of climatic conditions on chemical composition of the seed are still quite scarce.[8]

Studies conducted in Costa Rica at altitudes ranging from 900 to 1450 m describe a significant positive effect on biochemical composition (5-caffeoylquinic acid content and fatty acid concentration) of coffee beans of Caturra variety at higher altitudes. However, this effect was not observed for others hybrids of arabica coffee cultivated in El Salvador, Costa Rica, and Honduras, for altitudes from 700 to 1600 m.[4] Since microclimatic parameters were not registered, it was not possible to conclude if the effects of altitude were related to the temperature gradient or to other edaphic and climatic variables.

A more rigorous study relating the main climatic variables (temperature, rainfall, total irradiance, and potential evapotranspiration) and the biochemical composition of green coffee beans was carried out on Réunion island, near Madagascar.[8] Samples from 16 arabica coffee plots were taken in this region because of their particular characteristics, such as rich, homogeneous volcanic soil, different microclimates within short distances, and a dense network of meteorological stations. All the plots were sown in the same

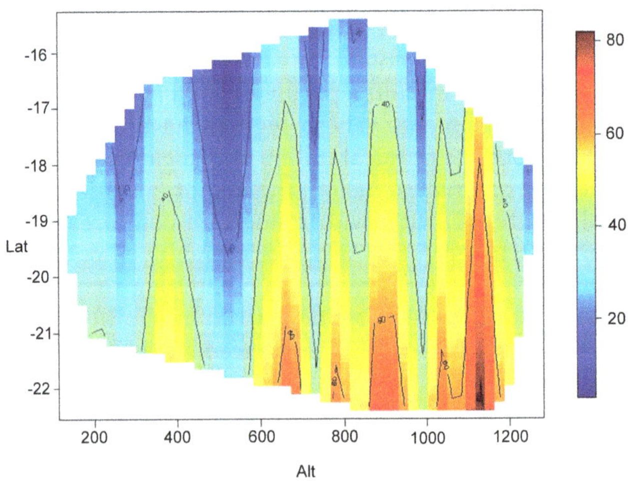

Figure 16.1 Surface of the samples relating the beverage quality score, altitude, and latitude. The greatest sensory scores are associated with the interaction between altitudes and latitude. The greater the latitude, the lower the requirement for high altitudes to achieve the best scores.[41]

year, with the same cultivar and crop management practices. Variations in lipids, chlorogenic acids, caffeine, and sugar content were quantified. However, the results continue to show contradictions since the content of some compounds is dependent on average temperature during seed development while other compounds are not influenced by climate.[8] This reveals the complexity of finding consistent relationships between edaphic/climatic variables, biochemical constituents, and improvement in the aroma and flavor of the coffee beverage.

Later studies carried out in the same location describe that, among the climatic factors, mean air temperature during seed development highly influenced sensory profile.[46] Positive quality features, such as acidity, fruitiness, and quality flavor were correlated with coffees produced in colder microclimates. In contrast, coffees grown under higher temperature conditions had lower acidity, low aromatic quality, and green, earthy notes. This study suggested that climate change, with an accentuated increase in mean temperatures in mountainous tropical areas, will have a substantially negative impact on the quality of highland coffees.

Another approach for identifying the relationships between coffee quality and environmental characteristics proposes studying information contained on the labels of specialty coffees from determined regions. A study of this nature was carried out in Colombia for the Cauca and Nariño regions. The results indicate significant differences in the environments studied and a high relationship between environmental characteristics (such as number of dry months, annual rainfall, and daily temperature range) and beverage quality.[47]

Calculation of the sum of potential evapotranspiration may also be used to improve understanding of the relationships between climate, coffee bean maturation, and the quality of the arabica coffee beverage. Studies carried out in some locations in Brazil showed that the best climatic conditions for natural production of specialty beverages were found at properties located at 1050 m altitude, with high water deficit and low temperatures at the time of harvest. Nevertheless, these indicators are more related to the interruption of fermentation processes that impair the beverage. These results do not provide information about the influence of climatic parameters on the quality of the coffee beverage.[10]

16.2.2 Ecological and Socio-environmental Benefits Associated with the Presence of Vegetation in Areas Planted to Coffee

There are diverse ecological and socio-environmental benefits associated with the presence of vegetation in coffee-growing areas. Ecosystem services such as pollination, pest control, climate regulation, and carbon capture are more noticeable on coffee farms that have fragments of native vegetation or that adopt the shaded coffee management system, or both.[48–52]

Growing shaded coffee has diminished in various regions of the world, and it is associated mainly with large farms. In contrast, most of the shaded coffee management occurs in small farms, which have high diversity index as shown by studies conducted in Central America[53] and India.[54] The contribution of shaded coffee to conservation of biodiversity is noteworthy.

It is known that diverse environmental factors, such as altitude and temperature, influence coffee beverage quality. Nevertheless, the quality of a product is not restricted to its organoleptic aspects. Although scientific reports on the presence of native forest fragment on coffee sensory quality remains scarce, other intangible values, such as fair trade and environmental preservation, are prized by the final consumers. Although some ecosystem services are well known by coffee producers,[55] many other services are still poorly understood due to their indirect effects and the difficulty of being measured.[56]

The preservation of natural plant resources has a much broader impact since these resources minimize soil degradation and favor preservation of water resources. In mountainous regions, the preservation of areas with natural plant growth is even more urgent since these areas also appear on steep slopes.[57]

Studies carried out in mountainous coffee-growing regions of Brazil[57] indicate that the natural vegetation in the area is quite fragmented due to the frequent changes in land use that characterizes these regions. Nevertheless, the high number of small fragments making up the remaining vegetation is of utmost importance. These small areas of natural vegetation are fundamental in the landscape, even though fragmented. These are areas that may favor the continuity of biodiversity and make for overlapping of hotspots of biodiversity in service to the coffee-growing regions.[58] In these mountainous areas of coffee growing, natural vegetation occupies 26.5% of the total area, greater than that recommended by Brazilian environmental legislation.[59] Situations such as these, common to many countries, provide a great environmental service.

An example of the importance of ecosystem services[60] indicates that native vegetation harbors predatory mites, natural enemies of pest-mites that occur in the coffee crop. This occurrence allows the development of ecological management programs with areas of natural vegetation and adjacent coffee-growing agroecosystems.

16.3 Genotype and Coffee Quality

The synthesis, degradation, and accumulation of chemical compounds in green coffee beans, precursors of the flavor and aroma of the beverage, depend on the interaction between genotype and environment.[2,4–6,61,62] Although various studies have been carried out seeking to understand the isolated effect of the factors that determine coffee quality, few of them correlate the influence of the genetic composition of different arabica coffee cultivars with the quality of the coffee beverage.[63]

The subject of coffee quality has been debated for years, and will certainly continue to be studied in the coming decades. Concepts regarding quality are broad and may be understood according to their contextualization and from different perspectives. From the point of view of "product quality", unlike the concept related to "consumer preference", it is necessary to understand that the intrinsic quality of coffee corresponds to all properties that a normal coffee bean may have in terms of its chemical composition and physical and physiological characteristics in accordance with the genetic constitution of the plant, of its interaction with the environment, and manner of post-harvest processing. These properties are manifested in a sensory manner after roasting and preparation of the beverage.

Certain characteristics of coffee quality are relatively stable and inherent to the genotype, and may serve as an indication of species or cultivars. Thus, genotype assumes a relevant role in determination of beverage flavor and aroma. Even when grown under the same environmental conditions, different cultivars may produce coffees with marked differences in overall beverage quality, with different flavor and aroma scores.[64] It is understood that when a cultivar has a genetic predisposition for expressing distinct flavors and aromas in the beverage, it will continue to be recognized by its characteristic flavor and aroma even if there are variations in the intensity of determined sensory attributes in response to environmental variations.

Currently there are a large number of arabica coffee cultivars used in the world. In Brazil alone, 128 arabica coffee cultivars are registered for commercial growing.[65] Nevertheless, most of the cultivars are derived from a narrow genetic base, resulting in great genetic similarity among the main arabica coffee cultivars dispersed throughout the world.[66,67] In the case of Brazil, with the largest production of coffee in the world, it is acknowledged that only four distinct genotypes (Típica, Bourbon, Arabusta, and Híbrido de Timor) contributed to formation of the genetic base of commercial cultivars. Thus, even with the genotype assuming a relevant role in the determination of beverage flavor and aroma, the genetic similarity among the commercial cultivars is so extensive that it impedes any type of differentiation, leading to a situation in which the most outstanding sensory perceptions are conditioned to the region of origin of the coffee and the "local know-how" (environment and processing), rather than the genotype.[67]

The possibility of developing new varieties adapted to different environments and growing systems, making accurate recommendation of productive, high quality cultivars viable, is of great importance to ensure sustainable coffee growing. Some approaches regarding the quality of the arabica coffee beverage indicate that the flavor and aroma are organoleptic characteristics determined essentially by the genotype, serving as a selection criterion for improving the quality of the arabica coffee beverage.[68,69] Coffee aroma, acidity, and flavor seem to be more affected by the genotype than by the environment, indicating that it is possible to improve sensory quality through genetic breeding.[70] On the other hand, astringency, bitterness, and body of the beverage are considerably affected by the environment, indicating low efficiency of genetic selection for these characteristics.[71]

The effects of genotype are also perceived in plant yield and in some physical characteristics of the coffee beans, such as color, shape, and size. These characteristics are direct consequences of the physiological response of the cultivar to the growing environment.[72] Although genetic selection of new arabica coffee cultivars is still undertaken, based primarily on traditional agronomic and technological parameters (yield and size of the coffee bean), it is necessary to understand that the size or shape of the beans, or both, are not always good indicators of beverage quality. It is erroneous to believe that cultivars that produce larger beans are of better quality. Up to now, there has not been sufficient scientific evidence to maintain any significant positive correlation between physical characteristics of the beans and the quality of the arabica coffee beverage.[72] Therefore, it is quite difficult to use physical characteristics as a selection criterion for cultivars with a view toward improving sensory quality. Even though the physical aspect of the coffee, especially that relates to coffee bean size and appearance, is a relevant factor in the sale and assessment of green coffee, beverage quality is certainly the main determinant of the price of specialty coffee, and thus it should be considered as a selection criterion in genetic breeding programs.[73,74]

As for coffee bean size, it should be noted that there are genotypes conditioned to produce very small beans, such as the cultivar Ibairi IAC 4761,[75] and others that produce very large beans, such as the cultivar Maragogipe.[76] Surprisingly, in spite of its reduced bean size, the cultivar Ibairi IAC 4761 is considered a reference for high beverage quality, with a differentiated sensory profile.[77] This reinforces that the physical characteristics of the beans, intrinsic to the genotypes, are permanent and may not be changed according to the environment. However, changes in the size of the beans as a result of climate challenges, such as reduction in bean size in response to prolonged dry periods, are temporary and should not be confounded with an effect of genotype.

The genetic component has been receiving more and more attention from breeding programs for the production of specialty coffees, with new patterns of flavor and aroma. Nevertheless, although the influence of genetic features on the final quality of the coffee beverage is clear, the attention of researchers with a focus on improving quality is quite recent.[78] An important source of sensory variability may be found in coffees coming from non-commercial genotypes of *Coffea arabica* L. from African countries. Studies undertaken in populations of wild accessions and varieties coming from Ethiopia, Sudan, Tanzania, and India indicate the existence of different types of plants, with variations in type of branching, caffeine content, and rust tolerance, confirming the high genetic variability of germplasm banks.[79] In these collections, it is also possible to find variability for qualitative factors related to color, size, and shape of the beans, and also to the sensory profile of the beverage. This offers new perspectives for genetic breeding programs with a view toward selection of new cultivars with superior beverage quality. In these programs, the genetic constitution of the plants determines the manifestation of distinct flavor and aroma among genotypes, which may be used as a differentiating characteristic of cultivars.[63,80]

Studies developed at IAC (Instituto Agronômico de Campinas – Brazil) described the sensory profile of the coffee beverage of some wild accessions coming from Ethiopia, under natural processing (Table 16.1). The results indicate high quality and sensory variability among the accessions analyzed. In addition, a relevant factor for consideration is that the environment where these accessions are grown is of low altitude and high mean annual temperature, showing that it is possible to produce coffees with high sensory quality even in an environment considered (in classical terms) unfavorable to quality. Furthermore, they show the strength of the genetic factor in determination of the quality of the coffee beverage.[41]

The description of sensory profile of wild accessions is an indispensable tool for characterization of plants in the pre-selection of coffee trees in breeding programs with a view toward beverage quality. The superiority of the beverage from the wild accessions as compared to the commercial Brazilian cultivars may be observed in Table 16.1. Both the variety "Catuaí Vermelho IAC 81" and the variety "Catuaí Amarelo IAC 62", used as a control, had an overall score of less than 80 points, not being classified as specialty coffees.

The coffee tree is a perennial plant, and genetic breeding by the genealogical method may take up to 30 years to arrive at a new cultivar. In contrast, there are techniques that allow the incorporation of qualitative characteristics of wild accessions in commercial cultivars in a much shorter period of time. Thus, the development of cultivars through hybridization constitutes an alternative for obtaining new cultivars, in which early selection of F1 hybrids may mean reduction of the time period for such. It is noteworthy that the use of non-traditional varieties for technological applications that result in progress in coffee growing, in addition to making it possible to add

Table 16.1 Sensory profile of some wild accessions coming from Ethiopia and commercial varieties under natural processing, Campinas, SP, 2009.[a]

Genotype/origin	Score	Description of the sensory profile
Ethiopia/Gojjan	86	Fruity, caramel, winey, raisin
Ethiopia/Kaffa	85	Mild floral, fruity, earthy, winey
Ethiopia/Illubabor	84	Mild floral, chocolate, molasses
India/BA	84	Chocolate, caramel, earthy, winey
Ethiopia/Harar	84	Chocolate, fruity, grape, cedar, complex
Ethiopia/Híbrido	84	Chocolate, fruity, caramel, honey, cedar
Ethiopia/Harar	83	Intense floral, fruity, honey, raisins
Ethiopia/Shoa	81	Earthy, woody
Brazil/Catuaí IAC81	78	Good sweetness, caramel, immature, astringent
Brazil/Catuaí IAC62	77	Strong bitterness, rubber, unpleasant flavor

[a]Source: Unpublished data from research (IAC).

value to specialty coffees, will be able to stimulate greater investments in the characterization, use, and maintenance of rare plants, thus avoiding their disappearance.

16.3.1 The Case of Yellow Bourbon

The Yellow Bourbon is currently considered as one of the commercial cultivars of greatest genetic potential for the production of specialty coffees. Due to the sweet flavor and the characteristic aroma it confers to the beverage, this cultivar has drawn the interest of coffee growers in different countries.[74] Several attempts have been made in Brazil and around the world to recover the use of older cultivars with high potential for beverage quality, both through the demands of international buyers and through signs that cultivars like Yellow Bourbon produce differentiated coffees, with particular flavors and aromas that are highly desired by consumers.[80,81]

With the advancement of genetic breeding, the lines IAC 662 of Red Bourbon and lines IACJ2, IACJ9, IACJ10, IACJ19, IACJ20, IACJ22, and IACJ24 of Yellow Bourbon were selected. These lines have plants that are highly susceptible to rust and of medium to high plant height, with red or yellow fruit, early maturation, seeds with average sieve 16, and excellent beverage quality. They are recommended for planting in tropical regions above 1000 meters, for the production of specialty coffees.[82] Studies carried out with diverse lines of Yellow Bourbon indicated the existence of genetic variability for diverse agronomic and technological factors, confirming that the genetic constitution of the plants has a quantitative and qualitative effect on some physico-chemical characteristics of the coffee beans, impressing distinct sensory profiles on the beverage from each cultivar.[63,83]

16.3.2 Beverage Quality of Rust Resistant Cultivars

Though the use of rust resistant cultivars is considered a great advance in the modernization of coffee growing, with some agronomic advantages in relation to susceptible cultivars, there are sectors of the production chain that attribute lower beverage quality to resistant cultivars due to the genetic background inherited from *C. canephora* Pierre through Timor hybrid. Some lines of *C. arabica*, descendants of Timor hybrid,[83] resistant to rust, may in fact have inferior beverage quality due to the introgression of *C. canephora* Pierre genes. However, the low quality was associated with the low sucrose content and high content of chlorogenic acids, which are known not to depend exclusively on the genetic factor.[84]

In studies on genetic similarity,[85] it was observed that the cultivar Ruiru 11, descendent of Timor hybrid, was grouped together with the robusta variety of *C. canephora* Pierre. Considering that beyond genetic composition other factors interfere in determination of sensory quality of coffee,[86] caution is recommended when interpreting data referring to the beverage quality of rust resistant cultivars with introgression of genes from *C. canephora* Pierre, because not all these cultivars exhibit problems with beverage quality. This fact has generated discussions in various countries that seek to develop strategies to combat the economic damage caused by rust. If, on the one hand, studies conducted in Indonesia observed that Timor hybrid showed beverage quality inferior to that of cultivar S795,[87] indicating problems in various sensory attributes, in Brazil there is no concise information that allows a technical–scientific approach to the qualitative performance of arabica coffee cultivars resistant to rust. On the other hand, various technical–scientific reports describe the Castillo cultivar, resistant to rust, as having sensory quality similar to the Caturra variety, traditionally grown for production of specialty coffees.

16.4 Post-harvest Processing and Coffee Quality

16.4.1 Brief History on Post-harvest Methods Nomenclature and Proposal for a New One

Newly harvested lots of coffee may consist of fruits at different stages of maturity, as well as impurities and foreign matters. The uniformity of the fruits and the quantity of impurities depend on the harvesting method and the agronomic conditions of the crop.

Post-harvest includes the operations following harvest up to industrialization of the coffee. The purpose is to eliminate impurities and foreign matter from the harvested coffee and separate it into lots with similar characteristics according to moisture content, the size and shape of the beans, and fruit maturity.

Traditionally, coffee processing is classified as dry process and wet process, also known as "natural" coffee and "washed" coffee, respectively (see Chapter 1).[88] This distinction arose at the beginning of the 18th century in equatorial regions where continual rainfall during the harvest period almost always resulted in inferior quality coffees. Mechanical removal of the exocarp, followed by removal of the mesocarp through natural fermentation, allowed faster drying, avoiding deterioration processes in coffee. Nevertheless, much more than simply avoiding coffee deterioration, the use of this new method resulted in a clear and consistent flavor difference, which was easily perceived between the processes. This method was adopted by various countries throughout the world and this distinction has been used since then. Although, the natural coffee was historically associated to low quality it is not completely correct to be assumed considering the current technologies available to produce natural coffee with high quality.

At the end of the 20th century, new coffee processing methods were developed, and it is not possible to fit them into one category or another. Methods that remove the skin mechanically but that maintain the mucilage intact (method used in Brazil and known as pulped natural), methods that remove the skin mechanically with low use of water (method used in Colombia, Central America, and Brazil and known as mechanically demucilaged coffee), and methods that remove the skin mechanically while removing the parchment with the seed still wet (method used in Indonesia and known as wet hulled coffee) have arisen as processing options. In addition, the terminologies that define the forms of coffee processing have undergone the influence of cultural and regional values. Depending on the producer country, there are many variations in the names given to the different types of post-harvest processes. In addition to being confusing to have various names for the methods variations, they often do not provide an accurate technical description of the process. Such is the case, for example, in the process called "semi-washed" (see Chapter 2).

Not only the fact of applying water or not in post-harvest procedures is important when using different terminologies to classify the existing methods. Another important factor to be considered is the remaining presence of tissues (exocarp and/or mesocarp) adhered to coffee seed after processing. A number of studies provide evidence that the remainder of such tissues after processing as well as the events that occur within the seed during processing promote differences in coffee flavor.[89–93]

In order to assist in inclusion of new processing technologies an alternative nomenclature is proposed considering the above factors:[89–93]

"Fruit-dried coffee": Dry coffees with the exocarp intact. This includes coffees harvested partially or completely dried on the tree, as well as coffees that were harvested mature and dried in a drying yard or on a screen. This category includes the coffees known as "dry process" or "natural" coffees.

"Pulp-dried coffee": These are coffees dried without the skin (exocarp), while maintaining the mesocarp (mucilage) intact. Every method that results in coffee with any remainder of mucilage during drying would fit within this category. This category includes what many call "honey process" or "pulped natural" coffee.

"Parchment-dried coffee": These are the coffees that had the exocarp and the mesocarp removed before drying. The coffee is dried with its parchment (endocarp) intact. The mucilage may be removed mechanically or through natural fermentation. This category includes coffees known as "washed" and "wet process" and "semi-washed".

"Seed-dried coffee": These are coffees that had all the fruit tissues removed (exocarp, mesocarp, and endocarp) before being completely dried. In this case, the exocarp and the mesocarp are removed and the seeds are partially dried with the presence of the parchment. However, the parchment is removed and the final phase of drying occurs only with the silverskin (perisperm) intact. This category includes the coffees known in Indonesia as Giling Basah, or "wet-hulled" coffees.

16.4.2 Influence of Processing on Coffee Quality

In addition to environmental and genetic factors and other factors related to management practices of the coffee crop, the differences found in the flavor and aroma of the coffee beverage are directly associated with physical, chemical, and physiological changes that occur in coffee beans during processing.[89–94]

In general, reports that compare differences in quality between fruit-dried coffee and parchment-dried coffee attribute greater body and lower acidity to fruit-dried coffees, while parchment-dried coffees are described as having a more accentuated aroma and a light and pleasurable acidity.[95–97] In addition, fruit-dried coffees are frequently described as having lower quality compared to coffees produced through the parchment-dried methods that used natural fermentation for removal of mucilage.[61,95–97]

In spite of the numerous physiological and biochemical events that occur in coffees during post-harvest and that may result in the characteristics that distinguish them, the traditional explanation for these differences makes reference to deterioration processes frequently associated with fruit-dried coffee, to the absence of care at the time of harvest and drying. However, such assertions are not sufficient to explain the differences in the quality of specialty coffees processed under controlled conditions, regardless of the processing method.[88,98–101]

In addition to the effect that the presence of the exocarp has on final coffee quality, several issues emerge in regard to the effect of the presence of the mesocarp (mucilage) during drying, and the method used to remove it. Studies conducted in Colombia did not reveal differences in the quality of parchment-dried coffee, regardless of the type of mucilage removal, *i.e.*, by natural fermentation or by mechanical means.[95]

The presence of the mesocarp (mucilage) in some processing methods has been used to explain variations in the sensory profile of the coffee beverage, such as variations in sweetness. Even today, there are hypotheses that variations in coffee sweetness are associated with the quantity of sucrose that moves from the mesocarp to the seed during drying. Nevertheless, there is not sufficient scientific evidence to prove the occurrence of this phenomenon.

Removal of the exocarp favors the embryo germination process.[89] It is believed that there is differentiated germination in pulped seeds because of the removal of inhibitors in the exocarp and mesocarp. Physiological and biochemical analyses have been used to confirm the differences in metabolic activity of seeds among the different processing methods. The expression of specific germination enzymes and the reactivation of cell division due to accumulation of β-tubulin may be used to study these differences.[102–105] Greater values of isocitrate lyase (ICL) enzyme activity, as well as greater accumulation of β-tubulin, are found in parchment-dried coffee seeds compared to fruit-dried coffee seeds, thus indicating greater germination and physiological activity in pulped seeds.[93]

The coffee pulping would trigger diverse reactions related to germination, such as mobilization of reserves, resulting in different metabolic profiles compared to fruit-dried coffee.[102] The breakdown products from sucrose hydrolysis may be consumed or accumulated depending on the coffee metabolism. If, on the one hand, greater activation of the metabolism of parchment-dried coffees results in consumption of glucose and fructose, on the other hand, the lower physiological activity of the embryo in fruit-dried coffees results in accumulation of these sugars. In addition to greater glucose and fructose contents in fruit-dried coffees, differences in the content of several other compounds have been described when compared to other processing methods. These variations occur both in the species *Coffea arabica* L. and in the species *Coffea canephora* Pierre. Analysis of the chemical composition of *C. canephora* coffee, robusta variety, of lots processed as parchment-dried coffee indicate lower contents of free carbohydrates (fructose and glucose), organic acids (quinic and oxalic acid), minerals (K^+, Ca^{2+}, Mg^{2+}, Cu^{2+}), and trigonelline, and greater contents of chlorogenic acids and polysaccharides of the cell wall (arabinogalactans and mannans) and lipids.[100] Sensory analysis of the samples studied showed that in comparison to fruit-dried coffee, parchment-dried coffee exhibited lower astringency, bitterness, rubbery and woody flavor, and a slightly more fruity and acidic flavor. It is believed that the variations in chemical composition are derived from degradation mechanisms, *i.e.*, free glucose and fructose resulting in hydrolysis of polysaccharides; quinic acid, stemming from the breakdown of chlorogenic acids; and phosphoric acid, from the degradation of phospholipids. Greater drying time and the associated lower rate of water removal in fruit-dried coffee are factors which may be considered responsible for the occurrence of these reactions, observed with greater intensity in fruit-dried coffee in comparison to parchment-dried coffee.[100]

The accumulation of free amino acids in green arabica coffee beans, *e.g.*, γ-aminobutyric acid, varies depending on the form of processing.[93] The γ-aminobutyric acid is accumulated in greater amount during fruit-dried coffee processing compared to parchment-dried coffee processing. These results reinforce the evidence that, in addition to degradation mechanisms, different metabolic reactions may occur in coffee fruits during post-harvest, and the extent of these reactions depends on the processing method.[89] In plants, γ-aminobutyric acid is formed from glutamic acid α-decarboxylase, and its accumulation represents a stress reaction.[106] This stress may stem from drying of the processed coffee in different manners. In processing of fruit-dried coffee, the time available for the occurrence of the decarboxylase reaction is greater, as a result of the lower drying rate that occurs in intact fruit compared to drying of parchment-dried coffee.

16.5 Spatial Distribution and Relationship Between Quality, Environment, Genotype, and Processing: Case Study of Specialty Coffees from the Mantiqueira de Minas Region, Brazil

To assess the influence of genotype, environment, and processing on coffee beverage quality, specialty coffees from the southern part of the state of Minas Gerais in Brazil were investigated. The Mantiqueira de Minas region, shown in Figure 16.2 is an important specialty coffee producing area and a geographical indication for coffee in Brazil, whose high quality coffees have been obtaining increasing recognition worldwide.

This region has a territorial extension of approximately 6300 km^2, with altitudes ranging from 812 to 2252 meters (m). Mean annual temperature is 17.9 °C, with a mean minimum temperature of 13 °C and a mean maximum temperature of 18.5 °C. Mean annual rainfall is 1665 mm.[42] Geographic information systems and remote sensing were used to characterize the landscape, the climate and land use/land cover and produce valuable spatial information for understanding the region's agroecosystem.[108–110] A geographic database was created to store all the geoinformation generated. Land use was mapped using RapidEye and Spot images from 2007 to 2013. A digital elevation model was generated from digital topographic information, which was then used to produce altitude and relief thematic maps. These maps are presented in Figure 16.3.

In 60% of the total area, altitudes reach up to 1000 m. Areas with altitudes between 1000 and 1100 m represent 15%; and areas with altitudes above 1100 m represent 25% of the entire region. The relief was also characterized (Figure 16.3) according to slope gradient ranges defined by Embrapa.[111] Approximately 40% of the landscape was classified as strongly sloping, followed by sloping (33%) and 9% as steeply sloping or mountainous relief. This limits the use of agricultural machinery and imposes a high degree of difficulty to coffee management and harvest. The region is

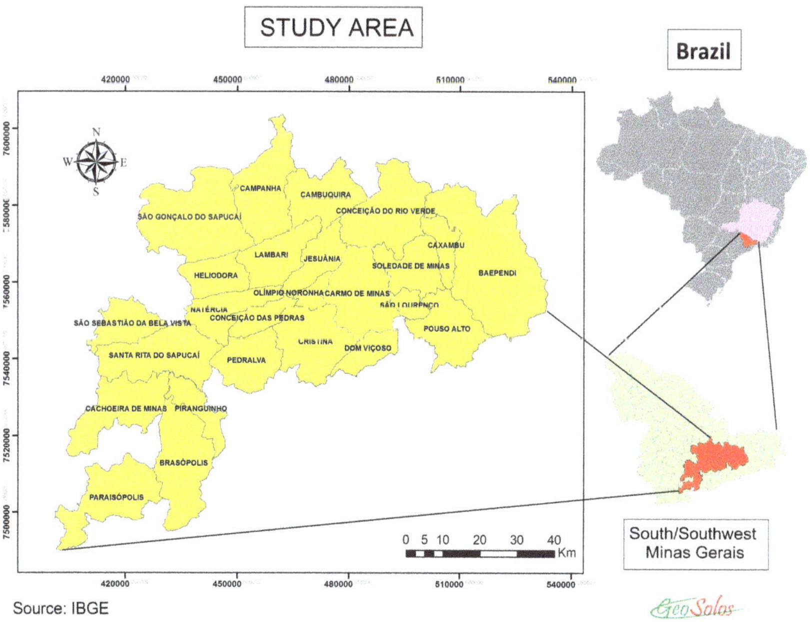

Figure 16.2 Location of the municipalities that comprise the Mantiqueira de Minas region in the state of Minas Gerais, Brazil. Source: DO BRASIL.[107]

(a)

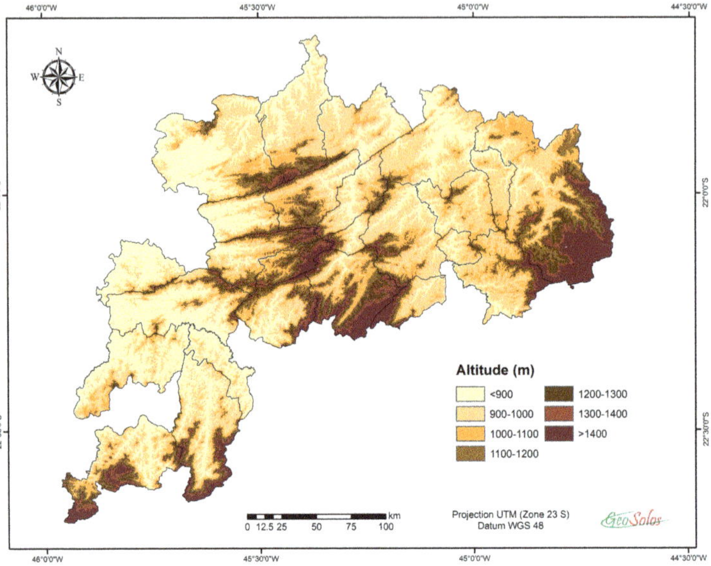

(b)

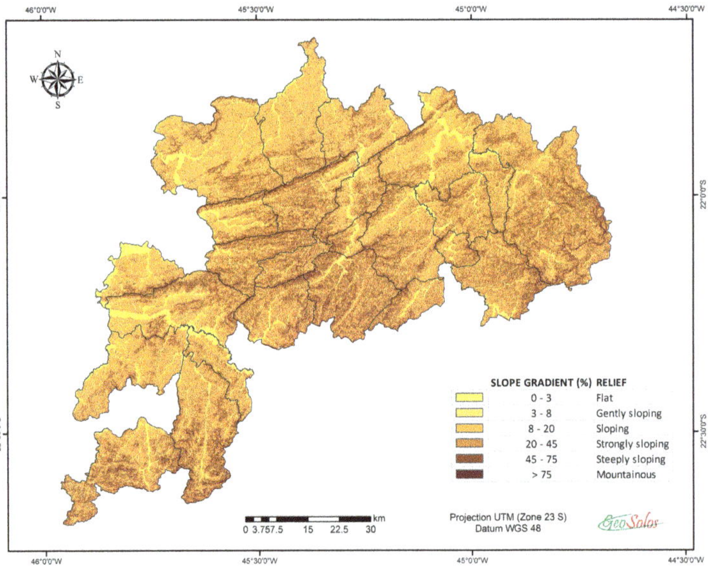

Figure 16.3 Altitudes (a) and relief (b) of Mantiqueira de Minas region, Minas Gerais, Brazil.

characterized by mountainside coffee growing, where most of the harvest is done manually.

To better characterize the land occupied by coffee plantations in the regional landscape, the altitude and land use maps were overlaid using the geographic information system. Half of the coffee fields are located below 1000 m. Coffee plantations in altitudes between 1100 and 400 m represent approximately 25% of the total coffee lands.

The characterization of the geographic space allowed the region to be divided into homogeneous environments for sample collection. An experimental area, representative of the regional landscape, was selected. This study area was stratified into three altitude classes (below 1000 m, from 1000 to 1200 m, and above 1200 m) and two groups of slope direction (sunnier faces: NE, N, NW, W and shaded faces: E, SE, S, SW), resulting in the combination of six environmental conditions. For *in situ* meteorological characterization, automatic weather stations were set up in each environment. Coffee samples of two genotypes were collected in each environment: Yellow Bourbon (representing genotypes with yellow fruits) and Acaiá (representing genotypes with red fruits). For all the combinations of environment and genotype, three replications were collected and then processed using two processing methods: fruit-dried coffee and mechanically demucilaged parchment-dried coffee. Seventy-two samples per crop season were collected.

Harvest was carried out manually and selectively, collecting only mature fruits. The fruits were separated by density, and only the denser ones were used. After hydraulic separation, another manual selection was made to ensure that the samples consisted only of mature fruits. Drying, milling, and storing were done carefully, according to the technical recommendations for production of specialty coffees.[88]

Sensory analyses[112] were carried out to assess the interaction of genotype, environment, and processing on coffee quality.

Different statistical techniques of multivariate data analysis were used, including logistic regression and correspondence analysis.

Logistic regression[113] was used to predict and explain the geographic distribution of coffee quality according the final coffee scores, altitude, latitude, and longitude. The distribution and definition of coffee quality depend on the environment in which each coffee variety is produced (Figure 16.4). However, it is not possible to describe the predominant sensory profile.

The results of correspondence analysis[114] seek to characterize the environment related to a determined coffee sensory profile capable of differentiating it from the other coffees produced in a given region of study.

As an exploratory analysis, several diagrams are generated for the purpose of finding the best correspondence among the data studied. Initially, altitudes and final scores categorized with whole values were considered. In Figure 16.5, it may be seen that the categories of slope direction and crop season variables do not exhibit a cluster profile and, moreover, that the confidence ellipses for the categories of these variables overlap, thus indicating that such categories do not exhibit significant differences among themselves. Thus, it may be affirmed for this case study that neither slope directions nor crop seasons contribute to definition of the coffee sensory profile, regardless of the genotype, altitude, and processing method.

Analyzing the categories of the altitude and final score variables in Figure 16.6, it is possible to observe the formation of three different groups. The samples further to the right of the correspondence diagram exhibit scores

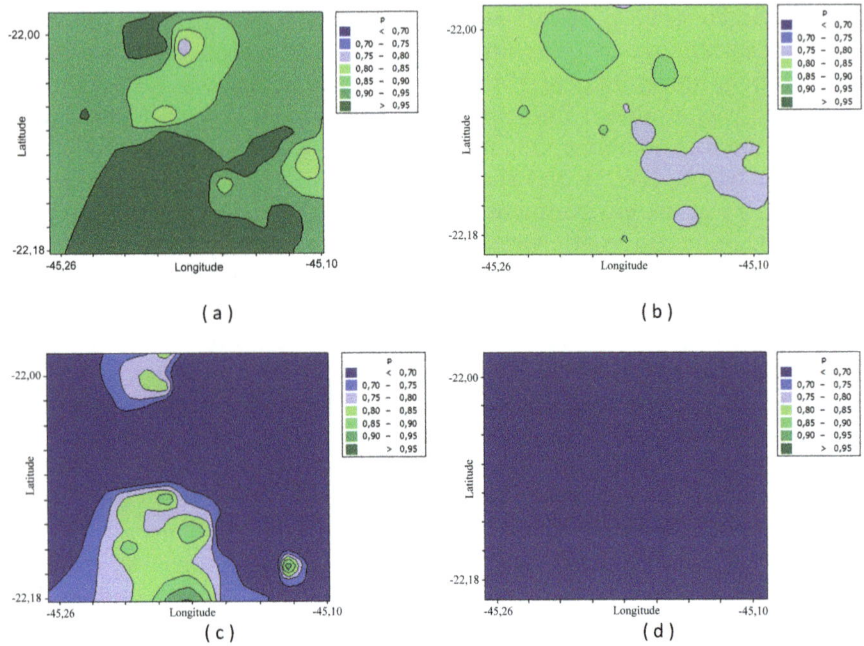

Figure 16.4 Contour diagrams considering the event of interest: (a) yellow fruits with scores ≥80; (b) red fruits with scores ≥80; (c) yellow fruits with scores ≥84; (d) red fruits with scores ≥84.

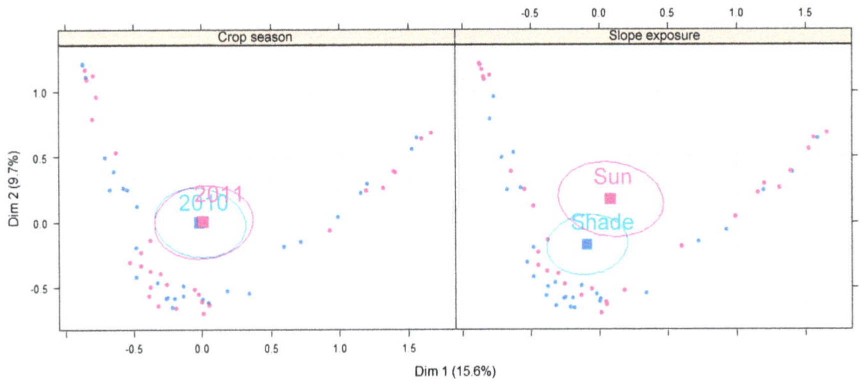

Figure 16.5 Correspondence diagrams of the categories of crop season (2010 and 2011) and slope direction (sun and shade) variables.

less than 82. Scores from 83 to 88 clustered in the lower left part, whereas scores above 88 were in the upper left part (Figure 16.6). For the altitude diagram it is possible to observe similar clusters. The group of samples located to the right on the correspondence diagram mostly exhibit altitudes less than 1050 m, whereas samples from altitudes greater than 1050 m are clustered

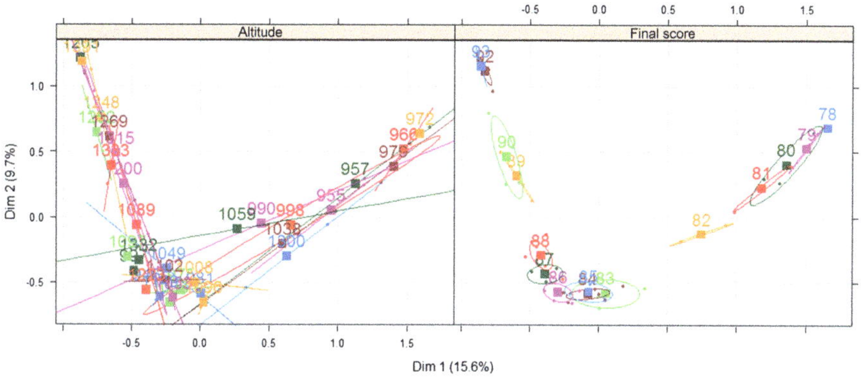

Figure 16.6 Correspondence diagrams for the categories of altitude and final score variables.

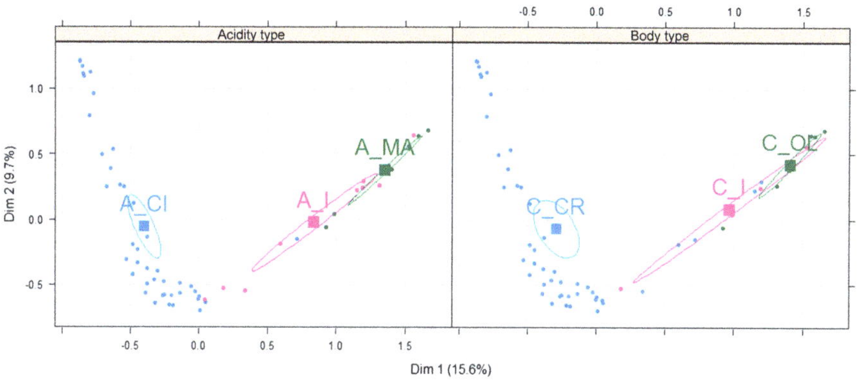

Figure 16.7 Correspondence diagrams of the categories of the acidity type and body type variables.

more to the left of the diagram. In this first analysis, it is possible to identify the relationship between altitudes above 1050 m and scores greater than 82. However, these characteristics do not define the product characteristics associated with the geographic environment.

In Figure 16.7, blue color points indicate samples that have citric acidity; green color points, samples with malic acidity; and pink color points, samples with undefined acidity. The confidence ellipses for malic and undefined acidity are overlapping, indicating that the samples belonging to these two groups are not significantly different. However, the confidence ellipse of the group of samples with citric acidity is not interposed with any other. Based on these observations and on the arrangement of the samples in the correspondence diagrams of Figure 16.7, it is possible to assert that altitudes above 1050 m produce coffees with scores greater than 82 and citric acidity. Body type is likewise identified as creamy.

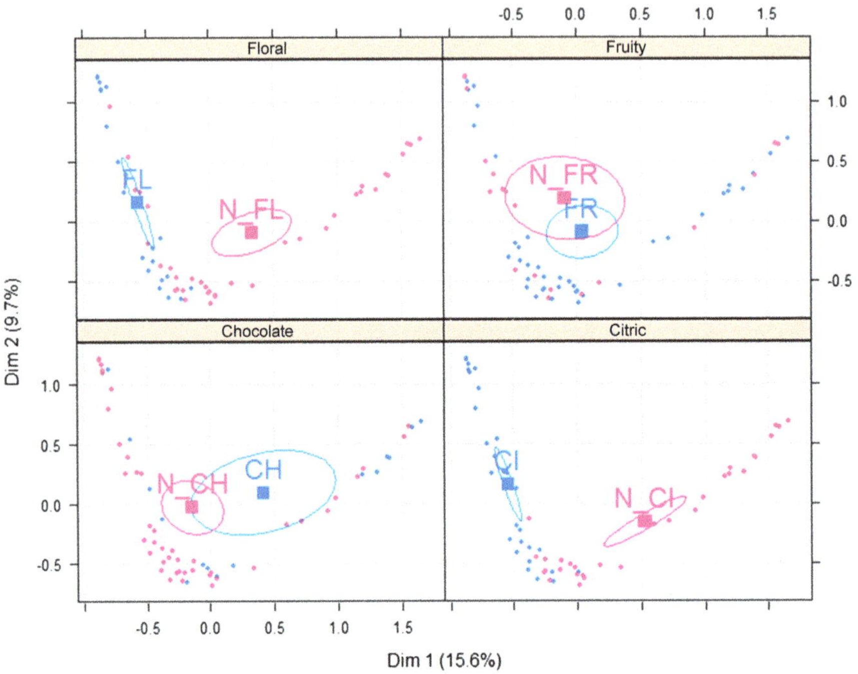

Figure 16.8 Correspondence diagrams of the categories of the flavor type variable (floral, fruity, chocolate, and citric).

In Figure 16.8, it can be seen that the coffee samples analyzed do not show significant differences between the presence or absence of the fruity and chocolate flavors. Though the citric and floral flavors exhibit significant differences, the clusters formed by the samples for the categories of these variables show behavior different from the behaviors shown for the altitude and final score variables.

Based on this information, we sought to find the best adjustment/fit to characterize the coffee through different combinations of the binary variables derived from the altitude and final score variables. In this case, processing has a marked effect, interacting with genotype and environment.

The best result for the mechanically demucilaged Yellow Bourbon genotype was found for altitudes greater than 1100 m. From these results, it is possible to observe that coffees grown at altitudes above 1100 m have final scores above 85, citric acidity, creamy body, floral/citric flavor, medium/high acidity and body intensities, and high sweetness intensity.

However, the best fit for the mechanically demucilaged Acaiá genotype was found for altitudes above 1100 m, with scores above 86 points. In this case, the predominant sensory profile was described as citric acidity, creamy body, floral/citric flavor, medium/high acidity and body intensities, and high sweetness intensity.

The best fit found for the natural processed Yellow Bourbon genotype was for altitudes above 1100 m and final scores above 86. In this case, the predominant sensory profile was citric acidity, creamy body, fruity/citric flavor, and high acidity, body, and sweetness intensities. However, it was not possible to find a sensory profile with significant correspondence with altitude for the natural processed Acaiá genotype.

Among the climate variables studied for this region, air temperature was the parameter that showed the greatest differences among the altitude ranges studied.[115] The mean air temperatures were 21.6 °C below 1000 m altitude and 19.7 °C above 1000 m. The greatest differences of temperature among the altitudes occur in the hottest months, with values ranging from 2.4 °C to 2.7 °C.

However, the new fact, contrary to what many believe, is that the greatest values of thermal amplitudes are related to the occurrence of coffee with lower beverage quality. In this case, the greatest thermal amplitudes were found in the altitude range below 1000 m. The lowest mean values for coffee sensory quality were found in this altitude range. Several reasons may be given for this occurrence. Among the various possibilities, the seasonal variation of ecophysiological and metabolic properties of arabica coffee stands out. This variation is dependent on altitude. The physiological differences indicate that the plants at lower altitudes underwent greater oxidative stress.[116] Both the greater mean temperatures and the greater thermal amplitude may be related to this fact. The lower antioxidant activity in the plants grown under thermal stress conditions is related to increased formation of reactive oxygen species, which contributes to lower beverage quality.[116,117]

16.6 Concluding Remarks

This chapter began with a question: What are the reasons behind the quality of specialty coffees? Coffee quality is directly influenced by environment, genotype, and processing techniques. The review and results presented in this chapter show the complexity of the relations between these factors and demonstrate more than the effect of each factor in itself; it is the interaction amongst them that really counts in the expression of coffee beverage quality. The interaction between the environment and genotype changes the chemistry of the raw coffee bean. Then, post-harvest processing changes again the chemical compounds in green beans, which are precursors of the flavor and aroma. Finally, the chemistry of the coffee bean is once more changed during roasting and preparation of the beverage. Quality is a result of multiple sources that affect the coffee bean like waves, which carry within themselves many other possible variations, like smaller noises. This demonstrates the difficulty of finding consistent relationships between edaphic/climatic variables, biochemical constituents, and improvement in the aroma and flavor of the coffee beverage. Nevertheless, small parts of this complex puzzle have been assembled. Although the favorable effect of increasing altitudes on the quality of arabica coffees is a consensus in studies conducted in different countries, the results presented here for the Mantiqueira de Minas region show that it is possible to fingerprint this altitude, *i.e.* to establish the exact height at which the beneficial effect occurs.

The subject of coffee quality has been increasingly debated in recent years, and will certainly continue to be studied in the coming decades. The intriguing question asked in the beginning of the text still remains to be fully answered, and will probably challenge researchers in the years to come.

References

1. H. V. Amorin, L. C. Scoton, A. de Castilho, F. P. Gomes and E. Malavolta, in *Anais XXI ESTUDOS SÔBRE A ALIMENTAÇÃO MINERAL DO CAFEEIRO*, ed. E. S. A. L. de Queiroz, Piracicaba, 1967, pp. 139–152.
2. H. V. de Amorim, L. C. Scoton, A. de Castilho, F. P. Gomes and E. Malavolta, in *Anais XXI ESTUDOS SÔBRE A ALIMENTAÇÃO MINERAL DO CAFEEIRO*, Anais da Escola Superior de Agricultura Luiz de Queiroz, Piracicaba, 1973, pp. 323–333.
3. A. Farah, T. de Paulis, L. C. Trugo and P. R. Martin, *J. Agric. Food Chem.*, 2005, **53**, 1505–1513.
4. B. Bertrand, P. Vaast, E. Alpizar, H. Etienne, F. Davrieux and P. Charmetant, *Tree Physiol.*, 2006, **26**, 1239–1248.
5. D. Villarreal, A. Laffargue, H. Posada, B. Bertrand, P. Lashermes and S. Dussert, *J. Agric. Food Chem.*, 2009, **57**, 11321–11327.
6. E. Alpizar and B. Bertrand, in *International Conference in Coffee Science*, ed. ASIC, Bangalore, 2004, CD-ROM.
7. J. Avelino, B. Barboza, J. C. Araya, C. Fonseca, F. Davrieux, B. Guyot and C. Cilas, *J. Sci. Food Agric.*, 2005, **85**, 1869–1876.
8. T. Joët, A. Laffargue, F. Descroix, S. Doulbeau, B. Bertrand, A. De Kochko and S. Dussert, *Food Chem.*, 2010, **118**, 693–701.
9. A. P. Camargo, R. Santinato and J. G. Cortez, in *Congresso Brasileiro de Pesquisas Cafeeiras*, ed. I. B. de Café, Araxá, 1992, pp. 70–74.
10. J. G. Cortez, *Inf. Agropecuário - Qual. do Café*, 1997, vol. 18, pp. 27–31.
11. H. M. R. Alves, M. M. L. Volpato, T. G. C. Vieira, F. M. Borém and J. N. Barbosa, *Inf. Agropecuário*, 2011, vol. 32, pp. 1–12.
12. A. R. Pereira, A. P. Camargo and M. B. P. Camargo, *Agrometeorogia de cafezais no Brasil*, Instituto Agronômico de Campinas, Campinas, 2008.
13. C. H. S. Carvalho, *Cultivares de café: origem, características e recomendações*, Embrapa Café, Brasília, 2008.
14. J. M. Gimase, W. M. Thagana, D. T. Kirubi, E. K. Gichuru and C. W. Kathurima, *Afr. J. Food Sci.*, 2014, **8**, 456–464.
15. C. A. Krug and A. J. T. Mendes, *Bragantia*, 1941, **1**, 189–203.
16. E. J. L. Meireles, M. M. L. Volpato, H. M. R. Alves and T. G. C. Vieira, *Informe Agropecuário*, Geotecnologias, Belo Horizonte, 2007, vol. 28(241), pp. 50–57.

17. A. R. Pereira, A. P. Camargo and M. B. P. Camargo, *Agrometeorologia de Cafezais no Brasil*, Instituto Agronômico, Campinas, 2008, p. 127.
18. A. J. A. Meirelles, P. D. D. O. Paiva, M. I. De Oliveira and T. S. Tavares, *Cienc. Agrotecnol.*, 2007, **31**, 1884–1887.
19. F. M. DaMatta and A. B. Rena, in *Tecnologias de Produção de Café com Qualidade*, ed. L. Zambolin, Universidade Federal de Viçosa, Viçosa, 2001, pp. 65–100.
20. A. P. Camargo, *Pesqui. Agropecu. Bras.*, 1985, **20**, 831–839.
21. J. A. Bieto and M. Talon, *Madrid: Interamericana*, 1996, 537–553.
22. A. P. Camargo and A. R. Pereira, *Agrometeorology of the Coffee Crop*, World Meteorological Organization, Geneva, Switzerland, 1994.
23. R. A. Thomaziello, E. D. Oliveira, J. A. Toledo Filho and T. E. Costa, Bol. Técnico CATI, 1999, **193**, 52–55.
24. G. C. Sediyama, J. César, F. De Melo, A. Rosa, A. Ribeiro, M. H. Costa and P. J. Hamakawa, *Rev. Bras. Agrometeorol.*, 2001, **9**, 501–509.
25. H. S. Pinto and J. Zullo Junior, *Rev. Bras. Agrometeorol.*, 2001, **9**, 495–500.
26. J. E. Drinnan and C. M. Menzel, *J. Hortic. Sci.*, 1995, **70**, 25–34.
27. Â. P. De Camargo and M. B. P. De Camargo, *Bragantia*, 2001, **60**, 65–68.
28. C. Alègre, *Agron. Trop.*, 1959, **14**, 23–58.
29. A. B. Rena and M. Maestri, in *Ecofisiologia do Cafeeiro*, ed. P. R. C. Castro, S. O. Ferreira and T. Yamada, Piracicaba, 1987, pp. 119–147.
30. J. B. Matiello, *O café: do cultivo ao consumo*, Globo, São Paulo, Coleção do agricultor, Grãos, 1991.
31. M. B. P. Camargo, G. S. Rolim and M. A. Santos, *Inf. Agropecuário* - Geotecnologias, 2007, vol. 28, pp. 58–65.
32. A. P. Camargo, in *Recursos naturais, meio ambiente e poluição: contribuição de um ciclo de debates.*, ed. IBGE, IBGE, Rio de Janeiro, 1977, pp. 68–76.
33. O. Ovalle-Rivera, P. Läderach, C. Bunn, M. Obersteiner and G. Schroth, *PLoS One*, 2015, **10**, e0124155.
34. E. D. Assad, H. S. Pinto, J. Zullo and A. M. Helminsk Ávila, *Pesqui. Agropecu. Bras.*, 2004, **39**, 1057–1064.
35. A. B. Rena and M. Maestri, Ecofisiologia do cafeeiro, in *Ecofisiologia da Produção Agrícola*, ed. P. R. C. Castro, S. O. Ferreira and T. Yamada, Potafos, Piracicaba, pp. 119–147.
36. D.-J. Kwon, H.-J. Jeong, H. Moon, H.-N. Kim, J.-H. Cho, J.-E. Lee, K. S. Hong and Y.-S. Hong, *Food Res. Int.*, 2015, **67**, 175–182.
37. P. Vaast, B. Bertrand, J.-J. Perriot, B. Guyot and M. Génard, Fruit thinning and shade improve bean characteristics and beverage quality of coffee (*Coffea arabica* L.) under optimal conditions, *J. Sci. Food Agric.*, 2006, **86**, 197–204.
38. C. E. B. Serrano and J. J. C. Castrillón, *Cenicafé*, 2002, **53**, 119–131.
39. F. Decazy, J. Avelino, B. Guyot, J. J. Perriot, C. Pineda and C. Cilas, *J. Food Sci.*, 2003, **68**, 2356–2361.
40. B. Guyot, D. Gueule, J. C. Manez, J. J. Perriot, J. Giron and L. Villain, *Plant., Rech., Dev.*, 1996, **3**(4), 272–283.

41. J. Barbosa, F. Borém, H. Alves, M. Cirillo, M. Sartori and C. Ducatti, *J. Agric. Sci.*, 2014, **6**, 55–64.
42. B. G. Laviola, H. E. P. Martinez, L. C. C. Salomão, C. D. Cruz and S. M. Mendonça, *Rev. Bras. Cienc. Solo*, 2007, **31**, 1451–1462.
43. J. O. Ayoade, *Introdução à climatologia para os trópicos*, Bertrand Brasil, Rio de Janeiro, 9th edn, 2003.
44. J. N. Barbosa, F. M. Borem, M. A. Cirillo, M. R. Malta, A. A. Alvarenga and H. M. R. Alves, *J. Agric. Sci.*, 2012, **4**, 181–190.
45. L. G. Carvalho, in *Zoneamento econômico ecológico de estado de Minas Gerais: componente geofísico e biótico*, ed. J. R. Scolforo, L. M. T. Carvalho and A. D. Oliveira, Editora UFLA, Lavras, 2007, CD-ROM.
46. B. Bertrand, R. Boulanger, S. Dussert, F. Ribeyre, L. Berthiot, F. Descroix and T. Joët, *Food Chem.*, 2012, **135**, 2575–2583.
47. T. Oberthür, P. Läderach, H. Posada, M. J. Fisher, L. F. Samper, J. Illera, L. Collet, E. Moreno, R. Alarcón, A. Villegas, H. Usma, C. Perez and A. Jarvis, *Food Policy*, 2011, **36**, 783–794.
48. P. Donald, *Conserv. Biol.*, 2004, **18**, 17–37.
49. T. Tscharntke, Y. Clough, S. A. Bhagwat, D. Buchori, H. Faust, D. Hertel, D. Hölscher, J. Juhrbandt, M. Kessler, I. Perfecto, C. Scherber, G. Schroth, E. Veldkamp and T. C. Wanger, *J. Appl. Ecol.*, 2011, **48**, 619–629.
50. M. de Beenhouwer, R. Aerts and O. Honnay, *Agric., Ecosyst. Environ.*, 2013, **175**, 1–7.
51. I. Perfecto, R. A. Rice, R. Greenberg and M. E. VanderVoort, *Bioscience*, 1996, **46**, 598–608.
52. S. M. Philpott, W. J. Arendt, I. Armbrecht, P. Bichier, T. V. Diestch, C. Gordon, R. Greenberg, I. Perfecto, R. Reynoso-Santos, L. Soto-Pinto, C. Tejeda-Cruz, G. Williams-Linera, J. Valenzuela and J. M. Zolotoff, *Conserv. Biol.*, 2008, **22**, 1093–1105.
53. V. E. Méndez, C. M. Bacon, M. Olson, K. S. Morris and A. Shattuck, *Prof. Geogr.*, 2010, **62**, 357–376.
54. S. A. Bhagwat, C. G. Kushalappa, P. H. Williams and N. D. Brown, *Conserv. Biol.*, 2005, **19**, 1853–1862.
55. C. R. Cerdán, M. C. Rebolledo, G. Soto, B. Rapidel and F. L. Sinclair, *Agric. Syst.*, 2012, **110**, 119–130.
56. E. M. Bennett, G. D. Peterson and L. J. Gordon, *Ecol. Lett.*, 2009, **12**, 1394–1404.
57. L. Zanella, R. A. T. Borém, C. G. Souza, H. M. R. Alves and F. M. Borém, Atlantic forest fragmentation analysis and landscape restoration management scenarios, *Nat. Conservacao.*, 2012, **10**, 57–63.
58. J. Sousa, A. L. Luz, F. N. Sousa, M. Cassama, A. Dabo, F. Dafa and M. Bivar Abrantes, *Agroecol. Sustain. Food Syst.*, 2015, **39**, 134–154.
59. *BRASIL, Presidência da República. Lei 12.651, de 25 de maio de 2012, Brasília*, 2012, Disponível em: http://www.planalto.gov.br/ccivil_03/_ato2011-2014/2012/lei/l12651.htm.
60. E. A. Silva, P. R. Reis, M. S. Zacarias and P. P. Marafeli, *Cienc. Agrotecnol.*, 2010, **34**, 1146–1153.

61. J. H. da Silva Taveira, F. M. Borém, L. P. Figueiredo, N. Reis, A. S. Franca, S. A. Harding and C. J. Tsai, *Food Res. Int.*, 2014, **61**, 75–82.
62. T. C. Villela, Universidade Federal de Lavras, 2002.
63. G. S. Giomo, J. C. Mistro, L. C. Fazuoli and E. S. Mantovani, in *VII Simpósio de Pesquisa dos Cafés do Brasil*, Consórcio Pesquisa Café, Araxá, 2011, CD-ROM.
64. H. P. Medina-Filho, M. P. Maluf, R. Bordignon, O. Guerreiro-Filho and L. C. Fazuoli, *Acta Hortic.*, 2007, **745**, 351–368.
65. *BRASIL, Ministério da Agricultura, Pecuária e Abastecimento. Lei 10.711, de 05 de agosto de 2003, Brasília*, 2003, Disponível em: http://www.agricultura.gov.br/vegetal/registros-autorizacoes/registro/registro-nacional-cultivares.
66. T. Sera, Crop breed, *Appl. Biotechnol.*, 2001, **1**, 179–199.
67. L. C. Fazuoli, H. P. Medina-Filho, W. Gonçalves, O. Guerreiro-Filho and M. B. Silvarolla, in *O estado da arte de tecnologias na produção de café*, ed. L. Zambolim, Editora UFV, Viçosa, 1st edn, 2002, pp. 63–215.
68. D. Yigzaw, University of The Free State, 2005.
69. A. Tessema, S. Alamerew, T. Kufa and W. Garedew, *Int. J. Plant Breed. Genet.*, 2011, **5**, 302–316.
70. G. Bekele, Alemaya University, 2009.
71. O. Kitila, S. Alamerew, T. Kufa and W. Garedew, *Int. J. Agric. Res.*, 2011, **6**, 537–549.
72. C. W. Kathurima, B. M. Gichimu, G. M. Kenji, S. M. Muhoho and R. Boulanger, *Afr. J. Food Sci.*, 2009, **3**, 365–371.
73. C. O. Agwanda, P. Baradat, A. B. Eskes, C. Cilas and A. Charrier, *Euphytica*, 2003, **131**, 1–14.
74. G. S. Giomo and F. M. Borem, *Inf. Agropecuário*, 2011, vol. 32, pp. 7–16.
75. C. H. S. Carvalho, *Cultivares de café: origem, características e recomendações*, Embrapa Café, Brasília, 2008, p. 334.
76. J. E. T. Mendes, Ensaio de variedades de cafeeiro II, *Bragantia*, 1949, **9**, 1–21.
77. L. C. Fazuoli, M. B. Silvarolla, T. J. G. Salva, O. G. Filho, H. P. M. Filho and W. Gonçalves, *O Agron.*, Campinas, 2007, **59**, 12–15.
78. C. Montagnon, P. Marraccini and B. Bertrand, 2012, 1–29.
79. A. Carvalho, L. C. Mônaco and H. J. Scaranari, *Bragantia*, 1962, **21**, 215–239.
80. G. S. Giomo, F. M. Borem, R. Saath, J. C. Mistro, L. P. Figueiredo, F. C. Ribeiro, S. P. Pereira and M. R. Bernardi, in *24th International Conference on Coffee Science*, ed. Association for Science and Information on Coffee, San José, Costa Rica, 2012, CD-ROM.
81. L. P. Figueiredo, F. M. Borém, M. Â. Cirillo, F. C. Ribeiro, G. S. Giomo and T. D. J. G. Salva, *J. Agric. Sci.*, 2013, **5**, 87–98.
82. O. Guerreiro-Filho, L. C. Fazuoli and A. T. E. Aguiar, Cultivares de *Coffea arabica* selecionadas pelo IAC: características botânicas, tecnológicas, agronômicas e descritores mínimos, http://www.infobibos.com/artigos/2006_2/cultivares_cafe/index.htm, 2006.

83. G. S. Giomo, M. B. Silvarolla, R. Saath and A. de C. Iobbi, in *VII Simpósio de Pesquisa dos Cafés do Brasil*, Consórcio Pesquisa Café, Araxá, 2011, CD-ROM.
84. B. Bertrand, B. Guyot, F. Anthony and P. Lashermes, *Theor. Appl. Genet.*, 2003, **107**, 387–394.
85. C. W. Kathurima, *Int. J. Biodiversity Conserv.*, 2012, **4**, 39–46.
86. H. A. M. Van Der Vossen, *Exp. Agric.*, 2009, **45**, 323.
87. S. Mawardi and R. Hulip, in *International Scientific Conference on Coffee Science, Kyoto-Japan*, 1995, pp. 637–644.
88. F. M. Borém, E. P. Isquierdo and J. H. S. Taveira, Coffee processing, in *Handbook of Coffee Post-harvest Technology*, ed. F. M. Borém. (Org.), Norcross, Georgia: Gin Press, 1st edn, 2014, vol. 1, pp. 49–68.
89. G. Bytof, S.-E. Knopp, P. Schieberle, I. Teutsch and D. Selmar, *Eur. Food Res. Technol.*, 2005, **220**, 245–250.
90. G. Bytof, S.-E. Knopp, D. Kramer, B. Breitenstein, J. H. W. Bergervoet, S. P. C. Groot and D. Selmar, *Ann. Bot.*, 2007, **100**, 61–66.
91. G. S. Duarte, A. A. Pereira and A. Farah, *Food Chem.*, 2010, **118**, 851–855.
92. S. Knopp, G. Bytof and D. Selmar, *Eur. Food Res. Technol.*, 2005, **223**, 195–201.
93. D. Selmar, G. Bytof, S.-E. Knopp and B. Breitenstein, *Plant Biol.*, 2006, **8**, 260–264.
94. E. C. Dias, F. M. Borém, R. G. F. A. Pereira and M. C. Guerreiro, *Eur. Food Res. Technol.*, 2011, **234**, 25–32.
95. G. I. P. Quintero, *Cenicafé*, 1996, **47**, 85–90.
96. J. C. Vincent, in *Coffee: Chemistry*, Elsevier, London/New York, 1985, pp. 1–33.
97. R. Wilbaux, *Agricultural Engineering*, FAO, Rome, Italy, 1963.
98. M. N. Clifford and J. R. Ramirez-Martinez, *Food Chem.*, 1991, **40**, 191–200.
99. M. N. Clifford, *Food Chem.*, 1991, **40**, 35–42.
100. V. Leloup, C. Gancel, R. Liardon, A. Rytz and A. Pithon, in *International Conference in Coffee Science*, ed. Association for Science and Information on Coffee, Bangalore, 2004, p. CD-ROM.
101. P. Mazzafera and R. P. Purcino, in *International Conference in Coffee Science*, ed. Association for Science and Information on Coffee, Bangalore, 2004, CD-ROM.
102. J. H. da S. Taveira, Universidade Federal de Lavras, 2014.
103. J. Z. Zhang, M. Gomez-Pedrozo, C. S. Baden and J. J. Harada, *Mol. Gen. Genet.*, 1993, **238**, 177–184.
104. S. J. Reynolds and S. M. Smith, *Plant Mol. Biol.*, 1995, **27**, 487–497.
105. J. M. Vázquez-Ramos and M. P. Sánches, *Seed Sci. Res.*, 2003, **13**, 113–130.
106. A. M. Kinnersley and F. J. Turano, *Crit. Rev. Plant Sci.*, 2000, **19**, 479–509.
107. DO BRASIL, *Divisão Territorial, Divisão Territorial do Brasil e Limites Territoriais*, Instituto Brasileiro de Geografia e Estatística (IBGE) (1 de julho de 2008), Página visitada em, vol. 11, 2008.

108. F. M. V. Carvalho, P. de Marco Júnior and L. G. Ferreira, The Cerrado into-pieces: habitat fragmentation as a function of landscape use in the savannas of central Brazil, *Biol. Conserv.*, 2009, **142**(7), 1392–1403.
109. M. C. Ribeiro, *et al.*, The Brazilian Atlantic Forest: how much is left, and how is the remaining forest distributed? Implications for conservation, *Biol. Conserv.*, 2009, **142**(6), 1141–1153.
110. T. Bernardes, *et al.*, Avaliação da acurácia do mapeamento do uso da terra no complexo Serra Negra, Patrocínio, MG, por interpretação visual e classificação automática de imagens Landsat, In: *Simpósio Brasileiro de Sensoriamento Remoto, 13, 2007, Florianópolis. Anais*, INPE, Florianópolis, 2007, pp. 1219–1225.
111. Empresa Brasileira de Pesquisa Agropecuária – EMBRAPA, *Sistema brasileiro de classificação de solos*, Brasília, Serviço de Produção de Informação, 1999, p. 412.
112. R. T. Lingle, *The Coffee Cupper's Handbook: Systematic Guide to the Sensory Evaluation of Coffee's Flavor*, Long Beach, 4th edn, 2011.
113. D. Kleinbaum, L. Kupper, A. Nizam and E. Rosenberg, *Applied Regression Analysis and Other Multivariable Methods*, Cengage Learning, 2013.
114. J.-P. Benzécri, *Correspondence Analysis Handbook*, Dekker, New York, 1992.
115. M. P. Luz, Universidade Federal de Lavras, 2014.
116. H. R. O. Silveira, Universidade Federal de Lavras, 2014.
117. M. O. Santos, Universidade Federal de Lavras, 2012.

CHAPTER 17

Coffee Certification

CARLOS HENRIQUE JORGE BRANDO

P&A Ltda – Praça Rio Branco, 13 – E.S. do Pinhal – SP, CEP 13990-000, Brazil
*E-mail: peamarketing@peamarketing.com.br

17.1 Introduction

Coffee certification is a broad concept that applies to how it is grown and traded – organic, fair trade, sustainable, *etc.* – where it is grown – origin – and how it is perceived in the cup – quality. As the concept evolved over time it reacted to new demands such as traceability from farm to cup and the need to reach small growers at reasonable costs. The responses were chain-of-custody certification and verification, respectively.

The process of coffee certification requires a code or standard governing how it should be produced and traded, where it should be grown or what its desirable organoleptic features should be. Next, a third-party independent auditing system should be in place to check that actual practices or features conform to standards. Finally, a way should be devised to inform consumers about the process, for example, a label on the package.

Since the process of certification involves costs, there is a major question on how they should be paid for and allocated along the supply chain. The most obvious reply to this question is a price premium that not only covers costs but also leaves behind a reward for the grower to adopt the certification system and to do everything that the code requires. Feasible as this can be in niche markets – for example, specialty coffees – it becomes more complicated as certification expands into mainstream markets and may indeed become a cost of production item if required by all markets.

Coffee: Production, Quality and Chemistry
Edited by Adriana Farah

Published by the Royal Society of Chemistry, www.rsc.org

17.2 The Focus of Certification: Grower or Consumer?

In spite of the fact that some certification systems had their roots in coffee producing countries, *e.g.*, UTZ Certified and Fairtrade, the inspiration came from the more advanced importing countries and consuming markets that were worried about the environment, quality of life and working conditions and quality in the cup itself. There is no doubt that the focus of improving working conditions is on the growing side but the question becomes more complex regarding environment and quality of life. Whose quality of life – consumers' or growers' or both? – is the consumer who demands and pays a premium for environmentally friendly coffee worried about? Does a consumer who pays a premium for organic coffee know that coffee unlike the vegetables he or she buys and consumes does not carry residues of the agrochemicals used in its production and that in buying organic coffee the benefits go directly to growers' health and only indirectly to the consumer? Theoretical and not very relevant as these questions may sound, they play an important role in the growers' decision to seek any certification, especially when price premiums are small or do not exist, which seems to be increasingly the case with certification/verification of sustainability.

The willingness of growers to certify quality or origin is attached to the price premium that they get that in turn depends on the ability of that specific quality or origin to command a market premium and the supply chain to transfer a part of that premium to growers. In the case of sustainability, the reward to growers can go beyond price premium and be attached to the use of better farming practices that in turn lead to greater productivity, lower production costs and larger incomes. This being the case, growers should embrace sustainable practices because there are economic and business reasons for doing it. But do they need to certify their use in the absence of a price premium? This remains for the market to answer in coming years.

17.3 Certification, Verification and Others

The difference between certification and verification of sustainability has to do with the way auditing takes place. Because the costs of auditing become disproportionately high for small growers, the practice has developed of self-assessment by growers themselves with actual auditing of a sample of a few growers belonging to the same group. This is known as verification. The credibility of the process should be ensured by peer pressure within the group since a failure in the audit by a grower chosen as part of the sample can disqualify the whole group.

Auditing may not be required in the case of Geographical Indication (GI) – certification of origin – because the identification of where coffee is grown is in itself the proof of origin. However, auditing may be required in more sophisticated GI schemes that involve an assurance of quality.

Certification of cup quality of green coffee lots may involve pre-shipment evaluation by qualified cuppers rather than auditing schemes. Quality certification of finished coffee products offered to consumers requires sampling at point of sale and cupping by independent auditors.

17.4 Sustainability

Sustainability in coffee growing is defined as the ability to produce today without compromising the ability of future generations to produce.

Sustainability is a wider, more comprehensive concept than organic or fair trade but the “umbrella” of sustainability now includes the two even though some organic or fair trade coffee may not be necessarily sustainable (Table 17.1).

Greater detail about the sustainability schemes above may be obtained from SCAA's 2010 Sustainable Coffee Certification Matrix (http://www.scaa.org/PDF/Sustainable%20Coffee%20Certifications%20Comparison%20Matrix%202010.pdf) and in the sites listed below:

- Fairtrade: http://www.fairtrade.net
- Rainforest Alliance: http://www.rainforest-alliance.org
- UTZ Certified: http://www.utzcertified.org
- 4C: https://www.cas-veri.com/

As the sustainable coffee certifications above developed and expanded in the last two decades, they not only developed a clientele of their own but also came to be recognized by the market by their main features and appeal. The schemes do compete with each other but are often identified by the market on the basis of their origins, areas of concentration or appeal to clients, namely, Fairtrade with social, Rainforest Alliance with environment, UTZ Certified with better farm management and 4C as a baseline entry level standard. This is obviously an oversimplification but it does convey a message in a world so sensitive to branding.

The Brazilian state of Minas Gerais created its own coffee sustainability scheme called *Certifica Minas Café* that reaches almost 1500 growers who produce a total of 1.8 million bags of certified coffee. Successful in the production side, the scheme is still toiling with the marketing of the coffees it certifies. As a way to improve access to markets abroad, benchmarking of Certifica Minas Café has been accomplished against UTZ Certified. The case of Certifica Minas Café is a good example of the need to work on both the supply and demand sides of certification to be successful.

A few companies have their own codes of conduct regarding sustainability that are applied only to the coffees they buy. The best known cases are Nespresso's AAA Sustainable Quality and Starbucks' C.A.F.E. Practices. Table 17.2 summarizes the volumes of sustainable coffee certified and marketed in the world from 2010 to 2014. The coffee sector is very dynamic. 4C has undergone major changes since 2016, including the move from a

Table 17.1 Comparative overview of sustainability schemes for coffee. Extracted from the Third Edition of the International Trade Centre Coffee Exporters' Guide, summarizing the certification schemes most widely used then.

Aspect	Organic	Fairtrade	Rainforest Alliance	UTZ Certified	Common Code 4C
Premium	No assured premium paid – it varies considerably from market to market (but 15 to 20 cts was paid in some countries in 2011, if double certified with Fairtrade then it gets an automatic 20 cts premium)	Fixed premium always assured (but overall level of demand not always in tandem with production)	No assured premium (but 5 to 8 cts range was common in 2011)	No assured premium paid (but 2 to 5 cts was common in 2011)	No assured premium (but may be paid in certain circumstances if seller/buyer so agree)
Yield and quality	Short-term impact on yields may be negative; possibly positive impact on quality	Only indirect (and possibly positive) impact on yields and quality (through higher income, thus increased possibility of purchasing inputs and hiring labor)	Potentially negative yield impact; positive impact on quality	Possibly positive but limited	Possibly positive through improved farming and processing methods
Labor inputs	Higher labor inputs	Higher labor inputs linked to collective processes such as coordination, meetings *etc.*	Higher labor inputs	Moderately higher labor inputs	Moderately higher labor inputs
Other income impacts	Possibility of selling other organic products from the farm; income diversification	Possible indirect impact through wider trade networking offering possibility of selling other Fairtrade products	Possibility of selling forest by-products and fruit	Increasing visibility of UTZ may improve conditions of trade	Over time improved conditions of trade may be possible

(*continued*)

Table 17.1 (*continued*)

Aspect	Organic	Fairtrade	Rainforest Alliance	UTZ Certified	Common Code 4C
Market access, networking	Access to well-established and reliable market	Access to well-established, reliable market; technical assistance from Fairtrade importers	Buyers and markets increasing steadily	Number of buyers and markets increasing steadily	Potentially easier access to large segment of the mainstream market
Extension, credit	Possibly more effective extension from field staff supported by NGOs and some buyers, but limited support from public system	Access to trade financing and traditional credit sources due to Fairtrade membership and improved financial position of cooperatives	More effective agro-forestry extension from supportive NGOs, but limited support from public system	Potentially better extension services from supportive NGOs and some buyers, but limited support from public extension services	Potential support from 4C-support platform and participating buyers; limited support from public extension services
Organizational capacity; community impact	Potential increase in mutual support among farmers to solve farming management problems	Increased organizational capacity of participating farmers; access to training; better organizational ability to serve members; community projects	Mutual support amongst farmers for forest management	Strengthening organizational capabilities (if registration is done *via* farmer groups rather than as individuals)	Strengthening of organizational capabilities through potential assistance from 4C-support platform; access to training
Environment	Potential adoption of new farming techniques to improve soil fertility as well as drought and erosion resilience	Limited environmental benefits	Improved biodiversity and agro-ecological conditions; enhancement of soil fertility	Limited environmental benefits through the gradual elimination of inappropriate farming and processing methods	Limited environmental benefits through the gradual elimination of inappropriate farming and processing methods
Risk, planning capabilities	Risk reduction through reduced external inputs; no mono-cropping; improved soil resilience; planning may improve	Better planning for coffee production, personal and household needs; guaranteed price reduces risk	Reduced pest management and social risk; planning may improve	Potential for some reduced pest management and social risk; planning may improve	Better planning and reduced risk through improved market access may be possible

membership association to a private company. Rainforest Alliance and UTZ Certified decided to merge in 2017 and started to operate together in 2018.

17.4.1 Niche and Mainstream Markets

A comparison of the figures in Table 17.2 with world production of about 150 million bags (9 million metric tons) shows that the total volume of sustainable coffee marketed is still only 13%, which means it still qualifies as a niche market. A closer analysis of the volumes indicates that, at that time, the production of verified 4C compliant coffee was growing much faster than certified Rainforest Alliance, UTZ Certified and Fairtrade coffees. The same comparison for marketed volume showed an even stronger growth of the 4C baseline standard. Together this could indicate that there was a larger chance that verified coffees would reach the mainstream market whereas certified coffees would remain in the realm of niche markets. If 4C is considered a baseline verification standard from which growers can move into certification, a scenario could develop in which certified coffee could extrapolate the specialty coffee market into differentiated coffees with verification remaining the option for commercial coffees. However, the volume of 4C verified coffee subsequently fell for several reasons that are not worth describing here.

Since the price premiums for sustainable coffees are defined by the market, they have fallen as the volumes of verified and certified coffees produced increased. Fairtrade is the exception because of its pre-established price premium that in itself restricts growth to the availability of buyers wishing to pay the price premium. As the mainstream coffee market adopts sustainable practices the tendency is indeed for price premiums to fall and sustainability may remain attractive to growers as a way to access specific markets. In the long run sustainability may simply become a requirement without price premiums or preferential access to markets. Long-term trends aside, it seems unlikely that certification and verification as they exist today will reach the mainstream market. That is why the institutions mentioned above are changing and trying, along with others, *e.g.*: the Global Coffee Platform, to redesign the way they work along the lines of new concepts that are being developed like landscape approaches and sustainable coffee regions.

17.4.2 Benefits to Growers and the Role of Government

What would be the reason for growers to adopt sustainable practices in the absence of price premiums or preferential access to markets? There is growing evidence of a strong business case that recommends the adoption of sustainable practices on grounds that growers become more efficient, lower their costs and increase their profits when they adopt the good agricultural practices that are embedded in all sustainability codes. The challenge is to demonstrate to growers that it is first of all in their own benefit to become sustainable. This being the case, it should be the role of governments, cooperatives and associations – all of whom have a stake in the business of growing coffee – to promote the implementation of sustainable coffee practices.

Table 17.2 Production and received volumes of verified/certified coffee.[a]

	2010		2011		2012		2013		2014	
Standard	Produced	Marketed	Produced	Marketed	Produced	Marketed	Produced	Marketed	Produced	Marketed
Rainforest Alliance	219 337	114 884	260 702	129 864	377 757	139 856	454 962	167 967	456 042	184 599
UTZ Certified	394 003	120 994	476 903	136 752	715 648	188 096	726 591	224 028	729 918	258 867
Fairtrade International	358 000	104 216	393 000	115 574	396 641	133 000	473 600	142 400	521 387[b]	148 157[b]
4C Association	646 496	23 167	906 348	48 617	1 785 078	152 000	2 359 868	453 899	2 871 682	596 983
Total (tons)	1 617 809	363 254	2 036 953	430 807	3 275 124	612 952	4 015 021	988 294	4 579 029	1 188 606

[a]Source: 4C 2014 Annual Report

[b]Fairtrade 2014: estimated

Public extension services should have as a priority to promote sustainable practices and there should be no preference for any specific sustainability code if the service is provided by government. Countries may even adopt and promote a basic sustainability code that, if the grower so desires, can be used as a stepping stone into one of the verification or certification systems described above.

The Sustainable Coffee Program of IDH, the Sustainable Trade Initiative, now Global Coffee Platform (GCP), has launched the concept of a Coffee Sustainability Curriculum as the basis for public extension services to disseminate and promote sustainability with freedom for growers to decide whether to verify or certify and, if yes, which standard to use. Brazil and Vietnam have already developed such curricula and other countries are currently developing it. Figure 17.1 was extracted from the materials used in the initial approach to growers about the Brazilian Coffee Sustainability Curriculum.

17.4.3 Labels or Not?

Is there a future for certification labels in final products? Such labels may perhaps only survive for specialty coffees with the bulk of the coffees marketed having their sustainability granted by the manufacturers and brands themselves. These in turn will rely on a provider of sustainability services that will certify or verify that the coffee is produced in a sustainable way and guarantee this to the manufacturers of the final product whose brand will have the concept of sustainability embedded in itself.

17.4.4 Traceability

When price premiums are involved there should be a strong concern that what is paid by the buyer effectively trickles down through the coffee supply chain and reaches the grower. Verification or certification should therefore include the chain-of-custody from the farms to the buyers who pay the premiums. Likewise, buyers who pay premiums must have guarantees that the coffees that they buy are indeed the ones that were verified or certified at farm level. This brings up the concept of traceability of sustainable coffees from farm to cup, an area where there is a lot of development yet to take place. In spite of modern digital technology and bar and QR codes, only a small percentage of the coffees in Table 17.2 are effectively traced from production to consumption.

17.4.5 Sustainable Coffee Content

Last but not least, there is often an assumption that a package or cup of sustainable coffee contains 100% sustainable coffee. This may not be always the case and more transparency in the communication of sustainability is required. The actual percentage of sustainable coffee is more often stated in the sustainability code or procedures than on the package itself. It may not necessarily be 100% but a given minimum percentage in which case either

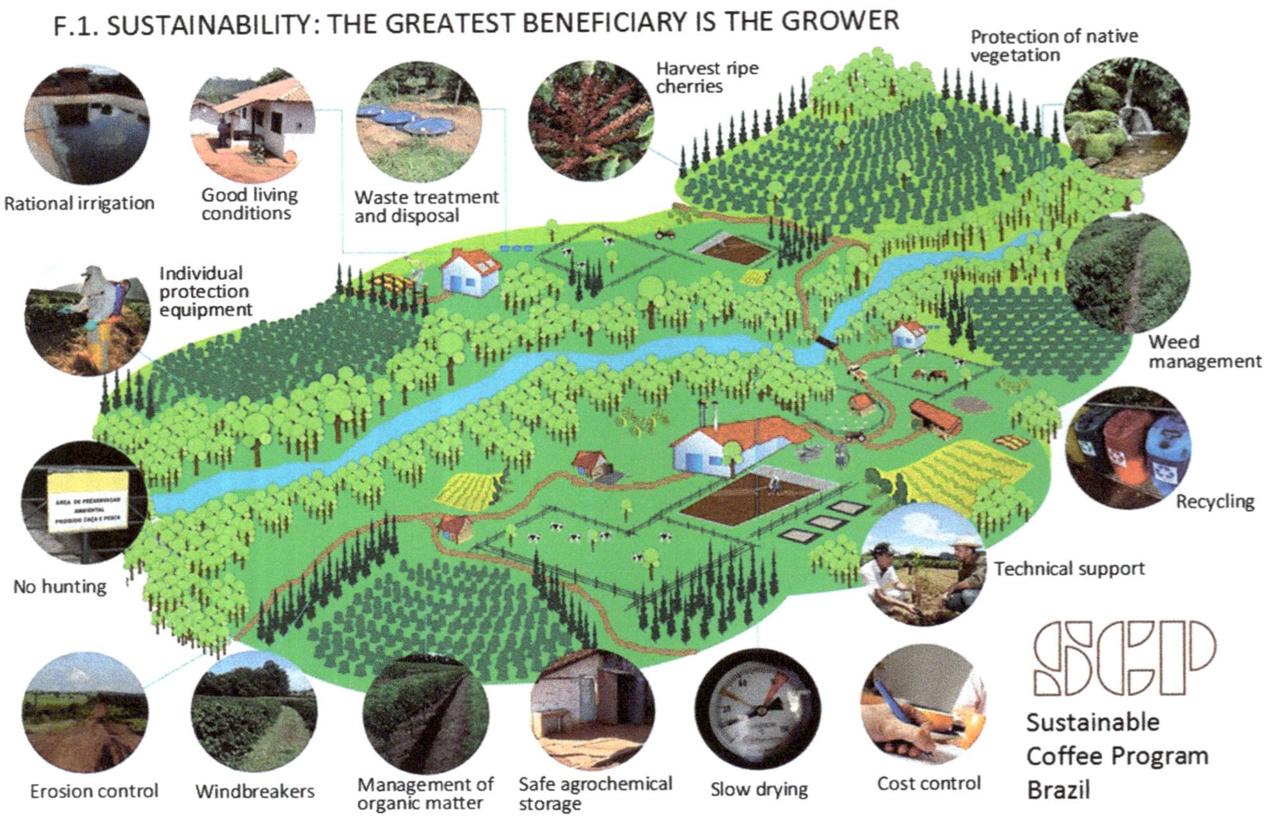

Figure 17.1 Sustainability: the greatest beneficiary is the grower. Reproduced from ref. 1 with permission from the International Trade Centre.

such percentage should be clearly printed or "contains sustainable coffee" should be stated making it clear that not all coffee in the package is sustainable. The least rigorous concept in accounting for sustainable content is mass balance. Still little used in coffee but prevalent with some other products, it means that the company selling the product buys raw material produced in a sustainable way but the sustainable product may not necessarily be in that specific retail package.

17.5 Origin

The association of organoleptic or cup features and origins has existed as long as the coffee business has existed. The market has traditionally associated specific coffee tastes and flavors with the region or country where coffee is grown, as shown by expressions like Costa Rica Tarrazu, Ethiopia Yirgacheffe or Indonesia Sumatra, that not only indicate the origin of the coffee but also what general or emblematic cup features to expect. Coffee has been traded and blended for centuries using origins; grading and cupping features were progressively added and became more sophisticated and complex over time. Certification of origin came up much later and is rather recent in the coffee business.

Certification of origin is much more straight-forward than that of sustainability. It simply requires a proof of where the coffee has been produced that is inherent to the legal and fiscal invoicing system used in most coffee-producing countries whereby the original sales invoice states the place where the coffee was produced, *i.e.*, the farm of origin.

Geographical Indication can of course be much more sophisticated than this if it associates origin with minimum or specific quality standards in which case a process to evaluate and certify quality has to be included. This has been partially addressed in the chapter about coffee grading and may be complemented by organoleptic evaluation, *i.e.*, cup tasting, which will be covered next.

Geographical Indication is regulated by specific legislation issued by the countries that use it. This legislation may be the same for several products or specific for coffee. In most cases the legislation requires that a specific representative body of the origin or region is created and this body establishes the criteria to be applied in accordance with the legislation and including minimum or specific qualities if desired or required.

17.6 Quality

Quality is generally described by the combination of grading (screen size and defect count) and sensorial evaluation, *i.e.*, cupping. Whereas grading is a rather exact activity, cupping has a subjective content that is much more difficult to standardize. Most coffee is traded without any formal certification of quality. Instead there are quality descriptions that are known to and accepted by seller and buyer. They are different for each origin but known to the trade.

Table 17.3 Total score quality classification.

Total score quality classification		
90–100	Outstanding	Specialty
85–89.99	Excellent	
80–84.99	Very good	
<80.0	Below specialty quality	Not specialty

Usually the buyer requires a given quality that the seller delivers or the seller offers a quality that the buyer accepts. Samples are kept at origin after shipment and if there is disagreement between buyer and seller these samples are used for arbitration against those drawn in the arrival country according to established procedures and carried out by bodies agreed upon by both parties, for example, the Green Coffee Association in the US.

There have been attempts to standardize cupping and to make it a more consistent and reliable tool across countries. The most successful initiative has been SCA's Cupping Protocols whose scores, found in Table 17.3, are today widely used for specialty coffees.

The scores in Table 17.3 have been incorporated into trading practices but no wide-spread certification system accompanies it. In other words, the scores are used in the same way country descriptions have always been, as explained above.

The Coffee Quality Institute (USA) has created the Q Coffee System that uses the SCA scores and three qualified Q Graders to certify Arabicas, Robustas and blends. The system is being used for specialty coffees coming from several countries.

A totally different quality certification approach has been used within Brazil to support its program to promote local coffee consumption. In continuation of the Coffee Purity Program, that roasters could choose to join and that checked the purity of coffees sold, the Brazilian Coffee Roasters Association – ABIC's Coffee Quality Program set three quality standards, Gourmet, Superior and Traditional that roaster members can voluntarily adopt. Participating roasters qualify and register their quality control departments that self-certify their coffees that are labeled accordingly. ABIC's independent auditors monitor these qualities by means of random sampling at point-of-sale and quality analysis against the program's cupping standards.

Reference

1. The Coffee Exporters' Guide, *International Trade Centre*, 3rd edn, 2012, 60.

Part III

Coffee Chemistry

Section I

Natural Coffee Compounds and Derivatives

CHAPTER 18

Proteins of Coffee Beans: Recent Advances

PAULO MAZZAFERA*, FLÁVIA SCHIMPL AND EDUARDO KIYOTA

Department of Plant Biology, Institute of Biology, State University of Campinas – UNICAMP, Cidade Universitária Zeferino Vaz, Rua Monteiro Lobato 255, CEP, 13083-862, Campinas, SP, Brazil
*E-mail: pmazza@unicamp.br

18.1 Introduction

Coffea arabica was the first species of the genus *Coffea* to be commercially cultivated and therefore became the reference for beverage quality.

Coffee beverage quality is a result of a balance between a series of factors, actions and decisions, starting in the field during cultivation and continuing to bean storage, processing and roasting. Both non-genetic and genetic sources of variation have been shown to affect beverage quality.[1] Soil fertility, fertilization, climate, altitude, plant health, the stage of fruit maturation at harvest, post-harvest processing and genetic background are all examples of key factors known to influence coffee beverage quality.[1–3] However, the consequence of all of these sources of variation is the final chemical composition of the beans, which will dramatically change during roasting, when aroma and flavor develop.

The complexity of coffee chemistry is best demonstrated by the number of volatile compounds identified in the aroma, which is over a thousand.[4,5]

Coffee: Production, Quality and Chemistry
Edited by Adriana Farah

Published by the Royal Society of Chemistry, www.rsc.org

Sugars, amino acids, peptides, lipids, trigonelline, chlorogenic acids, organic acids and carotenoids are thought to influence coffee quality.[6,7] It is also well known that amino acids, peptides and proteins play an important role in coffee aroma production[6,8–10] because they interact with sugars during roasting by the Maillard reaction.[11]

Over the past 50 years, several reports on the protein and/or amino acid content and profile of coffee beans have consistently confirmed the importance of these compounds for coffee flavor;[9,10,12–14] however, despite the improvement of the techniques and scientific approaches used in these studies, all of these studies failed to clearly identify their relative importance (proportion and concentration). A wide range of protein contents in *C. arabica* and *C. canephora* beans, from 10–15%, has been reported in the literature, but this variation results from the methods used to estimate protein content.[6,15] If corrections are made for caffeine and trigonelline, an average value of approximately 11.5% dry basis is found, and no significant difference seems to exist between *C. arabica* and *C. canephora* beans.[6,15] One third of the total coffee bean protein is thought to bind to arabinogalactans in the cell wall, but there is no detailed information regarding the nature of these proteins. It is probable that most are enzymes, such as polyphenol oxidase and peroxidases, while the largest fraction of proteins remains in the cytoplasm.[15]

The contents of free amino acids are much lower than protein, ranging from 0.15–2.5% in *C. arabica* and *C. canephora* varieties.[6,15] Despite that, they are the single most important group of compounds related with the development of coffee flavor.[6] *C. canephora* varieties have the higher levels of all free amino acids when compared with *C. arabica*, except for glutamate.[6]

Several other reports have described the investigation of the redox status of the coffee beans and its relationship to proteins and coffee quality. Particular attention has been given to chlorogenic acids, which represent approximately 5–7.5% of the compounds in *C. arabica*, and may reach even higher levels in the coffee beans of other *Coffea* species.[6] Early studies showed that the enzyme polyphenol oxidase oxidizes these phenolic acids to quinones, which in turn react with proteins, affecting the final bean quality and consequently the beverage.[9,13,16] Recent modeling studies have shown that *C. arabica* proteins are more susceptible to interactions with chlorogenic acids and have confirmed that polyphenol oxidase activity plays a crucial role in the process.[17]

Although the importance of the Maillard reaction in the determination of food quality, including the quality of coffee, is well known, proteins have been neglected in genetic studies aiming to select/breed coffee for quality.[1,18] Caffeine, trigonelline, sucrose and chlorogenic acids are the most frequently reported compounds in studies, presumably because they are easier to quantify and do not have the chemical complexity of proteins. A simple way to understand this property is to observe the complexity of a two-dimensional gel electrophoresis of coffee seed proteins.[8,9,19]

Based on the data of Thaler and Gaigl,[20] Clifford[6] called attention to the fact that the amino acid profile of the water-soluble proteins (albumin fraction) of coffee beans is similar to the profile of the total proteins. This similarity would be expected because the most abundant protein in coffee beans is an 11S-globulin seed storage protein.[8,21,22]

It must also be considered that not only amino acid type but also carbohydrates play a role in the flavors and aroma developed in the Maillard reaction.[23] Sucrose is the main soluble sugar in coffee beans, but fructose and glucose may be present in significant amounts depending on the stage of fruit maturity.[3] Low-molecular-weight carbohydrates, such as raffinose, stachyose, ribose, mannose and galactose, have also been reported in coffee beans.[15,24] Sucrose content may vary greatly in coffee beans, and values between 4.6 and 8.6 have been reported for *C. arabica* beans.[15] Coffee beans are also rich in polysaccharides, which represent 40–50% of the dry weight;[25] on average, the polysaccharides consist of 50% of mannans, 30% arabinogalactan, 15% cellulose and 5% pectin.[24] Low-molecular-weight sugars are progressively destroyed during roasting; however, the formation of monosaccharides from polysaccharides is not believed to occur at significant rates, although it is known that they undergo changes to form more complex structures.[25] Like low-molecular-weight sugars, intact polysaccharides and these complex carbohydrates may react with free amino acids, peptides and proteins. Due to such complexity, it is therefore not possible to determine the precise role of each carbohydrate and amino acid in the development of coffee flavor and aroma, although we know they play an important role.

Several reports about the participation of proteins in the development of coffee flavor and aroma have been published and have been reviewed by Clifford[6,15] and De Castro and Marraccini.[26] However, except for a few publications on the biochemistry and molecular control of the formation of the 11S reserve protein[8,21,22] and reserve proteins in different coffee species,[27] peptides and proteases,[28] very few studies have been conducted on coffee proteins and their relationship to cup quality. Some studies on specific enzymes have been published, but they are not essential to the development of quality.[16,29,30] Two exceptions were published in 2003: Montavon *et al.*[6] reported on the evolution of protein profiles with maturation and related it to cup quality, and Montavon *et al.*[15] reported on the changes in coffee protein profiles during roasting. To our knowledge, Homma[31] wrote the last review that dedicated a chapter to coffee amino acids and proteins, but it is less than two full pages. Although mentioning the importance of protein to the formation of aroma, the book by Flament[32] does not have a chapter dedicated to proteins. The effect of proteins on the formation of volatile compounds during coffee roasting was recently studied by Hwang *et al.*[33] A simple roasting apparatus using a mineral oil bath was set up that allowed the authors to conclude that protein increased sucrose degradation during roasting and that a greater amount of protein produced a roasted coffee with a lighter color.

In this chapter special attention will be given to the recent literature published after the reviews of Clifford.[6,15] Most of the reports are in some way related to the 11S reserve protein. The role of proteases will also be discussed.

18.2 The 11S Seed Storage Protein of Coffee

The 11S seed storage protein was cloned from *C. arabica* beans and has been well characterized.[8,21,22] Previous reports have shown that two main bands were predominant in SDS-PAGE electrophoresis,[12,13,34,35] and Luthe[36] suggested that they were the α- and β-subunits of a legumin-like protein.

Acuña *et al.*[21] isolated an 11S globulin from seeds of *C. arabica* variety Colombia with an apparent molecular weight of 55 kDa, which under denaturating conditions (SDS-PAGE) was cleaved into two polypeptides of 22 kDa (β-arm) and 33 kDa (α-arm). Molecular characterization showed that the coffee legumin had a putative legumin box.

Rogers *et al.*[8] used 2D-electrophoresis and amino acid microsequencing for a better characterization of the 11S coffee protein extracted from *C. arabica* variety Caturra. They sequenced several abundant polypeptides, which were identified as α- and β-subunits, suggesting that the protein had different isoforms. The comparison of these polypeptides with those described by Acuña *et al.*[21] supported the presence of isoforms because 23 amino acids differed between the sequences. The similarity with 11S storage proteins from other plants led Rogers *et al.*[8] to suggest that this protein has a storage function in coffee. This finding was further confirmed by Shimizu and Mazzafera,[37] who showed that the levels of α- and β-subunits declined during germination in seeds of *C. arabica*.

According to Rogers *et al.*,[8,15] the 11S protein accounts for approximately 45% of the total soluble protein in coffee endosperm, but the two subunits were also detected in 2D-electrophoresis of proteins found in embryos. Because it is the predominant protein in the endosperm, the deduced amino acid composition of the cloned 11S coding sequence confirmed previous results that sulfur-containing amino acids (methionine and cysteine) were present in low amounts in the total protein of coffee beans. Glutamate/glutamine and aspartate/asparagine are the main amino acids of the coffee 11S protein as well as of the total soluble protein fraction of coffee beans. A specific promoter controls the expression of the 11S gene in the coffee bean.[22] Rogers *et al.*[8] found that cysteine and methionine represented only 1.1 and 0.2, respectively, of the 20 amino acids coded by *csp1*. Thaler and Gaigl[20] found that methionine was the least abundant (2.38%) amino acid in the total protein and albumin fraction of Brazilian arabica coffee. Cysteine was not analyzed. It has been suggested that two cysteine residues are involved in the formation of the disulfide bridge between the α- and β-subunits of the 11S coffee protein.[8] Because sulfur-containing amino acids are important for beverage taste and aroma development during coffee roasting, examining the genetic variability in 11S proteins to isolate the ones with higher contents of cysteine and methionine would be useful.[8]

18.3 A Family of 11S Proteins in *Coffea*

The two-dimensional electrophoretic profiles of proteins isolated from *C. arabica* and *C. canephora* variety Robusta beans are different.[8] Not only were other peptides identified as precursors but corresponding groups of α- and β-arms also appeared in the profile. They had slightly higher masses and more acidic pIs. Curiously, not all beans of the Robusta variety showed these protein characteristics.

Baú *et al.*[27] used SDS mono-dimensional electrophoresis to examine the proteins of several varieties of *C. arabica* and other species of *Coffea*. Three plants from each of the Typica, Maragogipe, Catuaí, Mundo Novo, Cera and Caturra varieties of *C. arabica* were studied and showed very similar protein profiles, in which two bands at 37.6 ± 3.3 and 20.9 ± 1.4 kDa were predominant. Catuaí and Caturra are dwarf varieties, Maragogipe has large beans, Cera has yellow beans and Mundo Novo is a mutation of the variety Bourbon, which descends from Typica.[38–42]

Baú *et al.*[27] analyzed three plants of *C. eugenioides* and *C. canephora* variety Kouillou (or Conillon) and one plant each of *C. racemosa*, *C. bukobensis*, *C. kapakata*, *C. liberica*, *C. dewevrei*, *C. stenophylla*, *C. salvatrix*, *C. klainii* and *C. laurentii* (Figure 18.1). In addition to variation in the protein contents, from 4.7% in *C. racemosa* to 9.1% in *C. liberica*, a large variation was observed in the protein profiles, *i.e.*, both the mass of the two subunits and the number of subunits. For example, of the three *C. eugenioides* plants, plant 2 (lane 16) had an additional α-subunit. *C. canephora* var. Kouillou plant 70/9 (lane 7) clearly had additional α- and β-subunits. The *C. racemosa* (lane 2) profile was the most different; it had an intense band of approximately 12 kDa. Additionally, although a band was detected at the position of the α-arm, no band was observed at the apparent mass of the β-arm.

Bandil *et al.*[43] produced 2D-electrophoretic profiles of *C. arabica* beans at two different developmental stages: stage 1, cell division and elongation of the perisperm; and stage 2, early growth of the endosperm progressively replacing the perisperm. Only six proteins were differentially expressed during stage 2; four of them were related to energy metabolism, one was identified as thaumatin and one as the 11S.

18.4 2S Protein in *Coffea*

Lin *et al.*[44] identified a gene coding for a 2S-type protein as the dominant transcript in fruits of *C. canephora*. The second most highly expressed gene was related to the 11S reserve protein. In this study, an EST database was built on libraries from leaves, pericarps, young cherries with mixed seed and pericarp, middle-stage seeds and late-stage seeds. Curiously, the 2S gene showed transient expression during fruit maturation; 21 ESTs were found in the early stage, 1037 in the middle stage and 139 in the late stage, which accounted for 98.2% of the whole database (1197 of 1219 ESTs). However, the 11S gene showed the highest transcript numbers during the late stage (412)

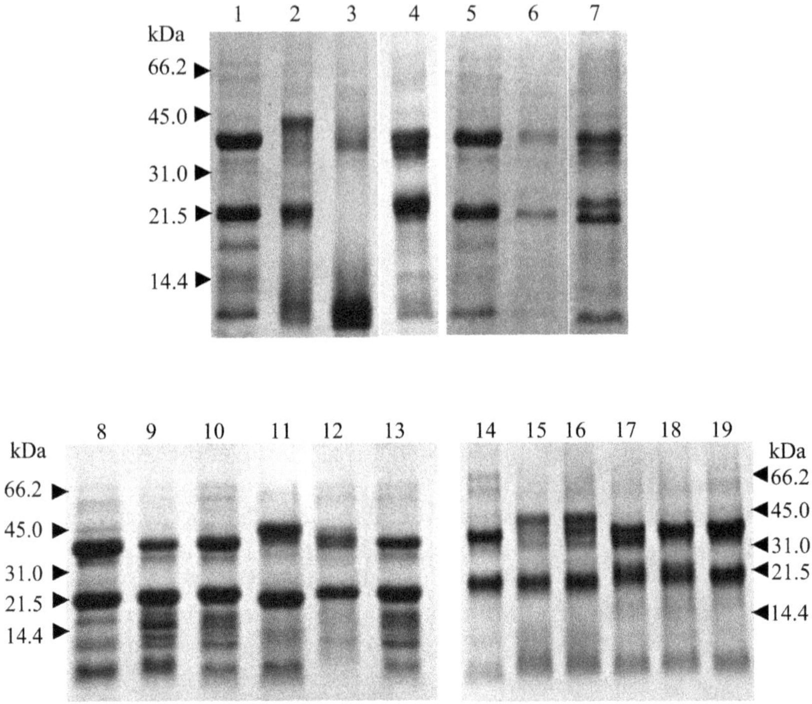

Figure 18.1 SDS-PAGE of protein seed extracts from coffee species: *C. arabica* var. Typica (lanes 1, 5, 8, 19); *C. eugenioides* (plant 1 – lane 2, plant 2 – lane 16, plant 3 – lane 15); *C. racemosa* (lane 3); *C. bukobensis* (lane 4); *C. kapakata* (lane 6); *C. canephora* var. Kouillou (70/9 – lane 7, 70/10 – lane 18, 70/14 – lane 17); *C. liberica* (lane 9); *C. dewevrei* (lane 10); *C. stenophylla* (lane 11); *C. salvatrix* (lane 12); *C. klainii* (lane 13); *C. laurenti* (lane 14). Molecular markers: phosphorylase b (97.4 kDa), BSA (66.2 kDa), egg albumin (45 kDa), carbonic anhydrase (31 kDa), trypsin inhibitor (21.5 kDa), lysozyme (14.4 kDa). This figure was reproduced with permission of the Sociedade Brasileira de Fisiologia Vegetal from S.M.T. Baú, P. Mazzafera and L.G. Santoro, *Rev. Bras. Fisiol. Veg.*, 2001, **13**, 33.

compared with the early (28) and middle stages (244). The 2S gene showed the highest similarity to the 2S proteins of tomato, sesame, sunflower and Brazil nut, in decreasing order.

Mondego *et al.*[45] combined 8 EST libraries from the study with *C. canephora* carried out by Lin *et al.*[44] with 43 libraries of *C. arabica* from the Brazilian Coffee Genome Project.[46] The sources for the Brazilian project were mainly two *C. arabica* cultivars (Catuai and Mundo Novo, with the exception of germinating seeds from cv. Rubi) and one *C. canephora* genotype (Conillon). In the study by Lin *et al.*[44] the sources were five different genotypes of *C. canephora*. Like in *C. canephora*, a 2S gene homolog to the 2S protein of sesame was highly expressed in *C. arabica*. It was the fifth most highly expressed gene;

it was found in four libraries and was represented by 1271 ESTs. The 11S gene was not among the 20 most expressed genes in *C. arabica*. Unfortunately, the Brazilian Coffee Genome Project did not have libraries specifically dedicated to fruits that discriminated their developmental stages. The libraries FR1 and FR2 were described as flower buds + pinhead fruits + fruits at different stages in Mundo Novo (libraries FR1 and FR2). Most of the ESTs related to the 2S protein were found in these libraries.

C. racemosa was also studied in the Brazilian EST project.[46] Two libraries from fruits were described as "FR4 – fruits" and "FV2 – stages 1, 2 and 3". There is no further information about the ESTs of *C. racemosa*, but a search of the publicly available Electronic Northerns of the Brazilian EST Project (http://bioinfo03.ibi.unicamp.br/cafe/) shows that the 2S gene was among the 20 most expressed genes and showed homology with the 2S protein of sesame. The 11S gene was not among the genes found in the ESTs of *C. racemosa*. Other more highly expressed genes were mostly related to processes such as defense (chitinases, chaperones, lipid transfer proteins), photosynthesis (Rubisco small unit, chlorophyll a/b binding protein), redox control (catechol oxidase), phosphate metabolism (acid phosphatase was the most highly expressed gene) and water transport (aquaporins).

The sesame 2S reserve protein has a molecular mass ranging from 13–15 kDa,[47] which is close to the mass of the main protein in the electrophoretic profile of *C. racemosa* (Figure 18.1 – from Baú *et al.*[27]) and of the translated open reading frame of the nucleotide sequence found in the Brazilian EST project (Mazzafera, unpublished data). The analysis of the sequences in this EST database suggested that two isoforms are present in *C. racemosa* (Mazzafera, unpublished data), which is in agreement with the results found by Coelho *et al.*[48] These authors purified legumin-like proteins from the mature seeds of *C. arabica* and *C. racemosa* in two chromatographic steps and separated two fractions from each species, named CaL-1 and CaL-2, and CrL-1 and CrL-2, respectively. Denaturing electrophoresis revealed that CaL-1 was composed of two prominent bands of 33 and 24 kDa. CaL-2, CrL-1 and CrL-2 showed a single protein band with an apparent molecular mass of 14 kDa. However, the amino acid compositions of CaL-2, CrL-1 and CrL-2 were quite different. Because Coelho *et al.*[48] observed a statistically significant reduction in the larval development and increase in mortality of *Callosobruchus maculatus* in dose–response experiments using an artificial diet, they suggested that the 2S legumin-like protein in *C. racemosa* might be involved in the resistance of this species to the coffee berry borer *Hypothenemus hampei*.[49] In this aspect, the susceptibility of *C. arabica* to this insect might be explained by the different amino acid sequence of its 2S protein when compared with the 2S protein of *C. racemosa* and the low amounts found in the seeds. Additionally, the EST study by Lin *et al.*[44] showed a decrease in 2S transcripts during the maturation of *C. canephora* beans, which are also susceptible to *H. hampei*, and an increase in 11S transcripts.

18.5 Peptides and Proteases

The composition of free amino acids has been investigated in several types of coffee, and their role as aroma precursors is well known. Peptides have not received the same attention. Ludwig *et al.*[28] found that on average, peptides were approximately 0.59% of the dry mass in several *C. arabica* samples and 0.43% in *C. canephora*. Samples freed of amino acids and proteins were separated by 2D-PAGE, allowing the identification of 30 peptides with molecular weights ranging from 4 to 10 kDa. *C. arabica* and *C. canephora* beans differed in their 2D-electrophoresis profiles and like the total protein and the 11S reserve proteins, methionine was the least abundant amino acid of the peptides. Glutamic acid/glutamine, aspartic acid/asparagine and glycine were the most abundant amino acids in both coffee types. Peptides were melted with dextrose in a model roasting experiment using the sensory impressions of five persons and their role in aroma formation in coffee was confirmed. Ludwig *et al.*[28] also assayed coffee protein extracts for proteolytic activity using activity staining following gel electrophoresis and found at least five to seven activity bands in all of the coffee samples. Their activities were not decreased by serine and pepsin protease inhibitors, but iodoacetamide caused a clear reduction. A very important observation was the fact that proteolytic activity is stable for months in stored coffee beans. However, it is interesting to mention that the amino acid profiles of beans stored for 1 year do not significantly change either the total content of protein amino acids or the concentration of most individual amino acids, except that there was a decrease in glutamine.[50] Thus, it appears that proteolytic activity has to be triggered because the 11S protein of coffee is degraded during germination.[37]

Montavon *et al.*[9] observed that the aerobic incubation of powdered beans, but not anoxic incubation, favored the fragmentation or digestion of the 11S reserve protein and the release of free amino acids. Under aerobic conditions the α-subunit of the 11S storage protein promptly disappeared from the extracts, whereas the β-arm was partially preserved. Although these results suggest an oxidative process rather than a proteolytic digestion, the authors argued that oxidation might alter protein conformation and increase the sensitivity toward active proteases.

Lepeley *et al.*[51] isolated the gene sequences of two cysteine proteases from *C. canephora* beans and showed that their expression increased with fruit maturation. However, the expression level of *CcCP1* was 600% higher than the expression level of *CcCP4*. Abreu *et al.*[52] searched the Brazilian EST project database for protease sequences. They found 41 genes (contigs) from all libraries made from *C. arabica* tissues, and based on their protein domains, 34 were identified as cysteine proteases, six as aspartic proteases and only one as a serine protease. Fourteen genes were exclusive to fruit libraries, and three cysteine proteases (*CaCP1*, *CaCP4* and *CaCP23*) and one aspartic protease (*CaAP2*) were chosen for further studies. The expression profiles of *CaCP4*, *CaCP23* and *CaAP2* indicated that they were more highly expressed in immature fruits than mature fruits, whereas the opposite trend was observed for *CaCP1*.[52] Phylogenetic trees built with the translated sequences showed that *Ca*CP1 and *Ca*CP4 were highly similar to *Cc*CP1 and *Cc*CP4, respectively. *Ca*CP1, *Ca*CP4 and *Ca*CP23 were more closely related to proteases from other species than to each other. *Ca*AP2 was highly similar to *Cc*AP2, which was isolated from *C. canephora*,[53] and both were completely different from *Cc*AP1.

The predominance of cysteine protease sequences in the Brazilian EST project database and that they are most highly expressed in immature fruits are in agreement with activity assays.[52] *In gel* activity assays showed that the highest activity was found in protein extracts obtained from immature beans, and strong inhibition of this activity was detected after iodoacetamide treatment, supporting the results obtained by Ludwig *et al.*[28] Interestingly, Abreu *et al.*[52] also detected serine protease activity, which was inhibited by phenylmethylsulfonyl fluoride (PMSF), although they had not identified any serine protease genes in their search for proteases in the coffee EST database. The characteristics of the varieties used in these studies, the developmental stage examined and the location where the fruits were harvested were used as arguments to explain this inconsistency.

Abreu *et al.*[52] collected samples from two different locations in Brazil and, in addition to protease activity and related gene expression, they also assayed the protein and free amino acids contents and produced an amino acid profile. Samples were collected from Adamantina and Mococa in São Paulo State, which have marked differences in temperature and produce coffee with distinct qualities. Mococa is a cooler location and produces better coffee quality. The overall results indicated that higher temperatures were related to lower quality and high protease activity. The results are corroborated by a previous report where the influence of irrigation was investigated in the same locations.[54] Principal component analysis (PCA) of the contents of sucrose, reducing sugars, starch, total phenolics, chlorogenic acid (5-caffeoylquinic acid), caffeine, lipids, proteins, free amino acids and nitrogen, and the activities of proteases and polyphenol oxidase in *C. arabica* plants subjected to four irrigation treatments showed that temperature rather than irrigation was the main factor affecting quality.

18.6 Does Coffee Have Bioactive Proteins and Peptides?

Bioactive proteins and peptides have been investigated in many food sources regarding their physiological significance beyond their role as nitrogen source. They can be found in several food sources like milk, corn, fish, wheat, rice, soybean, *etc.*[55,56] and a wide range of physiological effects – positive and adverse – have been indicated.[57] The information on the bioactive role of proteins and peptides and their nutritional impact increased considerably in the last years.[58] Because of their size, intact proteins are not absorbed in the human gastrointestinal tract limiting their physiological action. However, peptides also present in other food sources and those resulting from protein digestion are smaller and more likely to be absorbed, enabling a different array of bioactivities.[58] Unfortunately, the knowledge regarding the relationships between functionality and structure properties is still scant.

There is no information regarding bioactivity of proteins and peptides in coffee. Farah[59] listed several bioactive components of coffee, including caffeine, trigonelline, chlorogenic acids, terpenes, fibers, lipids, minerals, carbohydrates and other compounds, but proteins and peptides were not included. Coffee is roasted before brewing, altering significantly the protein profile.[9,10] Under the temperatures used to roast coffee not only proteins are destroyed, entirely or partially, but peptides are released as observed for the 11S coffee storage protein.[10]

Based on the global perspective of "foodomics" – which encompass several "omic" studies – protein "digestomics" has been proposed as a new platform to study proteins and peptides functionality in food.[60] Coffee research should make profit from this initiative. A recent book explored comprehensively the benefits and risks of coffee components but no chapter was dedicated to proteins and peptides.[61] Thus, it is clear that there is a lacuna in the research of coffee proteins and that the knowledge generated by the proteomic studies cited in this chapter should be expanded and this effort joined with health studies.

18.7 Conclusion

A large number of reports have underlined the importance of proteins, peptides and amino acids in the formation of aroma and flavor in roasted coffee beans, and this fact is indisputable. However, the information is sparse and has mostly been obtained using mature beans. Association of the available genomic information[44–46] with data from a well-planned proteome experiment should be used as a strategy to identify proteins in the beans. Ideally, such an experiment should be carried out using *C. arabica* beans from varieties that display marked differences in cup quality or even from the same variety cultivated in at least two locations where the quality is affected by climatic conditions, as suggested by studies carried out by Abreu *et al.*[52] and Silva *et al.*[54] Unfortunately, the proteome study carried out by Bandil *et al.*[43] analyzed only two developmental stages. Therefore, a proposed study should consider beans from fruits at different stages near maturation and the effects of storage under optimal and non-optimal conditions, thus provoking decay of quality. Protease activity should also be followed.

Such an experiment might not answer all of the questions about the participation of proteins such as the 11S protein, which represents 45% of the total protein of the bean, in the development of coffee aroma and flavor but would certainly provide a solid base for linking the available information and reveal new directions for coffee quality research. The 2S protein should also be considered because the 2S genes are highly expressed during coffee bean maturation. We are also aware that carbohydrates are key players in this scenario and perhaps a parallel study should also be carried out.

In addition, an effort should be made to address the physiological effects of coffee proteins and peptides. Moderate consumption of coffee has been suggested to bring benefits to human health and, like many other foods, part of this reward might be related to these compounds.

Acknowledgements

The author thanks the Brazilian National Research Council (Conselho Nacional de Desenvolvimento Científico e Tecnológico-CNPq) for a research fellowship.

References

1. T. Leroy, F. Ribeyre, B. Bertrand, P. Charmetant, M. Dufour, C. Montagnon, P. Marraccini and D. Pot, *Braz. J. Plant Physiol.*, 2006, **18**, 229.
2. F. M. Da Matta and J. D. C. Ramalho, *Braz. J. Plant Physiol.*, 2006, **18**, 55.
3. C. Geromel, L. P. Ferreira, S. M. C. Guerreiro, A. A. Cavalari, D. Pot, L. F. P. Pereira, T. Leroy, L. G. Vieira, P. Mazzafera and P. Marraccini, *J. Exp. Bot.*, 2006, **55**, 3243.
4. S. K. Dart and H. E. Nursten, in *Coffee: Chemistry*, ed. R. J. Clarke and R. Macrae, Elsevier Applied Science, London, 1985, p. 223.
5. I. Flament, in *Volatile Compounds in Foods and Beverages*, ed. H. Maarse, Dekker, New York, 1991, p. 617.
6. M. N. Clifford, in *Coffee: Chemistry*, ed. R. J. Clarke and R. Macrae, Elsevier Applied Science, London, 1985, p. 153.
7. W. Holscher and H. Steinhart, in *Thermally Generated Flavors*, ed. T. Parliament, M. Morello and R. McGorrin, ACS Symposium Series 543; American Chemical Society, Washington, DC, 1992, p. 207.
8. W. J. Rogers, G. Bézard, A. Deshayes, I. Meyer, V. Pétiard and P. Marraccini, *Plant Physiol. Biochem.*, 1999, **37**, 261.
9. P. Montavon, E. Duruz, G. Rumo and G. Pratz, *J. Agric. Food Chem.*, 2003, **51**, 2328.
10. P. Montavon, A.-F. Mauron and E. Duruz, *J. Agric. Food Chem.*, 2003, **51**, 2335.
11. J. Kerler, C. Winkel, T. Davidek and I. Blank, in *Food Flavour Technology*, ed. A. J. Taylor and R. S. T. Linforth, Blackwell Publishing Ltd., Oxford, 2nd edn, 2010, p. 51.
12. H. V. Amorim and R. V. Josephson, *J. Food Sci.*, 1975, **40**, 1179.
13. H. V. Amorim and V. L. Amorim, in *Enzymes in Food and Beverages Processing*, ed. R. L. Ory and A. J. St Angelo, ACS Symposium Series, 47, American Chemical Society, 1977, p. 27.

14. U. Arnold and E. Ludwig, *Z. Lebensm.-Unters. Forsch.*, 1996, **203**, 379.
15. M. N. Clifford, in *Coffee: Botany, Biochemistry and Production of Beans and Beverage*, ed. M. N. Clifford and K. C. Wilson, AVI Publishing Company, Inc., Westport, Connecticut, 1985, p. 305.
16. P. Mazzafera and S. P. Robinson, *Phytochemistry*, 2000, **55**, 285.
17. M. Ali, T. Homann, J. Kreisel, M. Khalil, R. Puhlmann, H.-P. Kruse and H. Rawel, *J. Agric. Food Chem.*, 2012, **60**, 11601.
18. T. Leroy, F. Bellis, H. Legnate, E. Kananura, G. Gonzales, L. F. Pereira, A. C. Andrade, P. Charmetant, C. Montagnon, P. Cubry, P. Marraccini, D. Pot and A. de Kochko, *Tree Genet. Genomes*, 2011, **7**, 781.
19. M. T. Gil-Agusti, N. Campostrini, L. Zolla, C. Ciambella, C. Invernizzi and P. G. Righetti, *Proteomics*, 2005, **5**, 710.
20. H. Thaler and R. Gaigl, *Z. Lebensm.-Unters. Forsch.*, 1963, **120**, 357.
21. R. Acuña, R. Bassüner, V. Beillinson, H. Cortina, G. Cadena-Gómez, V. Montes and N. C. Nielsen, *Physiol. Plant.*, 1999, **105**, 122.
22. P. Marraccini, A. Deshayes, V. Pétiard and W. J. Rogers, *Plant Physiol. Biochem.*, 1999, **37**, 273.
23. M. C. Lane and H. E. Nursten, in *The Maillard Reaction in Food and Nutrition*, ed. G. R. Waller and M. S. Feather, ACS Symposium Series 215, American Chemical Society, Washington, DC, 1983, p. 141.
24. A. G. W. Bradbury, in *Coffee: Recent Developments*, ed. R. J. Clarke and O. G. Vitzthum, Blackwell Science, Oxford, 2001, p. 1.
25. L. C. Trugo, in *Coffee Chemistry*, ed. R. Clarke and R. Macrae, Elsevier Applied Science, London, 1985, p. 83.
26. R. De Castro and P. Marraccini, *Braz. J. Plant Physiol.*, 2006, **18**, 175.
27. S. M. T. Baú, P. Mazzafera and L. G. Santoro, *Rev. Bras. Fisiol. Veg.*, 2001, **13**, 33.
28. E. Ludwig, U. Lipke, U. Raczek and A. Jäger, *Eur. Food Res. Technol.*, 2000, **211**, 111.
29. P. Marraccini, W. J. Rogers, V. Caillet, A. Desahyes, D. Granato, F. Lausanne, S. Lechat, D. Pridmore and V. Pétiard, *Plant Physiol. Biochem.*, 2005, **43**, 909.
30. M. G. Cotta, L. M. Barros, J. D. de Almeida, F. de Lamotte, E. A. Barbosa, N. G. Vieira, G. S. Alves, F. Vinecky, A. C. Andrade and P. Marraccini, *Plant Mol. Biol.*, 2014, **85**, 11.
31. S. Homma, in *Coffee: Recent Developments*, ed. R. J. Clarke and O. G. Vitzthum, Blackwell Science, Oxford, 2001, p. 50.
32. I. Flament, *Coffee Flavor Chemistry*, John Wiley & Sons, Inc., New York, 2001, p. 424.
33. C.-F. Hwang, C.-C. Chen and C.-T. Ho, *Int. J. Food Sci. Technol.*, 2012, **47**, 2117.
34. M. Centi-Grossi, C. Tassi-Micco and V. Silano, *Phytochemistry*, 1969, **8**, 1749.
35. H. Bade and H. Stegemann, *J. Agron. Crop Sci.*, 1982, **151**, 89.
36. D. S. Luthe, *Plant Mol. Biol. Rep.*, 1992, **10**, 254.
37. M. M. Shimizu and P. Mazzafera, *Braz. Arch. Biol. Technol.*, 2000, **43**, 259.
38. A. Carvalho, *Coffee Tea Ind. Flavor Field*, 1958, **81**, 30.

39. A. Carvalho, *Documentos IAC*, 1993, **34**, 1.
40. A. Carvalho, C. A. Krug, J. E. T. Mendes, H. Antunes Filho, H. Moraes, J. Aloisi Sobrinho, M. V. Moraes and T. R. Rocha, *Bragantia*, 1952, **12**, 97.
41. A. Carvalho, H. P. Medina Filho, L. C. Fazuoli, O. Guerreiro Filho and M. M. A. Lima, *Rev. Bras. Genet.*, 1991, **14**, 135.
42. A. Carvalho, H. P. Medina Filho, O. Guerreiro Filho and L. C. Fazuoli, *Ciênc. Cult.*, 1991, **43**, 32.
43. G. B. Bandil, R. M. Etto, C. W. Galvão, H. J. O. Ramos, E. M. Souza, F. O. Pedrosa, D. F. S. Chaves, L. F. Huergo and R. A. Ayub, *Genet. Mol. Res.*, 2013, **12**, 5102.
44. C. Lin, L. A. Mueller, J. McCarthy, D. Crouzillat, V. Pétiard and S. D. Tanksley, *Theor. Appl. Genet.*, 2005, **112**, 114.
45. J. M. Mondego, R. O. Vidal, M. F. Carazzolle, E. K. Tokuda, L. P. Parizzi, G. G. Costa, L. F. Pereira, A. C. Andrade, C. A. Colombo, L. G. Vieira, G. A. Pereira and Brazilian Coffee Genome Project Consortium, *BMC Plant Biol.*, 2011, **11**, 30.
46. L. G. E. Vieira, A. C. Andrade, C. A. Colombo, A. H. D. A. Moraes, A. Metha, A. C. D. Oliveira, C. A. Labate, C. L. Marino, C. D. B. Monteiro-Vitorello, D. D. C. Monte, E. Giglioti, E. T. Kimura, E. Romano, E. E. Kuramae, E. G. M. Lemos, E. R. P. D. Almeida, E. C. Jorge, E. V. S. Albuquerque, F. R. D. Silva, F. Vinecky, H. E. Sawazak, H. F. A. Dorry, H. Carrer, I. N. Abreu, J. A. N. Batista, J. B. Teixeira, J. P. Kitajima, K. G. Xavier, L. M. D. Lima, L. E. A. D. Camargo, L. F. P. Pereira, L. L. Coutinho, M. V. F. Lemos, M. R. Romano, M. A. Machado, M. M. D. C. Costa, M. F. G. D. Sá, M. H. S. Goldman, M. I. T. Ferro, M. L. P. Tinoco, M. C. Oliveira, M.-A. Van Sluys, M. M. Shimizu, M. P. Maluf, M. T. S. D. Eira, O. Guerreiro Filho, P. Arruda, P. Mazzafera, P. D. S. C. Mariani, R. L. B. C. D. Oliveira, R. Harakava, S. F. Balbao, S. M. Tsai, S. M. Z. D. Mauro, S. N. Santos, W. J. Siqueira, G. G. L. Costa, E. F. Formighier, M. F. Carazzolle and G. A. G. Pereira, *Braz. J. Plant Physiol.*, 2006, **18**, 95.
47. E. Onsaard, *Int. Food Res. J.*, 2012, **19**, 1287.
48. M. B. Coelho, M. L. Macedo, S. Marangoni, D. S. Silva, I. Cesarino and P. Mazzafera, *J. Agric. Food Chem.*, 2010, **58**, 3050.
49. O. Guerreiro Filho and P. Mazzafera, *J. Agric. Food Chem.*, 2003, **51**, 6987.
50. D. Selmar, G. Bytof and S.-E. Knopp, *Ann. Bot.*, 2008, **101**, 31.
51. M. Lepelley, M. Amor, N. Martineau, G. Cheminade, V. Caillet and J. Mccarthy, *BMC Plant Biol.*, 2012, **12**, 31.
52. H. M. C. Abreu, P. M. Nobile, M. M. Shimizu, P. Y. Yamamoto, E. A. Silva, C. A. Colombo and P. Mazzafera, *Braz. J. Bot.*, 2012, **35**, 357.
53. J. McCarthy, M. B. Amor, V. Pétiard, S. D. Tanksley and C. Lin, *US Pat.* 2007/0022489 A1, 2007.
54. E. A. Silva, P. Mazzafera, O. Brunini, E. Sakai, F. B. Arruda, L. H. C. Mattoso, C. R. L. Carvalho and R. C. M. Pires, *Braz. J. Plant Physiol.*, 2005, **17**, 229.
55. D. D. Kitts and K. Weiler, *Curr. Pharm. Des.*, 2003, **9**, 1309.
56. D. Agyei, *Recent Pat. Food, Nutr. Agric.*, 2015, **7**, 100.

57. Y. Mine, E. Li-Chan and B. Jiang, in *Bioactive Proteins and Peptides as Functional Foods and Nutraceuticals*, ed. Y. Mine, E. Li-Chan and B. Jiang, Blackwell Publishing Ltd. and Institute of Food Technologists, Chicago, 2010, p. 3.
58. B. Walther and R. Sieber, *Int. J. Vitam. Nutr. Res.*, 2011, **81**, 181.
59. A. Farah, in *Coffee: Emerging Health Effects and Disease Prevention*, ed. Y.-F. Chu, John Wiley & Sons, Inc., 2012, p. 21.
60. G. Picariello, G. Mamone, C. Nitride, F. Addeo and P. Ferrantia, *TrAC, Trends Anal. Chem.*, 2013, **52**, 120.
61. V. Preedy, *Coffee in Health and Disease Prevention*, Academic Press, San Diego, 2015, p. 1046.

CHAPTER 19

Polysaccharides and Other Carbohydrates

JOANA SIMÕES[a], ANA S. P. MOREIRA[a,b], CLÁUDIA P. PASSOS*[a,c], FERNANDO M. NUNES[d], M. ROSÁRIO M. DOMINGUES[a,b] AND MANUEL A. COIMBRA[a,c]

[a]QOPNA, Department of Chemistry, University of Aveiro, 3810-193 Aveiro, Portugal; [b]CESAM, Department of Chemistry, University of Aveiro, 3810-193 Aveiro, Portugal; [c]REQUIMTE/LAQV, Department of Chemistry, University of Aveiro, 3810-193 Aveiro, Portugal; [d]CQ-VR, Chemistry Research Centre, Department of Chemistry, University of Trás-os-Montes e Alto Douro, 5001- 801 Vila Real, Portugal
*E-mail: cpassos@ua.pt

19.1 Introduction

Polysaccharides are the predominant carbohydrates present in green coffee beans, accounting for about 50% of their dry weight. Galactomannans, type II arabinogalactans, and cellulose are the most abundant ones. The other carbohydrates are mainly sucrose, which accounts for 4–8%, with small amounts of oligosaccharides such as raffinose and stachyose and monosaccharides. During roasting, sucrose is completely degraded, as well as the other low molecular weight carbohydrates. The galactomannans and arabinogalactans structures are also changed during the roasting process, namely by the occurrence of depolymerization, debranching, dehydration, isomerization, and polymerization (*via* non-enzymatic transglycosylation) reactions. The occurrence of non-enzymatic transglycosylation reactions involving galactomannans and arabinogalactans

Coffee: Production, Quality and Chemistry
Edited by Adriana Farah

Published by the Royal Society of Chemistry, www.rsc.org

results in arabinogalactan–galactomannan hybrid polysaccharides. During the roasting of the coffee beans, galactomannans and arabinogalactans, as well as sucrose, are involved in the formation of melanoidins, the high molecular weight nitrogenous brown compounds that have been associated with health benefits of coffee consumption. In this chapter, the structural features and abundance of green and roasted coffee carbohydrates and their extractability to the brew are presented, paving the way for the presentation and discussion of melanoidins in Chapter 29.

19.2 Green Coffee Polysaccharides and Other Carbohydrates

Carbohydrates account for 59–61% of the dry weight of green coffee beans for the most commercialized coffee species, *Coffea arabica* (arabica) and *Coffea canephora* (robusta). Polysaccharides are the most abundant carbohydrates (50–54%), followed by sucrose (4–8%).[1] Besides sucrose, which is the major oligosaccharide (>90%), small amounts of raffinose (≤0.9%) and stachyose (≤0.13%) are present in green coffee beans. Also, minor amounts of monosaccharides, including fructose, glucose, galactose, and arabinose, are present, accounting for 0.2–0.5% of their dry weight.[2]

The whole polysaccharide fraction of green coffee beans is mainly composed of mannose, followed by galactose, glucose, and arabinose (Table 19.1). These monosaccharide residues are the building blocks of three main polysaccharides present in green coffee beans: galactomannans, type II arabinogalactans, and cellulose. They comprise, respectively, 22%, 14–17%, and 8% of the beans' dry weight.[1,3] The different values given for type II arabinogalactans are associated with the bean species, respectively, 14% for arabica and 17% for robusta beans.[3]

Only a small fraction of the coffee polysaccharides (1.85%) can be extracted with hot water from green coffee beans, with arabinogalactans being more water soluble than galactomannans.[4] Galactomannans extracted with hot water represent 24% of the polysaccharides present in green coffee beans, whereas arabinogalactans represent 57–69%.[5,6] The hot water extractable green coffee galactomannans are composed of a main backbone of (β1 → 4)-linked D-mannose residues, some of them substituted at *O*-6 by single residues of α-D-galactose or L-arabinose and at *O*-2 and/or *O*-3 by acetyl groups

Table 19.1 Monosaccharides composition (mol%) of the polysaccharides of arabica and robusta green beans. Adapted from A. Illy, E. Illy, R. Macrae, M. Petracco, M. R. Sondahl, S. Valussi and R. Viani, *Espresso Coffee: The Chemistry of Quality*, Academic Press, London, 1995.[1]

Monosaccharide	Arabica	Robusta
Arabinose	9.9	10.0
Mannose	49.3	45.7
Galactose	24.6	27.2
Glucose	16.1	17.0

(Figure.19.1A). Mannose residues can be singly acetylated or di-acetylated and the acetyl groups can occur in consecutive residues. Also, (β1 → 4)-linked D-glucose residues are components of the mannan backbone.[7] The higher water solubility of coffee arabinogalactans is related to their higher degree of branching. The hot water extractable green coffee arabinogalactans are composed of a main backbone of (β1 → 3)-linked D-galactose residues, some of them substituted at *O*-6 with short chains of (β1 → 6)-linked D-galactose residues (Figure 19.1B). The galactose residues of these short chains are substituted with various combinations of α-L-arabinose, α-L-rhamnose, and β-D-glucuronic acid residues.[8] The arabinogalactans of green coffee beans are covalently linked to proteins, forming arabinogalactan-proteins (AGPs).[9]

19.3 Roasting-induced Changes

The roasting of green coffee beans is the most relevant step in coffee processing. The characteristic aroma, taste, and colour of the coffee brew, prepared by hot water extraction from roasted and ground coffee beans, are to a great

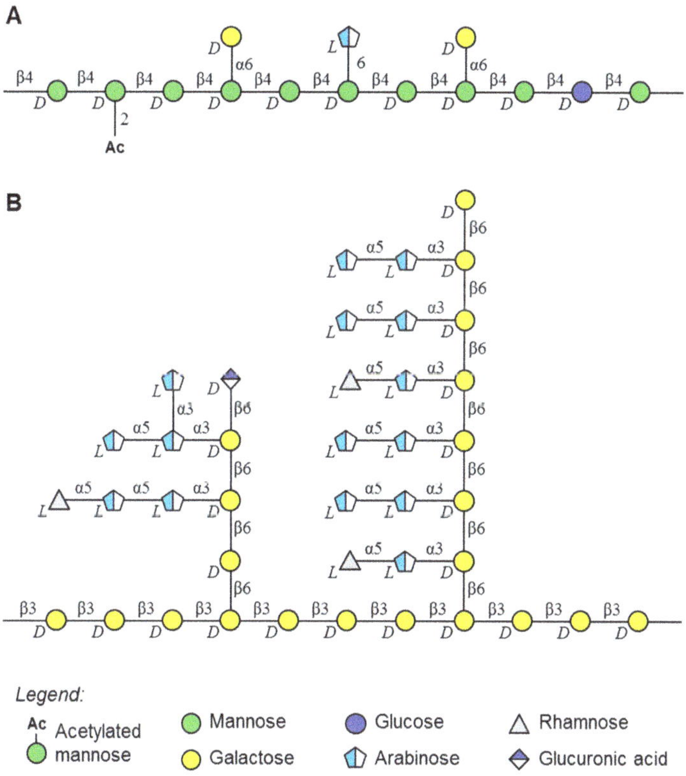

Figure 19.1 Illustration of the main structural features of galactomannans (A) and arabinogalactans (B) isolated by hot water extraction of green coffee beans. Reproduced from ref. 18, with permission from The Royal Society of Chemistry.

extent determined by the roasting process.[1] During this treatment, green coffee beans undergo several physical changes, including the colour change from green to brown, the increase in bean volume, and the appearance of larger micropores in the cell walls.[1,10] Also, several chemical reactions take place inside the beans, changing the structure of their components, including the carbohydrates.

19.3.1 Structural Changes of Carbohydrates

Sucrose, the main low molecular weight carbohydrate in green coffee beans, is nearly completely degraded during roasting.[1]

The polysaccharide content of arabica coffee beans, depending on the degree of roast, is 12–40% lower than the amount in green beans.[11] Arabinogalactans are the most vulnerable of the coffee polysaccharides to degradation, affecting mainly the arabinose residues present as side chains.[4,6,11] Also, depolymerization of arabinogalactans occurs during roasting, with a decrease in molecular weight, even after a light roast.[11]

Although coffee bean galactomannans are known to be more resistant than arabinogalactans,[12] the galactomannans also undergo structural modifications during the roasting of the coffee beans. Roasted coffee galactomannans are modified at the reducing end by the occurrence of caramelization, isomerization, oxidation, decarboxylation, and Maillard reactions. This was observed by the occurrence of 1,6-β-anhydromannose, mannonic acid, arabinonic acid, and Amadori compounds, among other modifications, at the reducing end of oligosaccharides obtained by β-*endo*-mannanase hydrolysis of hot water soluble roasted coffee galactomannans (Figure 19.2).[13] Similar to the arabinogalactans, galactomannans undergo depolymerization and debranching during the coffee roasting.[4–6,14]

Because roasted coffee polysaccharides have complex and diverse structural features, the understanding of the modifications induced by roasting on their structure has been carried out using model oligosaccharides to mimic galactomannans and arabinogalactans, dry heated at 200 °C for several periods of time.

The dry heating of (β1 → 4)-mannotriose (Man_3), an oligosaccharide with a degree of polymerization (DP) of 3 structurally related to galactomannans, led to the formation of a large range of new compounds (Figure 19.3), resulting from depolymerization, dehydration, and formation of oligosaccharides having a higher DP in comparison with Man_3.[15] The occurrence of oligosaccharides with DP4 to DP8, identified by electrospray mass spectrometry, showed that these compounds should be formed *via* non-enzymatic transglycosylation reactions. Similarly, the dry heating of (α1 → 5)-arabinotriose (Ara_3), an oligosaccharide structurally related to the arabinose side chains of arabinogalactans, promoted the formation of oligosaccharides with a higher DP in comparison with Ara_3.[16] These outcomes support the possibility of occurrence of non-enzymatic transglycosylation reactions during coffee roasting involving the same type of polysaccharides.

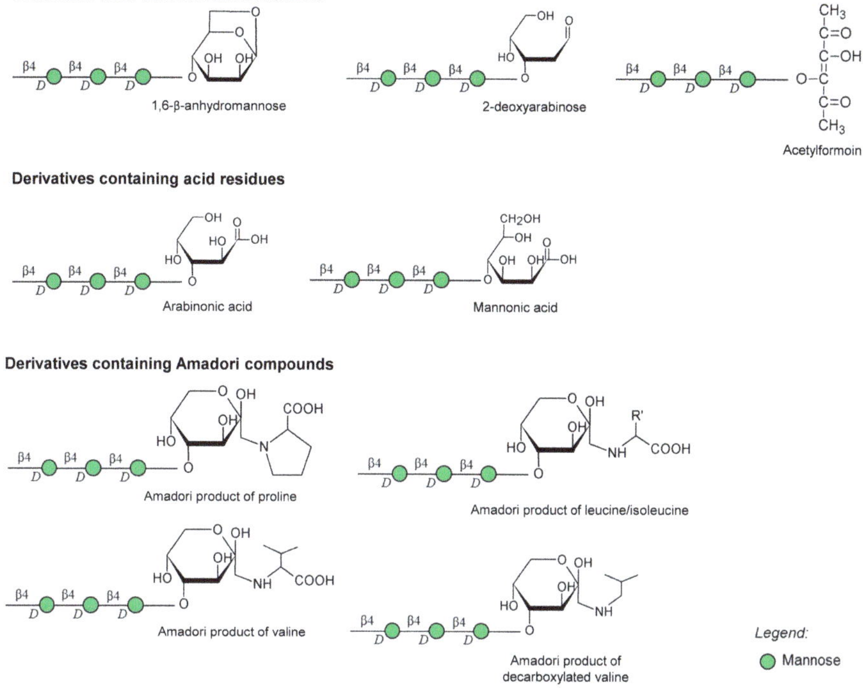

Figure 19.2 Structural modifications identified at the reducing end of roasted coffee galactomannans. Adapted from F. M. Nunes, A. Reis, M. R. M. Domingues and M. A. Coimbra, *J. Agric. Food Chem.*, 2006, **54**, 3428. Copyright 2006 American Chemical Society.[13]

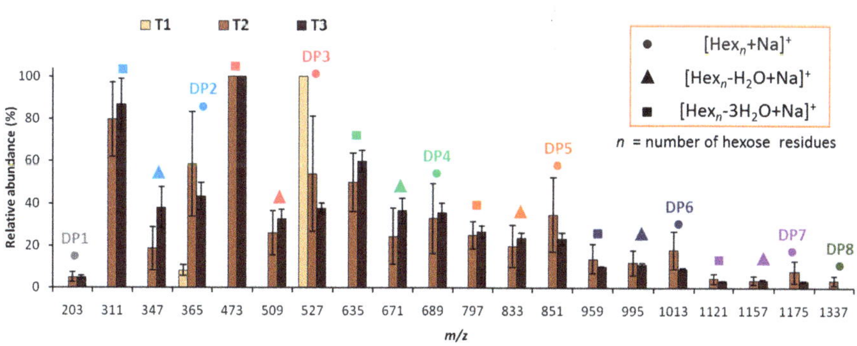

Figure 19.3 Relative abundances of [M + Na]$^+$ ions identified for Man$_3$ after heating from room temperature to 200 °C and maintained at 200 °C for 0 min (T1), 30 min (T2), and 60 min (T3). Adapted with permission from A. S. P. Moreira, M. A. Coimbra, F. M. Nunes, J. Simões and M. R. M. Domingues, *J. Agric. Food Chem.*, 2011, **59**, 10078. Copyright 2011 American Chemical Society.[15]

Dry heated mixtures of Man_3 and Ara_3 showed the possibility of the occurrence of non-enzymatic transglycosylation reactions between galactomannans and arabinose side chains of arabinogalactans during coffee roasting.[17] The dry thermal processing of these mixtures promotes the formation of oligo- and polysaccharides composed by pentose and hexose residues with a degree of polymerization up to 18 residues, including arabinan–mannan hybrid polysaccharides. Also, new types of glycosidic linkages of mannose and arabinose, as well as terminally and 5-linked ribose, possibly derived from arabinose isomerization, are formed after thermal processing. Xylose and lyxose are also formed from mannose (Figure 19.4).

The occurrence of non-enzymatic transglycosylation reactions has been also observed in roasted coffee samples, supporting the formation of arabinogalactan–galactomannan hybrid polysaccharides during coffee roasting (Figure 19.5).[17] These hybrid structures formed between galactomannans and arabinogalactans can be incorporated into melanoidin structures.[18,19]

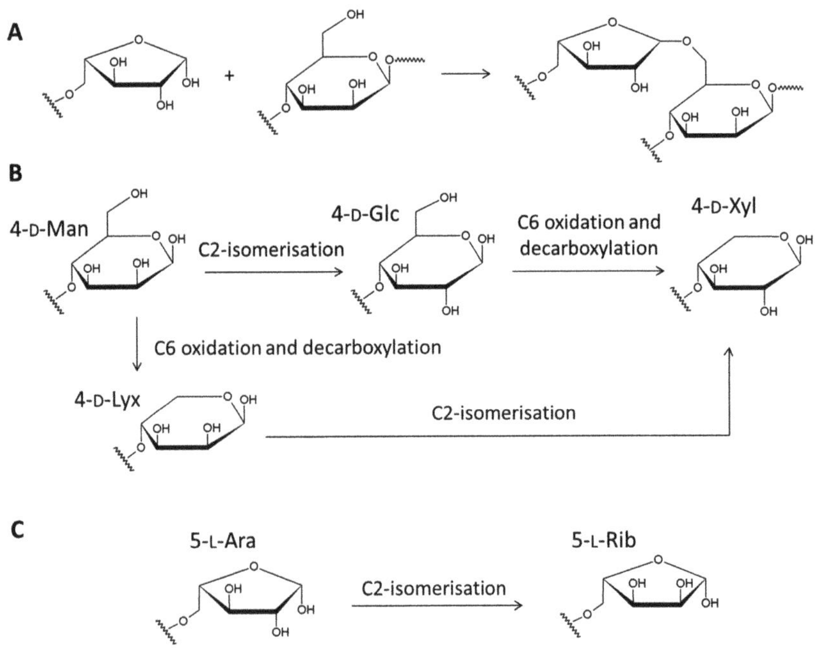

Figure 19.4 Reactions proposed to occur during the roasting of the carbohydrates, originating new carbohydrate structures. (A) Transglycosylation reaction involving arabinose (Ara) and mannose (Man) residues forming a 6-Man linkage; (B) reactions of isomerization and oxidative decarboxylation resulting in the formation of glucose (Glc), xylose (Xyl) and lyxose (Lyx); (C) reaction of isomerization of arabinose with concomitant formation of ribose (Rib). Reproduced from A. S. P. Moreira, J. Simões, A. T. Pereira, C. P. Passos, F. M. Nunes, M. R. M. Domingues and M. A. Coimbra, Transglycosylation reactions between galactomannans and arabinogalactans during dry thermal treatment, *Carbohydr. Polym.*, **112**, 48. Copyright 2014, with permission from Elsevier.[17]

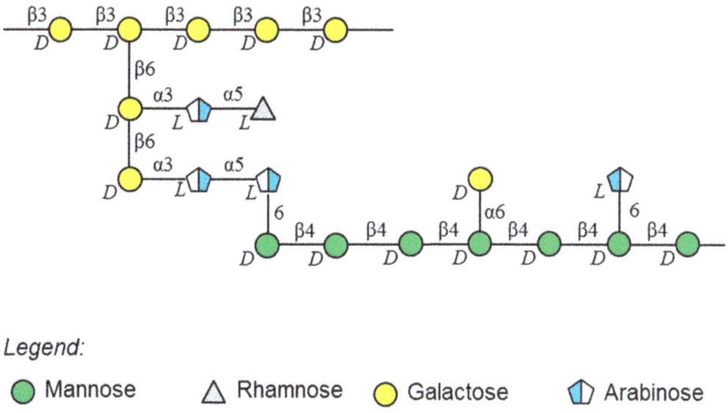

Figure 19.5 Proposed structure of arabinogalactan–galactomannan hybrid polysaccharides formed during coffee roasting.

Despite the exact structures of coffee melanoidins remaining unclear, these compounds have been associated with health benefits of coffee consumption (see Chapter 29).

19.3.2 Differences in Thermal Stability of Coffee Galactomannans and Arabinogalactans

Galactomannans and type II arabinogalactans show a different stability during coffee roasting (Figure 19.6). Isothermals at 160, 180, 200, 220, and 240 °C during 3 h allow evaluating the long-term thermal resistance of both coffee polysaccharides.[20]

During the isothermals of coffee galactomannans (Figure 19.6A), an initial mass loss stage from 0 to 10 min occurs (approximately 6–8%), attributed to the loss of sorbed water. After these mass losses, no variation occurs at temperatures below 200 °C during the 3 h, and variations of 4% and 10% occur for roasting at 220 °C and 240 °C, respectively.

Taking into account that galactomannans are thermally stable at temperatures lower than or equal to 200 °C during long-term exposures up to 3 h, and the coffee roasting process is usually performed at temperatures around 200 °C, it is expectable that during the roasting of the coffee beans the galactomannans keep their thermal stability. Even though there is no apparent weight loss at 200 °C up to 3 h, structural changes of the spent coffee grounds (SCG) galactomannans occurs.

As observed for galactomannans thermogravimetric profiles, the different isothermal weight loss curves of coffee arabinogalactans (Figure 19.6B) have also two main mass loss stages. The first one occurs from 0 to 6 min (5% for all isotherms, except for 220 °C and 240 °C), attributed to the loss of water. At 200 °C a slight decrease of the amount of water lost happens, due to its consumption in hydrolysis reactions. This effect is highly evident at 220 °C when only one-third of the water was lost in the first minute of heating. At 220 °C, the mass loss was not as well defined as the previous ones, due to the

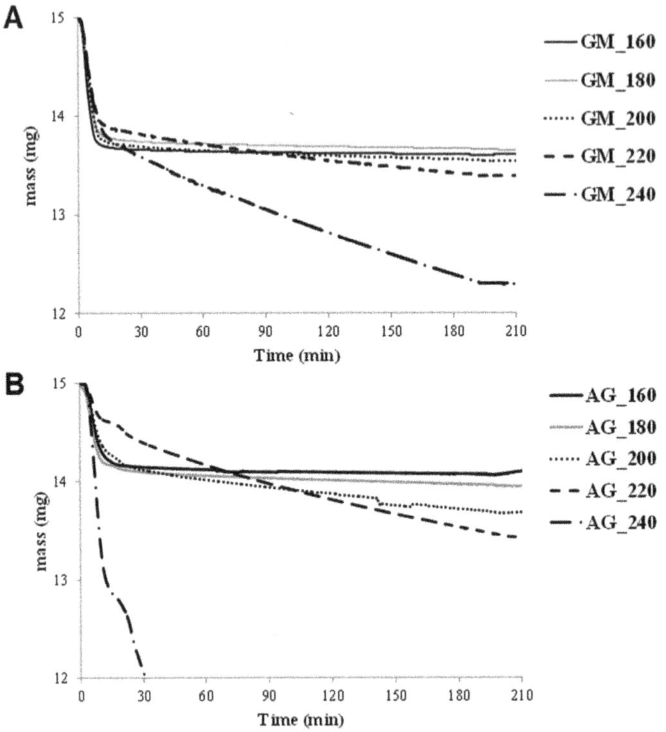

Figure 19.6 Thermograms at different isothermals (160, 180, 200, 220, and 240 °C) of (A) coffee galactomannans, (B) coffee arabinogalactans. Reproduced from J. Simões, É. Maricato, F. M. Nunes, M. R. Domingues and M. A. Coimbra, Thermal stability of spent coffee ground polysaccharides: Galactomannans and arabinogalactans, *Carbohydr. Polym.*, **101**, 256. Copyright 2014, with permission from Elsevier.[20]

simultaneous degradation of the polysaccharides at this stage. At 240 °C a higher extent of mass loss (23%) was observed in the first 30 min of roasting. The mass loss that occurs from 6 to 210 min for all temperatures may be attributed to polysaccharide decomposition. This loss is approximately 6% for the experiments performed at 160 °C and 180 °C, 9% and 10% for the experiments performed at 200 °C and 220 °C, and 46% for the experiments performed at 240 °C. Coffee arabinogalactans are thermally stable at temperatures lower than or equal to 180 °C during long-term exposures up to 3 h, which is a temperature lower than the 200 °C observed for coffee galactomannans. Taking into account that the coffee roasting process is usually performed at temperatures around 200 °C, it is expectable that during the coffee roasting the arabinogalactans are degraded, contrarily to the galactomannans that are more resistant to temperature.

The activation energy (E_a) of thermal decomposition (Figure 19.7) is calculated based on thermograms obtained by the isothermal thermogravimetric analysis, using the plot $\ln(m)$ *vs.* $1/T$ of the equivalent equation: $\ln(m) = -(E_a/R)/T + \text{constant}$, where m is the slope obtained in different thermograms, T

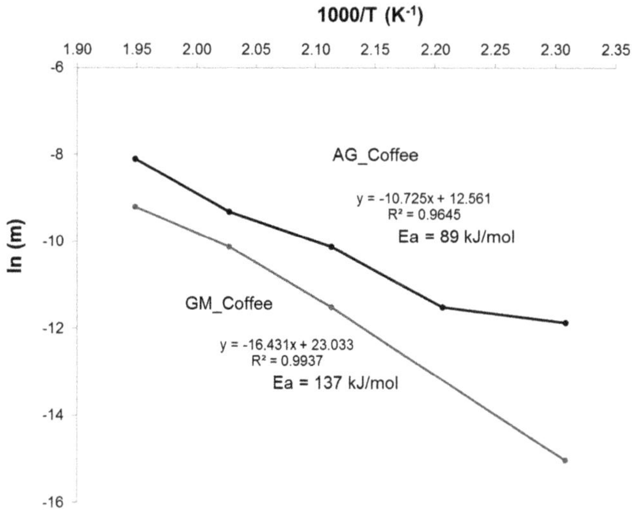

Figure 19.7 Determination of activation energy of thermal decomposition of coffee galactomannans and coffee arabinogalactans. Reproduced from J. Simões, É. Maricato, F. M. Nunes, M. R. Domingues and M. A. Coimbra, Thermal stability of spent coffee ground polysaccharides: Galactomannans and arabinogalactans, *Carbohydr. Polym.*, **101**, 256. Copyright 2014, with permission from Elsevier.[20]

is the temperature documented on the thermogram, in K, and R is the gas constant (8.314×10^{-3} kJ K^{-1} mol^{-1}). The E_a of thermal decomposition estimated for coffee galactomannans is 137 kJ mol^{-1}, and for the coffee arabinogalactans the E_a is 89 kJ mol^{-1}. This difference is due to the higher thermal resistance of coffee galactomannans than coffee arabinogalactans. The E_a of coffee arabinogalactans is slightly higher than the E_a value of other arabinogalactans, namely Gum Arabic arabinogalactans (71 kJ mol^{-1}),[20] due to the higher thermal resistance that could have been conferred during the roasting process to the less branched polysaccharides left.

19.3.3 Changes in Cell Walls and Extractability of Coffee Polysaccharides

The roasting process improves the amount of coffee polysaccharides extracted in relation to the green coffee beans, namely those extracted with hot water,[5] changing also their relative content, depending on the degree of roasting.[4,6] In contrast to green coffee beans, the polysaccharides extracted with hot water from roasted coffee beans are mainly galactomannans.[5,12] The changes in the hot water extractability of galactomannans and arabinogalactans are due to modification of the cell wall of coffee beans during the roasting process and also due to changes in the polysaccharide structures. Since arabica beans cell walls are more degraded during the roasting and contain more pores than roasted robusta,[21] arabica beans have a higher extraction yield of polysaccharides.

Although the roasting process changes the amount of hot extractable polysaccharides, the majority of the galactomannans and arabinogalactans remain associated to the cellulose-rich cell wall matrix. Nevertheless, the roasting of the coffee beans also promotes the extraction of the cell wall polysaccharides when solutions of NaOH are used.[14,22] The changes in the cell walls porosity are a key factor in the extraction process of roasted ground coffee. The scanning electron microscopy (SEM) image (Figure 19.8) of the roasted ground coffee beans before espresso preparation (Figure 19.8A) shows thick cell walls (light surfaces) when compared to the diameter of cell lumen (dark circular shape structures). The thickness of the cell walls remains high in the spent coffee grounds (SCG) obtained after espresso coffee preparation (Figure 19.8B). The SEM image of the residue left after roasting of SCG and extraction with NaOH (Figure 19.8C) shows the appearance of better defined and hollowed structures, possibly cell walls. A similar effect occurs after five consecutive microwave-assisted extractions using water as the solvent used to obtain polysaccharides from SCG (Figure 19.8D).[23,24]

Espresso coffee, prepared by short percolation of hot water at high pressure through a layer of roasted and ground coffee, is a worldwide established form of coffee brew consumption. The espresso coffee polysaccharides, mainly galactomannans and arabinogalactans, have an important role in

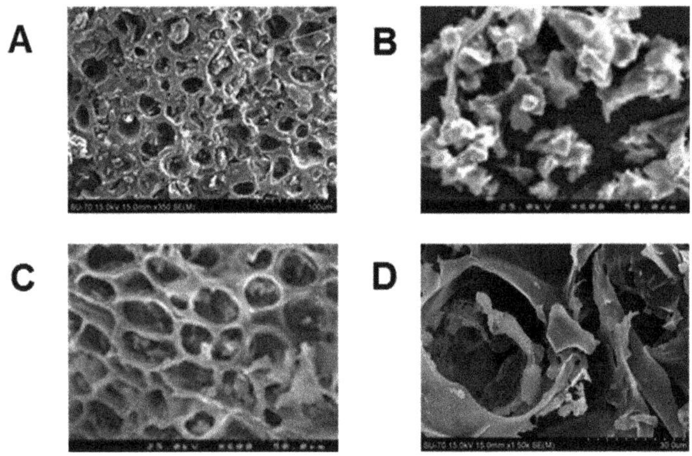

Figure 19.8 Scanning electron microscopy (SEM) images: (A) roasted ground coffee (before espresso preparation); (B) spent coffee grounds (SCG) obtained after espresso preparation; (C) residue left after roasting and hot alkali extraction of SCG; (D) residue left after five consecutive microwave-assisted extractions of SCG. Adapted from J. Simões, F. M. Nunes, M. R. Domingues and M. A. Coimbra, Extractability and structure of spent coffee ground polysaccharides by roasting pre-treatments, *Carbohydr. Polym.*, **97**(1), 81–89. Copyright 2013, with permission from Elsevier[22] and C. P. Passos, A. S. P. Moreira, M. R. M. Domingues, D. V. Evtuguin and M. A. Coimbra, Sequential microwave superheated water extraction of mannans from spent coffee grounds, *Carbohydr. Polym.*, **103**, 333. Copyright 2014, with permission from Elsevier.[24]

the foam stability of the brew, which is a relevant quality attribute. Table 19.2 shows how the total solids and carbohydrate content in espresso coffee is affected by the degree of roast (DR) and the beans species from different geographical origins. While the total extracted solids in espresso coffee is positively correlated with the increase of the DR, the carbohydrate content, mainly composed by galactomannans and arabinogalactans, achieves a maximum at medium roasting conditions, independently of the source, with higher carbohydrate yield for arabica coffee.[25]

Single-dose capsule systems offer consumers convenience for homemade espresso coffee preparation, with a diversified amount of blends, reproducible extraction conditions, and guarantee of coffee powder protection against moisture and oxidation processes.[26] These espresso coffees contain a total amount of solids between 1.15 and 1.24 g per cup (Table 19.3),[27] which is similar to the amount of total solids of regular espresso coffees when normalized for the same amount of coffee powder and volume of water used.[28] The carbohydrate content in espresso coffees prepared from single-dose capsule systems ranges between 110 and 285 g per cup, accounting for 11 to 24% of the total solids (Table 19.3). The carbohydrate content is significantly lower in decaffeinated blend, indicating that the decaffeination process may also contribute to differences in the carbohydrates extractability.[27]

19.4 Conclusions

Coffee carbohydrates have been extensively studied, leading to progress in the understanding of the complex chemistry of the coffee. This involves the structural features of coffee carbohydrates as well as their modifications by roasting. During the roasting process, the sucrose, the main low molecular weight carbohydrate in green coffee beans, is nearly completely degraded, whereas the polysaccharides are decreased by 12–40%. Arabinogalactans and galactomannans, the most abundant coffee polysaccharides, suffer

Table 19.2 Total solids extracted with relation to the content of carbohydrates in espresso coffee as a function of degree of roast (DR) for two coffee origins. Adapted from F. M. Nunes, M. A. Coimbra, A. C. Duarte, and I. Delgadillo, *J. Agric. Food Chem.*, 1997, **45**, 3238.[25]

Coffee origin	DR (%)	Total solids (g)	Carbohydrates (mg)	Carbohydrates (% of total solids)
Brazil (arabica)	4.0	3.78	442.4	11.7
	5.5	4.16	583.0	14.0
	6.6	4.25	777.4	18.3
	9.7	4.74	947.9	20.0
	12.1	4.95	826.2	16.7
	14.1	5.01	716.9	14.3
Uganda (robusta)	3.1	3.69	526.9	14.3
	5.7	4.38	609.3	13.9
	7.6	4.95	757.2	15.3
	9.5	5.05	706.3	14.0
	15.5	5.14	714.8	13.9

Table 19.3 Total solids extracted with relation to the content of carbohydrates in different blends (1–7) of single-dose espresso coffee capsules. Adapted from G. R. Lopes, A. S. Ferreira, M. Pinto, C. P. Passos, E. Coelho, C. Rodrigues, C. Figueira, S. M. Rocha, F. M. Nunes and M.A. Coimbra, *Food Res. Int.*, 2016, **89**, Part 2, 989.[27]

Blend	Total solids (g)	Carbohydrates (mg)	Carbohydrates (% of total solids)
1[a]	0.99	110	11.1
2	1.21	171	14.1
3	1.20	285	23.8
4	1.15	268	23.3
5[b]	1.24	181	14.6
6[c]	1.20	211	17.6
7[d]	1.23	216	17.5

[a]Decaffeinated coffee.
[b]Selected source: Jamaica.
[c]Selected source: Tanzania.
[d]Selected source: Ethiopia.

structural modifications, namely depolymerization, which increase their extractability, as well as the formation of new linkages by non-enzymatic transglycosylation reactions, leading to chimeric polysaccharides. These polysaccharides deserve attention in order to understand their properties and possible health benefits associated with coffee, as well as their role in the formation of melanoidins.

Acknowledgements

Thanks are due to Fundação para a Ciência e a Tecnologia (FCT)/MCTES for financial support to the research units QOPNA (FCT UID/QUI/00062/2013) and CESAM (UID/AMB/50017 - POCI-01-0145-FEDER-007638) at University of Aveiro and CQ-VR at UTAD Vila Real (PEst-OE/QUI/UI0616/2014) through national funds (PIDDAC) and, where applicable, co-financed by the FEDER, within the PT2020 Partnership Agreement and Compete 2020. Thanks are also due to FCT for funding the Portuguese Mass Spectrometry Network, RNEM (LISBOA-52201-0145-FEDER-402-022125), and the individual grants to Joana Simões (FRH/BPD/90447/2012), Cláudia Passos (SFRH/BDP/107881/2015), and Ana Moreira (SFRH/BD/80553/2011).

References

1. A. Illy, E. Illy, R. Macrae, M. Petracco, M. R. Sondahl, S. Valussi and R. Viani, *Espresso Coffee: The Chemistry of Quality*, Academic Press, London, 1995.
2. H.-D. Belitz, W. Grosch and P. Schieberle, *Food Chemistry*, Springer, 4th edn, Berlin, 2009.
3. A. G. W. Bradbury and D. J. Halliday, *J. Agric. Food Chem.*, 1990, **38**, 389.
4. F. M. Nunes and M. A. Coimbra, *J. Agric. Food Chem.*, 2002, **50**, 7046.

5. F. M. Nunes and M. A. Coimbra, *J. Agric. Food Chem.*, 2001, **49**, 1773.
6. F. M. Nunes and M. A. Coimbra, *J. Agric. Food Chem.*, 2002, **50**, 1429.
7. F. M. Nunes, M. R. Domingues and M. A. Coimbra, *Carbohydr. Res.*, 2005, **340**, 1689.
8. F. M. Nunes, A. Reis, A. M. S. Silva, M. R. M. Domingues and M. A. Coimbra, *Phytochemistry*, 2008, **69**, 1573.
9. R. J. Redgwell, D. Curti, M. Fischer, P. Nicolas and L. B. Fay, *Carbohydr. Res.*, 2002, **337**, 239.
10. S. Schenker, S. Handschin, B. Frey, R. Perren and F. Escher, *J. Food Sci.*, 2000, **65**, 452.
11. R. J. Redgwell, V. Trovato, D. Curti and M. Fischer, *Carbohydr. Res.*, 2002, **337**, 421.
12. A. Oosterveld, A. G. J. Voragen and H. A. Schols, *Carbohydr. Polym.*, 2003, **54**, 183.
13. F. M. Nunes, A. Reis, M. R. M. Domingues and M. A. Coimbra, *J. Agric. Food Chem.*, 2006, **54**, 3428.
14. A. Oosterveld, J. S. Harmsen, A. G. J. Voragen and H. A. Schols, *Carbohydr. Polym.*, 2003, **52**, 285.
15. A. S. P. Moreira, M. A. Coimbra, F. M. Nunes, J. Simões and M. R. M. Domingues, *J. Agric. Food Chem.*, 2011, **59**, 10078.
16. A. S. P. Moreira, M. A. Coimbra, F. M. Nunes and M. R. M. Domingues, *Food Chem.*, 2013, **138**, 2291.
17. A. S. P. Moreira, J. Simões, A. T. Pereira, C. P. Passos, F. M. Nunes, M. R. M. Domingues and M. A. Coimbra, *Carbohydr. Polym.*, 2014, **112**, 48.
18. A. S. P. Moreira, F. M. Nunes, M. R. Domingues and M. A. Coimbra, *Food Funct.*, 2012, **3**, 903.
19. F. M. Nunes and M. A. Coimbra, *J. Agric. Food Chem.*, 2007, **55**, 3967.
20. J. Simões, É. Maricato, F. M. Nunes, M. R. Domingues and M. A. Coimbra, *Carbohydr. Polym.*, 2014, **101**, 256.
21. C. Gutiérrez, M. D. Ortolá, A. Chiralt and P. Fito, *Análisis por Meb de la Porosidad del Café Tostado*, ASIC- Association for Science and Information on Coffee, Paris, 1993, pp. 661–671.
22. J. Simões, F. M. Nunes, M. R. Domingues and M. A. Coimbra, *Carbohydr. Polym.*, 2013, **97**, 81.
23. C. P. Passos and M. A. Coimbra, *Carbohydr. Polym.*, 2013, **94**, 626.
24. C. P. Passos, A. S. P. Moreira, M. R. M. Domingues, D. V. Evtuguin and M. A. Coimbra, *Carbohydr. Polym.*, 2014, **103**, 333.
25. F. M. Nunes, M. A. Coimbra, A. C. Duarteand and I. Delgadillo, *J. Agric. Food Chem.*, 1997, **45**, 3238.
26. A. Parenti, L. Guerrini, P. Masella, S. Spinelli, L. Calamai and P. Spugnoli, *J. Food Eng.*, 2014, **121**, 112.
27. G. R. Lopes, A. S. Ferreira, M. Pinto, C. P. Passos, E. Coelho, C. Rodrigues, C. Figueira, S. M. Rocha, F. M. Nunes and M. A. Coimbra, *Food Res. Int.*, 2016, **89**(part 2), 989.
28. D. Gniechwitz, N. Reichardt, M. Blaut, H. Steinhart and M. Bunzel, *J. Agric. Food Chem.*, 2007, **56**, 11027.

CHAPTER 20

Lipids

K. SPEER* AND I. KÖLLING-SPEER

Food Chemistry, Technical University of Dresden, 01062 Dresden, Bergstraße 66, Germany
*E-mail: karl.speer@chemie.tu-dresden.de

20.1 Introduction

The coffee species with the greatest commercial value are *Coffea arabica* L. and *Coffea canephora* Pierre var. *robusta*, commonly known as "arabica" and "robusta" coffees. There are important differences between these two species, both in their organoleptic properties and in their chemical composition. The lipid content of green arabica coffee beans averages some 15% whilst robusta coffees contain much less, about 10%. Most of the lipids, the so-called coffee oil, are located in the endosperm of the green coffee beans; only a small amount, the coffee wax, is located on the outer layer of the beans. The relatively large unsaponifiable fraction is rich in diterpenes of the kaurane family, mainly cafestol, kahweol, and 16-*O*-methylcafestol, which have gained more and more interest in recent years because of their different physiological effects (see *Coffee: Consumption and Health Implications*). Furthermore, 16-*O*-methylcafestol is the reliable indicator for the presence of robusta beans in coffee blends. Among the sterols, which are also part of the unsaponifiable fraction, various desmethyl-, methyl-, and dimethylsterols have been identified. The mean percentage composition of the lipid fraction of green coffee is presented in Figure 20.1.

Coffee: Production, Quality and Chemistry
Edited by Adriana Farah

Published by the Royal Society of Chemistry, www.rsc.org

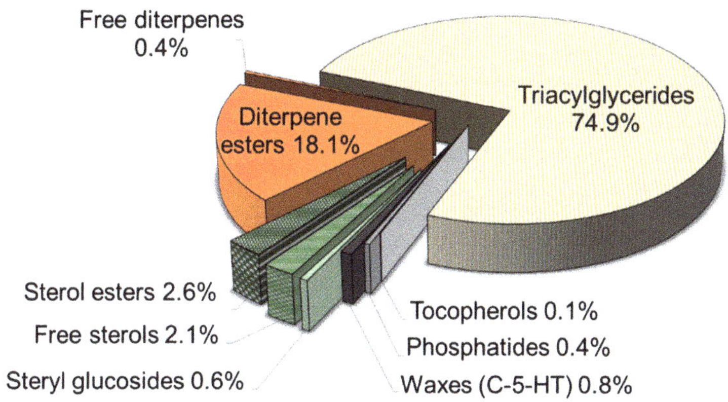

Figure 20.1 Composition of the lipid fraction of green coffee.

20.2 Coffee Oil

20.2.1 Total Oil Content

The lipid content of green arabica coffee averages some 15% on a dry basis, whilst robusta coffee contains much less, around 10%. Most of the lipids, the coffee oil, are located in the endosperm of the coffee bean, while only a small amount, the coffee wax, is in the outer layer. The yield of crude coffee oil is a result not only of the composition of the beans but also of the extraction conditions, the choice of solvent, and the particular particle size. Due to the great differences in yield, the term "coffee oil" needs to be defined more explicitly. Values for the oil content in roasted coffees are higher compared with those in the green coffees from which they were derived, due to the overall dry matter content loss on roasting, which varies according to the level of the roasting degree. The loss in actual lipid matter, however, is low.[1]

Due to its high content of diterpenes, green coffee oil is unsuitable for human consumption. Nevertheless thanks to the emollient properties provided by its fatty acids and diterpenes and its capacity to block the damaging rays of the sun, which have proved harmful to human skin, green coffee oil has been employed widely in the cosmetics industry for years.[2] Coffee oil from defective green beans has also been proposed as a possible raw material for biodiesel production.[3]

Obtaining coffee oil for an in-depth study of its chemical composition necessitates direct solvent extraction without acid treatment. According to Picard *et al.*,[4] several authors used diethyl ether, petroleum ether with different boiling point ranges, *n*-hexane, and a mixture of diethyl ether and *n*-hexane for this. Their results were not homogeneous since, depending on the selected solvent, some more polar and more non-lipid substances such as caffeine were extracted partially as well.

Folstar *et al.*[5] demonstrated that the obtainable yield in solvent extraction depends on the particle size to which the coffee is ultimately ground.

Speer[6] extracted ground coffees of a particle size between 0.25 and 0.63 mm and used tertiary butyl methyl ether as extraction solvent, avoiding the dangerous diethyl ether. This method was adopted as the DIN method 10779 (ref. 7) and is described in the following: roasted coffee beans are coarsely ground in a regular coffee mill and passed through a 0.63 mm sieve. Five grams of the sieved material are then powdered together with sodium sulfate in a mortar and extracted with tertiary butyl methyl ether in a Soxhlet (4 h), siphoning 6–7 times per hour. The solvent is evaporated and the residue is dried to constant weight (105 °C). Longer extraction times (6, 8, or 10 hours) do not increase the lipid content. For green coffee beans, grinding in the mill is prepared together with dry ice. As a result, higher yields for polar substances, such as free diterpenes, were obtained. An extraction using supercritical carbon dioxide proposed by de Azevedo *et al.*[8] seems to be rather convenient for the analysis of diterpenes.

20.3 Fatty Acids

20.3.1 Total Fatty Acids and Fatty Acids in Triacylglycerides

For the most part, the fatty acids are present in the combined state; approx. 75% are esterified with glycerol in the triacylglycerides (TAG), some 20% are esterified with diterpenes, a small proportion is to be found in the sterol esters and in the so-called carboxylic acid-5-hydroxytryptamides (C-5HT). Triacylglycerides are molecular species composed of palmitic (P), linoleic (L), stearic (S), and oleic (O) acids with PLL, PLP, PLO + SLL, LLL, OLL, OLO, POP, and SOS being the typical TAGs species identified thus far. Moreover, it is of interest to underline relevant differences between arabica and robusta species in terms of TAG composition and concentration. As such, arabica samples have a higher concentration of LLL; characteristic percentages are 6.1% for arabica and 3.7% for robusta. However, arabica is characterized by a lower content of OLO, which is 1.1% compared with the 2.3% in robusta. This is also true for PLO + SLL. In fact, arabica has a lower amount of PLO + SLL, which is 13.8% compared with the 16.4% of robusta.[9]

Folstar *et al.*,[5] Picard *et al.*,[4] and Speer *et al.*[10] analyzed the fatty acids in triacylglycerides, sterol esters as well as in diterpene esters. For the separation of the different lipid classes, Folstar *et al.*[5] used a Florisil column. Speer *et al.*[10] isolated the triacylglycerides by means of gel permeation chromatography, transesterified them with potassium methylate, and chromatographed the methylated fatty acids using a 60 m fused silica capillary column coated with RTX 2330. There were no significant differences between the total fatty acids and the fatty acids in the triacylglycerides.[11] It should be pointed out that in arabica and robusta coffees the roasting process increases the trans fatty acid levels, specifically the contents of $C_{18:2ct}$ and $C_{18:2tc}$.[12]

20.3.2 Free Fatty Acids

The presence of free fatty acids (FFA) in coffee was described by various authors.[13–16] Yet, all these data were expressed by acid value, a common but indirect determination procedure used in the analysis of fat. In the case of coffee, this titration method is only very approximate, for it includes not only the free fatty acids themselves but other acid compounds as well. Therefore, Speer *et al.*[10] developed a method for determining the free fatty acids directly. Using the gel chromatographic system with BioBeads S-X3 mentioned above, the coffee lipids, extracted with tertiary butyl methyl ether, can be divided into three individual fractions: a fraction with the triacylglycerides, a fraction containing the diterpene fatty acid esters, and one with the free fatty acids. The latter were converted with BF_3/methanol and determined as methyl esters by capillary gas chromatography.

Nine free fatty acids were detected; they are very uniformly distributed in the robusta and arabica coffees, respectively. In both coffee species, the main fatty acids were $C_{18:2}$ and C_{16}. It was also possible to detect large proportions of C_{18}, $C_{18:1}$, C_{20}, and C_{22}, whereas there were no more than traces of C_{14}, $C_{18:3}$, and C_{24}. Differences between arabica and robusta only become visible when their stearic acid and oleic acid contents are compared on the chromatograms (Figure 20.2). While the proportion of stearic acid is noticeably smaller than that of oleic acid in the robustas, the percentages of these two acids in the arabica coffees are almost equal. The ratio stearic acid/oleic acid may give a first indication of robusta in coffee blends.

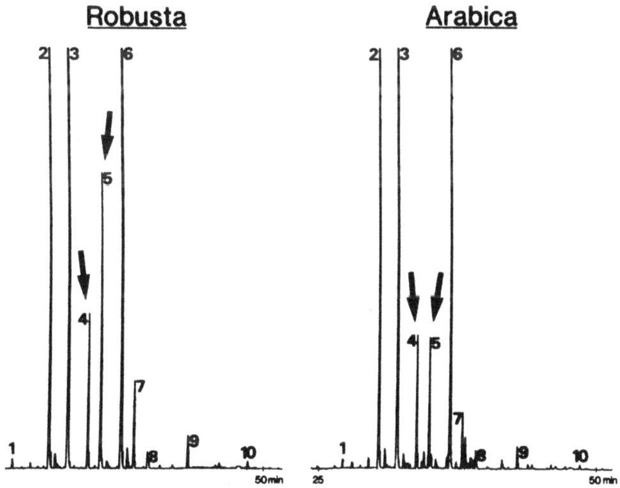

Figure 20.2 Gas chromatography (GC) chromatograms of the methylated free fatty acids (IS = internal standard heptadecanoic acid ethyl ester). Reproduced from ref. 142 with permission from John Wiley and Sons. Copyright 2001 Blackwell Science Ltd.

Martin *et al.*[17] used the Principal Components Analysis (PCA) and the Linear Discriminant Analysis (LDA) as their fundamental tools for processing the data obtained from gas chromatographic tests. PCA allows for separating arabica and robusta species by considering oleic, linoleic, and myristic acids. Rui Alves *et al.*[18] analyzed the fatty acid profile to determine the authenticity of green and roasted coffee varieties.

20.4 Diterpenes in the Lipid Fraction of Robusta and Arabica Coffees

The main diterpenes in coffee are pentacyclic diterpene alcohols based on the kauran skeleton. Working groups under Bengis and Anderson,[19] Chakravorty *et al.*,[20,21] Wettstein *et al.*,[22] Haworth and Johnstone,[23] and Finnegan and Djerassi[24] elucidated the structure of two of the coffee diterpenes, namely kahweol and cafestol. Both are sensitive to acids, heat, and light, and especially kahweol is unstable in purified form. In 1989, 16-*O*-methylcafestol (16-OMC) was isolated from robusta coffee beans, and its structure was elucidated by NMR, GC/MS, X-ray analysis of crystals, and synthesis.[25,26] With 16-*O*-methylkahweol (16-OMK), another O-methyl diterpene was identified in robusta coffee beans.[11] The structural formulae of these diterpenes are assembled in Figure 20.3.

Arabica coffees contain cafestol and kahweol, and robusta coffees contain cafestol, small amounts of kahweol (less than 0.1 g kg^{-1} dry weight), and, additionally, 16-OMC, which was determined only in robusta coffee beans[26,27] (Figures 20.4 and 20.5). The absence of 16-OMC in arabica coffee beans was confirmed by Frega *et al.*,[28] White,[29] Trouche *et al.*,[30] Kamm *et al.*,[31] D'Amelio *et al.*,[32] and more recently by Monakhova *et al.*[33] Due to its stability even during the roasting process, 16-OMC has become the ideal quality characteristic for the reliable detection of robusta in arabica coffee blends.[27]

It should be mentioned here that although 16-OMC was not detectable in arabica coffee beans, it was found in other parts of the arabica coffee plant, for instance in the leaves.[34]

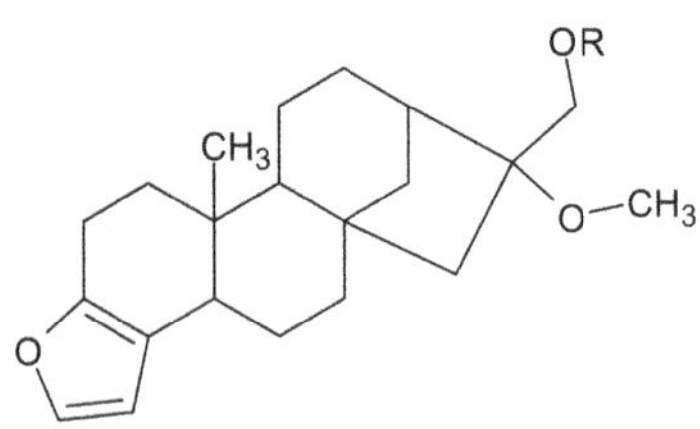

Figure 20.3 Structural formulae of the diterpenes. Reproduced from ref. 142 with permission from John Wiley and Sons. Copyright 2001 Blackwell Science Ltd.

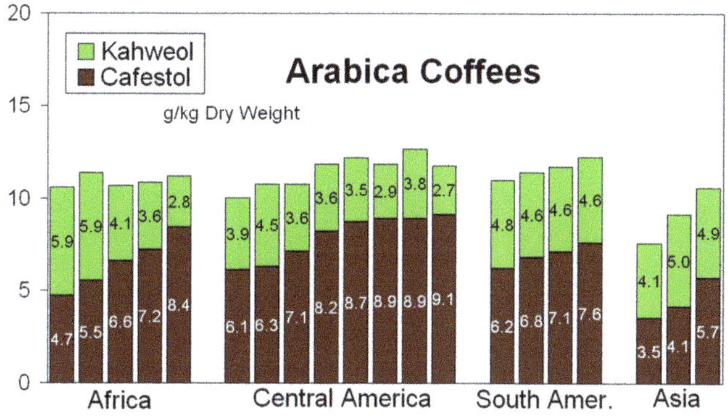

Figure 20.4 Contents of cafestol and kahweol in arabica coffees. Reproduced from ref. 11 with permission from *The Brazilian Journal of Plant Physiology*.

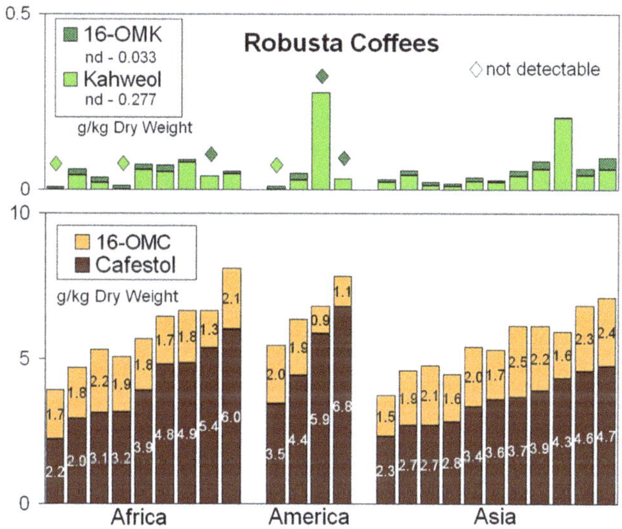

Figure 20.5 Contents of cafestol, 16-OMC, kahweol, and 16-OMK in robusta coffees. Reproduced from ref. 11 with permission from *The Brazilian Journal of Plant Physiology*.

16-OMK, which was clearly identified using different spectroscopic methods, was determined in both green and roasted beans of various robusta coffees.[11] These findings stand in contrast to the statement by de Roos *et al.*,[35] who described it as present exclusively in *Coffea stenophylla*. The diterpenes mentioned are unique chemicals to coffee beans.

The diterpenes cafestol, kahweol, 16-OMC, and 16-OMK are mainly esterified with various fatty acids, and only 2–3% occur in the free form. In order to

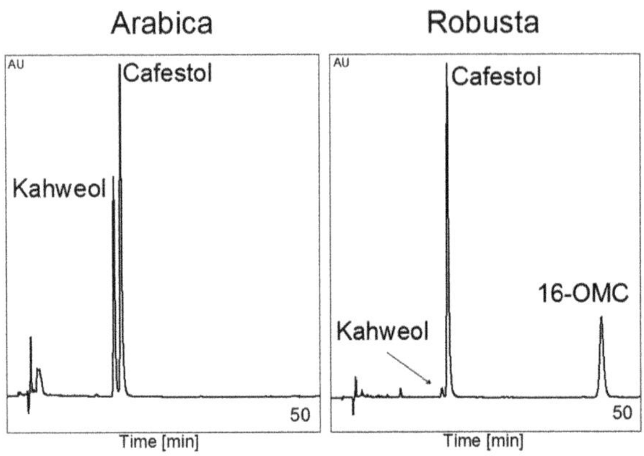

Figure 20.6 HPLC chromatograms of green arabica and robusta coffees. Reproduced from ref. 142 with permission from John Wiley and Sons. Copyright 2001 Blackwell Science Ltd.

analyze the total diterpene amount, the ground coffee beans or the extracted coffee oil thereof was saponified. Then the diterpenes were determined in the unsaponifiable matter by means of GC[26,28] or directly by RP-HPLC with acetonitrile/water as eluent (Nackunstz and Maier, 1987,[36] Speer, 1989,[6] see Figure 20.6). When analyzing the free diterpenes by means of GC or GC/MS without derivatization, a cold-on-column injector should be used instead of a split/splitless injector to avoid artefacts.[37]

In Germany, a validated method for the 16-OMC determination in roasted coffee was published as the DIN method 10779 (ref. 7) of the German Institute for Standardization. This DIN method based on the HPLC method by Speer[6] allows for detecting robusta parts under two percent in blends with arabica coffees.

The validated DIN method 10779 (ref. 7) realizes good results. However, the method is time-consuming and requires organic solvents in higher amounts, especially due to the first step, the extraction of the coffee oil. Therefore, the coffee samples were saponified directly without previous lipid extraction. Applying this method, 16-OMC is even quantifiable in instant coffee.[38] For further acceleration, the liquid–liquid extraction is replaced by an SPE using kieselguhr (absorbent diatom earth) (Figure. 20.7).

Faster results were obtained when analyzing the amount of the 16-OMC-esters in the coffee sample directly *via* the 16-OMC-palmitate using HPLC or TLC.[39,40] The 16-OMC-palmitate constantly represents approx. 50% of the total 16-OMC-esters and is therefore suitable for the qualification and quantification of the robusta parts in coffee blends (Figure 20.8). Assuming a mean total 16-OMC content of 1.7 g kg^{-1} (free and esterified) analyzed with the DIN method 10779 (ref. 7) for numerous robusta coffees, an admixture of less than 2% robusta is detectable.

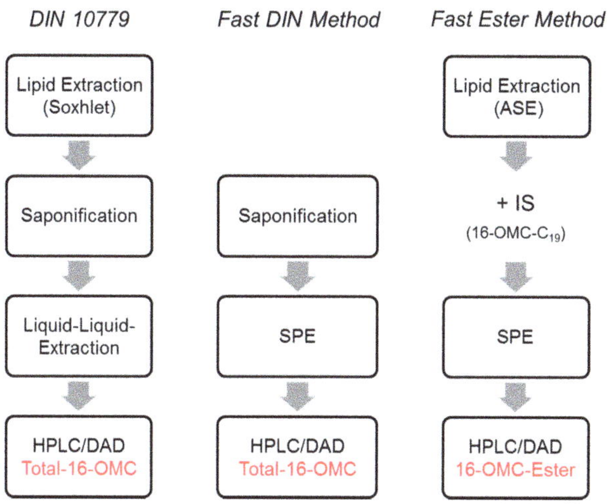

Figure 20.7 Three different methods for analyzing the 16-OMC content in roasted coffees (DIN Method 10779,[7] Fast DIN Method,[39] and Fast Diterpene Ester Method[39]).

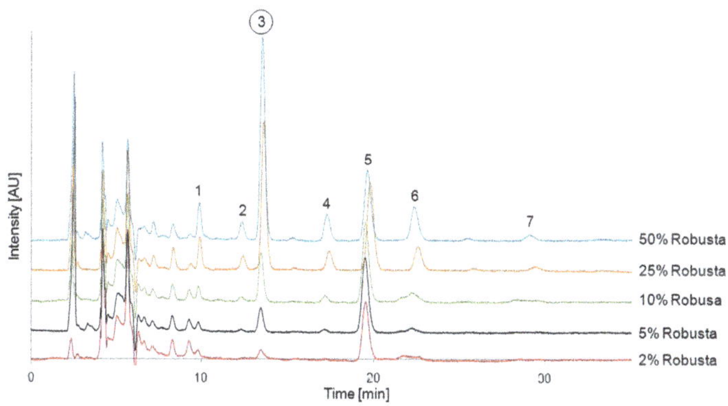

Figure 20.8 HPLC chromatograms of the 16-OMC esters of different arabica/robustamixtures.(1)16-OMC-$C_{18:2}$,(2)16-OMC-$C_{18:1}$,(3)16-OMC-C_{16},(4)16-OMC-C_{18}, (5) 16-OMC-C_{19} (IS), (6) 16-OMC-C_{20}, (7) 16-OMC-C_{22}.

Comparable to the DIN method,[7] contents of up to 2% are determinable. Taking into account that the small amounts of free 16-OMC will not be incorporated, the obtained results are significant. In comparison to the DIN method, where the analysis requires 2 to 3 days, approx. only 3–4 hours are necessary. Apart from saving time, the consumption of solvents could be reduced considerably.

Other different methods were applied for the analysis of robusta parts in arabica: an on-line LC-GC method was developed by Kamm *et al.*[31]

Furthermore, using NMR, green and roasted coffees can be analyzed directly without any clean-up.[32,33] Contents of up to 2% and 3% and higher could be quantified; as a consequence, the evidence of 16-OMC in instant coffees is not possible.

Several studies dealt with Near-Infrared Spectroscopy (NIR); differences between the two coffee species, arabica and robusta, became obvious. However, coffee is a combination of numerous organic molecules in small and large amounts such as caffeine, chlorogenic acids and their derivatives, polysaccharides, proteins, and lipids. It is therefore difficult to resolve the effect of every molecule on the NIR spectrum. So NIR may become a valuable tool for the coffee industry, the results however will not be able to hold up in court as here defined substances have to be analyzed.[41–43]

Likewise, Raman spectroscopy was carried out for discrimination of green arabica and robusta coffee beans by Keidel *et al.*,[44] and Wermelinger *et al.*[45] The latter obtained extracts of robusta and arabica coffees using a Soxhlet system. The solutes were then analyzed by means of Raman spectroscopy without further clean-up. The intensity ratio between two Raman peaks, one characteristic for kahweol and one characteristic for fatty acids, allows for the determination of the robusta content in a given mixture. The detection limit for a model system in the presented study was in the range between 4.9 and 7.5%.

20.4.1 Free Diterpenes

In their free form, the diterpenes cafestol, kahweol, and 16-OMC occur only as minor components in coffee oil. Quantifying them requires an effective method for the separation from the major compounds of the lipid fraction, namely the diterpenes esters and triacylglycerides, which interfere with the analysis. Using gel permeation chromatography, the fraction of the free diterpenes could be analyzed by subsequent RP-HPLC.[46,47] The proportions of the free diterpenes with the total content of each are usually smaller than 3.5%. Only in some robusta coffees, traces of kahweol were detected as expected due to the small amounts of total kahweol, which is smaller than 0.1 g kg^{-1} dry weight.

The amounts of the free diterpenes as well as of the free fatty acids were influenced by the storage conditions of the green coffee beans. While cold and dry storage increased the content only slightly, higher levels of up to 16% of the total cafestol content were determined in not well-dried coffees stored both at 5 °C and at 40 °C. The reason for this was the activity of the enzyme lipase. Cold temperatures and low water contents inhibited this enzyme reversibly.[48] For its detection, coffees of different ages were analyzed with a lipase test (Roche Company for clinical diagnostics) modified by the authors.[48] The test was based on the hydrolyzation of a specific substrate, the 1,2-*O*-dilauryl-*rac*-*glycero*-3-glutaric acid-resorufin

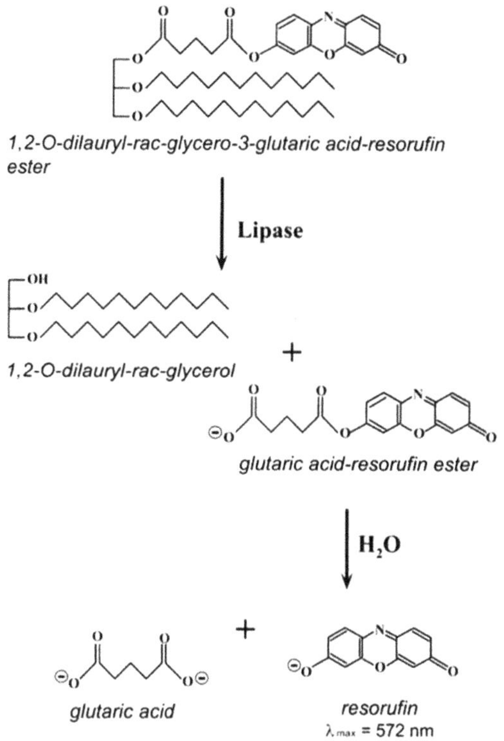

Figure 20.9 Scheme of lipase test. Reproduced from ref. 49 with permission from John Wiley and Sons, © WILEY-VCH Verlag GmbH, 69451 Weinheim, 2000.

ester (Figure 20.9). After separation from the diacylglycerides, the rest of the molecule dissociated spontaneously in an alkaline medium into glutaric acid and resorufin. The latter can be detected photometrically at λ = 572 nm. The intensity of the red-colored dye is proportional to the activity of the lipase.

First, ground coffee was extracted with a sodium ascorbate solution. For measuring the activity of the enzyme, an aliquot of this solution was mixed with an alkaline buffer solution and the color reagent. After measuring the extinction, the mixture was incubated for 6 h at 37 °C, and the extinction was measured again. The activity of the lipase was proved when the extinction of the coffee sample was higher than the extinction of the blank value. This way, the activity of lipase could be determined for fresh coffees as well as for coffees stored up to 10 years.

Steaming coffee beans prior to the roasting process reduces the content of the individual free diterpenes depending on the chosen steaming parameters.[50,51] By intensive steaming (120 min at 2 bars), the free kahweol was completely degraded. If both the steamed and the unsteamed coffee are available, the free kahweol content will be a helpful tool for the assessment of steamed coffee.[52]

20.4.2 Diterpene Fatty Acid Esters

Until 1987, only a few esters of cafestol and kahweol with different fatty acids were reported.[5,53–55] The team under Speer identified a number of further esters of 16-OMC.[58,59] Using gel chromatography, the diterpene esters were isolated together with sterol esters, the latter were removed by using solid phase extraction on silica cartridges. For robusta, one fraction was achieved containing the cafestol and kahweol esters and a second one containing the 16-OMC esters; for arabica, accordingly, only one fraction was achieved. The subsequent analysis by RP-HPLC with acetonitrile/isopropanol as eluent permits the determination of the individual esters. Figure 20.10 shows the semi-preparative chromatogram of the cafestol esters of a robusta coffee sample, Figure 20.11 the chromatogram of the 16-OMC esters of the same sample. In Figure 20.12, Table 20.1 the cafestol and kahweol esters of an arabica coffee are presented.

Cafestol esters with fatty acids such as C_{14}, C_{16}, C_{18}, $C_{18:1}$, $C_{18:2}$, $C_{18:3}$, C_{20}, C_{22}, C_{24} were identified as well as esters with the fatty acid $C_{20:1}$ and some odd-numbered fatty acids such as C_{17}, C_{19}, C_{21}, and C_{23}. These data were confirmed for the fatty acids with 16-OMC, and they seem to be valid also for kahweol.[60,61]

The individual diterpene esters were present irregularly in the coffee oil. The odd-numbered fatty acid esters were minor components, whereas the diterpenes esterified with palmitic, linoleic, oleic, stearic, arachidic, and behenic acid existed in larger amounts,[56–58] making up a sum of nearly 98% of the respective diterpenes.

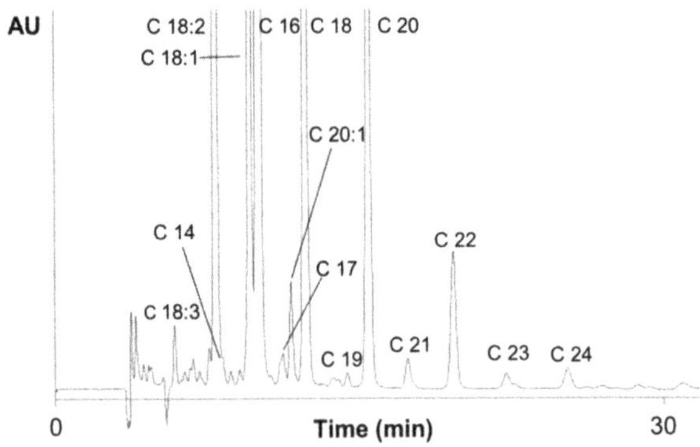

Figure 20.10 HPLC chromatogram of the cafestol fatty acid esters of a robusta coffee. Conditions: column 250 × 4 mm, Nucleosil 120-3 C_{18}, eluent: acetonitrile/isopropanol (60 : 40), detection: UV 220 nm. Reproduced from ref. 142 with permission from John Wiley and Sons. Copyright 2001 Blackwell Science Ltd.

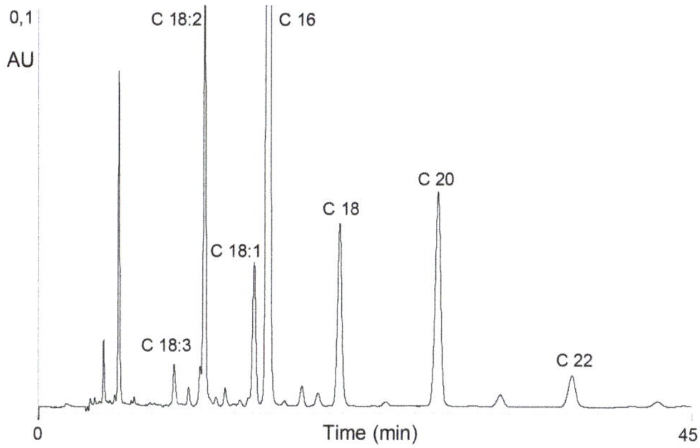

Figure 20.11 HPLC chromatogram of the 16-*O*-methylcafestol esters of a robusta coffee. Conditions: column 250 × 4 mm, Nucleosil 120-3 C_{18}, eluent: acetonitrile/isopropanol (60 : 40), detection: UV 220 nm. Reproduced from ref. 60 with permission from Taylor & Francis Ltd (http://www.tandfonline.com).

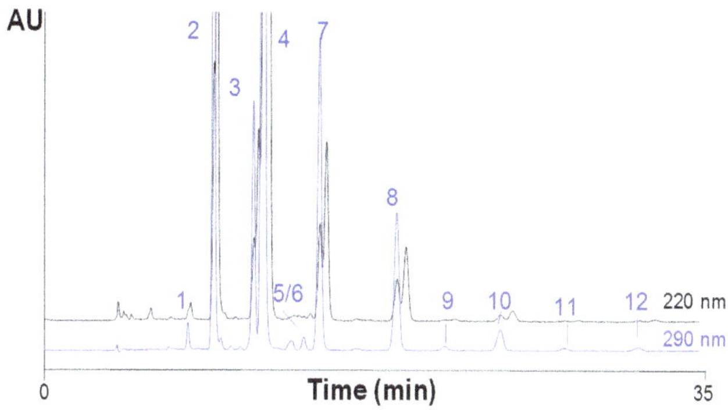

Figure 20.12 HPLC chromatogram of the cafestol (220 nm) and kahweol (290 nm) esters of an arabica coffee. Conditions: column 250 × 4 mm, Nucleosil 120-3 C_{18}, eluent: acetonitrile/isopropanol (70 : 30), for numbers for kahweol esters see Table 20.1.

20.4.3 Synthesis of Diterpene Esters

For quantitative investigations, standard substances are necessary. However, free diterpenes are very expensive, and diterpene esters are not commercially available. Therefore, standard substances have either to be isolated from the coffee oil or to be synthesized. For this purpose, the free diterpenes can be extracted from the coffee together with the other lipid components and isolated upon saponification.

The isolation of the diterpene esters from the coffee oil can be more difficult than the synthesis process. Depending on the quantitative occurrence of the individual compounds, this can be very expensive.

Table 20.1 Kahweol fatty acid esters identified by LC/MS/DAD.

Peak no.	Kahweol ester	$[M + H]^+$	$[M + H - H_2O]^+$
1*	Kahweol linolenate	575	557
2	Kahweol linoleate	577	559
3	Kahweol oleate	579	561
4	Kahweol palmitate	553	535
5*	Kahweol heptadecanate	567	549
6*	Kahweol gadoleinate	607	589
7	Kahweol stearate	581	563
8	Kahweol arachidate	609	591
9*	Kahweol henicosanate	623	605
10*	Kahweol behenate	637	619
11*	Kahweol tricosanate	651	633
12*	Kahweol lignocerate	665	647

Therefore, methods for the synthesis of cafestol esters and 16-OMC esters were developed. Because of the sensitivity of diterpenes to acids, a mild procedure is required for esterification. The reaction in the alkaline environment is favorable according to the Einhorn variant. Using pyridine as catalyst, alcohols are converted with carboxylic acid chlorides since the reaction with these chlorides proceeds much more easily and with higher yields than with the respective acids. A technique for esterifying cafestol was first described by Kaufmann and Hamsagar.[13] Cafestol was converted with fatty acid chlorides in a solvent mixture of benzene and pyridine (3 + 1, v/v), under cooling of the batch. For the individual esters they obtained yields between 45% (cafestol arachidate and cafestol linolate) and 55% (cafestol palmitate and cafestol stearate). If for toxicological reasons the benzene was dispensed by toluene, smaller yields were obtained. After little modification with respect to the quantities of chemicals used, different cafestol esters were produced according to the method described by Kaufmann and Hamsagar.[53]

Due to the sterically hindered free hydroxyl group, the synthesis of the 16-OMC esters needs a stronger catalyst, namely 4-dimethylaminopyridine. With this substance, acylation reactions proceed much faster and with greater yields than with pyridine as catalyst. With *n*-hexane and pyridine (5 + 1, v/v) as solvent, the synthesis of eight different 16-OMC esters was successfully carried out.[58]

After the reaction, the formed diterpene ester as well as the free fatty acid and the unconverted diterpene are present in the reaction vessel. By liquid–liquid extraction between *t*-butyl methyl ether and a sodium carbonate solution it is possible to separate the free fatty acid. Subsequently, the diterpene can be separated from the ester by means of solid-phase extraction on a silica-gel column. After transferring the sample onto the column, first, the ester is eluted with a mixture of *n*-hexane and ethyl acetate (9 + 1, v/v). Then, the unconverted diterpene can be eluted with ethyl acetate and reused for further syntheses. The yields of the different diterpene esters achieved by the syntheses are compiled in Table 20.2. The yields are directly related to the

Table 20.2 Yields of synthesized diterpene esters (purity >97%)..[58,60]

Fatty acid	16-*O*-Methylcafestol esters (mol%)	Cafestol esters (mol%)
C14	—	81.0
C16	80.6	77.7
C18	61.2	74.8
C18:1	75.3	51.0
C18:2	74.8	50.0
C18:3	58.5	—
C20	81.6	67.5
C22	81.0	49.6

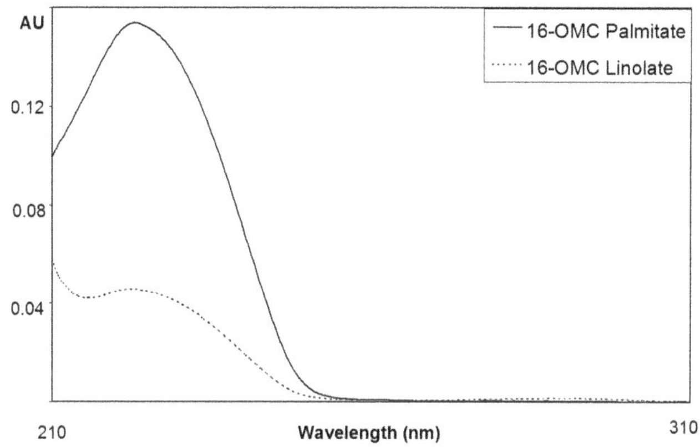

Figure 20.13 UV spectra of two 16-*O*-methylcafestol esters. Reproduced from ref. 60 with permission from Taylor & Francis Ltd (http://www.tandfonline.com).

applied quantity of the diterpene. The purity of the esters was determined by means of mass spectrometry as well as by HPLC in connection with a diode-array detector.

Moreover, the molar extinction coefficients were determined; they were very different and seemed to depend on the structure of the fatty acid (see Figure 20.13). As a consequence, for precise quantification, a number of individual esters are necessary.

20.4.4 Other Diterpene Compounds

20.4.4.1 Atractylosides

A further important group of diterpene derivatives found in coffee is the class of the atractylosides, which are mainly present as glycosides (Obermann and Spiteller, 1976, Maier and Wewetzer, 1978, Aeschbach *et al.*, 1982, Bradbury and Balzer, 1999).[62–65] Total bean glycoside contents varied between approx. 1 and 5 g kg^{-1}. On thermal treatment, decarboxylation to the corresponding atractyligenin glycosides occurred rapidly. The atractyligenin glycosides

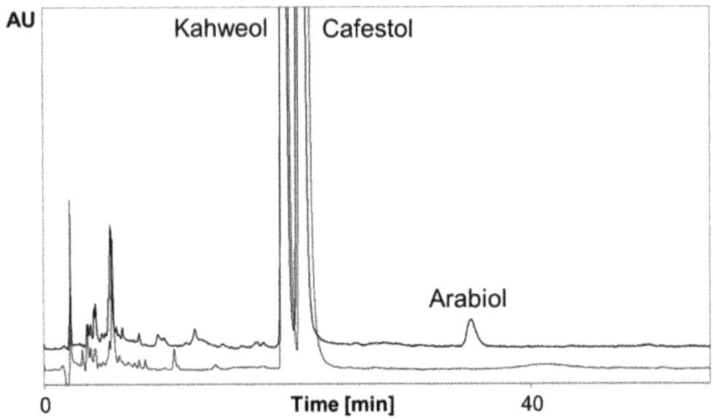

Figure 20.14 HPLC chromatograms of the fresh and stored arabica coffee.

were relatively stable to roasting, the content being reduced by about half under conditions of dark roast.[65]

20.4.4.2 *Arabiol*

With arabiol, a further diterpene derivative was identified in wet-processed green arabica coffee beans from Colombia. These coffee beans were stored for 18 months under the various conditions of critical factors such as elected moisture levels (6.2%, 11.5%, 13.8%) and temperatures (12 °C, 25 °C, 40 °C) in special bags under a defined atmosphere (oxygen content: 2%, 5%, normal) (Speer and Kölling-Speer, 2006).[11] In contrast to the fresh coffee beans, a further signal was detectable in the chromatogram (Figure 20.14). The structure of the new substance was determined by ^{1}H and ^{13}C NMR spectroscopy, in which the exact assignment of all proton and carbon signals was carried out by 2D NMR techniques COSY, HSQC, HMBC, and NOESY. In contrast to cafestol, the new compound indicates a different number and arrangement of aromatic or olefinic structures. The ^{1}H/^{13}C-correlated HSQC spectra prove six CH= groups in the aromatic region, and the NOESY spectrum offers the proton orientation. Using the correlations 12/13-H, 13/2-H, 2/1-H, and 4/3-H, the coordination of the linkage for these protons became obvious. In combining the data from mass spectrometry and NMR, the structural formula shown in Figure 20.15 can be postulated. The IUPAC-name for this substance is 8-(hydroxymethyl)-11b-methyl-4b,5,6,7,8,9,10,11,11a,11b-decahydro-6a,9-methanocyclohepta[*a*]phenanthren-8-ol (according to the software of ACD/Labs™ IUPAC Name Educators Pack 14.5 Version 6.0 from 2002). As a trivial name the authors of this chapter proposed arabiol.[66]

20.4.4.3 *Coffeadiol*

During the isolation of cafestol as a standard substance, a new compound was detected. By using silica-gel column chromatography and a following semi-preparative RP-HPLC the substance was isolated. Structure elucidation

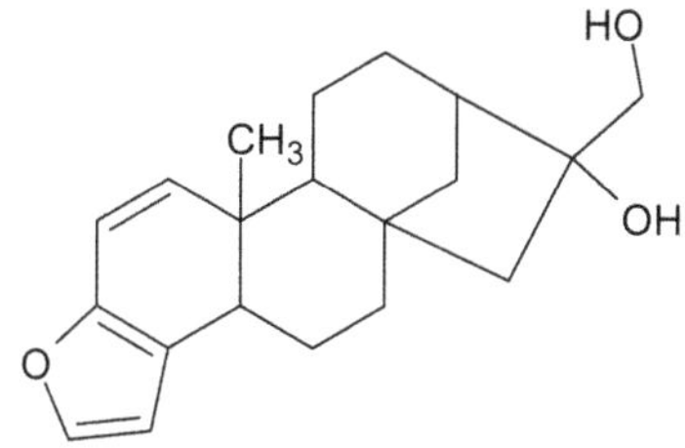

Figure 20.15 Structural formula of arabiol. Reproduced from ref. 11 with permission from *The Brazilian Journal of Plant Physiology*.

Table 20.3 ^{13}C and ^{1}H NMR chemical shifts in ppm of coffeadiol.

Position	δ (C)	δ (H)
		α/β
1	47.4	—
2	35.3	1.67/1.20
3	26.9	2.72/1.63
4	126.6	—
5	129.6	—
6	32.1	1.75/1.92
7	22.7	1.32; 1.67[a]
8	23.6	1.32/1.59
9	44.2	2.08
10	40.6	—
11	32.3	1.32/1.14
12	24.2	1.52; 1.54[a]
13	44.4	1.94
14	82.6	—
15	44.7	2.27/1.32
16	38.7	1.67; 1.70[a]
17	19.5	1.61
18	15.9	0.85
19	66.0	3.68 and 3.78
14-(OH)		1.56
19-(OH)		1.94

[a]Assignment of α/β protons uncertain.

was carried out by NMR spectroscopy (^{1}H-NMR, ^{13}C-NMR, DEPT, COSY, HSQC, HMBC, NOESY), mass spectroscopy, and X-ray crystal structure analysis.

The high-resolution EI mass spectrum displayed a molecular ion peak at *m*/*z* 290.225 accounting for the molecular formula $C_{19}H_{30}O_2$. Three more intense fragment peaks occurred at *m*/*z* 272.215 ($C_{19}H_{28}O$), *m*/*z* 259.206 ($C_{18}H_{27}O$), and *m*/*z* 241.195 ($C_{18}H_{25}$), originating from consecutive losses of water and hydroxymethyl.

The ^{13}C-NMR spectrum (Table 20.3) indicates that the new compound has a total of 19 carbon atoms. The carbon atoms were determined from the DEPT spectrum as two methyls, two methines, ten methylenes, two quaternary carbon atoms (sp^2), and three quaternary carbon atoms (sp^3). The presence of one double bond between two quaternary carbon atoms was confirmed by

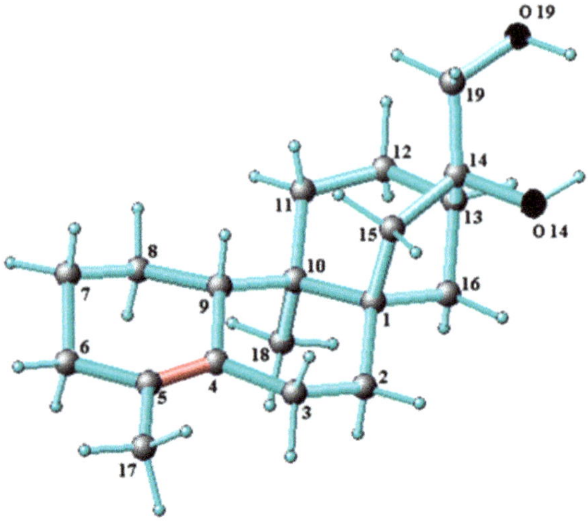

Figure 20.16 X-ray structure of coffeadiol.

^{13}C-NMR (δ = 129.6 and 126.6 ppm). The HMBC spectrum shows intensive cross-peaks between the protons of one methyl group (δ = 1.612 ppm), and the two quaternary carbon atoms of the double bond C4 and C5 and the C6, as well. In consequence, this methyl group is situated at C5. The NOESY spectrum shows cross-peaks between the methyl protons (δ = 0.85 ppm) and the protons H8α, H2α, H11α, und H16α. This can only be interpreted by locating the methyl group at the carbon atom C10. If it were located at carbon atom C9 like in cafestol, there should not be a NOE-peak of C10 with the proton H16α.

The crystal structure was analyzed by X-ray single crystal method to confirm that the methyl group C18 is really bound to another carbon in comparison to the known diterpenes.

The supposed structure could be confirmed (Figure 20.16), and the analysis indicates the presence of an H bridged dimer. The most important differences of the new structure in comparison to cafestol are the absence of the furan ring and the location of one methyl group.

14-(Hydroxymethyl)-5,10-dimethyltetracyclo-[11.2.1.0^{1,10}.0^{4,9}]hexadec-4-en-14-ol was obtained as light yellow crystals (m.p. 110 °C). As a trivial name, coffeadiol was proposed (Kurt and Speer, 1999).[67]

20.4.5 Diterpenes in the Lipid Fraction of Roasted Coffees

During the roasting process, a number of new diterpene compounds are formed. With dehydrocafestol and dehydrokahweol, two decomposition products from cafestol and kahweol were identified in roasted coffee (Figure 20.17). The amounts of both compounds increase with the rising roasting temperatures but also depend on the contents of cafestol and kahweol in the green coffee.[27,68,69]

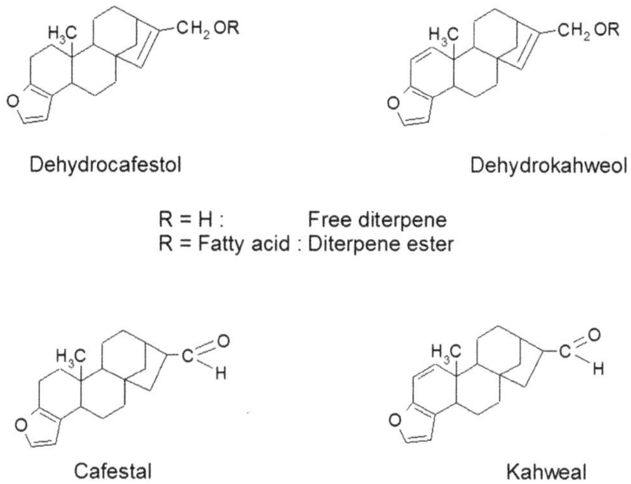

Figure 20.17 Structural formulae of decomposition products of cafestol and kahweol. Reproduced from ref. 142 with permission from John Wiley and Sons. Copyright 2001 Blackwell Science Ltd.

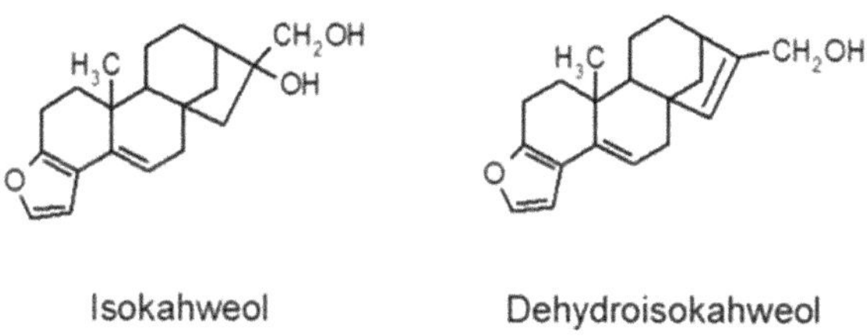

Figure 20.18 Structural formulae of isokahweol and of dehydroisokahweol.

Searching for an objective characteristic for the roasting degree of coffees, the cafestol/dehydrocafestol ratio is suitable; thus, a ratio of 25–40 describes a well-roasted coffee, whereas a ratio of up to 15 describes a strongly roasted coffee. More strongly roasted espresso coffees, however, have a ratio of 10–15.[69]

Cafestal and kahweal (Figure 20.17) are two further new degradation products of cafestol and kahweol detected in commercial roasted coffees.[70] In 2005, with isokahweol and dehydroisokahweol (Figure 20.18), and in 2008, with secokahweol (Figure 20.19), three new diterpenes were elucidated in roasted coffees by means of high-resolution mass spectrometry and different NMR-experiments.[71,72] A typical HPLC chromatogram of a strong-roasted coffee sample is presented in Figure 20.20. Using new columns and for example UHPLC the analysis could be run within 6 minutes without losing separation efficiency (Figures 20.21 and 20.22).[73,74]

The roasting stability of the fatty acid esters of the three diterpenes is quite different. Examinations of the 16-OMC esters have shown that they

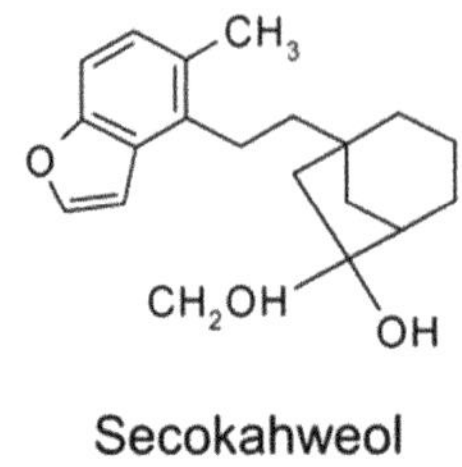

Figure 20.19 Structural formula of secokahweol.

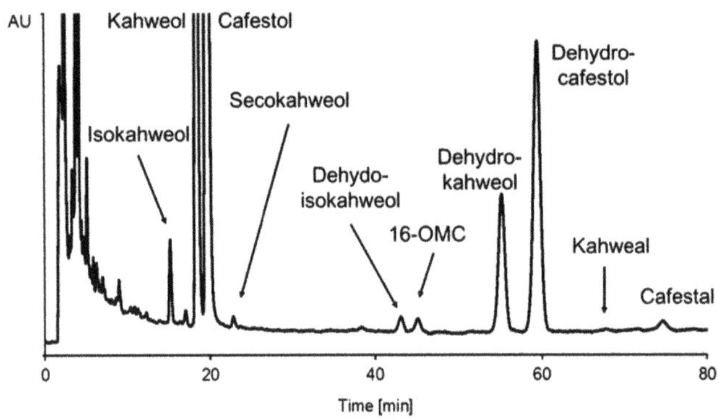

Figure 20.20 HPLC chromatogram of a dark-roasted arabica coffee with 2% robusta.

are clearly stable during roasting in spite of different roasting temperatures so that both the content and the proportional distribution for the diterpene esters are almost the same.[11]

In contrast, the contents of the cafestol and kahweol esters decrease depending on the roasting temperature, whereas the distribution of the esters remains nearly the same.[58,59] Kurzrock *et al.*[75] demonstrated that cafestol was dehydrated within the fatty acid esters as well. In model experiments by heating cafestol palmitate and cafestol linoleate, the corresponding dehydrocafestol esters, which were also identified in roasted coffee, were obtained.[75]

20.4.6 Diterpenes in Coffee Beverages

The sensory quality of coffee beverages can be negatively affected by various parameters. Taking into account that dried green beans are stored up to three years prior to roasting, the storage conditions are a key factor influencing the development of off-flavors in the coffee beverage. Negative aroma changes are described as "moldy" and "earthy" as well as "raw-old" and "woody". Although not all proveniences are affected by the latter, the conditions applied for raw coffee storage are assumed to produce this type

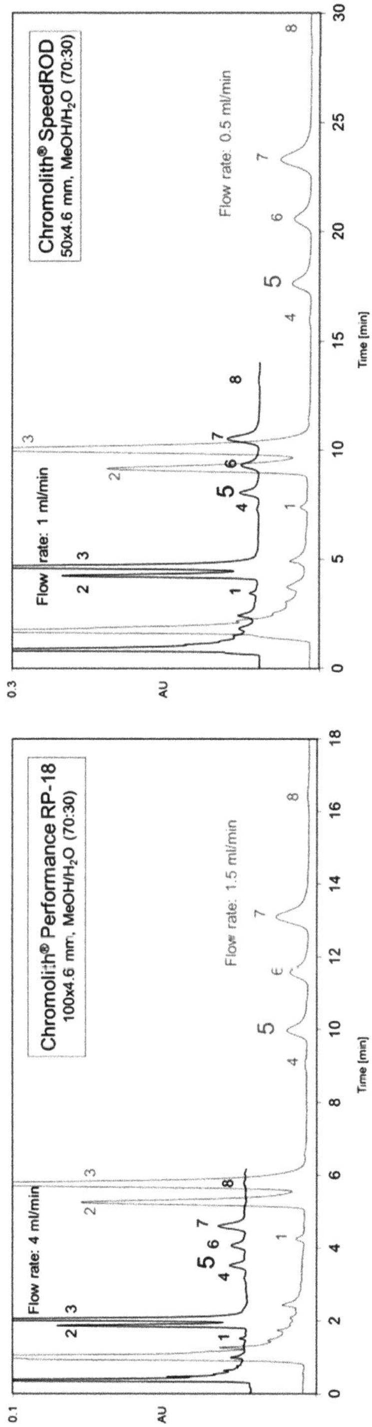

Figure 20.21 HPLC of an arabica/robusta blend using different columns; (1) isokahweol, (2) kahweol, (3) cafestol, (4) dehydroisokahweol, (5) 16-OMC, (6) dehydrokahweol, (7) dehydrocafestol, (8) cafestal.

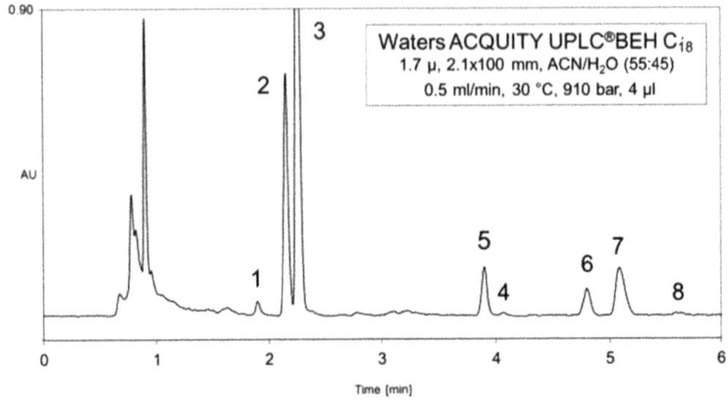

Figure 20.22 UHPLC of an arabica/robusta blend. (1) isokahweol, (2) kahweol, (3) cafestol, (4) dehydroisokahweol, (5) 16-OMC, (6) dehydrokahweol, (7) dehydrocafestol, (8) cafestal.

of off-flavor. In particular, the lipid fraction is suggested as a precursor for off-flavor development.[48]

Several studies reported that through the consumption of especially prepared coffee, the serum cholesterol level might increase. Because of the high prevalence of high serum LDL cholesterol levels in countries where unfiltered coffees were used, it was inferred that this effect was caused by the lipids present in the coffee brew, which, although poorly soluble in water, might be incorporated in the brew depending on the method of infusion. Initially, triacylglycerides were said to be responsible for this effect, but in more recent years, it has been established that the diterpenes, especially cafestol and kahweol, both in the free form and as palmitate esters influenced the serum cholesterol level.[76–80] Other diterpenes have not been tested yet.

Furthermore, a substantial number of scientific publications exist where positive effects of diterpenes were reported.[81–84] Lam *et al.*[81] has shown that cafestol stimulates the glutathione-*S*-transferase activity through which the decomposition of xenobiotica is accelerated. Other authors reported that cafestol and kahweol protect against aflatoxin B1-induced genotoxicity.[82–84]

Therefore, investigations into the presence of diterpenes in differently prepared coffee beverages are of great interest.[85–90]

Using the example of 16-OMC esters, Sehat *et al.*[86] showed that lipophilic diterpene esters flow into the coffee infusion and are even detectable in instant coffee granules. The amount in the beverage strongly depends on the method of preparation and is directly related to the amount of total lipids in the coffee brew. With filtered coffee prepared in a common household coffee machine, the total amount of lipids was less than 0.2% of the lipids in the coffee powder. In contrast, when preparing an espresso, between 2 and 4% of the lipids, depending on the vending machine and the chosen parameters, and thereby the diterpenes as well, scattered from the finely ground espresso coffee into the coffee brew. When coffee was prepared Scandinavian style, it contained even up to 22% of the lipids in the coffee powder. The proportional distribution of diterpenes in the coffee beverage was nearly identical to the distribution in the roasted coffee powder.

In 2007, the Consumentenbond[91] (consumer federation) of the Netherlands reported that coffee prepared in vending machines might increase the serum cholesterol level. Due to the negative physiological effects, conditions to reduce these components in the coffee brew are of interest. As a result of the increasing market share of vending machines, the aim of the authors' research was to study the cafestol content in differently prepared espresso coffees (Buchmann *et al.*, 2010).[90]

Commercially roasted arabica coffee was ground using an industrial centrifugal mill (VTA6S, Stahwert Mühlenbau, Germany). The particle size distribution was determined employing different sieves (90–1000 µm) in the sieve shaker AS 200 tap, Retsch, Germany. The coffee brews were prepared using a vending machine (espresso unit). In addition to the grind size (see below), the variable preparation parameters were the water temperature (80, 95, and 110 °C), the amount of coffee (6.5, 7.5, and 8.5 g), and the amount of water (40, 65, and 90 ml). The fresh coffee brews were lyophilized immediately. Then the dry residues were saponified, and after the isolation of the unsaponifiable matter, cafestol was quantified by HPLC with UV-detection.

To ensure comparable conditions, a single charge of a roasted arabica coffee was used throughout this study. The cafestol content was determined at 0.46 g 100 g^{-1}. The roasted arabica coffee beans were ground with a centrifugal mill at three settings. The obtained charges were analyzed regarding the particle size distribution. Obviously there were only slight differences between finely and medium-ground coffees. The main particle size was between 315 and 500 µm. Medium-ground coffees differ only in particle sizes between 500 and 630 µm. In contrast, coarsely ground coffee consisted primarily of particles larger than 700 µm (Figure 20.23).

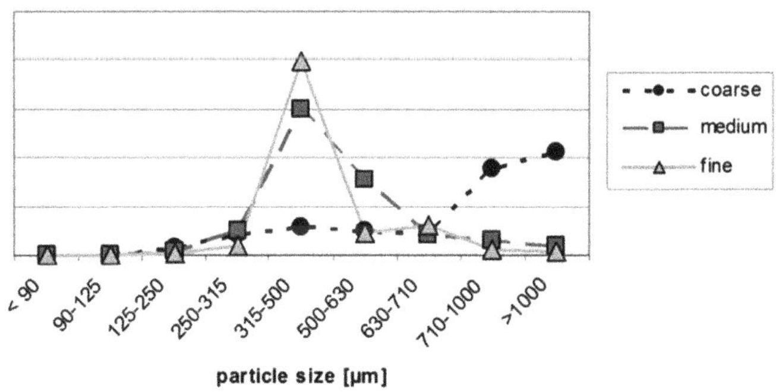

Figure 20.23 Particle size distribution of the ground coffee.

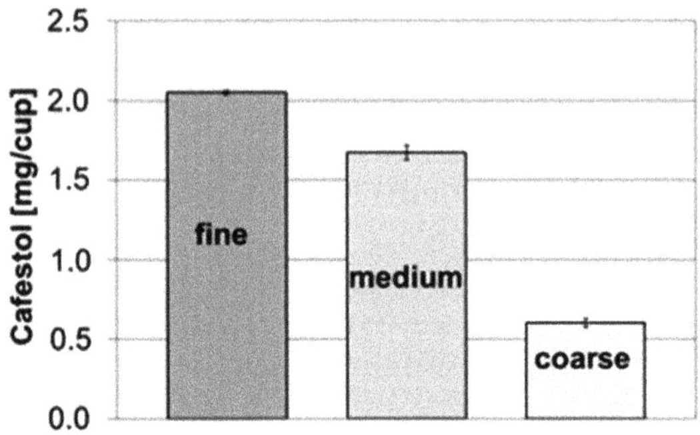

Figure 20.24 Impact of grinding grade on the cafestol content in espresso.

The impact of the grind size was studied on an espresso beverage prepared with 7.5 g ground roast coffee and 65 ml water at 95 °C. The finer the particle size, the higher was the extraction yield (Figure 20.24). Regarding the cafestol content, especially the difference between the medium and the coarse grinding grade was remarkable. A decrease from 2 mg per cup to 0.6 mg per cup was observed. However, for espresso, the coarsely ground coffee was inapplicable. Therefore, the following studies were carried out with the grinding grade medium and fine.

Statistical screening designs were used to optimize the extraction parameters and evaluate the main effects, interaction effects, and quadratic effects. After determining the significant factors, the optimal operation conditions were attained by using a Box–Behnken design.

In Figure 20.25 the 3D response surface plot for the medium grind size is presented. With the increase in the amount of coffee, the cafestol yield rose almost linearly. Especially with a simultaneous increase in water, the extraction yields of cafestol reached a maximum in the espresso brew.

As expected, the finely ground coffee led to an overall higher cafestol content. Due to the enlarged surface of fine coffee powder, the extraction yield increased (Figure 20.26).

The highest extractions yields of 2.7 mg cafestol were reached with 8.5 g coffee and 90 ml water. In contrast to the medium grinding grade with an increasing water quantity and a constant amount of coffee, only a slight flattening in the run of the cafestol level was observed.

It was shown that in addition to the particle size, the coffee/water ratio was the most important factor influencing the cafestol contents in the brew. By varying the preparation parameters of the espresso beverage, the cafestol content was reduced from 2.6 mg per cup (10 g coffee powder, medium ground, 180 mL of water, at 110 °C) to 0.9 mg per cup (8 g coffee powder, coarse ground, 70 mL of water, at 80 °C). However, a coffee drink with a low cafestol content and, consequently, a low lipid content will not delight the senses of the coffee consumer. Thus, a compromise needs to be found between the cafestol content and the sensory impression.

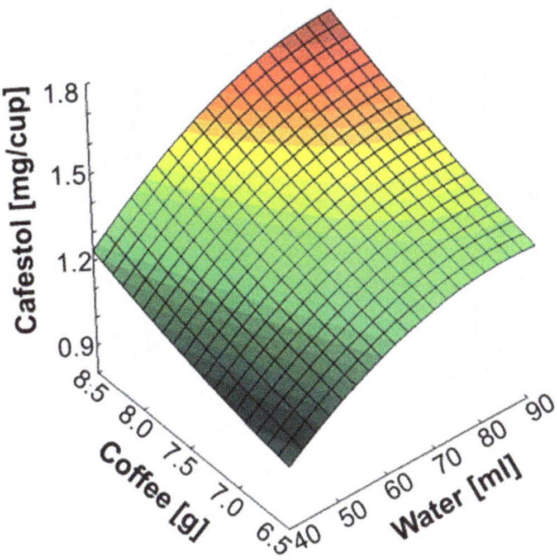

Figure 20.25 3D response surface plot for cafestol contents in brews prepared with medium ground coffee at 95 °C.

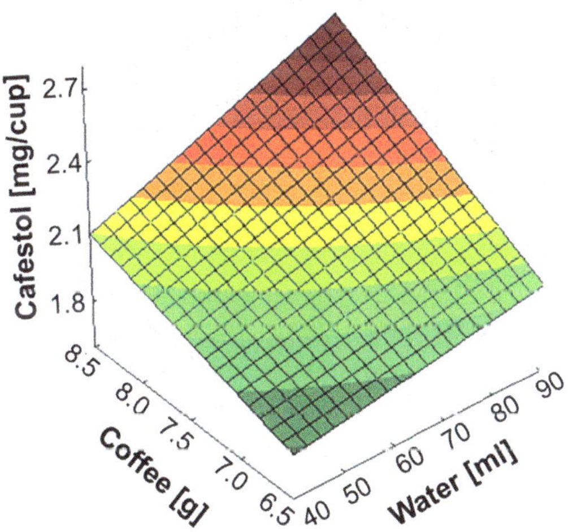

Figure 20.26 3D response surface plot for cafestol contents in brews prepared with finely ground coffee at 95 °C.

20.5 Sterols

Over the past years, phytosterols have achieved more importance due to their beneficial properties. It was reported that the phytosterols reduce the cholesterol level and have also demonstrated anti-carcinogenic properties. Furthermore, an immune modulating effect was described.[92,93]

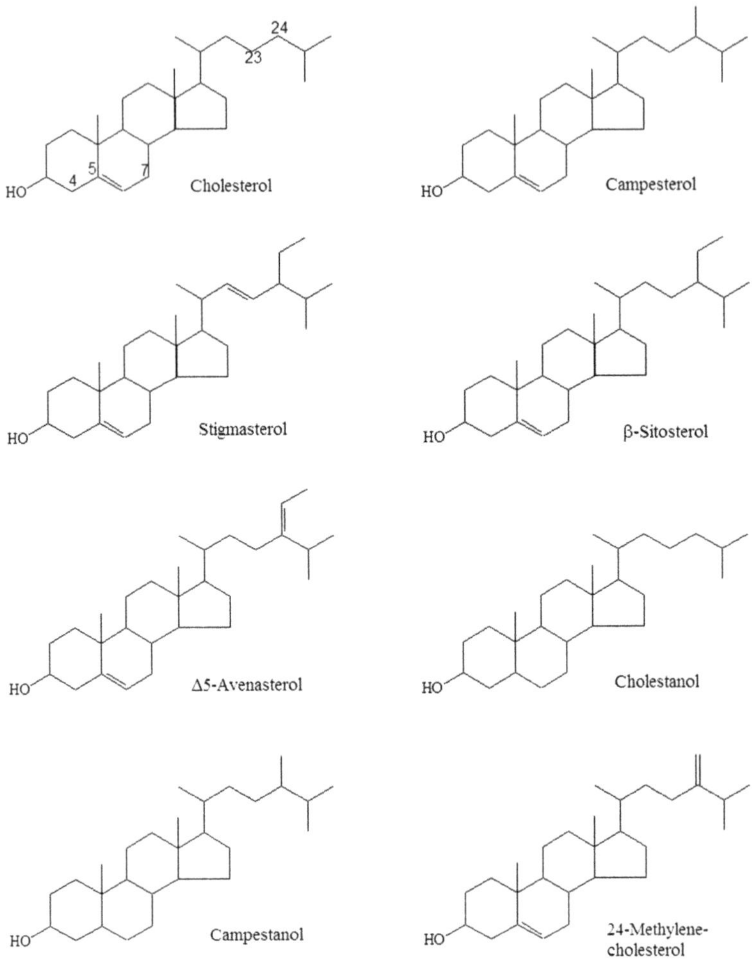

Figure 20.27 The structures of 4-desmethylsterols. Reproduced from ref. 142 with permission from John Wiley and Sons. Copyright 2001 Blackwell Science Ltd.

Coffee oil contained a number of sterols, which are typical for other seed oils as well. In addition to 4-desmethylsterols, various 4-methyl- and 4,4-dimethylsterols have been identified. The structural formulae of the main sterols are presented in Figures 20.27–20.29.

Sterols in coffee oil were described both in free (around 40%) and in esterified form (around 60%) (Picard *et al.*, 1984).[4] The total amount was determined in the unsaponifiable matter as TMS-derivatives by means of GC or GC/MS. Often, a fractionation in desmethyl, 4-methyl- and 4,4-dimethylsterols using TLC, HPLC, or silica gel cartridges was carried out (Alcaide *et al.*, 1971, Itoh *et al.*, 1973, Picard *et al.*, 1984, Horstmann and Montag, 1986, Homberg and Bielefeld, 1989, Speer, 1989).[4,6,94–98] The desmethylsterols represent 90% of

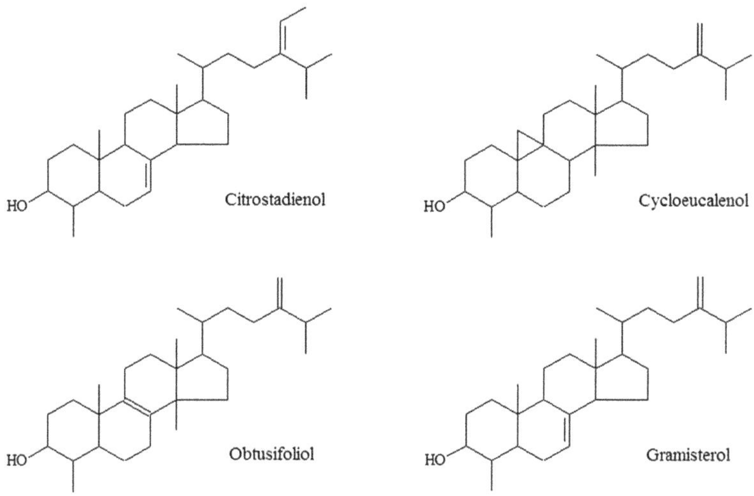

Figure 20.28 The structures of 4-methylsterols. Reproduced from ref. 142 with permission from John Wiley and Sons. Copyright 2001 Blackwell Science Ltd.

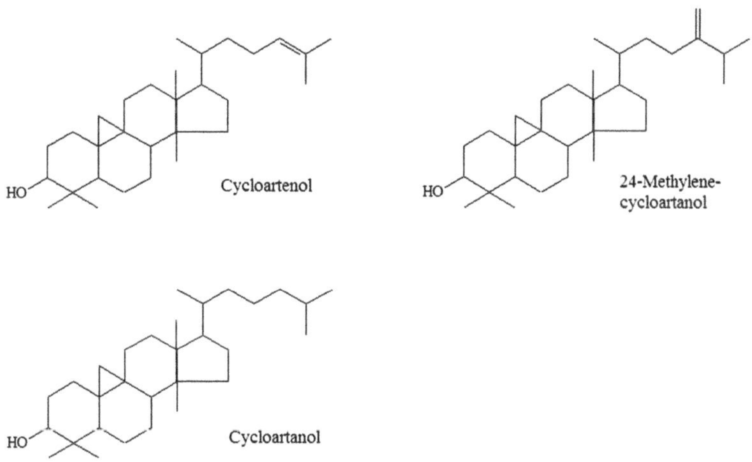

Figure 20.29 The structures of 4,4-dimethylsterols. Reproduced from ref. 142 with permission from John Wiley and Sons. Copyright 2001 Blackwell Science Ltd.

the total sterol fraction, which ranged from 1.5% to 2.4% of the lipids (Picard *et al.*, 1984);[4] with 5.4% an even higher content was reported by Nagasampagi *et al.* (1971).[99] The distribution of the main desmethylsterols in different robusta und arabica coffee samples is clarified in Table 20.4. The main sterol is β-sitosterol with about 50%, followed by stigmasterol, campesterol, and Δ5-avenasterol, whereby the latter, which occurs in much higher amounts in robusta than in arabica coffee, is suitable for coffee blend studies due to its

Table 20.4 Distribution (%) of desmethylsterols in arabica and robusta green coffees (30 samples).[104]

			Mean value	
Sterols	Arabica	Robusta	Arabica	Robusta
Cholesterol	0.2–0.4	0.1–0.3	0.3	0.2
Campesterol	14.7–17.0	15.5–18.8	15.8	16.9
Stigmasterol	20.5–23.8	20.0–26.7	21.9	23.1
β-Sitosterol	46.7–53.8	40.6–50.7	51.6	45.4
Δ5-Avenasterol	1.6–4.1	5.1–12.6	2.7	9.1
Campestanol	0.2–0.6	0.1–0.3	0.4	0.2
24-Methylenecholesterol	0.0–0.4	1.5–2.4	0.2	1.9
Sitostanol	1.4–2.8	0.5–1.2	2.0	0.8
Δ7-Stigmastenol	0.9–4.5	0.1–0.8	2.2	0.2
Δ7-Avenasterol	1.2–2.1	0.2–0.6	1.5	0.4
Δ7-Campesterol	0.4–1.2	0.1–0.6	0.6	0.2
Δ5,23-Stigmastadienol	0.2–0.5	0.1–2.0	0.3	0.5
Δ5,24-Stigmastadienol	0.0–0.4	0.0–0.3	0.1	0.0
Clerosterol	0.2–0.8	0.5–1.0	0.5	0.7

high stability even during the roasting process (Duplatre *et al.*, 1984, Saltor *et al.*, 1989, Frega *et al.*, 1994, Valdenebro *et al.*, 1999, Dussert *et al.*, 2008).[28,100–103] Robusta beans contain five-fold higher amounts of Δ5-avenasterol than arabica (1.7 g 100 g^{-1} *vs.* 350 mg 10 kg^{-1}, Kamm *et al.*, 2002).[31] However, due to their varying natural contents, they are qualified for determining robusta parts in arabica coffee mixtures only from approx. 20% onward (see diterpene section).

Picard *et al.*[4] separated the free sterols and the sterol fatty acid esters by means of column chromatography (aluminium oxide). Analyzing 12 robusta coffee samples, they noticed a changed distribution for the main sterols in comparison to the total sterols. For the free sterols, the order is stigmasterol > β-sitosterol > campesterol, for the sterol esters it is β-sitosterol > campesterol > stimasterol. Furthermore, Picard *et al.* studied the individual fatty acids in the sterol esters. C_{18}, C_{16}, and $C_{18:1}$ were the main compounds with a proportional distribution similar to that reported in triglycerides.

Cholesterol, campesterol, stigmasterol, β-sitosterol, stigmastanol, Δ5-avenasterol, Δ7-stigmastenol, Δ7-avenasterol, citrostadienol, gramisterol, cycloartenol, and traces of 24-methylenecycloartenol were identified and quantified in different coffee infusions, Scandinavian style coffee, espresso, and filtered coffee.[105]

The presence of sterol glucosides in food and dietary supplements was proved by Phillips *et al.*[106] Recently, sterol glucosides were isolated and elucidated in green arabica and robusta coffee beans for the first time by Buchmann *et al.*[107] Figure 20.30 shows the modified analysis scheme first developed by Oelschlägel *et al.*[108] for free and esterified sterols in pumpkin oils. Three sterol fractions were obtained, the first containing sterol fatty acid esters, the second free sterols, and the third sterol glucosides. The sterol

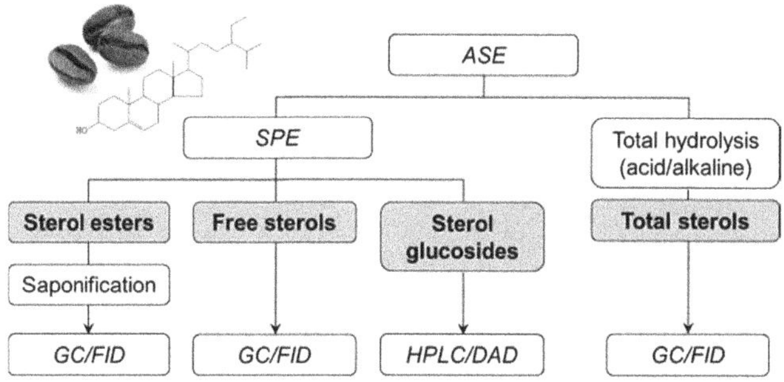

Figure 20.30 Analysis scheme for the three sterol fractions and the total sterol content.

fatty acid esters were analyzed after saponification and derivatization, the free sterols after silylation by GC/FID and then quantified *via* β-sitosterol. The separation of the sterol glucosides was accomplished on a Phenomenex Synergi 4 μ Fusion 250 × 3 mm equipped with a diode array detector. The quantification was carried out either *via* β-sitosterol or stigmasterol, depending on the number of double bounds in the molecule.

Typical GC chromatograms of the sterol ester fraction obtained from a green arabica coffee and a green robusta coffee as well are shown as trimethylsilyl (TMS) derivatives in Figure 20.31. The HPLC chromatograms of the subsequent sterol glucosides are presented in Figure 20.32.

The important ESI mass spectrum data of the identified sterol glucosides are listed in Table 20.5.

Figure 20.33 shows the ESI mass spectra for the detected sterol glucosides quantified by HPLC. Between 8% and 12% of the total sterols were sterol glucosides. Consequently, the sterol fatty acid esters represent the main fraction, followed by the free sterols (Figure 20.34). The percentage distribution of the three groups for the main individual sterols in green coffees is presented in Figure 20.35. It becomes obvious that the distribution of stigmasterol differs from that of β-sitosterol and campesterol. The individual contents of the four sterol glucosides are represented in Figure 20.36.

20.6 Tocopherols

The presence of tocopherols in coffee oil was described for the first time by Folstar *et al.*[109]; α-tocopherol was clearly identified, β- and γ-tocopherol, not separated by TLC and GC, were considered as one group (Figure 20.37). Cros *et al.*[110] also determined β- and γ-tocopherol in sum by HPLC. Folstar *et al.*[109] found concentrations of α-tocopherol of 89–188 mg kg^{-1} oil, for β- + γ-tocopherol 252–530 mg kg^{-1} oil.

In 1988, Aoyama *et al.*[111] analyzed α-, β-, and γ-tocopherols in different varieties of coffee beans, the total content being about 5.5–6.9 mg 100 g^{-1} coffee. Ogawa *et al.*[112] determined the contents of tocopherols in 14 green coffee beans, their roasted beans and infusions, and in 38 instant coffees by HPLC.

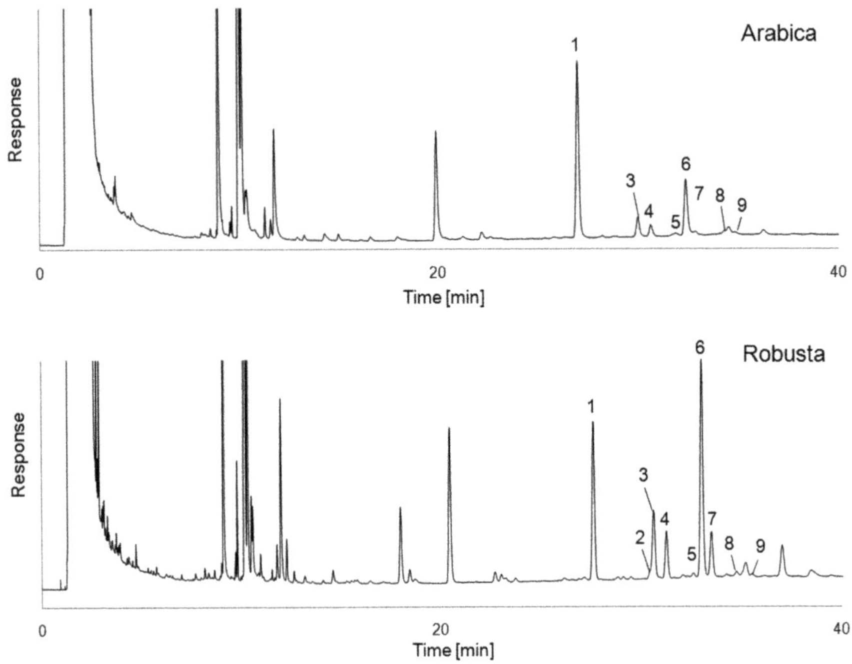

Figure 20.31 GC chromatograms of the sterol esters, after saponification and silylation in arabica green coffee and robusta green coffee: (1) cholesterol (internal standard), (2) 24-methylencholesterol, (3) campesterol, (4) stigmasterol, (5) clerosterol, (6) β-sitosterol, (7) Δ5-avenasterol and stigmastanol, (8) Δ7-stigmastenol, (9) Δ7-avenasterol.

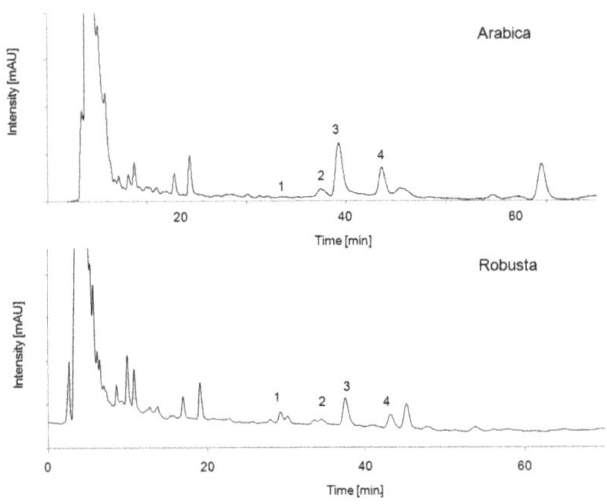

Figure 20.32 HPLC chromatograms of sterol glucosides (λ = 195 nm) in arabica green coffee and robusta green coffee: (1) Δ5-avenasterol glucoside, (2) campesterol glucoside, (3) stigmasterol glucoside, (4) β-sitosterol glucoside.

Table 20.5 Measurement of sterol glucosides by ESI-LC/MS.

Sterol glucosides	Molecular masses [g mol^{-1}]	Fragment ions negative [M − H + 60]$^-$	[M − H]$^-$	Fragment ions positive [M + H − 162 − 18]$^+$
Δ5-Avenasterol glucoside	574	633	573	395
Campesterol glucoside	562	621	561	383
Stigmasterol glucoside	574	633	573	395
β-Sitosterol glucoside	576	635	575	397

The maximum of total tocopherols in the green coffee beans was 15.7 mg 100 g^{-1} and the average was 11.9 mg 100 g^{-1}. The contents of α- and β-tocopherol were 2.3–4.5 and 3.2–11.4 mg 100 g^{-1}, respectively. γ- and δ-Tocopherol were not found. Roasting slightly diminishes the content of α-, β-tocopherol, and total tocopherols to 79–100%, 84–100%, and 83–99%, respectively.

Using GC-MS and HPLC as well, γ-tocopherol was detected by Speer and Kölling-Speer in some robusta coffees (Figure 20.38).[113] Less comprehensible are the results by González *et al.*[114] as they found higher amounts of γ-tocopherol in roasted coffees than in green coffees.

20.7 Coffee Wax

A thin waxy layer covers the surface of green coffee beans. Coffee wax is generally defined as the material obtained by extraction from coffee beans, using chlorinated organic solvents. The amount of the surface wax is about 0.2–0.3 g 100 g^{-1} of green coffee beans. The main constituents of the coffee wax are the so-called carboxylic acid-5-hydroxytryptamides (C-5HT). This substance group, amides of serotonin (5-hydroxytryptamine, 5HT) and fatty acids of different chain lengths, was first introduced by Wurziger and coworkers.[115,116] They isolated and characterized three 5HT with arachidic (C_{20}), behenic (C_{22}), and lignoceric acid (C_{24}) (Figure 20.39). Later on, Folstar described stearic acid-5HT as well as 20-hydroxy-arachidic- and 22-hydroxy-behenic acid-5HT.[117,118]

Kurzrock *et al.*[119] introduced two carboxylic acid-5HT with the odd-numbered fatty acids henicosanoic (C21) and tricosanoic acid (C23) at the 20th International Conference on Coffee Science held in Bangalore 11–15 October 2004. Later these results were confirmed by Lang and Hofmann (2005).[120] Furthermore, apart from palmitic (C16), eicosenoic (20 : 1)-, and octadecadienoic acid (C18 : 2)-5-hydroxytryptamide, the hydroxy derivatives of the two unsaturated 5-hydroxytryptamides (C20 : 1 and C22 : 1) were described by Hinkel and Speer for the first time.[121]

Several working groups developed analytical methods for determining the contents of C-5HT in green, roasted, and differently treated coffees. In

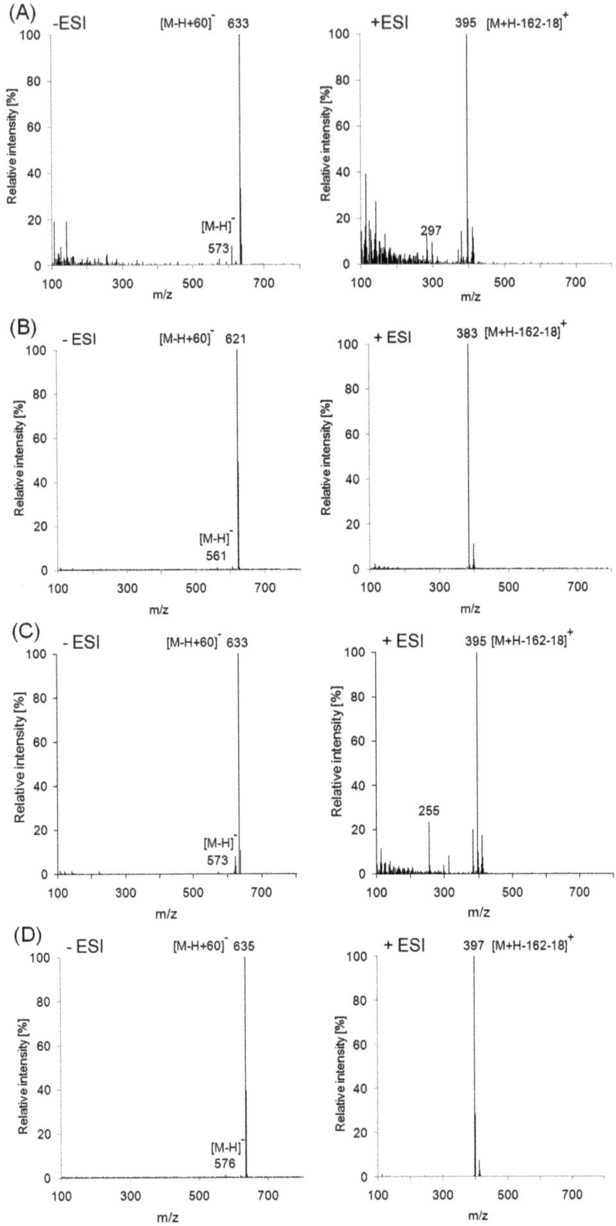

Figure 20.33 ESI mass spectra of (A) Δ5-avenasterol glucoside, (B) campesterol glucoside, (C) stigmasterol glucoside, and (D) β-sitosterol glucoside (left: negative ESI, right: positive ESI mode).

the beginning, an analysis was carried out by thin layer chromatography with spectral photometric or densitometric determination,[122–124] followed by liquid chromatography with UV detection at 278 nm.[117,125,126] In addition to the analysis by HPLC with fluorescence detection[121,127] at an excitation

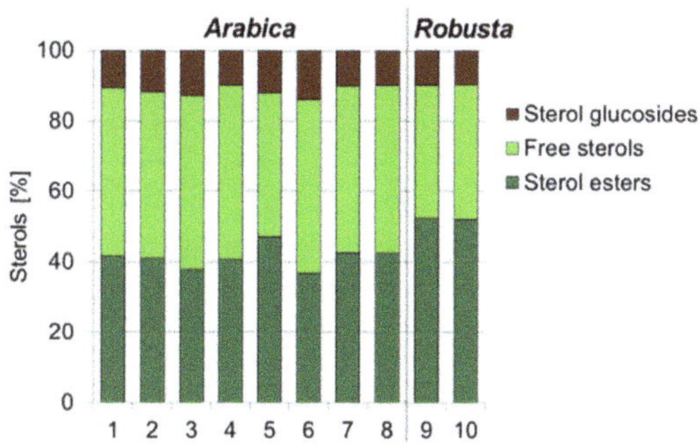

Figure 20.34 Percentage of the sterol groups in relation to the total sterols.

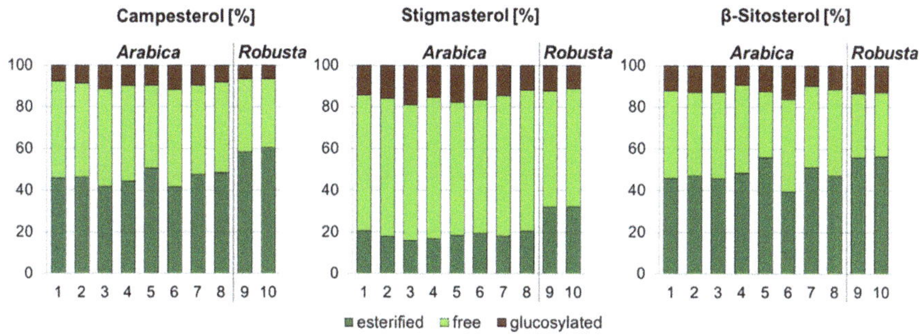

Figure 20.35 Distribution of the main individual sterols in green arabica and robusta coffees.

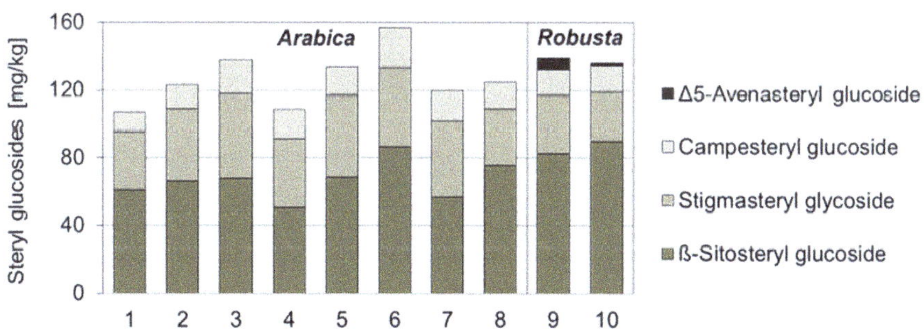

Figure 20.36 Contents of the main sterol glucosides in different green coffees.

wavelength of 280 nm and an emission wavelength of 330 nm, LC-MS/MS-methods were also described.[119–121]

In Figure 20.40 a typical HPLC chromatogram is shown of the C-5HT in a robusta coffee. The ground beans were extracted by using the accelerated solvent extraction (ASE), and before injection the extract was purified by solid phase extraction (SPE) (Hinkel and Speer, 2005).[121] The recovery rate of this method was about 90%, the limit of detection 2.5 mg kg^{-1}, and the variation

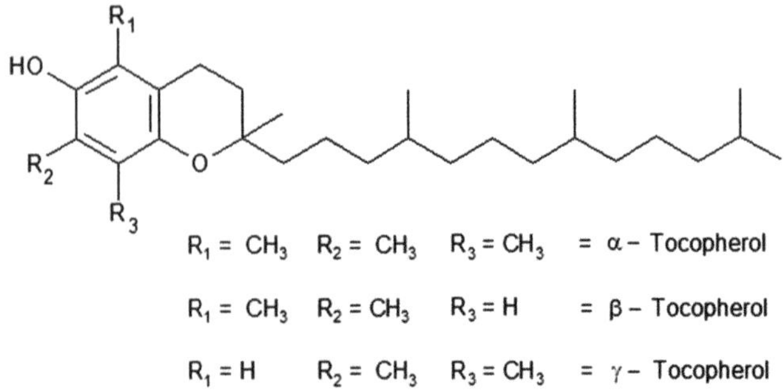

Figure 20.37 Structural formulae of tocopherols. Reproduced from ref. 142 with permission from John Wiley and Sons. Copyright 2001 Blackwell Science Ltd.

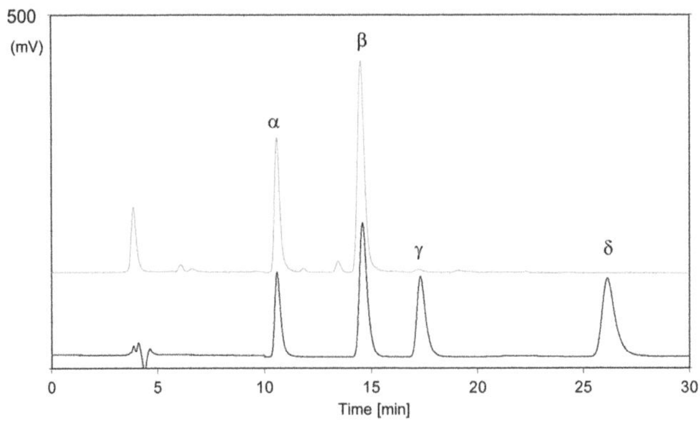

Figure 20.38 HPLC chromatogram of the tocopherols of a robusta coffee (upper chromatogram) in comparison to a standard chromatogram.

coefficients were below 5%. The C-5HT contents were calculated on the basis of synthesized standards (DE-Patent 102008025893 A1)[128] and, to allow the comparison of different treated samples, all the values are based on dry weight.

C-5HT with hydroxy and unsaturated fatty acids eluted in the first 15 minutes of the chromatogram, whereas those with saturated fatty acids have retention times greater than 18 minutes. Arachidic acid-5-hydroxytryptamide and behenic acid-5-hydroxytryptamide dominate, the other amides are only minor components.

In Figure 20.41, the total contents, including the standard deviations of analyzed green coffee samples separated into robusta and arabica, are presented. Concerning the total content of C-5HT, the amount in arabica coffees was nearly twice the value of that in the robusta coffees. The differences between the two species also became apparent in regard to the distribution of the C-5HT amounts in the different coffees. Whereas about 90% of the robusta coffees analyzed had C-5HT contents below 1.20 mg kg^{-1}, 90% of the arabica coffees showed values beyond 1.40 mg kg^{-1}. Comparing the peak area ratios of different

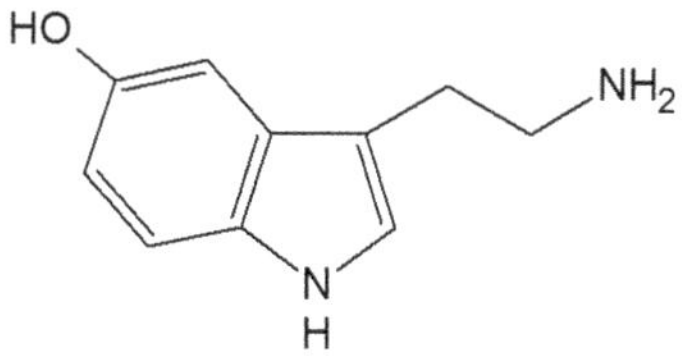

R =	Abbreviated form	References
• C16	16-5HT	Hinkel et al. (2005, 2009)
• C18	18-5HT	Folstar et al. (1979, 1980)
• C18:2	18:2-5HT	Hinkel et al. (2005, 2009)
• C20	20-5HT	Wurziger et al. (1969)
• C20:1	20:1-5HT	Hinkel et al. (2005, 2009)
• C20-OH	20-OH-5HT	Folstar et al. (1979, 1980)
• C20:1-OH	20:1-OH-5HT	Hinkel et al. (2005, 2009)
• C21	21-5HT	Kurzrock et al. (2004)
• C22	22-5HT	Wurziger et al. (1969)
• C22-OH	22-OH-5HT	Folstar et al. (1979, 1980)
• C22:1-OH	22:1-OH-5HT	Hinkel et al. (2005, 2009)
• C23	23-5HT	Kurzrock et al. (2004)
• C24	24-5HT	Wurziger et al. (1969)

Figure 20.39 Structural formulae of carboxylic acid-5-hydroxytryptamides (C-5HT).

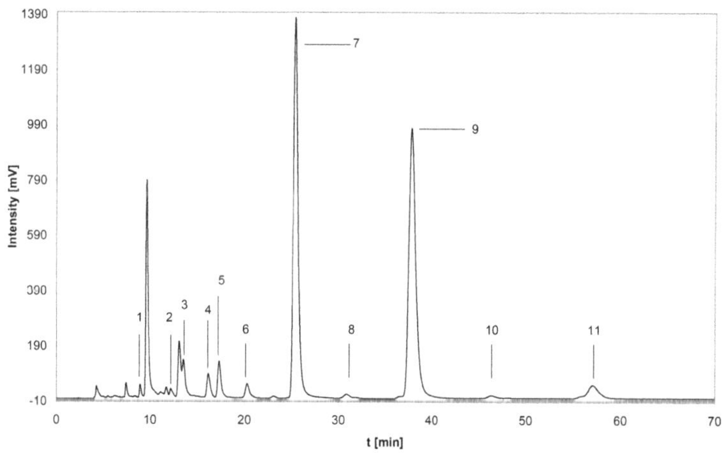

Figure 20.40 HPLC chromatogram of carboxylic acid-5-hydroxytryptamides (C-5HT) in robusta. **1:** $C_{18:2}$-5HT, **2:** C_{16}-5HT, **3:** C_{20}-Hydroxy-5HT, **4:** $C_{20:1}$-5HT, **5:** C_{18}-5HT, **6:** C_{22}-Hydroxy-5HT, **7:** C_{20}-5HT, **8:** C_{21}-5HT, **9:** C_{22}-5HT, **10:** C_{23}-5HT, **11:** C_{24}-5HT. Reproduced from ref. 11 with permission from *The Brazilian Journal of Plant Physiology*.

C-5HT signals by referring to the peak area of the arachic-5HT signal, another difference between arabica and robusta coffees can be observed. In robusta samples the amounts of arachic and behenic acid-5HT are similar, resulting in a peak area ratio of 1; by contrast, in arabica coffee the amount of arachic acid-5HT is two-fold higher so that the ratio of the two signals reaches a factor of 2.

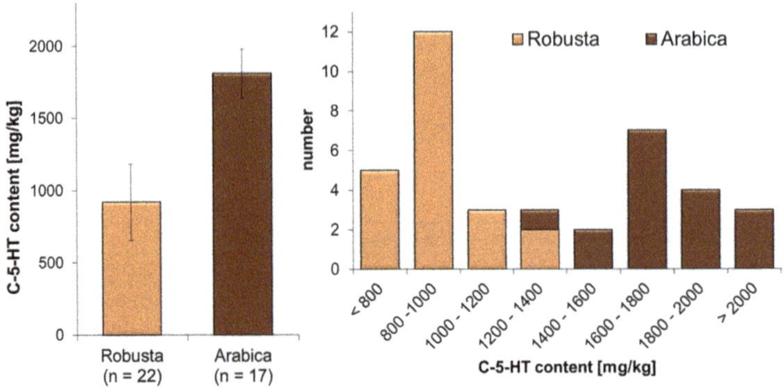

Figure 20.41 Mean values and contents of C-5HT in green arabica and robusta coffees.

The removal of the waxy layer of the coffee beans by technological treatment such as polishing, dewaxing, steaming, or decaffeinating results in a more digestible coffee brew.[125,129,131] Hence, in 1933, a first steaming method was developed to minimize any irritating effects the coffee brew has on certain coffee drinkers.[132] In the course of time, this method was improved repeatedly.[133–135]

Even though the C-5HT are the main constituents of the coffee wax, it is unlikely that they are solely responsible for the undesirable effects of untreated coffee. One reason for this assumption is their poor water solubility (2.3 mg l^{-1}), another is the absence in the percolated coffee brew made from untreated beans.[130] Fehlau and Netter,[136] studying the influence of coffee infusions on the gastric mucosa in rats, came to a similar conclusion.

Steaming the coffee beans resulted in a partial removal of the waxy layer and thereby of the C-5HT.[137] When comparing the total C-5HT content of non-treated and treated beans, the loss of C-5HT proves equivalent to 30%. However, the influence of the steam treatment on the different C-5HT is evident. The reduction of the hydroxy fatty C-5HT was about 40% while the unsaturated C-5HT, in contrast, were only reduced at a rate of about 20%. A variation of the processing parameters (time, temperature) leads to no additional reduction over 30%. This means that independent of the parameters, there is always the same reduction of approx. 30% of the total amount of the tryptamides.

As there is a wide variety of the C-5HT contents in coffees, the declaration “treated” or “untreated” to our knowledge cannot be stated by only looking at the total amount of C-5HT in the samples. Other parameters such as the free content of kahweol and cafestol must be taken into account.

Another important processing method for coffee is the decaffeination process. The effects of different decaffeination methods were analyzed and the results recorded in Figure 20.42. Whereas one CO_2 method – A – leads to a reduction of about 20%, another CO_2 method – B – decreases the total C-5HT content by 60%. The greatest effect on the total amount of C-5HT was noticed

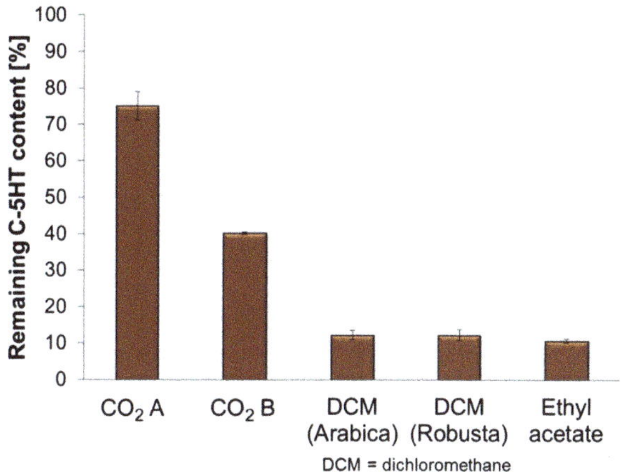

Figure 20.42 Effect of different decaffeinating processes on the amounts of C-5HT.

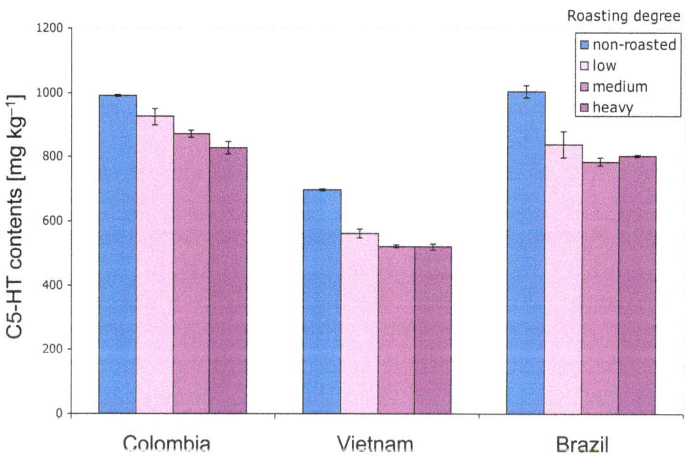

Figure 20.43 Effect of roasting on C-5HT.

when performing decaffeination with dichloromethane and ethyl acetate. A decrease of up to 85% was observed.

Furthermore, C-5HT is partially decomposed by roasting.[125,138] Figure 20.43 shows the results of two arabica coffees and one robusta coffee roasted at the roasting levels low, medium, and dark. The reduction of the C-5HT content is less distinctive than when steaming the coffee beans. Depending on the roasting degree, there is a decomposition of 7% at a low roasting level, 12% at medium roasting conditions, and approximately a 17% reduction in the dark roasted Columbia. The roasting process had the greatest influence on the group of C-5HT with hydroxy fatty acids as it is comparable to the steaming of the beans. Even at mild roasting conditions, the amount of this group is reduced by about 20% and, in contrast to the other groups, remains at this level when the roasting degree is increased.

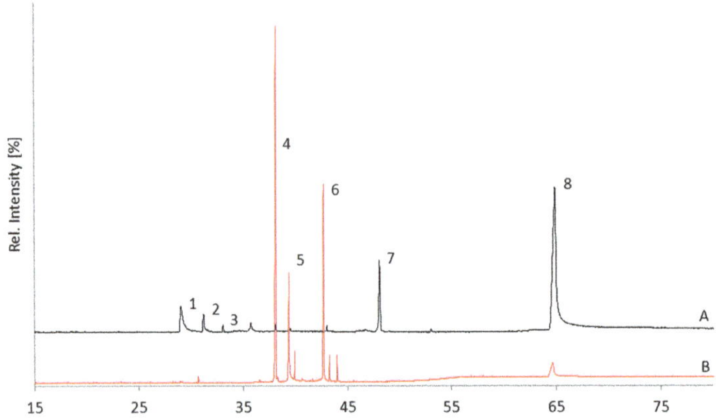

Figure 20.44 Degradation products of C_{16}-5HT (8) by roasting at 200 °C, 30 min (A) and 90 min (B); 5-hydroxyindole (1), 3-methyl-5-hydroxyindole (2), 3-ethyl-5-hydroxyindole (3), octadecanamid (4), octadecanenitril (5), and octadecanoic acid (6), serotonin (7).

Viani and Horman[139] proposed pathways for the thermal decomposition of carboxylic acid-5HT, however, only some of the supposed components were confirmed by Zahm *et al.*[140] (Figure 20.44).

Comparing them to commercial standard compounds and mass spectra from literature, the main non-volatile degradation products were identified as serotonin, 5-hydroxyindole (5HI), 3-methyl-5-hydroxyindole (5-hydroxyskatol, 5HS), and 3-ethyl-5-hydroxyindole.[141] Due to the increased thermal load, longer roasting times led to a further pyrolysis of the compounds mentioned and at the same time to an increase in other compounds identified as octadecanamide, octadecanenitrile, and octadecanoic acid. Roasting experiments with C-5HT consisting of unsaturated and long chain fatty acids (C_{21} and $C_{20:1}$) showed corresponding degradation compounds.[141]

20.7.1 Pyrolysis/GC-MS Experiments

Synthesized C-5HT was used for the pyrolysis study.[141] In Figure 20.45, the Py/GC-MS chromatogram of C_{18}-5HT shows homologue rows of the alkanes and alkenes (C_6–C_{17}) and the already identified degradation products 5-hydroxyindole (1), 3-methyl-5-hydroxyindole (2), 3-ethyl-5-hydroxyindole (3), and serotonin (4). Furthermore, octadecanonitrile (5), octadecanoic acid (6), and octadecanoamide (7) were detected. Likewise, the corresponding degradation products were analyzed by pyrolysis of $C_{20:1}$-5HT.

With regard to the results of the roasting trials, a most probable degradation pathway of the C-5HT is shown in Figure 20.46. The C-5HT will first degrade *via* α-cleavage to serotonin, 5-hydroxyindole, 3-methyl-5-hydroxyindole, 3-ethyl-5-hydroxyindole, and to the corresponding fatty acid, fatty acid nitrile, and fatty acid amide. An increased thermal load causes pyrolysis

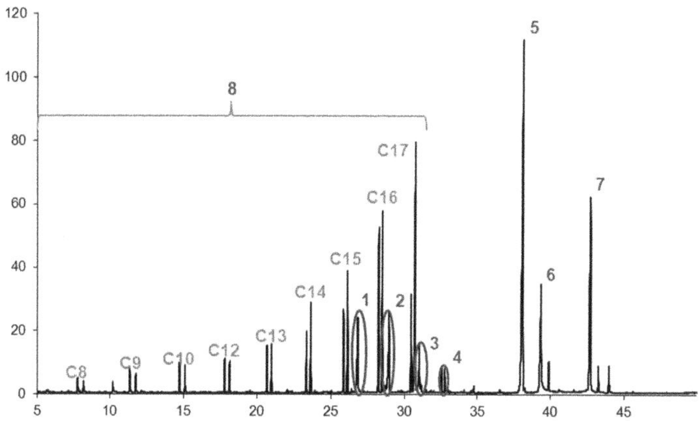

Figure 20.45 Py/GC-MS chromatogram of C_{18}-5HT hydroxyindole (1), 3-methyl-5-hydroxyindol (2), 3-ethyl-5-hydroxyindol (3), serotonin (4), octadecanonitrile (5), octadecanoic acid (6), octadecanoamide (7), homologue rows of alkanes and alkenes (8).

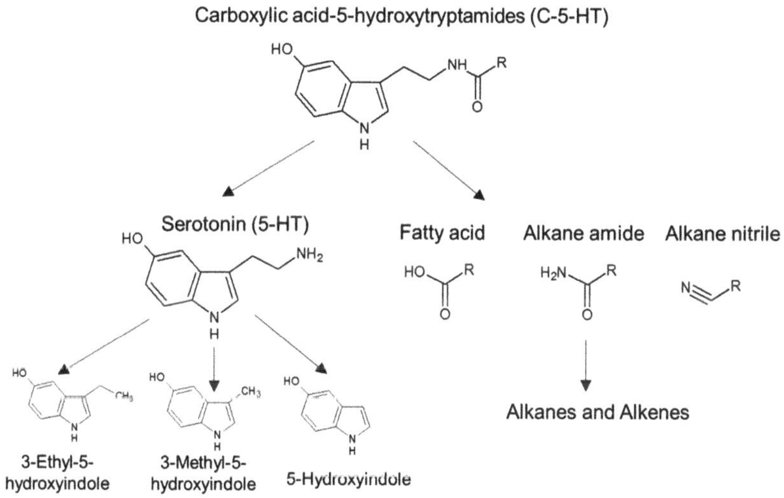

Figure 20.46 Pathway of C-5HT degradation.[141]

of the indol frame and the formation of alkanes and alkenes. However, no compounds with indene or with indole-frame could be detected as Viani and Hormann[139] had assumed.

Acknowledgements

Thanks are due to Fundação para a Ciência e a Tecnologia (FCT)/MCTES for financial support to the research units QOPNA (FCT UID/QUI/00062/2013) and CESAM (UID/AMB/50017 - POCI-01-0145-FEDER-007638) at University of Aveiro and CQ-VR at UTAD Vila Real (PEst-OE/QUI/UI0616/2014) through national funds (PIDDAC) and, where applicable, co-financed by the FEDER, within the PT2020 Partnership Agreement and Compete 2020. Thanks are also due to FCT for funding the Portuguese Mass Spectrometry Network, RNEM (LISBOA-52201-0145-FEDER-402-022125), and the individual grants to Joana Simões (FRH/BPD/90447/2012), Cláudia Passos (SFRH/BDP/107881/2015), and Ana Moreira (SFRH/BD/80553/2011).

References

1. O. G. Vitzthum, Chemie und Bearbeitung des Kaffees, In: *Kaffee und Coffein*, ed. O. Eichler, Springer Verlag, Berlin-Heidelberg-New York, 2nd edn., 1976, pp. 3–64.
2. T. A. L. Wagemaker, C. R. L. Carvalho, N. B. Maia, S. R. Baggio and O. Guerriero Filho, Sun protection factor, content and composition of lipid fraction of green coffee beans, *Ind. Crops Prod.*, 2011, **33**, 469–473.
3. G. Knothe and J. A. Kenar, Determination of fatty acid profile by 1H NMR spectroscopy, *Eur. J. Lipid Sci. Technol.*, 2004, **106**, 88–96.
4. H. Picard, B. Guyot and J.-C. Vincent, Étude des composés stéroliques de l'huile de café coffea canephora, *Café, Cacao, Thé*, 1984, **28**, 47–62.
5. P. Folstar, W. Pilnik, J. G. de Heus and H. C. van der Plas, The composition of fatty acids in coffee oil and wax, *Lebensm.-Wiss. Technol.*, 1975, **8**, 286–288.
6. K. Speer, 16-O-Methylcafestol – ein neues Diterpen im Kaffee – Methoden zur Bestimmung des 16-O-Methylcafestols in Rohkaffee und in behandelten Kaffees, *Z. Lebensm.-Unters. Forsch.*, 1989, **189**, 326–330.
7. DIN 10779, *Analysis of Coffee and Coffee Products – Determination of 16-O-Methylcafestol Content of Roasted Coffee – HPLC-Method*, Deutsches Institut für Normung e. V., Berlin, 1999.
8. A. B. A. de Azevedo, T. G. Kieckbush, A. K. Tashima, R. S. Mohamed, P. Mazzafera and S. A. B. Vieira de Melo, Extraction of green coffee oil using supercritical carbon dioxide, *J. Supercrit. Fluids*, 2008, **44**, 186–192.
9. G. González, F. Pablos, M. J. Martin, M. León-Camacho and M. S. Valdenebro, HPLC analysis of tocopherols and triglycerides in coffee and their use as authentication parameters, *Food Chem.*, 2001, **73**, 93–101.
10. K. Speer, N. Sehat and A. Montag, Fatty acids in coffee, *Proc. 15th ASIC Coll.*, ASIC, Paris, 1993, pp. 583–592.
11. K. Speer and I. Kölling-Speer, The lipid fraction of the coffee bean, *Braz. J. Plant Physiol.*, 2006, **18**, 201–216.
12. S. Casal, M. B. Oliveira and M. A. Ferreira, Discrimination of *Coffea arabica* and *Coffea canephora* var. *robusta* beans by their fatty acid composition, In: *Proceedings of Euro Food Chem IX, Interlaken, Switzerland*, ed. R. Amadò and R. Battaglia, 1997, vol. 3, p. 685.

13. H. P. Kaufmann and R. S. Hamsagar, Zur Kenntnis der Lipoide der Kaffeebohne. II. Die Veränderung der Lipoide bei der Kaffee-Röstung, *Fette, Seifen, Anstrichm.*, 1962, **64**, 734–738.
14. C. Calzolari and E. Cerma, Sulle sostanze grasse del caffè, *Riv. Ital. Sostanze Grasse*, 1963, **40**, 176–180.
15. A. Carisano and L. Gariboldi, Gas chromatographic examination of the fatty acids of coffee oil, *J. Sci. Food Agric.*, 1964, **15**, 619–622.
16. P. Wajda and D. Walczyk, Relationship between acid value of extracted fatty matter and age of green coffee beans, *J. Sci. Food Agric.*, 1978, **29**, 377–380.
17. M. J. Martin, F. Pablos, A. G. González, M. S. Valdenebro and M. León-Camacho, Fatty acid profiles as discriminant parameters for coffee varieties differentiation, *Talanta*, 2001, **54**, 291–297.
18. M. Rui Alves, S. Casal, M. B. P. P. Oliveira and M. A. Ferreira, Contribution of FA profile obtained by high-resolution GC/chemometric techniques to the authenticity of green and roasted coffee varieties, *J. Am. Oil Chem. Soc.*, 2003, **80**, 511–517.
19. R. O. Bengis and R. J. Anderson, The chemistry of the coffee bean. I. Concerning the unsaponifiable matter of the coffee bean oil. Extraction and properties of kahweol, *J. Biol. Chem.*, 1932, **97**, 99–113.
20. P. N. Chakravorty, M. M. Wesner and R. H. Levin, Cafesterol II, *J. Am. Chem. Soc.*, 1943, **65**, 929–932.
21. P. N. Chakravorty, R. H. Levin, M. M. Wesner and G. Reed, Cafesterol III, *J. Am. Chem. Soc.*, 1943, **65**, 1325–1328.
22. A. Wettstein, M. Spillmann and K. Miescher, Zur Konstitution des Cafesterols 6, *Mitt. Helv. Chim. Acta*, 1945, **28**, 1004–1013.
23. R. D. Haworth and R. A. W. Johnstone, Cafestol. Part II, *J. Chem. Soc.*, 1957, 1492–1496.
24. R. A. Finnegan and C. Djerassi, Terpenoids. XLV. Further studies on the structure and absolute configuration of cafestol, *J. Am. Chem. Soc.*, 1960, **82**, 4342–4344.
25. K. Speer and P. Mischnick, 16-O-Methylcafestol – ein neues Diterpen im Kaffee – Entdeckung und Identifizierung, *Z. Lebensm.-Unters. Forsch.*, 1989, **189**, 219–222.
26. K. Speer and P. Mischnick-Lübbecke, 16-O-Methylcafestol – ein neues Diterpen im Kaffee, *Lebensmittelchemie*, 1989, **43**, 43.
27. K. Speer, R. Tewis and A. Montag, A new roasting component in coffee, *Proc. 14th ASIC Coll.*, ASIC, Paris, 1991, pp. 615–621.
28. N. Frega, F. Bocci and G. Lercker, High resolution gas chromatographic method for determination of Robusta coffee in commercial blends, *J. High Resolut. Chromatogr.*, 1994, **17**, 303–307.
29. D. R. White, Coffee adulteration and a multivariate approach to quality control, *Proc. 16th ASIC Coll.*, ASIC, Paris, 1995, pp. 259–266.
30. M.-D. Trouche, M. Derbesy and J. Estienne, Identification of Robusta and Arabica species on the basis of 16-O-methylcafestol, *Ann. Falsif. Expert. Chim.*, 1997, **90**, 121–132.

31. W. Kamm, F. Dionisi, L.-B. Fay, C. Hischenhuber, H.-G. Schmarr and K.-H. Engel, Rapid and simultaneous analysis of 16-O-methylcafestol and sterols as markers for assessment of green coffee bean authenticity by on-line LC-GC, *J. Am. Oil Chem. Soc.*, 2002, **79**, 1109–1113.
32. N. D'Amelio, E. De Angelis, L. Navarini, E. Schievano and S. Mammi, Green coffee oil analysis by high-resolution nuclear magnetic resonance spectroscopy, *Talanta*, 2013, **110**, 118–127.
33. Y. B. Monakhova, W. Ruge, T. Kuballa, M. Ilse, O. Winkelmann, B. Diehl, F. Thomas and D. W. Lachenmeier, Rapid approach to identify the presence of Arabica and Robusta species in coffee using 1H NMR spectroscopy, *Food Chem.*, 2015, **182**, 178–184.
34. I. Kölling-Speer and K. Speer, Diterpenes in coffee leaves, *Proc. 17th ASIC Coll.*, ASIC, Paris, 1997, pp. 150–154.
35. B. de Roos, G. van der Weg, R. Urgert, P. van de Bovenkamp, A. Charrier and M. B. Katan, Levels of cafestol, kahweol, and related diterpenoids in wild species of the coffee plant Coffea, *J. Agric. Food Chem.*, 1997, **45**, 3065–3069.
36. B. Nackunstz and H. G. Maier, Diterpenoide im Kaffee. III. Cafestol und Kahweol, *Z. Lebensm.-Unters. Forsch.*, 1987, **184**, 494–499.
37. I. Kölling-Speer, T. Kurzrock, and K. Speer, Is the GC analysis of cafestol and kahweol a problem? *Proc. 21st ASIC Coll.*, ASIC, Paris, 2006, pp. 194–196.
38. K. Speer, T. Kurzrock, and I. Kölling-Speer, 16-O-Methylcafestol in the quality control of instant coffees, *Proc. 20th ASIC Coll.*, ASIC, Paris, 2004, pp. 302–304.
39. S. Buchmann, G. Küchenmeister, I. Kölling-Speer, and K. Speer, Fast method for the quantification of 16-O-methylcafestol in roasted coffee, *Proc. 23rd ASIC Coll.*, ASIC, Paris, 2010, pp. 235–238.
40. S. Buchmann, I. Kölling-Speer and K. Speer, TLC screening for the detection of Robusta admixtures to Arabica coffee, *CBS*, 2012, **109**, 2–4.
41. G. Downey, J. Boussion and D. Beauchêne, Authentication of whole and ground coffee beans by near infrared reflectance spectroscopy, *J. Near Infrared Spectrosc.*, 1994, **2**, 85–92.
42. I. Esteban-Diez, J. M. Gonzalez-Saiz, C. Saenz-Gonzalez and C. Pizarro, Coffee varietal differentiation based on near infrared spectroscopy, *Talanta*, 2007, **71**, 221–229.
43. C. Pizarro, I. Esteban-Diez and J. M. Gonzalez-Saiz, Mixture resolution according to the percentage of robusta variety in order to detect adulteration in roasted coffee by near infrared spectroscopy, *Anal. Chim. Acta*, 2007, **585**, 266–276.
44. A. Keidel, D. von Stetten, C. Rodrigues, C. Máguas and P. Hildebrandt, Discrimination of green Arabica and Robusta coffee beans by Raman spectroscopy, *J. Agric. Food Chem.*, 2010, **58**, 11187–11192.
45. T. Wermelinger, L. D'Ambrosio, B. Klopprogge and C. Yeretzian, Quantification of the Robusta fraction in a coffee blend via Raman spectroscopy: proof of principle, *J. Agric. Food Chem.*, 2011, **59**, 9074–9079.

46. K. Speer, R. Tewis and A. Montag, 16-O-Methylcafestol – ein neues Diterpen im Kaffee – Freies und gebundenes 16-O-Methylcafestol, *Z. Lebensm.-Unters. Forsch.*, 1991, **192**, 451–454.
47. I. Kölling-Speer, S. Strohschneider and K. Speer, Determination of free diterpenes in green and roasted coffees, *J. High Resolut. Chromatogr.*, 1999, **22**, 43–46.
48. T. Kurzrock, I. Kölling-Speer, and K. Speer, Effects of controlled storage on the lipid fraction of green Arabica coffee beans, *Proc. 20th ASIC Coll.*, ASIC, Paris, 2004, pp. 150–154.
49. F. Beisson, A. Tiss, C. Rivière and R. Verger, Methods for lipase detection and assay: a critical review, *Eur. J. Lipid Sci. Technol.*, 2000, **102**, 133–153.
50. K. Speer and A. Kurt, Effects of steam treatment on diterpenes. *Proc. 19th ASIC Coll.*, ASIC, Paris, 2001.
51. A. Kurt and K. Speer, Freie Diterpene in gedämpften Roh- und Röstkaffees – Optimierung der Bestimmungsmethode und erste Ergebnisse, *Dtsch. Lebensm.-Rundsch.*, 2001, **97**, 373–376.
52. T. Kurzrock and K. Speer, Identification of kahweol fatty acid esters in an Arabica coffee by LC/MS, *Proc. 19th ASIC Coll.*, ASIC, Paris, 2001.
53. H. P. Kaufmann and R. S. Hamsagar, Zur Kenntnis der Lipoide der Kaffeebohne. I. Über Fettsäure-Ester des Cafestols, *Fette, Seifen, Anstrichm.*, 1962, **64**, 206–213.
54. P. Folstar, Lipids, In: *Coffee Chemistry*, ed. R. J. Clarke, and R. Macrae, Elsevier Applied Science Publishers, London, vol. 1, 1985, pp. 203–222.
55. B. C. Pettitt Jr, Identification of the diterpene esters in Arabica and Canephora coffees, *J. Agric. Food Chem.*, 1987, **35**, 549–551.
56. K. Speer, 16-O-methylcafestol – a new diterpene in coffee; the fatty acid esters of 16-O-methylcafestol, In: *Proc. Euro Food Chem. VI*, ed. W. Baltes, T. Eklund, R. Fenwick, W. Pfannhauser, A. Ruiter, and H.-P. Thier, Behr's Verlag, Hamburg, vol. 1, 1991, pp. 338–342.
57. K. Speer, Fatty acid esters of 16-O-methylcafestol, *Proc. 16th ASIC Coll.*, ASIC, Paris, 1995, pp. 224–231.
58. T. Kurzrock and K. Speer, Fatty acid esters of cafestol, *Proc. 17th ASIC Coll.*, ASIC, Paris, 1997, pp. 133–140.
59. T. Kurzrock and K. Speer, Identification of cafestol fatty acid esters, In: *Proc. Euro Food Chem. IX*, ed. R. Amadò and R. Battaglia, Interlaken, Switzerland, vol. 3, 1997, pp. 659–663.
60. T. Kurzrock and K. Speer, Diterpenes and diterpene esters in coffee, *Food Rev. Int.*, 2001, **17**, 433–450.
61. T. Kurzrock and K. Speer, Identification of kahweol fatty acid esters in arabica coffee by means of LC/MS, *J. Sep. Sci.*, 2001, **24**, 843–848.
62. H. Obermann and G. Spiteller, Die Strukturen der "Kaffee-Atractyloside", *Chem. Ber.*, 1976, **109**, 3450–3461.
63. H. G. Maier and H. Wewetzer, Bestimmung von Diterpen-Glykosiden im Bohnenkaffee, *Z. Lebensm.-Unters. Forsch.*, 1978, **167**, 105–107.
64. R. Aeschbach, A. Kusy and H. G. Maier, Diterpenoide in Kaffee, *Z. Lebensm.-Unters. Forsch.*, 1982, **175**, 337–341.

65. A. G. W. Bradbury and H. H. Balzer, Carboxyatractyligenin and atractyligenin glycosides in coffee, *Proc. 18th ASIC Coll.*, ASIC, Paris, 1999, pp. 71–77.
66. I. Kölling-Speer, T. Kurzrock, M. Gruner, and K. Speer, Formation of a new substance in the lipid fraction during the storage of green coffee beans, In: *Proceedings of Euro Food Chem XIII*, ed. T. Eklund, M. Schwarz, H. Steinhart, H.-P. Thier, and P. Winterhalter, Hamburg, Germany, vol. 2, 2005, 491–494.
67. A. Kurt and K. Speer, A new component in the lipid fraction of coffee. In: *Proceedings of Euro Food Chem X,* 22–24 September, Budapest, Hungary, vol. 3, 1999, pp. 882–886. ISBN 963420 615 8 III.
68. R. Tewis, A. Montag, and K. Speer, Dehydrocafestol and dehydrokahweol – two new roasting components in coffee, *Proc. 15th ASIC Coll.*, ASIC, Paris, 1993, pp. 880–883.
69. I. Kölling-Speer, A. Kurt, N. Thu, and K. Speer, Cafestol and dehydrocafestol in roasted coffee, *Proc. 17th ASIC Coll.*, ASIC, Paris, 1997, pp. 201–204.
70. A. Hruschka and K. Speer, Cafestal in coffee, In: *Proceedings of Euro Food Chem IX*, ed. R. Amadò and R. Battaglia, Interlaken, Switzerland, vol. 3, 1997, pp. 655–658.
71. I. Kölling-Speer, M. Nickol, and K. Speer, Two new diterpenes in roasted coffee, *Proc. 20th ASIC Coll.*, ASIC, Paris, 2004, pp. 271–275.
72. I. Kölling-Speer, S. Schumann, M. Gruner, and K. Speer, Secokahweol – a new diterpene degradation product in roasted coffee, *Proc. 22nd ASIC Coll.*, ASIC, Paris, 2008, pp. 496–499.
73. I. Kölling-Speer, 16-O-methylcafestol determination in coffee using different HPLC columns and eluents, *Proc. 21st ASIC Coll.*, ASIC, Paris, 2006, pp. 205–208.
74. I. Kölling-Speer, Diterpene determination in coffee using fast LC-systems of different manufacturers, *Proc. 22nd ASIC Coll.*, ASIC, Paris, 2008, pp. 488–491.
75. T. Kurzrock, I. Kölling-Speer, and K. Speer, Identification of dehydrocafestol fatty acid esters in coffee, In: *Proc. 20th Int. Symposium on Capillary Chromatogr.*, ed. P. Sandra and A. J. Rackstraw, No 27. Naxos Software Solutions, M. Schaefer, Schriesheim, Germany, 1998.
76. A. A. Bak and D. E. Grobbee, The effect on serum cholesterol levels of coffee brewed by filtering or boiling, *N. Engl. J. Med.*, 1989, **321**, 1432–1437.
77. M. P. Weusten Van der Wouw, M. B. Katan, R. Viani, A. C. Huggett, R. Liardon, P. G. Lund-Larsen, D. S. Thelle, I. Ahola, A. Aro, S. Meyboom and A. C. Beynen, Identity of the cholesterol-raising factor from boiled coffee and its effect on liver function enzymes, *J. Lipid Res.*, 1994, **35**, 721–733.
78. R. P. Mensink, W. J. Lebbink, I. E. Lobbezoo, M. P. Weusten Van der Wouv, P. L. Zock and M. B. Katan, Diterpene composition of oils from Arabica and Robusta coffee beans and their effects on serum lipids in man, *J. Intern. Med.*, 1995, **237**, 543–550.

79. B. de Roos and M. B. Katan, Possible mechanisms underlying the cholesterol-raising effect of the coffee diterpene cafestol, *Curr. Opin. Lipidol.*, 1999, **10**, 41–45.
80. S. T. van Cruchten, L. H. de Haan, P. P. Mulder, C. Kunne, M. V. Boekschoten, M. B. Katan, J. M. Aarts and R. F. Witkamp, The role of epoxidation and electrophile-responsive element-regulated gene transcription in the potentially beneficial and harmful effects of the coffee components cafestol and kahweol, *J. Nutr. Biochem.*, 2010, **21**, 757–763.
81. L. K. Lam, V. L. Sparnis and L. W. Wattenberg, Isolation and identification of kahweol palmitate and cafestol palmitate as active constituents of green coffee beans that enhance glutathione S-transferase activity in the mouse, *Cancer Res.*, 1982, **42**, 1193–1198.
82. E. G. Miller, A. P. Gonzales-Sanders, A. M. Couvillion, W. H. Binnie, G. I. Sunahara, and R. Bertholet, Inhibition of oral carcinogenesis by roasted coffee beans and roasted coffee bean fractions, *Proc. 15th ASIC Coll.*, ASIC, Paris, 1993, pp. 420–425.
83. C. Cavin, D. Holzhauser, A. Constable, A. C. Huggett and B. Schilter, The coffee-specific diterpenes cafestol and kahweol protect against aflatoxin B1-induced genotoxicity through a dual mechanism, *Carcinogenesis*, 1998, **19**, 1369–1375.
84. F. Ferk, W. W. Huber, B. Grasl-Kraupp, K. Speer, S. Buchmann, R. Bohacek, M. Mišík, L. Edelbauer and S. Knasmüller, Protective effects of coffee against induction of DNA damage and pre-neoplastic foci by aflatoxin B_1, *Mol. Nutr. Food Res.*, 2014, **58**, 229–238.
85. W. M. Ratnayake, R. Hollywood, E. O'Grady and B. Starvric, Lipid content and composition of coffee brews prepared by different methods, *Food Chem. Toxicol.*, 1993, **31**, 263–269.
86. N. Sehat, A. Montag, and K. Speer, Lipids in the coffee brew, *Proc. 15th ASIC Coll.*, ASIC, Paris, 1993, pp. 869–872.
87. R. Urgert, G. van der Weg, T. G. Kosmeijer-Schuil, P. van der Bovenkamp, R. Hovenier and M. B. Katan, Levels of the cholesterol-elevating diterpenes cafestol and kahweol in various coffee brews, *J. Agric. Food Chem.*, 1995, **43**, 2167–2172.
88. G. Gross, E. Jaccaud and A. C. Huggett, Analysis of the content of the diterpenes cafestol and kahweol in coffee brews, *Food Chem. Toxicol.*, 1997, **35**, 547–554.
89. N. Naidoo, C. Chen, S. A. Rebello, K. Speer, E. S. Tai, J. Lee, S. Buchmann, I. Koelling-Speer and R. M. van Dam, Cholesterol-raising diterpenes in types of coffee commonly consumed in Singapore, Indonesia and India and associations with blood lipids: a survey and cross sectional study, *Nutr. J.*, 2011, **10**, 48.
90. S. Buchmann, A. Zahm, I. Kölling-Speer, and K. Speer, Lipids in coffee brews—impact of grind size, water temperature, and coffee/water ratio on cafestol and the carboxylic acid-5-hydroxytryptamides, *Proc. 23rd ASIC Coll.*, ASIC, Paris, 2010, pp. 101–109.

91. Consumentenbond, Automatenkoffie verhoogt cholesterol, *Gezondgids*, November 2007, 8–11.
92. W. H. Ling and P. J. Jones, Dietary phytosterols: a review of metabolism, benefits and side effects, *Life Sci.*, 1995, **57**, 195–206.
93. R. A. Moreau, B. D. Whitaker and K. B. Hicks, Phytosterols, phytostanols, and their conjugates in foods: structural diversity, quantitative analysis, and health-promoting uses, *Prog. Lipid Res.*, 2002, **41**, 457–500.
94. A. Alcaide, M. Devys, M. Barbier, H. P. Kaufmann and A. K. Sen Gupta, Triterpene and sterols of coffee oil, *Phytochemistry*, 1971, **10**, 209–210.
95. T. Itoh, T. Tamura and T. Matsumoto, Sterol composition of 19 vegetable oils, *J. Am. Oil Chem. Soc.*, 1973, **50**, 122–125.
96. T. Itoh, T. Tamura and T. Matsumoto, Methylsterol compositions of 19 vegetable oils, *J. Am. Oil Chem. Soc.*, 1973, **50**, 300–303.
97. P. Horstmann and A. Montag, Neue Methoden zur schnellen Isolierung von Sterinen aus Fettmatrices, *Fette, Seifen, Anstrichm.*, 1986, **88**, 262–264.
98. E. Homberg and B. Bielefeld, Einfluß von Minorbestandteilen des Unverseifbaren auf die Sterinanalyse, *Fat Sci. Technol.*, 1989, **91**, 105–108.
99. B. A. Nagasampagi, J. W. Rowe, R. Simpson and L. J. Goad, Sterols of coffee, *Phytochemistry*, 1971, **10**, 1101–1107.
100. A. Duplatre, C. Tisse and J. Estienne, Contribution á l'identification des espèces arabica et robusta par étude de la fraction stérolique, *Ann. Falsif. Expert. Chim.*, 1984, **828**, 259–270.
101. M. Saltor, A. Duplatre and J. Boatella, Identification of coffee species by their sterolic fraction, *An. Bromatol.*, 1989, **41**, 1–8.
102. M. S. Valdenebro, M. León-Camacho, F. Pablos, A. G. González and M. J. Martín, Determination of the Arabica/Robusta composition of roasted coffee according to their sterolic content, *Analyst*, 1999, **124**, 999–1002.
103. S. Dussert, A. Laffargue, A. de Kochko and T. Joët, Effectiveness of the fatty acid and sterol composition of seeds for the chemotaxonomy of *Coffea* subgenus *Coffea*, *Phytochemistry*, 2008, **69**, 2950–2960.
104. C. Mariani and E. Fedeli, Sterols of coffee grain of Arabica and Robusta species, *Riv. Ital. Sostanze Grasse*, 1991, **68**, 111–115.
105. K. Speer, C. Grasse, I. Kölling-Speer, *Sterine in Roh- und Röstkaffees sowie im Kaffeegetränk*. Diploma Thesis C. Grasse, 1996.
106. K. M. Phillips, D. M. Ruggio and M. Ashraf-Khorassani, Analysis of steryl glucosides in foods and dietary supplements by solid-phase extraction and gas chromatography, *J. Food Lipids*, 2005, **12**, 124–140.
107. S. Buchmann, H. Borch, I. Kölling-Speer, and K. Speer, Steryl glucosides in the coffee plant, *Proc. 24th ASIC Coll.*, ASIC, Paris, 2012, 256–260.
108. S. Oelschlägel, C. Menzel and K. Speer, Phytosterols and steryl esters in diverse Cucurbita, Cucumis and Citrullus seed oils, *Lipid Technol.*, 2012, **24**, 181–184.
109. P. Folstar, H. C. van der Plas, W. Pilnik and J. G. de Heus, Tocopherols in the unsaponifiable matter of coffee bean oil, *J. Agric. Food Chem.*, 1977, **25**, 283–285.

110. E. Cros, G. Fourny and J. C. Vincent, Tocopherols in coffee. High-pressure liquid chromatographic determination, *Proc. 11th ASIC Coll.*, ASIC, Paris, 1985, pp. 263–271.
111. M. Aoyama, T. Maruyama, H. Kanematsu, I. Niiya, M. Tsukamoto, S. Tokairin and T. Matsumoto, Studies on the improvement of antioxidant effect of tocopherols. XVII. Synergistic effect of extracted components from coffee beans, *Yukagaku*, 1988, **37**, 606–612.
112. M. Ogawa, C. Kamiya and Y. Iida, Contents of tocopherols in coffee beans, coffee infusions and instant coffee, *Nippon Shokuhin Kogyo Gakkaishi*, 1989, **36**, 490–494.
113. K. Speer and I. Kölling-Speer, Lipids, In: *Coffee Recent Developments*, ed. R. J. Clarke and O. G. Vitzthum, Blackwell Science, 2001, pp. 33–49.
114. A. G. González, F. Pablos, M. J. Martın, M. León-Camacho and M. S. Valdenebro, HPLC analysis of tocopherols and triglycerides in coffee and their use as authentication parameters, *Food Chem.*, 2001, **73**, 93–101.
115. G. Dickhaut, *Über phenolische Substanzen in Kaffee und deren analytische Auswertbarkeit zur Kaffeewachsbestimmung*, PhD Thesis, University of Hamburg, 1966.
116. U. Harms and J. Wurziger, Carboxylic acid 5-hydroxytryptamides in coffee beans, *Z. Lebensm.-Unters. Forsch.*, 1968, **138**, 75–80.
117. P. Folstar, H. C. van der Plas, W. Pilnik, H. A. Schols and P. Melger, Liquid chromatographic analysis of Nb-alkanoyl-hydroxy-tryptamide (C-5HT) in green coffee beans, *J. Agric. Food Chem.*, 1979, **27**, 12–15.
118. P. Folstar, H. A. Schols, H. C. van der Plas, W. Pilnik, C. A. Landheer and A. van Veldhuizen, New tryptamine derivatives isolated from wax of green coffee beans, *J. Agric. Food Chem.*, 1980, **28**, 872–874.
119. T. Kurzrock, I. Koelling-Speer and K. Speer, Chromatography of carbonic acid-5-hydroxytryptamides, *Proc. 20th ASIC Coll.*, ASIC, Paris, 2005, pp. 305–308.
120. R. Lang and T. Hofmann, A versatile method for the quantitative determination of N-alkanoyl-5-hydroxytryptamides in roasted coffee, *Eur. Food Res. Technol.*, 2005, **220**, 638–643.
121. C. Hinkel and K. Speer, New carboxylic acid-5-hydroxytryptamides in coffee wax, In: *Proceedings of Euro Food Chem. XIII*, ed. T. Eklund, M. Schwarz, H. Steinhart, H.-P. Thier and P. Winterhalter, Hamburg, Germany, 2005, vol. 2, pp. 487–490.
122. O. Culmsee, Methode zur quantitativen Bestimmung der Carbonsäure-5-hydroxy-tryptamide im Kaffee, *Dtsch. Lebensm.-Rundsch.*, 1975, **71**, 425–427.
123. P. Kummer and E. Bürgin, Neue Erkenntnisse zur quantitativen Bestimmung der Carbonsäure-5-hydroxytryptamide in Kaffee, *Mitt. Geb. Lebensmittelunters. Hyg.*, 1976, **67**, 212–215.
124. A. Studer and H. Traitler, Quantitative HPTLC determination of 5-hydroxytryptamides of carboxylic acids and tryptamines in food products, *J. High Resol. Chromatogr. Chromatogr. Commun.*, 1982, **5**, 581–582.

125. H. R. Hunziker and A. Miserez, Bestimmung der 5-Hydroxytryptamide in Kaffee mittels Hochdruck-Flüssigkeitschromatographie, *Mitt. Geb. Lebensmittelunters. Hyg.*, 1979, **70**, 142–152.
126. M. Kele and R. Ohmacht, Determination of serotonin released from coffee wax by liquid chromatography, *J. Chromatogr.*, 1996, **730**, 59–62.
127. A. Laganà, L. Curini, S. de Angelis Curtis and A. Marino, Rapid liquid chromatographic analysis of carboxylic acid 5-hydroxytryptamides in coffee, *Chromatographia*, 1989, **28**, 593–596.
128. C. Hinkel and K. Speer, *Verfahren zur Herstellung von Fettsäureamiden mit gesättigten, ungesättigten oder Hydroxy-Fettsäuren*, *DE Pat.*, DE102008025893 A1, 2008.
129. J. Wurziger, Carbonsäuretryptamide oder ätherlösliche Extraktstoffe um Nachweis und zur Beurteilung von bearbeiteten bekömmlichen Röstkaffees, *Kaffee- und Tee-Markt*, 1972, **22**, 3–11.
130. G. H. D. van der Stegen, The effect of dewaxing of green coffee on the coffee brew, *Food Chem.*, 1979, **4**, 23–29.
131. R. Corinaldesi, R. de Giorgio, V. Stanghellini, G. F. Paparo, A. Paternico, P. Sataguida, C. Ghidini, M. R. Maccarini and L. Barbara, Effect of the removal of coffee waxes on gastric acid secretion and serum gastrin levels in healthy volunteers, *Curr. Ther. Res.*, 1989, **46**, 13–18.
132. P. Lendrich, E. Wemmering and O. Lendrich, *Verfahren zum Verbessern von Kaffee*, *DR Pat.*, DE576515 C, 1933.
133. W. Roselius, O. Vitzthum and P. Hubert, *Process for the Removal of Undesirable Irritants from Raw Coffee Beans*, *US Pat.* US3770456 A, 1971.
134. P. Werkhoff, *Verfahren zum Entkoffeinieren von Rohkaffee*, *DE Pat.*, DE 2853169, 1980.
135. H. Seidlitz and E. Lack, *Verfahren zum Behandeln von Kaffee*, *EP* 0247999 A2, 1987.
136. R. Fehlau and K. J. Netter, The influence of untreated and treated coffee and carboxylic acid hydroxytryptamides on the gastric mucosa of the rat, *Z. Gastroenterol.*, 1990, **28**, 234–238.
137. C. Hinkel and K. Speer, A contribution of the occurrence and contents of carboxylic acid 5 hydroxytryptamides (C-5-HT) in green and processed coffee beans, *Proc. 21st ASIC Coll.*, ASIC, Paris, 2006, pp. 133–142.
138. G. H. D. van der Stegen and P. J. Noomen, Mass balance of carboxy-5-hydroxytryptamides (C-5-HT) in regular and treated coffee, *Lebensm.-Wiss. Technol.*, 1977, **10**, 321–323.
139. R. Viani and I. Horman, Thermal behavior of trigonelline, *J. Food Sci.*, 1974, **39**, 1216–1217.
140. A. Zahm, C. Hinkel and K. Speer, Carboxylic acid-5-hydroxytryptamides (C-5-HT) in coffee brews, *Proc. 22nd ASIC Coll.*, ASIC, Paris, 2008, pp. 479–483.
141. A. Zahm, K. Speer, Identification of C5-HT degradation products using pyrolysis-GC/MS, *Proc. 23rd ASIC Coll.*, ASIC, Paris, 2010, pp. 244–247.
142. K. Speer and I. Kölling-Speer, Chemistry 1: Non-volatile compounds, in *Coffee: Recent Developments*, ed. R. J. Clarke, O. G. Vitzthum, Blackwell Science Ltd, 2008, Ch 1, pp. 33–46.

CHAPTER 21

Minerals

CARMEN MARINO DONANGELO

Escuela de Nutrición, Universidad de la República, Uruguay
*E-mail: cmarino@nutricion.edu.uy

21.1 Introduction

Mineral elements are natural constituents of coffee beans as a result of biological incorporation during fruit formation in the developing coffee plant, mainly as essential nutrients (*e.g.* P, K, Mg, S, Ca, Fe, Zn, Cu, B) but also as non-essential and toxic contaminants (*e.g.* Pb, Cd, Al, Sb). The mineral composition of green coffee beans reflects the botanical source (species, cultivar) and the environmental growing conditions (geographical location, type and chemical composition of soil, agronomic practice such as crop rotation, and fertilization strategy). Moreover, harvesting techniques and industrial processing of coffee may affect the mineral composition of green coffee beans. Finally, the different forms of preparation of the coffee infusion affect the final mineral content of the beverage. In fact, determination of the mineral composition of green coffee and coffee products with adequate elemental analytical methods and data mining techniques have been used for authentication and discrimination of the geographical origin of coffee samples, and for detection and discrimination of different species and contaminants in coffee blends. Differences in the mineral composition of the beverages may affect their contribution to the total mineral intake from the diet, and may also impact taste and flavor thus affecting consumer acceptance.

Coffee: Production, Quality and Chemistry
Edited by Adriana Farah

Published by the Royal Society of Chemistry, www.rsc.org

In this chapter, literature studies on mineral composition of coffee and coffee products are reviewed with special attention to analytical aspects, factors affecting composition during processing of the coffee beans and preparation of the beverage, and the potential contribution of coffee to the dietary mineral intake of the consumer.

21.2 Methods of Analysis

The total mineral content of green coffee and coffee products can be roughly estimated by measuring ash, the solid residue obtained after dry incineration of the sample at 550–580 °C for several hours until constant weight. The ash content of green coffee is in the range of 3.0–4.5% and increases slightly in roasted coffee (4–5%) due to the reduced moisture content during roasting. The ash content of instant coffee is in a higher range (9–10%) due to the high water solubility of mineral salts during the production of instant coffee. The ash content is a simple and useful parameter for standardization of coffee processing in the industry and for detection of unwanted coffee substitutes (adulterants) added to instant coffee.[1]

The analysis of individual mineral components in coffee and coffee products requires specific methods and complex techniques. Although colorimetric methods have been used for the analysis of certain elements in coffee such as P,[2] elemental analysis is usually done using instrumental methods based on atomic spectroscopy techniques.[3,4] These techniques include atomic absorption spectrometry (AAS, flame, or graphite furnace), inductively coupled plasma atomic emission spectroscopy (ICP-AES), also referred to as inductively coupled plasma optical emission spectrometry (ICP-OES), and inductively coupled plasma mass spectrometry (ICP-MS). Instrumental neutron activation analysis (INNA) has also been used for elemental analysis of coffee although this instrumentation is less widely available. Some of these techniques (ICP-OES, ICP-MS, INNA) have very low detection limits, high sensitivity, and the capability for multi-element detection thus being of choice when aiming at elemental fingerprinting of coffee products in association with data mining techniques.[4] In all cases, adequate preparation of representative samples is required prior to analysis in order to ensure that all minerals are completely solubilized, that there are no mineral losses, and that matrix effects are reduced.[3] Sample preparation is usually done by dry ashing followed by dissolution with a strong acid or, more commonly, by wet digestion in open or pressurized closed vessels with the addition of strong acids and/or oxidant reagents. Spiking and recovery tests are often carried out to verify the accuracy of the results and to study potential matrix effects. Quality control of elemental analysis is usually done using one or more certified reference materials prepared from a food matrix different than coffee such as wheat flour, peach leaves, or apple leaves. A candidate for certified reference material from organic green coffee has been developed and shown to be homogeneous for Ca, Co, Cs, K, and Sc but not for other elements.[5]

Overall, more than 50 mineral elements have been determined in coffee, including Al, As, B, Ba, Be Br, Bi, Ca, Cd, Ce, Cl, Co, Cr, Cs, Cu, Dy, Eu, Er, Gd, Ge, Fe, Hg, In, K, La, Lu, Mg, Mn, Mo, Na, Ni, P, Pb, Rb, S, Sb, Sc, Se, Sm, Sn, Sr, Yb, W, U, Ta, Tb, Th, Tl, Zr, and Zn.[3,4] Based on the order of magnitude of concentration in the coffee bean and biological relevance for the coffee plant, mineral elements in coffee can be divided into major (*e.g.* Ca, K, Mg, Na, S, P) and minor (*e.g.* Cu, Fe, Zn, Mn) mineral nutrients, and in trace non-essential (*e.g.* Ba, La, and Sc) and toxic (*e.g.* Cd, Hg, and Pb) elements.

21.3 Minerals in Green and Roasted Coffee Beans

21.3.1 Green Coffee

In a pioneer study, applying several complementary analytical techniques (INNA, graphite furnace AAS, flame AAS, and combustion elemental analysis), Krivan *et al.*[6] quantified 22 elements in green coffee samples from eight different countries (Colombia, Costa Rica, Cuba, El Salvador, Mexico, Nicaragua, Panamá, and Papua New Guinea). Elements were ranked into five groups according to concentration: % level, or g 100 g^{-1} (C, H, K, *n*); ‰ level, or g kg^{-1} (Ca, Mg); ppm level, or μg g^{-1} (Ba, Cu, Fe, Mn, Na, Rb, Sr, Zn); ppb level, or μg kg^{-1} (Br, Co, Cr, Cs, La), and sub-ppb level, or ng kg^{-1} (Sc). An attempt was made to differentiate the origin of the green coffee samples by elemental composition, and Mn was found to be best suited for this purpose. Differences in Mn concentration of sets of coffee samples from different countries were highly significant in 84% of the cases. The main reason for this was thought to be differences in soil composition.

A recent study using high-resolution continuum source atomic absorption spectrometry (HR-CS-AAS) confirmed that coffee beans from different countries and different continents may differ in the content of specific minerals.[7] Coffees originated from Central America (Honduras, Cuba, Mexico, Guatemala) had the lowest content of Ca, Mg, K, P, and Fe compared to those originated from South America (Colombia, Brazil), Africa (Kenya, Mussulo, Ethiopia), Asia (China, India, Timor), and Oceania (Papua New Guinea). The highest Mn content was found in coffee from South America, and the highest Na and K contents were found in coffee from South America and Asia. Samples from South America had, on average, the highest concentrations of all the analyzed elements, except for Ca. The P content allowed discrimination between coffees from Africa, Central America, and Asia. Applying canonical discrimination analysis, Mn and Ca were found to be the best chemical descriptors for continental and country origin. The variation in mineral concentration of coffee grown in different regions of the world possibly reflects differences in type of soil (pH, content of organic and inorganic matter, drainage status), climate conditions, and agricultural practice such as pattern of crop rotation and use of fertilizers with different chemical composition.

The metal content of green coffee has also been shown to differentiate coffee species[8,9] and genotypes.[10] In a study in Brazil, of 11 metal elements analyzed by ICP-AES in different varieties of green *Coffea arabica* L. (or arabica coffee) and *Coffea canephora* Pierre species (popularly known as robusta),[8] P, Mn, and Cu concentrations were found to present the largest differences between *C. arabica* and *C. canephora* varieties. The contents of P and Cu were higher for *C. canephora* whereas the content of Mn was higher for *C. arabica.* In another study, measuring eight elements by ICP-AES in different genotypes of a *C. arabica* variety from Colombia,[10] P, N, and Mg concentrations, but not those of other minerals, differed significantly between genotypes. Location where samples were grown, rather than genotype, was the main determinant of differences in mineral composition of the green coffee samples studied.

The process of conversion from conventional to organic agriculture has been evaluated in terms of possible effects on metal content in Brazilian coffee.[4,11] The concentrations of K, Na, Ca, Mg, Cu, Fe, Mn, Cd, Zn, and Pb determined by AAS were measured in soil and in plant coffee tissues (leaf and fruit-green coffee) from two farms in Bahia (Brazil) moving from conventional to organic agriculture.[11] Coffee samples from both farms had relatively high levels of Cd, Zn, and Cu (although below the limits specified by the Brazilian Food Legislation) highly correlated to the available metal fraction in the soil. These results were interpreted to mean that the high amount of organic matter used in organic agriculture favor the bioavailability of soil metal ions for plant uptake.[11] More recently, the contents of 38 elements determined by ICP-MS were compared between 20 organic and 34 conventional green coffee samples.[4] In general, higher levels were found in the conventional samples. Considering essential elements, concentrations were quite similar between conventional and organic samples for Co, Cr, Cu, Mn, Sr, and Zn, but were higher for Mo and Se in the conventional samples. In relation to toxic and potentially toxic elements, conventional coffee was found higher in Al, U, Ce, La, Th, Yb, and Zr. On the other hand, Cs, Tl, and W were higher in the organic samples.

21.3.2 Ground Roasted Coffee

Most mineral elements are stable under the heating conditions (temperature/time) used during roasting of coffee beans. On the other hand, since the water content of coffee beans is reduced during roasting, mineral concentrations are usually higher in roasted compared to green coffee.[1] Differences in mineral composition of coffee beans due to species, environmental, and agronomic factors remain evident in roasted coffee samples.

Mineral composition, determined by flame AAS and flame AES, was analyzed in green (n = 30) and roasted (n = 5) Brazilian coffee samples.[12] Using principal component and hierarchical cluster analysis it was shown that Na, K, Ca, Cu, Mg, and Fe were the principal elements with contents discriminating between green and roasted samples.

The metal content of roasted coffee has been used to characterize *C. arabica* and *C. canephora* beans and to estimate the percentage contribution of these species (mixture resolution) in coffee blends.[13] In this study, the contents of Ba, Ca, Cu, Fe, K, Mg, Mn, Na, P, Sr, and Zn were determined by ICP-AES in roasted coffee samples of *C. arabica* (n = 9) and *C. canephora* (n = 9) species and in coffee blends (n = 12). Principal component and cluster analysis indicated that P, Mn, and Cu were the most discriminating variables between species. Using partial least squares regression, the relative content of each species in the coffee blends was estimated with a prediction error of about 7%.

The contents of Ca, Cu, Fe, K, Mg, Mn, Na, and Zn, determined by ICP-OES, were used to characterize the geographical origin of roasted coffee beans in Mexico.[14] All samples (n = 51) were 100% arabica. Using forward stepwise linear discriminant analysis it was possible to differentiate coffee origin, with Ca, K, Mn, Mg, Na, and Zn as the best chemical descriptors. Results were improved applying a multilayer perceptron artificial neural networks model that allowed differentiation of the geographical origin of Mexican roasted coffees with 93% prediction ability and 98% specificity.

21.3.3 Instant Coffee

In general, concentrations of mineral elements are higher in soluble (instant) coffees than in regular roasted and ground coffee powder, due to the industrial procedures used to obtain the soluble coffee solids that favor extraction of water soluble mineral salts.[1] Also, since instant coffee is usually obtained from blends with a high proportion of *C. canephora*, the mineral profile of instant coffee tends to reflect that of *C. canephora* varieties. As for roasted coffees, environmental and processing variables may affect the mineral composition of soluble coffees, and several studies aimed at discriminating instant coffees based on these factors.

The analysis of mineral nutrients and toxic elements (Na, K, Mg, Al, P, S, Ca, Mn, Fe, Ni, Cu, Zn, Cd, Sb, Pb, Cr, and Sn) in 21 samples of Brazilian soluble coffee by ICP-AES allowed for clustering by similarities in mineral composition using principal component and agglomerate hierarchical analysis.[15] Variations in composition were attributed to differences in the process of industrial production and/or in factors during cultivation of the coffee plant (type of soil and use of fertilizers).

In a study done in market coffee samples in Poland, with metal content determined by flame AAS,[16] it was shown by multivariate methods that 12 elements (Co, Mn, Fe, Cr, Ni, Zn, Cu, Ca, Mg, K, Na, and P) were able to differentiate chemometrically different types of coffee, distinguishing *C. arabica* from *C. canephora* species, and ground from instant coffee.

Commercial instant coffees and coffee substitutes (containing barley, malt, chicory, and/or rye and 0–66% ground roasted coffee) from Oporto (Portugal) (n = 49) were compared in terms of nine elements (Ca, Mg, K, Na, P, Fe, Mn, Cr, and Ni) analyzed by high-resolution continuum source

Table 21.1 Concentration range of selected major and minor minerals elements in green, ground roasted, and instant coffee.[a]

	Type of coffee		
Element	Green	Ground roasted	Instant
	mg g^{-1}		
K	12.1–21.4	11.4–29.1	18.7–61.5
P	1.41–2.20	0.19–4.03	0.35–4.29
Ca	0.79–1.87	0.49–2.20	0.11–2.65
Mg	0.14–2.30	0.75–3.10	0.84–5.54
Na	<0.01–0.12	<0.01–1.47	0.03–6.67
	μg g^{-1}		
Fe	24.8–108	12.0–617	6.3–451
Mn	13.4–57.7	6.6–320	3.6–495
Cu	5.5–76.9	0.4–30.1	0.30–12.9
Zn	2.9–61.3	1.2–803	1.8–15.0
B	0.3–19.5	0.6–17.5	13.3–21.3
Ba	1.6–10.2	0.8–10.1	3.4–3.9
Al	3.8–118	3.0–200	<0.5–233
Cr	<0.08–1.80	0.02–1.29	<0.01–52
Co	0.02–0.62	0.06–1.90	0.04–14.2
Cd	<0.01–0.75	<0.01 to <0.1	0.02–0.03
Pb	<0.01–0.29	0.02 to <2.6	0.09–0.91

[a]Source: ref. 2–6 and 9–16.

flame and graphite furnace AAS.[2] Plain instant coffees had higher mineral contents than coffee substitutes except for Na and Cr. Coffee substitutes containing 20–66% coffee had intermediate amounts of minerals, and lower quantities were found in substitutes containing no coffee, particularly those with barley.

Table 21.1 summarizes the information from published studies on the concentration range of selected major and minor mineral elements in green, roasted, and instant coffee powders.

21.4 Minerals in Coffee Beverages

Coffee beverages are prepared using various techniques that usually involve either brewing (mug, drip, Turkish, espresso) ground coffee or dissolving instant coffee in hot water, thus releasing mineral elements into the infusion. The extent of this release (leachability or extractability) varies with the type of coffee, the brewing method, and the specific mineral element. In general, coffee beverages have lower mineral concentrations than the respective powdered coffee from which they are prepared, but variability in mineral content of the beverage is typically affected by the same factors that influence the mineral composition of coffee beans. Most studies indicate that the method of preparation of the coffee beverage is a significant factor affecting the concentration of minerals in the infusion.

For beverages prepared from roasted and ground coffees, the extraction efficiencies are generally higher for K (71–88%), intermediate for Cr (51–63%), P (45–46%), Na (36–48%), Mn (19–52%), Co (26–50%), Ni (29–45%), Ca (10–65%), Mg (6–55%), Al (2–64%), and Zn (1–24%), and lowest for Fe (4–16%) and Cu (2–12%).[3,16,17] Differences in extractability between elements may be due to the nature and strength of the complexes their ions form with constituents of the coffee matrix, such as polyphenols, caffeine, and other compounds. For example, Fe, Cu, Zn, and Mn tend to form strong covalent bonds within the coffee matrix whereas K and Na are possibly attached by electrostatic bonds, thus more easily released into the coffee beverage.[3] The large variability in the extraction efficiencies reported by different studies may reflect differences in the quality of the coffee used as raw material (species/variety, origin, processing) and in the method (mug, drip, Turkish, espresso) and conditions (water temperature, brewing time, particle size) of preparation of the beverage.[3,17,19,20] It appears that highest amounts of minerals are released from the coffee powder to the beverage in the first 10 min of brewing[17,19] and that the satisfactory extraction efficiencies for Ca, Cu, Fe, Mg, Mn, and Zn are achieved using drip filter coffee machine, boiled methods, and pressure methods such as Italian mocha and an espresso machine.[17,18]

In the case of instant coffee beverages, it is generally assumed that all of the coffee powder is dissolved in water and therefore the mineral content of the beverage can be calculated from the mineral concentration in the instant coffee powder used for the preparation of the beverage.[2,3,21]

Drinking green coffee infusions is becoming popular worldwide due to potential health benefits and there is an increased scientific interest in describing the chemical composition and bioactivity of these beverages. In a recent study,[22] the content of Ca, Cu, Fe, Mg, and Mn was determined by high-resolution continuum source flame AAS in green coffee brews with the purpose to evaluate the effect of the preparation method (mug, Turkish, and dripped) and the relationship with antioxidant (radical scavenger) activity. Mug coffees were prepared by infusing for 10 min coffee powder (at 1%, powder/water) in hot water (90 °C), followed by filtration. Turkish coffees were prepared by heating the infusion of ground or whole bean (at 1%, powder/water) for about 10 min to boil, followed by filtration. The dripped coffees were prepared by percolating 90 °C water through green coffee powder (at 1%) followed by filtration. It was found that Ca and Mg had the highest leachabilities (75% and 70%, respectively) independent of the form of preparation of the brew. The extraction of all elements in the beverages was increased using the dripper or Turkish method as compared to the mug preparation. The antioxidant activity of the beverages correlated positively with their Ca concentration.

Table 21.2 summarizes the information from published studies on the concentration range of selected major and minor mineral elements in coffee beverages.

Table 21.2 Concentration range of selected major and minor mineral elements in coffee beverages.[a]

Element	Type of coffee beverage Roasted brewed	Instant	Turkish
	$mg\ g^{-1}$		
K	12–61.5	18–61.5	0.43–0.73
Ca	0.42–1.42	0.21–2.7	—
Mg	3.1–9.5	0.8–5.5	0.11–0.17
Na	0.02–2.9	0.03–3.5	0.19–0.34
	$\mu g\ g^{-1}$		
Fe	4.6–108	18.5–108	—
Mn	5.8–40	9.3–40	1.4–4.1
Cu	0.4–3.2	0.3–1.2	—
Zn	4.1–29	1.8–10.7	3.8–25
B	3.3–11.8	—	—
Al	1.2–9.6	—	—
Cr	<0.003–0.23	0.02–0.13	—
Co	0.15–0.80	0.2–0.8	—
Ni	<0.002–1.4	0.31–5.9	—
Cd	0.005–0.019	—	—
Pb	0.23–0.31	—	—

[a]Source: ref. 2, 3, 16–18, 20, and 26.

21.5 Contribution of Coffee to Dietary Mineral Intake

Coffee consumption is a source of essential and non-essential mineral elements from the diet. In fact, the contribution of coffee to dietary mineral intake was estimated in several studies.[2,7,16,23,24] Obviously, this contribution depends on the mineral concentrations in the coffee beverages habitually consumed, which are quite variable depending on several factors already discussed, and on the habitual amounts of coffee consumed, which vary among populations.

In an early study in New Zealand,[23] daily intakes of tea and coffee were compared as sources of Na, K, Ca, Cu, Mg, Mn, and Zn in the national diet. Coffee was found to be a better source of K than tea, and to contribute significantly to Mn intake. However, both beverages were considered in general of low nutritional significance as sources of mineral nutrients in the diet.

In a study in India,[21] five varieties of instant coffee powder were analyzed for Ca, Cr, Cd, Cu, Fe, K, Mg, Ni, Sr, Zn, and Pb, and based on habitual coffee consumption, the daily mineral intakes through ingestion of coffee were estimated. It was concluded that the contribution of coffee to the total dietary intake of metals in the population studied was ≤1% for most elements, except for Fe (1.7%) and for Cr and Ni which were around 3%. Similarly, in a study in Brazil[24] assessing the daily intake of trace elements (Al, Cd, Cr, Cu, Mn, Ni, Pb, U, and Zn) due to consumption of foodstuffs by adults of Rio de Janeiro, it was estimated that coffee contributed to 8% of the total intake of Cr, but for other minerals the contribution of coffee to total intake was much lower.

Estimates have been made of the potential contribution of coffee intake to cover the RDA according to European recommendations for specific mineral nutrients from ground,[16] instant,[2,16] and espresso[7] coffee. These estimates indicate that in general the contribution of coffee to recommended mineral nutrient intakes is low, in most cases lower than 5% although it may reach 10–15%. Moreover, estimates indicate that the consumption of coffee in usual amounts do not exceed the tolerance limits of ingestion of toxic metals, such as Pb and Cd.[16]

For infusions prepared from ground coffee in Poland,[16] the consumption of 300 ml (2 cups) was estimated to supply, on average, 4.5% of RDA for Mg, 3.5% for K, 2.8% for Mn, 2.4% for Cr, 1.9% for P, 0.32–0.43% for Ca and Na, 0.26–0.33% for Cu and Fe, 0.13% for Zn, and 2.6–15.6% for Ni. Higher average percentages were estimated from the ingestion of 300 ml infusions prepared from instant coffee: 12.3% for Mg, 8.9% for K, 8.6% for Mn, 4.9% for Cr, 7.4% for P, 1.6% for Ca, 2.5% for Na, 0.32% for Cu, 2.9% for Fe, 0.35% for Zn, and 4.9–29.7% for Ni.

In a study in Portugal,[2] two cups of instant coffee (4 g) were estimated to supply 9.5% of RDA for K, 5.2% for Mg, 4.4% for Mn, 3.5% for Ni, 2.2% for P, 1.5% for Fe, 0.5% for Cr, 0.4% for Ca, and 0.2% for Na. In a more recent study also in Portugal,[7] one cup of espresso coffee beverage (from 5–6 g ground espresso coffee) was estimated to provide 5.2–7.0% of RDA for K, 2.8–7.2% for P, 1.4–2.2% for Mg, 1.4–1.9% for Mn. 0.14–0.28% for Ca, 0.07–0.15% for Fe, and <0.02% for Na.

Since coffee is one of the most consumed food products in Brazil, the fortification of ground roasted coffee with iron, zinc, and calcium was evaluated[25] as a possible means for increasing micronutrient intake in the population. Different mineral salts and methods of preparation of the beverage for varieties of *C. arabica and C. canephora* species were compared. Iron bis-glycine chelate, zinc lactate, and calcium lactate were the salts with higher extractability, especially when using an espresso machine, on average 88%, 96%, and 87%, respectively. Recoveries were between 57 and 65% when using an electric coffee maker with paper filter, and slightly lower with nylon filter. These results suggest that ground roasted coffee may be a promising food vehicle for fortification with mineral nutrients.

21.6 Conclusions

The mineral composition of coffee has been extensively studied from the plant fruit to the final coffee products in different countries from different continents. It has been clearly established that the individual mineral contents of coffee and coffee products are highly variable reflecting the influence of complex and interacting factors acting from the cultivation of the coffee plant to the final consumption of the coffee beverages. Major factors identified include botanical origin (species, genotype), environmental/geographical and agronomic growing conditions, industrial processing, and different techniques of beverage preparation. Determination of the mineral composition of coffee and coffee products using advanced chemometric methods is emerging as a useful tool for authentication of the geographical origin of coffee and for discrimination of species and components in coffee blends.

References

1. L. C. Trugo, Coffee/Analysis of Coffee Products, in *Encyclopedia of Food Sciences and Nutrition*, ed. L. C. Trugo and P. M. Finglas, Academic Press—Elsevier Science, London, UK, 2nd edn, 2003.
2. M. Oliveira, S. Casal, S. Morais, C. Alves, F. Dias, S. Ramos, E. Mendes, C. Delerue-Matos and B. P. P. Oliveira, Intra- and interspecific mineral composition variability of commercial coffees and coffee substitutes, Contribution to mineral intake, *Food Chem.*, 2012, **130**, 702–709.
3. P. Pohl, E. Stelmach, M. Welna and A. Szymczycha-Madeja, Determination of the elemental composition of coffee using instrumental methods, *Food Aanal. Methods*, 2013, **6**, 598–613.
4. R. M. Barbosa, B. L. Batista, R. M. Varrique, V. Coelho, A. D. Campiglia and F. Barbosa Jr, The use of advanced chemometric techniques and trace element levels for controlling the authenticity of organic coffee, *Food Res. Int.*, 2014, **61**, 246–251.
5. F. S. Tagliaferro, E. A. de Nadai Fernandes and M. A. Bacchi, Quality assessment of organic coffee beans for the preparation of a candidate reference material, *J. Radioanal. Nucl. Chem.*, 2006, **269**, 371–375.
6. V. Krivan, P. Barth and A. F. Morales, Multielement analysis of green coffee and its possible use for the determination of origin, *Microchim. Acta*, 1993, **110**, 217–236.
7. M. Oliveira, S. Ramos, C. Delerue-Matos and S. Morais, Espresso beverages of pure origin coffee: Mineral characterization, contribution for mineral intake and geographical discrimination, *Food Chem.*, 2015, **177**, 330–338.
8. M. J. Martin, F. Pablos and A. G. Gonzalez, Characterization of green coffee varieties according to their metal content, *Anal. Chim. Acta*, 1998a, **358**, 177–183.
9. M. J. Martin, F. Pablos and A. G. Gonzalez, Discrimination between arabica and robusta green coffee varieties according to their chemical composition, *Talanta*, 1998b, **46**, 1259–1264.
10. B. Bertrand, D. Villarreal, A. Laffargue, H. Posada, P. Lashermes and S. Dussert, Comparison of the effectiveness of fatty acids, chlorogenic acids, and elements for the chemometric discrimination of coffee (*Coffea arabica* L.) varieties and growing origins, *J. Agric. Food Chem.*, 2008, **56**, 2273–2280.

11. J. S. Dos Santos, M. L. P. dos Santos, M. M. Conti, S. N. dos Santos and E. de Oliveira, Evaluation of some metals in Brazilian coffees cultivated during the process of conversion from conventional to organic agriculture, *Food Chem.*, 2009, **115**, 1405–1410.
12. V. R. M. Filho, W. L. Polito and J. A. G. Neto, Comparative studies of the sample decomposition of green and roasted coffee for determination of nutrients and data exploratory analysis, *J. Braz. Chem. Soc.*, 2007, **18**, 47–53.
13. M. J. Martin, F. Pablos and A. G. Gonzalez, Characterization of arabica and robusta roasted coffee varieties and mixture resolution according to their mineral content, *Food Chem.*, 1999, **66**, 365–370.
14. R. Muñiz-Valencia, J. M. Jurado, S. G. Ceballos-Magaña, A. Alcázar and J. Hernández-Dıaz, Characterization of Mexican coffee according to mineral contents by means of multilayer perceptrons artificial neural networks, *J. Food Compos. Anal.*, 2014, **34**, 7–11.
15. E. J. Dos Santos and E. de Oliveira, Determination of mineral nutrients and toxic elements in Brazilian soluble coffee by ICP-AES, *J. Food Compos. Anal.*, 2001, **14**, 523–531.
16. M. Grembecka, E. Malinowska and P. Szefer, Differentiation of market coffee and its infusions in view of their mineral composition, *Sci. Total Environ.*, 2007, **383**, 59–69.
17. E. Stelmach, P. Pohl and A. Szymczycha-Madeja, The suitability of the simplified method of the analysis of coffee infusions on the content of Ca, Cu, Fe, Mg, Mn and Zn and the study of the effect of preparation conditions on the leachability of elements into the coffee brew, *Food Chem.*, 2013, **141**, 1956–1961.
18. L. L. Costa, C. M. Donangelo, C. L. P. Silviera and A. Farah, Mineral content in Brazilian Arabica coffee brewed by different methods, 23rd ASIC. International Conference on Coffee Science, Indonesia, Bali, 2010.
19. A. M. Cozmuta, A. Conea, A. Peter, L. Mihaly-Cozmuta and C. Nicula, Study the influence of work parameters on the leachability of mineral elements into the coffee infusion, *Carpathian J. Food Sci. Technol.*, 2013, **5**, 23–35.
20. Ö. Özdestan, Evaluation of bioactive amine and mineral levels in Turkish coffee, *Food Res. Int.*, 2014, **61**, 167–175.
21. B. Suseela, S. Bhalke, A. V. Kumar, R. M. Tripathi and V. N. Sastry, Daily intake of trace metals through coffee consumption in India, *Food Addit. Contam.*, 2001, **18**, 115–120.
22. E. Stelmach, P. Pohl and A. Szymczycha-Madeja, The content of Ca, Cu, Fe, Mg and Mn and antioxidant activity of green coffee brews, *Food Chem.*, 2015, **182**, 302–308.
23. M. E. Gillies and J. A. Birkbeck, Tea and coffee as sources of some minerals in the New Zealand diet, *Am. J. Clin. Nutr.*, 1983, **38**, 936–942.
24. E. E. Santos, D. C. Lauria and C. L. Porto da Silveira, Assessment of daily intake of trace elements due to consumption of foodstuffs by adult inhabitants of Rio de Janeiro city, *Sci. Total Environ.*, 2004, **327**, 69–79.

25. N. M. Barros, T. Saint Pierre, L. Costa, V. Calado, C. M. Donangelo and A. Farah, Fortification of ground roast coffees with iron, zinc, and calcium: mineral recovery in the beverage, 25th ASIC. International Conference on Coffee Science, Quindío, Colombia, 2014.
26. J. H. Zaidi, I. Fatima, M. Arif and I. H. Qureshi, Determination of trace elements in coffee beans and instant coffee of various origins by INAA, *J. Radioanal. Nucl. Chem.*, 2006, **267**, 109–112.

CHAPTER 22

Organic Acids

ADRIANA FARAH* AND ÂNGELA GALVAN DE LIMA

Laboratório de Química e Bioatividade de Alimentos & Núcleo de Pesquisa em Café Luiz Carlos Trugo. Instituto de Nutrição, Universidade Federal do Rio de Janeiro, 21941-902, Rio de Janeiro, Brazil
*E-mail: afarah@nutricao.ufrj.br

22.1 Introduction

The taste sensations experienced when drinking a coffee are the consequence of a specific balance between the concentration of the various volatile and non-volatile compounds present in coffee, being technically reported as a multimodal or synesthetic perception, including taste, smell, touch, sound, and vision, where the stimulation of one sense causes a perception in another sense.[1] One of the primary taste sensations in coffee is acidity, as a positive sensation, or sourness, as a negative sensation, which has a direct influence on the sensory quality of the beverage, together with aroma and bitter and sweet compounds.[2] These attributes together with other taste attributes are currently known to be perceived in many places on the tongue rather than in a specific region.[3]

The main compounds responsible for acidity in coffee are non-volatile organic acids,[4,5] but low-molecular-mass organic acids also contribute to acidity as well as to the formation of aroma and flavor of coffee beverage due to their volatility.[2] Organic acids contribute specific types and intensity of acidity in different ways, depending on the sensory characteristics, concentration in the beverage, and also on their strengths determined by their dissociation constant (K_a) value, which is used to describe a tendency of compounds or ions

Coffee: Production, Quality and Chemistry
Edited by Adriana Farah

Published by the Royal Society of Chemistry, www.rsc.org

to dissociate. K_a is most commonly expressed as its negative logarithm ($pK_a = -\log K_a$), resulting in smaller numbers which are simpler and easier to remember. The lower the pK_a value, the stronger the acid, *i.e.* the lower the pH at which the molecule dissociates to give protons and anions in the solution.[6]

Organic acids are formed in the coffee fruits during maturation and roasting, having sucrose and other carbohydrates as the main precursor.[5,7,8] Genetic aspects, growing conditions such as altitude, agricultural methods, post-harvest treatment, roast degree, and brewing method may lead to changes in the final brew's acidity.[9–11]

The purpose of this chapter is to present a brief overview of the organic acids contained in the green and roasted beans and beverages of *Coffea arabica* and *Coffea canephora* species, the most cultivated and commercialized coffee species in the world. The chapter also approaches the analytical methods used to determine their contents in coffee and discusses their implication in flavor, cup quality, and health.

22.2 Coffee Organic Acids

According to Woodman,[12] the organic acids found in coffee may be classified into four groups: aliphatic, alicyclic, chlorogenic acids, and phenolic acids (Table 22.1). Although this classification has been suggested decades ago, it is still one of the most accepted in the literature, but some authors also include other acidic compounds such as amino acids and fatty acids in their organic acids list. To date, about 18 aliphatic acids[2,8,13–19] have been identified in coffee beans and beverages. Among them, some of the most important for the beverage's flavor are citric, malic, acetic, lactic, and formic acids.[4,5,18] The chlorogenic acids are in fact esters of *trans*-hydroxycinnamic acids (phenolic acids), mainly caffeic, ferulic, and *p*-coumaric acids with quinic acid (alicyclic acid), and there are more than 100 major and minor compounds identified (for details on the chemistry of chlorogenic acids, see Chapters 24 and 25), but because of their large amount in coffee and contribution to the beverage's perceived acidity (although secondary[20]) they can be included among organic acids. Small amounts of unesterified quinic acid, hydroxycinnamic acids, and other phenolic acids also contribute acidity to coffee beverage.[11,21] Phosphoric acid, which is actually an inorganic compound present in very low concentrations in coffee, has also been incorporated into the list of coffee acids because its very low pK_a may influence the perceived acidity of the brew.[15,22,23]

Table 22.1 presents the main organic acids identified in green and roasted coffees.

22.2.1 Methods Used for Determination of Acidity and Organic Acids Content in Coffee

Total organic acids determination traditionally involves the use of titulometric methods but it can also be determined by colorimetric methods. The determination of individual acids involves chromatographic, capillary

Table 22.1 The main organic acids and phosphoric acid identified in coffee and their pK_a values, organized in alphabetical order.

Organic acid	pK_a value (μmol 100 mL^{-1})	Chemical formula	Chemical structure[26]	Molar mass (g mol^{-1})[26]
Aliphatic acids				
Acetic acid	4.56[24]	$C_2H_4O_2$	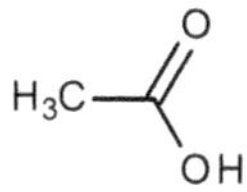	60.05
Ascorbic acid	4.03[24]	$C_6H_8O_6$	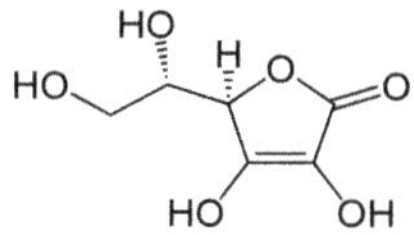	176.12
Citric acid	2.87[24]	$C_6H_8O_7$	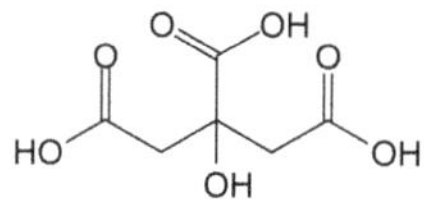	192.124
Citraconic acid	2.20[24]	$C_5H_6O_4$	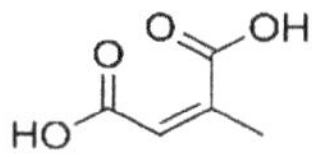	130.10
Formic acid	3.55[24]	CH_2O_2	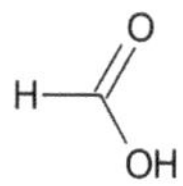	46.02
Fumaric acid	2.85[24]	$C_4H_4O_4$	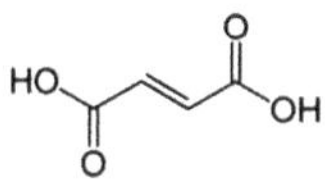	116.07
Glycolic acid	3.52[24]	$C_2H_4O_3$	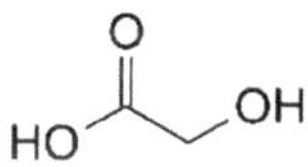	76.05
Isocitric acid	3.02[24]	$C_6H_8O_7$	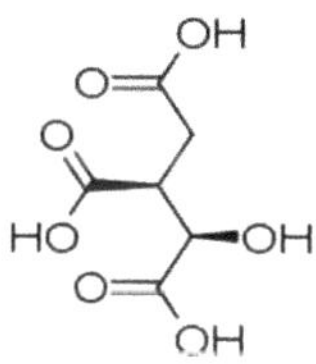	192.12
Lactic acid	3.66[1]	$C_3H_6O_3$	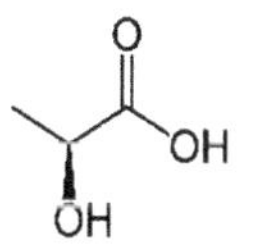	90.08
Maleic acid	1.75[24]	$C_4H_4O_4$	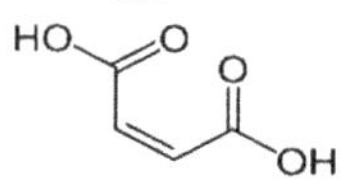	116.07
Malic acid	3.24[24]	$C_4H_6O_5$	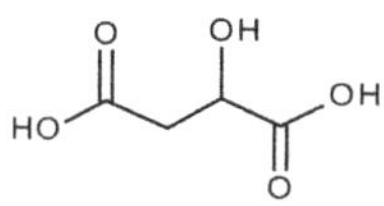	134.09
Mesaconic acid	2.61[24]	$C_5H_6O_4$	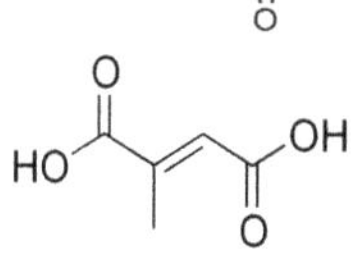	131.00
Oxalic acid	1.04[24]	$C_2H_2O_4$	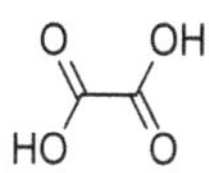	90.03

(*continued*)

Table 22.1 *(continued)*

Organic acid	pK_a value (µmol 100 mL^{-1})	Chemical formula	Chemical structure[26]	Molar mass (g mol^{-1})[26]
Pyroglutamic acid	3.61 [24]	$C_5H_7NO_3$	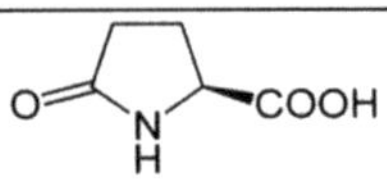	129.11
Pyruvic acid	2.50 [24]	$C_3H_4O_3$	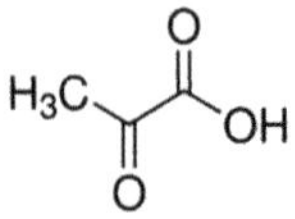	88.06
Propionic acid	4.67 [24]	$C_3H_6O_2$	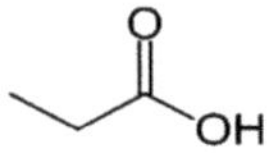	74.08
Succinic acid	4.00 [24]	$C_4H_6O_4$	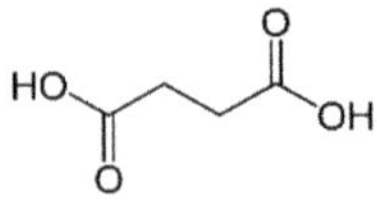	118.09
Tartaric acid	2.82 [24]	$C_4H_6O_6$	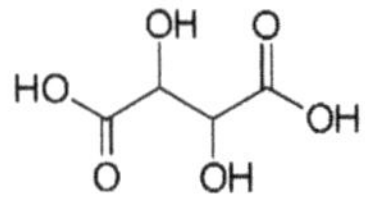	150.09
Alicyclic acids				
Quinic acid	3.4 [25]	$C_7H_{12}O_6$	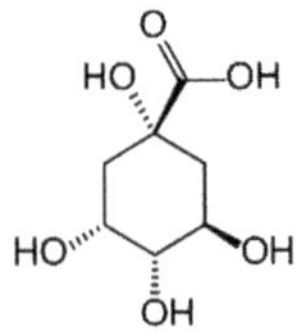	192.17
Phenolic acids				
Caffeic acid	4.62 [25]	$C_9H_8O_4$	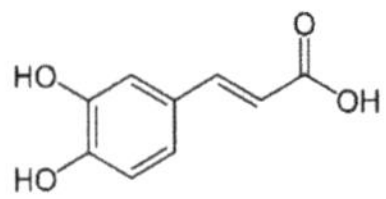	180.16
Ferulic acid	4.58 [25]	$C_{10}H_{10}O_4$	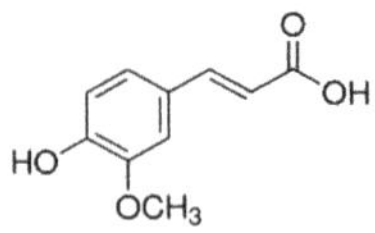	194.18
p-coumaric acid	4.4 [25]	CO_3	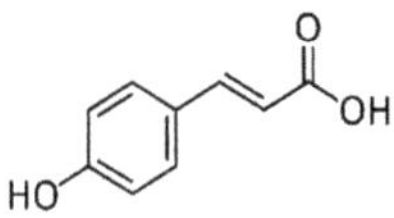	164.05
Esthers				
Chlorogenic acids	See Chapter 24 for chemical characteristics			
Inorganic acids				
Phosphoric acid	2.15 [15]	H_3O_4P	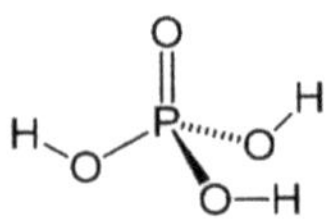	97.99

electrophoretic, or spectrophotometric methods, but enzymatic methods can also be used. They will be covered below.

pH and titratable acidity – The relationship between pH and acid content has been established long ago and is well explained in many physical chemistry text books. There are potentiometric and optical methods for determining the pH value. Optical methods like pH indicator paper strips involve photometric and visual analyses. pH-dependent color changes are produced by acidic or basic halochromic substances (indicators). The pH value is estimated by means of a visual comparison of the color against a color scale. However, the precision is only sufficient to provide a rough estimate and for more precision there has to be different strips for given pH ranges. Also, if the solution is opaque or if it is already colored, like coffee, the reading will probably be affected. Liquid kits for specific pH ranges using the same principle of measurement are commercially available.

The potentiometric methods relate to the measurement of electrical voltages on pH-sensitive electrodes. This method (also called pH meter) uses the electrical potential of a pH-sensitive electrode (usually a glass electrode). The voltmeter measures the electronic impedance in the electrode and displays pH units instead of volts. The pH meter typically has to be calibrated for each use with standard liquid solutions (buffers) of known pH. This is a sensitive and selective sensor and measures from pH 0 to 14.

The determination of pH, and titratable acidity, of whole coffee extracts will produce data resulting from all the present acidic components. This will include not only the aliphatic acids but also phenolic acids and amino acids, both free and protein-bound.[12,27] Conversely, the contribution of these components to the final pH, or titratable acidity, will be counteracted by the presence of both inorganic and organic bases forming salts.[12] Tomoda *et al.*[28] reported values obtained by titrating to both pH 7.0 and 8.2. According to Woodman,[12] at 8.2 and higher values, the phenolic hydroxy groups begin to ionize and the results give greater emphasis to the chlorogenic acids, whereas at the end-points of pH 6.5 to 7.0 only the carboxylic groups of all the organic acids will be titrated, therefore giving a better measure of total acidity. A further factor which produces considerable error in the determination is the buffering action of other components, for example proteins. Thus, there is no clear "pH jump" as the end-point is reached but rather a slow change over a considerable addition of alkali. This precludes the use of pH indicators, which usually require a pH change of some 1–2 pH units to show a clear color change, and this is further complicated by the natural color of the coffee extracts.[12]

Although the beverage pH is a good indicator of acidity of a simple solution and can be directly related to titratable acidity, this is not always observed in coffee extracts. In a study performed by Gloess *et al.*,[29] the same coffee samples have been extracted by nine methods to obtain coffee brews which were evaluated for pH, titratable acidity and perceived acidity. A total of 40 mL of coffee brew was titrated with 0.1 M NaOH at 20 °C to (i) a pH of 6.6 and (ii) a pH of 8.0. A considerable difference was observed among methods and the espresso extraction achieved the highest score for flavor acidity while presenting the highest pH value and titratable acidity (3–12 mL 0.1 M NaOH to pH 8 per 50 mL of coffee brew) and therefore only titratable acidity reflected sensory results. No correlation between pH or titratable acidity and perceived acidity in flavor or after-sensation was found. Espresso Moka extraction achieved the highest score for flavor acidity while being the least acidic coffee brew (highest pH value). In the same way, no correlation was found between pH and titratable acidity. The authors justified that many acids which exist in a coffee brew are not (fully) deprotonated at the pH of the brew, and thus do not contribute to the brew's pH value, but are measured during titration with a base. They also mentioned the possibility of interference of other sensory attributes, like bitterness and temperature, since, in general, acidity is more pronounced at lower temperatures of the coffee brew.[29] On the same note, Rodrigues *et al.*[30] analyzed coffee organic acids by liquid chromatography as well as the sensory acidity in several *C. arabica* and *C. canephora* samples. While in *C. canephora* samples the total amount of organic acids correlated well with sensory acidity, no correlation was found in the case of the arabica brews.

In contrast to the findings in these referred studies, Fuse *et al.*[31] found positive correlations between pH value of Guatemalan coffee brews and titratable acidity, as well as between pH and the perceived sourness intensity of the brew. The range of titratable acidity was 15–30 mL 0.1 M NaOH to pH 8.2 per 50 mL of coffee brew.

Thus, neither titratable acidity nor pH seem to reflect directly the organic acids content of coffee, but they can be good indicators of acidity.

Spectrophotometric methods – These are based on the reaction of the organic acid with a certain substance, giving rise to the formation of a compound or colored complex that is measured at a certain wavelength. In order to avoid interferences, organic acids are isolated by precipitation, ionic exchange resins, *etc.* These methods lack specificity and therefore they are not advisable for individual acids, unless derivatization reactions are used, which is time consuming and costly.[32,33]

Enzymatic methods – The enzymatic methods have been mainly used for quantification of malic, lactic, and citric acids in juices and wine,[33,34] but they have also been applied to other foods,[25] including coffee,[35] in which malic, lactic, citric, and pyruvic acid were determined. These methods are based on the measurement of changes in the absorbance of the reduced form of coenzymes which absorb in a certain wavelength region (usually reading is performed at 340 nm).[33] These methods are highly specific because the determination of L- and D-isomers of some acids is possible and easy. Nonetheless, such methods usually require a specific kit for each individual acid, which increases the cost of analysis in addition to being time consuming.[33,36]

Chromatographic methods – Ion chromatography, high performance liquid chromatography (HPLC), and gas chromatography (GC) techniques are used for analyses of organic acids in different types of foods. According to

Shukla *et al.*[37] and Svecová *et al.*,[38] methods using HPLC, GC, and related techniques associated or not with mass spectrometry are currently among the most used for determination of individual organic acids in coffee and in other beverages, presenting advantages such as high selectivity, sensitivity, and reliability. Drawbacks are the high cost of this equipment and related costs such as equipment maintenance and acquisition of high purity standards, which generally increases commercial value. In addition, they often require previous sample preparation, and, in some cases, extraction. Furthermore, purification of the analytes is necessary, consuming time, and further increasing the analytical cost.[39]

Below are examples of studies that determined organic acids content using chromatography.

Jhan *et al.*[40] compared the performances of methods using HPLC-UV and GC-FID for the quantification of eight organic acids (citric, malic, oxalic, succinic, fumaric, glutaric, tartaric, and quinic) in green *C. arabica* samples. An ion exchange column, phosphoric acid 1% and UV detection at 210 nm were applied for HPLC analysis. For GC-Flame Ionization Detector (FID) analysis, ethanolic extract was purified on two ion exchange columns and trimethylsilyl derivatives were prepared; for chromatography, a DB-1 column and oven temperature gradient from 100 °C to 280 °C were used. The authors concluded that the proposed HPLC method was simple, reliable, and therefore more suitable for the analysis of organic acids compared to the use of purification of the crude extract on ion exchange columns prior to HPLC analysis. They also concluded that GC was unsuitable for analysis of a large number of samples, since it involves very laborious methods.

Alcázar *et al.*[18] quantified five organic acids (acetic, malic, ascorbic, citric, and succinic acids) and inorganic anions in green and roasted *C. arabica* and *C. canephora* samples using ion chromatography/HPLC. An anion-exchange column kept at 40 °C, isocratic elution with phthalate/phthalic acid (0.6 mM) and 4% (v/v) acetonitrile, and a conductivity detector were used.

Rodrigues *et al.*[30] used Solid Phase Microextraction (SPME) prior to HPLC-UV separation to analyze seven organic acids (citric, malic, acetic, formic, pyruvic, succinic, and quinic) in *C. arabica* and *C. canephora* beverages. For chromatographic analysis, an isocratic system was used. Mobile phase was 25 mM phosphate buffer, pH 2.4, methanol 1%; UV detection was performed at 210 nm.

Electrophoretic methods – An alternative to the chromatographic methods for analysis of organic acids in coffee is capillary electrophoresis (CE), which has been previously used as a clean-up procedure associated with ultrafiltration prior to GC/MS analysis.[16] CE has been used for the determination of organic acids in very different matrices, including coffee and other food samples.[41–44] Trying to respond to the lack of rapid and simple analytical methods for short-chain organic acids in complex matrices, Weers *et al.*,[15] developed a method using electrophoresis.[2] A UV-absorbing buffer system was set at a pH at which the acids were essentially fully ionized. According to the authors, the method was adequate for the investigation of low-molecular-mass organic acids in complex aqueous samples like coffee,[45] with high efficiency and minimum sample pre-treatment.

A comprehensive review of CE methods developed for the measurement of short-chain organic acids and inorganic anions in a wide variety of matrices has been first published by Galli *et al.*,[44] with no inclusion of coffee samples. A year later, Galli and Barbas[2] validated a CE method for the analysis of short-chain organic acids and applied it to various coffee extracts of different origins and brands available in the market. Seventeen organic acids were detected and quantified in roasted and soluble (lyophilized) coffees. The authors concluded that the validated method was simple, fast, reliable, and suitable for application in a large number of samples, showing important differences in some of these compounds due to the coffee processing.

22.3 Organic Acids in Green Coffee

The first significant contributions found on the quantification of organic acids in green coffee date back to the 1950s.[46,47] By the end of the 1980s,[12,48] besides chlorogenic and quinic acids, most of the important organic acids in green coffee had already been identified. To date, eight aliphatic acids have been identified in green coffee (citric, malic, acetic, ascorbic, succinic, lactic, oxalic, and tartaric acids), in addition to the non-aliphatic acids (chlorogenic, quinic, phenolic acids, and phosphoric acid). Citric and malic acids represent a significant portion (more than 50%) of the total aliphatic acids content in green coffee and are important for the development of perceived acidity.[8,11,18,19]

Acids may correspond to as much as 11–12% of the dry weight of green coffee, including the esters chlorogenic acids, which account for 4–9% of this percentage, depending on species and other factors,[5,11] and phosphoric acid. In addition to genetic factors, the degree of acidity in *C. arabica* and *C. canephora* species may also differ with origin, climate conditions (related to geographic location and altitude), harvest and post-harvest methods applied.[2,5,30,49] *C. arabica* has notably higher acidity than *C. canephora*.[18,50] Although the content of chlorogenic acids is about 70% higher in *C. canephora* than in *C. arabica*, their contribution is not as important as that of the sum of aliphatic acids and phosphoric acid. Belitz *et al.*[51] reported that the sum of some of the major non-volatile aliphatic acids (citric, malic, and quinic acids) in green *C. arabica* and *C. canephora* beans corresponded to 2–2.9% and 1.3–2.2% (dry weight) of total beans' composition. Such difference does not seem to be genetically established. Usually, *C. arabica* trees grow in higher altitude and in lower temperatures compared to *C. canephora* species and the plant development in high altitude regions and in volcanic soils has been associated with high acidity in brews and this contributes to differences in final acidity.[10,52,53] Climate changes may also interfere with the formation and maturation of coffee fruits,[54,55] which allows differences in the final chemical composition and sensory profile of the beans, including acidity.

The degree of fruit maturation plays a major role in this aspect, since during maturation, organic acids profile is changed. For example, the degradation of sucrose and other sugars present in the bean forms aliphatic acids such as formic acid, acetic acid, glycolic acid and lactic acid as maturation progresses and acidity tends to increase gradually, elevating the quality of the beverage.[9,49] The contents of quinic and malic acids decrease while that of citric acid increases[9] to the point that it can double the concentration of malic acid, in both green *C. arabica* and *C. canephora* seeds.[18] Therefore, the acids contents and profile in defective beans related to the degree of maturation will be different from those in seeds from cherry fruits.

Fermentation during post-harvest (depulping) operations can also increase acidity[56,57] and affect quality, either positively or negatively.[58–60] Wet depulping methods tend to increase acidity in a positive way compared to traditional dry depulping.[61] The acids produced during fermentation may penetrate the husks of the coffee bean changing beans composition and beverage pH.[62–64] In the case of natural post-harvest processing (drying followed by depulping), undesired fermentation may occur during drying with production of butyric, acetic, and propionic acids, which can damage the final quality of the beverage, conferring an onion flavor when present in concentrations higher than 1 mg mL^{-1}.[65,66]

Bórem *et al.*,[67] studied the effect of drying temperature on the quality of natural and pulped coffees subjected to air circulation at temperatures of 40 °C and 60 °C during drying. Titratable acidity increased gradually with temperature elevation, in spite of the post-harvest process applied. According to the authors, drying at 60 °C may alter seeds' cell membrane structures leading to degradation and changes in seeds' chemical components and consequently to an increase in acidity.[67,68]

Table 22.2 presents values reported in the literature for green *C. arabica* and *C. canephora* seeds.

22.4 Organic Acids in Ground Roasted Coffees

The presence of different acids and their content in green beans are not necessarily indicators neither of the acids profile nor or of the quality of roasted beans. The role of roasting and brewing is pivotal in flavor and cup quality development. Coffee roasting is a method of exposing the green coffee beans to heat (at a certain range of temperatures and for a certain time) that, after drying of the beans, generates color and flavor through a sequence of chemical reactions and physical changes that occur simultaneously producing hundreds of chemical compounds. Although it appears to be simple in terms of processing conditions, the chemistry underlying the flavor development is highly complex and not completely understood.[11] Roasting temperatures range commonly from about 210 °C to 240 °C for 12–30 min[11,69] in conventional drum roasters, and for much shorter times (*e.g.* 8–12 min) in fluidized bed and similar types of roasters, in which the circulation of hot air in the roasting chamber increases heat exchange and accelerates the process. The roasting

Table 22.2 Organic acids identified in green coffee beans by different techniques and authors.[a]

Organic acids content (mg 100 g^{-1})						
	C. arabica				*C. canephora*	
	HPLC[18 b]	HPLC[19 c]	HPLC[8 d]	CE[15]	HPLC[18]	CE[15]
Acetic	58	60	80	29	40	15
Ascorbic	337	—	—	—	308	—
Citric	852	1160	1360	1311	608	1350
Citraconic	—	—	—	—	—	—
Formic	—	—	—	—	—	18
Fumaric	—	—	—	—	—	—
Glycolic	—	—	—	—	—	—
Isocitric	—	—	—	—	—	—
Lactic	—	60	80	—	—	—
Maleic	—	—	—	—	—	—
Malic	414	520	510	402	295	247
Mesaconic	—	—	—	—	—	—
Oxalic	—	—	50	—	—	—
Pyrogutamic	—	—	—	—	—	—
Propionic	—	—	—	—	—	13
Succinic	—	170	—	—	—	13
Tartaric	—	10	—	—	—	—
Quinic	—	250	—	687	—	47
Phosphoric	154	—	—	145	177	279

[a]Note: HPLC – High Performance Liquid Chromatography; CE – Capillary Electrophoresis.
[b]Mean of five samples (dry base).
[c]Average levels, Yellow Bourbon (dry base).
[d]Yellow Bourbon (dry base).

profile influences coffee flavor in general, including the correct balance between bitterness and acidity.

Throughout roasting, part of the alicyclic and aliphatic acids existing in green coffee are degraded,[9,12,18,22,30] for example, part of malic acid may form fumaric and maleic acids, while the content of citric acid may remain intact[12,30,70,71] or be degraded to form citracomic, glutaric, itaconic, mesaconic, and succinic acids.[15,70] Ascorbic acid is also degraded in the process.[18]

Despite the partial loss of acids, additional acids are generated from carbohydrates in the initial phase of the process. In the beginning of roasting, the brew's acidity tends to increase as levels of aliphatic acids rise.[5,12,72,73] Formic and acetic acids' yields increase up to medium roast degree and then begin to fall as the process continues. The formation of these acids is considered a key factor related to the higher acidity and lower pH obtained at light to medium roasting levels compared to dark roasts.[5,30,70]

Sucrose represents almost the total amount of free sugars in green coffee beans, and it has been for a long time the focus of studies involving coffee quality as it is an important precursor of taste and aroma compounds.[5,9,53,74] During roasting, sucrose is degraded and in addition to participating in reactions such as Maillard and caramelization, it is partly converted to a variety of organic acids.[5] The decrease in sucrose content during roasting may reach 98%.[12,75] *C. arabica* species contains almost twice (7 to 11 g per 100 g) the concentration of sucrose compared to *C. canephora* species (3 to 7 g per 100 g). This helps to explain the higher acidity of roasted *C. arabica* seeds compared to *C. canephora* seeds.[76] Polysaccharides and additional compounds also contribute to formation of organic acids.[77,78]

Yields of the acids which are less volatile and presumably not breaking down products of the polymeric carbohydrates fraction, for example, glycolic, quinic, and lactic acids, tend to level off at the higher roasts.[15] The content of free quinic and caffeic acids slightly increases due to chlorogenic acids degradation, although not in a proportional way, since they participate in other simultaneous reactions such as lactonization producing pleasant bitter compounds and incorporation into melanoidins. As roasting progresses, chlorogenic acids and lactones break down to form aroma compounds, like 4-vinyl guaiacol (positive clove note), guaiacol (spicy, vanilla, meat, smoky, and sometimes medicinal notes), and phenol[79] (phenolic compounds appear to make a positive contribution to aroma at about threshold concentration, but tend towards an unpleasant note, as their concentration rises in coffee and other foods).[11,70,71,80] 4-Vinylcatechol then oligomerizes to give multiply hydroxylated phenylindanes, which yield a "lingering, harsh type of bitter sensation" typical of over-roasted coffee.[81] In any case, chlorogenic acids do not contribute to the brew's acidity as importantly as the sum of aliphatic acids and phosphoric acid.

In particular, pyroglutamic acid was identified in roasted coffee by Galli and Barbas.[2] This is an amino acid that naturally occurs in vegetables and presents special interest since it was generally affirmed that no free amino acids were still present in coffee after roasting. Furthermore, pyroglutamate is known to have a number of remarkable cognitive enhancing effects, which could be also related to the properties of coffee traditionally associated with caffeine.[2]

The roasting parameters seem to interfere in the formation and degradation of acids and in the acidity results but to date there is no clear correlation between these factors.[82] Wang and Lim[72] and Santos *et al.*[73] obtained maximum titratable acidity at lower roasting temperatures (about 200–210 °C). In agreement with these results, roasting at lower temperatures and for a longer period of time to obtain the same color degree increased degradation of chlorogenic acids and possibly the production of free quinic acids compared to higher temperatures for shorter duration.[83,84] As the roaster type and roasting method affect the acidity of the beverage, roasters usually create methods (roasting profiles) to try to bring out the best mixture of naturally occurring acids and additional flavor compounds that can result in uniqueness to the brew's acidity.

The average total content of aliphatic acids in ground roasted coffees vary mostly with roasting method/degree and analytical method. Values for *C. arabica* range from 0.7% to 1.6%., for *C. canephora* species from 0.4% to 1.6%,[15,18] and in commercial blends from 2.1% to 4.1%.[2,13,14]

Table 22.3 Organic acids identified in roasted coffee beans by different techniques and authors.[a]

	Organic acids content (mg 100 g^{-1})						
	C. arabica		*C. canephora*		*Commercial blend*		
	HPLC[18][b]	CE[15][c]	HPLC[18][b]	CE[15][d]	HPLC[14]	CE[2]	GC[13]
Acetic	174	498	178	325	360–550	730	515
Ascorbic	Nd	—	Nd	—	—	—	—
Citric	522	634	460	838	430–700	1010	672
Citraconic	—	—	—	—	—	26	54
Formic	—	228	—	161	180–250	270	201
Fumaric	—	—	—	—	—	20.6	12
Glycolic	—	202	—	130	170–490	169	132
Isocitric	—	—	—	—	—	—	—
Lactic	—	130	—	77	0.0–180	72	88
Maleic	—	—	—	—	—	10.2	9
Malic	174	200	50	145	100–390	220	216
Mesaconic	—	—	—	—	—	7.7	8
Oxalic	—	—	—	—	—	25.6	—
Pyrogutamic	—	—	—	—	—	85	87
Propionic	—	—	—	—	—	6	—
Succinic	—	33	—	45	—	18	7
Quinic	—	912	—	1319	890–1500	940	793
Phosphoric	214	218	370	595	—	—	—

[a]Note: HPLC – High Performance Liquid Chromatography; CE – Capillary Electrophoresis; GC – Gas Chromatography; Nd – not detected.
[b] Mean of five samples (dry base).
[c] Roast loss (6.4%).
[d] Roast loss (6.5%).

The contents of organic acids reported for roasted *C. arabica* and *C. canephora* seeds are presented in Table 22.3.

22.5 Organic Acids in Brewed and Soluble Coffees

During the brew preparation, soluble (and part of the insoluble) components are extracted from roasted coffee beans with pre-boiling, boiling, or, most recently, cold water, using several different methods. The acidity of coffee beverage is not only dependent on the factors that lead to variations in ground roast composition but also it highly depends on the grid, water temperature, water pressure, and duration of extraction, in addition to other subtleties of each used method for beverage preparation.[85]

Brews prepared from medium or light medium ground roasted *C. arabica* seeds tend to be slightly more acidic (pH 4.3–5.2) than those from *C. canephora* seeds (pH 5.3–5.8).[4,85–87] Additionally, the perceived acidity may be even higher for *C. arabica* than titratable acidity and pH indicate.[30]

Regarding roasting degree, according to Yate and Tuo,[82] the drink from light medium roasted coffee brews (average weight loss of 14%) have an acidity described as fine, well-developed, and easily perceived by the consumer, while the heavily roasted coffee brews (average weight loss of 18%) are a little acidic (for example, pH 5.0–6.3) or acidic in such a way that bitter taste is highlighted in the drink.

The methods of preparation using hot water tend to extract a higher percentage of organic acids and phosphoric acid, as well as chlorogenic acids, while cold brewing decreases extraction.[88,89] The International Coffee Organization (ICO)[90] compared the concentration of nine acids in brews extracted for 5 min. with water at 70 °C, 94 °C, and 100 °C, from finely ground coffee. Extraction of fatty acids was better at 100 °C, chlorogenic acids were similarly (better) extracted with water at 94 °C and 100 °C, quinic, lactic, acetic, citric, and malic acids were better extracted at 94 °C, decreasing at 100 °C, possibly by degradation and/or volatilization. Phosphoric acid extraction did not seem to be affected by water temperature. When brews were extracted at 94 °C from course, fine, and extra fine ground coffees, only the concentration of acetic acid was higher in coarsely ground coffee, possibly due to volatilization during grinding of fine and extra fine coffees. Time of extraction was also evaluated. Brews were extracted for 1, 5, and 14 minutes, from finely ground coffees at 94 °C. Five minutes was the best extraction time for lactic, citric, malic, chlorogenic, and palmitic acids. Time did not seem to affect quinic and acetic acids, as well as fatty acids extraction.[90]

In a study conducted by Parenti *et al.*,[91] three extraction procedures were used for espresso coffee preparation: the bar method (the traditional way to make espresso coffee), a simple commercial capsule method, and a more advanced commercial capsule method, primarily designed to increase the sensory attributes related to the colloidal state of the beverage.[92] The differences were assessed in terms of sensory characteristics, physical parameters, and key extracted volatile compounds. The results showed that with the same batch of roasted coffee, the same water and the same operative settings, the espresso coffee prepared in a professional machine with a pressure of 9 bar showed lower acidity than that obtained using capsules, which resulted in an unusually high acidity.[91]

The storage time and conditions of roasted beans may also affect the sensory quality of the beverage. Acidity, sour taste, bitterness, and aroma are recognized by tasters as the most important attributes indicating the evolution of the sensory quality of coffee upon storage.[93] Lipid oxidation is one of the chemical reactions that contribute to quality decay during storage.[94,95] The formation of sensory perceivable acid compounds is also one of the prevalent events occurring during coffee staling.[70] Kreuml *et al.*[96] observed an increase in the brew's sourness when ground roasted *C. arabica* and *C. canephora* was stored under vacuum for 18 months, which was about 60% higher in comparison to the same coffee beverage prepared from freshly roasted beans. In robusta coffee, the perceived sourness doubled at the end of the storage compared to the preparation from freshly roasted beans. Bitterness also increased. However, the bitter taste intensity in robusta coffee beverages was much higher even at the first determination of the unstored brew. Despite all the changes in chemical composition and flavor during storage, they are not always perceived by consumers.[93]

Regarding soluble or instant coffees, there have been few specific studies reporting on the loss of aliphatic acids during spray or freeze-drying of commercial extracts. According to Woodman[12] no losses would be expected in the non-volatile acids, but for the volatile ones, the loss would depend very markedly on drying conditions, initial concentrations in the extract, and so forth. Using model systems, Gero and Smyrl[97] showed loss of acetic acid varying between 40% and 75% during freeze-drying. Both increase and decrease in acetic acid content have been observed in other studies.[12,13] There are only a few reports on organic acids content in commercial soluble coffees, and recent data are even more scant. Data available for soluble coffees commercialized almost 60 years ago[98] reported average 4.18% of aliphatic acids (sum of citric, acetic, malic, lactic, pyruvic).[12] Galli and Barbas[2] reported average content of 3.38% (sum of acetic, ascorbic, citric, citraconic, formic, fumaric, glycolic, isocitric, lactic, maleic, malic, mesaconic, oxalic, pyrogutamic, succinic, and quinic) in soluble coffees from Colombia. The aliphatic acids content (sum of citric, malic, lactic, piruvic, and acetic) evaluated in a number of spray-dried and freeze-dried commercial soluble coffees marketed in the USA, Germany, France, and Switzerland in 1977 was reported to vary between 1.5% and 4.7%.[35]

Angelucci *et al.*[99] provided pH values from 16 commercial soluble coffees manufactured in Brazil, ranging widely from 4.75 to 5.45 (average 4.9). In general, pH values for soluble coffee tend to be somewhat lower than those for domestic brews for the corresponding blend of roasted coffee,[12] due to technology used to increase soluble solids extraction but, currently, potassium hydroxide can be added to the extract in order to decrease acidity (personal information from industry). Angelucci *et al.*[99] also reported the phosphate ion contents in soluble coffees averaging around 0.1%, with one sample at 0.17%, which led to a considerable increase in titratable acidity.

Engelhardt and Maier[13] estimated the contributions of different organic acids to total acidity in ground roasted and soluble coffees (by titration to pH 8 after cation exchange of the coffee solutions). According to the authors the contributions of citric, acetic, formic, chlorogenic, quinic, malic, phosphoric, and pyrrolidonecarboxylic acids were estimated to account for about 67% and 72% of ground roasted and soluble coffees acidity, respectively. Citric acid (12.2% in roast coffee/10.7% in soluble coffee), acetic acid (11.2%/8.8%), and the high molecular weight acids (8%/9%) contributed importantly to the total acidity. Contributions of chlorogenic acids (9%/4.8%), formic acid (5.3%/4.6%), quinic acid (4.7%/5.9%), malic acid (3.9%/3%), and phosphoric acid (2.5%/5.2%) were also highlighted. The contribution of malic acid was surprisingly low in this study. The authors highlighted that a notable difference between roast and soluble coffee was found for phosphoric acid and pyroglutamate (pyrrolidonecarboxylic acid) (0.7%/1.9% contribution,

Table 22.4 Organic acids identified in coffee beverages and soluble coffee by different techniques and authors.

	Organic acids content					
	C. arabica		*C. canephora*		Soluble powder	
	GC[143] (mg 100 g^{-1} dm)	HPLC[30 a] (mg 100 mL^{-1})	GC[143] (mg 100 g^{-1} dm)	HPLC[30 a] (mg 100 mL^{-1})	HPLC[144] (mg 100 g^{-1})	CE[2] (mg 100 g^{-1})
Acetic	—	27–229	—	30–131	4356–6242	1440
Ascorbic	—	—	—	—	—	—
Citric	35	6.5–50	75	33–40	729–1707	1440
Citraconic	—	—	—	—	—	82
Formic	—	Nd–40	—	Nd–16	—	810
Fumaric	7.2	—	23	—	53.4–79.30	114
Glycolic	—	—	—	—	—	400
Isocitric	—	—	—	—	—	1830
Lactic	78	—	88	—	—	210
Maleic	3.8	—	54	—	—	46
Malic	15	Nd–0.4	26	Nd–0.6	—	570
Mesaconic	—	—	—	—	—	41
Oxalic	35	—	49	—	—	17
Pyro-gutamic	—	—	—	—	2701–6050	880
Pyruvic	58	Nd–4	85	Nd	—	—
Propionic	—	—	—	—	—	—
Succinic	5.5	Nd–140	6.7	29–143	Nd	80
Quinic	270	Nd	710	Nd–8.4	1555–5092	4400
Phosphoric	—	—	—	—	—	—

[a]Mean of ten *C. arabica* and ten *C. canephora* brews; dm – dry matter; Nd - not detected.

respectively) and concluded that those two acids are formed or released from *e.g.* their esters in higher amounts than other acids during the production of soluble coffee. Pyroglutamate is greatly increased in the lyophilized (soluble) coffee; the increase is probably related to some stage of the industrial process.[13]

The contents of organic acids reported for samples of soluble coffee and coffee brew are presented in Table 22.4.

22.6 Contribution of Organic Acids to Perceived Acidity and Cup Quality

Acceptance of acidic foods in general is probably an inherited feature and might reflect the fact that acidic foods are generally better preserved, from a nutritional as well as a safety point of view, compared with foods showing neutral or even slightly alkaline characteristics. In most cases, a food's acidity reflects the presence of organic anions, such as malate, citrate, or lactate, together with inorganic anions (for example, phosphate).[100]

Accordingly, the acidity or sourness has (together with aroma and bitterness) always been recognized as an important attribute for sensory quality and consumer acceptance of coffee brews.[2] However, acidity is not always indicative of good quality, or of the content of acids due to various reasons including their different pK_as and unknown interactions with the taste buds of the tongue.[22] Like with bitterness, the perceived acidity may also differ between individuals, since it depends on the receptors located in the individual's tongue. According to Clifford[101] the reaction of the acid with the receptor is a titration, and thus very similar to the process used in measuring titratable acidity. In fact, Lugaz *et al.*[102] performed an extensive study on the perception of acid taste in general as a function of saliva flow rates, pH, and chemical composition of the acid and concluded that the titratable acidity rather than the pH was associated to the acidic taste. This was also observed by Kampmann and Maier,[103] Maier, Balcke, and Thies,[104] and Engelhardt and Maier.[13] However, some studies did not find any correlation between pH value and the perceived acidity,[29,105] nor between titratable acidity and perceived acidity.[106,107]

Acids are not only responsible for perceived acidity but also contribute flavor in different ways and intensities. For example, citric acid has a fresh and short-lived tartness[25] and its flavor recalls lemon and orange. When associated with phosphoric acid it may have notes of grapefruit. Phosphoric acid alone tends to impart sweet notes and quinic acid a "clean" finish; when submitted to dry heat quinic acid is partly lactonized forming bitter compounds.[5,11,20,108] Malic acid is present in many fruits but it is responsible for almost all of the organic acids in apple, imparting fruity, apple, and sometimes pear, peaches, or plum notes to the beverage; it has a smooth lingering taste, and tartness is not as sharp as that of citric acid, although it is longer lasting.[25] Malic acid is used in the food industry to enhance flavor, mostly in combination with citric acid.[109] Lactic acid conveys a buttery note.[110,111]

Moreover, depending on both the nature and predominance, acidity may be desirable or undesirable.[110,112] The desirable acidity contributes to the vivacity and freshness of the coffee beverage, increasing the sweetness perception and, in some cases, providing a dry fruit taste. An over-expressed acidity, on the other hand, may be unpleasant, and related to unusual tastes in coffee drinks.[110,112] Acidity resulting from citric and malic acids, for example, is usually desirable and can indicate good quality;[40,49,90,114] on the other hand, acetic acid resembles fermentation. In small amounts acetic acid contributes to global flavor and can impart a pleasant sharpness to the beverage but in higher concentrations it may be very unpleasant.[110] In fact, not only acetic, but also oxalic,[8] propionic, and butyric[65,115] acids tend to affect quality negatively. Similarly, tartaric acid may produce a sour taste at high concentrations, but at low levels it can present grape-like or winey notes, which are related to its presence in grapes. These acids often originate from excessive or inappropriate fruit fermentation during post-harvest processing as previously mentioned in this chapter, and can be also produced as the fruits overripe and ferment on the tree as is the case with black defective beans. Some of these undesired acids can also be produced during roasting.[5] In addition to objectionable acids, black and sour defective beans (which greatly affect cup quality) seem to have more acidic characteristics than regular beans[116] and immature defective beans were reported to have lower acidity, since acidity increases with maturation.[116,117]

After brewing, perceived acidity tends to increase in spite of temperature. Chemical components continue to react mutually in the cup and hydrolysis of week organic acids is one of them, contributing to dramatic changes until the beverage is consumed.[81,118] Keeping brewed coffee on a hot plate may cause hydrolysis of chlorogenic acid, producing sensory changes,[11] but a relation of such changes to staleness does not seem to be scientifically established.

As previously mentioned, seeds of *C. arabica* and *C. canephora* species have different chemical compositions and sensory characteristics. If picked and processed appropriately, in addition to having richer aroma, *C. arabica* has higher (desired) perceived (desirable) acidity compared to *C. canephora*[30] (despite the content of organic acids), and this may be one of several reasons to increase its commercial value.[30] *C. canephora* usually has lower sensory quality and is often used in blends with *C. arabica* to give "body" to the beverage. However, as pronounced acidity is not always a determinant of quality, *C. canephora* beans from cherry fruits that are carefully harvested and processed can produce better quality beverages than arabica blends containing a high percentage of defective beans and peculiar flavor, being, like arabica coffees, found in gourmet markets of different countries.

22.7 Coffee Organic Acids and Health

Organic acids might be of interest to health since certain acids may lead to lowering of the postprandial blood glucose and insulin responses,[119] maintain the liver in a healthy condition, and help in the digestion process.[120] They and their derivatives can also act as antioxidants synergists. Citric acid is an example of a synergist.[121,122] It improves the stability and activity of primary antioxidants and lipids by acidifying the medium.[121,123,124] Besides chlorogenic acids whose health effects are discussed in *Coffee: Consumption and Health Implications*, citric acid appears to be the most important aliphatic acid in coffee related to health. It is naturally concentrated in citrus fruits (about 5% of lemon and 2% of orange composition) and even though in coffee the concentration is about half of that in orange, it is worth knowing some of its potential effects.

In addition to enhancing antioxidant activity of other compounds, in high concentrations citric acid showed antioxidant and anti-inflammatory activities in brain tissue.[125] Bryland *et al.*[126] demonstrated that citric acid can improve endothelial function by reducing inflammatory markers and by decreasing neutrophil diapedesis in hyperglycemia. In other studies, citrate decreased lipid peroxidation and downregulated inflammation.[127–129] The use of citric acid in the treatment of toxic and inflammatory conditions of the brain and liver tissues has been suggested.[125] Moreover, oral administration of citric acid has been shown to reduce induced hepatocellular injury in rats.[130] Citric acid (like malic acid) is also an antimicrobial agent against psychrophilic and mesophilic microorganisms.[111,131,132]

Pyroglutamate – Although in very small concentrations in coffee, pyroglutamate is an intermediate of the gamma-glutamyl cycle of glutathione synthesis and degradation. This amino acid is formed during coffee roasting and occurs naturally in vegetables. It is known to have a number of remarkable cognitive enhancing effects and according to Galli and Barbas[2] this could also be related to some of the properties of coffee traditionally associated with caffeine. Pyroglutamate is greatly increased in lyophilized soluble coffee and can be found in the market as a nutraceutical. Its actual effect when ingested in very small amounts like in coffee is still to be evaluated.

Oxalic acid – Oxalic acid is a common component of many plant foods, including nuts, fruits, vegetables, grains, and legumes, and it is present in minimal amounts in coffee. Consumption of large amounts can cause some pathologies, like generation of stones in the kidneys besides bone decalcification;[133,134] in the case of coffee, the amount would not be enough to produce that effect and oral intake of citric acid, the main aliphatic acid in coffee, has been suggested as an effective means of treating oxalate stones.[135–137]

Coffee acids and gastro-esophageal reflux – Coffee acidity does not affect everyone's health in the same way. If on one hand some acids and derivatives contained in the beverage have antioxidant, anti-inflammatory activities,[11] on the other hand some people are sensitive to coffee's acidity. For those, the consumption of acidic foods and beverages often results in physical discomfort in the form of indigestion and heartburn among other discomforts, especially those who suffer from digestive tract disorders, such as acid reflux or ulcers.[85]

Gastro-esophageal reflux, also called heartburn, is caused by the reflux of gastric fluid to the esophagus because of the low pressure in the sphincter muscle at the junction of the stomach and esophagus. A number of people suffering from this condition also mention that coffee may be included in the list of food products causing the complaint. Although some studies have been trying to investigate this, it is still unclear what may be the role of coffee consumption in reflux. It has, for example, been reported that a few compounds in coffee as well as the hot temperature at which it is drunk stimulate the production of gastric juice, which is very acidic when first produced in the stomach (pH about 1.5).[138,138a] Chlorogenic acids, Nβ-alkanoil-5-hydroxytryptamides (C5HTs) from coffee wax, and caffeine, are some of the main compounds thought to promote this effect.[85,138,139] The stimulating effect of caffeine to produce gastric acid secretion seems to be linked to bitter receptors both in the oral cavity and in the stomach, possibility due to induction of gastrine production, like for other bitter compounds.[138]

Additionally, it has been hypothesized that roasting products of chlorogenic acid such as pyrogallol can, like C5HTs, "irritate" the gastric mucosa.[85,140] Besides the stimulation of gastric juice, coffee consumption is believed to cause muscle contraction impairment (relaxation effect) on the lower-esophageal sphincter of some individuals, promoting heartburn in these sensitive individuals. This effect has been attributed to caffeine and to hot temperatures.[85,141] The coffee brew's pH is mildly acidic, commonly fluctuating between 5.8 and 5.5 in medium light roasted *C. canephora beans*, and can vary commonly between 4.3 and 4.8 in fresh light roasted acidic *C. arabica* coffees, with higher pH (about 5.0–6.3) being more usual in dark roasted blends. In soluble coffees, pH tends to vary between 4.7 and 5.2. This is much higher than for the gastric juice pH or, for example, the pHs of apple juice (pH 4.3–3.3) or citric juices (pH 2.3–3.3).[4,85,113]

Therefore, based on the knowledge currently available, the stimulation of gastric juice production by hot temperature and a few compounds, along with the relaxation of the esophageal sphincter, seems to be the most likely cause for heartburn, although some studies support the contribution of foods' acidity to heartburn.[85,142]

Some pre-roasting technological methods have been developed aiming to decrease heartburn, although no clinical studies have yet proved that these treatments are effective against the disease in humans. Reducing coffee wax, and thus C5HTs, can be obtained by applying steam-treatment, either as a stand-alone process, or as part of the water or supercritical carbon dioxide decaffeination methodologies (which in addition to caffeine decreases chlorogenic acids and other organic acids content, especially the water method).[85]

22.8 Concluding Remarks

In conclusion, although coffee organic acids composition has been studied for a long time, there are some aspects that still remain unclear and need more investigation, such as the real contribution of each acid and of acids association to global coffee acidity, the relationship between pH, titratable acidity, and perceived acidity by individual consumers, as well as coffee acids effects on health. As technological processes have changed considerably in the last decade in the coffee production industry, and coffee quality has been considerably refined, new compositional data related to processing and quality are needed for individual acids.

More mechanistic and clinical studies are necessary to prove the involvement of specific coffee compounds in gastro-esophageal reflux and improved conditions in their absence. Considering the most recent results on the stimulatory effects of caffeine on gastric acid secretion, additional bitter coffee compounds such as the chlorogenic acids lactones should also be investigated.

Acknowledgement

The authors would like to thank the National Research Council of Brazil – CNPq and the Research Support Foundation of Rio de Janeiro – FAPERJ (Brazil) for scholarships.

References

1. M. Auvray and C. Spence, *Conscious. Cognit.*, 2008, **17**, 1016–1031.
2. V. Galli and C. Barbas, *J. Chromatogr. A*, 2004, **1032**, 299–304.
3. A. Huang, X. Chen, M. Hoon, J. Chandrashekar, W. Guo, D. Tränkner, N. Ryba and C. Zuker, *Nature*, 2006, **442**, 934–938.
4. O. G. Vitzthum and P. Werkhoff, *J. Agric. Food Chem.*, 1975, **23**, 510–516.
5. M. Ginz, H. H. Balzer, A. G. W Bradbury and H. G. Maier, *Eur. Food Res. Technol.*, 2000, **211**, 404–410.
6. M. Clifford, What factors determine the intensity of coffee's sensory attributes, *Tea Coffee Trade J.*, 1989, **8**, 35–39.
7. D. J. Lyman, R. Benck, S. Dell, S. Merle and J. Murray-Wijelath, *J. Agric. Food Chem.*, 2003, **51**, 3268–3272.
8. F. M. Borém, L. P. Figueiredo, F. C. Ribeiro, J. H. S. Taveira, G. S. Giomo and T. J. G. Salva, *Afr. J. Agric. Res.*, 2016, **11**, 709–717.
9. W. J. Rogers, S. Michaux, M. Bastin and P. Bucheli, *Plant Sci.*, 1999, **149**, 115–123.
10. J. Avelino, B. Barboza, J. C. Araya, C. Fonseca, F. Davrieux, B. Guyot and C. Cilas, *J. Sci. Food Agric.*, 2005, **85**, 1869–1876.
11. A. Farah, in *Coffee: Emerging Health Effects and Disease Prevention*, ed. Y. F. Chu, Wiley-Blackwell, USA, 1st edn, 2012, vol. 1, ch. 2, pp. 21–58.
12. J. S. Woodman, in *Coffee: Chemistry*, ed. R. J. Clarke and R. Macrae, Elsevier, Applied Science Publishers, London, 1st edn, 1985, vol. 1, ch. 8, pp. 266–289.
13. U. H. Engelhardt and H. G. Maier, *Z. Lebensm.-Unters. Forsch.*, 1985, **181**, 20–23 apud J. S. Woodman, in *Coffee: Chemistry*, ed. R. J. Clarke and R. Macrae, Elsevier, Applied Science Publishers, London, 1st edn, 1985, vol. 1, ch. 8, pp. 266–289.
14. G. H. D. van der Stegen, J. Duijin, in *Proceedings of The12th ASIC Colloquium*, Montreux, 1987, p. 238, apud V. Galli and C. Barbas, *J. Chromatogr. A*, 2004, **1032**, 299–304.
15. M. Weers, H. Balzer, A. Bradbury and O. G. Vitzthum, *presented in part at the 16th Int. Sci. Coll. Coffee*, ASIC, Kyoto, 1995, pp. 218–223.
16. F. Bähre and G. H. Maier, *Fresenius' J. Anal. Chem.*, 1996, **355**, 190–193.
17. - F. Bähre, *Dissertation*, Technical University of Braunschweig, Germany, 1997 apud H. H. Balzer, in *Coffee Recent Developments*, ed. R. J. Clarke and O. G. Vitzthum, Blackwell Science, Berlin, 1st edn, 2001, vol. 1, ch. 2, pp. 18–30.
18. A. Alcázar, P. L. Fernández-Cáceres, M. J. Martín, F. Pablos and A. G. Gonzál, *Talanta*, 2003, **61**, 95–101.

19. D. E. Ribeiro, *PhD thesis*, Federal University of Lavras, 2017.
20. I. Esteban-Díez, J. M. González-Sáiz and C. Pizarro, *Anal. Chim. Acta*, 2004, **514**, 57–67.
21. M. B. S. Scholz, V. R. G. Figueiredo, J. V. N. Silva and C. S. G. Kitzberger, *Coffee Sci.*, 2011, **6**, 245–255.
22. H. G. Maier, *The Acids of Coffee, presented in part at the 12th Int. Sci. Coll. Coffee*, ASIC, Paris, France, 1987, pp. 229–237.
23. M. N. Clifford and T. Kazi, *Food Chem.*, 1987, **26**, 59–69.
24. L. Bhattacharyya and J. S. Rohrer, in *Applications of Ion Chromatography for Pharmaceutical and Biological Products*, John Wiley & Sons, Canada, 2012, Appendix 1, pp. 449–453.
25. M. M. Theron and J. F. R. Lues, *Food Rev. Int.*, 2007, **23**, 141–158.
26. Royal Society of Chemistry, http://www.rsc.org/, accessed October 2017.
27. R. S. Shallenberger, *Food Chem.*, 1996, **56**, 209–214.
28. G. Tomoda, J. Matsuyama, K. Hiramoto and K. Izu, *Jpn. Soc. Food Sci. Technol.*, 1977, **24**, 286–290, apud J. S. Woodman, in *Coffee: Chemistry*, ed. R. J. Clarke and R. Macrae, Elsevier, Applied Science Publishers, London, 1st edn, 1985, vol. 1, ch. 8, pp. 266–289.
29. A. N. Gloess, B. Schönbächer, B. Klopprogge, L. D´Ambrosio, K. Chatelain, A. Bongartz, A. Strittmatter, M. Rast and C. Yeretzian, *Eur. Food Res. Technol.*, 2013, **236**, 607–627.
30. C. I. Rodrigues, L. Marta, R. Maia, M. Miranda, M. Ribeirinho and C. Máguas, *J. Food Compos. Anal.*, 2007, **20**, 440–448.
31. T. Fuse, F. Kusu and K. Takamura, *J. Agric. Food Chem.*, 1997, **45**, 2124–2127.
32. M. González and V. González, in *Food Analysis by HPLC*, ed. L. M. L. Nollet and F. Toldrá, Taylor & Francis Group, New York, NY, 3rd edn, 2012, vol. 1, ch. 11, pp. 443–466.
33. I. Mato, S. Suárez-Luque and J. F. Huidobro, *Food Res. Int.*, 2005, **38**, 1175–1188.
34. H. U. Bergmeyer, in *Methods of Enzymatic Analysis*, ed. J. Bergmeyer and M. Grabl, Verlag. Chemie., Weinheim, Germany, 3rd edn, 1985, vol. 8, pp. 78–85.
35. M. B. Blanc, G. E. Davis, J. M. Parchet and R. Viani, *J. Agric. Food Chem.*, 1977, **37**, 926–930, apud J. S. Woodman, in *Coffee: Chemistry*, ed. R. J. Clarke and R. Macrae, Elsevier, Applied Science Publishers, London, 1st edn, 1985, vol. 1, ch. 8, pp. 266–289.
36. G. Saccani, S. Gherardi, A. Trifiró, C. Soresi Bordini, M. Calza and C. Freddi, *J. Chromatogr. A*, 1995, **706**, 395–403.
37. S. Shukla, H. K. Park, J. K. Kim and M. Kim, *Food Chem. Toxicol.*, 2010, **48**, 1191–1195.
38. B. Svecová, M. Bordovská, D. Kalvachová and T. Hájek, *J. Food Compos. Anal.*, 2015, **38**, 80–88.
39. H. Cen and Y. He, *Trends Food Sci. Technol.*, 2007, **18**, 72–83.
40. G. M. Jham, A. S. Fernandes, C. F. Garcia and A. A. Silva, *Phytochem. Anal.*, 2002, **13**, 99–104.

41. C. Barbas, N. Adeva, R. Aguilar, M. Rosillo, T. Rubio and M. Castro, *Clin. Chem.*, 1998, **44**, 1340–1342.
42. A. García, C. Barbas, R. Aguilar and M. Castro, *Clin. Chem.*, 1998, **44**, 1905–1911.
43. A. García and C. Barbas, *Clin. Chem. Lab. Med.*, 2003, **41**, 755–761.
44. V. Galli, A. García, L. Saavedra and C. Barbas, *Electrophoresis*, 2003, **24**, 1951–1981.
45. L. Saavedra, A. García and C. Barbas, *J. Chromatogr. A*, 2000, **881**, 395–401.
46. A. F. Mabrouk and F. E. Deatherage, *Food Technol.*, 1956, **10**, 194–197, apud J. S. Woodman, in *Coffee: Chemistry*, ed. R. J. Clarke and R. Macrae, Elsevier, Applied Science Publishers, London, 1st edn, 1985, vol. 1, ch. 8, pp. 266–289.
47. C. Lentner and F. E. Deatherage, *Food Res.*, 1959, **24**, 483–492, apud J. S. Woodman, in *Coffee: Chemistry*, ed. R. J. Clarke and R. Macrae, Elsevier, Applied Science Publishers, London, 1st edn, 1985, vol. 1, ch. 8, pp. 266–289.
48. A. Scholze and H. G. Maier, *Z. Lebensm.-Unters. Forsch.*, 1984, **178**, 5–8.
49. C. J. Pimenta, *PhD thesis*, Federal University of Lavras, 2001.
50. J. N. Wintgens, *Coffee: Growing, Processing, Sustainable Production*, Wiley, Weinheim, 2nd edn, 2009.
51. H.-D. Belitz, W. Grosch and P. Schieberle, in *Food Chem.*, ed. H.-D. Belitz, W. Grosch and P. Schieberle, Springer, Berlin, 4th edn, 2009, vol. 21, pp. 938–951.
52. B. Guyot, D. Gueule, J. C. Manez, J. J. Perriot, J. Giron and L. Villain, *Plant., Rech., Dev.*, 1996, **3**, 272–280.
53. F. Decazy, J. Avelino, B. Guyot, J. J. Perriot, C. Pineda and C. Cilas, *J. Food Sci.*, 2003, **68**, 2356–2361.
54. P. Montavon, E. Duruz, G. Rumo and G. Pratz, *J. Agric. Food Chem.*, 2003, **51**, 2328–2334.
55. R. Dal Molin, M. Andreott, A. R. Reis, E. Furlani Junior, G. C. Braga and M. B. S. Scholz, *Acta Sci., Agron.*, 2008, **30**, 353–358.
56. E. E. Miya, R. S. Garruti, M. A. Chaib, E. A. I. Figueiredo and I. Shirose, *Colet. Inst. Tecnol. Aliment.*, 1973, **5**, 417–432.
57. V. D. Carvalho, S. M. Chalfoun, S. J. R. Chagas, N. Botrel and E. S. G. Juste Junior, *Pesqui. Agropecu. Bras.*, 1994, **29**, 449–454.
58. P. T. G. Guimarães, F. D. Nogueira, P. C. Lima, M. J. C. L. Guimarães and A. A. A. Pozza, *Agricultural Report*, 2002, **23**, 63–81.
59. V. Leloup, C. Gancel, R. Liardon, A. Rytz and A. Pithon, *presented in part at the 20th Int. Conf. on Coffee Sci.*, ASIC, 2004, pp. 93–101.
60. A. S. Franca, L. S. Oliveira, J. C. F. Mendonça and X. A. Silva, *Food Chem.*, 2005, **90**, 84–89.
61. M. Kleinwächter, G. Bytof and D. Selmar, in *Coffee in Health and Disease Prevention*, ed. V. R. Preedy, Kin's College London, London, U.K., 1st edn, 2015, vol. 1, ch. 9, pp. 73–81.

62. A. D. Ferreira, G. R. Carvalho, J. C. Rezende, C. E. Botelho, R. M. Rezende and A. M. Carvalho, *Brazilian Agricultural Research*, 2013, **48**, 388–394.
63. P. S. Nigam and A. Singh, in *Encyclopedia of Food Microbiology*, ed. C. A. Batt and M. Luo, 2014, vol. 2, pp. 485–492.
64. K. Koskei, M. Patrick and M. Simon, *Afr. J. Food Sci.*, 2015, **9**, 230–236.
65. C. I. Lopez, E. Bautista, E. Moreno and E. Dentan, *presented in part at the ASIC, 13th Int. Sci. Coll. Coffee*, ASIC, Colombia, 1989, pp. 373–383.
66. C. F. Silva, D. M. Vilela, C. C. de Souza, W. F. Duarte, D. R. Dias and R. F. Schwan, *World J. Microbiol. Biotechnol.*, 2013, **29**, 235–247.
67. F. M. Borém, P. C. Coradi, R. Saath and J. A. Oliveira, *Science and Agrotechnology*, 2008, **32**, 1609–1615.
68. C. C. Pereira, P. D. Oliveira, E. P. Isquierdo, F. M. Borém, J. H. S. Taveira and G. E. Alves, *presented in part at the VII Brazilian Coffee Research Symposium*, 2011.
69. L. C. Mendes, H. C. Menezes, M. Aparecida and A. P. Silva, *Food Qual. Prefer.*, 2001, **12**, 153–162.
70. H. H. Balzer, in *Coffee Recent Developments*, ed. R. J. Clarke and O. G. Vitzthum, Blackwell Science, Berlin, 1st edn, 2001, vol. 1, ch. 2, pp. 18–30.
71. C. S. G. Kitzberger, *PhD thesis*, State University of Londrina, 2012.
72. N. Wang and L. T. Lim, *J. Agric. Food Chem.*, 2012, **60**, 5446–5453.
73. J. R. Santos, M. Lopo, A. O. S. S. Rangel and J. A. Lopes, *Food Control*, 2016, **60**, 408–415.
74. B. Bertrand, B. Guyot, P. Anthony and P. Lashermes, *Theor. Appl. Genet.*, 2003, **107**, 387–395.
75. L. C. Trugo and R. Macrae, *Analyst*, 1984, **109**, 263–266.
76. L. M. V. Lopes, R. G. F. A. Pereira, A. N. G. Mendes, E. R. Vilela and V. D. Carvalho, *Brazilian Journal of Storage*, 2000, **1**, 3–8.
77. M. Fischer, S. Reimann, V. Trovato and R. J. Redgwell, *Carbohydr. Res.*, 2001, **330**, 93–101.
78. F. Wei, K. Furihata, M. Koda, F. Hu, T. Miyakawa and M. Tanokura, *J. Agric. Food Chem.*, 2012, **60**, 1005–1012.
79. R. Dorfner, T. Ferge, A. Kettrup, R. Zimmermann and C. Yeretzian, *J. Agric. Food Chem.*, 2003, **51**, 5768–5773.
80. A. Farah, T. Paulis, L. C. Trugo and P. R. Martin, *J. Agric. Food Chem.*, 2005, **53**, 1505–1513.
81. O. Frank, S. Blumberg, C. Kunert, G. Zehentbauer and T. Hofmann, *J. Agric. Food Chem.*, 2007, **55**, 1945–1954.
82. D. K. Yate and S. Tuo, *presented in part at the 16th Int. Sci. Coll. Coffee*, ASIC, Kyoto, 1995, pp. 886–901.
83. V. Leloup, A. Louvrier and R. Liardon, *presented in part at the 16th Int. Sci. Coll. Coffee*, ASIC, Kyoto, 1995, pp. 192–198.
84. V. D. Nagaraju, C. T. Murthy, K. Ramalakshmi and P. N. S. Rao, *J. Food Eng.*, 1997, **31**, 263–270.
85. A. Farah, in *Archieving Sustainable Cultivation of Coffee*, ed. P. Lashermes, Cambridge, 1st edn, 2017, vol. 1, ch. 14, pp. 1–53.

86. M. Dalla Rosa, M. C. Nicoli and C. R. Lerici, *Ind. Aliment.*, 1986, **25**, 629–633.
87. M. C. Nicoli, M. Anese, L. Manzocco and C. R. Lerici, *Lebensm.-Wiss. Technol.*, 1997, **30**, 292–297.
88. A. Peters, *Brewing Makes the Difference, presented in part at the 14th Int. Conf. on Coffee Sci.*, ASIC, Paris, France, 1991, pp. 97–106.
89. C. Severini. G. G. Pinnavaia, S. Pizzirani, M. C. Nicoli and C. R. Lerici, *presented in part at the 15th Int. Sci. Coll. Coffee*, ASIC, Montpellier, 1993, pp. 601–606.
90. International Coffee Organization, (ICO), *Sensory Evaluation of Coffee: Technical Unit Quality Series*, ICO, 1991, vol. 9, pp. 209–243.
91. A. Parenti, L. Guerrini, P. Masella, S. Spinelli, L. Calamai and P. Spugnoli, *J. Food Eng.*, 2014, **121**, 112–117.
92. L. Navarini, S. Colomban, V. Lonzarich, D. Rivetti, G. Brollo and F. Suggi-Liverani, *presented in part at the 22th Int. Conf. on Coffee Sci.*, ASIC, 2008, pp. 296–309.
93. L. Manzocco and C. Lagazio, *Food Qual. Prefer.*, 2009, **20**, 24–29.
94. A. T. Toci, V. J. M. F. Neto, A. G. Torres and A. Farah, *LWT-Food Sci. Technol.*, 2013, **50**, 581–590.
95. M. Y. Rendón, T. J. G. Salva and N. Bragagnolo, *Food Chem.*, 2013, **147**, 279–286.
96. M. T. L. Kreuml, D. Majchrzak, B. Ploeder and J. Koenig, *Food Sci. Nut.*, 2013, **1**, 267–272.
97. L. Gero and T. G. Smyrl, *J. Food Sci.*, 1982, **47**, 954–957.
98. J. Schormuller, W. Brandenburg and H. Langner, 1961 *Quoted by Blanc, 1977*, apud J. S. Woodman, in *Coffee: Chemistry*, ed. R. J. Clarke and R. Macrae, Elsevier, Applied Science Publishers, London, 1st edn, 1985, vol. 1, ch. 8, pp. 266–289.
99. E. Angelucci, Y. Yokomizo, R. M. Moraes, R. B. Campos, E. E. Miya, I. B. Figueiredo, *6th. Int. Colloq. Chem. Coffee*, ASIC, Bogotá, 1974, pp. 178–183, apud J. S. Woodman, in *Coffee: Chemistry*, ed. R. J. Clarke and R. Macrae, Elsevier, Applied Science Publishers, London, 1st edn, 1985, vol. 1, ch. 8, pp. 266–289.
100. C. Demigné, h. Sabboh, C. Puel and C. Rémésy, *Nutr. Res. Rev.*, 2004, **17**, 249–258.
101. M. N. Clifford, in *Coffee: Botany, Biochemistry and Production of Beans and Beverage*, ed. M. N. Clifford and K. C. Willson, Croom Helm, New York, 1985, vol. 1, ch. 13, pp. 305–374.
102. O. Lugaz, A. M. Pillias, N. Boireau-Ducept and A. Faurion, *Chem. Senses*, 2005, **30**, 89–103.
103. B. Kampmann and H. G. Maier, *Z. Lebensm.-Unters. Forsch.*, 1982, **175**, 333–336.
104. H. G. Maier, C. Balcke and F. C. Thies, *Lebensmittelchem. Gerichtl. Chem.*, 1983, **37**, 81–83.
105. S. Andueza, M. A. Vila, M. P. Peña and C. Cid, *J. Sci. Food Agric.*, 2007, **87**, 586–592.

106. A. Voilley, F. Sauvageot and D. Simatos, *J. Food Process. Preserv.*, 1981, **5**, 135–143.
107. J. H. S. Taveira, F. M. Borém, S. D. V. F. da Rosa, P. D. Oliveira, G. S. Giomo, E. P. Isquierdo and V. A. Fortunato, *Afr. J. Agric. Res.*, 2015, **10**, 1457–1466.
108. O. Frank, G. Zehentbauer and T. Hofmann, *Eur. Food Res. Technol.*, 2006, **222**, 492–508.
109. B. Taylor, in *Carbonated Soft Drinks: Formulation and Manufacture*, ed. D. Steen and P. R. Ashurst, Blackwell Publishing, Oxford, UK, 2006, pp. 48–86.
110. T. R. Lingle, *The Coffee Cupper's Handbook: Systematic Guide to the Sensory Evaluation of Coffee's Flavor*, Long Beach, California, USA, 4th edn, 2011, p. 66.
111. M. Strafford, in *Encyclopedia of Food Microbiology*, ed. R. K. Robinson, Academic Press, Inc., New York, N.Y., 1999, pp. 1729–1737.
112. A. Illy and R. Viani, *Espresso Coffee: The Chemistry of Quality*. London, San Diego, Academic Press, 2nd edn, 2005, vol. 1, p. 339.
113. M. Sivetz and N. W. Desrosier, *Physical and Chemical Aspects of Coffee*, Coffee Technology, Westport, 1979, ch. 13, pp. 527–575.
114. H. E. P. Martinez, Y. Poltronieri, A. Farah and D. Perrone, *Rev. Ceres*, 2013, **60**, 293–299.
115. C. J. Pimenta and E. R. Vilela, *Cienc. Agrotec., Lavras*, 2003, **27**, 1315–1320.
116. R. G. F. A. Pereira, *PhD thesis*, Federal University of Lavras, 1997.
117. J. Arcila-Pulgarin and G. Valência-Aristizábal, *Cenicafé, Chinchiná*, 1975, **26**, 55–71.
118. M. Petracco, in *Coffee Recent Developments*, ed. R. J. Clarke and O. G. Vitzthum, Blackwell Science, Berlin, 1st edn, 2001, vol. 1, ch. 7, pp. 140–164.
119. M. Suni, M. Nyman, N. Eriksson, L. Björk and I. Björck, *J. Sci. Food Agric.*, 2000, **80**, 1538–1544.
120. S. L. Soumya and R. N. Bindu, *Int. J. Pharm. Sci.*, 2014, **6**, 572–577.
121. J. Pokorny and J. Korczak, in *Antioxidants in Food: Practical Application*, ed. J. Pokorny, N. Yanishlieva and M. Gordon, Woodhead Publishing Ltd and CRC Press LLC, Cambridge, 2001, ch. 13, pp. 311–327.
122. G. C. Yen, P. D. Duh and H. L. Tsai, *Food Chem.*, 2002, **79**, 307–313.
123. D. F. Buck, in *Antioxidants*, ed. J. Smith, Springer, New York, 1991, ch. 1, pp. 1–46.
124. D. Kregiel, *BioMed Res. Int.*, 2015, 1–15, DOI: 10.1155/2015/128697.
125. O. M. E. Abdel-Salam, E. R. Youness, N. A. Mohammed, S. M. Y. Morsy, E. A. Omara and A. A. Sleem, *J. Med. Food*, 2014, **17**, 588–598.
126. A. Bryland, A. Wieslander, O. Carlsson, T. Hellmark and G. Godaly, *Diabetes Vasc. Dis. Res.*, 2012, **9**, 42–51.
127. L. Gabutti, N. Ferrari, G. Mombelli, F. Keller and C. Marone, *J. Nephrol.*, 2004, **17**, 819–825.

128. M. Gritters, M. P. Grooteman, M. Schoorl, M. Schoorl, P. C. M. Bartels, P. G. Scheffer, T. Teerlink, C. G. Schalkwijk, M. Spreeuwenberg and M. J. Nubé, *Nephrol., Dial., Transplant.*, 2006, **21**, 153–159.
129. K. Tiranathanagul, O. Jeamsujitwimol, P. Susantitaphong, N. Kijkriengkraikul, A. Leelahavanichkul, N. Srisawat, K. Praditpornsilpa and S. Eiam-Ong, *Ther. Apheresis Dial.*, 2011, **15**, 556–564.
130. O. M. E. Abdel-Salam, A. A. Sleen and N. M. Shaffie, *Excli. J.*, 2009, **8**, 41–49.
131. M. K. Nielsen and N. Arneborg, *Food Microbiol.*, 2007, **24**, 101–105.
132. D. Rico, A. B. Martín-Diana, J. M. Barat and C. Barry-Ryan, *Trends Food Sci. Technol.*, 2007, **18**, 373–386.
133. R. Siener, R. Hönow, A. Seidler, S. Voss and A. Hesse, *Food Chem.*, 2006, **98**, 220–224.
134. C. O. Lopes, G. V. Dessimoni, M. C. Silva, G. Vieira and N. A. V. D. Pinto, *Aliment. Nutr.*, 2009, **20**, 669–675.
135. D. E. Kang, R. L. Sur, G. E. Haleblian, N. J. Fitzsimons, K. M. Borawski and G. M. Preminger, *J. Urol.*, 2007, **177**, 1358–1362.
136. G. E. Haleblian, V. A. Leitao, S. A. Pierre, M. R. Robinson, D. M. Albala, A. A. Ribeiro and G. M. Preminger, *J. Endourol.*, 2008, **22**, 1359–1366.
137. K. L. Penniston, S. Y. Nakada, R. P. Holmes and D. G. Assimos, *J. Endourol.*, 2008, **22**, 567–570.
138. K. I. Liszt, J. P. Ley, B. Lieder, M. Behrens, V. Stöger, A. Reiner, C. M. Hochkogler, E. Köck, A. Marchiori, J. Hans, S. Widder, G. Krammer, G. J. Sanger, M. M. Somoza, W. Meyerhof and V. Somoza, *PNAS*, 2017, E6260–E6269, DOI: 10.1073/pnas.1703728114; a. C. Yawen, *et al.* Consumption of hot beverages and foods and the risk of esophageal cancer: a meta-analysis of observational studies, *BMC Cancer*, 2015, **15**(n. 1), 1.
139. R. Fehlau and K. J. Netter, *Z. Gastroenterol.*, 1990, **28**, 234–238.
140. A. Darboven, *Food Sci. Technol.*, 1997, **8**(10), 348.
141. P. Terry, J. Lagergren, A. Wolk and O. Nyren, *Nutr. Cancer*, 2000, **38**, 186–191.
142. M. Feldman and C. Barnett, *Gastroenterology*, 1995, **108**, 125–131.
143. J. R. Feldman, W. S. Ryder and J. T. Kung, *J. Agric. Food Chem.*, 1969, **17**, 733–739.
144. R. López-Froilán, E. Ramírez-Moreno, N. S. Podio, M. L. Pérez-Rodríguez, M. Cámara, M. V. Baroni, D. A. Wunderlin and M. C. Sánchez-Mata, *Food Funct.*, 2016, **7**, 2706–2711.

CHAPTER 23

Caffeine and Minor Methylxanthines in Coffee

JULIANA DE PAULA LIMA AND ADRIANA FARAH*

Laboratório de Química e Bioatividade de Alimentos, Instituto de Nutrição, Universidade Federal do Rio de Janeiro, Av. Carlos Chagas Filho, 373, CCS, Bl. J, Rio de Janeiro, 21941-902, Brazil
*E-mail: afarah@nutricao.ufrj.br

23.1 Introduction

Methylxanthines are known to be alkaloids because they are secondary plant metabolites derived from purine nucleotides with a heterocyclic nitrogen atom (definition of true alkaloids).[1,2] Some authors, however, call them pseudo-alkaloids, since they do not have the incorporation of an amino acid in their biosynthesis.[1]

Present in nearly 100 species, methylxanthines are in 13 orders of the plant kingdom.[3] In nature, they are found mainly in *Coffea* seeds and leaves, *Camelia sinensis* and maté (*Ilex paraguariensis*) leaves, cocoa (*Theobroma cacao*) seeds, guaraná (*Paulinia cupana*) seeds, kola (*Cola* sp.)[1,4] nuts, and coca leaves (*Erythroxylon coca*). Caffeine and theobromine are also found in a number of commercial non-alcoholic beverages, powders, capsules, and in association with therapeutic drugs. Caffeine is the most abundant methylxanthine in foods, followed by theobromine and theophylline, with the last being present in very low concentrations.[5]

Coffee: Production, Quality and Chemistry
Edited by Adriana Farah

Published by the Royal Society of Chemistry, www.rsc.org

In this chapter, we briefly present the chemical characteristics, contents, and distribution of methylxanthines in coffee seeds and beverages, including low-caffeine varieties. The impact of seed processing on these contents is also discussed.

23.2 Chemical Characterization of Methylxanthines

Methylxanthines are heterocyclic organic compounds. They have a purine base called xanthine, consisting of a pyrimidine ring fused to an imidazole ring[1] (Figure 23.1).

Although there are reports on the use of coffee at least since the 9th century, caffeine and other methylxanthines were only understood as substances and as components of food matrixes in the middle of the 19th century.

Johann Wolfgang von Goethe (1749–1832), one of the most important writers and philosophers of Germany, was an advisor of the University of Jena and an active stimulator of the scientific progress. An autodidact on natural sciences, he was always surrounded by specialists to support his studies. One of them was the experimentalist Johann Wolfgang Döbereiner (1780–1849), who was a chemistry and technology professor at the University of Jena. In 1818, Ferdinand Runge (1795–1867), a young researcher, arrived in Jena and was soon introduced to Döbereiner. After being told about Runge's experiences in isolating the active substances of some plants and about his findings on the medicinal plant Belladona, Döbereiner scheduled a meeting with Goethe and Runge that occurred on October 3rd, 1819.[6] Runge was interviewed by Goethe, who passed on to him a box with unroasted coffee seeds, which had been given to him by a Greek researcher as a uniqueness. Shortly after, Runge isolated caffeine from coffee seeds and called it first "Kaffein", which later became "caffeine" in English, and was included in the medical vocabulary in 1823.[7]

In 1826, Theodore von Martius isolated guaranine from the guaraná plant and in 1827 Pflanzenstoff zuerst von Oudry found theine in tea. Later, in 1838, all of these substances were proven to be the same chemical compound.[8] In 1843, caffeine was found in maté prepared from *Ilex paraguariensis*

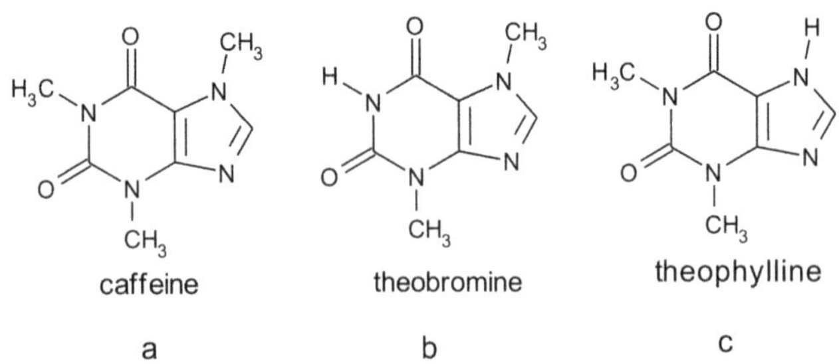

Figure 23.1 Chemical structures of caffeine (a), theobromine (b), and theophylline (c).

and in 1865 in kola nuts. Theobromine was first identified in 1842, in cocoa seeds, by Voskresensky. The molecule was named after the cocoa plant (*Theobroma cacao*), referring to the name given to it by Cortés, in "the food of Gods".[9]

The structure of caffeine was first proposed in 1875 by Ludwig Medicus (1847–1915) in his professorial dissertation. Unusual for that time, he deduced the structure from the already known pure compound.[8] The chemical structure of caffeine and other methylxanthines was validated in 1882 by Hermann Emil Fischer (1852–1919), who published a series of studies on purine synthesis that were explicitly cited in his Nobel Prize in Chemistry in 1902. The study of caffeine was one of his first major efforts, published when he was 29–30 years of age.[10–12] He used oxidation with moist chlorine, and found that caffeine had a similar heterocyclic skeleton to uric acid. He then found that it was a trimethylxanthine, but he struggled for some time to clarify the structure of xanthine. It was only when Fischer turned to a synthetic approach to the structure that he finally realized that the xanthine structure proposed by Ludwig Medicus prior to Fischer, in 1895, was the correct one.[13]

Fisher also identified theophylline as a minor component in tea in 1888, when he established its structure and devised a synthetic pathway.[14] He also synthesized theobromine for the first time in 1882. Already at the beginning of the 20th century, theophylline was introduced in the clinic, first as a diuretic, and later in the treatment of asthma.[15]

The main biosynthetic and catabolic pathways of caffeine and other methylxanthines were not fully established until the beginning of the century by Kato *et al.*[16,17] when highly purified caffeine synthase was obtained from tea leaves and a gene encoding the enzyme was cloned. This finding facilitated molecular studies on the regulation of caffeine production mainly with coffee plants.[18] The synthesis of caffeine in the coffee plant is detailed in Chapter 4.

Methylxanthines are colorless at room temperature, odorless, and bitter.[19] Regarding solubility, they have different behaviors.[1] Caffeine dissolves well in boiling water and its solubility is increased by the addition of acids and formation of complexes such as benzoate, citrate, and salicylate at high temperatures – 1% (w/v) at 15 °C and 10% at 60 °C.[20] However, it shows better solubility in chloroform at room temperature.[1] Theobromine is generally much less soluble than caffeine, but it can quickly dissolve in aqueous acidic or basic extract. Theophylline has intermediate solubility between caffeine and the theobromine.[1]

In aqueous solution, at physiological pH, caffeine is a non-ionized compound. Its melting point is 234–239 °C and the temperature of sublimation at atmospheric pressure is 178–180 °C, which is a lower temperature of sublimation than for theobromine (290–296 °C) and theophylline (269–272 °C).[1]

Reports on methylxanthines ultraviolet (UV) absorption region differ slightly in the literature. A reasonable interval between 250 and 280 nm can be concluded,[1] although wavelength limits of 243 and 302 nm have been observed for caffeine.[21] But in a general way, it can be accepted that caffeine is well absorbed between 271 and 275 nm,[22] with maximum absorption (λ_{max}) in aqueous solution at 272/273 nm.[21]

23.3 Analysis of Methylxanthines

Several analytical methods have been proposed for determination of methylxanthines in foods. Gravimetry was the first method developed for the analysis of methylxanthines, especially caffeine, in food products.[23] The technique consisted of extraction of the product with water or ethanol, cleaning the filtrate with magnesium and subsequent extraction with chloroform. However, this method was time-consuming and not all interferers were removed, which caused an overestimation of methylxanthines contents.[24]

The absorption of electromagnetic radiation in the UV region by caffeine was described in the early 20th century by Hartley.[25] Ishler *et al.*[26] developed a low-cost spectrophotometric analytical method in coffee products. The quantification of caffeine was based on the absorbance analysis at the wavelength of 272 nm. This method is faster, simpler, and more accurate than gravimetric methods but the results can still be overestimated by the presence of matrix interferers usually associated with food products.[24,27] The method is especially not recommended when used for simultaneous determination of individual methylxanthines given the high spectral overlap of these compounds[24,27] combined with the occurrence of very low amounts or lack of theobromine and theophylline relative to caffeine contents in most foods.

The High Performance Liquid Chromatography (HPLC) technique was used in methylxanthine analysis for the first time in the early 1970s.[28] Ion exchange chromatography was also used to separate caffeine from other organic compounds in coffee.[29] The use of HPLC for determination of caffeine was an important qualitative leap in terms of parameters of precision, accuracy, and speed[24] also allowing multi-analysis of methylxanthines. The development of stationary phases, containing microparticles (5 μm), and the use of the gradient system to separate caffeine from other methylxanthines allowed better resolution in a shorter period of time.[24] However, in coffee matrix, HPLC isocratic systems have been the most commonly used technique for determination and quantification of these compounds in coffee, using a reverse-phase column and a mixture of methanol and water as the mobile phase, which, depending on the column, can separate well all three methylxanthines.[30–32]

Methods for HPLC analysis evaluating simultaneously compounds different from caffeine were develop in the 1980s. The first study of this kind was published by Trugo *et al.*[33] and analyzed trigonelline and methylxanthines in soluble coffees and other beverages (tea, cocoa drink, and cola beverages) in a short run. The authors used saturated aqueous solution of basic lead acetate to clarify the samples, since Carrez solutions, which are often used for clarification of coffee extracts, tend to precipitate trigonelline. The precipitate was removed by simple filtration, obtaining extracts that were much clearer than those obtained by treatment with organic solvent.[24]

In some cases, however, when only one methylxanthine is present in the matrix, UV-Vis spectrophotometric determination may be preferred because it is possible to obtain rapidly high accuracy and reproducibility from small amounts of samples using a relatively simple and inexpensive procedure when compared to chromatographic techniques.[30] With the development of UV/Diode Array Detector (DAD), the multi-analysis of methylxanthines improved in terms of precision and accuracy, since each alkaloid was analyzed at its maximum absorption wavelength.[34,35]

Liquid chromatography mass spectrometry (LC-MS) is one of the preferred methods for determining caffeine at low levels and/or in very complex matrices. The three advantages of LC-MS over conventional HPLC methods can be represented by 3 "Ss": sensitivity, selectivity, and speed. However, due to the high cost, LC-MS techniques are not economically justifiable for the analysis of caffeine in major sources of this compound, such as coffee.[36] Zhu *et al.*[37] established a methodology for analyzing multiple bioactive compounds in tea and tea extracts, including caffeine and minor methylxanthines, by LC-DAD-ESI-MS (liquid chromatography – diode array detector – electrospray ionization – mass spectrometry). The proposed method provided more qualitative and quantitative information compared with general HPLC in the analysis of multi-components in tea. Complex extraction and sample pretreatment were also unnecessary.[36]

In recent years, as an alternative method to wet chemical methods,[24,38] the spectroscopy technique in the infra-red region presents very interesting results when applied in the pharmaceutical and food areas, specially the use of Fourier Transform Infra-red Spectroscopy (FTIR).[39] The method consists in the dissolution of the active principles in chloroform, followed by filtration of the solutions to remove the excipients.[24]

Ohnsmann *et al.*[40] used the infra-red technique to determine caffeine in tea leaf extracts. The method was based on the extraction with ammonia and chloroform and on the direct determination of the caffeine content in chloroform extracts, using the absorbance measurements associated with the peak heights and external calibration. The time required to perform the analysis of a single sample was relatively short (about 15 min).

Chemometrics is a valuable mathematical tool that, in combination with different chemical methods, enables the analysis of a large number of variables in a single sample.[24] Principal Components Analysis (PCA) has been most commonly used among chemometric methods to discriminate samples with different chemical patterns. In the study by Briandet *et al.*,[38] analyses of caffeine were performed by FTIR, followed by PCA to discriminate *C. arabica* and *C. canephora* species in lyophilized dry instant coffee blends.

23.4 Contents of Caffeine and Minor Methylxanthines in Coffee and Coffee Products

23.4.1 Content of Methylxanthines in Regular Green Coffee

Caffeine function in coffee plant and seeds seems to be basically related to protection. The "chemical defense theory" proposes that caffeine in young leaves, fruits, and flower buds acts to protect soft tissues from predators such as insect larvae[41] and beetles.[42] The "allelopathic theory" proposes that caffeine in seed coats is released into the soil and inhibits the germination of other seeds.[43] The steps on biosynthesis, catabolism, and the biological role of purine alkaloids in *Coffea* sp. are covered in detail in Chapter 4 of this book.

Caffeine, theobromine, and theophylline have been identified in coffee seeds,[44,45] with the last being found only in *C. canephora* species.[45,46] However, caffeine is by far the main alkaloid in green coffee seeds, with contents corresponding to more than 99% of total methylxanthines.[45,46]

The genetic variation is a key factor in controlling methylxanthines contents in green coffee seeds, which varies greatly, especially between species and, to a lesser extent, varieties.[47] The range of caffeine contents reported for green *C. arabica* L. vary in the literature between 0.7 and 1.7 g 100 g^{-1} (dry matter - dm),[21,32,45,46,48–60] while those for *C. canephora* vary in the range between 1.4 and 3.3 g 100 g^{-1} (dm).[32,45,46,49,50,53,57,58,60–64] However, common values are in the narrower range between 1.0 and 1.2 g 100 g^{-1} for *C. arabica*[21,50,52–57,59] and between 1.7 and 2.1 g 100 g^{-1} for *C. canephora*.[32,46,49,57,61,62]

In general, *C. canephora* seeds contain 40–50% more caffeine than those of *C. arabica* species,[65] with higher variability observed in *C. canephora* seeds.[63] *C. arabica* presents a more homogeneous composition independent of their geographical origin, due to low genetic diversity characteristic of the species.[45,66] The lower caffeine content in *C. arabica* species makes it more vulnerable to phytopathogens as well as to biological and mechanical stress than *C. canephora*.[67]

Caffeine seems to form a complex with chlorogenic acids, more specifically caffeoylquinic acids[68] in a 1:1 molecular ratio.[69] They may also bind with other purine alkaloids, such as theophylline and theobromine.[70] There is a lack of information on these complexes as well as on the complexation of methylxanthines with other subclasses of chlorogenic acids. Additional complexes may be formed, but at a lower rate, with caffeoylquinic acid precursors such as quinic or caffeic acids.[71] All of these compound combinations could explain the close correlation observed between chlorogenic acids and caffeine contents within the subgenus *Coffea*.[72]

The effect of environment, agricultural practices, including use of fertilizers, and post-harvest processing on methylxanthines contents seems to be less important than genetic aspects, except in the case of secondary processing such as decaffeination.[48,54,56]

Regarding post-harvest processing, after fruits harvesting, green coffee seeds are obtained by one of several different methods from which the most commonly known are called dry, wet, and semi-dry processing[73,74] (see Chapters 2 and 16 for details). Although all methods aim at removing the fruit flesh of the coffee cherry, they do it in different ways and this will affect the final contents of some compounds in the seeds.[75] Duarte *et al.*[56] compared the caffeine contents in coffee seeds processed by the wet and semi-dry methods and did not observe significant differences between both methods, in agreement with results from Balylaya and Clifford,[76] for wet and dry methods and from Leloup *et al.*[77] for dry and semi-dry methods, possibly due to the stability of caffeine in coffee.

Reports on the contents of theobromine in coffee seeds are scarce. They tend to be from 100 to 200 times lower than caffeine's content. For example, a study by Mehari *et al.*[59] investigated the methylxanthines composition of 99 green coffee samples from 8 arabica varieties (Harar, Jimma, Kaffa, Wollega, Sidama, Yirgachefe, Benishangul, and Finoteselam) from the major producing regions of Ethiopia, using HPLC-DAD. Theobromine contents varied from 4.8 to 9.4 mg 100 g^{-1} (dm), while caffeine contents varied from 0.9 to 1.4 g 100 g^{-1} (dm). The varieties from the east (Harar region) did not contain theobromine, suggesting geographical influence on contents. Based on this hypothesis, the authors suggested that the absence of this alkaloid may be used to discriminate coffees originated in that particular region. In the study, theophylline was not detected.

Due to low concentrations found in coffee, like with theobromine, reports on theophylline are scarce. Values reported for green *C. canephora* species range from 0.01 to 0.8 mg 100 g^{-1} (dm).[45,46] Reports on the theobromine content in *C. arabica* species and in roasted seeds in general were not found.

Caffeine data from seeds derived from fruits with different maturation degrees are somewhat conflicting. Clifford and Kazi[46] and Franca *et al.*[78] reported a small variation in caffeine content in arabica coffee during seed development, with the range of 1.39–1.42 and 1.23–1.36 g 100 g^{-1}, for both studies respectively. Toci and Farah,[79] on the other hand, observed significant differences among seeds from different degrees of maturation. Green and sour defects (from immature fruits), regular seeds (from mature fruits), and black seeds (from over-matured fruits) were evaluated. The highest caffeine contents were found in green (immature) (0.94 to 1.40 g 100 g^{-1}) and in control seeds (0.91 to 1.42 g 100 g^{-1}). Sour seeds (0.85 to 1.03 g 100 g^{-1}) and black seeds (0.42 to 0.94 g 100 g^{-1}) followed. The lower contents of caffeine found in black and sour seeds are in agreement with the differences in chlorogenic acids observed by Farah *et al.*,[80] although differences in caffeine content are smaller than those in chlorogenic acids. Using HPLC, Mazzafera *et al.*[50] also evaluated the levels of caffeine, theobromine, and theophylline in aqueous extracts of endosperm (seeds) from immature and mature fruits of *C. arabica* and *C. canephora* and five other *Coffea* species. The authors observed higher contents of caffeine in the seeds from immature fruits (1.1 and 3.4 mg 100 g^{-1} for *C. arabica* and *C. canephora*, respectively) compared with the seeds from mature fruits (1.0 and 2.4 mg 100 g^{-1}, respectively). Theobromine content varied from 0.03 to 0.08 mg 100 g^{-1} in seeds from immature fruits and from 0.032 to 0.033 mg 100 g^{-1} in seeds from mature fruits. Small amounts of theophylline were found in most immature endosperm extracts of *C. arabica* (0.05 to 0.2 mg 100 g^{-1}) and *C. canephora* (0.1 mg 100 g^{-1}).[50]

Despite the fact that caffeine contributes bitterness to the cup, the relationship between the content of caffeine in green seed and sensory quality of the beverage is not clear.

Data from a few studies investigating the contents of caffeine, theobromine, and theophylline in green *C. arabica* and *C. canephora* seeds are presented in Tables 23.1 and 23.2, respectively.

23.4.2 Contents of Methylxanthines in Regular Roasted Coffee

The contents of caffeine and other methylxanthines are not significantly altered during coffee roasting due to their thermal stability, but small losses may occur owing to sublimation. In terms of percentage composition, an increase in the caffeine content may be observed due to the loss of thermolabile compounds.[74]

The range of caffeine contents reported for *C. arabica* in roasted coffee seeds varies in the literature between 0.7 and 1.5 g 100 g^{-1} (dm)[21,32,35,58,78] while that for *C. canephora* varies between 1.8 and 2.5 g 100 g^{-1} (dm).[35,58,62]

Data from a few studies investigating the contents of caffeine, theobromine and theophylline in roasted *C. arabica* and *C. canephora* seeds and in commercial blends are presented in Table 23.3.

23.4.3 Contents of Methylxanthines in Coffee Brews

Caffeine contents in coffee are closely related to its stimulating properties as well as to about 10% or less of the bitterness of the beverages.[49,53,54] Past studies have suggested that there may be a mild association between caffeine content in *C. arabica* seeds and cup quality,[54,78,85] although the minor contribution of caffeine to the beverage's bitterness has led Ramalakshmi *et al.*[64] to conclude that this alkaloid is not responsible for differences in beverage quality.

The final content of caffeine and minor methylxanthines ingested by coffee drinkers depends on many factors such as blend composition (which depends on genetics, edaphoclimatic conditions, and post-harvest method), the type of roaster, the roasting conditions, grid, and the method used to brew the coffee, which varies dramatically according to the social and cultural habits of each country.[86]

Table 23.1 Summary data from studies investigating the contents of caffeine and theobromine in green *Coffea arabica* seeds.[a]

Coffee cultivar	Origin	Samples (*n*)	Method	Content range of caffeine (g 100 g^{-1})	Mean content of caffeine (g 100 g^{-1})	Mean content of theobromine (mg 100 g^{-1})	Reference
C. arabica cv. Maragogipe	Ivory Coast	1	HPLC-UV	Nr	1.4	Nd	46
C. arabica	Nr	4	HPLC-UV	0.9–1.4	Nr	Nr	49
C. arabica cv. Mundo Novo	Brazil	2	HPLC-UV	Nr	1.1	0.0003	50
C. arabica cv. Red Caturra, cv. Red Bourbon	Nr	2	HPLC-UV	0.7/0.8	Nr	Nr	51
C. arabica cv. Catuaí	Nr	3	Spectrophotometer	0.8–1.3	1.1	Nr	52
C. arabica	Africa	38	HPLC-DAD	0.9–1.6	1.2	Nr	53
C. arabica	Brazil	5	HPLC-DAD	0.9–1.2	1.0	Nr	54
C. arabica	Ethiopia	4	UV/vis spectrometer	0.9–1.1	1.1	Nr	21
C. arabica cv. Mundo Novo	Brazil	1	LC-MS	0.8–0.9	0.9	Nr	32
C. arabica	America	35	HPLC-DAD	1.0–1.7	1.5	Nr	45
C. arabica	Africa	10	HPLC-DAD	1.2–1.6	1.4	Nr	45
C. arabica	Asia	5	HPLC-DAD	1.4–1.6	1.5	Nr	45
C. arabica cv. Mundo Novo, cv. Red Catuaí, cv. Red Bourbon	Brazil	3	HPLC-DAD	1.0–1.1	1.1	Nr	55
C. arabica cv. Yellow Bourbon, cv. Red Catuaí, cv. Rubi, cv. Topazio	Brazil	5	HPLC-DAD	1.0–1.4	1.2	Nr	56
C. arabica	Brazil	5	HPLC-DAD	1.1–1.3	1.2	Nr	57
C. arabica cv. Minas, cv. Cioccolatato	Croatia	6	HPLC-DAD	0.6–1.2	0.9	Nr	58
C. arabica	Ethiopia	99	HPLC-DAD	0.8–1.3	1.1	0.0048–0.0094	59
C. arabica	Honduras	8	LC-MS	0.8–1.1	0.9	Nr	60
C. arabica	Kenya	8	LC-MS	0.8–1.1	0.9	Nr	60

[a]HPLC: High Performance Liquid Chromatography; DAD: diode array detector; UV: ultra violet; vis: visible; LCMS: Liquid Chromatography–Mass Spectrometry. Nr, not reported; Nd, not detected.

Table 23.2 Summary data from studies investigating the contents of methylxanthines in green *Coffea canephora* seeds.[a]

Coffee cultivar	Origin	Samples (*n*)	Method	Content range of caffeine (g 100 g^{-1})	Mean content of caffeine (g 100 g^{-1})	Mean content of theobromine (g 100 g^{-1})	Mean content of theophylline (g 100 g^{-1})	Reference
C. canephora cv. robusta	Ivory Coast	2	HPLC-UV	Nr	2.1	0.01	Nd	46
C. canephora	Nr	2	HPLC-UV	2.1/2.7	Nr	Nr	Nr	49
C. canephora cv. conilon	Brazil	1	HPLC-UV	Nr	2.4	0.0003	Nd	50
C. canephora cv. robusta	Ivory Coast	3	HPLC-UV	2.7–2.8	Nr	Nr	Nr	61
C. canephora cv. robusta	Uganda	3	HPLC-UV	1.7–3.2	Nr	Nr	Nr	61
C. canephora cv. robusta	Ivory Coast	1	HPLC-UV	2.2–3.2	Nr	Nr	Nr	62
C. canephora	Nr	2	HPLC-DAD	1.5/3.3	2.4	Nr	Nr	63
C. canephora cv. robusta	India	2	Spectrophotometry	1.6/1.8	Nr	Nr	Nr	64
C. canephora cv. robusta	Africa	40	HPLC-DAD	1.5–3.3	2.5	Nr	Nr	53
C. canephora cv. conilon	Brazil	2	LC-MS	1.6–1.7	1.7	Nr	Nr	32
C. canephora	Brazil	3	HPLC-DAD	1.9–2.0	2.0	Nr	Nr	57
C. canephora cv. robusta	Africa	17	HPLC-DAD	2.0–2.8	2.5	Nr	0.01–0.007	45
C. canephora cv. robusta	Asia	39	HPLC-DAD	2.3–3.1	2.7	Nr	0.01–0.007	45
C. canephora cv. robusta	Croatia	2	HPLC-DAD	1.9/2.1	1.9	Nr	Nr	58
C. canephora	Brazil	1	LC-MS	Nr	1.2	Nr	Nr	60
C. canephora	Uganda	2	LC-MS	1.8/1.9	Nr	Nr	Nr	60
C. canephora	Vietnam	2	LC-MS	1.6–1.7	Nr	Nr	Nr	60

[a]HPLC: High Performance Liquid Chromatography; DAD: diode array detector; UV: ultra violet; vis: visible; LCMS: Liquid Chromatography - Mass Spectrometry; Nr, not reported; Nd, not detected.

Table 23.3 Summary data from studies investigating the contents of caffeine, in roasted *Coffea arabica*, *C. canephora* seeds and commercial blends.[a]

Coffee cultivar	Origin	Samples (*n*)	Method	Content range of caffeine (g 100 g^{-1})	Mean content of caffeine (g 100 g^{-1})	Reference
Coffea arabica						
C. arabica	From several geographic origins	9	HPLC-DAD	1.2–1.6	1.3	35
C. arabica	Brazil	4	HPLC-DAD	0.7–1.1	0.9	78
C. arabica	Ethiopia	4	UV/vis spectrometer	0.9–1.1	1.0	21
C. arabica	Hawaii	1	LC-MS	Nr	0.8	32
C. arabica	Brazil	3	LC-MS	0.8–0.9	Nr	32
C. arabica	Colombia	1	LC-MS	Nr	0.8	32
C. arabica cv. Minas, cv. Cioccolatato	Croatia	6	HPLC-DAD	0.8–1.4	1.3	58
C. arabica	USA	2	HPLC-DAD	0.9/1.2	1.1	Unpublished
Coffea canephora						
C. canephora cv. robusta	From several geographic origins	20	HPLC-DAD	2.0–2.6	2.3	35
C. canephora cv. robusta	Croatia	2	HPLC-DAD	1.8/2.5	2.2	58
C. canephora cv. robusta	Ivory Coast	1	HPLC-UV	1.8–2.0	2.0	62
Commercial blends of *C. arabica* or *C. arabica* and *C. canephora*						
Blend	Brazil	10	HPLC-UV	0.8–1.4	Nr	81
Blend	USA	6	HPLC-UV	1.0–1.6	1.3	82
Nr	Nr	3	HPLC-UV	1.4–1.9	1.7	83
Blend	Africa	1	LC-MS	Nr	1.4	32
Blend	Brazil	40	HPLC-DAD	0.8–1.6	1.2	84
Arabica blends	USA	36	HPLC-DAD	0.8–1.4	1.2	Unpublished

[a]HPLC: High Performance Liquid Chromatography; DAD: diode array detector; UV: ultra violet; vis: visible; LCMS: Liquid Chromatography - Mass Spectrometry. Nr, not reported.

Caffeine is well extracted by the most common hot brewing methods, from 75% to nearly 100%.[74,87]. Percolation and decoction methods tend to extract more than infusion methods which have low extraction power,[65] but not only the method is important. During brewing, the percentage of caffeine extracted will depend on the proportion of water to powder, the water temperature, the size of the particle and the duration of the brewing process.[74,82] Caffeine seems to also be more easily extracted from dark roasted coffees. The type of filter may also alter the final result.[74]

As with seeds, in coffee brews, caffeine is the most abundant methylxanthine. Based on data from over 2000 coffee drinks analyzed in the UK, Burg[89] suggested mean values of 55 mg of caffeine per 100 mL cup of percolated coffee, 39 mg 100 mL^{-1} of reconstituted instant coffee, and 2 mg 100 mL^{-1} in decaffeinated coffee. In another study that analyzed the caffeine content of 20 different specialty coffees purchased at coffee shops in the United States, the amount of caffeine in brewed coffee ranged from 76 to 112 mg 240 mL^{-1}.[88] In addition, it was found that the caffeine content of the same type of coffee purchased in the same store on six separate occasions ranged from 130 to 282 mg 240 mL^{-1}. Results from the analyses of seven decaffeinated coffee beverages were equal to or lower than 17.7 mg 473 mL^{-1}.[88] Data from studies investigating the contents of caffeine in coffee beverages prepared by different methods are presented in Table 23.4.

In general, the contents of theobromine and theophylline have not been within the detection limits in most studies analyzing coffee brews for methylxanthines, but using HPLC-DAD, Rodrigues and Bragagnolo[95] evaluated the contents of methylxanthines in coffee brews from roasted seeds and identified 10.9 to 12.9 mg of theobromine per 100 g dry extract and 3.7 to 10.2 mg of theophylline per 100 g dry extract.

Instant coffee production typically involves treating ground-roast coffee with hot water and high pressure to extract the water-soluble compounds, followed by drying. The instant coffee production is covered in detail in Chapter 11. While ground-roasted coffee generally consists of *C. arabica* species alone or of a high percentage of *C. arabica*, in various countries *C. canephora* seeds are often used at a high percentage or alone in blends designated for instant coffee production because they yield a higher amount of soluble solids in the brew. So this explains the higher caffeine levels observed in reconstituted soluble coffee purchased in some countries.[74] Data from studies investigating the contents of caffeine in instant coffee are presented in Table 23.5.

23.4.4 Content of Methylxanthines in Decaffeinated and Low-Caffeine Coffees

Decaffeination is a process that significantly influences the final caffeine concentration and can be performed for all products that originally contain the alkaloid.[65] Demand for decaffeinated coffee has increased gradually since the

Table 23.4 Summary data from studies investigating the contents of caffeine in coffee beverages.[a]

Coffee sp.	Origin	Samples (n)	Amount of powder to water volume (%)	Water temperature/brewing time	Analytical method	Mean content of caffeine (mg 100 mL^{-1})	Reference
Infusion (coffee bag)							
Arabica	Japan	2	10	95 °C/5 min	HPLC-DAD	43	Unpublished
French press							
Arabica	Guatemala	2	8	98 °C/5 min	HPLC-DAD	118	90
Robusta	Vietnam	2	8	98 °C/5 min	HPLC-DAD	156	90
Blends	Spain – local market	3	8	98 °C/5 min	HPLC-DAD	136	91
Arabica	USA	1	7	90 °C/5 min	HPLC-DAD	75	Unpublished
Cold brewing							
Arabica	USA	1	7	95 °C followed by 10 °C	HPLC-DAD	52	Unpublished
Arabica	USA	1	7	10 °C	HPLC-DAD	51	Unpublished
Electric dripper							
Arabica	Guatemala	1	6	90 °C/6 min	HPLC-DAD	100	90
Robusta	Vietnam	1	6	90 °C/6 min	HPLC-DAD	220	90
Blends	Spain – local market	3	6	90 °C/7 min	HPLC-DAD	110	91
Commercial blends	Brazil	40	10	95 °C/4 min	HPLC-DAD	110	84
Arabica	USA	1	7	95 °C/4 min	HPLC-DAD	76	Unpublished
Manual drip							
Nr	UK – coffee shops	Nr	Nr	Nr	Nr	55	89
Nr	USA – local market	26	Nr	Nr	GLC	51	92
	Nr	Nr	Nr	Nr	Nr	52.1	93
Nr	USA – coffee shops	14	Nr	Nr	GC	45.7	88
Nr	USA – coffee shops	27	Nr	Nr	Nr	65.1	94
Nr	USA – local market	7	3	Nr	HPLC-UV	131	82

(*continued*)

Table 23.4 (*continued*)

Coffee sp.	Origin	Samples (n)	Amount of powder to water volume (%)	Water temperature/ brewing time	Analytical method	Mean content of caffeine (mg 100 mL^{-1})	Reference
Blends	Brazil	8	10	92.3 °C/—	HPLC-DAD	45	95
Blends	Brazil – local market	40	10	90 °C/4 min	HPLC-DAD	56.7	84
Moka							
Blends	Spain – local market	3	8	93 °C/10 min	HPLC-DAD	156	91
Robusta	Vietnam	2	8	93 °C/10 min	HPLC-DAD	226	90
Arabica	USA	1	7	95 °C/10 min	HPLC-DAD	92	Unpublished
Espresso							
Nr	USA	Nr	Nr	Nr	Nr	123.2	93
Nr	USA – coffee shops	6	Nr	Nr	GC	96.8	88
Nr	USA – coffee shops	20	Nr	Nr	Nr	250	94
Blends	Australia – coffee shops	97	Nr	Nr	MECC	247	96
Blends	Scotland – coffee shops	20	Nr	Nr	HPLC-DAD	360	97
Blends	Croatia – local coffee manufacturer	13	14	95–97 °C	HPLC-DAD	280.2	98
Arabica	Guatemala	1	17.5	90 °C/24 s	HPLC-DAD	100	90
Robusta	Vietnam	1	17.5	90 °C/24 s	HPLC-DAD	177	90
Blends	Brazil – local market	40	10	90 °C/30 s	HPLC-DAD	196.4	84
Arabica	Spain – local market	1	18.7	96 °C/21 s	HPLC-DAD	209	99
Robusta	Spain – local market	1	18.7	96 °C/21 s	HPLC-DAD	288	99
Blends	Spain – local market	3	17.5	90 °C/30 s	HPLC-DAD	192	91
Arabica	USA	1	7	90 °C/28 s	HPLC-DAD	82	Unpublished
Boiled							
Arabica	USA	1	7	95 °C/5 min	HPLC-DAD	85	Unpublished

Turkish coffee							
Blends	Croatia – local coffee manufacturer	13	14	95–97 °C	HPLC-DAD	248	98
Arabica	USA	1	7	95 °C/4 min	HPLC-DAD	85	Unpublished
Soluble or instant coffee beverages							
Nr	UK – coffee shops	Nr	Nr	Nr	Nr	39	89
Nr	USA – local market	26	Nr	Nr	GLC	44	92
Nr	USA – coffee shops	20	Nr	Nr	Nr	39	94
Blends	Brazil – local market	40	2	90 °C	HPLC-DAD	169	84
Nr	Nr	2	1	Nr	HPLC-UV	212	27
Nr	Nr	2	2.2	Hot water	HPLC-UV	74	100
Nr	Nr	2	1	Hot water	GC	246	101
Ready to drink coffee beverages							
Blend	Japan	4	—	—	HPLC-DAD	57	Unpublished

[a]HPLC: High Performance Liquid Chromatography; DAD: diode array detector; UV: ultra violet; GC: Gas chromatography; GLC: Gas Liquid Chromatography; MECC: Micellar electrokinetic capillary chromatography; Nr, not reported.

Table 23.5 Summary data from studies investigating the contents of caffeine in instant coffees.[a]

Coffee sp.	Origin	Samples (n)	Method	Content range of caffeine (g 100 g^{-1})	Mean content of caffeine (g 100 g^{-1})	Reference
Blend	Brazil	9	HPLC-DAD	1.6–3.2	2.5	102
Blend	Brazil	2	LC-MS	2.0–2.2	2.1	32
Nr	Kenya	2	HPLC-DAD	1.6–3.4	2.7	103
Arabica	Brazil	3	HPLC-DAD	2.8–4.1	3.4	104
Robusta	Brazil	3	HPLC-DAD	3.9–5.8	5.0	104
Nr	Nr	2	HPLC-UV	4.5–4.8	Nr	83
Nr	Nr	5	HPLC-DAD	2.2–3.9	2.8	105

[a]HPLC: High Performance Liquid Chromatography; DAD: diode array detector; UV: ultra violet; LCMS: Liquid Chromatography - Mass Spectrometry. Nr, not reported.

early 1970s. Worldwide sales of "decaf" coffee have achieved a 15% share of the market in the last decade.[106] The process can be performed using different methods and solvents (see Chapter 7 for details) and often leaves small amounts of caffeine in the seeds.[74]

According to various national and international regulations, the decaffeination process should lower the methylxanthine concentration by 97–99.9% in ground roasted coffee and 97% in instant coffee[107–110] without a significant influence on other natural compounds.[65] A more restricted norm is obligatory in European Community countries[96] where decaffeinated coffee has to be 99.9% alkaloid free.[65,107] Ranges of 0.3–0.5 mg caffeine 100 g^{-1} in decaffeinated ground roasted[82] and of 0.7–0.9 mg 100 g^{-1} in instant decaffeinated coffee[83] have been reported.

Values from 0.0 to 13.9 mg 100 mL^{-1} of caffeine have been reported for brews from decaffeinated ground roasted coffee,[30,89,111] while values from 0.9 to 1.2 mg 100 mL^{-1} have been reported for reconstituted instant decaffeinated coffee.[27,30] It is noteworthy that for a heavy coffee drinker, the ingestion of multiple servings of decaffeinated beverages could result in caffeine doses equivalent to a caffeinated beverage.[111]

Low caffeine and caffeine-free species have been found in African nature due mainly to the low biosynthetic activity of purine alkaloids, producing possibly extremely weak *N*-methyltransferase reactions in caffeine biosynthesis.[112] Coffee cultivars combining high cup quality with a low caffeine content may provide a less expensive and "ecofriendly" (referring to decaffeination methods using organic solvents) alternative to meet the demand for decaffeinated coffee.[112] Seeds of *C. arabica* var. laurina, for example, have about 50% lower caffeine content than other *C. arabica* plants (0.5–0.65%, dry matter weight).[49,52,113] This variety has been among the most sensorially acceptable low caffeine varieties and therefore it has been cultivated around the world and used to compose low caffeine gourmet blends. However, productivity is lower than in commercial plants with higher caffeine content and they are more susceptible to diseases,[52] which implies a higher market value than regular gourmet coffee.

Silvarolla *et al.*[114] reported the existence of naturally decaffeinated mutant plants in the progeny of *C. arabica* accessions from Ethiopia, which they called AC1, AC2, and AC3 in honor of the Brazilian geneticist Alcides Carvalho. The seeds of those plants had very low caffeine content (mean caffeine content 0.07% dry weight), but significant amounts of theobromine (about 0.6%), which has a milder stimulating effect (about one-third) compared to caffeine.[115,116]

Other low-caffeine species include: *C. eugenioides* (0.4–0.5%), *C. charrieriana* (0.0–0.4%), *Mascarocoffea* sp. (0.7%), and *C. recemosa* (0.8–0.9%).[47,51,63,65,114,117]

Attempts to use genetic engineering to produce transgenic low-caffeine coffees have had to date only limited success. Seeds of *C. charrieriana* were introduced on the market as the first natural caffeine-free coffee. In this plant, caffeine has been replaced by theobromine due to genetic modification. This feature could be transferred to other plants by crossing it with other *Coffea* plants.[65]

Ogita *et al.*[118] reported the construction of transgenic low-caffeine *C. canephora* in which expression of the gene encoding caffeine synthase (CaDXMT1) and theobromine synthase (CaMXMT1) were repressed by RNA interference (RNAi), but the caffeine content of the resulting seeds was variable, reaching a reduction of up to 70%.[118,119] *C. arabica* has a superior flavor compared to that of *C. canephora* coffee but low-caffeine transgenic *C. arabica* seeds have not been produced yet. When this is achieved, it is likely to have major commercial implications.[120]

More recently, Nagai *et al.*[121] reported production of a new low-caffeine tetraploid interspecific hybrid developed in Madagascar from *C. eugenioides*, *C. canephora*, and *C. arabica*, with green seeds containing 0.3 to 0.4 g of caffeine 100 g^{-1} and no detectable theobromine.

23.5 Concluding Remarks

In the present review, the contribution of genetic variation, especially regarding species, for caffeine and other methylxanthines content in coffee seeds was highlighted, while post-harvesting processing and roasting do not seem to affect greatly its contents. Large variation on reported contents of caffeine in brews is observed due to cultural habits, which are reflected in the blend composition, preparation methods, and proportion of powder/water. Lower amounts of caffeine in coffee can be obtained by decaffeination processes, but naturally low amounts have been found in African species different from *C. arabica* and *C. canephora*. The search for development of resistant and well-accepted "caffeine-free" seeds for individuals who are sensitive to the stimulating effect of caffeine is an agricultural trend.

Acknowledgements

The authors would like to acknowledge the scholarships from National Research Council – CNPq and Research Support Foundation of Rio de Janeiro – FAPERJ (Brazil).

References

1. S. M. Tarka and W. J. Hurst, in *Caffeine*, ed. G. A. Spiller, CRC Press, Florida, 1998, p. 12.
2. K. G. Zulak, D. K. Liscome, H. Ashihara and P. J. Facchini, in *Plant Secondary Metabolites: Occurrence Structure, and Role in the Human Diet*, ed. A. Crozier, M. N. Clifford and H. Ashihara, Blackwell, Oxford, 2006, p. 1.
3. H. Ashihara and A. Crozier, *Adv. Bot. Res.*, 1999, **30**, 117.
4. T. M. Chow and N. L. Benowitz, *Comp. Biochem. Physiol.*, 1994, **109**(2), 173.
5. R. A. Pauwels, *Am. J. Respir. Crit. Care Med.*, 2001, **163**(2), 1256.

6. J. W. von Goethe and F. W. von Biedermann, in *Hauswirtschaftlichen Briefen*, ed. J. W. von Goethe and F. W. von Biedermann, G. Gespräche, Nachträge, Leipzig, 1896, p. 89.
7. A. Weinberg and B. K. Bealer, in *The World of Caffeine*, ed. B. A. Weinberg and B. K. Bealer, Routledge, New York, 2001, p. 394.
8. R. Meusinger, *Anal. Bioanal. Chem.*, 2009, **393**, 1381.
9. M. J. Arnaud, in *Methylxanthines*, ed. B. B. Fredholm, Springer, New York, 2011, p. 33.
10. E. Fischer, *Ber. Dtsch. Chem. Ges.*, 1881, **14**, 1905.
11. E. Fischer, *Ber. Dtsch. Chem. Ges.*, 1881, **14**, 637.
12. E. Fischer, *Ber. Dtsch. Chem. Ges.*, 1882, **15**, 29.
13. E. Fischer, *Ber. Dtsch. Chem. Ges.*, 1897, **30**, 549.
14. B. B. Fredholm, in *Methylxanthines*, ed. B. B. Fredholm, Springer, New York, 2011, p. 549.
15. H. H. Salter, in *On Asthma: Its Pathology and Treatment*. John Churchill, London, 1860.
16. M. Kato, K. Mizuno, T. Fujimura, M. Iwama, M. Irie, A. Crozier and H. Ashihara, *Plant Physiol.*, 1999, **120**(2), 579.
17. M. Kato, K. Mizuno, A. Crozier, T. Fujimura and H. Ashihara, *Nature*, 2000, **406**(6799), 956.
18. S. Ogita, H. Uefuji, Y. Yamaguchi, N. Koizumi and H. Sano, *Nature*, 2003, **423**, 823.
19. O. G. Vitzthum and P. Werkhoff, *Z. Lebensm.-Unters. Forsch.*, 1976, **160**, 277.
20. M. J. Arnaud, in *Caffeine: Encyclopedia of Human Nutrition*, ed. B. Caballero, L. Allen and A. Prentice, Elsevier Academic Press, Cambridge, 2005, p. 247.
21. A. Belay, K. Ture, M. Redi and A. Asfaw, *Food Chem.*, 2008, **108**, 310.
22. E. R. Holiday, *Biochem. J.*, 1930, **24**, 619.
23. Association of Official Agricultural Chemists in *Official and Tentative Methods of Analysis*, ed. George Banta Publishing, Washington, 1945.
24. C. A. B. De Maria and R. F. A. Moreira, *Quim. Nova*, 2007, **30**, 99.
25. W. N. Hartley, *J. Chem. Soc., Trans.*, 1905, **87**, 1796.
26. N. H. Ishler, T. P. Finucane and E. Borker, *Anal. Chem.*, 1948, **20**, 1162.
27. L. López-Martínez, P. L. López-de-Alba, R. García-Campos and L. M. De León-Rodríguez, *Anal. Chim. Acta*, 2003, **493**, 83.
28. S. R. Brown, P. A. Cann and N. W. Read, *Gut*, 1990, **31**, 450.
29. E. Murgia, P. Richards and H. F. Walton, *J. Chromatogr.*, 1973, **87**, 523.
30. M. S. Bispo, M. C. C. Veloso, H. L. C. Pinheiro, R. F. S de Oliveira, J. O. N. Reis and J. B. de Andrade, *J. Chromatogr. Sci.*, 2002, **40**, 45.
31. A. Farah, T. de Paulis, L. C. Trugo and P. R. Martin, *J. Agric. Food Chem.*, 2006, **54**(2), 374.
32. D. Perrone, C. M. Donangelo and A. Farah, *Food Chem.*, 2008, **110**, 1030.
33. L. C. Trugo, R. Macrae and J. Dick, *J. Sci. Food Agric.*, 1983, **34**, 300.
34. S. Casal, M. B. Oliveira and M. A. Ferreira, *Food Chem.*, 1998, **21**, 3187.

35. S. Casal, M. B. Oliveira and M. A. Ferreira, *J. Agric. Food Chem.*, 2000, **48**, 3420.
36. D. Perrone and A. Farah, in *Caffeine: Chemistry, Analysis, Function and Effects*, ed. V. Preedy, Royal Society of Chemistry, Cambridge, 2012, p. 97.
37. X. Zhu, B. Chen, M. Ma, X. Luo, F. Zhang, S. Yao, Z. Wan, D. Yang and H. Hang, *J. Pharm. Biomed. Anal.*, 2004, **34**, 695.
38. R. Briandet, E. K. Kemsley and R. H. Wilson, *J. Agric. Food Chem.*, 1996, **44**, 170.
39. Z. Bouhsain, S. Garrigues and M. de la Guardia, *Analyst*, 1997, **122**, 441.
40. J. Ohnsmann, G. Quintas, S. Garrigues and M. de la Guardia, *Anal. Bional. Chem.*, 2002, **374**, 561.
41. J. B. Harborne, in *Introduction to Ecological Biochemistry*, ed. J. B. Harborne, Elsevier Academic Press, Cambridge, 1993, p. 1.
42. P. Hewavitharanage, S. Karunaratne and N. S. Kumar, *Phytochemistry*, 1999, **51**, 35.
43. G. R. Waller, *Biol. Plant.*, 1989, **31**, 418.
44. C. W. Huck, W. G. Uggeabichler and G. K. Bonn, *Anal. Chim. Acta*, 2005, **538**, 195.
45. R. M. Alonso-Salces, F. Serra, F. Reniero and A. Rolyheberger, *J. Agric. Food Chem.*, 2009, **57**(10), 4224.
46. M. N. Clifford and T. Kazi, *Food Chem.*, 1987, **26**, 59.
47. P. Mazzafera and A. Carvalho, *Euphytica*, 1992, **59**, 55.
48. R. Macrae, in *Coffee Chemistry*, ed. R. J. Clarke and R. Macrae, Elsevier, New York, 1985, p. 115.
49. M. N. Clifford, in *Coffee: Botany, Biochemistry and Production of Beans and Beverage*, ed. M. N. Clifford and K. C. Wilson, Springer, New York, 1985. p. 305.
50. P. Mazzafera, A. Crozier and A. C. Magalhães, *Phytochemistry*, 1991, **30**, 3913.
51. M. N. Clifford and J. R. Ramirez-Martinez, *Food Chem.*, 1991, **40**, 35.
52. P. Mazzafera, A. Carvalho, L. C. Fazuoli and H. P. Medina Filho, *Turrialba*, 1992, **42**, 231.
53. C. L. Ky, J. Louarn, B. Guyot, S. Dussert, S. Hamon and M. Noirot, *Food Chem.*, 2001, **75**, 223.
54. A. Farah, M. C. Monteiro, V. Calado, A. S. Franca and L. C. Trugo, *Food Chem.*, 2006, **98**, 373.
55. P. Mazzafera and M. B. Silvarolla, *Seed Sci. Res.*, 2010, **20**(3), 163.
56. G. S. Duarte, A. A. Pereira and A. Farah, *Food Chem.*, 2010, **118**, 851.
57. R. M. N. de Souza and M. T. Benassi, *J. Braz. Chem. Soc.*, 2012, **23**, 1347.
58. I. Hečimović, A. Belščak-Cvitanović, D. Horžić and D. Komes, *Food Chem.*, 2011, **129**(3), 991.
59. B. Mehari, M. Redi-Abshiro and B. S. Chandravanshi, *Food Anal. Methods*, 2016, **9**, 1627.
60. O. Babova, A. Occhipinti and M. E. Maffei, *Phytochemistry*, 2016, **123**, 33.
61. M. J. Martín, F. Pablos and A. G. Gonzalez, *Talanta*, 1998, **46**, 1259.

62. E. Nebesny and G. Budryn, *Eur. Food Res. Technol.*, 2003, **217**, 157.
63. C. Campa, S. Doulbeau, S. Dussert, S. Hamon and M. Noirot, *Food Chem.*, 2005, **93**, 135.
64. K. Ramalakshmi, R. Kubra and L. J. M. Rao, *J. Food Sci.*, 2007, **72**(5), 333.
65. F. Burdan, in *Coffee in Health and Disease Prevention*, ed. V. R. Preedy, Elsevier, Cambridge, 2015, p. 823.
66. C. Bertrand, M. Noirot, S. Doulbeau, A. de Kochko, S. Hamon and C. Campa, *Plant Sci.*, 2003, **165**, 1355.
67. N. Fujimori and H. Ashihara, *Phytochemistry*, 1994, **36**, 1359.
68. S. Payen, *Ann. Phys.*, 1846, **60**, 286.
69. E. Sondheimer, C. D. Szymanski and J. W. Corse, *J. Agric. Food Chem.*, 1961, **9**, 146.
70. A. W. Kappeler, T. W. Baumann and H. Greutert, *Proc. Int. Congr. ASIC*, 1987, **12**, 247.
71. F. Anthony, M. Noirot and M. N. Clifford, *Genet. Resour. Crop Evol.*, 1993, **40**, 61.
72. I. Horman and R. Viani, *J. Food Sci.*, 1972, **37**, 925.
73. A. A. Teixeira, C. H. Brando, R. A. Thomaziello, R. Teixeira and R. Drying, in *Espresso Coffee: The Science of Quality*, ed. A. Illy and R. Viani, Elsevier Academic Press, San Diego, 1995, p. 91.
74. A. Farah, in *Coffee: Emerging Health Effects and Disease Prevention*, ed. Y. F. Chu, IFT press, New York, 2012, p. 21.
75. L. C. Trugo and R. Macrae, *Analyst*, 1984, **109**, 263.
76. K. J. Balyaya and M. N. Clifford, in *Proceedings of the 16th of the International Conference on Coffee Science*, 1995, p. 316.
77. V. Leloup, A. Louvrier and R. Liardon, in *Proceedings of the 16th of the International Conference on Coffee Science*, 1995, p. 192.
78. A. S. Franca, L. S. Oliveira, C. F. Mendonça and X. A. Silva, *Food Chem.*, 2005, **90**, 89.
79. A. T. Toci and A. Farah, *Proceedings of the 23th of the International Conference on Coffee Science*, 2011, p. 260.
80. A. Farah, M. C. Monteiro and L. C. Trugo, *Proceedings of the 21th of the International Conference on Coffee Science*, 2007, p. 240.
81. M. C. Monteiro and L. C. Trugo, *Quim. Nova*, 2005, **28**, 637.
82. K. Fujioka and T. Shibamoto, *Food Chem.*, 2008, **106**, 217.
83. J. R. Santos and A. O. S. S. Rangel, *Anal. Chim. Acta*, 2012, **715**, 57.
84. J. P. Lima and A. Farah, *Proceedings of the 25th of the International Conference on Coffee Science*, 2015.
85. A. Farah, Doctorate Thesis, Instituto de Química, Universidade Federal do Rio de Janeiro, Brasil, Distribuição nos grãos, influência sobre a qualidade da bebida e biodisponbilidade dos ácidos clorogênicos do café, 2004.
86. A. Farah, in *Functional and Speciality Beverage Technology*, ed. P. Paquim, CRC Press, London, 2009, p. 370.
87. A. D'Amicis and R. Viani, in *Caffeine, Coffee and Health*, ed. S. Garattini, Raven Press, New York, 1993, p. 17.

88. R. R. McCusker, B. A. Goldberger and E. J. Cone, *J. Anal. Toxicol.*, 2003, **27**, 520.
89. A. W. Burg, *Tea Coffee Trade J.*, 1975, **147**, 40.
90. J. Bravo, I. Juaniz, C. Monente, B. Caemmerer, L. W. Kroh, M. P. De Peña and C. Cid, *J. Agric. Food Chem.*, 2012, **60**, 12565.
91. I. Lopez-Galilea, M. P. de Peña and C. Cid, *J. Agric. Food Chem.*, 2007, **55**, 6110.
92. R. M. Gilbert, J. A. Marshman, M. Schwieder and R. Berg, *Can. Med. Assoc. J.*, 1976, **114**(3), 205.
93. J. J. Barone and H. R. Roberts, *Food Chem. Toxicol.*, 1996, **34**(1), 119.
94. Center for Science in the Public Interest, <<http://www.cspinet.org/new/cafchart.htm >>, last acessed January 2017.
95. N. P. Rodrigues and N. Bragagnolo, *J. Food Compos. Anal.*, 2013, **32**, 105.
96. B. Desbrow, R. Hughes, M. Leveritt and P. Scheelings, *Food Chem. Toxicol.*, 2007, **45**, 1588.
97. T. W. M. Crozier, A. Stalmach, M. E. J. Lean and A. Crozier, *Food Funct.*, 2012, **3**, 30.
98. T. Niseteo, D. Komes, A. Belščak-Cvitanović, D. Horžić and M. Budeč, *Food Chem.*, 2012, **134**, 1870.
99. L. Maeztu, S. Andueza, C. Ibañez, M. P. de Peña, J. Bello and C. Cid, *J. Agric. Food Chem.*, 2001, **49**, 4743.
100. N. M. D. Aragao, M. C. C. Veloso, M. S. Bispo, S. L. C. Ferreira and J. B. Andrade, *Talanta*, 2005, **67**, 1007.
101. H. Sereshti and S. Samadi, *Food Chem.*, 2014, **158**, 8.
102. M. Nogueira and L. C. Trugo, *Cienc. Tecnol. Aliment.*, 2003, **23**(2), 296.
103. H. N. Wanyika, E. G. Gatebe, L. M. Gitu, E. K. Ngumba and C. W. Maritim, *Afr. J. Food Sci.*, 2010, **4**(6), 353.
104. J. A. Vignoli, D. G. Bassoli and M. T. Benassi, *Food Chem.*, 2011, **124**, 863.
105. G. Del Campo, I. Berregi, R. Caracena and J. Zuriarrain, *Talanta*, 2010, **81**, 367.
106. National Coffee Association, <<http://www.ncausa.org>>, last acessed January 2017.
107. European Coffee Federation, <<http://www.ecf-coffee.org>>, last acessed January 2017.
108. International Coffee Organization, <<http://www.ico.org>>, last acessed January 2017.
109. United States Department of Agriculture, <<https://ndb.nal.usda.gov/ndb>>, last acessed January 2017.
110. Agência Nacional de Vigilância Sanitária, <<http://www.anvisa.gov.br>>, last acessed January 2017.
111. R. R. McCusker, E. A. Goldberger and J. Cone, *J. Anal. Toxicol.*, 2006, **30**, 611.
112. H. Ashihara, S. Ogita and A. Crozier, in *Plant Metabolism and Biotechnology*, ed. H. Ashihara, A. Crozier and A. Komamine, John Wiley & Sons Ltd, Chichester, 2011, p. 163.
113. A. Carvalho, J. S. Tango and L. C. Monaco, *Nature*, 1965, **205**, 314.
114. M. B. Silvarolla, P. Mazzafera and L. C. Fazuoli, *Nature*, 2004, **429**, 826.

115. M. U. Eteng, E. U. Eyong, E. O. Akpanyong, M. A. Agiang and C. Y. Aremu, *Plant Foods Hum. Nutr.*, 1997, **51**, 231.
116. G. K. Mumford, S. M. Evans, B. J. Kaminski, K. L. Preston, C. A. Sannerud, K. Silverman and R. R. Griffiths, *Psychopharmacology*, 1994, **115**, 1.
117. J. Hamon, M. Renner, M. Jamei, A. Lukas, A. Kopp-Schneider and F. Y. Bois, *Toxicol. In Vitro*, 2015, **30**, 203.
118. S. Ogita, H. Uefuji, Y. Yamaguchi, N. Koizumi and H. Sano, *Nature*, 2003, **423**, 823.
119. S. Ogita, H. Uefuji, M. Morimoto and H. Sano, *Plant Mol. Biol.*, 2004, **54**, 931.
120. B. B. Fredholm, in *Methylxanthines*, ed. B. B. Fredholm, Springer, New York, 2011, p. 1.
121. C. Nagai, J. J. Rakotomalala, R. Katahira, Y. Li, K. Yamagata and H. Ashihara, *Euphytica*, 2008, **164**, 133.

CHAPTER 24

Chlorogenic Acids

MARIUS FEBI MATEI, LEE SEUNG-HUN AND NIKOLAI KUHNERT*

Department of Life Sciences and Chemistry, Jacobs University Bremen, Campusring 8, 28759 Bremen, Germany
*E-mail: n.kuhnert@jacobs-university.de

24.1 Introduction – Chlorogenic Acids and Hydroxycinnamates

Chlorogenic acids (CGAs) are the most abundant secondary metabolites in both green and roasted coffee beans. In green beans they account for around 10% of the dry weight. With 142 million coffee bags sold in 2014 (60 kg a bag), it can be estimated that human consumption of CGAs from coffee alone is in the order of 500 000–900 000 t per annum. Chemically, we refer to CGAs as esters of hydroxycinnamates (or hydroxycinnamic acids) with quinic acid. The name chlorogenic acid comes from the Greek χλωρός (light green) and -γένος (a suffix meaning 'give rise to'), because of the green colour produced when CGAs are oxidised.

Hydroxycinnamates form a subclass of polyphenolic natural products. The interest in dietary hydroxycinnamates has expanded rapidly in the last 5–10 years. Numerous epidemiological studies have frequently linked the consumption of a diet rich in hydroxycinnamates, in particular CGAs, with numerous beneficial health effects,[1–5] which have been in many cases substantiated by *in vitro*, *in vivo* and human intervention studies. Further

Coffee: Production, Quality and Chemistry
Edited by Adriana Farah

Published by the Royal Society of Chemistry, www.rsc.org

interest in this ubiquitous class of compounds is motivated by their attractive sensory and organoleptic properties.[6]

Hydroxycinnamic acids are secondary metabolites, found ubiquitously in plants including most fruits and vegetables relevant to the human diet. They can occur in their free form with caffeic (**1**), ferulic (**5**), sinapic (**4**) and *p*-coumaric acid (**2**) being the most widespread examples (structures in Figure 24.1). Other minor cinnamic acid derivatives such as isoferulic acid (**6**), dimethoxycinnamic (**7**) or trimethoxycinnamic acid (**8**) have also been reported in plants.[7–10]

Hydroxycinnamates may be conjugated to many molecules, with conjugates to (−)-quinic acid (QA) being to present knowledge the most widespread compounds in the human diet. Other conjugates, of compounds with limited distribution found in only a few species, or frequently found only as minor components in plants, include: (1) esters of (−)-quinic acid (chlorogenic acids); (2) esters of hydroxyl acids such as shikimic, tartaric, galactaric, glucaric, gluconic, malic, dihydrocaffeic, hydroxycitric, phenylpyruvic, methoxyaldaric, lactic and tartronic acid; (3) glycosides; (4) amides of biogenic amines and amino acids including aromatic amino acids, glycine, spermidine, spermine, choline and anthranilic acid; (5) esters of carbohydrates and polyols including mono-saccharides, sugar alcohols including glycerol, inositols, and glycosides of anthocyanins, flavanols and diterpenes; (6) esters of lipids including sterols.[11]

CGAs form a class of compounds in the comprehensive family of hydroxycinnamates. CGAs are classically referred to as a class of esters afforded from *trans*-cinnamic acids and quinic acid, the latter bearing either axial substitution positions (hydroxyl groups on carbons 1 and 3) or equatorial ones (hydroxyls on carbons 4 and 5).[12,13] Most commonly encountered *trans*-cinnamic acids moieties are of caffeic, *p*-coumaric, and ferulic acids[13–15] but sinapic acid and dimethoxycinnamic acid are also present in certain plant species.[9,16–18] In the IUPAC system (−)-quinic acid is defined as 1L-1(OH),3,4/5-tetrahydroxycyclohexane carboxylic acid, but Eliel and Ramirez recommend 1α,3*R*,4α,5*R*-tetrahydroxycyclohexane carboxylic acid.[19]

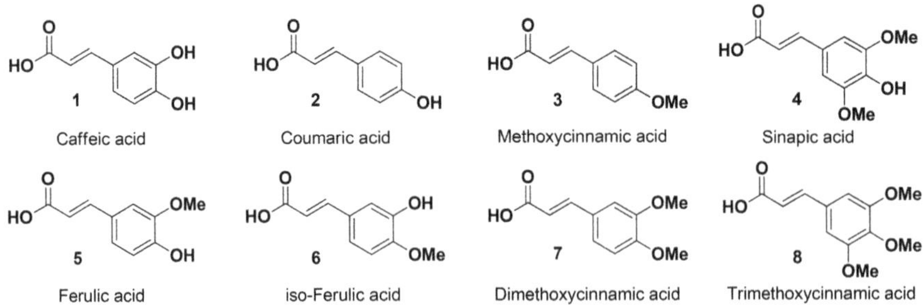

Figure 24.1 Selected structures of typical hydroxycinnamic acids present in the human diet.

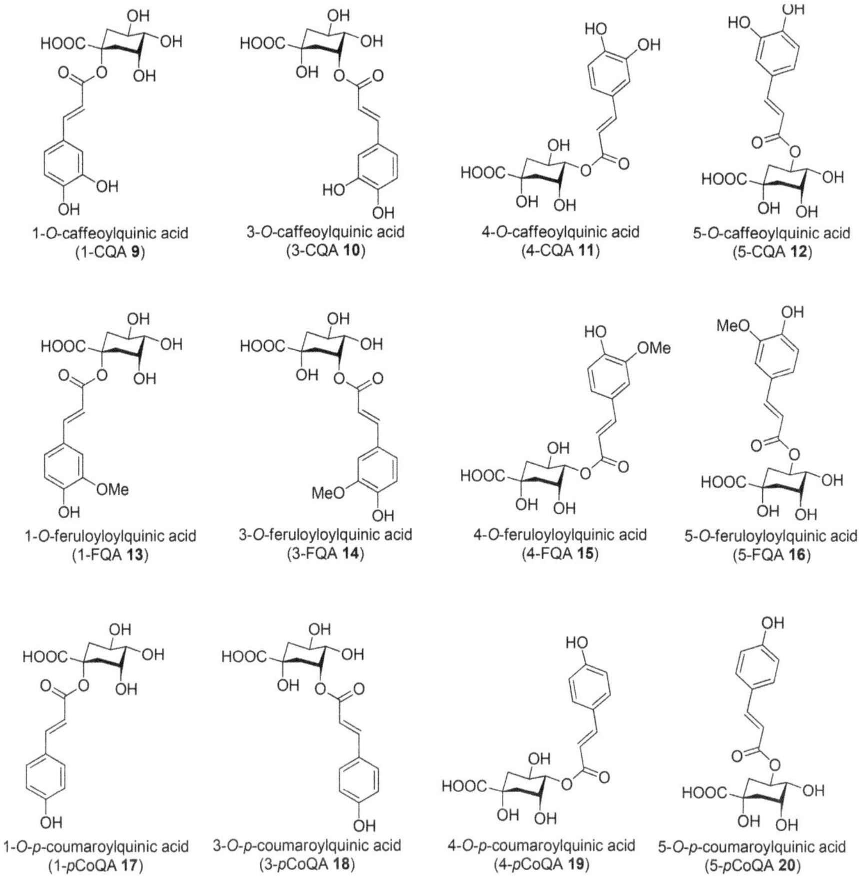

Figure 24.2 Representative chemical structures of mono-acylated CGAs.

Representative structures of both mono- and di-acylated CGAs are shown in Figures 24.2 and 24.3.

The CGAs show a variety of biological activities like antioxidant, anti-inflammatory, anti-HIV, anti-HBV, radical scavenging, inhibit mutagenesis and carcinogenesis, and are considered to be beneficial to human health.[20–24] The metabolism of CGAs in mammals has been poorly studied so far and relatively little is known[25] though their importance as non-nutritive food constituents (*ortho*-diphenyl: antioxidative power,[26] radical scavengers[27]) cannot be challenged. It should not come as a surprise that such metabolic studies are still limited since isotopically labelled CGAs are not readily available; also, the first efficient synthesis of a compound in this class was only published in 2001,[28] though analogues of chlorogenic acids were reported earlier by Hemmerle and colleagues.[20]

A classical subdivision of CGAs discriminates between compounds of this family based on the identity, number and position of the acyl residues.

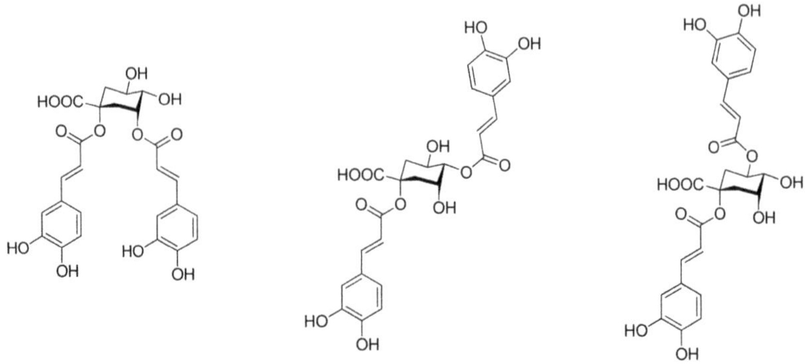

1,3-di-*O*-caffeoylquinic acid (1,3-diCQA **21**)

1,4-di-*O*-caffeoylquinic acid (1,4-diCQA **22**)

1,5-di-*O*-caffeoylquinic acid (1,5-diCQA **23**)

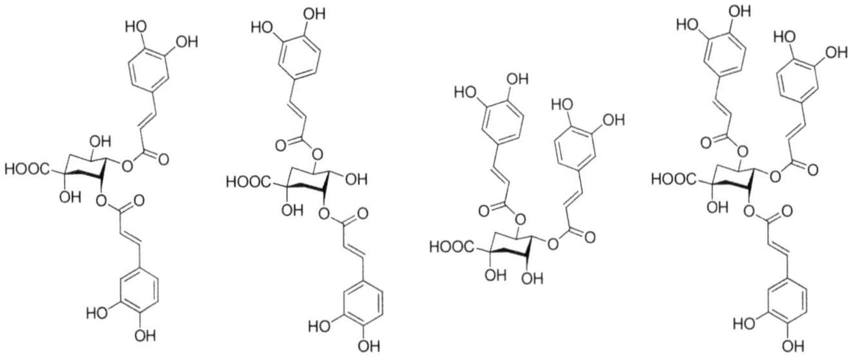

3,4-di-*O*-caffeoylquinic acid (3,4-diCQA **24**)

3,5-di-*O*-caffeoylquinic acid (3,5-diCQA **25**)

4,5-di-*O*-caffeoylquinic acid (4,5-diCQA **26**)

3,4,5-tri-*O*-caffeoylquinic acid (3,4,5-diCQA **27**)

Figure 24.3 Representative chemical structures of non-mixed di- and tri-acylated CGAs.

Clifford divided them into: (a) mono-esters of caffeic acid (*e.g.*, caffeoylquinic acids or CQAs, *p*-coumaroylquinic acids or pCoQAs, feruloylquinic acids or FQAs); (b) di-esters, tri-esters and a single tetra-ester of caffeic acid; (c) mixed di-esters of caffeic acid and either ferulic acid (*e.g.*, caffeoylferuloylquinic acids) or sinapic acid (*e.g.*, caffeoylsinapoylquinic acids); (d) mixed esters derived from permutations of one to three caffeic acid moieties with one to two residues of a dicarboxylic acid (*e.g.*, oxalic, succinic, glutaric).[11]

CGAs are known as common secondary plant metabolites and coffee, tea and potatoes as well as many vegetables and fruits are known to display particularly high content of such esters. Coffee beans and commercial coffee products are usually the main source of CGAs in the human diet and it was reported that esterification never occurs at position 1 of the quinic acid but only at positions 3, 4 and 5 in green coffee.[12] Food processing, however, can produce the remaining isomer at position 1 and may partially convert *trans* to *cis* isomers among other transformations.

24.2 Chlorogenic Acids and Derivatives: Analysis and Structure Elucidation

Analytical methods for efficient screening and unambiguous identification of dietary natural products and their chemical transformation products form the root of understanding the impact of diet on human health. Once such analytical methods have been developed and established, dietary materials can be screened for novel compounds, structures can be assigned, novel compounds can be quantified, their dietary burden determined, their biological activity tested, their impact on human health estimated and, after pharmacokinetic evaluation, scientifically sound dietary advice to the consumer can be provided.

While figures for an average dietary intake exist in particular for CGAs, for other hydroxycinnamate derivatives figures need to be regarded as sketchy. The average daily human intake of CGAs varies depending on the author of the study between 1 g per day per human and 2.5 g per day per human. A cup of coffee (200 mL serving) contains on average 200 mg CGA per serving (cup of 200 ml). To an extent, the discrepancies between such reported figures are a result of the individual diets and consumer preferences. However, they are inevitably a consequence of the limited knowledge of this important class of compounds as well. Numbers on exact quantities of CGAs in dietary material are relatively rare and little is known about statistical variances between the CGA content in different plant species and varieties.

Moreover, available analytical data are in many cases obsolete, obtained through methods based on derivatisation followed by colorimetry, quantifying only one particular class of CGAs *e.g.*, caffeoyl or feruloyl esters, resulting in a possible underestimation of real CGAs content. Secondly, agricultural practice has changed dramatically over the last decade with many new varieties of fruit and vegetables introduced to the market and older varieties consequently disappearing from the market for patent reasons. Therefore it is likely that both CGA acid profiles and quantities within these new varieties have changed dramatically in comparison to the published data. With a dietary intake figure of around 2 g per day, combined with bioavailability data suggesting a high level of absorption, hydroxycinnamates must be considered as the most relevant secondary plant metabolites for the human diet.

Prior to 2003 only 40 different CGA derivatives were known in the human diet. Since 2003, more than 300 novel chlorogenic acid derivatives have been identified in the human diet by other research groups and more than 200 novel derivatives by the Kuhnert group alone. In 2003 Clifford and Kuhnert published a novel method for the identification and structure elucidation of regioisomeric CGA derivatives (hydroxycinnamate esters of quinic acid). This approach revolutionised CGA identification and structure elucidation by introducing a method based on tandem mass spectrometry, allowing assignment of regiochemistry based exclusively on fragment spectra.[12]

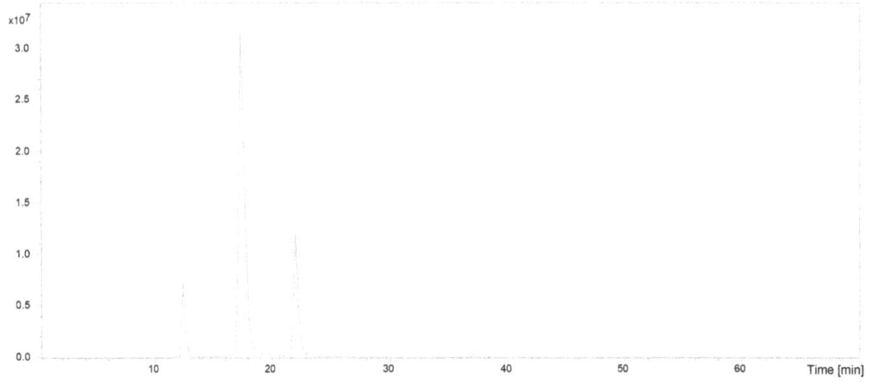

Figure 24.4 Extracted ion chromatogram in negative ion mode (EIC) at *m*/*z* 353 (caffeoyl quinic acids) of Brazilian green coffee blend extract.

The method has moreover the advantage that CGAs do not need to be isolated but can be identified and their structure elucidated directly from analytical LC-tandem-MS runs, even if present as minor components or as chromatographically close eluting compounds. Figure 24.4 shows a typical extracted ion chromatogram of caffeoylquinic acids in green coffee beans with three peaks corresponding to 3-CQA, 4-CQA and 5-CQA. Due to the diagnostic differences in the tandem MS fragment spectra, a consistent and predictive structure diagnostic hierarchical key for CGA structure elucidation has been established, which allows reliable determination of CGA regiochemistry even for minor component CGAs from tandem MS data exclusively, superior to structure elucidation methods by Nuclear Magnetic Resonance (NMR).

The basis of these differences in tandem MS spectra was rationalised in terms of different hydrogen bonding arrays found in gas phase ions of regio-isomeric CGAs inducing distinct pathways for fragmentation. This hydrogen bonding in turns leads to activation of the ester carbonyl group and to cleavage of the carbonyl C–O bond. The degree of activation is different in all regioisomers hence cleavage of the C–O bond occurs at different energies leading to significant differences in the fragment spectra. Figure 24.5 shows characteristic MS^2 spectra of the key monocaffeoyl quinic acids in coffee. Tandem MS data for the most common CGAs are further summarised in Table 24.1. More details can be found in the literature. An additional feature of the tandem spectra is the occurrence of characteristic CGA fragment ions in MS^n spectra at *m*/*z* 191.1 and 173.1. By creating extracted ion chromatograms in all MS^n modes, it is straightforward to search fragment ions corresponding to any CGA derivative.

This method was recently extended to the structure elucidation of regio-isomeric shikimic acid esters.[17] Furthermore, to illustrate the power of the method, all four regioisomers of feruloylquinic acid (FQA) and isoferu-loylquinic acid (iFQA) were synthesised in a co-authored study and then the

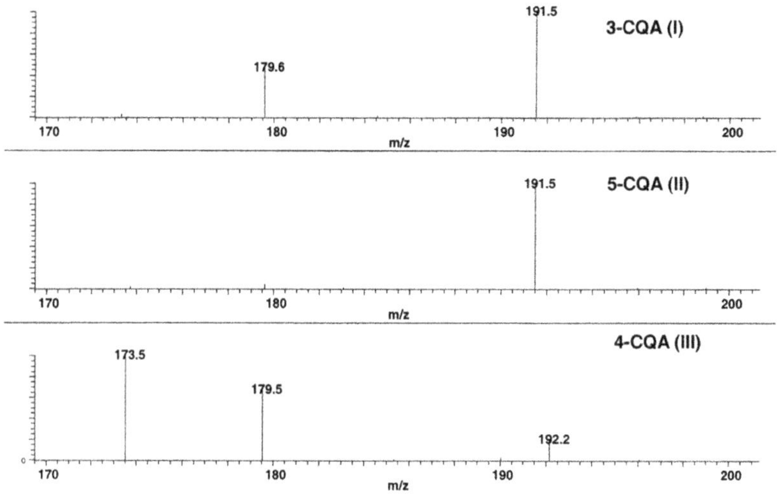

Figure 24.5 MS^2 spectra of three different region-isomers of CQAs.

total of eight regioisomers was resolved by LC/MS/MS (chromatogram shown in Figure 24.6).[10] All structures of the derivatives could be unambiguously identified to the regioisomeric level. Therefore it is definitely worth exploring whether such diagnostic differences in tandem MS spectra can as well be observed and hence successfully employed for structure elucidation of structurally related derivatives including other cyclohexenes, lactones, oxidation products or carbohydrate trans-esterification products.

The cinnamate chromophore of CGAs is responsible for a typical UV absorption at roughly 320 nm. λ_{max} values of selected chlorogenic acids as well as data for molar extinction coefficients (ε) were recently summarised as shown in Table 24.2.[33] Such data are especially valuable in HPLC analysis, which employs UV detectors such as DAD for quantification purposes.

In the same study, the authors observe that the formation of phenolate anions by CGAs in alkaline media is responsible for a bathochromic shift; $\log E$ progressively increases with the increasing degree of acyl substitution.[33] UV-VIS spectroscopy is additionally able to discriminate between *cis* and *trans* isomers of CGAs since in the absorption spectra the *cis* isomer typically displays a red-shifted shoulder.[34]

For quantification of hydroxycinnamates and CGAs from dietary sources most of the published information was based on UV-VIS spectroscopy. HPLC-MS techniques have gained popularity recently for quantification purposes. HPLC-MS allows quantification of CGAs with the help of appropriate reference standards. The samples to be analysed by this approach should give MS spectra with few signals, no co-elution of compounds and no signal overlap. These might appear as demanding conditions to be met at times but for the quantification of the more complex samples, single ion monitoring in tandem MS is an effective approach.

Table 24.1 MS and tandem MS data in negative ion mode of selected key chlorogenic acids in green coffee bean.

		MS[1]	MS[2]						MS[3]					
			Base peak	Secondary peak				Base peak	Secondary peak					
RT [min]	Compd	Parent ion	*m/z*	*m/z*	int	*m/z*	int	*m/z*	*m/z*	int	*m/z*	int	*m/z*	int
12.3	3-CQA	352.81	190.6	178.7	43	134.8	10	132.7	134.7	12	133.6	7		
17.4	5-CQA	352.78	190.6	214.6	4	178.7	2	126.7	172.6	100	85	96	93	64
20.7	3-FQA	366.84	192.6	133.8	11	193.4	8	133.7	148.6	29	190.6	4	116.8	3
22	4-CQA	352.81	172.7	178.6	58	190.6	20	92.9	110.8	75	71.2	55	154.6	34
24.1	5-*p*CoQA	336.84	190.6	162.8	5	172.7	1	172.6	126.7	94	85	72	92.9	56
27	5-FQA	366.79	190.6	192.6	7	172.7	2	172.6	126.7	87	85	75	92.9	50
27.6	4-*p*CoQA	336.85	172.6	162.7	6	173.5	5	92.9	110.8	77	71.1	42	59.4	25
30.1	4-FQA	366.87	172.6	190.6	25	192.6	14	92.9	110.8	69	71.3	30	154.6	27
37	3,5-diCQA	514.81	352.8	190.7	8	178.7	4	190.6	178.7	38	134.9	6	172.9	3
38.3	3,4-diCQA	514.82	352.8	172.7	23	178.7	18	172.7	178.6	87	190.6	70	134.8	12
40.4	4,5-diCQA	514.81	352.8	172.7	16	202.7	15	172.7	178.6	68	190.6	33	134.9	7
41	3-C-4-*p*CoQA	498.83	336.8	352.7	31	162.7	17	162.6	172.7	59	334.7	14	190.6	6
42.4	3-F-5-CQA	528.85	366.8	352.9	49	367.6	15	190.6	178.6	43	350.8	16	134.7	11
43	3-*p*Co-4-CQA	498.83	352.8	336.8	49	318.8	44	172.6	178.6	52	190.6	20	134.7	6
43.9	3-C-4-FQA	528.83	366.8	172.7	51	352.8	23	172.6	192.6	36	133.7	5	173.5	4
44.5	4-*p*Co-5-CQA	498.83	336.8	172.6	25	337.7	14	172.6	162.6	27	334.8	8		
45.7	4-C-5-FQA	528.85	366.8	352.8	57	172.7	22	172.6	178.6	45	190.6	25	350.7	6

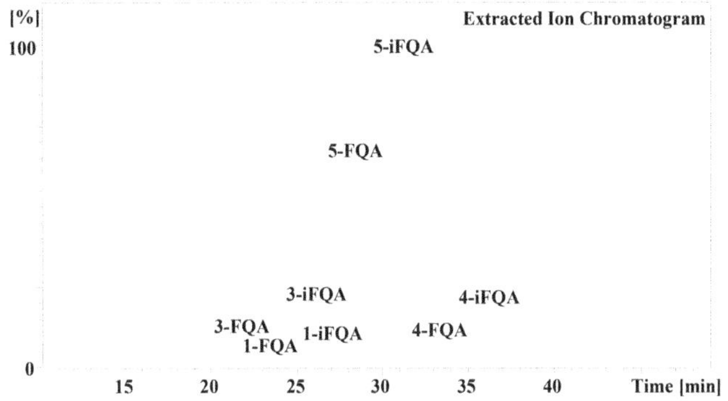

Figure 24.6 HPLC chromatogram (EIC at *m/z* 367.1) of mixtures of regioisomers of FQAs and iFQAs. Reproduced from ref. 10 with permission from John Wiley & Sons. Copyright 2011.

Table 24.2 Characteristic UV-VIS data of selected chlorogenic acids.[33]

CGA	$\lambda_{max}{}^{b}$ [nm]	log E^{b}
1-CQA **9**	330	4.26
3-CQA **10**	330	4.22
4-CQA **11**	330	4.26
5-CQA **12**	327	4.29
1-FQA **13**	325	4.27
3-FQA **14**	325	4.28
4-FQA **15**	325	4.29
5-FQA **16**	325	4.29
1-*p*CoQA **17**	375	4.31
3-*p*CoQA **18**	315 (310)[a]	4.30
4-*p*CoQA **19**	315	4.32
5-*p*CoQA **20**	315 (310)	4.31
1,3-diCQA **21**	325	4.50
1,4-diCQA **22**	327	4.53
1,5-diCQA **23**	327	4.50
3,4-diCQA **24**	330	4.53
3,5-diCQA **25**	329	4.55
4,5-diCQA **26**	330	4.52
3,4,5-triCQA **27**	327	4.75

[a]Two reported values in literature.
[b]All reported values in ethanol.

Before HPLC-MS mainly replaced the quantification techniques based on UV-VIS spectroscopy, quantification was performed either by HPLC-UV-VIS or by derivatisation followed by UV-VIS spectroscopy. The first approach (HPLC-UV-VIS) has the advantage of being capable of quantifying individual CGAs while the second was used to calculate the total amount of hydroxycinnamates or CGAs. Table 24.2 summarises useful values.

In HPLC-UV-VIS the HPLC instrument effects chromatographic separation and the UV-VIS detector coupled to the HPLC system then quantifies the individual chromatographically well-resolved components. Since the maximum absorption of the cinnamate chromophore is at around 320 nm as shown above, the quantification is done at this wavelength value based on the previously obtained calibration curves.[33,35] An important share of the quantification data on total content of hydroxycinnamates and CGAs in various samples was afforded through colour derivatisation reactions. The method is not without limitations and the numbers generated in this manner should be viewed critically.

By this approach, a coloured product which is at the end quantified by UV-VIS spectroscopy is the result of a reaction between a crude dietary extract and a reagent which reacts specifically with a targeted hydroxycinnamate residue. Examples of reagents typically employed to effect this transformation are periodate, molybdate or thiobarbituric acid. But this method can result in either an over- or an under-estimation of the CGAs content. An over-estimation of the CGAs may be due to the presence of additional (poly)phenolic dietary constituents derivatised by this method while CGAs with different substituents from those being targeted (and thus not being detected) would produce an under-estimation of the total content.[33,35,36]

24.3 Chlorogenic Acids Derivatives in Food Processing

Diets particularly rich in chlorogenic acids and therefore relevant to human health comprise coffee, tea and plants from the Solanaceae family including potatoes, tomatoes and eggplants. All these plants are being processed by heat treatment prior to consumption. Coffee is roasted and potatoes and eggplants are always, and tomatoes frequently, baked, fried, cooked, steamed or roasted. In this processes the secondary plant metabolites undergo significant chemical changes and new products are formed. Coffee and the above-mentioned plants prompt further investigation for clarifying the chemical fate of hydroxycinnamate derivatives in food processing.

For instance, an HPLC-MS measurement on green coffee beans reveals the presence of around 80 different chlorogenic acids, which upon roasting increases to around 200 derivatives.[16,29,30] Figure 24.7 shows chromatograms of green and roasted coffee beans to illustrate the increase of complexity of the sample after roasting. This number can be estimated by identifying typical cinnamate absorptions in the UV traces of the chromatogram at 320 nm and by studying extracted ion chromatograms in all MS^n modes indicating fragmentation patterns characteristic to CGA derivatives, typically fragment ions at *m/z* 191.1 and 173.1. Two typical chromatograms are shown in Figure 24.6.[31]

There is a limited number of potential CGAs isomers produced by plants genuinely as secondary metabolites; in contrast to this, the number of

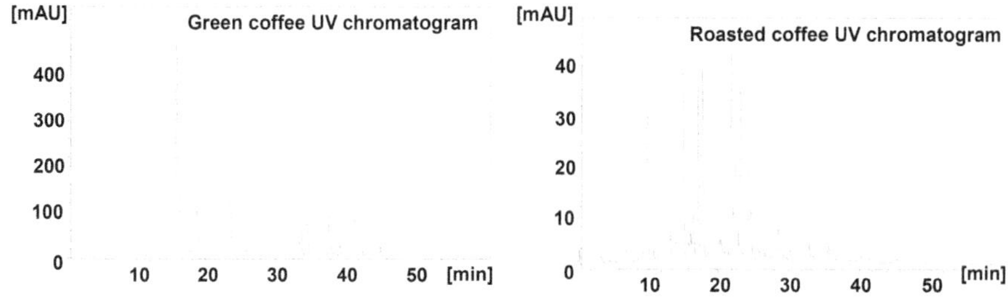

Figure 24.7 UV chromatograms of green and roasted coffee at λ_{max} = 320 nm.

possible regio- and stereoisomers which could be chemically or enzymatically generated by food processing and metabolism has the potential to be large. Food processing (roasting, cooking, baking, frying, steaming, microwaving, fermenting) of many dietary plants can potentially enrich the profile of the CGAs and their derivatives in a given food, which are then available for human consumption and potentially human metabolism.

Among the transformations of CGAs during food processing, hydrolysis can release cinnamic acids from conjugates, which can further be decarboxylated by heat or microorganisms to yield various alkyl phenols. Not much attention has so far been given to the possibility of formation of novel transformation products triggered by food processing. The first reports on this topic identified the grape reaction product 2-*S*-glutathionylcaftaric acid as well as 2,5-di-*S*-glutathionylcaftaric acid formed from caftaric acid (caffeoyl-tartaric acid) during wine making.[37–39] Adducts generated from the interaction of anthocyanins with vinyl phenol resulted from *p*-coumarate decarboxylation were detected in red wine.

During coffee roasting caffeic acid (**1**) is partially converted to tetrahydroxy-phenylindans.[40–42] As suggested by Engelhardt and co-workers, the resulting hydrolysed quinic acid residue from coffee roasting can be subsequently converted to the full theoretical complement of quinic acid and quinic acid lactone (quinide) diastereomers (Figure 24.8).[43] There are eight possible stereoisomers for quinic acid, out of which there are four *meso* forms and two pairs of enantiomers.

Quinides (including 1,5-quinides) can be the products of the intra-molecular water elimination processes from the parent unhydrolysed caffeoylquinic acids and feruloylquinic acids; additional diastereomers of the original acids or of the newly formed lactones might also be generated.[44,45] The elevated temperatures employed in the roasting process can result in a breakage of the C–C bonds in CGAs and their derivatives. Roasting times as short as 5 minutes can alter significantly the CGAs profile: the levels of both 3-CQA (**10**) and 4-CQA (**11**) were observed to double their original values at the expense of the more abundant 5-CQA (**9**), whose level decreased substantially. Feruloylquinic acids analogues behaved similarly in the 5 minute roasting experiment. Besides the obvious isomerisation processes, partial hydrolysis of di-acylated CQAs to mono-acylated CQAs was also

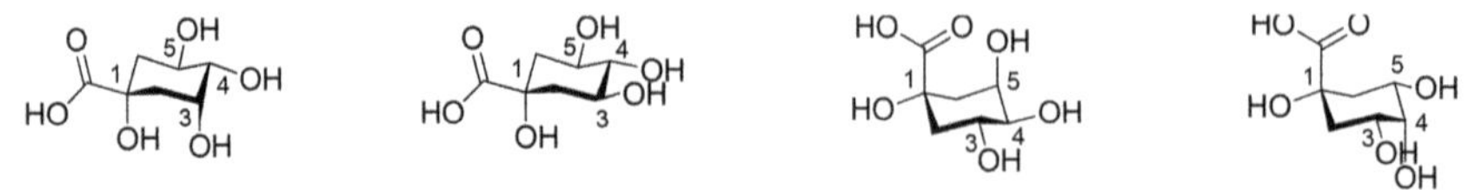

(-)-quinic acid **28** | *muco*-quinic acid **29** C3 inverted | (-)-*epi*-quinic acid **30** C4 inverted | *cis*-quinic acid **31** C5 inverted

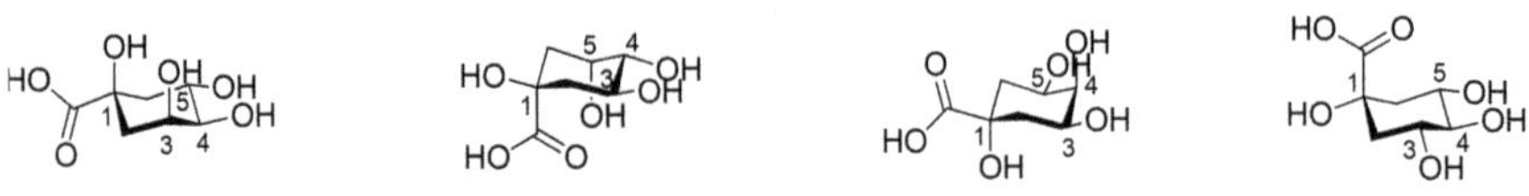

(+)-quinic acid **32** | (+)-*epi*-quinic acid **33** | *neo*-quinic acid **34** C3 and C4 inverted | *scyllo*-quinic acid **35** C4 and C5 inverted

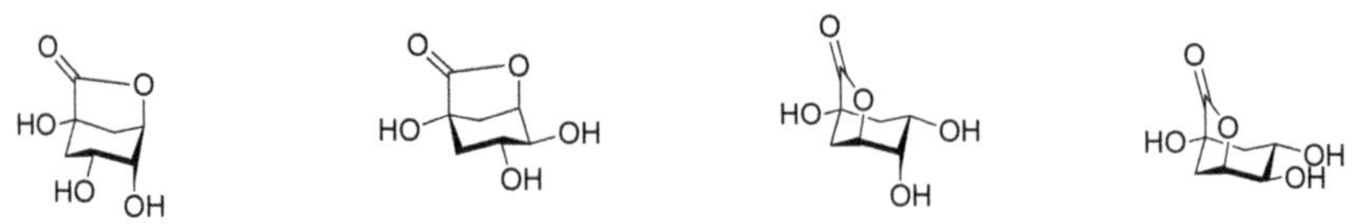

γ-quinide **36** | *epi*-γ-quinide **37** | *muco*-γ-quinide **38** | *neo*-γ-quinide **39**

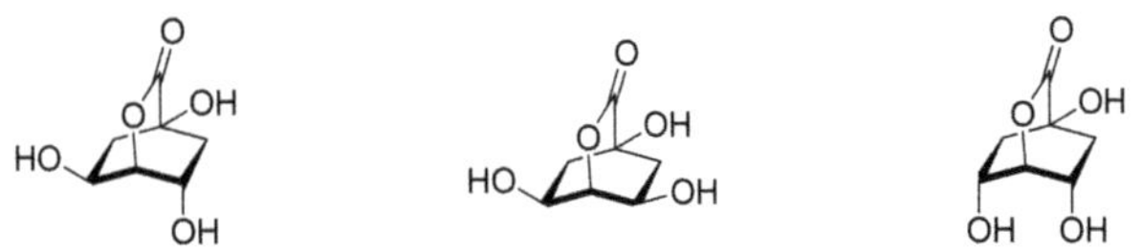

epi-δ-quinide **40** | *neo*-δ-quinide **41** | *scyllo*-δ-quinide **42**

Figure 24.8 Chemical structures of stereoisomers of quinic acid and quinic acid lactones.

observed,[42] where a similar behaviour to that during the roasting of coffee was observed: migration of acyl from C5 to C3 positions of the quinic acid moiety.[44]

Only selected efforts have been made to elucidate the structure of such reaction products. For example, the group of Farah reported on the presence of chlorogenic acid lactones in roasted coffee formed by dehydration of the parent chlorogenic acid,[29,32] whereas the Kuhnert group reported on the presence of alternative shikimic acid dehydration products in roasted maté tea.[17]

Nevertheless, a systematic work addressing the fate of hydroxycinnamate derivatives in food processing is lacking at the moment and, as can clearly be seen from the chromatograms above, there are many more unknown structures present in processed food. Being part of the common human diet such derivatives are to humans of foremost importance and relevance and urgent clarification of their structures and properties must be set as a priority.

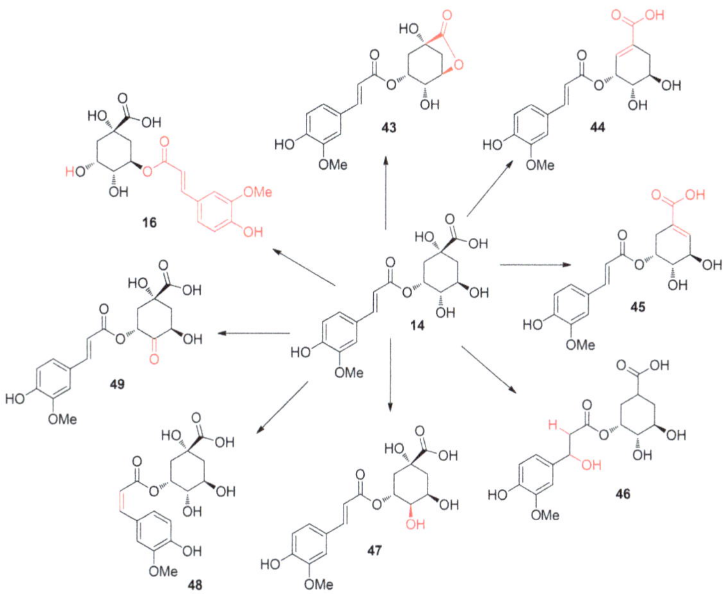

Figure 24.9 Chemical fate of chlorogenic acids at elevated temperatures. Reproduced from ref. 45 with permission from the Royal Society of Chemistry.

From the above arguments it becomes obvious that a large proportion of the chlorogenic acids in food are converted by heat treatment into novel structures. Considering the chemical fate of chlorogenic acids at elevated temperatures, a reaction diagram can be developed and a hypothesis on possible reaction pathways introduced. Figure 24.9 represents such a diagram with examples of one compound per specific transformation. Therefore, it is hereby proposed that at elevated temperatures, neglecting all further chemical species present in the plant, chlorogenic acids (*e.g.*, (**14**)) can dehydrate to form lactones (**43**) or cyclohexene derivatives (**44**) including shikimic acid derivatives (**45**), can add water (**46**), can undergo epimerisation (**47**), can undergo thermal *trans–cis* isomerisation (**48**), oxidation (**49**) or can undergo acyl migration (**16**) (either intra- or inter-molecular with other CGAs or alcohols present in the human diet *e.g.*, carbohydrates).

Underestimated solubility differences for the real coffee consumption – Most quantitative analyses of CGAs use alcoholic solvents for extractions of beans. However, for human consumption water is used as the extracting solvent leading to discrepancies between measured total CGA content in a bean and dietary exposure. Although we consume all the coffee beans in some brewing methods like Turkish coffee, we generally extract only 0.5–40% of the water-soluble part of coffee for drinking. Therefore, there were always some differences between research results and real consuming compounds. We could confirm that there is solubility differences between water and 70% methanol as expected. Table 24.3 shows the results in

Table 24.3 Relative ratios against 5-CQA (100%) of major compounds in water and 70% methanol extracted coffees.

Compound	Green/water	Green/MeOH	Roasted/water	Roasted/MeOH
5-FQA	12.78	27.09	22.75	36.08
3,5-diCQA	5.47	10.18	2.10	4.42
3,4-diCQA	2.85	8.29	2.71	6.61
4,5-diCQA	4.83	17.05	2.49	8.86
Methyl-CQA	—	—	0.77	10.84
CQAL 25.2	—	—	27.34	39.28
CQAL 28.9	—	—	24.23	26.92

chlorogenic acid concentration if alcoholic extraction is compared to hot water extraction. We can find that there are significant differences caused by solvent difference.

24.4 Intake of Chlorogenic Acids and Derivatives

The main dietary source of CGAs and their derivatives is coffee for most consumers.[11] Figures of ingested CGAs per serving vary between 20 mg and 675 mg depending on several parameters such as the consumed volume, the concentration of the brew and the type of the roast. Simple math indicates that a figure of more than 1 g CGAs per day is a realistic estimation for regular coffee consumers, with an intake of 4–5 servings daily.[11,46] A study performed by Crozier in Glasgow about the CGAs content in espresso coffee samples purchased from 21 coffee shops confirmed the above estimation. The values for the total CGAs dosage per serving varied in a wide range (almost 20-fold), from a minimum of 24 mg per cup to a maximum quantity of 423 mg per cup.[47]

The higher end of this reported interval (423 mg or 1195 μmol per cup) is substantially superior to what other sources of CGAs could offer: 25 mg (72 μmol) per 200 mL serving of cloudy apple juice[48] or 34 mg (96 μmol) per 200 mL serving of apple smoothie.[49] Although there are other rich sources of CGAs (*e.g.*, globe artichoke with 268 mg or 762 μmol per 100 g),[50] they are very rarely consumed in a high enough quantity to contest coffee as the leading source of CGAs for the human diet. With a content of 94–111 mg (270–320 μmol) CGAs per 200 mL cup volume,[51] maté represents an important source in certain populations, particularly in South America, given the daily number of servings for regular consumers. Table 24.4 summarises the reported CGAs content in different relevant products.[52]

Coffee represents the dominant source of dietary CGAs and was used in most human studies on CGAs absorption and metabolism. Besides coffee, there are numerous dietary sources of CGAs with varying relevant content. Their compositional data are summarised in Table 24.4; however, caution should be exercised when using such data since the composition of the cultivars can be of significant difference.[52]

Table 24.4 CGAs content in different food products (ref. 52).[a]

Product	Serving	Dose µmol	Dose mg	Reference
Espresso coffee	27 mL	68	24	Crozier *et al.*[47]
	52 mL	1195	422	Crozier *et al.*[47]
Instant coffee	200 mL	412	146	Stalmach *et al.*[53]
	200 mL	384	136	Stalmach *et al.*[54]
Cloudy apple juice	200 mL	72	25	Kahle *et al.*[48]
Apple smoothie	200 mL	96	34	Hagl *et al.*[49]
Globe artichoke	100 g	762	268	Pandino *et al.*[50]
Maté	200 mL	270–320	107–133	Clifford and Ramìrez-Martìnez[51]

[a]Data expressed as mean values (n = 3); in all instances the SE was <10% of the mean value.

24.5 Final Considerations

From the data presented it can be clearly appreciated that our knowledge of the occurrence and relevance of CGAs has increased substantially over the last decade. While 15 years ago around 100 derivatives were described in the literature this number has grown to more than 500 derivatives that have been described to date. This fact can certainly be traced down to the major improvements in analytical techniques, which today not only allow structural elucidation of minor metabolites but also the use of targeted tandem MS experiments to perform a simplified search for CGAs due to their characteristic fragment ions. It is worth noting that CGA chemistry has been a main driving force behind the development of tandem MS techniques able to unambiguously assign natural product stereo- and regiochemistry contributing to a more general use of tandem mass spectrometry in reliable chemical structure elucidation.

References

1. G. Czok, Coffee and health, *Z. Ernahrungswiss.*, 1977, **16**, 248–255.
2. J. V. Higdon and B. Frei, Coffee and health: a review of recent human research, *Crit. Rev. Food Sci. Nutr.*, 2006, **46**, 101–123.
3. R. M. van Dam, Coffee consumption and risk of type 2 diabetes, cardiovascular diseases, and cancer, *Appl. Physiol., Nutr., Metab.*, 2008, **33**, 1269–1283.
4. A. Crozier, I. B. Jaganath and M. N. Clifford, Dietary phenolics: chemistry, bioavailability and effects on health, *Nat. Prod. Rep.*, 2009, **26**, 1001–1043.
5. A. Tavani and C. La Vecchia, Coffee and cancer: a review of epidemiological studies, 1990-1999, *Eur. J. Cancer Prev.*, 2000, **9**, 241–256.
6. O. Frank, S. Blumberg, C. Kunert, G. Zehentbauer and T. Hofmann, Structure determination and sensory analysis of bitter-tasting 4-vinylcatechol oligomers and their identification in roasted coffee by means of LC-MS/MS, *J. Agric. Food Chem.*, 2007, **55**, 1945–1954.

7. M. N. Clifford, S. Marks, S. Knight and N. Kuhnert, Characterization by LC-MS^n of four new classes of *p*-coumaric acid-containing diacyl chlorogenic acids in green coffee beans, *J. Agric. Food Chem.*, 2006, **54**, 4095–4101.
8. Y. Maruta, J. Kawabata and R. Niki, Antioxidative caffeoylquinic acid derivatives in the roots of burdock (*Arctium lappa* L.), *J. Agric. Food Chem.*, 1995, **43**, 2592–2595.
9. R. Jaiswal, M. A. Patras, P. J. Eravuchira and N. Kuhnert, Profile and characterization of the chlorogenic acids in green Robusta coffee beans by LC-MS^n: identification of seven new classes of compounds, *J. Agric. Food Chem.*, 2010, **58**, 8722–8737.
10. R. Jaiswal, M. F. Matei, F. Ullrich and N. Kuhnert, How to distinguish between cinnamoylshikimate esters and chlorogenic acid lactones by liquid chromatography-tandem mass spectrometry, *J. Mass Spectrom.*, 2011, **46**, 933–942.
11. M. N. Clifford, Chlorogenic acids and other cinnamates - nature, occurrence and dietary burden, *J. Sci. Food Agric.*, 1999, **79**, 362–372.
12. M. N. Clifford, K. L. Johnston, S. Knight and N. Kuhnert, Hierarchical scheme for LC-MS^n identification of chlorogenic acids, *J. Agric. Food Chem.*, 2003, **51**, 2900–2911.
13. M. N. Clifford, S. Knight and N. Kuhnert, Discriminating between the six isomers of dicaffeoylquinic acid by LC-MS^n, *J. Agric. Food Chem.*, 2005, **53**, 3821–3832.
14. IUPAC Commission on the Nomenclature of Organic Chemistry (CNOC) and IUPAC-IUB Commission on Biochemical Nomenclature (CBN), Nomenclature of cyclitols, Recommendations, 1973, *Biochem. J.*, 1976, **153**, 23–31.
15. M. N. Clifford, Chlorogenic acids and other cinnamates - nature, occurrence, dietary burden, absorption and metabolism, *J. Sci. Food Agric.*, 2000, **80**, 1033–1043.
16. M. N. Clifford, S. Knight, B. Surucu and N. Kuhnert, Characterization by LC-MS^n of Four New Classes of Chlorogenic Acids in Green Coffee Beans: Dimethoxycinnamoylquinic Acids, Diferuloylquinic Acids, Caffeoyl-dimethoxycinnamoylquinic Acids, and Feruloyl-dimethoxycinnamoylquinic Acids, *J. Agric. Food Chem.*, 2006, **54**, 1957–1969.
17. R. Jaiswal, T. Sovdat, F. Vivan and N. Kuhnert, Profiling and characterization by LC-MS^n of the chlorogenic acids and hydroxycinnamoylshikimate esters in maté (*Ilex paraguariensis*), *J. Agric. Food Chem.*, 2010, **58**, 5471–5484.
18. R. Jaiswal and N. Kuhnert, Hierarchical scheme for liquid chromatography/multi-stage spectrometric identification of 3,4,5-triacyl chlorogenic acids in green Robusta coffee beans, *Rapid Commun. Mass Spectrom.*, 2010, **24**, 2283–2294.
19. E. L. Eliel and M. B. Ramirez, (−)-Quinic acid: configurational (stereochemical) descriptors, *Tetrahedron: Asymmetry*, 1997, **8**, 3551–3554.

20. H. Hemmerle, H. J. Burger, P. Below, G. Schubert, R. Rippel, P. W. Schindler, E. Paulus and A. W. Herling, Chlorogenic Acid and Synthetic Chlorogenic Acid Derivatives: Novel Inhibitors of Hepatic Glucose-6-phosphate Translocase, *J. Med. Chem.*, 1997, **40**, 137–145.
21. S. Gorzalczany, V. Sulsen, F. Redko, C. Vescina, L. Muschietti, V. Martino and C. Acevedo, Choleretic and antispasmodic effects of *Lippia integrifolia* aqueous extract, *Rev. Bras. Farmacogn.*, 2008, **18**, 16–20.
22. M. H. Kweon, H. J. Hwang and H. C. Sung, Identification and Antioxidant Activity of Novel Chlorogenic Acid Derivatives from Bamboo (*Phyllostachys edulis*), *J. Agric. Food Chem.*, 2001, **49**, 4646–4655.
23. H. C. Kwon, C. M. Jung, C. G. Shin, J. K. Lee, S. U. Choi, S. Y. Kim and K. R. Lee, A new caffeoyl quinic acid from *Aster scaber* and its inhibitory activity against human immunodeficiency virus-1 (HIV-1) integrase, *Chem. Pharm. Bull.*, 2000, **48**, 1796–1798.
24. G. F. Wang, L. P. Shi, Y. D. Ren, Q. F. Liu, H. F. Liu, R. J. Zhang, Z. Li, F. H. Zhu, P. L. He, W. Tang, P. Z. Tao, C. Li, W. M. Zhao and J. P. Zuo, Anti-hepatitis B virus activity of chlorogenic acid, quinic acid and caffeic acid in vivo and in vitro, *Antiviral Res.*, 2009, **83**, 186–190.
25. J. Kroll, H. M. Rawel and N. Seidelman, Physicochemical Properties and Susceptibility to Proteolytic Digestion of Myoglobin-Phenol Derivatives, *J. Agric. Food Chem.*, 2000, **48**, 1580–1587.
26. Y. Kono, S. Kashine, T. Yoneyama, Y. Sakemoto, Y. Matsui and H. Shibata, Iron chelation by chlorogenic acid as a natural antioxidant, *Biosci., Biotechnol., Biochem.*, 1998, **62**, 22–27.
27. H. Iwahashi, Some polyphenols inhibit the formation of pentyl radical and octanoic acid radical in the reaction mixture of linoleic acid hydroperoxide with ferrous ions, *Biochem. J.*, 2000, **346**, 265–273.
28. M. Sefkow, First Efficient Synthesis of Chlorogenic Acid, *Eur. J. Org. Chem.*, 2001, 1137–1141.
29. J. Shearer, A. Farah, T. de Paulis, D. P. Bracy, R. R. Pencek, T. E. Graham and D. H. Wasserman, Quinides of roasted coffee enhance insulin action in conscious rats, *J. Nutr.*, 2003, **133**, 3529–3532.
30. M. D. del Castillo, J. M. Ames and M. H. Gordon, Effect of roasting on the antioxidant activity of coffee brews, *J. Agric. Food Chem.*, 2002, **50**, 3698–3703.
31. R. Slimestad and M. Verheul, Review of flavonoids and other phenolics from fruits of different tomato (*Lycopersicon esculentum* Mill.) cultivars, *J. Sci. Food Agric.*, 2009, **89**, 1255–1270.
32. A. Farah, T. De Paulis, D. P. Moreira, L. C. Trugo and P. R. Martin, Chlorogenic acids and lactones in regular and water-decaffeinated Arabica coffees, *J. Agric. Food Chem.*, 2006, **54**, 374–381.
33. N. Kuhnert, H. Karaköse and R. Jaiswal, Analysis of Chlorogenic Acids and Other Hydroxycinnamates in Food, Plants, and Pharmacokinetic Studies, in *Handbook of Analysis of Active Compounds in Functional Foods*, ed. L. M. L. Nollet and F. Toldrae, CRC Press, Boca Raton, FL, 2012, pp. 461–510.

34. M. N. Clifford, J. Kirkpatrick, N. Kuhnert, H. Roozendaal and P. R. Salgado, LC-MS^n analysis of the *cis* isomers of chlorogenic acids, *Food Chem.*, 2008, **106**, 379–385.
35. H. Karaköse, R. Jaiswal and N. Kuhnert, Characterization and Quantification of Hydroxycinnamate Derivatives in *Stevia rebaudiana* Leaves by LC-MS^n, *J. Agric. Food Chem.*, 2011, **59**, 10143–10150.
36. M. N. Clifford, Chlorogenic acids, in *Coffee*, ed. R. J. Clarke and R. Macrae, Elsevier Applied Science Publishers, London, 1985, vol. 1, pp. 153–202.
37. V. F. Cheynier, E. K. Trousdale, V. L. Singleton, M. J. Salgues and R. Wylde, Characterization of 2-*S*-glutathionylcaftaric acid and its hydrolysis in relation to grape wines, *J. Agric. Food Chem.*, 1986, **34**, 217–221.
38. M. Salgues, V. Cheynier, Z. Gunata and R. Wylde, Oxidation of Grape Juice 2-*S*-Glutathionyl Caffeoyl Tartaric Acid by *Botrytis cinerea* Laccase and Characterization of a New Substance: 2,5-di-*S*-Glutathionyl Caffeoyl Tartaric Acid, *J. Food Sci.*, 1986, **51**, 1191–1194.
39. V. L. Singleton, M. Salgues, J. Zaya and E. Trousdale, Caftaric acid disappearance and conversion to products of enzymic oxidation in grape must and wine, *Am. J. Enol. Vitic.*, 1985, **36**, 50–56.
40. A. Morata, M. C. Gomez-Cordoves, F. Calderon and J. A. Suarez, Effects of pH, temperature and SO_2 on the formation of pyranoanthocyanins during red wine fermentation with two species of *Saccharomyces*, *Int. J. Food Microbiol.*, 2006, **106**, 123–129.
41. A. Morata, C. Gonzalez and J. A. Suarez-Lepe, Formation of vinylphenolic pyranoanthocyanins by selected yeasts fermenting red grape musts supplemented with hydroxycinnamic acids, *Int. J. Food Microbiol.*, 2007, **116**, 144–152.
42. A. Crozier, I. B. Jaganath and M. N. Clifford, Dietary phenolics: chemistry, bioavailability and effects on health, *Nat. Prod. Rep.*, 2009, **26**, 1001–1043.
43. C. Bennat, U. H. Engelhardt, A. Kiehne, F. M. Wirries and H. G. Maier, HPLC analysis of chlorogenic acid lactones in roasted coffee, *Z. Lebensm.-Unters. Forsch.*, 1994, **199**, 17–21.
44. M. N. Clifford, B. Kellard and G. G. Birch, Characterisation of chlorogenic acids by simultaneous isomerisation and transesterification with tetramethylammonium hydroxide, *Food Chem.*, 1989, **33**, 115–123.
45. R. Jaiswal, M. F. Matei, A. Golon, M. Witt and N. Kuhnert, How to distinguish between cinnamoylshikimate esters and chlorogenic acid lactones by liquid chromatography-tandem mass spectrometry, *Food Funct.*, 2012, **3**, 976–984.
46. A. Stalmach, W. Mullen, C. Nagai and A. Crozier, On-line HPLC analysis of the antioxidant activity of phenolic compounds in brewed, paper-filtered coffee, *Braz. J. Plant Physiol.*, 2006, **18**, 253–261.
47. T. W. M. Crozier, A. Stalmach, M. E. J. Lean and A. Crozier, Espresso coffees, caffeine and chlorogenic acid intake: potential health implications, *Food Funct.*, 2012, **3**, 30–33.

48. K. Kahle, W. Huemmer, M. Kempf, W. Scheppach, T. Erk and E. Richling, Polyphenols Are Intensively Metabolized in the Human Gastrointestinal Tract after Apple Juice Consumption, *J. Agric. Food Chem.*, 2007, **55**, 10605–10614.
49. S. Hagl, H. Deusser, B. Soyalan, C. Janzowski, F. Will, H. Dietrich, F. W. Albert, S. Rohner and E. Richling, Colonic availability of polyphenols and D-(−)-quinic acid after apple smoothie consumption, *Mol. Nutr. Food Res.*, 2011, **55**, 368–377.
50. G. Pandino, F. L. Courts, S. Lombardo, G. Mauromicale and G. Williamson, Caffeoylquinic Acids and Flavonoids in the Immature Inflorescence of Globe Artichoke, Wild Cardoon, and Cultivated Cardoon, *J. Agric. Food Chem.*, 2010, **58**, 1026–1031.
51. M. N. Clifford and J. R. Ramirez-Martinez, Chlorogenic acids and purine alkaloids contents of Maté (*Ilex paraguariensis*) leaf and beverage, *Food Chem.*, 1990, **35**, 13–21.
52. M. N. Clifford, I. B. Jaganath, I. A. Ludwig and A. Crozier, Chlorogenic acids and the acyl-quinic acids: Discovery, biosynthesis, bioavailability and bioactivity, *Nat. Prod. Rep.*, 2017, **34**, 1391–1421.
53. A. Stalmach, W. Mullen, D. Barron, K. Uchida, T. Yokota, C. Cavin, H. Steiling, G. Williamson and A. Crozier, Metabolite profiling of hydroxycinnamate derivatives in plasma and urine after the ingestion of coffee by humans: identification of biomarkers of coffee consumption, *Drug Metab. Dispos.*, 2009, **37**, 1749–1758.
54. A. Stalmach, H. Steiling, G. Williamson and A. Crozier, Bioavailability of chlorogenic acids following acute ingestion of coffee by humans with an ileostomy, *Arch. Biochem. Biophys.*, 2010, **501**, 98–105.

CHAPTER 25

Major Chlorogenic Acids' Contents and Distribution in Coffees

ADRIANA FARAH* AND JULIANA DE PAULA LIMA

Universidade Federal do Rio de Janeiro, Laboratório de Química e Bioatividade de Alimentos, Avenida Carlos Chagas Filho, 373, CCS, Bloco J, Rio de Janeiro, 5521, Brazil
*E-mail: afarah@nutricao.ufrj.br

25.1 Chlorogenic Acids Characterization

Chlorogenic acids, known as CGA, are the major phenolic compounds in coffee seeds. This family of compounds includes different molecules formed by the esterification of one molecule of (−)-quinic acid and one to three molecules of *trans*-hydroxycinnamic acids (*trans*-phenyl-3-propenoic acids, with different substitutions in the aromatic ring). *Trans*-hydroxycinnamic acids are the most relevant secondary plant metabolites in the human diet,[1] with caffeic acid (3,4-dihydroxycinnamic acid) being the most abundant in coffee, followed by ferulic acid (3-methoxy-4-hydroxycinnamic acid) and *p*-coumaric acid (4-hydroxycinnamic acid). After esterification, the main CGA classes found in *Coffea arabica* and *Coffea canephora* seeds are caffeoylquinic acids (CQA), dicaffeoylquinic acids (diCQA), feruloylquinic acids (FQA), and *p*-coumaroylquinic acids (*p*-CoQA), with at least three major isomers each, and caffeoyl-feruloyl-quinic acids (CFQA), with six isomers (Figure 25.1).

Coffee: Production, Quality and Chemistry
Edited by Adriana Farah

Published by the Royal Society of Chemistry, www.rsc.org

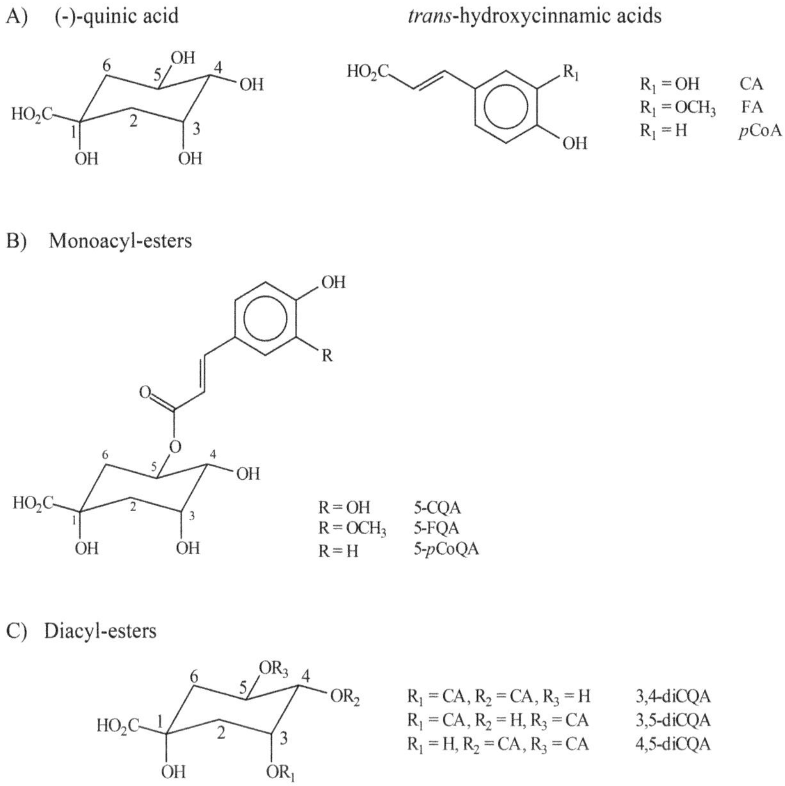

Figure 25.1 Major chlorogenic acids in coffee and their precursors. (A) The precursors. (B) Monoesters (exemplified by 5-acyl compounds). (C) Diesters of quinic acid with caffeic acid, and mixed esters. The authors adopted the IUPAC numbering system for chlorogenic acids. CA – caffeic acid; CQA – caffeoylquinic acid; FQA – feruloylquinic acid.[4,7]

Most recently, a series of minor CGA and related compounds have been identified in green *Coffea arabica* and/or *Coffea canephora* seeds, including acyl dicaffeoylquinic acids, dimethoxycinnamoylquinic acids, caffeoyl-dimethoxycinnamoylquinic acids, diferuloylquinic acids, feruloyl-dimethoxycinnamoylquinic, sinapoylquinic acids, sinapoyl-caffeoylquinic acids, sinapoyl-feruloylquinic acids, feruloyl-sinapoylquinic acids, and new minor *p*-coumaric acid-containing compounds.[2–5] (See Chapter 24 for information on CGA analysis and minor CGA chemical structures.) Conjugates with tryptophan, tyrosine, proteins, and caffeine have also been described in green and roasted coffees in the last two decades.[4–8]

25.2 Chlorogenic Acids Content in Green Coffee

Despite the large distribution of some CGA in the plant kingdom, green coffee is known as one of the main food sources of these compounds, with contents being equaled or exceeded only by the green leaves of *Ilex paraguariensis* (yerba mate).[9] Total CGA content in green coffee seeds may vary according to genetics – species and cultivar, degree of maturation, and, less importantly, agricultural practices, climate, and soil.[10,11] Literature values for total CGA in regular green coffee seeds on dry matter (dm) vary from about 2.1 to 7.8 g 100 g^{-1} for *Coffea arabica*[4,11–23] with 4 to 6 g 100 g^{-1} being a more common range. Reported values for *Coffea canephora* vary from 2.3 to 14.4 g 100 g^{-1} (dm) with 7 to 8 g 100 g^{-1} being more usual contents.[4,11–14,17,18,20–22,24–30] The lowest and highest values have been reported in just a few papers[17,22,25] and may be due to differences in analytical methodology and/or number of compounds considered.

Green coffee is safe for human consumption and because of its high antioxidant property and thermogenic effects among others, its dried aqueous extract has been used as nutraceuticals (23–40 g CGA 100 g^{-1})[31,unpublished data] or to formulate products like green and roasted soluble coffee blend (4.6 to 6.7 g CGA 100 g^{-1}).[32, unpublished data]

CQA account for approximately 70–85% of total CGA content, followed by diCQA (10–20%) and FQA (5–10%).[10,11,26,32,33] In particular, 5-CQA alone accounts for about 60% of total CGA in green coffee seeds; 3-CQA and 4-CQA each comprise approximately 10%.[10,11,32–34] The same tendency is observed for FQA isomers: 5-FQA is the major FQA isomer and accounts for about 5% of total CGA; 3-FQA and 4-FQA together comprise approximately 2%. Among diCQA, 3,5-diCQA represents, on average, 7% of total CGA; 3,4- and 4,5-together corresponds to about 6%.[14,15] Minor CGA and related compounds identified in green *C. arabica* and *C. canephora* seeds[2–5] correspond together to less than 1% of *C. canephora* seeds.[11,34]

CGA composition of coffee seeds varies considerably during fruit maturation. Its content increases in coffee seeds until fruits are completely formed and decreases as they ripen until senescence.[10,13,32] When coffee fruits are harvested by stripping the branches or mechanically, the different fruit maturation stages are reflected on raw coffee seeds appearance, chemical composition, and flavor and these seeds are called intrinsic defects. They are usually marketed with regular seeds and sold for lower prices. Inappropriate agricultural practices and careless management of fruits also produce defects. Allowing for variations between cultivars, average total CGA content of 7.3–8.7 g 100 g^{-1} (dm) has been reported for raw *C. arabica* immature seeds (originated mainly from immature fruits) and 7.7 g 100 g^{-1} (dm) for black-immature seeds (originated from use of high temperature during drying of immature seeds)[19] and therefore the amount of green defective seeds in a coffee blend may increase the total CGA content, depending on the amount in the blend. Sour defective seeds (generated by water paucity during fruits development or by abnormal fermentation of immature or ripe seeds) may also contain significant levels of total CGA (about 7.0 g 100 g^{-1}); lower contents have been reported for black defective seeds (originated from over-ripened cherries) (1.3 g 100 g^{-1}).[19] The distribution of CGA classes in coffee seeds does not tend to change along fruits maturation. However, from immature to black seeds, the contents of 5-acyl isomers tend to decrease while those of 3-acyl and 4-acyl isomers tend to increase suggesting acyl migration.[10,19]

After harvesting of coffee fruits, they are depulped/dehusked to obtain the green coffee seeds by one of the existing methods known as dry, wet, semi-dry, and other hybrid processes[35] (see Chapters 2 and 16 of this book). It is well known that post-harvest processing has pronounced effects on the chemical composition of coffee seeds, especially in water-soluble components like sugars, caffeine, trigonelline, and CGA when fruits are soaked in water and/or fermented.[11,16,36,37] Studies have reported higher CGA content in wet processed *C. canephora* seeds compared to dry processed seeds (Table 25.1).[16,37] This difference in CGA contents has been attributed to losses of other water-soluble components by lixiviation and fermentation in the wet method[38] in addition to slight degradation of CGA in dry processed seeds due to exposure to sun heat.[16] Regarding animal depulped coffees, neither differences in the contents nor significant changes in CGA isomers distribution have been observed in kopi luwak-depulped and jacu bird-depulped coffees, compared to the original seeds.[unpublished data]

CGA contents reported for a few green *C. arabica* and *C. canephora* samples are presented in Tables 25.1 and 25.2, respectively. Some hybrids present intermediate contents. Also, lower CGA contents have been observed in naturally low-caffeine and caffeine-free species native to East Africa.[7,32,39] The contents of CGA reported for a few hybrids and some species other than *C. arabica* and *C. canephora* are presented in Table 25.3.

Decaffeination may considerably change CGA content, decreasing or increasing it, depending on the type of solvent and method used for this purpose.[11] Losses of CGA during decaffeination with dichloromethane have been reported as 16% for *C. arabica* and 11% for *C. canephora*,[20] whereas 1.5% gain of CGA has been observed in green *C. arabica* seeds after water decaffeination.[15] On the other hand, a study using a different water decaffeination method reported average 20% loss of CGA in *C. arabica* samples.[24] A relative average gain of 1.2% in CGA content of green *C. arabica* samples has been reported when supercritical carbon dioxide was used.[24]

CGA contents reported for a few decaffeinated green *C. arabica* and *C. canephora* samples are presented in Table 25.4.

Coffee may be steam-treated before roasting to make it less "irritating" to the stomach. This process is popularly called "stomach friendly coffee".[44] The beneficial properties have been attributed to the reduction of CGA.[45] For more information on this subject, see Chapter 22 on organic acids. Gamma-irradiation has also been reported to reduce the contents of cinnamic acid derivatives in food products.[33]

Table 25.1 Contents of major chlorogenic acids (CGA) in green *Coffea arabica* cultivars from different origins.[a]

Coffee cultivar/type	Origin	*N*	3-CQA (mg 100 g^{-1})	4-CQA (mg 100 g^{-1})	5-CQA (mg 100 g^{-1})	3-FQA (mg 100 g^{-1})	4-FQA (mg 100 g^{-1})	5-FQA (mg 100 g^{-1})	3,4-diCQA (mg 100 g^{-1})	3,5-diCQA (mg 100 g^{-1})	4,5-diCQA (mg 100 g^{-1})	Total CGA (g 100 g^{-1})	Ref.
C. arabica cv. Maragogipe	Ivory Coast	1	Nr	Nr	Nr	Nr	Nr	Nr	Nr	Nr	Nr	6.9	12
C. arabica cv. Red Caturra	Venezuela	1	260.0	480.0	3920.0	Nr	Nr	330.0	100.0	300.0	260.0	5.6	13
C. arabica cv. Red Bourbon	Venezuela	1	240.0	520.0	4000.0	Nr	Nr	340.0	90.0	250.0	210	5.6	13
C. arabica cv. Mundo Novo	Brazil	1	667.2	807.5	3611.2	9.6	31.3	192.0	262.1	269.1	316.6	6.1	4
C. arabica cv. Red Catuaí	Brazil	1	618.0	770.9	3357.0	6.7	25.7	160.9	190.3	156.9	164.6	5.4	4
C. arabica cv. Bourbon	Brazil	1	483.3	543.5	3126.1	28.3	40.1	210.7	236.2	254.9	278.9	5.1	14
C. arabica cv. Longberry	Ethiopia	1	478.2	473.5	3605.8	22.9	34.5	234.9	234.3	263.5	342.6	5.5	14

C. arabica cv. Bourbon	Brazil	1	542.0	618.8	3062.0	28.8	66.6	200.9	168.7	232.8	167.0	5.1	15
C. arabica cv. Sumatra	Sumatra	1	543.6	647.9	3224.6	25.2	68.0	238.5	201.1	271.8	208.2	5.4	15
C. arabica cv. Sumatra	Sulawesi	1	436.7	620.8	4340.0	14.5	39.5	264.1	161.3	374.2	210.0	6,4	15
C. arabica cv. Herloom	Ethiopia	1	346.1	520.5	3662.7	21.2	49.3	296.6	146.7	362.7	200.4	5.6	15
C. arabica cv. Rubi	Brazil	1	445.9	617.9	3191.5	28.0	**	206.9**	238.0	442.3	248.5	5.2	16
C. arabica cv. Topazio	Brazil	1	536.0	689.4	3251.2	28.6	**	257.9**	238.3	423.2	225.8	5.4	16
C. arabica	Kenya	38	160.0–250.0	*	2610.0–3970.0*	Nr	Nr	80.0–220.0	80.0–170.0	170.0–360.0	180.0–380.0	3.4–4.8	17
C. arabica	Angola	3	Nr	Nr	Nr	Nr	Nr	Nr	Nr	Nr	Nr	5.6–7.8	18
C. arabica	Brazil	3	434.0–702.0	599.0–885.0	3995.0–4049.0	Nr	480.0–740.0	223.0–250.0	144.0–198.0	213.0–381.0	118.0–216.0	5.9–6.4	19
C. arabica	Brazil	1	830.0	840.0	3980.0	Nr	**	250.0**	260.0	300.0	230.0	6.7	20
C. arabica	Africa	10	149.0–251.1	254.9–346.8	2987.8–4532.5	10.5–24.2	7.9–23.4	167.5–240.6	63.0–94.2	146.3–370.1	65.5–145.4	3.4–5.4	21
C. arabica	Asia	5	221.4–398.3	337.9–482.5	3273.1–3661.4	21.4–30.5	19.3–30.7	149.3–263.1	96.1–129.1	164.4–347.5	85.4–164.1	3.8–5.5	21
C. arabica	Honduras	5	270.0–450.0	20.0–60.0	1540.0–1810.0	10.0–10.1	Nr	80.0–110.0	30.0–50.0	60.0–110.0	80.0–120.0	2.1–2.6	22

(*continued*)

Table 25.1 (*continued*)

Coffee cultivar/type	Origin	*N*	3-CQA (mg 100 g^{-1})	4-CQA (mg 100 g^{-1})	5-CQA (mg 100 g^{-1})	3-FQA (mg 100 g^{-1})	4-FQA (mg 100 g^{-1})	5-FQA (mg 100 g^{-1})	3,4-diCQA (mg 100 g^{-1})	3,5-diCQA (mg 100 g^{-1})	4,5-diCQA (mg 100 g^{-1})	Total CGA (g 100 g^{-1})	Ref.
C. arabica	Kenya	3	320.0–450.0	60.0–66.0	1115.0–1960.0	10.0–20.0	Nr	130.0–620.0	30.0–50.0	100.0–110.0	60.0–130.0	2.1–3.4	22
C. arabica	Peru	2	440.0/490.0	10.0/20.0	1620.0/1740.0	10.0/20.0	Nr	90.0/91.0	60.0/60.1	80.0/81.0	100.0/101.0	2.4/2.5	22
C. arabica	Colombia	2	Nr	Nr	Nr	Nr	Nr	Nr	Nr	Nr	Nr	6.1/6.2	23
Kopi luwak-depulped *C. arabica*	Indonesia	1	419.2	557.4	2932.6	1.1	59.6	180.4	159.2	237.0	187.4	4.8	Unpub.
Jacu bird-depulped *C. arabica*	Brazil	1	577.6	721.3	2510.4	0.8	68.8	305.5	97.8	222.6	178.7	4.7	Unpub.

[a]Note: 3-CQA: 3-caffeoylquinic acid; 4-CQA: 4-caffeoylquinic acid; 5-CQA: 5-caffeoylquinic acid; 3-FQA: 3-feruloylquinic acid; 4-FQA: 4-feruloylquinic acid; 5-FQA: 5-feruloylquinic acid; 3,4-diCQA: 3,4-dicaffeoylquinic acid; 3,5-diCQA: 3,5-dicaffeoylquinic acid; 4,5-diCQA: 4,5-dicaffeoylquinic acid. *N*: number of samples. Nr: not reported. *sum of 4-CQA and 5-CQA. **sum of 4-FQA and 5-FQA. Unpub.: Unpublished

Table 25.2 Contents of major chlorogenic acids (CGA) (mg 100 g^{-1}) in green *Coffea canephora* cultivars from different origins.[a]

Coffee cultivar	Origin	*N*	3-CQA (mg 100 g^{-1})	4-CQA (mg 100 g^{-1})	5-CQA (mg 100 g^{-1})	3-FQA (mg 100 g^{-1})	4-FQA (mg 100 g^{-1})	5-FQA (mg 100 g^{-1})	3,4-diCQA (mg 100 g^{-1})	3,5-diCQA (mg 100 g^{-1})	4,5-diCQA (mg 100 g^{-1})	Total CGA (g 100 g^{-1})	Ref.
C. canephora cv. robusta	Uganda	1	924.8	602.4	4243.3	34.1	57.8	379.2	423.5	400.7	511.5	7.6	14
C. canephora cv. robusta	Indonesia	1	575.8	821.8	4737.8	101.5	61.7	383.2	467.1	613.6	372.3	8.1	24
C. canephora cv. robusta	Africa	17	321.6–555.7	386.6–690.6	2162.1–3625.5	52.6–110.5	60.3–135.2	414.7–616.2	244.3–407.4	204.3–354.8	162.8–375.6	4.0–6.8	21
C. canephora cv. robusta	Asia-Oceania	39	265.6–568.6	367.0–663.8	2501.5–4290.3	43.3–104.5	58.1–125.7	501.5–827.1	264.9–403.9	201.8–441.0	201.7–371.4	4.8–7.8	21
C. canephora cv. robusta	Vietnam	2	380.0/400.0	10.0/30.0	1410.1/1520.0	20.0/20.1	Nr	130.0/140.0	80.0/90.0	100.0/130.0	120.0/150.0	2.3/3.6	22
C. canephora cv. conilon	Brazil	1	1230.0	1450.0	3180.0	Nr	**	720.0**	490.0	490.0	180.0	7.7	20
C. canephora cv. conilon	Brazil	1	1065.6	1277.2	4114.0	32.6	57.5	519.7	429.1	283.5	586.4	8.3	4
C. canephora	Venezuela	1	300.0	490.0	4560.0	Nr	Nr	790.0	440.0	560.0	50.0	7.2	13
C. canephora	Africa	8	430.0–1330.0	*	4680.0–8960.0*	Nr	Nr	620.0–2040.0	540.0–950.0	380.0–1130.0	410.0–1240.0	7.9–14.4	17
C. canephora	Colombia	3	Nr	Nr	1157.7–2740.8	Nr	Nd	243.6–476.3	201.2–314.0	264.6–419.1	155.0–256.7	2.0–4.2	25
C. canephora cv. robusta	Nr	1	Nr	Nr	Nr	Nr	Nr	Nr	Nr	Nr	Nr	6.9	26
C. canephora cv. robusta	Ivory Coast	2	Nr	Nr	Nr	Nr	Nr	Nr	Nr	Nr	Nr	8.4/9.5	12
C. canephora cv. robusta	Mexico	1	Nr	Nr	Nr	Nr	Nr	Nr	Nr	Nr	Nr	8.3	27

(*continued*)

Table 25.2 (*continued*)

Coffee cultivar	Origin	N	3-CQA (mg 100 g^{-1})	4-CQA (mg 100 g^{-1})	5-CQA (mg 100 g^{-1})	3-FQA (mg 100 g^{-1})	4-FQA (mg 100 g^{-1})	5-FQA (mg 100 g^{-1})	3,4-diCQA (mg 100 g^{-1})	3,5-diCQA (mg 100 g^{-1})	4,5-diCQA (mg 100 g^{-1})	Total CGA (g 100 g^{-1})	Ref.
C. canephora cv. robusta	Equador	1	Nr	Nr	Nr	Nr	Nr	Nr	Nr	Nr	Nr	7.8	27
C. canephora cv. robusta	Brazil	1	Nr	Nr	Nr	Nr	Nr	Nr	Nr	Nr	Nr	8.3	28
C. canephora cv. robusta	Angola	2	Nr	Nr	Nr	Nr	Nr	Nr	Nr	Nr	Nr	6.1/7.2	18
C. canephora cv. conilon	Brazil	6	Nr	Nr	Nr	Nr	Nr	Nr	Nr	Nr	Nr	7.8	28
C. canephora cv. conilon	Brazil	1	Nr	Nr	Nr	Nr	Nr	Nr	Nr	Nr	Nr	9.5	29
C. canephora cv. conilon	Brazil	1	Nr	Nr	Nr	Nr	Nr	Nr	Nr	Nr	Nr	10.1	30
C. canephora cv. kouillou	Tanzania	1	Nr	Nr	Nr	Nr	Nr	Nr	Nr	Nr	Nr	6.8	28
C. canephora cv. ugandae	Tanzania	1	Nr	Nr	Nr	Nr	Nr	Nr	Nr	Nr	Nr	7.7	28

[a]Note: 3-CQA: 3-caffeoylquinic acid; 4-CQA: 4-caffeoylquinic acid; 5-CQA: 5-caffeoylquinic acid; 3-FQA: 3-feruloylquinic acid; 4-FQA: 4-feruloylquinic acid; 5-FQA: 5-feruloylquinic acid; 3,4-diCQA: 3,4-dicaffeoylquinic acid; 3,5-diCQA: 3,5-dicaffeoylquinic acid; 4,5-diCQA: 4,5-dicaffeoylquinic acid. *N*: number of samples. Nr: not reported. Nd: not detected. *sum of 4-CQA and 5-CQA. **sum of 4-FQA and 5-FQA.

Table 25.3 Contents of major chlorogenic acids (CGA) in non-commercial green Coffea species and hybrids from different origins.[a]

Coffee cultivar	Origin	N	Total CQA (mg 100 g^{-1})	Total FQA (mg 100 g^{-1})	Total diCQA (mg 100 g^{-1})	Total CGA (g 100 g^{-1})	Ref.
Hybrids between Coffea arabica and other species							
Timor hybrid *(C. arabica x C. canephora)*	Venezuela	1	4800.0	300.0	600.0	5.7	13
Timor hybrid	Colombia	3	2135.2–2707.4	167.7–260.1	373.8–465.5	2.7–3.5	25
Catimor (Timor hybrid x *C. arabica* cv. Caturra)	Venezuela	1	5400.0	300.0	500.0	6.2	13
C. arabica cv. Caturra x Timor hybrid	Colombia	1	2416.7	282.5	536.5	3.2	25
C. arabica cv. Caturra x Timor hybrid	Nr	4	4826–5600	355.0–418.0	792.0–1000.0	6.0–7.1	40
C. arabica cv. Yellow Catuai x Timor hybrid	Brazil	10	4135.6–5482.6	273.6–348.2	632.5–925.7	5.0–6.7	16
C. arabica cv. Red Caturra x Timor hybrid	Brazil	2	4152.5/4196.0	292.9/308.2	928.3/932.0	5.3/5.5	16
Other Coffea species							
C. liberica	Ivory Coast	1	9370.0	530.0	730.0	10.6	12
C. liberica cv. Dewevrei	Ivory Coast	1	5390.0	480.0	1110.0	7.0	41
C. liberica cv. Dewevrei	Ivory Coast	1	Nr	Nr	Nr	8.0	42
C. liberica liberica	Nr	Nr	Nr	Nr	Nr	9.8	13
C. racemosa Lour.	Brazil	2	4360.0/4790.0	40.0/130.0	420.0/500.0	5.3/5.5	28
C. eugienoides	Brazil	1	4240.0	170.0	150.0	4.5	28
C. brevipes Hiern	Ivory Coast	1	4800.0	320.0	1280.0	6.4	28
C. salvatrix Swynn.	Brazil	Nr	1950.0	160.0	140.0	2.2	28
C. pseudozanguebariae	Ivory Coast	1	1570.0	60.0	120.0	1.4	28
C. pseudozanguebariae	Ivory Coast	1	Nr	Nr	Nr	1.7	43
C. perrieri Drake ex Jaarb	Madagascar	Nr	350.0	Nr	Nr	0.3	28

[a]Note: total CQA: total caffeoylquinic acid: 3-caffeoylquinic acid + 4-caffeoylquinic acid + 5-caffeoylquinic acid; total FQA: total feruloylquinic acid: 3-feruloylquinic acid + 4-feruloylquinic acid + 5-feruloylquinic acid; total diCQA: total dicaffeoylquinic acids: 3,4-dicaffeoylquinic acid + 3,5-dicaffeoylquinic acid + 4,5-dicaffeoylquinic acid. *N*: number of samples. Nr: not reported.

Table 25.4 Contents of major chlorogenic acids (CGA) in decaffeinated *C. arabica* and *C. canephora* cultivars by different methods.[a]

Coffee cultivar and origin	Decaffeination method	3-CQA (mg 100 g^{-1})	4-CQA (mg 100 g^{-1})	5-CQA (mg 100 g^{-1})	3-FQA (mg 100 g^{-1})	4-FQA (mg 100 g^{-1})	5-FQA (mg 100 g^{-1})	3,4-diCQA (mg 100 g^{-1})	3,5-diCQA (mg 100 g^{-1})	4,5-diCQA (mg 100 g^{-1})	3-CQL (mg 100 g^{-1})	4-CQL (mg 100 g^{-1})	Total CGA and CGL (g 100 g^{-1})	Ref.
C. arabica														
cv. Bourbon/ Brazil	Water	1672.3	1585.6	1931.5	90.5	120.8	113.1	222.1	185.9	206.6	10.1	Nd	6.0	15
cv. Sumatra/ Sumatra	Water	1713.8	1590.8	1979.3	90.6	152.3	150.2	271.7	219.6	274.4	8.9	Nd	6.3	15
cv. Sumatra/ Sulawesi	Water	1884.0	1996.7	2264.0	84.1	143.4	151.5	262.9	220.2	260.0	Nr	Nr	7.0	15
cv. Herloom/ Ethiopia	Water	1613.5	1742.0	1937.3	134.2	146.4	169.8	255.7	173.2	232.9	Nr	Nr	6.4	15
Brazil	DCM*	1340.0	1340.0	2229.0	Nr		130.0**	190.0	140.0	160.0	Nr	Nr	5.6	20
cv. Supremo/ Colombia	Water	1338.7	1388.3	2079.1	114.6	82.4	135.8	145.1	122.0	140.6	4.6	0.8	5.6	24
cv. Sidamo/ Ethiopia	Water	1517.4	1535.6	2100.7	117.1	83.1	125.9	165.6	133.4	149.8	2.1	1.2	6.0	24
Colombia	sCO_2***	1260.4	1433.5	2888.5	103.0	88.9	192.7	257.2	230.2	289.4	3.4	0.7	7.2	24
Costa Rica	Water	1325.6	718.5	1723.6	52.3	74.3	120.4	180.9	160.3	170.3	9.3	0.6	4.6	Unpub.
Peru	Water	1429.4	811.6	1727.9	58.3	71.3	100.2	160.4	165.5	200.8	10.1	0.5	4.8	Unpub.
C. canephora														
cv. conilon/ Brazil	DCM*	1810.0	1980.0	2230.0	Nr		290.0**	260.0	270.0	70.0	Nr	Nr	6.9	20
cv. Sumatra/ Indonesia	Water	1325.5	1384.0	2366.1	122.9	83.4	156.4	132.9	114.3	131.8	5.4	0.5	5.8	24
Indonesia	sCO_2***	1684.8	1952.2	2273.8	319.5	188.6	242.5	441.5	336.5	414.9	8.3	2.2	7.8	24

[a]Note: 3-CQA: 3-caffeoylquinic acid; 4-CQA: 4-caffeoylquinic acid; 5-CQA: 5-caffeoylquinic acid; 3-FQA: 3-feruloylquinic acid; 4-FQA: 4-feruloylquinic acid; 5-FQA: 5-feruloylquinic acid; 3,4-diCQA: 3,4-dicaffeoylquinic acid; 3,5-diCQA: 3,5-dicaffeoylquinic acid; 4,5-diCQA: 4,5-dicaffeoylquinic acid. Nr: not reported. Nd: not detected. *DCM: dichloromethane; **sum of 4-FQA and 5-FQA; ***sCO_2: supercritical carbone dioxide; Unpub: Unpublished.

25.3 Chlorogenic Acids Content in Roasted Coffee

CGA take part in the generation of color, flavor, and aroma of coffee during roasting.[10,46–48] Due to CGA thermal instability and vulnerability to the intense heat conditions (200–310 °C, commonly 210–230 °C), they undergo many changes, participating in Maillard and pyrolysis reactions. In the beginning of roasting, acyl migration is observed (the levels of substitutes in the 5-position of the quinic acid decrease substantially while levels of substitutes in the 3- and 4-positions increase in some cases to almost double their original levels)[10,32] in all main CGA classes. As roasting progresses, CGA can be lactonized or epimerized. Both CGA and 1,5-γ-quinolactones (CGL), which contribute to the characteristic mild and pleasant bitterness of coffee, are later degraded to form free organic acids and low molecular mass and volatile phenol derivatives, including pyrogallol, hydroxyhydroquinone, catechols, methylcatechols, ethylcatechols, vinylcathechols, guaiacols among a series of di- and trihydroxybenzenes[6,14,15,33,49–55] (Figure 25.2). It has also been reported that as roasting proceeds 4-vinylcatechol is oligomerized to produce a family of polyhydroxylated phenylindans which exhibits a harsh and lingering bitter taste profile.[56,57] Moreover, di- and trihydroxybenzenes, catechol, pyrogallol, 3-methylcatechol, and 4-methylcatechol were reported to react with furfuryl alcohol during roasting to give (furan-2-yl)methylated benzene diols and triols, another class of (unpleasant) bitter taste compounds derived from coffee CGA.[58] CGA and derivatives also participate in the formation of polymeric materials like melanoidins,[30,59] which explain at least part of the antioxidant activity of melanoidins in coffee brews.[60,61]

Total CGA content in ground roasted coffees varies considerably for the same species, depending mainly on the roasting degree. In general, the values reported for total CGA (including lactones) in regular roasted *Coffea arabica* seeds (dm) vary from 2.9 to 5.1 g% in light roasts,[4,14,24,62,63] from 1.7 to 3.5 g% in medium roasts,[20,24,62,63] and from 0.4 to 2.1 g% in dark roasts.[4,14,20,24,62,63] In regular *Coffea canephora* seeds, the contents (dm) usually vary from 5.9 to 7.1 g% in light roasts,[4,14,62] from 1.0 to 4.3 g% in medium roasts,[20,24,62] and from 0.4 to 1.7 g% in dark roasts.[4,14,20,62]

In commercial ground roasted coffees, total CGA content also ranges depending on the percentage of *C. arabica* and *C. canephora* seeds in the blend and on the roasting degree, with reported contents (including lactones) varying from 0.2 to 3.1 g 100 g^{-1}.[11,64–70, unpublished data]

In roasted coffees, CQA account for approximately 78% of total CGA content, followed by diCQA (5–8%) and FQA (5–7%). Breaking into CGA groups of isomers, 5-CQA alone accounts for about 40% of total CGA in seeds (dm); and 3-CQA and 4-CQA comprise approximately 16% and 20%, respectively. The same tendency is observed for FQA isomers: 5-FQA is the major FQA isomer in roasted coffee seeds and accounts for about 4% of total CGA; 3-FQA and 4-FQA together comprise approximately 3%. Among diCQA, 3,4-diCQA comprise, on average, 3% of total CGA; and 3,5- and 4,5-diCQA about 2% and 2.5%, respectively.[4,11,14,15,20,24]

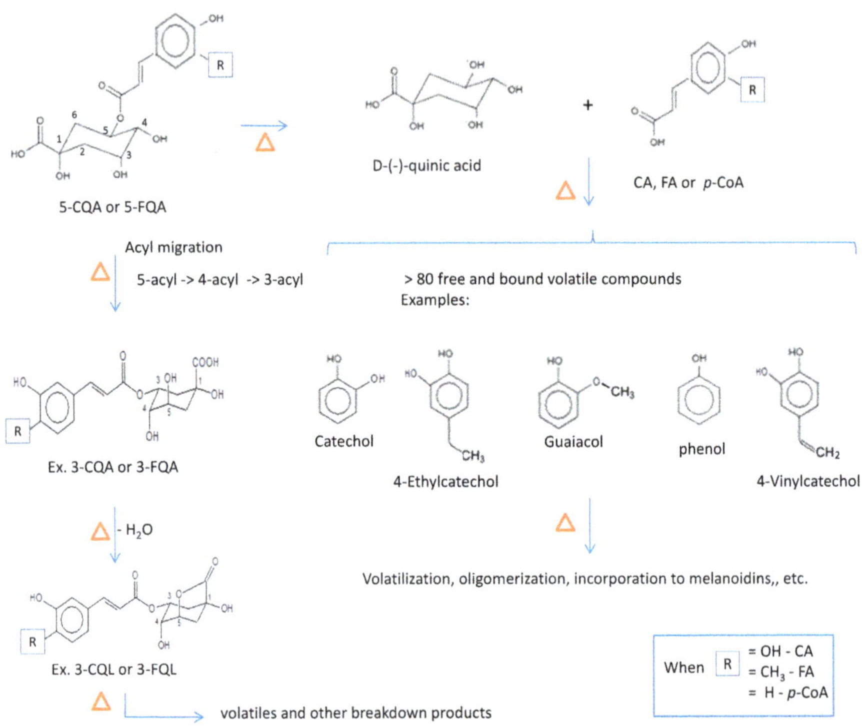

Figure 25.2 Acyl migration and formation of major 1,5-γ-quinolactones (exemplified by 3-caffeoylquinic-1,5-γ-lactone) and volatiles from chlorogenic acid (5-caffeoylquinic acid) or 5-feruloylquinic acid during roasting. Although under IUPAC rules the numbering system for the lactones is different from that for the acids, in order to avoid confusion, most reports in the literature have used for lactones the same numbering of the carbons as for the acid precursors. CQA – caffeoylquinic acid; CQL – caffeoylquinide; FQA – feruloylquinic acid; FQL – feruloylquinide; *p*-CoA.[14,77,78]

About 7–10% of CGA in *C. arabica* and 6–9% of *C. canephora* seeds were reported to be converted to 1,5-γ-quinolactones or 1,5-γ-quinides (CGL),[10] as they are commonly called (Figure 25.2). 3-Caffeoyl-1,5-quinide (3-CQL) and 4-caffeoyl-1,5-quinide (4-CQL) are the main CGA lactones in coffee[6,14,56,71–73] since CQA prevail among CGA and only those CGA compounds that lack a substitute in the 5-position are able to form a 1,5-γ-quinide. 1,5-γ-quinides derived from FQA, diCQA, and *p*-CoQA, in order of relevance and other minor quinides, are present in roasted coffee as well.[14,74,75] CGL levels peak most commonly from light medium to dark medium roasting degree, depending on roaster and roasting profile.[11,14,76] Average maximum levels vary from 380 to 430 mg 100 g^{-1} (dm).

Contents of CGA reported for roasted *C. arabica* and *C. canephora* seeds as well as commercial blends are presented in Tables 25.5, 25.6, and 25.7, respectively.

Table 25.5 Contents of major chlorogenic acids (CGA) and 1,5-γ-quinolactones (CGL) in roasted *Coffea arabica* seeds from different origins.[a]

Coffee cultivar and origin	RD	3-CQA (mg 100 g^{-1})	4-CQA (mg 100 g^{-1})	5-CQA (mg 100 g^{-1})	3-FQA (mg 100 g^{-1})	4-FQA (mg 100 g^{-1})	5-FQA (mg 100 g^{-1})	3,4-diCQA (mg 100 g^{-1})	3,5-diCQA (mg 100 g^{-1})	4,5-diCQA (mg 100 g^{-1})	3-CQL (mg 100 g^{-1})	4-CQL (mg 100 g^{-1})	Total CGA and CGL (g 100 g^{-1})	Ref.
C. arabica cv. Bourbon/Brazil	L	816.1	999.6	1995.8	47.8	68.0	109.1	119.7	88.6	127.8	160.2	92.1	4.6	14
C. arabica cv. Bourbon/Brazil	LM	458.5	590.4	1103.5	34.2	43.9	87.3	51.7	35.2	50.1	248.5	115.3	2.8	14
C. arabica cv. Bourbon/Brazil	D	122.8	155.8	244.3	9.7	19.9	22.1	7.4	5.1	9.2	95.7	53.1	0.7	14
C. arabica cv. Longberry/Ethiopia	L	804.7	933.0	1724.1	42.2	56.8	107.0	94.3	72.5	118.9	161.2	94.4	4.2	14
C. arabica cv. Longberry/Ethiopia	LM	343.8	494.8	815.7	41.1	42.2	68.4	40.9	36.3	54.1	210.5	116.0	2.3	14
C. arabica cv. Longberry/Ethiopia	D	110.1	156.2	218.6	9.2	17.0	14.5	3.4	5.1	6.1	94.5	47.9	0.7	14
C. arabica cv. Mundo Novo/Brazil	VL	1096.6	1344.7	2736.7	43.2	58.3	179.0	198.3	154.6	232.7	11.5	77.5	6.1	4
C. arabica cv. Mundo Novo/Brazil	L	720.8	868.4	1656.6	31.5	49.1	143.8	103.4	80.2	118.6	170.1	107.6	4.0	4

(*continued*)

Table 25.5 (*continued*)

Coffee cultivar and origin	RD	3-CQA (mg 100 g^{-1})	4-CQA (mg 100 g^{-1})	5-CQA (mg 100 g^{-1})	3-FQA (mg 100 g^{-1})	4-FQA (mg 100 g^{-1})	5-FQA (mg 100 g^{-1})	3,4-diCQA (mg 100 g^{-1})	3,5-diCQA (mg 100 g^{-1})	4,5-diCQA (mg 100 g^{-1})	3-CQL (mg 100 g^{-1})	4-CQL (mg 100 g^{-1})	Total CGA and CGL (g 100 g^{-1})	Ref.
C. arabica cv. Catuaí Vermelho/ Brazil	LM	495.3	592.6	1150.6	9.4	29.7	109.9	46.8	36.2	54.8	215.8	133.4	2.9	4
C. arabica cv. Catuaí Vermelho/ Brazil	D	328.0	393.4	761.4	7.1	21.4	77.4	26.2	20.2	29.7	172.1	107.6	1.9	4
C. arabica cv. Catuaí Vermelho/ Brazil	VD	70.4	82.4	118.0	Nd	7.0	18.0	Nd	1.2	4.4	47.7	26.6	0.3	4
C. arabica cv. Supremo/ Colombia	L	535.3	629.3	1127.3	44.5	36.2	71.0	57.3	44.8	57.2	173.8	120.5	2.9	24
C. arabica cv. Sidamo/ Ethiopia	M	450.6	503.0	859.9	32.2	32.8	61.5	34.6	25.6	30.7	193.7	134.3	2.4	24
C. arabica/ Guatemala	D	198.7	209.8	324.4	22.0	26.9	46.3	14.4	8.5	10.4	119.1	85.7	1.1	24
C. arabica/ Brazil	LM	1130.0	1120.0	2790.0	Nr		130.0**	160.0	120.0	160.0	Nr	Nr	5.6	20
C. arabica/ Brazil	M	350.0	440.0	700.0	Nr		90.0**	40.0	30.0	20.0	Nr	Nr	1.7	20
C. arabica/ Brazil	DM	230.0	320.0	410.0	Nr		70.0**	30.0	20.0	10.0	Nr	Nr	1.0	20

C. arabica/ Brazil	D	120.0	190.0	190.0	Nr		50.0**	Nd	Nd	Nd	Nr	Nr	0.6	20
C. arabica/ Brazil	L	987.0	1230.0	2210.0	70.0	Nr	162.0	166.0	130.0	175.0	Nr	Nr	5.1	62
C. arabica/ Brazil	M	594.0	814.0	1280.0	55.0	Nr	107.0	70.0	50.0	65.0	Nr	Nr	3.0	62
C. arabica/ Brazil	D	203.0	258.0	379.0	28.0	Nr	40.0	14.0	9.0	9.0	Nr	Nr	0.9	62
C. arabica/ Ethiopia	L	545.0	756.0	1511.0	Nr	128.0	56.0	65.0	56.0	92.0	Nr	Nr	3.2	63
C. arabica/ Nicaragua	M	350.0	490.0	819.0	Nr	100.0	30.0	33.0	18.0	44.0	Nr	Nr	1.9	63
C. arabica/ Panama	D	55.0	117.0	117.0	Nr	87.0	69.0	Nd	Nd	Nd	Nr	Nr	0.4	63

[a]Note: 3-CQA: 3-caffeoylquinic acid; 4-CQA: 4-caffeoylquinic acid; 5-CQA: 5-caffeoylquinic acid; 3-FQA: 3-feruloylquinic acid; 4-FQA: 4-feruloylquinic acid; 5-FQA: 5-feruloylquinic acid; 3,4-diCQA: 3,4-dicaffeoylquinic acid; 3,5-diCQA: 3,5-dicaffeoylquinic acid; 4,5-diCQA: 4,5-dicaffeoylquinic acid; 3-CQL: 3-caffeoylquinic-1,5 lactone; 4-CQL: 4-caffeoylquinic-1,5-lactone. RD: roasting degree: VL: very light; L: light; LM: light-medium; M: medium; DM: dark-medium; D: dark; VD: very dark. Nr: not reported. Nd: not detected. **sum of 4-FQA and 5-FQA.

Table 25.6 Contents of major chlorogenic acids (CGA) and 1,5-γ-quinolactones (CGL) (mg 100 g^{-1}) in roasted *Coffea canephora* seeds from different origins.[a]

Coffee cultivar and origin	RD	3-CQA (mg 100 g^{-1})	4-CQA (mg 100 g^{-1})	5-CQA (mg 100 g^{-1})	3-FQA (mg 100 g^{-1})	4-FQA (mg 100 g^{-1})	5-FQA (mg 100 g^{-1})	3,4-diCQA (mg 100 g^{-1})	3,5-diCQA (mg 100 g^{-1})	4,5-diCQA (mg 100 g^{-1})	3-CQL (mg 100 g^{-1})	4-CQL (mg 100 g^{-1})	Total CGA and CGL (g 100 g^{-1})	Ref.
C. canephora cv. robusta/Uganda	VL	1257.3	1481.0	3802.4	89.7	108.0	310.8	337.2	279.6	334.3	57.7	25.8	7.7	14
C. canephora cv. robusta/Uganda	L	1086.7	1342.9	2520.8	57.1	110.0	241.7	217.3	158.6	205.0	198.8	110.0	6.1	14
C. canephora cv. robusta/Uganda	LM	623.4	790.9	1347.1	79.2	86.1	169.8	81.8	61.4	91.9	253.6	138.8	4.2	14
C. canephora cv. robusta/Uganda	DM	334.0	308.3	448.0	40.8	70.1	73.4	23.1	18.0	19.0	165.4	125.9	2.4	14
C. canephora cv. robusta/Uganda	D	218.6	167.5	225.8	24.4	39.4	37.7	6.9	6.5	6.0	87.6	45.0	0.9	14
C. canephora cv. robusta/Uganda	VD	108.0	99.9	111.0	14.7	18.2	23.4	2.5	3.2	1.9	54.4	25.9	0.5	14
C. canephora cv. conilon/Brazil	VL	1308.4	1666.5	3175.5	115.3	139.0	411.1	269.6	195.8	503.5	153.2	111.4	7.5	4
C. canephora cv. conilon/Brazil	L	1064.6	1324.9	2460.5	108.1	125.3	362.6	181.2	138.8	370.9	210.4	147.2	6.0	4
C. canephora cv. conilon/Brazil	LM	711.0	894.4	1534.3	184.8	105.3	289.0	94.5	73.8	221.9	235.3	163.9	4.4	4

C. canephora cv. conilon/ Brazil	D	563.4	706.7	1145.1	155.3	89.5	223.5	61.1	48.5	163.4	218.3	146.8	3.4	4
C. canephora cv. conilon/ Brazil	VD	95.9	106.9	174.9	49.6	24.6	51.4	3.9	2.3	30.0	54.8	31.6	2.9	4
C. canephora cv. robusta/ Indonesia	L	499.1	570.7	944.0	41.9	38.6	69.4	42.8	31.0	38.2	182.0	126.8	2.6	24
C. canephora cv. robusta/ Indonesia	M	358.5	399.8	648.2	34.6	40.7	70.1	24.7	16.0	20.0	149.9	106.5	1.9	24
C. canephora cv. robusta/ Indonesia	D	204.2	212.6	322.3	24.5	31.2	51.5	13.2	7.4	8.3	114.7	80.0	1.1	24
C. canephora/ Brazil	LM	1410.0	1690.0	2590.0	Nr		410.0**	250.0	260.0	130.0	Nr	Nr	6.6	20
C. canephora/ Brazil	M	730.0	1040.0	1430.0	Nr		280.0**	110.0	110.0	80.0	Nr	Nr	3.7	20
C. canephora/ Brazil	DM	300.0	550.0	550.0	Nr		170.0**	30.0	30.0	40.0	Nr	Nr	1.6	20
C. canephora/ Brazil	D	150.0	300.0	260.0	Nr		110.0**	Nd	Nd	Nd	Nr	Nr	0.7	20
C. canephora/ India	L	1060.0	1310.0	2190.0	165.0	Nr	413.0	315.0	202.0	276.0	Nr	Nr	5.9	62
C. canephora/ India	M	777.0	1030.0	1580.0	142.0	Nr	305.0	180.0	116.0	160.0	Nr	Nr	4.3	62
C. canephora/ India	D	310.0	488.0	589.0	84.0	Nr	157.0	43.0	21.0	33.0	Nr	Nr	1.7	62

[a]Note: 3-CQA: 3-caffeoylquinic acid; 4-CQA: 4-caffeoylquinic acid; 5-CQA: 5-caffeoylquinic acid; 3-FQA: 3-feruloylquinic acid; 4-FQA: 4-feruloylquinic acid; 5-FQA: 5-feruloylquinic acid; 3,4-diCQA: 3,4-dicaffeoylquinic acid; 3,5-diCQA: 3,5-dicaffeoylquinic acid; 4,5-diCQA: 4,5-dicaffeoylquinic acid; 3-CQL: 3-caffeoylquinic-1,5-lactone; 4-CQL: 4-caffeoylquinic-1,5-lactone. RD: roasting degree: VL: very light; L: light; LM: light-medium; M: medium; DM: dark-medium; D: dark; VD: very dark. Nr: not reported. Nd: not detected. **sum of 4-FQA and 5-FQA.

Table 25.7 Chlorogenic acids (CGA) and 1,5-γ-quinolactones (CGL) contents (mg 100 g^{-1}) in commercial coffee blends from different countries.[a]

Coffee species or type/origin	*N*	RD	3-CQA (mg 100 g^{-1})	4-CQA (mg 100 g^{-1})	5-CQA (mg 100 g^{-1})	3-FQA (mg 100 g^{-1})	4-FQA (mg 100 g^{-1})	5-FQA (mg 100 g^{-1})	3,4-diCQA (mg 100 g^{-1})	3,5-diCQA (mg 100 g^{-1})	4,5-diCQA (mg 100 g^{-1})	3-CQL (mg 100 g^{-1})	4-CQL (mg 100 g^{-1})	Total CGA and CGL (g 100 g^{-1})	Ref.
Blend/Germany	12	Nr	439.0–554.0	584.0–731.0	247.0–996.0	Nr	108.0–176.0	108.0–173.0	51.0–86.0	36.0–52.0	68.0–99.0	172.0–233.0	74.0–110.0	1.8–3.2	64
Blend/Germany	5	Nr	Nr	Nr	Nr	Nr	Nr	Nr	Nr	Nr	Nr	78.0–186.0	77.0–158.0	0.1–0.3	65
Blend/Brazil	7	M	108.0–300.0	113.0–390.0	159.0–540.0	Nr	29.0–119.0	34.0–105.0	14.0–52.0	13.0–47.0	12.0–52.0	Nr	Nr	0.5–1.5	66
Blend/Brazil	3	D	75.0–285.0	74.0–377.0	114.0–500.0	Nr	32.0–73.0	24.0–71.0	15.0–49.0	13.0–33.0	12.0–34.0	Nr	Nr	0.4–1.4	66
Blend/Brazil	11	LM	88.9–371.7	117.1–524.8	151.2–843.1	Nr		49.5–199.5**	13.1–50.2	3.6–41.2	6.3–51.8	46.3–214.3	40.4–156.3	0.5–2.5	67
Arabica/Mexico	2	Nr	Nr	Nr	Nr	Nr	Nr	Nr	Nr	Nr	Nr	Nr	Nr	1.1/1.2	68
Robusta/Indonesia	3	D	107.2–265.1	110.9–203.1	371.3–498.3	Nr	22.0–41.8	33.2–101.5	11.0–45.2	1.4–41.3	1.2–41.1	31.2–80.4	16.6–28.4	0.8–1.5	69
Blend/Costa Rica	3	M	189.3–257.4	125.3–293.1	222.3–610.2	Nr	10.1–25.3	21.4–93.2	20.1–30.3	11.6–50.3	21.3–40.4	57.9–80.7	20.4–52.7	0.8–1.9	69
Blend/Colombia	3	D	112.3–201.1	157.9–273.1	341.4–567.4	Nr	11.5–20.8	28.5–40.3	1.0–30.3	2.3–33.1	1.5–22.2	32.5–56.3	24.7–49.1	0.9–1.6	69
Blends – USA/local market	7	Nr	132.0–395.0	144.0–4.56	213.0–706.0	12.0–129.0	21.0–84.0	27.0–89.0	11.0–47.0	1.0–67.0	11.0–73.0	Nr	Nr	0.8–1.7	70
Decaffeinated blends – USA/local market	5	Nr	45.0–342.0	51.0–378.0	82.0–623.0	4.0–12.0	9.0–70.0	0.01–0.1	3.0–40.0	2.0–27.0	1.0–41.0	Nr	Nr	0.2–1.5	70

Arabica – gourmet coffees/ local market	7	L	277.7–546.2	388.5–734.6	622.7–1304.6	0.9–2.8	70.5–185.9	80.7–151.6	29.9–65.9	20.3–70.9	7.6–66.0	81.1–152.9	58.6–124.8	1.6–3.4	Unpub.
Arabica – gourmet coffees/ local market	11	LM	85.3–549.8	120.2–669.3	169.0–1203.7	0.7–48.0	52.3–132.5	67.6–119.2	7.9–68.4	5.0–57.3	8.4–65.3	23.8–159.4	19.2–113.7	0.6–3.1	Unpub.
Arabica – gourmet coffees/ USA local market	3	DM	82.6–229.5	105.0–259.9	156.0–379.9	6.0–23.4	44.5–70.2	42.5–135.0	7.9–14.0	4.2–8.7	7.6–10.7	23.9–61.9	16.8–44.1	0.5–1.2	Unpub.
Blends/ USA local market	24	M	55.2–742.3	68.2–518.0	110.3–910.0	3.6–36.1	5.3–146.7	5.1–162.5	4.9–54.3	5.8–44.5	4.6–48.7	13.8–200.4	11.5–77.7	0.4–2.9	Unpub.
Blends/ USA local market	5	D	34.3–153.5	41.2–178.8	63.2–267.1	0.6–9.6	17.1–48.9	15.0–45.0	6.7–21.7	2.7–10.2	1.6–16.7	10.2–47.5	6.1–30.3	0.2–0.8	Unpub.

[a]Note: 3-CQA: 3-caffeoylquinic acid; 4-CQA: 4-caffeoylquinic acid; 5-CQA: 5-caffeoylquinic acid; 3-FQA: 3-feruloylquinic acid; 4-FQA: 4-feruloylquinic acid; 5-FQA: 5-feruloylquinic acid; 3,4di-CQA: 3,4 di-caffeoylquinic acid; 3,5di-CQA: 3,5 di-caffeoylquinic acid; 4,5di-CQA: 4,5 di-caffeoylquinic acid; 3-CQL: 3-caffeoylquinic-1,5-lactone; 4-CQL: 4-caffeoylquinic-1,5-lactone. *N*: number of samples. RD: roasting degree: L: light; LM: light-medium; M: medium; D: dark; DM: dark-medium. Nr: not reported.
**sum of 4-FQA and 5-FQA. Unpub.: unpublished

The type of roaster and speed at which coffee is roasted, related with time and temperature conditions (roasting profile), also affect the content of CGA in the coffee seeds. The higher the roasting speed to obtain one specific color degree, the lower the CGA loss tends to be.[32]

25.4 Contribution of Chlorogenic Acids to Cup-quality

The relationship of CGA with coffee cup quality is somewhat controversial and not completely understood. Phenolic compounds, including CGA, are reported to be essential contributors to coffee characteristic flavor.[78] CGA and derivatives contribute to the final acidity[46] and to the characteristic coffee bitterness.[77] As a result of Maillard and Strecker's reactions, coffee bitterness increases during roasting due to release of caffeic acid and formation of lactones, in addition to the derivatives responsible for flavor and aroma[79–81] (Figure 25.2). There are more than 80 CGA volatile derivatives catalogued.[78] At low concentrations, most of the major compounds are described as having warm, floral, balsamic, vanilla, carnation, clove, or anisic notes. However, when in high concentrations, many CGA derivatives are referred to as having a somewhat negative contribution to flavor with attributes such as tarry, smoky, woody, spicy, leathery, or medicinal[10,78] in addition to imparting undesirable astringency to the beverage.[80,82,83] Total CGA and especially CQA and FQA levels were reported to present inverse association with coffee quality, with higher CGA content observed in lower quality samples.[19,84] In one study,[19] the decrease in quality was accompanied by an increase in color intensity of raw beans which was attributed indirectly to the presence of reactive orthoquinones formed by the action of polyphenol oxidase over these compounds. Therefore, it was suggested that products of CGA oxidation in green coffee might be associated with cup quality decrease. This hypothesis has been strengthened by the following facts: (1) 5-CQA levels in seeds from mature coffee fruits were reported to be inversely associated with polyphenol oxidase activity in coffee beans;[84–87] (2) orthoquinones are reported to be responsible for the inhibition of the activity of polyphenol oxidase,[88] and (3) coffee cup quality has been directly associated to polyphenol oxidase activity.[89] Moreover, oxidation products of phenolic compounds have been suggested[89] as a possible cause for Rio-off-flavor (a negative phenolic note responsible for loss of cup quality). Amorin[90] also observed that 5-CQA levels above usual levels for a determined species contributed to a decrease in quality. Therefore, taking into account the well-known importance of CGA for flavor and aroma formation, there appears to be a limiting CGA level over which cup quality decreases. Lower CGA levels in green *C. arabica* also appear to explain its cup quality superiority when compared to *C. canephora*.[17,26,91] In dark roasted coffee, as previously mentioned, compounds which impact negative bitterness to the beverage are formed from CGA degradation products.[56,57]

25.5 Chlorogenic Acids Content in Coffee By-products

It is estimated that for every tonne of fresh roasted and ground coffee produced 0.5 tonne of pulp and 0.18 tonne of husk are produced.[92,93] Most of these by-products have been used for animal feeding, especially in coffee-producing countries where the disposal of wastes represent an important pollution problem.[10,94–97] In contrast to the considerable amount of studies performed on the chemical composition of green and roasted coffee seeds, there are relatively few studies concerning the chemical constituents of the other parts of the coffee plant, including pulp,[13,98,99] leaf,[43,100,101] silverskin,[95,102,103] and husk.[104,105]

Ramirez-Martinez[98] identified and quantified four CGA isomers in pulps from eight coffee cultivars. 5-CQA content ranged from 0.24 to 0.88 g 100 g^{-1} (dm), 3,4-diCQA content from 0.03 to 0.1 g 100 g^{-1} (dm), 3,5-diCQA from 0.09 to 0.45 g 100 g^{-1} (dm), and 4,5-diCQA from 0.03 to 0.1 g 100 g^{-1} (dm). As in the seeds, 5-CQA was the major CGA (42%). 3,4-diCQA, 3,5-diCQA, and 4,5-diCQA corresponded to 5.7, 19.3, and 4.4%, respectively. Later, Clifford and Ramirez-Martinez[13] identified and quantified seven CGA compounds (3-CQA, 4-CQA, 5-CQA, 5-FQA, 3,4-diCQA, 3,5-diCQA, 4,5-diCQA) in pulps of two *C. arabica* cultivars (*Red Caturra* and *Red Bourbon*) and one *C. canephora*. Total CGA contents (dm) were very similar in pulps of *C. arabica* cultivars: 1.01 g 100 g^{-1} in *Red Caturra* and 1.16 g 100 g^{-1} in *Red Bourbon*; and highest in *C. canephora* pulp, about 1.61 g 100 g^{-1} (dm).[13] Eight CGA compounds have been identified in methanolic extracts of *Coffea* leaves: (three CQA, two FQA, and three diCQA isomers).[100,101] Campa *et al.*[100] reported content (dm) of 0.1 g 100 g^{-1} for 3-CQA; 0.2 g 100 g^{-1} for 4-CQA; 3.2 g 100 g^{-1} for 5-CQA; 0.05 g 100 g^{-1} for 3,4-diCQA; 0.17 g 100 g^{-1} for 3,5-diCQA; 0.04 g 100 g^{-1} for 4,5-diCQA, and 0.12 g 100 g^{-1} for total FQA in coffee leaves of *C. arabica* cv. Laurina from Ethiopia. Mondolot *et al.*[101] identified five CGA compounds (5-CQA, 3,4-diCQA, 3,5-diCQA, 4,5-diCQA, and 5-FQA) in *C. canephora Pierre* from France. The authors reported that the oldest leaves presented lower CGA contents (dm) (4.02 mg 100 g^{-1}) when compared to the youngest ones (7.82 mg 100 g^{-1}).

In coffee silverskin (which is very light weight, eight CGA compounds have been identified: (three CQA, two FQA, and three diCQA isomers). Bresciani *et al.*[103] reported 432 mg 100 g^{-1} of total CQA (wet matter – wm); Regazzoni *et al.*[106] 130 mg 100 g^{-1} (dm), and Jiménez-Zamora *et al.*[107] 470 mg 100 g^{-1} (dm). Regarding total FQA: Bresciani *et al.*[103] reported 142 mg 100 g^{-1} (wm) and Regazzoni *et al.*[106] 274 mg 100 g^{-1} (dm). Regarding total diCQA, Regazzoni *et al.*[106] reported 55.8 mg 100 g^{-1} (dm) and Jiménez-Zamora *et al.*[107] content of 44 mg 100 g^{-1} (dm).

In arabica coffee husks from Brazil, eight CGA (3-CQA, 4-CQA, 5-CQA, 4-FQA, 5-FQA, 3,4-diCQA, 3,5-diCQA, and 4,5-diCQA) were quantified, totaling about 304 mg 100 g^{-1}, from which 51.4% corresponded to CQA, 12.1% to FQA, and 36.5% to diCQA.[unpublished data]

Spent coffee grounds are a by-product from the extraction of brewed coffee and production of soluble coffee[34,108,109] and therefore should have unextracted amounts of all CGA found in roasted and ground coffee,[96,107] which will depend on the beverage extraction efficiency. Annually, it is estimated that six million tonnes of spent coffee grounds are produced worldwide.[110] For every tonne of green coffee 650 kg of spent coffee are produced.[104] Contents from 0.3 to 1.4 g 100 g^{-1} (dm) of CQA and diCQA have been reported in spent coffee grounds.[107–109]

25.6 Conclusions

Genetic factors such as species and variety, the degree of maturation, and to some extent environmental conditions and agricultural practices are important determinants to green coffee beans composition and may affect the composition of the final beverage. Although the effect of different types of processing on coffee CGA composition needs to be further explored, considering the data available so far, of all studied coffee processing methods, roasting produces the most dramatic changes. However, up to medium or dark medium roast, the remaining CGA contents are still high. Contents in coffee beverage and estimates for daily CGA intake through coffee will be explored in the book *Coffee: Consumption and Health Implications*. Regarding cup quality, more studies are needed on the contribution of individual major and minor CGA and derivatives to coffee quality as well as on their interactive effects on taste buds.

References

1. G. Williamson and M. N. Clifford, *Biochem. Pharmacol.*, 2017, **139**, 24.
2. M. N. Clifford, S. Knight and N. Kuhnert, *Food Chem.*, 2005, **53**(10), 3821.
3. M. N. Clifford, S. Knight, B. Surucu and N. Kuhnert, *J. Agric. Food Chem.*, 2006, **54**, 1957.
4. D. Perrone, A. Farah, C. M. Donangelo, T. de Paulis and P. R. Martin, *Food Chem.*, 2008, **106**, 859.
5. R. Jaiswal, M. A. Patras, P. J. Eravuchira and N. Kuhnert, *J. Agric. Food Chem.*, 2010, **58**, 8722.
6. M. N. Clifford, *Food Chem.*, 1979, **4**, 63.
7. M. N. Clifford, *J. Sci. Food Agric.*, 1999, **79**, 362.
8. M. N. Clifford, S. Marks, S. Knight and N. Kuhnert, *J. Agric. Food Chem.*, 2006, **54**, 4095.
9. V. X. Marques and A. Farah, *Food Chem.*, 2009, **113**, 1370.
10. A. Farah and C. M. Donangelo, *Braz. J. Plant Physiol.*, 2006, **18**(1), 23.
11. A. Farah, in *Coffee: Emerging Health Effects and Disease Prevention*, ed. Y. F. Chu, IFT press, New York, 2012, p. 21.
12. M. N. Clifford and T. Kazi, *Food Chem.*, 1987, **26**, 59.

13. M. N. Clifford and J. R. Ramirez-Martinez, *Food Chem.*, 1991, **40**, 35.
14. A. Farah, T. De Paulis, L. C. Trugo and P. R. Martin, *J. Agric. Food Chem.*, 2005, **53**, 1105.
15. A. Farah, T. de Paulis, D. P. Moreira, L. C. Trugo and P. R. Martin, *J. Agric. Food Chem.*, 2006, **54**, 374.
16. G. S. Duarte, A. A. Pereira and A. Farah, *Food Chem.*, 2010, **118**, 851.
17. C. L. Ky, J. Louarn, B. Guyot, S. Dussert, S. Hamon and M. Noirot, *Food Chem.*, 2001, **75**, 223.
18. A. M. G. Correia, M. C. A. Leitão and M. N. Clifford, *Food Chem.*, 1995, **53**, 309.
19. A. Farah, M. C. Monteiro and L. C. Trugo, in *Proceedings of the International Conference on Coffee Science*, ASIC, 2006, p. 240.
20. A. Toci, A. Farah and L. C. Trugo, *Quim. Nova*, 2006, **29**(5), 965.
21. R. M. Alonso-Salces, F. Serra, F. Reniero and A. Rolyheberger, *J. Agric. Food Chem.*, 2009, **57**(10), 4224.
22. O. Babova, A. Occhipinti and M. E. Maffei, *Phytochemistry*, 2016, **123**, 33.
23. G. Baeza, B. Sarriá, L. Bravo and R. Mateos, *J. Agric. Food Chem.*, 2016, **64**(51), 9663.
24. A. Farah, D. Perrone, J. P. Fernandes and J. S. Kenny, in *Proceedings of the International Conference on Coffee Science*, ASIC, 2011, p. 367.
25. G. Guerrero, M. Suárez and G. Moreno, *J. Agric. Food Chem.*, 2001, **49**, 2454.
26. L. C. Trugo and R. Macrae, *Food Chem.*, 1984, **15**, 219.
27. M. N. Clifford and T. Jarvis, *Food Chem.*, 1988, **29**, 291.
28. F. Anthony, M. N. Clifford and M. Noirot, *Genet. Resour. Crop Evol.*, 1993, **40**, 61.
29. A. Farah, D. F. Neves, L. C. Trugo, A. Rosenthal and R. C. Modesta, in *Annals of the II Simpósio de PNP&D Embrapa Café*, 2001, p. 1144.
30. D. Perrone, R. Donangelo, C. M. Donangelo and A. Farah, *J. Agric. Food Chem.*, 2010, **58**, 12238.
31. A. Farah, M. Monteiro, C. M. Donangelo and S. Lafay, *J. Nutr.*, 2008, **138**, 2309.
32. A. Farah and G. Duarte, in*Coffee and Health Disease Prevention*, ed. V. R. Preedy, Elsevier, Cambridge, 2015, p. 789.
33. M. N. Clifford, *J. Sci. Food Agric.*, 2000, **80**, 1033.
34. Y. Narita and K. Inouye, in *Coffee and Health Disease Prevention*, ed. V. R. Preedy, Elsevier, Cambridge, 2015, p. 189.
35. A. A. Teixeira, C. H. J. Brando, R. A. Thomaziello and R. Teixiera, in *Espresso Coffee: The Science of Quality*, ed. A. Illy and R. Viani, Elsevier Academic Press, Italy, 1995, p. 53.
36. R. F. Smith, in *A History of Coffee*, ed. M. N. Clifford and K. C. Willson, Croom Helm, London Sydney, 1985, p. 1.
37. V. Leloup, C. Gancel, R. Liardon, A. Rytz and A. Pithon, in *Proceedings of the International Conference on Coffee Science*, ASIC, 2004.
38. A. E. Wootton, in *Proceedings of the International Conference on Coffee Science*, ASIC, 1974, p. 316.

39. M. N. Clifford, in *Coffee*, ed. R. J. Clarke and R. Macrae, Elsevier Applied Science Publications, London, UK, 1985.
40. B. Bertrand, D. Villarreal, A. Laffarque, H. Posada, P. Lashermes and S. Dussert, *J. Agric. Food Chem.*, 2008, **56**(6), 2273.
41. C. L. Ky, M. Noirot and S. Hamon, *J. Agric. Food Chem.*, 1997, **45**, 786.
42. C. L. Ky, J. Louarn, B. Guyoy, A. Charrier, S. Hamon and M. Noirot, *Theor. Appl. Genet.*, 1999, **98**, 628.
43. C. Campa, S. Doulbeau, S. Dussert, S. Hamon and M. Noirot, *Food Chem.*, 2005, **93**, 135.
44. H. Steinhart and A. Luger, in *Proceedings of the International Conference on Coffee Science*, ASIC, 1997, p. 155.
45. H. G. Maier in *Proceedings of the International Conference on Coffee Science*, ASIC, 1994, p. 567.
46. L. C. Trugo and R. Macrae, *Analyst*, 1984, **109**, 263.
47. R. F. A. Moreira, L. C. Trugo and C. A. B. Maria, *Quim. Nova*, 2000, **23**, 195.
48. P. Montavón, E. Duruz, G. Rumo and G. Pratz, *J. Agric. Food Chem.*, 2003, **51**, 1328.
49. R. Tressl, D. Bahri, H. Koeppler and A. Jensen, *Z. Lebensm.-Unters. Forsch.*, 1978, **167**, 111.
50. R. Dorfner, T. Ferge, A. Kettrup, R. Zimmermann and C. Yeretzian, *J. Agric. Food Chem.*, 2003, **51**, 5768.
51. J. W. Haffenden and V. A. Yaylayan, *J. Agric. Food Chem.*, 2005, **53**(12), 4938.
52. R. Lang, C. Müller and T. Hofmann, *J. Agric. Food Chem.*, 2006, **54**, 5755.
53. A. T. Toci and A. Farah, *Food Chem.*, 2008, **108**(3), 1133.
54. C. Petisca, T. Péreza-Palacios, A. Farah, O. Pinho and I. M. P. L. V. O. Ferreira, *Food Bioprod. Process.*, 2012, **91**, 233.
55. A. T. Toci and A. Farah, *Food Chem.*, 2014, **153**, 298.
56. O. Frank, S. Blumberg, C. Kunert, G. Zehentbauer and T. J. Hofmann, *J. Agric. Food Chem.*, 2007, **55**, 1945.
57. S. Blumberg, O. Frank and T. Hofmann, *J. Agric. Food Chem.*, 2010, **58**, 3720.
58. S. Kreppenhofer, O. Frank and T. Hofmann, *Food Chem.*, 2011, **126**, 441.
59. E. K. Bekedam, H. A. Schols, M. A. J. S. Van Boekel and G. J. Smit, *J. Agric. Food Chem.*, 2008, **56**, 2055.
60. M. D. Fuster, A. E. Mitchell, H. Ochi and T. Shibamoto, *J. Agric. Food Chem.*, 2000, **48**, 5600.
61. C. Delgado-Andrade and F. J. Morales, *J. Agric. Food Chem.*, 2005, **53**, 1403.
62. N. C. Bicho, A. E. Leitão, J. C. Ramalho and F. C. Lidon, *Eur. Food Res. Technol.*, 2001, **233**, 303.
63. J. K. Moon, H. S. Yoo and T. Shibamoto, *J. Agric. Food Chem.*, 2009, **57**, 5365.
64. K. Schrader, A. Kiehne, U. H. Engelhardt and H. G. Maier, *J. Sci. Food Agric.*, 1996, **71**, 392.

65. C. Bennat, U. H. Engelhardt, A. Kiehne, F. M. Wirries and H. G. Maier, *Z. Lebensm.-Unters. Forsch.*, 1994, **199**, 17.
66. M. C. Monteiro and L. C. Trugo, *Quim. Nova*, 2005, **28**(4), 637.
67. G. S. Duarte and A. Farah, in *Proceedings of the International Conference on Coffee Science*, ASIC, 2009.
68. L. M. Pérez-Hernández, K. Chávez-Quiroz, L. A. Medina-Juárez and N. G. Meza, *J. Mex. Chem. Soc.*, 2012, **56**(4), 430.
69. J. P. Lima, A. B. N. Martins, N. Marques, G. S. Duarte and A. Farah, in *Proceedings of the Simpósio Latino-Americano de Ciência de Alimentos (SLACA)*, 2013.
70. K. Fujioka and T. Shibamoto, *Food Chem.*, 2008, **106**, 217.
71. O. Frank, G. Zehentbauer and T. Hofmann, *Eur. Food Res. Technol.*, 2006, **222**, 492.
72. O. Frank, G. Zehentbauer and T. Hofmann, in *Flavour Science: Recent Advances and Trends*, ed. W. L. P. Bredie and M. A. Petersen, Elsevier, Academic Press, 2006, p.165.
73. O. Frank, S. Blumberg, G. Krümpel and T. Hofmann, *J. Agric. Food Chem.*, 2008, **56**(20), 9581.
74. B. M. Scholz and H. G. Maier, *Z. Lebensm.-Unters. Forsch.*, 1990, **190**, 132.
75. B. M. Scholz-Bottcher, E. Ludges and H. G. Maier, *Liebigs Ann. Chem.*, 1991, 1029.
76. R. Lang, E. F. Yagar, A. Wahl, A. Beusch, A. Dunkel, N. Dieminger, R. Eggers, G. Bytof, H. Stiebitz, I. Lantz and T. Hofmann, *J. Agric. Food Chem.*, 2013, **61**(49), 12123.
77. - L. C. Trugo, *Doctorate Thesis*, University of Reading, 1984.
78. I. Flament and Y. Bessière-Thomas, in *Coffee Flavor Chemistry*, ed. I. Flament and Y. Bessière-Thomas, John Wiley and Sons, LTD., New York, 2002, p. 410.
79. M. Ginz and U. H. Enhelhardt, in *Proceedings of the International Conference on Coffee Science*, ASIC, 1995.
80. P. S. Variyar, R. Ahmad, R. Bhat, Z. Niyas and A. Sharma, *J. Agric. Food Chem.*, 2003, **51**, 7945.
81. C. B. Maria, L. C. Trugo, R. F. A. Moreira and C. C. Werneck, *Food Chem.*, 1994, **50**, 141.
82. M. L. C. Carelli, C. R. O. Lopes and L. C. Monaco, *Turrialba*, 1974, **24**, 398.
83. M. N. Clifford and J. Wight, *J. Sci. Food Agric.*, 1976, **27**, 73.
84. - E. B. Silva, *Doctorate Thesis*, Federal University of Lavras, 1999.
85. H. V. Amorim and D. M. Silva, *Nature*, 1968, **219**, 381.
86. V. D. Carvalho, S. J. Chagas, S. M. Chalfon, N. Botrel and E. S. G. Juste-Junior, *Pesqui. Agropecu. Bras.*, 1994, **29**, 449.
87. P. Mazzafera, *Food Chem.*, 1999, **64**, 547.
88. P. Mazzafera and S. P. Robinson, *Phytochemistry*, 2000, **55**, 285.
89. H. V. Amorim, A. A. Teixeira, M. Melo, V. F. Cruz and E. Malavolta, *Turrialba*, 1975, **25**, 18.
90. - H. V. Amorim, *Doctorate Thesis*, São Paulo University, 1972.

91. C. Bertrand, M. Noirot, S. Doulbeau, A. Kochko, S. Hamon and C. Campa, *Plant. Sci.*, 2003, **165**, 1355.
92. S. Roussos, M. de los Angeles Aquiáhuatl and M. del Refugio Trejo-Hernández, *Appl. Microbiol. Biotechnol.*, 1995, **42**, 756.
93. B. Janissen and T. Huynh, *Resour., Conserv. Recycl.*, 2018, **128**, 110.
94. A. F. Barcelos, P. C. A. Paiva, J. R. O. Pérez, V. B. Santos and R. M. Cardoso, *Rev. Bras. Zootec.*, 2001, **30**(4), 1.
95. Y. Narita and K. Inouye, *Food Chem.*, 2012, **135**, 943.
96. P. Esquivel and V. M. Jimenez, *Food Res. Int.*, 2012, **46**, 488.
97. G. A. Figueroa, T. Homann and H. M. Rawel, *Austin Food. Sci.*, 2016, **1**(3), 1014.
98. J. R. Ramirez-Martinez, *J. Sci. Food Agric.*, 1988, **43**, 135.
99. A. Duangjai, N. Suphromb, J. Wungrathc, A. Ontawong, N. Nuengchamnong and A. Yosboonruange, *Integr. Med. Res.*, 2016, **5**, 324.
100. C. Campa, L. Mondolot, A. Rakotondravao, L. P. R. Bidel, A. Gargadennec, E. Couturon, P. La Fisca, J.-J. Rakotomalala, C. Jay-Allemand and A. P. Davis, *Ann. Bot.*, 2012, **110**, 595.
101. L. Mondolot, P. La Fisca, B. Buato, E. Talansier, A. de Kochk and C. Campa, *Ann. Bot.*, 2006, **98**, 33.
102. Y. Narita and K. Inouye, *Food Res. Int.*, 2014, **61**, 16.
103. L. Bresciani, L. Calani, R. Bruni, F. Brighenti and D. Del Rio, *Food Res. Int.*, 2014, **61**, 196.
104. S. Murthy and M. M. Naidu, *Resour., Conserv. Recycl.*, 2012, **66**, 45.
105. E. Bondesson, *Swedish University of Agricultural Sciences*, Swedish University of Agricultural Sciences, Uppsala, 2015.
106. L. Regazzoni, F. Saligari, C. Marinello, G. Rossoni, G. Aldini, M. Carini and M. Orioli, *J. Funct. Foods*, 2016, **20**, 472.
107. A. Jiménez-Zamora, S. Pastoriza and J. A. Rufián-Henares, *LWT--Food Sci. Technol.*, 2015, **61**, 12.
108. K. Ramalakshmi, L. J. M. Rao, Y. Takano-Ishikawa and M. Goto, *Food Chem.*, 2009, **115**, 79.
109. J. Bravo, I. Juániz, C. Monente, B. Caemmererb, L. W. Krohb, M. Paz De Peña and C. Cid, *J. Agric. Food Chem.*, 2012, **60**, 12565.
110. S. I. Mussatto, M. S. Ercília, S. M. Machado and J. A. Teixeira, *Food Bioprocess Technol.*, 2011, **4**, 661.

CHAPTER 26

Isoflavones, Lignans and Other Minor Polyphenols

LUCIANO NAVARINI*[a], SILVIA COLOMBAN[b], GIOVANNI CAPRIOLI[c] AND GIANNI SAGRATINI[c]

[a]illycaffè spa, via Flavia 110, 34147 Trieste, Italy; [b]Aromalab, illycaffè spa, AREA Science Park, Padriciano 99, 34149 Trieste, Italy; [c]School of Pharmacy, University of Camerino, via Sant'Agostino 1, 62032 Camerino, Italy
*E-mail: luciano.navarini@illy.com

26.1 Introduction

Among the chemical compounds present in a cup of coffee, caffeine is the most commonly recognized by consumers. In the last few years, chlorogenic acids have also become known, since these compounds are nominated by experts as major coffee polyphenols that play a relevant role in the antioxidant activity exerted by the drink. However, it is still not very well known, even by scientists, that a cup of espresso contains phytoestrogens. These are naturally occurring plant-derived substances promoting weak estrogenic actions in mammals and structurally similar to mammalian estrogen 17β-estradiol (Figure 26.1).[1]

There are several classes of phytoestrogens, for example, steroidal estrogens, found only in a few plants, and the more ubiquitous phenolic estrogens such as isoflavones, coumestans and lignans.[1] Other classes of phytoestrogens that have been reported in foods different from coffee include anthraquinones, chalcones, flavones, prenylflavonoids and

Coffee: Production, Quality and Chemistry
Edited by Adriana Farah

Published by the Royal Society of Chemistry, www.rsc.org

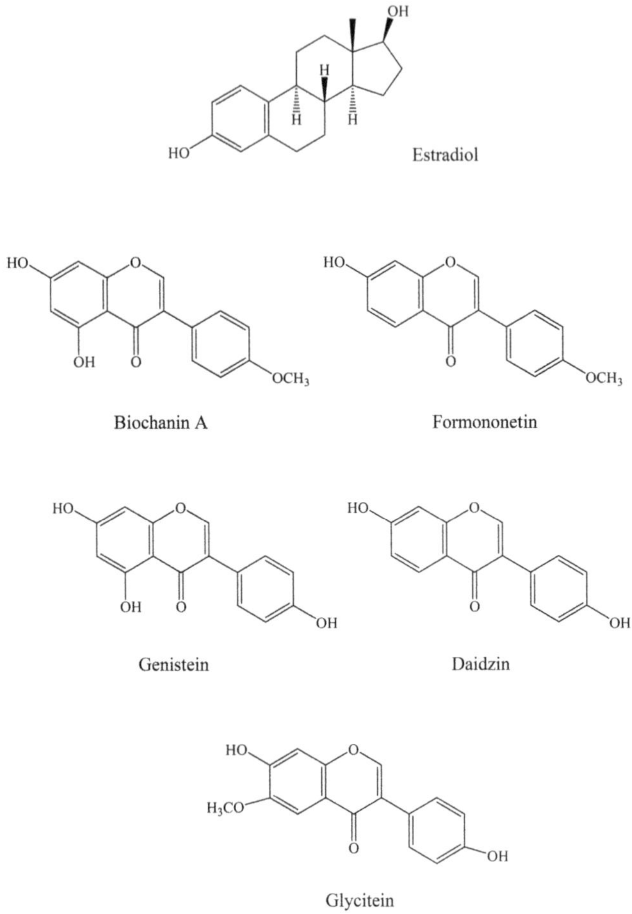

Figure 26.1 Chemical structures of 17β-estradiol and isoflavone aglycones.

saponins.[1] Isoflavones and lignans will be the focus of the present chapter. Some of these polyphenolic compounds have been detected in roasted coffee for the first time in 1998, by Mazur and collaborators.[2] In spite of the possible contribution offered by these compounds to health, the number of investigations dedicated to this class of coffee constituents so far is scarce. Isoflavones' and lignans' contents in coffee beverages described in the literature are low but highly variable, and this may reflect the lack of systematic studies aimed at investigating different raw materials and the effect of processing, brewing, sample preparation and analytical methods on the compositional data.

Regarding coffee lignans the situation is more critical, with fewer published scientific reports compared to those on coffee isoflavones. It has to be stressed that no studies on biological activity have been performed using isolated isoflavones and lignans from coffee, so far. However, most of them are structurally similar to compounds found in other foods. The present review is intended to raise the awareness of coffee scientists and consumers to the existence of these minor compounds in coffee which are potentially biologically active in humans, especially considering the high coffee intake in most countries. The presence of other minor polyphenols in coffee fruits and to a lower extent in seeds will also be discussed.

26.2 Chemistry

26.2.1 Isoflavones

Isoflavones are naturally occurring plant secondary metabolites having 1,2-diarylpropane structure (Figure 26.1). Over 60 members of this class have been identified[3] and although they are present in more than 300 kinds of plants, soy beans are the principal human dietary source.[3,4] Glycosidic-conjugate isoflavones represent the natural and biologically inactive form of these compounds. However, hydrolysis and the subsequent release of the aglycone (phenolic moiety) component are essential for the absorption of isoflavones in the digestive tract.[5] For this reason, aglycones are considered to be biologically active forms of isoflavones. In this subclass, the most thoroughly investigated and interesting compounds are genistein (5,7-dihydroxy-3-(4-hydroxyphenyl) chromen-4-one), daidzein (7-hydroxy-3-(4-hydroxyphenyl)chromen-4-one), glycitein (7-hydroxy-3-(4-hydroxyphenyl)-6-methoxychromen-4-one), biochanin A (5,7-dihydroxy-3-(4-methoxyphenyl)chromen-4-one) and formononetin (7-hydroxy-3-(4-methoxyphenyl)chromen-4-one)[6] (Figure 26.1). These compounds have been detected in roasted and instant coffees,[2,7–13] as well as in coffee beverage. Daidzin, glycitin and genistin are the glucosides of daidzein, glycitein, and genistein containing a *β*-glucoside group. Daidzin and genistin have been recently detected in spent coffee grounds (Figure 26.2).[14]

Coffee as a beverage, brewed from grounds and prepared with tap water, has been included in the USDA Database for the *Isoflavone Content of Selected Foods* section.[15] The mean values of total isoflavone contents in some selected foods are reported in Table 26.1. Interestingly, as far as we

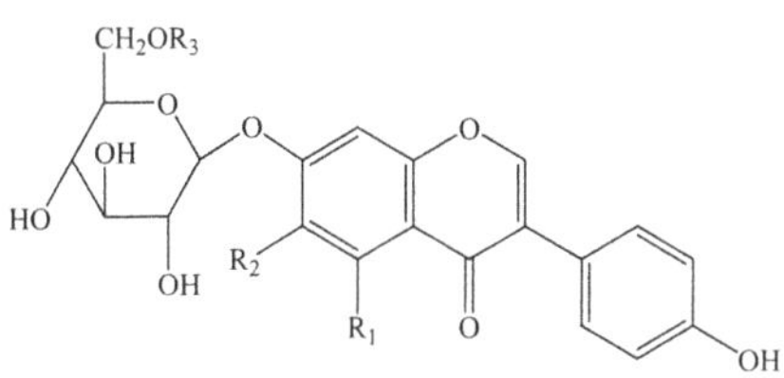

R_1	R_2	R_3	Compound
H	H	H	Daidzin
OH	H	H	Genistin
H	OCH_3	H	Glycitin
H	H	$COCH_3$	Acetyldaidzin
OH	H	$COCH_3$	Acetylgenistin
H	OCH_3	$COCH_3$	Acetylglycitin
H	H	$COCH_2COOH$	Malonyldaidzin
OH	H	$COCH_2COOH$	Malonylgenistin
H	OCH_3	$COCH_2COOH$	Malonylglycitin

Figure 26.2 Chemical structures of the most common isoflavone glucosides.

Table 26.1 Mean values of total isoflavone contents in some selected foods (wet weight basis). Data from ref. 15.

Food item	Total isoflavones (mg 100 g^{-1} of edible)
Dry roasted mature soybeans seeds	148.50
Soy protein drink	81.65
Soy yogurt	33.17
Different soy cheese varieties	6.02–25.72
Soybeans, green raw and boiled	48.95 and 17.92
Soybeans, mature seeds, sprouted, raw and steamed	34.39 and 12.50
Different breakfast cereals	0.01–93.90
Beans, raw and differently processed	0.02–0.59
Chickpeas, mature seeds, raw and boiled	0.38–0.02
Different fruits and fruit juices	0.01–0.08
Cereal grains and pasta	0.01–0.07
Coffee (beverage)	0.04

know, reports on the content of isoflavones in green coffee beans as well as in other *Coffea* plant organs are not available in the literature. Contents of isoflavones in roasted coffee reported in different research studies are presented in Section 26.4.

Acetyl daidzin, acetyl glycitin, acetyl genistin, malonyl daidzin, malonyl glycitin and malonyl genistin are conjugates of the glucosides having either an acetyl or a malonyl β-glucoside, which are normally present in soyfoods.[16] They have not been detected so far in coffee, however their presence cannot be excluded *a priori*, particularly in green coffee beans.

As thermal treatment is one of the most common food processing methods, changes in the content and distribution profile of isoflavones in soyfoods because of heat processing have been extensively studied. However very few of these studies have looked into the mechanism of isoflavones modifications during dry heating under conditions possibly similar to coffee roasting.[17–19]

It has been reported that when heated at 150 °C or higher, the conjugated glucosides are cleaved from the isoflavones to produce their corresponding aglycones.[19] The half-lives of major soy isoflavones daidzin, glycitin and genistin when dry heated at 200 °C were reported to be in the range of 5.8–6.0 min, whereas those of the corresponding aglycones resulted to be in the range 15.7–21.2 min.[19]

The higher thermostability of the aglycones has been observed also in other studies. In particular, glycitein and genistein were observed to start their degradation after 60 min at 200 °C while daidzein was stable until 90 min. At 215 °C, degradation started at 3, 7 and 15 min for glycitein, genistein and daidzein, respectively.[18] According to thermogravimetric measurements, genistein undergoes thermal degradation at an onset temperature of 275 °C.[20]

Unfortunately no similar studies have been carried out on isoflavones isolated from coffee, so far.

The involvement of isoflavones in Maillard reactions cannot be excluded *a priori*, since chlorogenic acids are supposed to be integrated in the melanoidins backbone. So far, development of Maillard reaction products during storage of soy protein isolates having isoflavones (genistein) as reactants has been just hypothesized.[21]

In view of the reported thermostability it is difficult to consider the pure aglycones as possible aroma precursors. However, this aspect has not yet been investigated from a coffee roasting point of view.

As far as coffee is concerned, only one study, by Alves *et al.*,[22] has focused on the effect of roasting on total isoflavones content. During the roasting of coffee beans (from light- to dark-roast degree), a decrease by approximately 30–40% of isoflavone levels has been observed. However, surprisingly, this degradation was not noted when analysing the corresponding brews. According to the authors, this was due to a significant increase in the extractability of the compounds from less to more intense roasting processes. This increase was more marked for formononetin, resulting in it being more extensively extracted, especially at dark roast, than other detected isoflavones.[22] Formononetin (a methylated derivate) is less polar than the other isoflavones, however, according to the authors, the observed higher extractability could be related to the existence of conjugated forms occurring in green coffee beans prior to the roasting process.[22]

26.2.2 Lignans

The term lignans, proposed by Haworth,[23] is used for a different class of plant phenols possessing a 2,3-dibenzylbutane structure. Lignans were identified in trees in the 19th century. They occur widely in vascular plants and are found in wooden parts, roots, leaves, flowers, fruits and seeds. Several hundred lignans have been identified. They occur either in free form or glycosidically linked to a wide variety of carbohydrates.[24]

Secoisolariciresinol and matairesinol are two lignan dimers identified in coffee.[2] Although they are not estrogenic by themselves, they are converted by the gut microflora to readily absorbed mammalian lignans (enterolignans), which are estrogenic: *trans*-2,3-bis(3-hydroxybenzyl)-γ-butyrolactone, commonly known as enterolactone, and 2,3-bis(3-hydroxybenzyl)-butane-1,4-diol, commonly known as enterodiol.[24] The mammalian-derived lignans differ from dietary phyto-lignans in possessing phenolic hydroxyl groups only in the meta position of the aromatic rings (see Figure 26.3). These compounds have been described as the major lignans present in serum, urine, bile and seminal fluids of humans and animals.[24] Recently, it has been found that other lignans, such as lariciresinol and pinoresinol also occur commonly in food products including coffee and they are precursors to mammalian lignans.[25] (See *Coffee: Consumption and Health Implications*, for more details.)

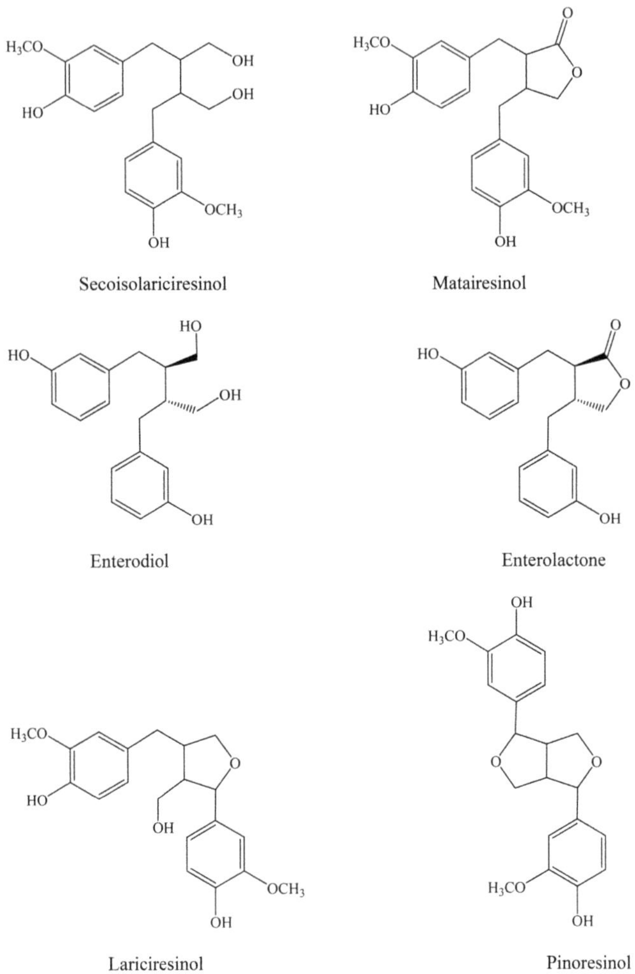

Figure 26.3 Chemical structures of plant and mammalian lignans.

The phyto-lignans, in addition to being present in tea and coffee, appear in higher amounts in flaxseed, sesame seeds and cereals but also in vegetable oils, legumes, wholegrain breads, fruits and vegetables.[2] In Table 26.2 the mean values of total lignan contents on the wet weight basis in some selected foods are reported.[25]

The effect of thermal heating on some lignans from major dietary sources have been recently investigated.[26] In flaxseeds, secoisolariciresinol, lariciresinol and pinoresinol, which are mainly present as esterified compounds, resulted to be stable if heated to 250 °C for 3.5 min. However, in other matrices like rye and sesame seeds, high roasting temperatures (250 °C) cause degradation of both aglycones and glycosides. This degradation is much less pronounced at 200 °C and begins after 10 min heating for rye and between 10

Table 26.2 Mean values of total lignan contents on the wet weight basis in some selected foods. Data from ref. 25.

Food item	Total lignans (mg 100 g^{-1} of edible)
Flaxseed	301.129
Sesame seed	39.348
Multi-grain bread	6.744
Wheat bread, wholegrain	0.121
Wheat wholemeal	0.210
Brassica vegetables	0.185–2.321
Spinach (frozen)	0.082
French bean	0.273
Pea (in jars)	0.034
Apricot	0.450
Strawberry	0.334
Orange	0.078
Olive oil (extra-virgin)	0.248
Tea (beverage)	0.039–0.077
Coffee (beverage)	0.019–0.031

and 20 min for flaxseeds. This finding demonstrated that the type of conjugation and the matrix influence the thermal stability of lignans.[26]

So far, no studies reported the effect of roasting on lignans as part of coffee components. This is probably related to the fact that as for isoflavones, no studies have reported their presence in green coffee beans as well as in other *Coffea* plant organs. The existence of bound forms that are released during roasting is possible.

26.3 Methods of Analysis

A high number of extraction procedures as well as many analytical methods for detection and quantitation of phytoestrogens and their metabolites in plant, plant derived products, foodstuffs and biological matrices have been reported.[27,28] Gas chromatography/mass spectrometry (GC/MS) methods as well as liquid chromatography (LC) coupled with different detectors (MS, UV, DAD, *etc.*) have been described.[4] For total aglycones determination, a hydrolytic step (chemical or enzymatic) is frequently used.

Reported recoveries % of isoflavones and lignans from food samples can be as low as 30%.[28] In the case of coffee samples, recovery % ranges from 48 to 129 according to the different methods.[11–13]

The first analytical method for the parallel determination of coffee lignans and isoflavones was developed by Mazur and collaborators.[2] In order to prepare the samples, after rehydration of the solid sample (diverse types of coffee including instant coffee) with distilled water or infusion preparation and addition of isoflavonoid internal standards, enzymatic hydrolysis with *Helix pomatia* extract was performed. After ethyl ether extraction and phase separation, the ethereal phase (containing mainly isoflavonoids compounds) was evaporated and preserved in methanol at −20 °C and then combined with the fraction containing mainly lignans. The ether in the aqueous phase was evaporated and this was followed by HCl hydrolysis (mainly of lignan conjugates), pH adjustment (to 3–5) and standards addition. The aqueous phase was then extracted with ethyl ether and the organic phase was evaporated to dryness. This fraction, containing mainly lignans, was combined with the isoflavonoid fraction in methanol. The sample was then submitted to ion-exchange chromatography process and the separated lignan and isoflavonoid fractions were converted to trimethylsilyl ethers; finally, derivatives were quantitatively measured using isotope dilution capillary GC-MS in the selected ion monitoring mode.[2]

A different approach was followed by Horn-Ross and collaborators.[7] In this case, the method was developed to determine phytoestrogens in a wide range of foods, including coffee. The sample was extracted with 80% aqueous methanol. Aliquots of the methanol extracts were evaporated to dryness, β-glucosidase in ammonium acetate buffer (pH 5.0) was added and the sample was incubated at 37 °C overnight. After addition of glacial acetic acid, the aglycones were recovered by extraction with ether. After evaporation to dryness, the extract was redissolved in 80% aqueous methanol and injected into the high-performance-liquid-chromatography-mass spectrometer (HPLC-MS). HPLC analysis was performed on a C8 reversed-phase column under isocratic conditions with 30% acetonitrile in 10 mM ammonium acetate. After chromatographic separation, the eluate stream was passed into the heated nebulizer of the atmospheric pressure chemical ionization (HN-APCI) interface of a triple quadrupole mass spectrometer. Multiple-reaction ion monitoring was carried out by selecting parent $[M-H]^-$ molecular ions and specific daughter ions formed by collision with 90% Argon ± 10% nitrogen gas. Parent/daughter ion pairs based on their *m*/*z* were selected as follows: for the isoflavones, daidzein (253/132), formononetin (267/208), genistein (269/133) and biochanin A (283/239), for the lignans, matairesinol (357/122), secoisolariciresinol (361/165) and the internal standard, 4-methylumbelliferone (175/122).[7]

Thompson and collaborators[8] reported a double extraction by using 70% methanol at 60–70 °C for 2 h, to extract phytoestrogens from a wide range of foods, including coffee. For lignan analysis, the sample passed through alkaline hydrolysis, a step of C18 solid-phase extraction (SPE), β-glucuronidase hydrolysis and another SPE step. Then the eluent was evaporated to dryness and the residue was reconstituted with methanol and stored at −20 °C until analysis in GC-MS. For isoflavone analysis, the method was the same as for lignan analysis, except for the alkaline hydrolysis step that was eliminated. When ready for the analysis, the internal standard solution (5α-androstane-3β, 17β-diol, 50 μg mL^{-1} methanol) was added to the above sample in methanol solution. The sample was then dried under nitrogen flow and, after adding silylating agent, it was incubated at 60 °C for 30 min. The silylating agent was then removed under nitrogen gas flow, and the trimethylsilylated sample was dissolved in hexane and injected in GC-MS. Analysis was performed in the scan ion monitoring mode and then two or three characteristic ions were chosen for each analyte or standard. The selected ions for each analyte were as follows: *m/z* 421.3, 346.3 and 436.3 for 5α-androstane-3β, 17β-diol; *m/z* 340.1 and 325.1 for formononetin; *m/z* 398.1 and 383.1 for daidzein; *m/z* 414.1 and 399.1 for genistein; *m/z* 560.3 and 209.1 for secoisolariciresinol; *m/z* 428.1 and 398.1 for glycitein; *m/z* 502.2 and 209.1 for matairesinol; *m/z* 576.3 and 486.2 for lariciresinol; *m/z* 502.2 and 487.2 for pinoresinol.[8]

Kuhnle and collaborators[9,29] developed an innovative analytical method based on automated SPE and LC-MS/MS for the analysis of phytoestrogens in food, including coffee. After liquid–solid extraction (LSE) (with 10% methanol in sodium acetate, 0.1%, pH 5) and enzymatic hydrolysis with a mixture of purified *H. pomatia* juice, cellulase and β-glucosidase in sodium acetate (0.1%, pH 5) containing internal standard, an SPE was performed by using Strata C18E cartridges (silica-based sorbent for extraction of hydrophobic molecules from aqueous and biological samples). Samples were separated on a diphenyl column kept at 50 °C, using a binary gradient formed by mobile phase A (40% ammonium acetate 0.1% at pH 4.8 and methanol) and mobile phase B (100% methanol): samples were analysed in the negative ion mode with a spray voltage of −4.3 kV and a spray temperature of 600 °C. Transitions used were different from those previously used.[7] Samples were quantified using the area ratio between analyte and internal standard. For analytes for which no $^{13}C_3$-labeled standard was available, $^{13}C_3$-enterolactone or $^{13}C_3$-enterodiol were used.[29]

Development and validation of an analytical method to specifically quantify three isoflavones (daidzein, genistein and formononetin) in coffee beans and coffee brews by HPLC with diode-array detection (DAD) was performed by Alves and collaborators in 2010.[10] In contrast to previous approaches, acid hydrolysis (methanol/3.4 M HCl (1 : 1), at 75 °C for 2.5 h under reflux, in the case of ground roasted coffee) to release aglycones was used for the first time for isoflavones quantification in coffee. After the hydrolysis step and neutralization, the hydrolysate was analysed without further purification apart from an ultra-centrifugation step. The chromatographic separation of the compounds was achieved with a C18 column operating at 40 °C. A gradient solvent system of formic acid 0.1% (A) and acetonitrile (B) was used. Daidzein, genistein and formononetin were identified in the chromatograms by comparing retention times and co-elution with authentic standards, and by their UV spectra characterized by two absorption maxima at 245–275 and 300–330 nm with low intensities in the latter range.[10] This method, according to the authors, offers higher recovery and it is less labour-intensive and time-consuming than those previously adopted.[10] However, it has been reported that LC methods with UV detection have questionable specificity for distinguishing compounds of very similar structures as they often exhibit fairly similar UV absorption characteristics.[28]

Moreover, the use of HPLC with conventional detection requires a typical purification step (*e.g.* SPE) of crude extracts to avoid possible co-elution of target analytes with matrix co-extracts.[28]

In a different paper, isoflavones from arabica roasted coffee have been analysed by a C18 column and HPLC equipped with a UV detector at 254 nm, but unfortunately no details about extraction and clean-up procedures have been provided.[14]

Recently, LC-tandem mass spectrometry methods have been proposed for analysis of coffee phytoestrogens.[11–13]

An LC-MS/MS method was developed by Sapozhnikova in 2014 for analysis of a wide range of polyphenolic compounds, including isoflavones and lignans, in liquid samples including coffee.[13] Ground coffee samples (described as premium dark and French roast) were brewed according to package directions, and liquid samples were subjected to analysis. As an extraction procedure, a "dilute and shoot" approach has been used (samples have been diluted ten times with 0.05% formic acid in water and filtered). C8 HPLC column and a binary gradient of methanol and 0.2 mM ammonium formate buffer were used for separation of selected polyphenolic compounds. Mass spectra were acquired in both electrospray ionization positive and negative modes to determine the most intense precursor ions and quantification was performed using genistein-d_4 as an internal standard.[13]

A different research group recently proposed a new HPLC-MS/MS method for the simultaneous determination of five isoflavones in espresso coffee.[11,12] After extraction with methanol and purification by using SPE C18, the sample was injected for analysis (triple quadrupole as mass analyser). Separation of the analytes was performed on a C18 analytical column by using acetonitrile (mobile phase A) and aqueous buffer solution containing formic acid 0.05% and ammonium formate 5 mM (mobile phase B) with gradient elution.[11,12]

26.4 Isoflavones Content in Coffee

In addition to the great variability in the extraction procedures and in the analytical methods to detect and to quantify isoflavones in coffee, this matrix is particularly challenging due to its great variability in terms of product types (ground roasted, instant powder, coffee brews, regular or decaffeinated), botanical species, blends, method of preparation and a number of other factors which can affect the chemical composition. These aspects, together with the low number of investigations performed so far, make it very difficult to draw an exhaustive picture. Since the first exploration[2] it was clear that coffee was characterized by inhomogeneous isoflavones content. In fact, Mazur and collaborators,[2] analysed six diverse types of coffee, including arabica instant among unspecified blends, but only one contained daidzein (10–66 μg 100 g^{-1}), two contained genistein (15–29 μg 100 g^{-1}) and three contained formononetin (72–78 μg 100 g^{-1}). This variability was evident in further studies. Alves and collaborators[10] confirmed the presence of daidzein, genistein and formononetin in all commercial roasted coffees analysed, obtaining remarkably higher isoflavones content (daidzein 620–1080 μg 100 g^{-1}; genistein 50–710 μg 100 g^{-1}; formononetin 1200–5800 μg 100 g^{-1}).

However, in one sample of arabica roasted coffee analysed by a different research group, only the aglycone daidzein (1480 μg 100 g^{-1}) was identified, together with a reasonable content of total glucosides (daidzin 160 μg g^{-1}; ononin 0.15 μg g^{-1}; genistin 0.02 μg g^{-1}),[14] and in another study evaluating unspecified coffee sample, only daidzein (50.3 μg 100 g^{-1}) and genistein (≤25 μg 100 g^{-1}) were detected, although biochanin A and formononetin were also investigated.[7]

Very few data have been reported as far as instant coffee powder is concerned. This product seems to present a remarkable variability in isoflavones' contents, going from regular (daidzein 153 μg 100 g^{-1}, genistein 594 μg 100 g^{-1}, formononetin 4 μg 100 g^{-1} and glycitein 162 μg 100 g^{-1}) to decaffeinated coffee (genistein 4 μg 100 g^{-1}; formononetin 1 μg 100 g^{-1}).[9] The low amounts in decaffeinated coffees are possibly derived from extraction of isoflavones with caffeine.

In the case of coffee brews, several published data refer to lack of details on preparation method and therefore it is extremely difficult to trace a conclusive view on the effect of such methods on isoflavones content. In one study[9] both regular and decaffeinated coffee infusions contained genistein, formononetin and biochanin A in concentrations lower than 10 μg L^{-1}, with no daidzein detection. In a different study, for regular coffee steeped in boiling water and filtered, 1 μg L^{-1} of both daidzein and genistein and 2 μg L^{-1} of formononetin were determined, while for decaffeinated coffee (same brand but no details on blend) with the same preparation, 1 μg L^{-1} of both daidzein and genistein, 2 μg L^{-1} of formononetin and 2 μg L^{-1} of glycitein were determined.[8] Very different concentrations have been reported for coffee beverages prepared by methods generally leading to low total soluble solids like filter and press-pot methods. In this case, daidzein (158–459 μg L^{-1}), genistein (below the quantification limit – 79 μg L^{-1}) and formononetin (1356–1644 μg L^{-1}) were found.[22] Coffee liquid samples prepared by using two different ground coffees and according to unspecified "package directions" were reported to contain higher concentrations of total isoflavones.[13] In particular, daidzein (3210–5200 μg L^{-1}), genistein (878–1380 μg L^{-1}) and formononetin (3–6 μg L^{-1}) together with two glucosides daidzin (788–2300 μg L^{-1}) and genistin (218–418 μg L^{-1}) were found. Similar concentrations but different distribution among aglycones were reported for coffee beverages prepared by espresso and moka methods. Contents in espresso brew were: daidzein (683 μg L^{-1}), genistein (230 μg L^{-1}) and formononetin (4594 μg L^{-1}), while in moka brew contents were: daidzein (937 μg L^{-1}), genistein (252 μg L^{-1}) and formononetin (4775 μg L^{-1}).[22]

Espresso coffee beverages have been the subject of a detailed investigation on isoflavones content.[22] The amount of isoflavones in a cup of 30 mL espresso coffee has been reported to range from about 40 μg (100% arabica) to about 240 μg (100% robusta). In addition to coffee species, roasting degree and cup volume have been indicated as factors affecting isoflavones content. For the latter, a long espresso (70 mL) contains double the amount of isoflavones compared with a short one (20 mL ristretto). Indeed, extraction percentages varied between 19, 16 and 31% in 20 mL espressos and 61, 50 and 72% in 70 mL ones for genistein, daidzein and formononetin, respectively. On average, a cup of 30 mL espresso regular coffee contains daidzein (17 μg), genistein (6 μg) and formononetin (77 μg) but no biochanin A.[30a] However, recent investigations on espresso coffee evaluating 100% arabica samples from different geographical origins (medium roast degree), reported different results, with much lower total isoflavone content (maximum of 0.11 μg for a 30 mL cup of espresso coffee) and with formononetin and biochanin A as the only aglycones detected.[11,12]

26.5 Lignans Content in Coffee

The great majority of the quantitative data on coffee lignans reported in the literature have been obtained from the same samples characterized in parallel for the content of isoflavones.[2,7–9] The six coffee samples characterized in the first analytical investigation on coffee phytoestrogens[2] contained only secoisolariciresinol, matairesinol being below the detection limit, in concentrations (3.93–7.16 mg kg^{-1}) higher than those of isoflavones. Similar figures were reported by another group when characterizing a single unspecified coffee sample.[7] In this case, matairesinol was not detected and 0.25 mg kg^{-1} of secoisolariciresinol was found.

As far as instant coffee is concerned, both regular and decaffeinated powders showed similar lignans quali-quantitative data, secoisolariciresinol and matairesinol being in the range 6.10–8.62 mg kg^{-1} and 0.32–0.58 mg kg^{-1}, respectively.[9]

In addition to secoisolariciresinol and matairesinol, more recent investigations on coffee beverages reported the presence of lariciresinol and pinoresinol,[8,25] lariciresinol being quantitatively similar to secoisolariciresinol which is confirmed in all cases to be the most abundant lignan in coffee. In Table 26.3, lignans content in coffee brews is reported.

Coffee liquid samples prepared by using two different ground coffees and according to unspecified "package directions" were reported to contain 97 and 135 μg L^{-1} of enterodiol as the single lignan detected.[13]

Differently from isoflavones, in the case of lignans there are no reports on the influence exerted by both raw material (in terms of coffee species and roasting degree) and preparation methods, on their content.

26.6 Other Flavonoids in Coffee

In addition to isoflavones and lignans, coffee beans contain small amounts of other flavonoids that are probably derived from flavonoids present in the coffee fruit pulp. Anthocyanidins such as cyanidins, pelargonidins and one peonidin have been identified in arabica coffee seeds.[30b] Typical flavonoids like catechins and flavonol glycosides typically found in tea or flavonols like quercetin and myricetin found in fruit juices and wines have not been

Table 26.3 Lignans contents in coffee brews.

Coffee brews from different preparation methods	Secoisolariciresinol ($\mu g\ L^{-1}$)	Matairesinol ($\mu g\ L^{-1}$)	Lariciresinol ($\mu g\ L^{-1}$)	Pinoresinol ($\mu g\ L^{-1}$)	Reference
Filtered coffee	92–161	0–7	90–131	4–15	25
Infusion	160	<10			9
Infusion (decaffeinated)	100	<10			9
Coffee steeped in boiling water and filtered	47	1	9	2	8
Coffee steeped in boiling water and filtered (decaffeinated)	34	2	11	1	8

detected in ground coffee (light, medium and dark roast), instant coffee and coffee brews.[31,32]

The increasing knowledge of the positive health effects of food polyphenols, as well as the awareness to look for sustainable processes, has prompted the need to explore some by-products of the coffee industry as a source of high-value chemicals including flavonoids. Coffee pulp is the first by-product obtained during processing and represents 29% dry-weight (about 40% fresh-weight) of the whole fruit.[33]

Ramirez-Colonel and collaborators,[34] by subjecting freeze-dried arabica coffee pulp powder to both direct or after solvent extractions thiolysis, quantified phenolic compounds by HPLC. Flavonoids resulted to be particularly abundant and flavan-3-ols, both monomers and proanthocyanidins, were detected in higher amounts. Differences in concentration of proanthocyanidins (about ten times more concentrated than flavan-3-ols monomers) were observed between fresh and 3 day-old (sun dried) coffee pulp. Further detailed characterization showed evidence that constitutive units were mainly epicatechin (Figure 26.4), representing more than 90% of the proanthocyanidin units, with average degrees of polymerization in the range of 3.8–9.1. Molecular size of oligomeric proanthocyanidins separated by normal-phase HPLC, and determined by matrix-assisted laser desorption ionization time-of-flight mass spectrometry (MALDI-TOF MS), confirmed the presence of oligomers of the flavan-3-ol (−)-epicatechin. Anthocyanidins were also found as a minor class of flavonoids present, however, the authors did not provide any structural details on these compounds.[34]

Anthocyanins from fresh coffee husks (outer skin and pulp) were the subject of more recent investigations.[35,36] Prata and Oliveira[35] investigated anthocyanin content and profile of fresh arabica coffee outer skin and pulp by both UV spectrophotometry and HPLC-DAD analysis. Independently on the coffee varieties studied (Mondo Novo, Icatu, Acaià, Catuai and Rubi), cyanidin

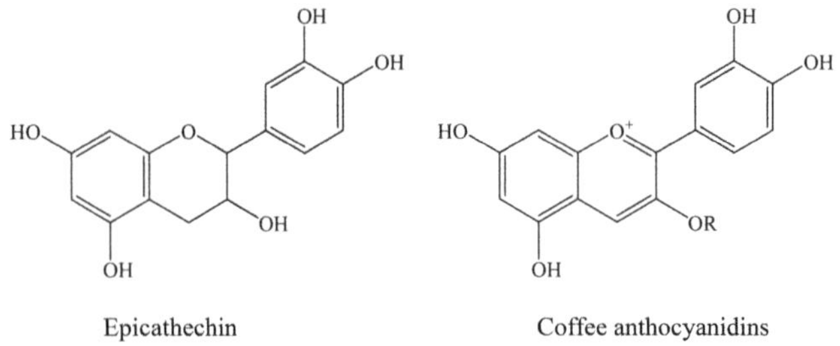

Figure 26.4 Chemical structure of flavonoids (R = rutenose – cyanidin-3-rutinoside or R = glucose – cyanidin-3-glucoside) present in coffee pulp.

3-rutinoside resulted to be the dominant anthocyanin in fresh coffee husk followed by a small amount of cyanidin 3-glycoside (see Figure 26.4).[35]

Similar results were obtained by Murthy and collaborators, who, in addition to chemically characterizing anthocyanins from fresh coffee pulp (no details on coffee species), performed a bioactivity characterization.[36] From a quantitative point of view, coffee pulp yielded 25 mg of monomeric anthocyanins 100 g^{-1} of fresh pulp on a dry weight basis whereas, from a functional point of view, the red colour of coffee peels was confirmed, by both ^{1}H and ^{13}C nuclear magnetic resonance spectroscopy, to be due to the presence of cyanidin 3-rutinoside.[36] Coffee anthocyanins have shown multiple biological effects resulting in effective α-glucosidase and α-amylase inhibitory activities. It was concluded that coffee skin/pulp are potential sources of colorants and bioactive ingredients to be used in formulated foods.[36]

26.7 Conclusions

Roasted coffee beans and instant coffee may represent a source of isoflavones and lignans, however, information on these coffee phytoestrogens is scarce and fragmentary. In view of the different extraction procedures and analytical methods, as well as the different coffee matrices characterized to date, it is very difficult to interpret the great variability in quali- and quantitative reported data. Contradictory data could derive from both methodological approaches as well as intrinsic characteristics of the analysed samples. So far, no data of these phenolic compounds in green coffee beans have been reported. It cannot be excluded *a priori* that the observed remarkable discrepancies could be ascribed to agronomical and post-harvest processes, degree of ripening or presence of variable amounts of intrinsic and extrinsic defects, storage conditions and, in general, the whole set of factors which can affect the green coffee chemical composition. In addition, roasting processes and brewing methods obviously contribute to increased variability in coffee phytoestrogen composition and content. In all cases, the confirmed presence of these well-known bioactive polyphenols in coffee must trigger studies to better understand their dietary contribution, as well as of other minor phenolic compounds, from coffee. The potential physiological implications are discussed in *Coffee: Consumption and Health Implications*.

References

1. A. L. Ososki and E. J. Kennelly, *Phytother. Res.*, 2003, **17**, 845.
2. W. Mazur, K. Wahala, S. Rasku, A. Salakka, T. Hase and H. Adlercreutz, *Br. J. Nutr.*, 1998, **79**(1), 37.
3. A. Cassidy, B. Hanley and R. M. Lamuela-Raventos, *J. Sci. Food Agric.*, 2000, **80**, 1044.
4. K.-P. Ko, *Asian Pac. J. Cancer Prev.*, 2014, **15**, 7001.
5. A. J. Day, M. S. Dupont, S. Ridley, M. J. C. Rhodes, M. R. A. Morgan and G. Wiliamsosm, *FEBS Lett.*, September 25 1998, **436**(1), 71.
6. W. Mazur, *Phytoestrogens: Occurrence in Foods, and Metabolism of Lignans in Man and Pigs*, University of Helsinki, Helsinki, 2000, Available from: http://ethesis.helsinki.fi/julkaisut/laa/kliin/vk/mazur/.
7. P. L. Horn-Ross, M. Lee, E. M. John and J. Koo, *Cancer Causes Control*, 2000, **11**, 299.
8. L. U. Thompson, B. A. Boucher, Z. Liu, M. Cotterchio and N. Kreiger, *Nutr. Cancer*, 2006, **54**(2), 184.
9. G. G. C. Kuhnle, C. Dell'Aquila, S. M. Aspinall, S. A. Runswick, A. A. Mulligan and S. A. Bingham, *J. Agric. Food Chem.*, 2008, **56**, 7311.
10. R. C. Alves, I. M. C. Almeida, S. Casal and M. B. P. P. Oliveira, *Food Chem.*, 2010, **122**, 914.
11. G. Sagratini, L. Navarini, M. Ricciutelli, E. Torregiani, S. Vittori and G. Caprioli, *Proceedings 4th MS Food DAY*, Foggia (Italy), October 7–9, 2015, 97.
12. G. Caprioli, L. Navarini, M. Cortese, M. Ricciutelli, E. Torregiani, S. Vittori and G. Sagratini, *J. Mass Spectrom.*, 2016, **51**, 698.
13. Y. Sapozhnikova, *Food Chem.*, 2014, **150**, 87.
14. Y. S. Kim, S. J. Yeom and D. K. Oh, *J. Agric. Food Chem.*, 2011, **59**, 11812.
15. United States Department of Agriculture (USDA), USDA database for the isoflavone content of selected foods, Release 2.1, November 2015.
16. T. Song, K. Barua, G. Buseman and P. A. Murphy, *Am. J. Clin. Nutr.*, 1998, **68**, 1474S.
17. Z. Xu, Q. Wu and S. Godber, *J. Agric. Food Chem.*, 2002, **50**, 7402.
18. S. Lee and J. Lee, *Food Chem.*, 2009, **112**, 316.
19. X. Yue, A. M. Abdallah and Z. Xu, *Int. J. Food Sci. Technol.*, 2010, **45**, 1878.
20. S. Buddhiranon and T. Kyu, *J. Phys. Chem.*, 2012, **116**, 7795.
21. C. G. A. Davies, F. M. Netto, N. Glassenap, C. M. Gallaher, T. P. Labuza and D. D. Gallaher, *J. Agric. Food Chem.*, 1998, **46**, 2485.
22. R. C. Alves, I. M. C. Almeida, S. Casal and M. B. P. P. Oliveira, *J. Agric. Food Chem.*, 2010, **58**, 3002.
23. R. D. Haworth, *Annu. Rep. Prog. Chem.*, 1936, **33**, 266.

24. S. M. Willför, A. I. Smeds and B. R. Holmbom, *J. Chromatogr. A*, 2006, **1112**, 64.
25. I. E. J. Milder, I. C. W. Arts, B. van de Putte, D. P. Venema and P. C. H. Hollman, *Br. J. Nutr.*, 2005, **93**, 393.
26. E. Gerstenmeyer, S. Reimer, E. Berghofer, H. Schwartz and G. Sontag, *Food Chem.*, 2013, **138**, 1847.
27. Q. Wu, M. Wang and J. E. Simon, *J. Chromatogr. B*, 2004, **812**, 325.
28. K. S. R. Raju, N. Kadian, I. Taneja and M. Wahajuddin, *Phytochem. Rev.*, 2015, **14**, 469.
29. G. G. C. Kuhnle, C. Dell'Aquila, Y.-L. Low, M. Kussmaul and S. A. Bingham, *Anal. Chem.*, 2007, **79**, 9234.
30. (a) R. C. Alves, C. M. D. Soares and M. B. P. P. Oliveira, *Coffee in Health and Disease Prevention*, ed. V. R. Preedy, Academic Press, London, 1st edn, 2015, ch. 16, pp. 143–148; (b) G. Mazza and E. Miniati, *Anthocyanins in Fruits, Vegetables and Grains*, CRC Press, Boca Raton, 1993.
31. C. Lakenbrink, S. Lapczynski, B. Maiwald and U. H. Engelhardt, *J. Agric. Food Chem.*, 2000, **48**, 2848.
32. M. G. L. Hertog, P. C. H. Hollman and B. van de Putte, *J. Agric. Food Chem.*, 1993, **41**, 1242.
33. P. S. Murthy and M. M. Naidu, *Resour., Conserv. Recycl.*, 2012, **66**, 45.
34. M. A. Ramirez-Coronel, N. Marnet, V. S. Kumar Kolli, S. Roussos, S. Guyot and C. Augur, *J. Agric. Food Chem.*, 2004, **52**, 1344.
35. E. R. B. A. Prata and L. S. Oliveira, *LWT*, 2007, **40**, 1555.
36. P. S. Murthy, M. R. Manjunatha, G. Sulochannama and M. M. Naidu, *Eur. J. Biol. Sci.*, 2012, **4**(1), 13.

CHAPTER 27

Trigonelline and Derivatives

ADRIANA FARAH*, THIAGO FERREIRA AND ANA CAROLINA VIEIRA

Universidade Federal do Rio de Janeiro, Laboratório de Química e Bioatividade de Alimentos, Avenida Carlos Chagas Filho, 373, CCS, Bloco J, Rio de Janeiro, 5521, Brazil
*E-mail: afarah@nutricao.ufrj.br

27.1 Introduction and Chemical Aspects

Trigonelline is a water-soluble alkaloid biologically derived in plants from the enzymatic methylation of the pyridine nucleotide of nicotinic acid by trigonelline synthase.[1,2] Trigonelline's chemical formula is $C_7H_7NO_2$, molecular weight 137.138 g mol^{-1}, melting point 230 to 233 °C, and IUPAC name 1-methylpyridin-1-ium-3-carboxylate[3] (Figure 27.1).

Trigonelline was first isolated in 1885 from the seeds of fenugreek (*Trigonella foenum-graecum* L.), one of the oldest existing medicinal plants originated in East Asia and Northern Africa.[4] Isolation from arabica coffee occurred in 1909.[5] This alkaloid is present in a wide variety of species in the plant kingdom.[6,7] In addition to being present in non-conventional plants, trigonelline is found in the seeds and other organs of various legume species including soy bean[7,8] and in rice, pumpkins, lettuce, cabbage, peas, orange, and coffee, with the last being the most important commercially available food source.[9] Trigonelline is also found in animal foods, including fish,

Coffee: Production, Quality and Chemistry
Edited by Adriana Farah

Published by the Royal Society of Chemistry, www.rsc.org

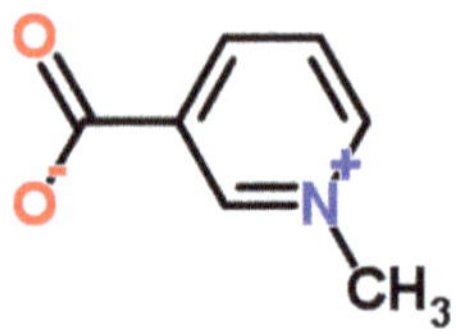

Figure 27.1 Molecular structure of trigonelline (IUPAC name: 1-methylpyridin-1-ium-3-carboxylate).[55]

shellfish, and the liver of several animals. In a study by Taguchi *et al.*,[10] the alkaloid was investigated in 104 plants and contents ranged from 0.05 g 100 g^{-1} in green asparagus to 0.23 g 100 g^{-1} in turban shell.

Among the functions of trigonelline in plants are: to serve as a nutrient source and as a bioactive substance for nyctinasty (a nastic movement such as the opening and closing of some flowers that is associated with diurnal changes of temperature or light intensity); to take part in cell cycle regulation and in signal transduction and detoxification of nicotinic acid.[7] The involvement of trigonelline in coffee plant metabolism is discussed in Chapter 4.

In raw coffee seeds, trigonelline is found in a similar degree of magnitude to caffeine, but differently from caffeine, due to its thermolability, during roasting part of its content is degraded generating both volatile and non-volatile compounds such as nicotinic acid and *N*-methylpyridinium.[11–13] In the last decade, trigonelline and its non-volatile derivatives have gained growing importance because of their potential contribution to health (see *Coffee: Consumption and Health Implications*). The content of trigonelline in green coffee seeds, the chemical changes that occur during roasting, and the reported contents in roasted coffee and brew are presented in this chapter.

27.2 Analysis of Trigonelline and Derivatives in Coffee

The first attempts to determine trigonelline in coffee products date to as early as 1930. These methods were based on trigonelline precipitation with iodine and titration techniques. Colorimetric methods and alkaline hydrolysis were used for quantitative purposes. However, they were unsatisfactory in terms of precision and accuracy.[14] An electrochemical method using reducible or oxidizable substances known as polarography (or polarographic analysis, or voltammetry) has also been applied to assess trigonelline in coffee.[15]

Currently, liquid chromatography associated with UV (HPLC-UV), diode array (HPLC-DAD), or mass spectrometry (LC-MS) detection are the most used techniques for the analysis of trigonelline and non-volatile derivatives, either individually[11] or together with other compounds.[12,16–18]

For the analysis of trigonelline alone, isocratic systems and reverse phase columns kept at 23–40 °C have been mostly used, with UV or DAD detectors set at 265–272 nm (lambda max is reported to be from 265 to 268 nm).[12,18] Additional compounds have been analyzed simultaneously.[1,12,18–22] Electrospray ionization mass spectrometry (ESI-MS) has been used for peaks confirmation, especially in simultaneous analyses.

Perrone *et al.*[12] performed simultaneous analyses of nicotinic acid (lambda max ~ 260 nm), caffeine, trigonelline, and sucrose. For compounds separation, a gradient of 0.3% aqueous formic acid and methanol was used. The ESI source was operated in the negative mode to generate the formic acid adduct of sucrose $(M + HCOO)^-$ and the mass spectrometer was operated in the single ion monitoring (SIM) positive mode to generate caffeine, trigonelline, and nicotinic acid $(M + H)^+$ ions.[12]

De Maria *et al.*[23] proposed a method for the simultaneous determination of total chlorogenic acids, trigonelline, and caffeine in green coffee using high performance gel filtration chromatography. The methodology was based on gradient elution with reverse-phase column. Readings were carried out at 272 nm.[23]

Lang *et al.*[13] performed quantitative analyses of trigonelline, *N*-methylpyridinium, caffeine, and caffeoylquinic acids in coffee. Ionization was in positive electrospray mode with ion spray voltage of 5.5 kV and temperature of the heater gas 550 °C. Optimization of ionization and fragmentation for sensitive MRM analysis was obtained by software-assisted ramping of the respective ion source and collision cell parameters. Trigonelline and *N*-methylpyridinium were analyzed by an isocratic HILIC-LC-MS/MS method based on stable isotope dilution. Caffeine and caffeoylquinic acids (as the sum of 3-, 4-, and 5-caffeoylquinic acids) were analyzed by HPLC-DAD, using a reverse-phase column and a gradient of ammonium formate buffer (0.25 M, pH 3.5) and methanol. Readings were performed at 272 and 324 nm, respectively.[13]

Vignoli *et al.*[21] performed simultaneous analyses of 5-caffeoylquinic acid, trigonelline, furfural, and 5-hydroxymethylfurfural (HMF) by HPLC-DAD. Compounds were eluted with a gradient using 5% acetic acid and acetonitrile as mobile phases. Trigonelline, caffeine, furfural, and HMF were detected at 272 nm, while 5-CQA was detected at 320 nm.[21]

Mehari *et al.*[22] analyzed trigonelline and methylxanthines (theobromine, caffeine, and theophylline) simultaneously in coffee by HPLC-DAD-MS. Chromatographic data for methylxanthines were collected at 272 nm for methylxanthines, and at 264 nm for trigonelline. DAD was set to collect UV-vis spectral data in the wavelength range 200–400 nm. MS was used for theobromine confirmation. Analyses were carried out under isocratic conditions using 90% deionized water (acidified as 0.1% aqueous formic acid) and 10% acetonitrile.[22]

Tran *et al.*[18] assessed trigonelline and caffeine contents in coffee. Detection was achieved using HPLC-DAD set at 265 nm for trigonelline and 273 nm for caffeine using gradient elution with phosphate buffer and methanol.[18]

Although trigonelline and derivates have been widely assessed by HPLC, gas chromatography has also been applied by Stennert and Maier[24] for the same purpose associated with a nitrogen–phosphorus-detector (NPD) (also known as a thermionic specific detector (TSD)). Helium was used as mobile phase with heating temperature slope from 50 to 195 °C at a rate of 15 °C min^{-1}, followed by 10 min isothermally at 195 °C.[24]

In addition to chromatographic techniques, proton nuclear magnetic resonance technique (^{1}H NMR) has also been applied in a few studies to determine trigonelline contents in coffee. Consonni *et al.*[25] used NMR-based method for simultaneous monitoring of different compounds followed by chemometrics to discriminate coffee samples from different origins. Trigonelline was among the main compounds serving this purpose. Campo *et al.*[26,27] used ^{1}H NMR spectrum for simultaneous determination of caffeine, formic acid, trigonelline, and HMF. No need for previous derivatization or treatment was observed.

Just as a complementary note, quantification of trigonelline using commercial standard should take into account that the alkaloid is available as trigonelline hydrochloride or similar forms and therefore calculation adjustments should be made, which will lower trigonelline content value.

For volatile derivatives, variations of gas chromatography associated with mass spectrometry (GC/MS) have been used in the last decades.[28–30] For information on the most recent techniques for analysis of volatile compounds, see Chapter 33 of this book.

27.3 Content of Trigonelline in Green Coffee Seeds

The content of trigonelline in green coffee seeds may be influenced by various intrinsic and extrinsic factors. Among important aspects are genetics (species and variety/cultivar), origin (climate/soil), and agricultural practices.[31,32] But excluding genetic variables and technological processes such as decaffeination, the analytical methods employed seem to influence the contents most importantly.[19]

Trigonelline contents in green coffee seeds vary widely within the same species. Reported values for *Coffea arabica* beans are in the range of 0.8 to 2.9 g 100 g^{-1} (dry weight basis – dw), with common values between 0.9 and 1.2 g 100 g^{-1}. For *Coffea canephora* species, reported values range from 0.5 to 3.1 (dw), with common values between 0.7 and 1.0 g 100 g^{-1}. Other coffee species have shown variable trigonelline amounts, from 0.3 to 2.3 g 100 g^{-1} (dw) with average values up to 1.2 g 100 g^{-1}.[12,18,19,31,33,34] The alkaloid is also distributed in other parts of the coffee plant, including leaves, stem, and root, varying between organs and with age. In a study by Zheng and Ashihara,[35] trigonelline concentration decreased with the age of tissues. Leaves collected from upper, middle, and lower parts of the plants showed trigonelline contents about 1.1, 1.0, and 0.8 g 100 g^{-1} (dw), respectively. Younger (upper) and older (lower) parts of stems showed concentrations about 2.5 and 0.6 g 100 g^{-1} (dw), respectively.

Regarding the content of trigonelline in Brazilian *C. arabica* seeds depulped by different post-harvest methods, average contents in seeds processed by the wet method were slightly higher (0.9 g 100 g^{-1}) when compared to the same seeds processed by the semi-dry method (0.8 g 100 g^{-1}).[20]

A few reported contents of trigonelline in green coffee seeds from different sources and analyzed by different methods are presented in Table 27.1.

Table 27.1 Contents of trigonelline reported for green coffee seeds from *Coffea arabica* and *Coffea canephora* species.[a]

Origin	Number of samples	Technique used for analysis	Content or content range and mean value (g 100 g^{-1}, dm)	Ref.
Coffea arabica				
Brazil	1	LC-MS	1.03	12
England	NR	HPLC-UV	1.31	36
Ethiopia	NR	HPLC-DAD	0.95–1.29 ($\bar{x}$ = 1.13)	19
Brazil	3	HPLC-UV	1.68–1.80 ($\bar{x}$ = 1.76)	33
Costa Rica	1		1.27	
Guatemala	1		1.66	
Colombia	1		1.92	
Nicaragua	1		1.24	
Honduras	2	Polarography	0.88, 1.01	15
Brazil	2		0.94, 0.96	
Mexico	1		0.83	
Kenya	1		0.85	
Bolivia	1		0.96	
NR	7	HPLC-UV	0.79–1.06 ($\bar{x}$ = 0.88)	37
Brazil	16	HPLC-UV	1.52–2.90 ($\bar{x}$ = 2.19)	38
Brazil	10	HPLC-DAD	($\bar{x}$ = 0.89)	34
Colombia	3	HPLC-UV	0.84–0.88 ($\bar{x}$ = 0.85)	39
Kenya	1		0.79	
East India	1		0.83	
Mocca	1		0.92	
Brazil	2		1.05–1.06	
NR	232	HPLC-DAD	0.80–1.38 ($\bar{x}$ = 1.11)	18
Brazil	17	HPLC-UV	0.80–1.40 ($\bar{x}$ = 0.90)	20
Ethiopia (East)	26	HPLC-DAD	1.06–1.32 ($\bar{x}$ = 1.19)	22
Ethiopia (Northwest)	6		1.15–1.29 ($\bar{x}$ = 1.22)	
Ethiopia (South)	49		1.01–1.31 ($\bar{x}$ = 1.15)	
Ethiopia (West)	18		0.98–1.20 ($\bar{x}$ = 1.10)	
Brazil	NR	HPLC-UV	0.96–1.23	40
Italy	4	HPLC-UV	1.0–1.6 ($\bar{x}$ = 1.10)	23
Brazil	5	HPLC-UV	0.96–1.34	11
Coffea canephora				
NR	2	HPLC-DAD	0.91, 1.24 ($\bar{x}$ = 1.08)	18
Ivory coast	NR	HPLC-UV	0.57–0.69	36
Uganda	3	HPLC-UV	0.96–1.12 ($\bar{x}$ = 1.06)	33
Indonesia	2		0.94, 1.24	
Ivory coast	1		0.91	
Thailand	1		1.34	
Brazil	NR	HPLC-UV	3.08	38
Ivory coast	4	HPLC-DAD	0.52–1.06 ($\bar{x}$ = 0.82)	19
NR	3	HPLC-UV	0.66–0.68	37
Cameroon	NR	HPLC-UV	0.66	39
India	NR		0.68	
Zaire	NR		0.66	
Karnataka	NR	HPLC-UV	0.98	32
Ivory coast	NR	HPVC-DAD	0.61	34

(continued)

Table 27.1 (*continued*)

Origin	Number of samples	Technique used for analysis	Content or content range and mean value (g 100 g^{-1}, dm)	Ref.
Other species				
C. pseudozanguebariae	NR	HPLC-UV	0.86–1.19 ($\bar{x}$= 1.02)	1
C. liberica var. dewevrei	NR	HPLC-UV	0.51–0.66 ($\bar{x}$= 0.57)	
C. liberica	NR	HPLC-UV	0.29	36
C. brevipes	1	HPLC-DAD	0.76	18
C. stenophylla	NR	HPLC-UV	1.89	38
C. salvatrix			1.21	
C. liberica			1.01	
C. racemosa			2.31	
C. brevipes			2.23	
C. liberica Koto	1	HPLC-DAD	0.30	19
C. brevipes	4		1.26–1.43 ($\bar{x}$= 1.35)	
C. congensis	4		0.91–1.12 ($\bar{x}$= 1.06)	
C. eugenioides	4		1.13–1.42 ($\bar{x}$= 1.33)	
C. heterocalyx	4		1.50–1.57 ($\bar{x}$= 1.53)	
C. humblotiana	4		0.71–0.88 ($\bar{x}$= 0.81)	
C. humilis	4		0.36–0.73 ($\bar{x}$= 0.52)	
C. kapakata	4		1.55–1.99 ($\bar{x}$= 1.77)	
C. liberica dewevrei	4		0.45–0.66 ($\bar{x}$= 0.56)	
C. liberica liberica	4		0.52–0.87 ($\bar{x}$= 0.67)	
C. pseudozanguebariae	4		0.87–1.16 ($\bar{x}$= 1.02)	
C. racemosa	4		0.75–1.23 ($\bar{x}$= 1.02)	
C. salvatrix	4		1.08–1.22 ($\bar{x}$= 1.16)	
C. pocsii	4		1.38–1.48 ($\bar{x}$= 1.45)	
C. stenophylla	4		0.98–1.26 ($\bar{x}$= 1.09)	
C. sp. Bakossi	4		0.50–0.88 ($\bar{x}$= 0.66)	
C. sp. Congo	4		1.35–1.92 ($\bar{x}$= 1.75)	
C. sp. Ngongo 2	4		0.97–1.19 ($\bar{x}$= 1.13)	
C. sp. Moloundou	4		1.39–1.87 ($\bar{x}$= 1.58)	
C. sp. Koumbala	4		1.09–1.34 ($\bar{x}$= 1.20)	
C. liberica var. Dewevrei	NR	HPLC-UV	1.37	38
C. dewevrei cv. Abeokutae	NR		0.88	
C. dewevrei cv. Dibowskii	NR		0.84	
C. dewevrei cv. Excelsa	NR		0.77	
C. stenophylla	NR		1.89	
C. salvatrix	NR		1.21	
C. liberica	NR		1.01	
C. racemosa	NR		2.31	
C. brevipes	NR		2.23	
C. arabica x C. liberica var. Dewevrei	NR		1.72	
C. racemosa x C. eugenioides	NR		1.92	

Table 27.1 (*continued*)

Origin	Number of samples	Technique used for analysis	Content or content range and mean value (g 100 g^{-1}, dm)	Ref.
C. sp. Excelsa	1	HPLC-UV	0.32	37
C. Stenophylla C.	1		0.51	
Liberica	1		0.28	
C. brevipes	1	HPLC-DAD	0.76	18

[a]NR – Not reported; dm - dry matter; HPLC - High performance liquid chromatography; DAD - Diod array detector.

27.4 Contents of Trigonelline, Nicotinic Acid, and Other Derivatives in Roasted Coffee Seeds

During coffee roasting, at temperatures approaching 200 °C, demethylation of trigonelline forms pyridine-3-carboxylic acid (under IUPAC name), or nicotinic acid, the main form of niacin, also called vitamin B_3, or vitamin PP,[41–45] with chemical formula $C_6H_5NO_2$ and molecular weight 123.11 g mol^{-1}. Very low amounts of nicotinamide ($C_6H_6N_2O$, 122.13 g mol^{-1}, IUPAC name pyridine-3-carboxamide), another form of niacin, are also formed.

Decarboxylation of trigonelline during roasting produces the ion *N*-methylpyridinium[46–54] with chemical formula $C_6H_8N^+$, molecular weight 94.134 g mol^{-1} and IUPAC name 1-methylpyridinium. The dietary occurrence of this compound is presumably restricted to coffee.[51] 1,2-, 1,3-, and 1,4-dimethylpyridinium ($C_7H_{10}N$, molecular mass 108.160 g mol^{-1}) are also produced as minor non-volatile derivatives.[48,48a,55] The importance of these compounds to coffee flavor and health are mostly unknown.

Regarding the volatile compounds formed during roasting, about 46% of trigonelline initial content was reported to form volatile compounds.[45] The main classes formed by trigonelline pyrolytic reactions are pyridines and pyrroles.[5,56,57] Among pyridines found in coffee are methylpyridines, ethylpyridines, 3-propylpyridine, 3-buthylpyridine, dimethylpyridines, 3-phenylpyridine, acetylpyridines, 2-propylpyridine, methylpiperidines, and 2-phenethylpiperidine. Examples of pyrroles are *N*-methylpyrrole, ethylpyrroles, pentylpyrroles, isobutylpyrroles, isopentylpyrroles, and dimethylpyrroles.[5,57]

Additional potentially important minor pyridinium compounds formed from trigonelline pyrolysis have been reported, such as methyl nicotinate or nicotinic acid methyl ester ($C_7H_7NO_2$, molecular weight 137.138 g mol^{-1} and IUPAC name methyl pyridine-3-carboxylate).[51,58]

Figure 27.2 presents the breakdown of trigonelline during roasting and formation of the main non-volatile and volatile classes of derivatives.

The loss of trigonelline and formation of derivatives is strongly dependent on temperature and time and therefore strongly associated with the degree

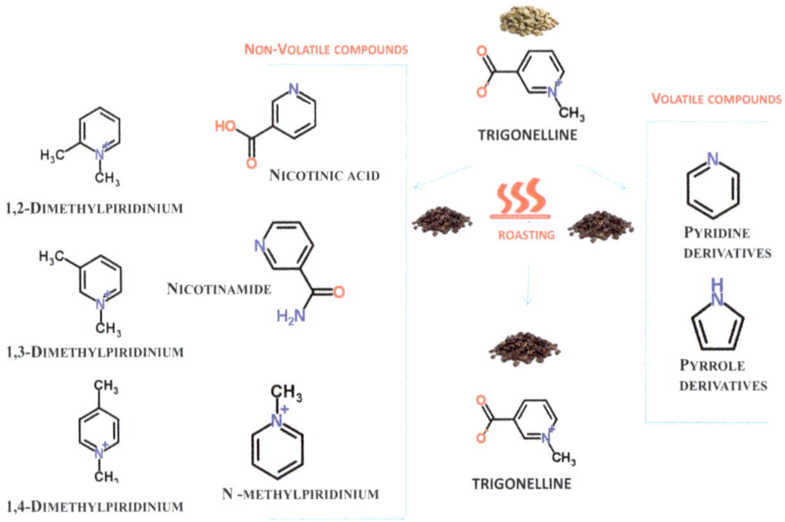

Figure 27.2 Breakdown of trigonelline and formation of non-volatile and volatile derivatives during roasting process.[12,13,46,48–54]

of roast.[45,59] The content of trigonelline decreases continuously during the roasting process, while those of *N*-methylpyridinium and nicotinic acid increase, with *N*-methylpyridinium as the major thermal product up to the dark-medium roast. From this point on towards the end of roast, the contents of both *N*-methylpyridinium and nicotinic acid decrease, probably due to further decomposition and/or interaction with other thermolytic products.[12,46,49,60] Trigonelline losses of 50–90% of the original content in green coffee have been reported after roasting.[11,61] In medium roasted coffees, however, trigonelline content is still relevant (270 to 663 mg 100 g^{-1}) (Table 27.2), which is interesting, due to the potential beneficial effects of trigonelline on health (see *Coffee: Consumption and Health Implications* for trigonelline related health aspects).

Regarding differences between species, reported contents in roasted coffee vary from 30 to 1100 mg 100 g^{-1} in *C. arabica* and from 60 to 683 mg 100 g^{-1} in *C. canephora* (considering light to dark roasts, dry weight, Table 27.2). Despite the initial lower contents in *C. canephora* green beans, the contents in both species may be similar in roasted *C. canephora* possibly due to differences in cell wall resistance during the roasting process. Contents of nicotinic acid in roasted and ground coffee (lab roasted and commercial) vary from 4.7 to 53 mg 100 g^{-1}. For *N*-methylpyridinium, they vary from 73 to 295 mg 100 g^{-1} and for 1,2-dimethylpyridinium reported values vary up to 2.5 mg 100 g^{-1}.[48]

The ratio of trigonelline to caffeine has been proposed for estimation of the roasting degree, decreasing from 0.86 in green beans to 0.15 in very dark roasted beans.[5,62]

Coffee type	Origin	Roast degree	Analytical technique	Content of trigonelline (mg 100 g^{-1}, dm)	Content of nicotinic acid (mg 100 g^{-1}, dm)	Content of N-methylpyridinium (mg 100 g^{-1}, dm)	Ref.
Coffea arabica							
NR	Brazil	NR	HPLC-DAD	550	11.4		34
Soft	Brazil	Light	HPLC-UV	1170			11
Soft	Brazil	Medium	HPLC-UV	390			11
Soft	Brazil	Dark	HPLC-UV	190			11
Soft	Brazil	Very dark	HPLC-UV	185			11
Hard	Brazil	Light	HPLC-UV	1300			11
Hard	Brazil	Medium	HPLC-UV	400			11
Hard	Brazil	Dark	HPLC-UV	50			11
Hard	Brazil	Very dark	HPLC-UV	30			11
Rio Zona	Brazil	Light	HPLC-UV	900			11
Rio Zona	Brazil	Medium	HPLC-UV	550			11
Rio Zona	Brazil	Dark	HPLC-UV	250			11
Rio Zona	Brazil	Very dark	HPLC-UV	90			11
NR	NR	Light	HILIC-LC-MS/MS	701		73.0	13
NR	NR	Dark	HILIC-LC-MS/MS	198		294.7	
NR	Brazil	Medium	HPLC-UV	570			40
NR	Brazil	Dark	HPLC-UV	40			
NR	Brazil	Medium	HPLC-UV	489	12.0		17
NR	Brazil	Dark	HPLC-UV	297	9.0		
NR	Germany	Dark	HILIC-LC-MS/MS	382		110.0	63
Coffea canephora							
Conilon	Brazil	Light	HPLC-UV	530			11
Conilon	Brazil	Medium	HPLC-UV	270			11
Conilon	Brazil	Dark	HPLC-UV	100			11
Conilon	Brazil	Very dark	HPLC-UV	60			11
Conilon	Brazil	Light	HPLC-UV	683	32.0		17
NR	Ivory coast	NR	HPLC-DAD	360	9.5		34

[a]dm: dry matter; NR: Not reported; HPLC - High performance liquid chromatography; DAD - Diod array detector; HILIC - hydrophilic interaction chromatography; MS - mass spectrometry

Differences in roasting profile affect trigonelline content. Trigonelline-rich coffee is produced by low-temperature roasting to prevent the degradation of trigonelline and it is commercially available in Japan.[7]

The contents of trigonelline and derivatives in lab-roasted *C. arabica* and *C. canephora* seeds and in commercial ground roasted blends are presented in Tables 27.2 and 27.3.

Table 27.3 Contents of trigonelline reported for commercial coffee blends.[a]

Coffee type	Origin	Roast degree	Analytical Technique	Content of trigonelline (mg 100 g^{-1}, dm)	Content of nicotinic acid (mg 100 g^{-1}, dm)	Ref.
Roasted and ground (commercial) coffee						
Regular blend	Spain	NR	HPLC-UV	590–850 ($n = 4$)		33
Torrefacto	Spain	NR	HPLC-UV	500–650 ($n = 3$)		33
Arabica	Colombia	NR	HPLC-DAD	430		65
Arabica	Hawaii	Dark	LC-MS	280	29.5	12
Arabica	Brazil	Moderately dark	LC-MS	527	30.4	12
Arabica	Brazil	Medium	LC-MS	641	20.3	12
Regular blend	Africa	Light medium	LC-MS	592	13.7	12
Arabica	Colombia	Light medium	LC-MS	683	9.6	12
Arabica	Brazil	Light medium	LC-MS	956	10.3	12
Decaf	Brazil	Medium	LC-MS	663	10.4	12
Decaf	Brazil	NR	HPLC-DAD-MS	2803–3301 ($n = 3$)	66.5–76.7 ($n = 3$)	65
Regular	Germany	NR	Polarography	490–650 ($n = 4$)		15
Decaf	Germany	NR	Polarography	690		15
Regular	Germany	NR	HPLC-UV	580–710 ($n = 11$)		37
Espresso blend	Germany	NR	HPLC-UV	430, 440 ($n = 2$)		37
Soluble coffee						
Regular	Brazil	Light medium	LC-MS	833, 917 ($n = 2$)	34.2, 36.3 ($n = 2$)	12
Regular	Japan	NR	HPLC-UV	302–976 ($n = 11$)	4.7–53.1 ($n = 11$)	45
Regular	England	NR	HPLC-UV	940–1640 ($n = 12$)		16
Decaf	England	NR	HPLC-UV	1120, 1210 ($n = 2$)		16
Decaf soluble	Germany	NR	Polarography	500		15

[a]NR: not reported; dm: dry matter; Decaf: decaffeinated; HPLC: High performance liquid chromatography; MS: mass spectrometry; DAD: diod array detector.

27.5 Content of Trigonelline, Nicotinic Acid, and Other Derivatives in Coffee Brew

The concentrations of trigonelline and derivatives in coffee brews vary, as for other soluble compounds, with the content in the roasted seeds, the method of extraction, the amount of powder to water, water hardness, filter, duration, and the pressure applied in the extraction process. Typical amounts of trigonelline in brews prepared from medium roasted coffee in a manual dripper with paper filter were reported to be about 40–50 mg 100 mL^{-1}.[12,13] Espresso regular coffees tend to contain higher amounts of trigonelline compared to other types of preparation methods. Despite the degradation during roasting of the seeds, espresso brews from dark roasted coffees may still contain reasonable amounts of trigonelline and *N*-methylpyridinium, for example, about 200 and 290 mg 100 g^{-1}, respectively.[13,31] In instant coffees, trigonelline contents decrease significantly with the roasting degree as in commercial sample.[45,64] Arai *et al.*[45] reported average values of 647 mg 100 g^{-1} ($n = 11$) for soluble coffees.

Data on the content of nicotinic acid in coffee brews are scarce. Perrone *et al.*[12] reported 10–30 mg nicotinic acid per 100 mL cup of paper-filtered manual-dripped brew from medium roasted coffee prepared at 10% (10 g ground coffee to 100 mL water). They also reported values of 34.2 and 36.3 mg 100 g^{-1} of nicotinic acid in two regular commercial instant coffees (powder) and 10.4 mg 100 g^{-1} in decaffeinated (by methanol chloride) ground roasted coffee.

27.6 Contribution of Trigonelline to Cup Quality

The contribution of trigonelline and volatile derivatives to global coffee flavor is still unclear and calls for thorough investigation. According to different authors, trigonelline itself seems to have little direct influence in flavor,[5] having a mild bitter taste, which corresponds to approximately one-quarter of caffeine's contribution to the brew's bitterness.[5,43] Its content in the green and light roasted seeds has been associated with higher cup quality when Brazilian samples from different cup qualities were evaluated.[11] In agreement with that, despite the very low amounts in coffee, the minor roasting product methyl nicotinate is a flavoring ingredient in the food industry (categorized by the Joint FAO/WHO Expert Committee on Food Additives (JECFA)),[65] having a low perception threshold. This compound is also present in guava fruit, papaya, strawberry, roasted filbert, roasted peanut, and in some alcoholic beverages like beer, grape brandy, and Bourbon vanilla, and therefore its contribution to coffee flavor should be investigated.[51,66] On the other hand, higher contents of trigonelline volatile derivatives, especially pyrroles, have been reported in low quality, defective beans compared to control beans.[29]

27.7 Concluding Remarks

A considerable variation is observed in the amount of trigonelline and derivatives reported for green and roasted ground coffees and for coffee brew. More important than variations between *C. arabica* and *C. canephora* species are those related to degree of roast, brewing, and analytical methods, especially the last. More studies investigating the amount of trigonelline as well as major and minor derivatives in different types of coffee and brews are needed. Their contribution to flavor and cup quality should also be more thoroughly investigated.

References

1. C. L. Ky, B. Guyot, J. Louarn, S. Hamon and M. Noirot, *Theor. Appl. Genet.*, 2001, **102**, 630.
2. G. Caprioli, M. Cortese, F. Maggi, C. Minnetti, L. Odello, G. Sagratini and S. Vittori, *Int. J. Food Sci. Nutr.*, 2014, **65**(4), 465.
3. Royal Society of Chemistry, *ChemSpider*, http://www.chemspider.com/Chemical-Structure.5369.html(last accessed January 2018).
4. E. Johns, *Ber. Dtsch. Chem. Ges.*, 1885, **18**, 2518.
5. I. Flament, *Coffee Flavor Chemistry*, John Wiley & Sons, West Sussex, England, 2002.
6. W. A. Tramontano, D. G. Lynn and L. S. Evans, *Phytochemistry*, 1983, **3**, 673.
7. H. Ashihara, I. A. Ludwig, R. Katahira, T. Yokota, T. Fujimura and A. Crozier, *Phytochem. Rev.*, 2015, **14**, 765.
8. J. Zhou, L. Chan and S. Zhou, *Curr. Med. Chem.*, 2012, **19**(21), 3523.
9. H. Ashihara, Y. Yin, R. Katahira, S. Watanabe, T. Mimura and H. Sasamoto, *Plant Physiol. Biochem.*, 2012, **60**, 190.
10. H. Taguchi, M. Sakaguchi and Y. Shimabayashi, *Vitamins*, 1986, **60**, 537, (in Japanese).
11. A. Farah, M. C. Monteiro, V. Calado, A. S. Franca and L. C. Trugo, *Food Chem.*, 2006, **98**, 373.
12. D. Perrone, C. M. Donangelo and A. Farah, *Food Chem.*, 2008, **110**, 1030.
13. R. Lang, E. F. Yagar, A. Wahl, A. Beusch, A. Dunkel, N. Dieminger, R. Eggers, G. Bytof, H. Stiebitz, I. Lantz and T. Hofmann, *J. Agric. Food Chem.*, 2013, **61**(49), 12123.
14. L. C. Trugo, PhD Thesis, University of Reading, England, 1984, p. 108.
15. U. Smola and G. Sontag, *Z. Lebensm.-Unters. Forsch.*, 1983, **177**, 114,(in German).
16. L. C. Trugo, R. Macrae and J. Dick, *J. Sci. Food Agric.*, 1983, **34**, 300.
17. S. T. Alves, R. C. E. Dias and M. T. Benassi, *Quim. Nova*, 2006, **29**(6), 1164,(in Portuguese).
18. H. T. M. Tran, C. A. C. Vargas, L. S. Lee, A. Furtado, H. Smyth and R. Henry, *Tree Genet. Genomes*, 2017, **13**, 54.

19. C. Campa, J. F. Ballester, S. Doulbeau, S. Dussert, S. Hamon and M. Noirot, *Food Chem.*, 2004, **88**, 39.
20. G. S. Duarte, A. A. Pereira and A. Farah, *Food Chem.*, 2010, **118**, 851.
21. J. A. Vignoli, M. C. Viegas, D. G. Bassoli and M. T. Benassi, *Food Res. Int.*, 2014, **61**, 279.
22. B. Mehari, M. Redi-Abshiro, B. S. Chandravanshi, M. Atlabachew, S. Combrinck and R. McCrindle, *Food Anal. Methods*, 2016, **9**, 1627.
23. C. A. B. De Maria, L. C. Trugo and R. F. A. Moreira, *Food Chem.*, 1995, **52**, 447.
24. A. Stennert and H. G. Maier, *Z. Lebensm.-Unters. Forsch.*, 1996, **202**, 45.
25. R. Consonni, L. R. Cagliani and C. Cogliati, *Talanta*, 2012, **88**, 420.
26. G. del Campo, I. Berregi, R. Caracena and J. Zuriarrain, *Talanta*, 2010, **81**, 367.
27. M. Jeszka-Skowron, A. Zgoła-Grzeskowiak and T. Grzekowiak, *Eur. Food Res. Technol.*, 2015, **240**, 19.
28. S. Risticevic, E. Carasek and J. Pawliszyna, *Anal. Chim. Acta*, 2008, **617**, 72.
29. A. T. Toci and A. Farah, *Food Chem.*, 2014, **153**, 298.
30. S. J. Leea, M. K. Kim and K. Lee, *Innovative Food Sci. Emerging Technol.*, 2017, **44**, 97.
31. A. Farah, in *Coffee: Emerging Health Effects and Disease Prevention*, ed. Y. Chu, John Wiley & Sons, Inc. Published by Blackwell Publishing, USA, 2012, vol. 1, p. 21.
32. V. Sridevi and P. Giridhar, *Journal of Food Studies*, 2013, **2**, 1.
33. M. J. Martin, F. Pablos, M. A. Bello and A. G. Gonzalez, *Fresenius. J. Anal. Chem.*, 1997, **357**, 357.
34. S. Casal, M. B. Oliveira and M. A. Ferreira, *J. Liq. Chromatogr. Relat. Technol.*, 1998, **21**(20), 3187.
35. X. Q. Zheng and H. Ashihara, *Plant Sci.*, 2004, **166**, 807.
36. M. N. Clifford and T. Kazi, *Food Chem.*, 1987, **26**, 69.
37. A. Stennert and H. G. Maier, *Z. Lebensm.-Unters. Forsch.*, 1994, **199**, 198.
38. P. Mazzafera, *Phytochemistry*, 1991, **30**(7), 2309.
39. A. Stennert and H. G. Maier, *Z. Lebensm.-Unters. Forsch.*, 1993, **196**, 430.
40. A. Toci and A. Farah, *Contents of chlorogenic acids, caffeine trigoneline and sucrose in Brazilian defective seeds*, ASIC, Paris, 2011, p. 260.
41. M. G. R. Cannell, in *Physiology of the Coffee Crop*, ed. M. N. Clifford and K. C. Wilson, AVI Publishing Company, Inc., CN, 1985, vol. 1, p. 108.
42. H. Taguchi, M. Sakaguchi and Y. Shimabayashi, *Vitamins*, 1986, **60**, 537.
43. L. C. Trugo and R. Macrae, *Food Chem.*, 1984, **15**, 219.
44. J. Adrian and R. Frangne, *Adv. Exp. Med. Biol.*, 1991, **289**, 49.
45. K. Arai, H. Terashima, S. Alzawa, A. Taga, A. Yamamoto, K. Tsutsumiuchi and S. Kodama, *Anal. Sci.*, 2015, **31**, 831.
46. R. H. Stadler, N. Varga, J. Hau, F. A. Vera and D. H. Welti, *J. Agric. Food Chem.*, 2002, **50**, 1192.
47. A. Crozier, H. Ashihara and F. Tomás-Barbéran, *Teas, Cocoa and Coffee: Plant Secondary Metabolites and Health*, 2011.

48. (a) A. Stalmach, M. N. Clifford, G. Williamson and A. Crozier, in *Teas, Cocoa and Coffee: Plant Secondary Metabolites and Health*, ed. A. Crozier, H. Ashihara and F. T. Barberán, Blackwell Publishing Ltd., USA, 2012, vol. 1, p. 143; (b) B. Kalaska, L. Piotrowski, A. Leszczynska, B. Michalowski, K. Kramkowski, T. Kaminski, J. Adamus, A. Marcinek, J. Gebicki, A. Mogielnicki and W. Buczko, *J. Agric. Food Chem.*, 2014, **62**, 2853.
49. F. Wei, K. Furihata, F. Hu, T. Miyakawa and M. Tanokura, *J. Agric. Food Chem.*, 2011, **59**, 9065.
50. R. Lang, A. Wahl, T. Stark and T. Hofmann, in *Recent Advances in the Analysis of Food and Flavors*, ed. S. Toth, et al., ACS Symposium Series; American Chemical Society: Washington, DC, 2012, vol. 1, p. 13.
51. A. Riedel, C. M. Hochkogler, R. Lang, G. Bytof, I. Lantz, T. Hofmann and V. Somoza, *Food Funct.*, 2014, **5**, 454.
52. V. Somoza, M. Lindenmeier, E. Wenzel, O. Frank, H. F. Erbersdobler and T. Hofmann, *J. Agric. Food Chem.*, 2003, **51**, 6861.
53. J. Gebicki, A. Marcinek, S. Chlopicki and J. Adamus, J. PCT International Application. WO 2008104920 A1, 2008.
54. R. Lang, A. Wahl, T. Skurk, E. F. Yagar, L. Schmiech, R. Eggers, H. Hauner and T. Hofmann, *Anal. Chem.*, 2010, **82**, 1486.
55. Royal Society of Chemistry, *ChemSpider*, http://www.chemspider.com/Chemical-Structure.63295.html(last accessed January 2018).
56. A. Toci and A. Farah, *Food Chem.*, 2008, **108**, 1133.
57. A. Toci and A. Farah, *Food Chem.*, 2014, **153**, 298.
58. R. Viani and I. Horman, *J. Food Sci.*, 1974, **39**, 1216.
59. X. Wu, K. Skog and M. Jagersta, *Mutat. Res.*, 1997, **391**, 171.
60. L. C. Trugo, in *Coffee*, ed. R. J. Clarke and R. Macrae, Elsevier Applied Science Publishers, 1985, vol. 1, p. 83.
61. A. S. Franca, L. S. Oliveira, C. F. J. Mendonça and X. A. Silva, *Food Chem.*, 2005, **90**, 89.
62. T. Nakabayashi and M. Masano, *Nippon Shokuhin Kogyo Gakkaishi*, 1986, **33**(10), 729,(in Japanese).
63. T. Bakuradze, R. Lang, T. Hofmann, G. Eisenbrand, D. Schipp, J. Galan and E. Richling, *Eur. J. Nutr.*, 2015, **54**, 149.
64. I. López-Galilea, M. P. de Peña and C. Cid, *Eur. Food Res. Technol.*, 2008, **227**, 141.
65. World Health Organization, Joint Expert Committee on Food Additives - JECFA, http://apps.who.int/foodadditivescontaminantsjecfadatabase/chemical.aspx?chemID=5328 (last accessed January 2018).
66. U.S. National Library of Medicine, National Center for Biotechnology Information, https://pubchem.ncbi.nlm.nih.gov/compound/methyl_nicotinate#section=Top (last accessed January 2018).

CHAPTER 28

Bioactive Amines

MARIA BEATRIZ A. GLORIA*[a,b] AND NICKI J. ENGESETH[b]

[a]Laboratório de Bioquímica de Alimentos, Departamento de Alimentos, Faculdade de Farmácia, Universidade Federal de Minas Gerais, Av. Antonio Carlos 6621, Belo Horizonte, MG, 31270-901, Brasil; [b]208 Bevier Hall, Food Science and Human Nutrition, University of Illinois, Urbana-Champaign, IL, 61801, USA
*E-mail: mbeatriz@ufmg.br

28.1 Introduction

Bioactive or biologically active amines fulfill several important metabolic and physiologic functions in living organisms. They have been known for centuries and still attract considerable interest in food, biomedical, and environmental research. They are formed during normal metabolic processes and are, therefore, present in foods. Bioactive amines play relevant roles in plant development and stress responses. The most common classification divides amines into polyamines and biogenic amines, due to their different roles in development and growth and in food quality and safety, respectively.[1] As a quality control tool, amines can be markers of authenticity and microbial contamination; attest the quality of raw materials, intermediates, and end products; monitor fermentation processes; and indicate freshness, spoilage, and ripening of foods. High concentrations of some amines can also affect the organoleptic characteristics of foods. Additionally, amines are relevant

Coffee: Production, Quality and Chemistry
Edited by Adriana Farah

Published by the Royal Society of Chemistry, www.rsc.org

in human health and prevention of several diseases.[1,2] However, at high concentrations they can pose a toxicological risk.

In this chapter, a brief review is provided on the chemistry of biologically active amines and also on the state of the art knowledge regarding the occurrence of amines in the whole coffee chain, from the plant to the ground roasted coffee available to consumers. The amines which are inherent to coffee as well as the factors which may affect their formation, accumulation, and degradation in coffee are presented. The possible impact of bioactive amines on human health either beneficially or adversely are addressed in *Coffee: Consumption and Health Implications*.

28.2 Chemical Characteristics of Coffee Bioactive Amines

Bioactive amines are aliphatic, cyclic, or heterocyclic organic bases of low molecular mass which express high biological activity. The common and chemical names, molecular formulas and masses and chemical structures of some bioactive amines relevant in coffee are summarized in Table 28.1.[1,3] The molecular masses are usually low, varying from 88.15 (putrescine) to 202.34 g mol^{-1} (spermine). Amines behave as cations, being protonated at physiological pH values.[1]

Even though bioactive amines can be classified by different means, they are typically grouped as polyamines and biogenic amines. The polyamines—spermine and spermidine—are ubiquitous to living cells and play important roles in cell division, inhibition of lipid oxidation, and response to stress; thus, they support life, health, and reproduction. The biogenic amines—putrescine, cadaverine, agmatine, histamine, tyramine, tryptamine, serotonin, melatonin, and phenylethylamine—are neuroactive or vasoactive, and can affect secretory activity. The biogenic amines can be further classified into indolamines (serotonin, tryptamine, and melatonin), imidazolamines (histamine), aliphatic amines (putrescine, cadaverine, agmatine), and aromatic amines (tyramine, phenylethylamine). They can be naturally present or can be formed by intentionally added or contaminating microorganisms, or by thermal treatment; therefore they can be used as a quality control criterion.[1,2,4]

Amines in plants can exist either free or conjugated, *i.e.*, bound to other food components. The conjugated amines can be bound to small molecules such as phenolic acids or to macromolecules, such as nucleic acids, proteins, and to the cell wall.[5] Amines conjugated to hydroxycinnamic acids (hydroxycinnamic acid amides) have been observed in some plants families.[5–7] Conjugated polyamines can accumulate in reproductive organs and seeds and can play roles in plant defense responses to pathogens. They are also associated with tolerance to abiotic stress, including drought, salinity, and mineral deficiency.[5–8]

28.3 Synthesis of Bioactive Amines

As indicated in Figure 28.1, there are several pathways for the synthesis of amines. The biogenic amines histamine, tyramine, tryptamine, phenylethylamine, agmatine, and cadaverine occur through decarboxylation of the precursor amino acids histidine, tyrosine, tryptophan, phenylalanine, arginine, and lysine, respectively.[1] In the synthesis of the indolamine serotonin, tryptophan is transformed by tryptophan-5-hydrolase into 5-hydroxytryptophan, which is decarboxylated by aromatic amino acid decarboxylase into serotonin (5-hydroxytryptamine). Also, tryptophan is decarboxylated by tryptophan decarboxylase into tryptamine, which can be transformed into serotonin by tryptamine-5-hydroxylase. Melatonin is synthesized from serotonin by *N*-acetyltransferase and *N*-acetylserotonin-*o*-methyltransferase, with *N*-acetylserotonin as an intermediate.[1,9–11] Prerequisites for the formation of biogenic amines are availability of free amino acids and the presence and activity of decarboxylases. In foods, free amino acids can be readily available or may be released from proteins as a result of proteolytic activity or thermal degradation. Decarboxylases can be naturally present but can also be provided by microorganisms which are naturally present, or are introduced purposely as starter cultures or by contamination before, during, and after processing.[1]

Polyamines synthesis is a more complex process (Figure 28.1). Ornithine is the main precursor of polyamines, producing putrescine, which is an obligate intermediate, by ornithine decarboxylase. However, arginine and citrulline can also be precursors of putrescine. Spermidine and spermine are formed by subsequent addition of aminopropyl moieties, derived from methionine through S-adenosylmethionine (SAM), to putrescine and to spermidine, respectively.[1,2,12]

28.4 Functions of Bioactive Amines in Plants

Bioactive amines participate in important metabolic and physiological functions in living organisms. In plants, polyamines are associated with many developmental processes, such as cell division, root formation, embryogenesis, organogenesis, floral initiation, leaf senescence, fruit development, and fruit ripening. Polyamines are also important in membrane stabilization by acting as free radical scavengers, inhibiting phospholipids movement and reducing lipoxygenase activity. Polyamines and ethylene synthesis share the same precursor – SAM – and, therefore, they have competitive effects on fruit development and ripening in many plants.[13,14] Polyamines are also involved in the control and modulation of stress tolerance in plants. They can increase plant survival under abiotic stresses, such as salt, drought, flooding, chilling, osmotic, acid, nutrient imbalance, radiation-induced oxidation, and heavy metal stress, as well as infection by pathogenic fungi and viruses.[2,4,6,7,12,15]

Table 28.1 Chemical structures and properties of the most common amines found in coffee.[1,3]

Amine	Chemical name	Chemical structure	Molecular formula	Molecular mass (g mol^{-1})
Polyamines				
Spermidine	*N*-(3-aminopropyl)-1,4-butane-diamine; *N*-(γ-amino-propyl) tetramethylenediamine	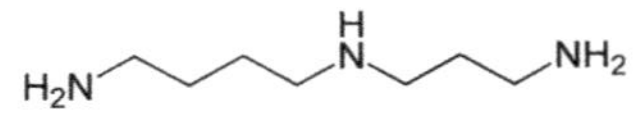	$C_7H_{19}N_3$	145.24
Spermine (gerontine, musculamine, neuridine)	*N*,*N*′-bis(3-aminopropyl)-1,4-butane-diamine; *N*,*N*′-bis(3-aminopropyl) tetramethylenediamine	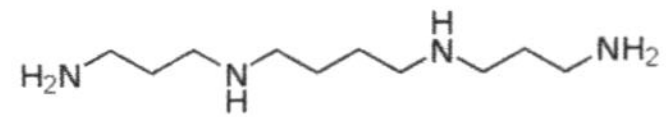	$C_{10}H_{26}N_4$	202.34
Biogenic amines				
Diamines				
Putrescine	1,4-Butane-diamine; tetramethylenediamine	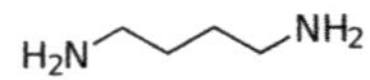	$C_4H_{12}N_2$	88.15
Cadaverine	1,5-Pentanediamine; pentamethylenediamine	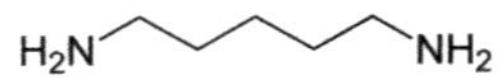	$C_5H_{14}N_2$	102.18
Indolamines				
Serotonin	5-Hydroxy-tryptamine; 5-hydroxy-3-(β-amino-ethyl) indol	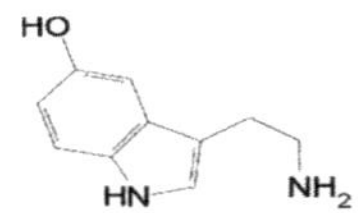	$C_{10}H_{12}N_2O$	176.21
Melatonin	*N*-acetyl-5-methoxytryptamine; 5-methoxy-*N*-acetyltryptamine	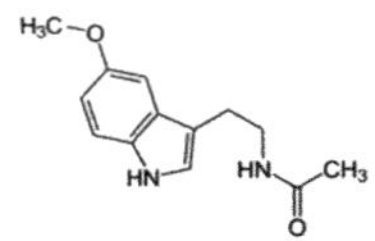	$C_{13}H_{16}N_2O_2$	232.28
Tryptamine	1H-indole-3-ethanamine; 3-(2-aminoethyl) indole	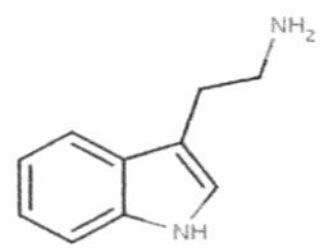	$C_{10}H_{12}N_2$	160.21

Tetramine				
Agmatine	4-(aminobutyl) guanidine; 1-amino-4-guanidobutane	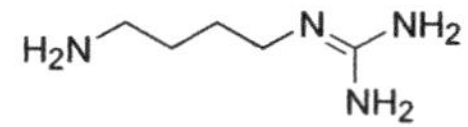	$C_5H_{14}N_4$	130.19
Aromatic amines				
Tyramine	4-(2-aminoethyl)phenol; 2-*p*-hydroxy phenyl-ethyl amine	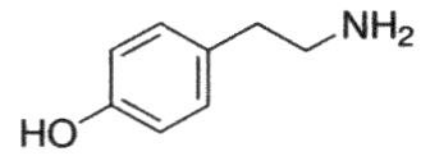	$C_8H_{11}NO$	137.18
Phenylethylamine	Benzene ethanoamine	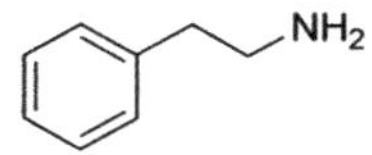	$C_8H_{11}N$	121.18
Imidazolamine				
Histamine	1H-imidazole-4-ethanamine; 2-(4-imidazolyl)-ethylamine	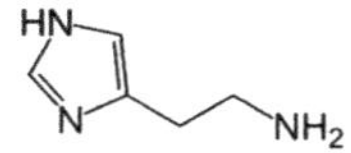	$C_5H_9N_3$	111.15

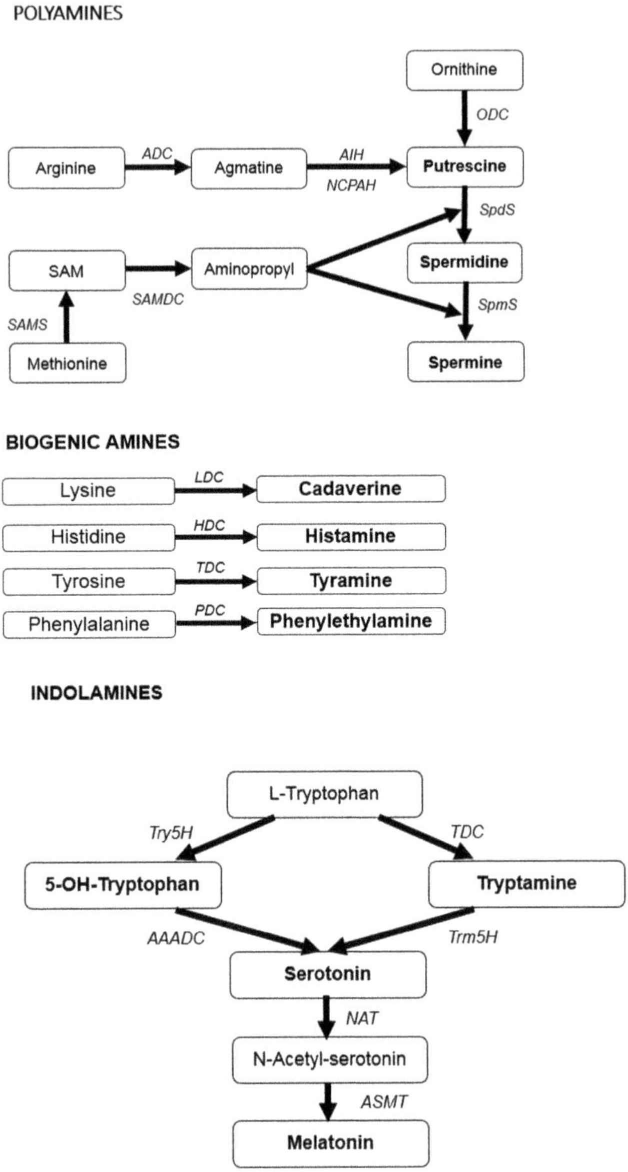

Figure 28.1 Synthesis of bioactive amines of relevance in coffee: polyamines, biogenic amines, and indolamines (polyamines: *ODC* – ornithine decarboxylase, *ADC* – arginine decarboxylase, *AIH* – agmatine iminohydrolase, *NCPAH* – *N*-carbamoylputrescine amidohydrolase, *SpdS* – spermidine synthase, *SpmS* – spermine synthase, *SAMS* – S-adenosylmethionine synthase, *SAMDC* – S-adenosylmethionine decarboxylase; biogenic amines: *LDC* – lysine decarboxylase, *HDC* – histidine decarboxylase, *TDC* – tyrosine decarboxylase, *PDC* – phenylalanine decarboxylase; indolamines: *Try5H* – tryptophan-5-hydroxylase, *TDC* – tryptophan decarboxylase, *AAADC* – aromatic L-amino acid decarboxylase, *Trm5H* – tryptamine-5-hydroxylase, *NAT* – *N*-acetyltransferase, *ASMT* – *N*-acetylserotonin-*o*-methyltransferase).[1,9–11]

Biogenic amines are also present in plants. Some are inherent to the plant, and others are synthesized as protection against predators. Putrescine is a required precursor in the biosynthetic pathway of the polyamines.[1,2] Agmatine in plants indicates another possible route for the synthesis of polyamines, even though the most common pathway is *via* ornithine (Figure 28.1).[1,9,11,12] Serotonin has been associated with various physiological functions in plants, among them, energy acquisition, xylem sap exudation, ion permeability, growth regulation, modulation of reproductive development, control of root and shoot organogenesis, maintenance of plant tissues, flowering, seed germination, ripening or maturation of plant organs, delay of senescence, seasonal cycles, response to light, protection from and adaptation to environmental changes, detoxification of excess ammonia, and resistance to pathogenic infections. Serotonin also shows antioxidant activity by scavenging reactive oxygen species. Moreover, it prevents the accumulation of tryptamine and maintains the reducing potential of cells, thereby delaying senescence.[16–18] Tryptamine is involved in auxin biosynthesis, *e.g.*, production of the plant growth hormone indole acetic acid. It can also be used as a precursor of a wide variety of alkaloids, including the tricyclic β-carbolines. Furthermore, it can deter insects.[1,16,19,20]

Biogenic amines conjugated to phenolic compounds are known to perform defensive functions in plants after wounds and pathogenic attack. They also show antimicrobial activity. For example, coumaroylserotonin and feruloylserotonin in rice have antifungal activity and are involved in plant defense responses against pathogen infection.[9,16,21] Conjugated polyamines can play roles in plant defense responses to pathogens. They are associated with tolerance to abiotic stress, including drought, salinity, and mineral deficiency.[5–8,14]

28.5 Methods for the Analysis of Bioactive Amines

The analysis of bioactive amines in coffee and foods is an analytical challenge due to diversity of chemical structures, high polarity, possible interactions with other components, and complexity of the coffee matrix. Also, they have low volatility, lack chromophores, and do not exhibit strong fluorescence. A comprehensive review has been published on methods for the chromatographic analysis of amines.[22] It provides detailed information on amines analysis, including extraction, clean-up, purification, concentration, derivatization, separation, and quantification. Sample preparation will depend on the method used for the separation and quantification of the amines. The most widely used method for quantification is based on high performance liquid chromatography – HPLC – with fluorescence or ultraviolet (UV) detection, after pre- or post-column derivatization of the amines. This technique is precise and sensitive. Gas chromatography has also been widely used for the analysis of amines in the clinical and environmental fields. This technique is simple. Although a derivatization step is also needed, it has high resolving power and sensitivity, short analysis time, and lower cost compared to HPLC.[22,23]

Most of the published studies on amines in coffee employed HPLC analysis. Extraction of amines from the coffee matrix is accomplished by means of 5% trichloroacetic acid;[8,24–31] however, other extractors have also been used, including perchloric acid[14] and borate buffer, pH 9.5.[32] Purification can be accomplished prior to derivatization and HPLC analysis with an ion-pair – bis(2-ethylhexyl) phosphate.[33] The most widely used HPLC method for amines analysis is ion-pair reversed phase separation followed by post-column derivatization with *o*-phthalaldehyde and fluorometric detection at 340 nm of excitation and 445 nm of emission.[8,25–28] This approach benefits from automatic online derivatization and no extra sample manipulation, avoiding imprecision and long sample preparations.[22] Reverse phase HPLC and UV detection at 254 nm has also been used after derivatization of the amines. The most commonly used derivatization reagents are dansyl chloride[8,29,31,32] and benzoyl chloride.[14,30]

The conjugated amines are first hydrolyzed under acid conditions prior to analysis of the free amines. Two fractions are obtained, the acid-soluble and acid-insoluble, being released or not, respectively, upon 12 M HCl hydrolysis for 18 h at 110 °C.[8,34]

28.6 Bioactive Amines During Coffee Growth and Development

During germination and fruit formation in both *C. arabica* and *C. canephora* species, putrescine was the only amine present at significant concentrations in every part of the developing seed: the hypocotyls and the seedling roots only contained putrescine; the cotyledonary leaf of *C. arabica* beans also contained spermidine and spermine, whereas the cotyledonary leaf of *C. canephora* contained spermidine but not spermine. Coffee flowers of both species contained spermine, spermidine, and cadaverine, but not putrescine; spermine was the prevalent amine followed by cadaverine and spermidine.[14]

The indolamines serotonin and melatonin were investigated in seedlings and in *in vitro* cultures of *C. canephora* and their concentrations varied, with high levels in callus and somatic embryos, but lower in further growth stages.[35]

Throughout fruit development, *C. arabica* beans contained putrescine, spermidine, spermine, cadaverine, and tryptamine. *C. canephora* beans contained the same amines and also tyramine. One of the varieties of *C. canephora* investigated did not contain cadaverine. Overall, spermidine, putrescine, and tryptamine increased continuously throughout development and ripening of coffee species; the concentrations of cadaverine increased early in development, followed by a decrease. The concentrations of spermine increased continuously in *C. canephora* but decreased in *C. arabica* during fruit ripening.[14]

28.7 Bioactive Amines in Green Coffee

Several studies have investigated the occurrence and concentration of free amines in green coffee (Table 28.2). The first amines identified in green coffee were the polyamines spermine and spermidine and the diamine putrescine. Putrescine was the prevalent amine, followed by spermidine and spermine in *C. arabica* var Mundo Novo.[24] Later on, it was observed that serotonin was the prevalent amine in green coffee followed by putrescine, spermidine, and spermine in *C. arabica* var Red Catuai.[25] Other researchers continued investigations on amines in coffee and from that time on, a larger number of amines (up to 11) have been identified in different coffee varieties and cultivars, as well as several commercial blends.

As summarized in Table 28.3, the types and occurrence of amines vary widely among studies.[8,14,17,24–29] Spermine, spermidine, and putrescine are always present, whereas other amines have been only detected in some samples. Results suggest that spermine, spermidine, and putrescine are inherent to green coffee. Agmatine and phenylethylamine were not detected in any sample. Cadaverine, histamine, and tyramine have been detected sporadically. Every evaluated sample contained either serotonin or tryptamine, or both. Melatonin has been only investigated in recent studies.[35,36]

Spermine, spermidine, putrescine, and tryptamines (indolamines) are typical of green coffee. The presence of the polyamines spermine and spermidine in coffee is expected as they are widely distributed in plants and play important roles in cellular division and growth. The presence of putrescine is also expected since it is a precursor of spermidine and spermine. The polyamines are important in plant development and growth and also in resistance to abiotic stress.[1,2,7]

The presence of indolamines such as tryptamines and derivatives (serotonin, tryptamine, melatonin, and N^{β}-alkanoyl-5-hydroxytryptamine), as well as their precursors (tryptophan and 5-hydroxytryptophan), has been reported in coffee wax, leaves, and beans[10,35–38] and also in many other plants.[39,40] As mentioned previously, they play important roles, serving as precursors of the plant growth hormone indole acetic acid and also the tricyclic β-carboline alkaloids harman and norharman.[11,36,41] Several studies *in vitro* and with other food items such as rice have indicated that indolamines play important roles in the protection of plants. They are potent radical scavengers, antioxidants, and radioprotective agents; they may act as phytoalexins; and they can also control cell wall strength as a mechanical barrier against pathogens and activate the intracellular signaling of immune responses.[9,40,42] In addition, melatonin can stabilize cellular redox status and protect tissues against reactive oxygen and nitrogen species which accumulate under stress. Melatonin can also improve phytoremediation efficiency of plants against different pollutants, seed germination, seedling/plant growth, and crop production.[11]

As indicated in Table 28.2, the contents of free amines in green coffee vary widely in the literature; however, putrescine is the most abundant amine in green coffee, followed by spermidine and spermine. Serotonin and

Table 28.2 Contents of free bioactive amines in green coffee.[b,c]

Coffee (n)	Amines (means ± standard deviation or minimum–maximum) in mg 100 g^{-1}										
	SPD	SPM	PUT	AGM	SRT	MLN	TRM	CAD	TYM	PHM	HIM
C. arabica Mundo Novo[a] [24]	1.56 ± 0.18	0.81 ± 0.18	4.26 ± 0.52	—	—	—	—	—	—	—	—
C. arabica Red Catuai[25]	0.60 ± 0.07	0.44 ± 0.08	1.03 ± 0.09	nd	1.13 ± 0.15	—	nd	nd	nd	nd	nd
C. arabica (11)[a] [8]	9.0 ± 3.0	5.5 ± 3.3	47.9 ± 11	—	2.5 ± 0.7	—	—	0.2 ± 0.2	0.2 ± 0.1	—	—
C. canephora (19)	8.3 ± 2.8	6.0 ± 3.8	11.1 ± 3.6	—	2.1 ± 0.8	—	—	0.4 ± 0.2	3.1 ± 4.6	—	—
C. arabica (Brazil)[26]	1.13–1.71	0.97–1.79	6.32–10.1	nd	nd	—	nd–0.3	nd	nd		nd–0.15
C. arabica (Brazil)[28]	1.99 ± 0.41	2.26 ± 0.45	7.65 ± 1.56	nd	0.17 ± 0.30	nd	nd	nd	nd		nd
C. canephora[27]	0.37 ± 0.06	0.15 ± 0.03	1.00 ± 0.31	nd	0.49 ± 0.08	—	0.07 ± 0.01	nd	0.34 ± 0.06	nd	nd
C. arabica[a] [14]	0.70 ± 0.02	0.65 ± 0.00	0.43 ± 0.03	—	—	—	0.11 ± 0.01	0.10 ± 0.00	nd	—	—
C. canephora (2)	0.48–0.92	0.79–0.82	1.21–1.56	—	—	—	0.10–0.22	0.08–0.14	0.08–0.11	—	—
C. arabica Topazio[29]	0.82 ± 0.08	0.72 ± 0.06	5.38 ± 0.26	—	0.15 ± 0.01	—	—	0.03 ± 0.00	0.04 ± 0.03	—	0.04 ± 0.00
C. arabica[a] [17]	—	—	—	—	1.25 ± 0.08	0.68 ± 0.04	—	—	—	—	—
C. canephora	—	—	—	—	1.05 ± 0.06	0.58 ± 0.08	—	—	—	—	—
C. canephora[a] [36]	—	—	—	—	9.8 ± 0.5	11.5 ± 0.6	—	—	—	—	—

[a]Dry weight basis.
[b]n = number of samples, – not investigated, nd – not detected.
[c]SPD – spermidine, SPM – spermine, PUT – putrescine, AGM – agmatine, SRT – serotonin, MLN – melatonin, CAD – cadaverine, TYM – tyramine, TRM – tryptamine, PHM – phenylethylamine, HIM – histamine.

Table 28.3 Contents of free and conjugated bioactive amines in green coffee.[b,c]

Coffee (n)	Amines (means ± standard deviation or minimum–maximum) in mg 100 g^{-1}							
	SPD	SPM	PUT	AGM	SRT	TRM	CAD	TYM
C. arabica (11)[a 8]								
Free	0.90 ± 0.30	0.55 ± 0.33	4.79 ± 1.10	—	0.25 ± 0.07	—	0.02 ± 0.02	0.02 ± 0.01
Acid-soluble	0.35 ± 0.27	0.20 ± 0.26	0.37 ± 0.28	—	0.08 ± 0.06	—	0.03 ± 0.02	0.02 ± 0.02
Acid insoluble	0.05 ± 0.02	0.58 ± 0.17	0.29 ± 0.08	—	nd	—	0.01 ± 0.01	0.04 ± 0.03
C. canephora (19)[a 8]								
Free	0.83 ± 0.28	0.60 ± 0.38	1.11 ± 0.36	—	0.21 ± 0.08	—	0.04 ± 0.02	0.31 ± 0.46
Acid-soluble	0.26 ± 0.30	0.36 ± 0.34	0.32 ± 0.24	—	0.12 ± 0.09	—	0.04 ± 0.03	0.24 ± 0.32
Acid insoluble	0.04 ± 0.02	0.52 ± 0.17	0.06 ± 0.03	—	nd	—	0.01 ± 0.01	0.09 ± 0.08
***C. canephora* green** (3)[a 27]								
Free	0.37 ± 0.02	0.15 ± 0.03	1.00 ± 0.31	nd	0.49 ± 0.08	0.07 ± 0.01	nd	0.34 ± 0.06
Acid-soluble	0.19 ± 0.07	0.25 ± 0.04	0.56 ± 0.24	nd	nd	nd	nd	nd

[a]Phenylethylamine and histamine were not detected in any sample.
[b]n = number of samples; – not investigated; nd – not detected.
[c]SPD – spermidine, SPM – spermine, PUT – putrescine, AGM – agmatine, SRT – serotonin, TRM – tryptamine, CAD – cadaverine, TYM – tyramine.

melatonin are present at intermediate concentrations whereas tryptamine, cadaverine, and histamine are present at low concentrations. In general, tyramine concentrations are higher in *C. canephora* compared to *C. arabica*.[8,14] Because of that, the amount of tyramine has been suggested[8] as a marker to discriminate between the two coffee species. However, as blending both species is a common practice in industry, it could be of limited value to discriminate these species in commercial coffee, because tyramine is present in both species.

In contrast to the wide availability of data regarding free amines, conjugated amines have been less investigated in coffee (Table 28.3), even though they have essential roles in modulation of plant responses to stress.[6,7] Amines in coffee are usually present at higher concentrations in the free form compared to conjugated forms, either acid-soluble or acid-insoluble[8,34] Conjugated spermine, spermidine, putrescine, cadaverine, and tyramine were detected in every sample analyzed; however, conjugated serotonin was only detected in the acid-soluble fraction. Serotonin conjugates in coffee bean wax have been described as alkanoyl derivatives, such as stearoyl, arachidoyl, behenoyl, and lignoceroyl.[37]

As can be seen in Tables 28.2 and 28 3, the occurrence and concentration of amines in green coffee vary widely among species, varieties, cultivation conditions, such as climate, soil type, mineral deficiency, stress, and physiological stage (ripening), processing and storage,[8] similar to what happens to other coffee chemical components. However, some conclusions can be drawn from literature data available. Among the 11 free amines studied in green coffee, putrescine, spermidine, spermine, and indolamines (serotonin and/or tryptamine) were detected in every sample, as both free and conjugated forms. Agmatine and phenylethylamine were not detected in any sample. Cadaverine and tyramine were present in a few samples at low concentrations, both free and conjugated. Histamine was found in the free but not the conjugated form. In recent studies, free melatonin was detected in green coffee, contributing to the pool of indolamines.[35,36]

28.8 Influence of Post-harvest Processing on Bioactive Amines in Coffee

Few studies were undertaken regarding the influence of coffee processing on bioactive amines and overall coffee quality. The influence of post-harvest processing methods (wet and dry processes) on amine concentrations in *C. arabica* was investigated.[8,29] Studies on amines in green coffee available commercially indicated that wet-processed coffee had higher free putrescine and tyramine and higher acid-soluble conjugated spermidine concentrations compared to dry-processed coffee. The higher concentrations of biogenic amines in wet-processed coffee were associated with fermentation, which takes place to remove the remaining adherent mucilage after pulping, incorporating the action of naturally present microorganisms and coffee enzymes.

Cluster analysis discriminated wet- from dry-processed coffees by the higher concentrations of free putrescine and spermine, and lower spermidine concentrations.[8] When seven free amines were analyzed in unripe *C. arabica* cv. Topazio submitted to dry- and wet-processing with either wet or dry resting periods, spermine and spermidine concentrations were lower in depulped compared to natural coffee and cadaverine concentrations were affected by dry- and wet-processing.[29] Based on these studies, it is clear that post-harvest processing affects the concentrations of amines, mainly spermine, spermidine, putrescine, histamine, and cadaverine. However, the impact of these changes on coffee quality is still to be determined.

28.9 Influence of Bean Quality on Bioactive Amines

One of the most important and commonly used criteria to evaluate the quality of coffee beans is cup quality (flavor), even though other characteristics are also important, such as presence of defects, bean size, color, shape, processing method, and crop year. As indicated in Table 28.4, no significant differences in concentrations of polyamines and putrescine were found in Brazilian green coffee samples classified as soft (high quality, low acidity, mellow sweetness, pleasant mouth feel) and rio (low quality, accentuated iodine, and medicine-like flavor).[24,26] However, higher concentrations of histamine and tryptamine were found in the coffee classified as rio compared to the one classified as soft.[26]

Comparison of good quality and defective green beans (Table 28.4) indicated some significant differences among them.[28] All defective beans

Table 28.4 Contents of free bioactive amines in *C. arabica* of different cup qualities and defective beans.[c,d]

	Amines (means ± standard deviation or minimum–maximum) in mg 100 g^{-1}					
Coffee type	SPD	SPM	PUT	SRT	TRM	HIM
Cup quality[a] [24]						
Soft	1.63 ± 0.26	0.83 ± 0.13	4.35 ± 0.72	—	—	—
Rio	1.50 ± 0.00	0.80 ± 0.08	4.18 ± 0.32	—	—	—
Cup quality[b] [26]						
Soft	1.13	1.16	6.62	nd	nd	nd
Rio	1.36	1.37	9.85	nd	0.3	0.15
Beans[b] [28]						
Good quality	1.99 ± 0.41	2.26 ± 0.45	7.65 ± 1.56	0.52	nd	nd
Black	0.42	<0.37	0.88 ± 0.16	0.39	nd	0.44 ± 0.08
Immature	1.59 ± 0.07	1.89 ± 0.17	5.39 ± 0.28	0.42	0.57 ± 0.06	0.69 ± 0.00
Sour	1.24 ± 0.36	1.74 ± 0.45	4.21 ± 1.09	nd	<0.37	1.02 ± 0.18

[a]Dry weight basis.
[b]Cadaverine, tyramine, phenylethylamine, and agmatine were not detected in any sample.
[c]*n* = number of samples, - not investigated, nd - not detected.
[d]SPD - spermidine, SPM - spermine, PUT - putrescine, SRT - serotonin, TRM - tryptamine, HIM - histamine.

contained histamine whereas sour and immature beans contained also tryptamine, which was not present in good quality coffee. Defective beans also contained lower putrescine concentrations. The concentrations of spermidine and spermine were different for black green beans, which had the most atypical amine profile compared to the other defective green coffees. Based on these studies, histamine could be a marker of coffee quality as it was only detected in defective compared to good quality green coffee beans.

28.10 Influence of Coffee Roasting on Bioactive Amines

Several studies[8,14,17,24,25,28,30,31] investigated the influence of roasting on the concentrations of free amines in coffee (Table 28.5). Different roasting conditions were used, with temperatures varying from 160 to 300 °C and times from 6 to 30 minutes. In general, the concentrations of free amines were significantly reduced during roasting at rates affected by the type of amine and roasting conditions. Changes in concentrations of some amines during roasting were also affected by their release from conjugated forms, which are more resistant to the roasting process.[5,34]

The formation of agmatine during roasting of *C. arabica*[25] and *C. canephora*[34] beans has been reported. Both studies suggested that agmatine was formed by thermal decarboxylation of the amino acid arginine. In different studies, serotonin[5,25,36] and melatonin[36] concentrations did not change during roasting, suggesting that indolamines are less sensitive to heat or that they can be formed during roasting. In fact, serotonin can be formed by decarboxylation of 5-hydroxytryptophan.[9]

The degradation of several amines, including spermidine, spermine, putrescine, histamine, and tryptamine during roasting takes place mainly during the drying stage, which takes place in initial roasting times.[26] This explains the significant loss of amines even under mild roasting conditions. The loss of amines during roasting occurs regardless of coffee quality.[26,28]

Amine concentrations in commercially available roasted coffee are presented in Table 28.5. Even though most amines are heat sensitive and, therefore, tend to degrade during roasting, high concentrations of putrescine ($\leq$26.1 mg 100 g^{-1}), cadaverine ($\leq$0.3 mg 100 g^{-1}), histamine ($\leq$16.6 mg 100 g^{-1}), tyramine ($\leq$16.1 mg 100 g^{-1}), and spermidine ($\leq$14.4 mg 100 g^{-1}) were found in commercial roasted coffee samples. Intermediate concentrations of spermine and phenylethylamine were also found.[30,31] Some of these amines reported in roasted coffee are not typical of green coffee, such as phenylethylamine. Other amines found are typical of poor quality green coffee, histamine for example, or of bacterial contamination, like cadaverine and putrescine. Since these amines are usually degraded during roasting, their occurrence in coffee is suggestive of contamination after roasting, during transportation or storage, or adulteration of coffee with other products; this deserves further investigation. Another possibility for the presence of phenylethylamine is its production as a result of lipid oxidation.[43]

Based on the information available regarding the fate of amines during roasting, only serotonin, melatonin, and agmatine would be expected to be present in the beverages. However, several amines known to be degraded during roasting, including spermine, spermidine, histamine, putrescine, and tryptamine were detected in the coffee beverages.[5,24,26,28] No information is available regarding the impact of roasting on other amines, such as tyramine and phenylethylamine. Some amines found in roasted coffee could result from a very mild roasting process or due to contamination of the beans after roasting, during commercialization and storage.

28.11 Other Factors Affecting Bioactive Amines in Coffee

The influence of the type of instant coffee and of the brewing or infusion method was investigated as indicated in Table 28.6. When comparing the concentration of amines in commercially available instant coffee including regular, regular organic, and decaffeinated coffee, the same amines were detected – serotonin, tyramine, and cadaverine – in the three types of instant coffee and no significant difference was found among them.[27] However, some organic instant coffee contained high cadaverine concentrations (up to 408 μg 50 mL^{-1} coffee beverage). Regular and decaffeinated instant coffee did not differ regarding amines,[27,31] although higher cadaverine levels were found in one brand of organic instant coffee.

28.12 Bioactive Amines in Coffee Beverages

Brewing and coffee preparation technique, including mocha, espresso, and instant coffees, significantly affected amines levels.[31] Mocha was the product with higher concentrations of amines; however, it is difficult to compare results because of the different coffees used in the preparation of these beverages. Turkish coffee, which is prepared by heating ground roasted coffee with ultra-pure water in a Turkish coffee machine, showed very high concentrations of serotonin, cadaverine, tyramine, tryptamine, and phenylethylamine, suggesting that this procedure extracts more amines from the coffee grains.[30] However, to be able to ascertain the influence of brewing processes, decaffeination, and other parameters on coffee components, the same coffee had to be used.

Concentrations of bioactive amines in 50 mL of coffee beverages, provided by or calculated from literature data,[17,27,30,31] are shown in Table 28.6. Several amines were detected in the beverages, with the exception of agmatine. Cadaverine was present in every evaluated beverage, at concentrations varying from 0.4 to 453 μg 50 mL^{-1}. Serotonin and tyramine were present in most of the samples at levels up to 677 and 985 μg 50 mL^{-1}, respectively.

Table 28.5 Contents of free bioactive amines in roasted coffee.[b,c]

Coffee (n)	Amines (mean ± standard deviation or minimum–maximum) in mg 100 g^{-1}										
	SPD	SPM	PUT	AGM	SRT	MLN	TRM	CAD	TYM	PHM	HIM
***C. arabica* Mundo Novo**[a 24]											
240 °C/10 min	nd	nd	0.20 ± 0.05	—	—	—	—	—	—	—	—
240 °C/12 min	nd	nd	nd	—	—	—	—	—	—	—	—
***C. arabica* Red Catuai**[25]											
300 °C/6 min	0.12 ± 0.02	nd	nd	nd	0.16 0.05	—	nd	nd	nd	nd	nd
300 °C/12 min	0.20 ± 0.01	nd	nd	0.12 ± 0.05	0.29 ± 0.22	—	nd	nd	nd	nd	nd
(160–220 °C/14 min)[a 8]											
***C. arabica* (11)**	0.1 ± 0.1	0.12 ± 0.06	0.10 ± 0.09	—	0.23 ± 0.12	—	—	nd	nd	—	—
***C. canephora* (19)**	0.1 ± 0.1	0.18 ± 0.08	0.13 ± 0.12	—	0.33 ± 0.08	—	—	nd	nd	—	—
(200 °C/30 min)[28]											
C. arabica	nd	nd	nd	nd	Traces	—	nd	nd	nd	nd	nd
C. canephora[a 14]	0.02 ± 0.00	0.01 ± 0.00	0.09 ± 0.00	—	—	—	nd	nd	0.02 ± 0.00	—	—
C. arabica[a 17]	—	—	—	—	0.87 ± 0.04	0.96 ± 0.08	—	—	—	—	—
C. canephora[a 17]	—	—	—	—	0.73 ± 0.05	0.80 ± 0.09	—	—	—	—	—
C. arabica											
Market (10)[30]	nd	nd	0.36–1.52	—	5.75–17.1	—	0.30–3.79	1.36–7.51	2.25–9.96	nd–2.28	nd
Market (9)[31]	4.97–15.2	nd–7.81	7.8–22.2	—	1.08–6.15	—	—	1.27–20.3	3.64–16.1	3.14–8.62	3.39–16.6
Pod or capsule[31]											
Regular (4)	10.1–17.4	nd–8.51	8.24–26.1	—	1.06–2.66	—	—	2.61–7.57	5.28–7.91	3.09–6.41	1.99–5.94
Decaffeinated (2)	7.19–9.27	2.87–5.47	1.27–5.12	—	1.42–2.81	—	—		3.67–4.51	2.72–3.96	1.39–4.29

[a]Dry weight basis.
[b]n = number of samples, – not investigated, nd – not detected.
[c]SPD – spermidine, SPM – spermine, PUT – putrescine, AGM – agmatine, SRT – serotonin, MLN – melatonin, TRM – tryptamine, CAD – cadaverine, TYM – tyramine, PHM – phenylethylamine, HIM – histamine.

Table 28.6 Contents of free bioactive amines (μg 50 mL^{-1}) in coffee beverages.[b,c]

Coffee beverage (*n*)	Amines (means ± standard deviation or minimum–maximum) in μg 50 mL^{-1}									
	SPD	SPM	PUT	SRT	MLN	TRM	CAD	TYM	PHM	HIM
C. arabica										
Turkish (10)[30]	nd	nd	25–77	186–677	—	nd–1009	73–453	nd–985	nd–250	nd
Instant[a 27]										
Regular (10)	nd	nd	nd	nd–7.8	—	nd	0.2–5.0	0.3–2.3	nd	nd
Regular organic (2)	nd	nd	nd	1.7–2.2	—	nd	0.9–408	0.3–0.6	nd	nd
Decaffeinated (6)	nd	nd	nd	nd–5.5	—	nd	0.4–6.0	1.3–4.9	nd	nd
C. arabica[17]	—	—	—	7.3 ± 0.6	3.9 ± 0.3	—	—	—	—	—
C. canephora[17]	—	—	—	4.0 ± 0.7	3.0 ± 0.6					
Mocha, regular (7)[31]	22.5–59	nd–97.5	31–113.5	10–45	—	—	14–92	32.5–94.5	43.5–90.5	30–81
Espresso[31]										
Regular (6)	27–90.5	nd–52.5	30.5–83	nd–29	—	—	15.5–30	12.5–59	11–52	11–26
Decaffeinated (2)	23–34.5	18.5	21.5–30	nd–15.5	—	—	9.5–14.5	14–15.5	10–27	19.5–42
Instant[31]										
Regular (2)	36–41.5	nd–27	42.5–46.5	nd–29.5	—	—	21–34.5	25.5–29	11.5–19	24–31
Decaffeinated (2)	30.5–38.5	nd–24	36.5–37	10.5–26.5	—	—	16.5–26	21.5–26	15–15.5	15.5–29.5

[a]Agmatine was not detected in any sample.
[b]*n* = number of samples; – not investigated; nd – not detected.
[c]SPD – spermidine, SPM – spermine, PUT – putrescine, AGM – agmatine, SRT – serotonin, MLN – melatonin, CAD – cadaverine, TYM – tyramine, TRM – tryptamine, PHM – phenylethylamine, HIM – histamine.

Melatonin was present in all the samples in which they were analyzed (3–4 μg 50 mL^{-1}). Tryptamine was only detected in Turkish coffee.[30] Spermidine, spermine, phenylethylamine, and histamine were detected sporadically. The profile and concentrations of amines varied widely among beverages, suggesting the influence of the type and origin of coffees and the brewing processes. Based on these results, amines from roasted coffee (Table 28.5) are extracted during the brewing process and therefore are present in the beverage.

28.13 Bioactive Amines as Markers of Coffee Quality

The presence or high concentrations of some amines can be reliable markers of quality. It can be useful to access coffee quality, help differ between defective and low quality beans, and also help determine coffee origin. Histamine is not usually present in good quality coffee, but it may be present in defective and in low quality green coffee beans;[26,28] therefore, it can help discriminate defective and other low quality green coffee. However, since histamine is degraded during roasting, it cannot be used as a quality indicator in roasted coffee.

The concentrations of putrescine and tyramine in green coffee can be used to help discriminate between coffee species and coffee origin. Putrescine was suggested as an index to discriminate green *C. arabica* from *C. canephora* coffees, as the former usually contain higher concentrations of putrescine; and tyramine content was proposed as an origin marker for green Angolan *C. canephora* beans.[8] Tyramine concentrations can also help discriminate between coffees species as they are present at higher concentrations in *C. canephora*. However, these descriptors can be used for green but not for roasted coffee due to changes in amines concentrations during roasting. The origins of coffee – Asia, South America, and Africa – were successfully discriminated combining several analytical data with multivariate analysis, among them protein, flavor profile, carbohydrate, monosaccharide, and the amines tryptamine and spermine; the concentrations of amines along with flavor profile could discriminate South American from Asian and African coffees by flavor components (high) and amines (low).[32]

The presence of high concentrations of some of the biogenic amines – putrescine, cadaverine, histamine, phenylethylamine, and tyramine – could be indicative of contamination of coffee beans after roasting. In addition, the presence of phenylethylamine could indicate lipid oxidation, which could be associated with coffee stored for a long time.[43] The presence of the polyamines spermine and spermidine and of the indolamines serotonin, tryptamine, and melatonin in coffee would be desirable due to their antioxidant activity, which would increase coffee bean shelf-life[1] and would also be beneficial to human health (see *Coffee: Consumption and Health Implications*). Histamine and tyramine concentrations in the coffee beverages can be associated with adverse effects to human health and this will be discussed in *Coffee: Consumption and Health Implications*.

Putrescine and cadaverine are reported to have unpleasant and putrid odors;[27] therefore, their presence at high concentrations in coffee could impact flavor in a negative way. Information regarding the sensory threshold of these amines in coffee is not available.

28.14 Concluding Remarks

Despite the important roles of bioactive amines in food quality and human health, few studies are available on coffee and much still needs to be ascertained. So far, it is clear that putrescine and serotonin are the major amines found in green coffee, followed by melatonin, spermidine, and spermine. These amines are present in coffee either free or conjugated to alkanoyl or phenolic acids, proteins, and nucleic acids.

The profile and concentrations of amines in coffee are affected by genetic factors like species (*C. arabica* x *C. canephora*), varieties or cultivars within the species, cultivation conditions, post-harvesting processing, bean quality (presence of defects), roasting temperature and time, and brewing process. Tyramine is present at higher concentrations in green *C. canephora*, and putrescine in *C. arabica*, which could help discriminate between these species. Histamine has been identified in low cup quality or defective coffee and could be used as a quality criterion for green coffee. Roasting can reduce the concentrations of amines in coffee, even though some can still be found in roasted coffee. Serotonin, melatonin, and spermidine are most resistant to roasting. They can also be released from conjugated forms. In addition, serotonin and agmatine can be thermally produced from precursors. Even though roasting degrades most of the amines, roasted samples from the market may contain some. Further studies are needed to ascertain their sources.

Acknowledgement

The authors acknowledge Coordenação de Pessoal de Nível Superior – CAPES (Brasília, DF, Brazil), Conselho Nacional de Desenvolvimento Científico e Tecnológico – CNPq (Brasília, DF, Brazil), and Fundação de Amparo a Pesquisa do Estado de Minas Gerais – FAPEMIG (Belo Horizonte, MG, Brasil) for the support.

References

1. M. B. A. Gloria, Bioactive amines, in *Handbook of Food Science, Technology and Engineering*, ed. H. Hui and L. L. Nollet, Taylor & Francis, 2006, vol. 1, p. 1.
2. P. Kalac, *Food Chem.*, 2014, **161**, 27.
3. https://pubchem.ncbi.nlm.nih.gov/, Last accessed December, 2015.
4. M. A. Ali, E. Poortvliet, R. Stromberg and A. Yngve, *Food Nutr. Res.*, 2011, **55**, 5455.

5. S. Casal, E. Mendes, M. B. P. P. Oliveira and M. A. Ferreira, *Electron. J. Environ., Agric. Food Chem.*, 2005, **4**(5), 1063.
6. Q. Fariduddin, P. Varshney, M. Yusuf and A. Ahmad, *J. Plant Interact.*, 2013, **8**(1), 1.
7. M. R. Pathak, J. A. Teixeira da Silva and S. H. Wani, *GM Crops Food: Biotechnol. Agric. Food Chain*, 2014, **5**(2), 87.
8. S. Casal, E. Mendes, M. R. Alves, R. C. Alves, M. B. P. P. Oliveira and M. A. Ferreira, *J. Agric. Food Chem.*, 2004, **52**, 6188.
9. T. Fujiwara, S. Maisonneuve, M. Isshiki, M. Mizutani, L. Chen, H. L. Wong, T. Kawasaki and K. Shimamoto, *J. Biol. Chem.*, 2010, **285**(15), 11308.
10. A. C. C. L. Martins and M. B. A. Gloria, *Food Chem.*, 2010, **118**, 529.
11. K. M. Janas and M. M. Posmyk, *Acta Physiol. Plant.*, 2013, **35**, 3285.
12. F. Sánchez-Jiménez, M. V. Ruiz-Pérez, J. L. Urdiales and M. A. Medina, *Br. J. Pharmacol.*, 2013, **170**, 4.
13. C. De La Pena, R. M. Galaz-Avalos and V. M. Loyola-Vargas, *Mol. Biotechnol.*, 2008, **39**(3), 215.
14. V. Sridevi, P. Giridhar and G. A. Ravishankar, *Acta Physiol. Plant.*, 2009, **31**, 757.
15. C. Sudhakar, G. Veeranagamallaiah, A. Nareshkumar, O. Sudhakarbabu, M. Sivakumar, M. Pandurangaiah, K. Kiranmai, *et al.*, *Plant Cell Rep.*, 2015, **34**, 141.
16. D. Ly, K. Kang, J.-Y. Choi, A. Ishihara, K. Back and S.-G. Lee, *J. Med. Food*, 2008, **11**(2), 385.
17. A. Ramakrishna, P. Giridhar and G. A. Ravishankar, *Plant Signaling Behav.*, 2011, **6**(6), 800.
18. L. A. E. Erland, C. E. Turi and P. K. Saxena, *Biotechnol. Adv.*, 2016, **34**(8), 1347.
19. W. Runguphan, J. J. Maresh and S. E. O'Connor, *PNAS*, 2009, **106**(33), 13673.
20. T. A. Smith, *Phytochemistry*, 1977, **16**, 171.
21. J. B. Park, *Phytomedicine*, 2008, **15**(12), 1093.
22. C. Fernandes and M. B. A. Gloria, Bioactive amines, in *Handbook of Food Analysis*, ed. L. L. Nollet and F. Toldra, CRC Press, Boca Raton, 2015, vol. 2, ch. 15, p. 301.
23. H. Kataoka, *J. Chromatogr. A*, 1996, **733**(1), 19.
24. H. V. Amorim, L. C. Basso, O. J. Crocomo and A. A. Teixeira, *J. Agric. Food Chem.*, 1977, **25**, 957.
25. M. P. G. Cirilo, A. F. S. Coelho, C. M. Araujo, F. R. B. Gonçalves, F. D. Nogueira and M. B. A. Gloria, *Food Chem.*, 2003, **82**, 397.
26. S. D. Oliveira, A. S. Franca, M. B. A. Gloria and M. L. A. Borges, *Food Chem.*, 2005, **90**, 287.
27. T. M. L. Silveira, E. Tavares and M. B. A. Gloria, *J. Food Compos. Anal.*, 2007, **20**, 451.
28. A. L. S. Vasconcelos, A. S. Franca, M. B. A. Gloria and J. C. F. Mendonca, *Food Chem.*, 2007, **101**, 26.

29. E. C. Dias, R. G. F. A. Pereira, F. M. Borém, E. Mendes, R. R. Lima, J. O. Fernandes and S. Casal, *J. Agric. Food Chem.*, 2012, **60**, 4120.
30. O. Özdestan, *Food Res. Int.*, 2014, **61**, 167.
31. D. Restuccia, U. G. Spizzirri, O. I. Parisi, G. Cirillo and N. Picci, *Food Chem.*, 2015, **175**, 143.
32. M.-Y. Choi, W. Choi, J. H. Park, J. Lim and S. W. Kwona, *Food Chem.*, 2010, **121**, 1260.
33. S. Casal, M. B. P. P. Oliveira and M. A. Ferreira, *J. Liq. Chromatogr. Relat. Technol.*, 2002, **25**(16), 2535.
34. T. M. L. Silveira, W. P. Evangelista, J. A. Vignoli, D. G. Bassoli and M. B. A. Gloria, *Documentos Embrapa: 6, VI Simpósio de Pesquisa dos Cafés do Brasil*, Vitoria, ES, 2010, p. 1.
35. A. Ramakrishma, P. Giridhart, K. U. Sankar and A. Ravishankar, *Acta Physiol. Plant.*, 2012, **34**, 393.
36. A. Ramakrishna, P. Giridhar, K. U. Sankar and G. A. Ravishankar, *J. Pineal Res.*, 2012, **52**, 470.
37. P. Folstar, H. C. van der Plas, W. Pilnik, H. A. Schols and P. Melger, *J. Agric. Food Chem.*, 1979, **27**(1), 12.
38. M. Kele and R. Ohmacht, *J. Chromatogr. A*, 1996, **730**, 59.
39. R. C. Adão and M. B. A. Gloria, *Food Chem.*, 2005, **90**, 705.
40. F. A. Badria, *J. Med. Food*, 2002, **5**(3), 153.
41. S. Casal, Neuroactive beta-carbolines norharman and harman in coffee, in *Coffee in Health and Disease Prevention*, ed. V. Preedy, 2015, ch. 82, p. 737.
42. A. Galano, D. X. Tan and R. J. Reiter, *J. Pineal Res.*, 2013, **54**, 245.
43. R. Zamora, R. M. Delgado and F. J. Hidalgo, *Food Res. Int.*, 2012, **46**, 321.

CHAPTER 29

Melanoidins

ANA S. P. MOREIRA*[a,b], JOANA SIMÕES[a], CLÁUDIA P. PASSOS[a,c], FERNANDO M. NUNES[d], M. ROSÁRIO M. DOMINGUES[a,b] AND MANUEL A. COIMBRA[a,c]

[a]QOPNA, Department of Chemistry, University of Aveiro, 3810-193 Aveiro, Portugal; [b]CESAM, Department of Chemistry, University of Aveiro, 3810-193 Aveiro, Portugal; [c]REQUIMTE/LAQV, Department of Chemistry, University of Aveiro, 3810-193 Aveiro, Portugal; [d]CQ-VR, Chemistry Research Centre, Department of Chemistry, University of Trás-os-Montes e Alto Douro, 5001-801 Vila Real, Portugal
*E-mail: ana.moreira@ua.pt

29.1 Introduction

Melanoidins are brown high molecular weight (HMW) compounds containing nitrogen, which are, by definition, end products of Maillard reaction.[1–3] They are formed during the roasting of green coffee beans, as well as during the heat processing of other food products like bread,[4] meat,[5] or malt.[6] However, coffee brews, prepared by hot water extraction from roasted and ground coffee beans, along with bread, are considered the main sources of melanoidins in the human diet.[4]

This chapter gives an overview on what is known about coffee brew melanoidins, namely about their structural components, possible formation routes, biological activities, and potential health impacts, as well as the strategies that have been used for their quantitation, isolation, and purification.

Coffee: Production, Quality and Chemistry
Edited by Adriana Farah

Published by the Royal Society of Chemistry, www.rsc.org

29.2 Strategies for Quantitation, Isolation, and Purification of Coffee Melanoidins

Despite several attempts that have been made to elucidate the structure of coffee brew melanoidins, they remain largely unknown. It is due to the extreme complexity and diversity of structures and the consequent difficulty in isolation of pure melanoidin fractions.[2]

As a result of the scarce information about their structures, melanoidins are usually quantified by difference, subtracting the total percentage of known compounds from 100 percent. In this way, they are estimated to account for up to around 25% (w/w) of the dry weight of roasted coffee beans.[7,8] Coffee brew, prepared from ground and roasted coffee beans by hot water extraction, is one of the main sources of melanoidins in the human diet.[4] Melanoidins are estimated to account for up to around 29% (w/w) of the dry weight of coffee brew.[8] Due to their contribution to coffee's characteristic brown color, the content of melanoidins in coffee brew (and in coffee brew fractions) has also been evaluated using color dilution analysis,[9,10] or by measuring the absorption near 400 nm, particularly at 405 nm.[11–17] A direct relation exists between the melanoidin content in coffee brews and the degree of roast of the coffee beans. This can be measured by the $K_{mix\,405\,nm}$ value, which is the specific extinction coefficient at 405 nm.[11] The melanoidin browning index (MBI), calculated by dividing the $K_{mix\,405\,nm}$ values by the relative amount of unknown material, was proposed for coffee melanoidin determination.[18] A direct relationship between the MBI and the molecular weight (MW) was observed when comparing instant coffee fractions of different MW recovered by sequential ultrafiltration, showing that the higher the MW, the higher the brown color of the compounds present.[19]

Figure 29.1 summarizes different approaches that have been employed for isolation and purification (if any) of coffee brew melanoidins. In all approaches, melanoidins are first isolated taking advantage of their HMW. Most commonly, the high molecular weight materials (HMWM) of coffee brews are isolated by dialysis,[9,20] diafiltration,[11,13,14] or ultrafiltration,[10,19,21–24] using membranes with a molecular weight cut-off (MWCO) ranging from 2 to 100 kDa. However, in addition to the HMW coffee brew melanoidins, which are estimated to account for 59% of all coffee brew melanoidins,[14] the existence of low and intermediate MW coffee brew melanoidins, recovered in the dialysate obtained by diafiltration using a membrane with a MWCO of 3 kDa, has been also reported.[11] The absence of a general consensus concerning the minimum MW of coffee melanoidins reflects the lack of knowledge on their structures, and the need for more studies focused on their structural characterization.

Even considering only the HMWM of coffee brews, beyond melanoidins, there are free proteins and polysaccharides, and thus additional purification procedures are required for their separation. Over the last decade, different melanoidin populations have been purified through approaches that

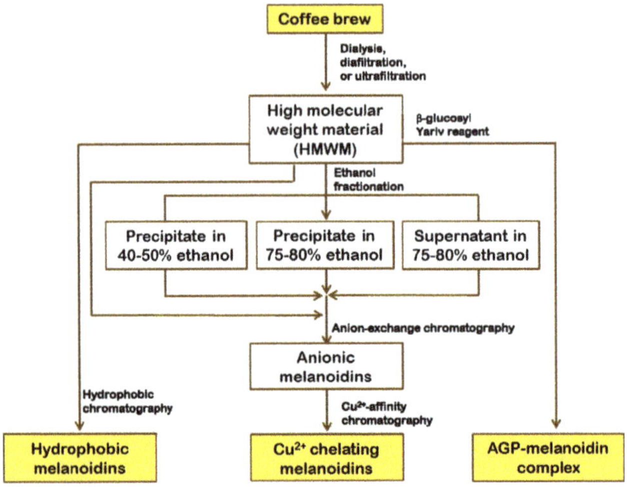

Figure 29.1 Most common experimental strategies used for isolation and purification of coffee melanoidins. Adapted from Springer and F. M. Nunes and M. A. Coimbra, Role of hydroxycinnamates in coffee melanoidin formation, *Phytochem. Rev.*, 2009, **9**, 171–185, Figures 5 and 6, © Springer Science+Business Media B.V. 2009. With kind permission from Springer Science and Business Media.[1]

exploit different physico-chemical properties, such as solubility, charge, metal chelating ability, and hydrophobicity.[9–11,14,17,24] As detailed throughout the following section, the chemical characterization of these purified coffee melanoidin fractions has given increasing evidences that polysaccharides, proteins, and chlorogenic acids are involved in their formation.[9–11,14,17]

29.3 Structural Components of Coffee Melanoidins

In the case of coffee, not only carbohydrates and proteins through the classical Maillard reaction are involved in the formation of melanoidins. The following paragraphs summarize what is known about the different structural components of coffee melanoidins, as well as the purification strategies that enable melanoidin fractions with different chemical compositions to be obtained.

Polysaccharides – The polysaccharides account for about 50% of the dry weight of green coffee beans, in which 22% comprise galactomannans and 14–17% type II arabinogalactans.[7,25] During the roasting process, galactomannans and arabinogalactans, by themselves, undergo structural modifications, namely depolymerization, debranching, and polymerization, among others (See Chapter 19).[26–31] Also, they react with each other by transglycosylation reactions, yielding hybrid polysaccharides.[32,33] Galactomannans and arabinogalactans are also involved in the formation of melanoidins.[9,11,14,20]

With the purpose of obtaining melanoidin fractions with different polysaccharide composition, the HMWM of coffee brews is fractionated by differential ethanol solubility (Figure 29.1). The richest fractions in galactomannans precipitate at an ethanol concentration of 40–50% (% by volume or weight) and the richest fractions in arabinogalactans precipitate at an ethanol concentration of 75–80%. The fractions that remain soluble at an ethanol concentration of 75–80% present the lowest carbohydrate content.[9,13,28,29,34]

Arabinogalactans, in the form of arabinogalactan-proteins (AGPs), are present in green and roasted coffee beans, as well as in green and roasted coffee brews.[14,35–39] Melanoidin fractions containing intact AGPs have been isolated from the HMWM of coffee brews by precipitation with β-glucosyl Yariv reagent, a reagent that selectively precipitates AGPs.[11,14] Instead of the white color of the AGP fraction recovered from green coffee,[40] a brown colored precipitate is obtained by the addition of the Yariv reagent to the HMWM of a roasted coffee brew.[14] Based on the $K_{\mathrm{mix}\ 405\ \mathrm{nm}}$ values, it was shown that the AGP fraction accounts for approximately half of the melanoidins present in the HMWM of coffee brews. However, as the amount of galactose and arabinose of the AGP fraction obtained by precipitation with the Yariv reagent was smaller than that observed in the HMWM, some of the green coffee bean AGPs may be transformed during roasting by losing their protein moiety.[14]

As shown in Figure 29.1, anionic melanoidin fractions have also been isolated by anion exchange chromatography on the fractions obtained by ethanol fractionation,[9] or on the HMWM isolated directly from the coffee brew.[14,17] The anionic character is not a ubiquitous property of coffee melanoidins as some do not contain it or contain it in a very low extent.[9,14] Also, due to the metal chelating capacity of coffee melanoidins,[41,42] a further fractionation of the anionic melanoidin fractions performed by copper affinity chromatography allows the isolation of anionic and chelating melanoidin fractions. This approach also showed that half of the anionic melanoidin fractions have chelating ability for immobilized copper ions.[9] Moreover, hydrophobic melanoidin fractions, mainly composed of arabinogalactan-derived fragments, have been isolated from the HMWM by gel permeation chromatography followed by hydrophobic interaction chromatography (Figure 29.1).[10]

Proteins – Proteins account for about 10% of the dry weight of green coffee beans.[7] During the roasting process, their relative content is decreased. Particularly, arginine, cysteine, lysine, and serine are the amino acids with the largest decrease.[7,13,43,44] As a consequence of structural modifications induced by roasting, sodium dodecyl sulfate-polyacrylamide gel electrophoresis patterns obtained under non-reducing conditions of the HMWM isolated from green and roasted coffee brews are clearly distinct. While green coffee brew presents a major protein band at 58 kDa and a second one at 38 kDa, roasted coffee brew presents a defined band with ≤14 kDa and a diffuse band with >200 kDa,[28,29,34] which is probably due to the involvement of proteins in melanoidins formation.

Different melanoidin fractions have also been isolated from the HMWM of coffee brews containing variable amounts of protein-like material, quantified by amino acid analysis after acid hydrolysis.[9,10,13] Comparing different melanoidin fractions obtained from ethanol fractionation of the HMWM, the fraction with the highest protein content remains soluble at a high ethanol concentration (75–80%), but all the fractions have a similar composition in terms of amino acid relative abundances. Alanine, aspartic acid/asparagine, glutamic acid/glutamine, and glycine are among the most abundant amino acids, whereas histidine, lysine, methionine, and tyrosine are the least abundant. Arginine was not found.[9,13] Furthermore, hydroxyproline, an amino acid found in high amounts in AGPs isolated from green coffee beans and green coffee brews,[35,36] was identified in melanoidin fractions containing arabinose and galactose residues, corroborating the existence of AGPs in coffee melanoidin structures.[9]

Although the intact amino acids account for most of the nitrogen present in the roasted coffee brew HMWM, as well as in the melanoidin fractions obtained from the HMWM by ethanol fractionation, non-amino acid nitrogen, also referred to as non-protein nitrogen (NPN), is also present. Comparing melanoidin fractions obtained by ethanol fractionation, the amount of NPN increases with the increase of their ethanol solubility.[13] As corroborated by the identification of Maillard reaction products derived from lysine, namely N^{ε}-(fructosyl)lysine, N^{ε}-(carboxymethyl)lysine, and N^{ε}-(carboxyethyl) lysine, in the HMWM and melanoidin fractions,[9] chemically modified amino acids formed during coffee roasting are also incorporated in melanoidin structures, contributing to their NPN content.

Chlorogenic acids – Various melanoidin fractions containing chlorogenic acids (or their derivatives) have also been isolated from the HMWM of coffee brews.[9,11,17] Chlorogenic acids account for up to 10% of the dry weight of green coffee beans.[7] Structurally, they consist of a quinic acid (QA) moiety esterified to one or more *trans*-cinnamic acids, such as caffeic, *p*-coumaric, and ferulic acids.[45] The most abundant chlorogenic acid in green coffee beans is 5-*O*-caffeoylquinic acid (5-CQA, following IUPAC numbering),[46] a caffeic acid (CA) ester.[47,48]

The presence of chlorogenic acids in the coffee brew HMWM and melanoidin fractions was suggested based on the quantification of phenolic compounds using the Folin–Ciocalteu reagent,[13,14] but this colorimetric assay is not specific for phenolic compounds.[49] The incorporation of chlorogenic acids in coffee melanoidins was also suggested by the detection of phenols after thermal degradation of the HMWM isolated from a coffee brew,[50] also identified after thermal treatment of ferulic acid.[51] However, this approach does not allow evidencing the type of interaction (covalent or non-covalent) between chlorogenic acids and other melanoidin components. The presence of covalently linked chlorogenic acids (or their derivatives) in the coffee brew HMWM and melanoidin fractions was first demonstrated by using alkaline fusion, an efficient method to release condensed phenolic structures.[9,42] The incorporation of covalently linked chlorogenic acids and derivatives was corroborated by subsequent studies using saponification and enzymatic treatments,[11,17,52] and other using a modified "in bean" coffee model enriched with a typical green coffee chlorogenic acid mixture.[53] These studies showed the presence of ester-linked phenolic and QA moieties, as well as intact chlorogenic acids incorporated *via* a CA moiety through mainly non-ester linkages.

Although there is strong evidence that chlorogenic acid derivatives are components of coffee melanoidins, it is not yet known how they are linked within the melanoidin structures. Based on studies demonstrating that the roasting-induced changes in coffee protein profiles are similar to those observed when oxidized chlorogenic acids react with green coffee proteins in model systems,[54,55] proteins were considered as a possible binding site for the chlorogenic acid derivatives.[1] Also, arabinose from arabinogalactan side chains was hypothesized as a possible binding site for the chlorogenic acid derivatives.[17] This hypothesis was proposed based on studies demonstrating that the arabinose residues are quite susceptible to degradation during roasting,[31,39] and a model study developed under simulated roasting conditions, in which the reaction of epicatechin with sugar fragments was reported.[56] In order to investigate this hypothesis, an equimolar mixture of 5-CQA and (α1→5)-L-arabinotriose, an oligosaccharide structurally related to arabinogalactan side chains, was submitted to dry thermal treatments, mimicking coffee roasting conditions. Compounds composed of one or two CQAs covalently linked with pentose (Pent) residues (1–12) were identified after thermal processing, along with compounds bearing a sugar moiety but composed exclusively of the QA or CA moiety of 5-CQA. Also, the formation of isomeric $Pent_n$CQA structures during thermal processing of the model mixture was demonstrated. Moreover, $Pent_{1\text{-}2}$CQA were identified in coffee samples. These results give evidence for a diversity of chlorogenic acid-arabinose hybrids formed during roasting, opening new perspectives for their identification in melanoidin structures.[57]

Sucrose – Sucrose is the major oligosaccharide in green coffee beans, but it is nearly completely degraded during the roasting process (see Chapter 19). The use of modified "in bean" green coffee models, produced by the incorporation of sucrose into the green beans, allowed the contribution of sucrose to coffee melanoidin formation to be shown, which is probably related to the formation of colored low molecular weight (LMW) structures attached to the polysaccharides and proteins or to protein fragments.[18]

29.4 Possible Formation Routes of Coffee Melanoidins

Although melanoidins are defined as end products of Maillard reaction, the different chemical compositions of melanoidin populations isolated from (roasted) coffee brews strongly suggest that the classical Maillard reaction mechanism is not the only mechanism operating during coffee melanoidin formation. About 40 years after the publication of the French scientist Louis Camille Maillard about the reaction between amino acids and sugars,[58] the first comprehensive scheme of the Maillard reaction was published in 1953.[59] Accordingly, this non-enzymatic browning reaction encompasses a network of various reactions between carbonyl compounds, especially reducing sugars, and compounds with a free amino group, such as amino acids, peptides and proteins, forming a variety of products. These products can be classified as early stage products, intermediate stage products, and last stage products (the melanoidins).

Based on model studies, mostly in solution, and in which the classical Maillard reaction mechanism was considered, three hypotheses on melanoidin formation have been described.[1,60] One hypothesis is that the melanoidins are formed by polymerization (*via* polycondensation reactions) of LMW Maillard reaction products, such as furans and pyrroles, formed in the advanced stages of the reaction.[61–63] Hoffmann[64–66] has suggested that the melanoidins are derived from cross-linking of LMW Maillard reaction products to proteins *via* reactive side chains of amino acids such as lysine, arginine, and cysteine. The third hypothesis is that the melanoidin skeleton is mainly built up of sugar degradation products, formed in the early stages of the Maillard reaction and polymerized through aldol-type condensation.[67–69] In particular, the first hypothesis was corroborated by analysis of the volatile compounds released upon the thermal degradation of coffee brew HMWM, showing the presence mostly of furans (65%) followed by carbonyl compounds (16%).[50] Also, based on a model study in which aqueous solutions of bovine serum albumin and glycolaldehyde were heated at 95 °C, the protein-bound 1,4-bis-(5-amino-5-carboxy-1-pentyl)pyrazinium radical cation (CROSSPY) was proposed as an intermediate in the formation of melanoidins. Effectively, the involvement of this radical in the formation of coffee melanoidins was corroborated by the detection of CROSSPY in roasted coffee beans.[70]

The involvement of other mechanisms, besides the classical Maillard reaction mechanism, in the formation of coffee melanoidins was also unequivocally demonstrated by the identification of covalently linked chlorogenic acids (and their derivatives) in coffee melanoidin fractions recovered from coffee brews (see Section 29.3). However, as already mentioned in this section, chlorogenic acids are not considered in the classical Maillard reaction mechanism.

In summary, the exact mechanisms involved in the formation of coffee melanoidins remain unclear. For this reason, a simplistic illustration of coffee melanoidin formation is presented in Figure 29.2. As was described in a previous section of this chapter, galactomannans, arabinogalactan-proteins, proteins, chlorogenic acids, and sucrose are known to be involved in coffee melanoidin formation. However, several questions remain open about their structures as the nature of the unknown material that can contribute up to 90% to the melanoidin weight, as well as the type of linkages between the

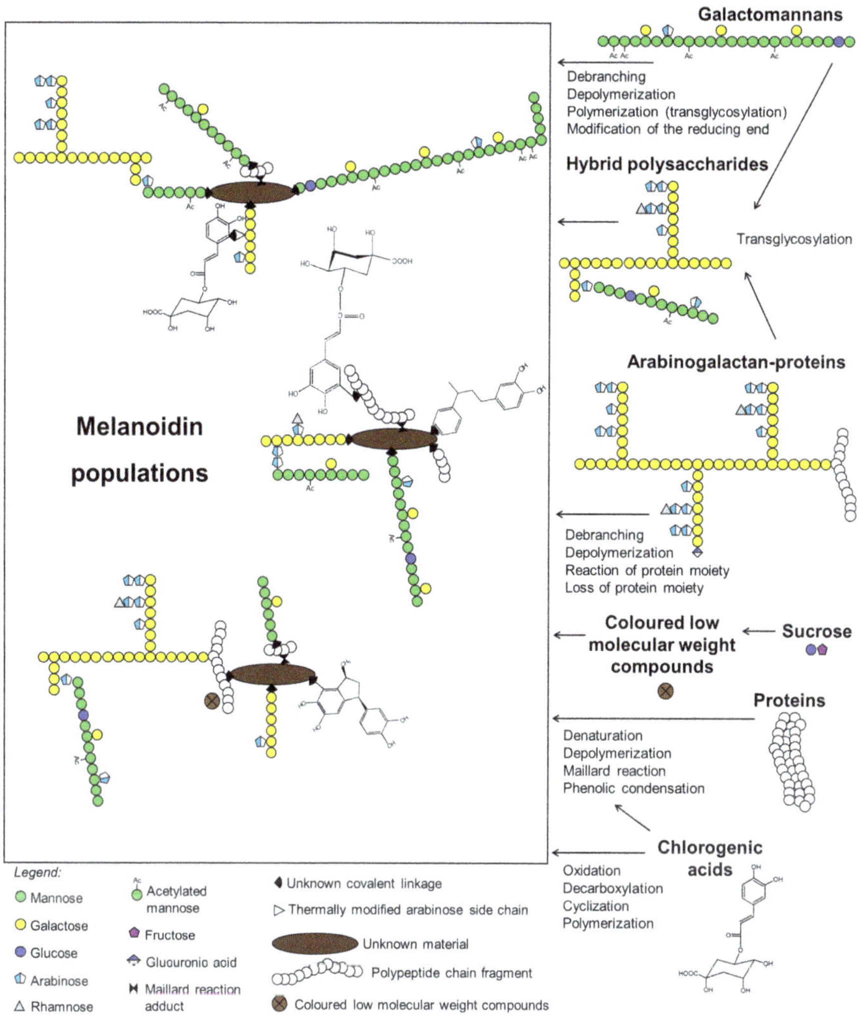

Figure 29.2 Simplified illustration of coffee melanoidin formation. Adapted from Springer and F. M. Nunes and M. A. Coimbra, Role of hydroxycinnamates in coffee melanoidin formation, *Phytochem. Rev.*, 2009, **9**, 171–185, Figures 5 and 6, © Springer Science+Business Media B.V. 2009. With kind permission from Springer Science and Business Media,[1] and from ref. 2 with permission from The Royal Society of Chemistry.

different compounds involved in the formation of coffee melanoidins. Thus, the structural characterization of coffee melanoidins remains a topic of great research interest, especially since several biological activities have been associated to these compounds.

29.5 Biological Activities and Potential Health Impacts of Coffee Melanoidins

Given that coffee brew is one of the main sources of melanoidins in the human diet,[4] biological activities and potential health impacts of coffee melanoidins are of great interest, as reflected by the increasing number of studies on this topic in the last years.

Despite the several biological activities that have been attributed to coffee melanoidins, including, among others, antioxidant, antimicrobial, anticariogenic, anti-inflammatory, antihypertensive, and antiglycative activities, the impacts of these activities (mostly based on *in vitro* studies) on human health remain unclear. Most of the studies reporting biological activities of coffee melanoidins were developed using the HMWM isolated from coffee brews without subsequent purification, simply denominating them as melanoidins. As coffee brew HMWM comprises other HMW compounds, namely polysaccharides and proteins, a definitive conclusion about the active principle responsible for the reported biological activities is not possible. On the other hand, little is known about the metabolic transit and biotransformation of melanoidins, either from coffee or from other foods.[71,72] This question is particularly relevant for those activities whose potential health effects rely on the presence of melanoidins in the bloodstream or their transportation through the blood to the organs, since melanoidins are by definition HMW compounds and their ability to cross the intestinal epithelial barrier has not yet been reported. In fact, coffee melanoidins appear to be largely resistant to digestion in the human gastrointestinal tract, as inferred by the LMW fraction recovered from coffee brew HMWM after simulated *in vitro* gastrointestinal enzymatic digestion, representing only 14% of the HMWM.[73,74]

The following paragraphs briefly present what is known to date about the biological activities attributed to coffee melanoidins. Depending on the data available, some considerations about their biological relevance are also presented, based on the estimated melanoidin concentration of 2 to 4 mg mL^{-1} in coffee brews,[4] and 0.25 to 1 mg mL^{-1} in colon of moderate and heavy consumers.[2] The concentration in colon was estimated considering the daily intake of coffee melanoidins ranging between 0.5 and 2.0 g for moderate and heavy consumers,[4] and the assumption that the colon accumulates its content over at least 24 h in a maximum volume of 2 L.[23]

Antioxidant activity – According to several studies, coffee melanoidins present *in vitro* antioxidant activity.[10,12,21,73,75] However, many of these studies were developed using the HMWM isolated from coffee brews without subsequent purification. On the other hand, LMW compounds released during overnight incubation of the HMWM in 2 M NaCl showed a higher antioxidant activity than that of the remaining polymeric material, suggesting that an important contribution to the overall antioxidant activity of the HMWM is given by the LMW compounds linked non-covalently to the melanoidin skeleton, such as chlorogenic acids.[21,73,75]

Regarding the influence of the degree of roast on the antioxidant activity of coffee brew melanoidins, different results have been presented.[12,21] The use of different assay procedures for testing antioxidant activity may contribute to the distinct results reported. Since there are a number of melanoidin populations, the use of different procedures for isolation of coffee brew melanoidins may also contribute to the distinct results. The different melanoidin populations may exhibit antioxidant capacities through different mechanisms. Although these mechanisms remain not fully elucidated, the antioxidant activity of coffee melanoidins has been related to their ability to scavenge free radical species and to their metal chelating capacity.[21,75] In terms of its implications on human health, the antioxidant activity of coffee melanoidins may be important in the protection against oxidative damage, namely by prevention of lipid peroxidation,[76,77] contributing to the prevention of age-related diseases, namely neurodegenerative and cardiovascular diseases, among others.[78]

Antimicrobial activity – The antimicrobial activity of coffee melanoidins (referring to the coffee brew HMWM isolated by ultrafiltration) was evaluated against bacterial strains of Gram-positive, like *Staphylococcus aureus*, and Gram-negative, like *Escherichia coli*. For all strains studied, the minimum inhibitory concentration (MIC), defined as the lowest concentration of melanoidin fractions that did not produce any detected cell growth, ranged between 2 and 10 mg mL^{-1}.[79,80] Gram-positive bacteria are more sensitive to the antimicrobial activity of coffee melanoidin fractions, showing lower MIC values (2–3 mg mL^{-1}) than Gram-negative bacteria (≥4 mg mL^{-1}). Considering these MIC values, and the estimated melanoidin concentration of 2 to 4 mg mL^{-1} in the coffee brews,[4] it can be inferred that all coffee brews should present antibacterial activity in the mouth against Gram-positive bacteria and possibly also against some Gram-negative. On the other hand, based on the estimated melanoidin concentration of 0.25 to 1 mg mL^{-1} in colon of moderate and heavy consumers,[2] it can be expected that the regular intake of coffee melanoidins has an inhibitory/regulatory effect on colon microflora, but does not prevent, as expected, the microbial growth. Regarding the effect of the degree of roast, a higher inhibitory bacterial growing activity occurs for the higher degree of roast. Also, the non-covalently linked compounds released from the HMWM after incubation in 2 M NaCl exert a lower antibacterial activity than that of the remaining HMWM.[81]

The antimicrobial capacity of coffee melanoidins has been associated to their metal chelating properties.[79,80] Also, three different mechanisms have been proposed for the antibacterial activity of coffee melanoidins. Accordingly, melanoidins may exert a bacteriostatic activity at low concentration by iron chelation from the culture medium. In the case of bacterial strains able to produce siderophores for iron acquisition, melanoidins may chelate the siderophore-Fe^{3+} complex, which could decrease the virulence of such pathogenic bacteria. Finally, coffee melanoidins may also exert a bactericide activity at high concentrations by removing Mg^{2+} cations from the outer membrane, promoting the disruption of the cell membrane and allowing the release of intracellular molecules. Regarding the distinct sensitiveness of Gram-negative and Gram-positive bacteria, it is assumed that the absence of the outer membrane in Gram-positive bacteria makes them more susceptible to the antibacterial activity of coffee melanoidins.[79]

Coffee melanoidins may react with salivary nitrite and thiocyanate in the gastric lumen, producing nitric oxide (NO). The formation of NO in the stomach may contribute to the inhibition of the microbial growth, as well as the regulation of mucosal flow, mucosal formation, and gastric mobility.[82]

Anticariogenic activity – Also related to their antibacterial activity, the potential anticariogenic activity of coffee melanoidins was reported based on the ability of HMWM isolated from a coffee brew, and melanoidin fractions recovered from the HMWM, to affect *Streptococcus mutans* sucrose-dependent adhesion to and detachment from saliva-coated hydroxyapatite beads and to inhibit biofilm formation on microliter plates. A concentration of 6 mg mL^{-1} of HMWM inhibits the *Streptococcus mutans* adhesion to the beads in 91% and abolishes the biofilm production.[83] Considering the estimated melanoidin concentration in coffee brews (2 to 4 mg mL^{-1}),[4] it can be expected that coffee brews promote dental caries protection. The anticariogenic activity of coffee melanoidins is also corroborated by the antiadhesive effect of coffee brews on the adhesive properties of *Streptococcus mutans*.[84]

Modulation of the bacterial colon population – As inferred by the decrease of the absorbance at 405 nm after *in vitro* fermentation of coffee brew HMW fractions (>100, 50–100, 10–50, and 3–10 kDa) with human fecal bacteria, melanoidins are degraded/modified in the human gut. For all coffee fractions, an increase of *Bacteroides-Prevotella* was also observed.[85] These results are in accordance with a human volunteer study demonstrating that the moderate consumption of an instant coffee produces an increase in the number of *Bacteroides-Prevotella* bacteria detected in feces.[86] In this *in vivo* study, the main bacteria increase was observed for *Bifidobacterium* spp., known for their probiotic effects.[87,88] However, it cannot be disclosed if this effect can be attributed only to the melanoidins because the melanoidins are present in mixtures with polysaccharides that are known to have prebiotic properties. Nevertheless, the capacity of coffee melanoidins to modulate the bacterial colon population is corroborated by model melanoidin solutions.[89,90]

Inhibition of matrix metalloproteases – The potential activity of coffee melanoidins as inhibitors of matrix metalloproteases (MMPs), endo-peptidases associated with tumor growth and metastasis,[91] was reported based on the ability of the HMWM isolated from (roasted) coffee brews to inhibit the *in vitro* activity of selected human MMPs, specifically MMP-1, MMP-2, and MMP-9 that are thought to be involved in the pathogenesis of colon cancer.[92] For all MMPs (and roasting times >10 min), the 50% inhibitory concentration (IC_{50}) values, *i.e.* the concentrations that produce 50% of inhibition, ranged between 0.2 and 1.1 mg mL^{-1}, whereas the HMWM from green coffee brew showed no significant inhibitory activity against any of the MMPs at concentrations up to 2.5 mg mL^{-1}.[23] The IC_{50} values obtained from roasted coffee brew are comparable to the estimated melanoidin concentration in the colon of moderate and heavy consumers (0.25–1 mg mL^{-1}), suggesting that the MMP inhibitory activity of coffee melanoidins may offer protective effects against colon cancer.

Antihypertensive activity – In respect to the modulation of other enzymatic systems, it has been reported that coffee melanoidins,[23,93] as well as Maillard reaction compounds obtained from model systems,[94] are capable of inhibiting the angiotensin-I converting enzyme (ACE). ACE is a key element of the renin-angiotensin system that regulates blood pressure, and thus ACE inhibitors are important for the treatment of hypertension.[95] The mechanism of action for potential ACE-inhibitory activity of melanoidins remains unclear, but different mechanisms have been suggested.[93,94] The ACE is a zinc-dependent enzyme,[95] and thus the inhibitory activity of melanoidins can come from their metal chelating properties. Also, melanoidins can act as ACE non-competitive inhibitors, bind to the enzyme in an area other than the active center, deform the enzyme, and hinder binding to the substrate.[96]

A concentration of 2 mg mL^{-1} of instant coffee HMWM inhibited the *in vitro* ACE activity by 37–45%.[93] This result is in accordance with another study showing that concentrations higher than 1.5 mg mL^{-1} of coffee brew HMWM are needed to inhibit *in vitro* ACE activity by 50%.[23] Based on the estimated intake of coffee melanoidins of 0.5 to 2.0 g per day,[4] their absorption, residence, and dilution in the blood, it can be estimated that the amount of melanoidins from coffee intake are far from reaching the required concentration to have an effect on ACE inhibition.

Anti-inflammatory activity – When liver samples from rats subjected to a high-fat diet for 3 months and ingestion of decaffeinated coffee or melanoidins (the HMWM isolated from the decaffeinated coffee) from the beginning of the second month were analyzed in comparison with samples from control rats drinking water, a reduction in the concentration of proinflammatory cytokines, such as tumor necrosis factor α (TNF-α) and interferon-γ (IFN-γ), and an increase of anti-inflammatory ones, such as interleukin-4 (IL-4), was observed, suggesting that coffee melanoidins may exert an anti-inflammatory activity, particularly in liver. Accordingly, in the same study, it was reported that the anti-inflammatory activity of coffee melanoidins may play a role in counteracting the progression of liver diseases, namely non-alcoholic steatohepatitis.[97] In agreement with this study, other studies conducted in mice reported the anti-inflammatory action of coffee brews.[98,99]

Antiglycative activity – The potential antiglycative activity of coffee melanoidins was reported based on the ability of the coffee brew HMWM to inhibit *in vitro* glycation of bovine serum albumin with glucose. A HMWM

Table 29.1 Summary of the biological activities attributed to coffee brew melanoidins.

Biological activity	References	Relevance[a]
Antioxidant activity	10, 12, 21, 73 and 75	—
Antimicrobial activity	79–81	✓
Anticariogenic activity	83 and 84	✓
Modulation of the bacterial colon population	85, 89 and 90	—
Inhibition of matrix metalloproteases	23	✓
Antihypertensive activity	23, 93 and 94	×
Anti-inflammatory activity	97	✓
Antiglycative activity	22	✓

[a]Presents (✓) or not (×) biological relevance considering the estimated melanoidin concentration in coffee brews and the daily intake; —, influenced by the contribution of other coffee components.

concentration of 274 μg mL^{-1} was able to inhibit 50% of the glycation (IC_{50}).[22] Although the proportion of melanoidins absorbed and present in the blood is not defined, based on the estimated intake of coffee melanoidins of 0.5 to 2.0 g per day,[4] it can be estimated that the concentrations of melanoidins in blood are in the range of the effective dose for a possible antiglycative activity of coffee brew. As advanced glycation end products are considered important mediators of diabetes complications, glycation inhibitors are of great interest due to their preventive or therapeutic potential.[100]

Summarizing what was reported in the subsections above, Table 29.1 shows the biological activities that have been associated to coffee melanoidins with the indication of their biological relevance considering the estimated melanoidin concentration in coffee brews and the daily intake.

29.6 Conclusions

During the roasting of green coffee beans, galactomannans, arabinogalactans, sucrose, proteins, and chlorogenic acids react and give origin to different populations of HMW brown compounds containing nitrogen, known as melanoidins. Despite their exact structures and mechanisms of formation remaining unclear, melanoidins have been associated to several biological activities with potential health benefits. However, more work is needed on biological activities of coffee melanoidins. As mentioned in the previous section, most of the studies undertaken to date have been performed using the HMWM isolated from coffee brews without subsequent purification, hampering a definitive conclusion about the active principle responsible for the biological activities. Also, future studies are needed to understand metabolism and biotransformation of coffee melanoidins, as well as the relation between the structure of different coffee melanoidin populations and their biological activities.

References

1. F. M. Nunes and M. A. Coimbra, *Phytochem. Rev.*, 2010, **9**, 171.
2. A. S. P. Moreira, F. M. Nunes, M. R. Domingues and M. A. Coimbra, *Food Funct.*, 2012, **3**, 903.
3. J. A. Rufián-Henares and S. Pastoriza, in *Coffee in Health and Disease Prevention*, ed. V. R. Preedy, Academic Press, San Diego, 2015, ch. 20, pp. 183–188.
4. V. Fogliano and F. J. Morales, *Food Funct.*, 2011, **2**, 117.
5. T. D. Obretenov, S. D. Ivanova, M. J. Kuntcheva and G. T. Somov, *J. Agric. Food Chem.*, 1993, **41**, 653.
6. V. Faist, M. Lindenmeier, C. Geisler, H. F. Erbersdobler and T. Hofmann, *J. Agric. Food Chem.*, 2002, **50**, 602.
7. A. Illy, E. Illy, R. Macrae, M. Petracco, M. R. Sondahl, S. Valussi and R. Viani, *Espresso Coffee: The Chemistry of Quality*, Academic Press, London, 1995.
8. H.-D. Belitz, W. Grosch and P. Schieberle, in *Food Chemistry*, ed. H.-D. Belitz, W. Grosch and P. Schieberle, Springer, Berlin, 4th edn, 2009, ch. 21, pp. 938–951.
9. F. M. Nunes and M. A. Coimbra, *J. Agric. Food Chem.*, 2007, **55**, 3967.
10. D. Gniechwitz, N. Reichardt, J. Ralph, M. Blaut, H. Steinhart and M. Bunzel, *J. Sci. Food Agric.*, 2008, **88**, 2153.
11. E. K. Bekedam, M. J. Loots, H. A. Schols, M. A. J. S. Van Boekel and G. Smit, *J. Agric. Food Chem.*, 2008, **56**, 7138.
12. R. C. Borrelli, A. Visconti, C. Mennella, M. Anese and V. Fogliano, *J. Agric. Food Chem.*, 2002, **50**, 6527.
13. E. K. Bekedam, H. A. Schols, M. A. J. S. van Boekel and G. Smit, *J. Agric. Food Chem.*, 2006, **54**, 7658.
14. E. K. Bekedam, M. P. F. C. De Laat, H. A. Schols, M. A. J. S. Van Boekel and G. Smit, *J. Agric. Food Chem.*, 2007, **55**, 761.
15. E. K. Bekedam, E. Roos, H. A. Schols, M. A. J. S. Van Boekel and G. Smit, *J. Agric. Food Chem.*, 2008, **56**, 4060.
16. T. Hofmann, M. Czerny, S. Calligaris and P. Schieberle, *J. Agric. Food Chem.*, 2001, **49**, 2382.
17. E. K. Bekedam, H. A. Schols, M. A. J. S. Van Boekel and G. Smit, *J. Agric. Food Chem.*, 2008, **56**, 2055.
18. F. M. Nunes, A. C. S. Cruz and M. A. Coimbra, *J. Agric. Food Chem.*, 2012, **60**, 8710.
19. C. P. Passos, M. R. Cepeda, S. S. Ferreira, F. M. Nunes, D. V. Evtuguin, P. Madureira, M. Vilanova and M. A. Coimbra, *Food Chem.*, 2014, **161**, 60.
20. D. Gniechwitz, N. Reichardt, E. Meiss, J. Ralph, H. Steinhart, M. Blaut and M. Bunzel, *J. Agric. Food Chem.*, 2008, **56**, 5960.
21. C. Delgado-Andrade, J. A. Rufián-Henares and F. J. Morales, *J. Agric. Food Chem.*, 2005, **53**, 7832.
22. E. Verzelloni, D. Tagliazucchi, D. Del Rio, L. Calani and A. Conte, *Food Chem.*, 2011, **124**, 1430.

23. L. M. de Marco, S. Fischer and T. Henle, *J. Agric. Food Chem.*, 2011, **59**, 11417.
24. C. Bartel, M. Mesias and F. J. Morales, *Food Res. Int.*, 2015, **67**, 356.
25. A. G. W. Bradbury and D. J. Halliday, *J. Agric. Food Chem.*, 1990, **38**, 389.
26. F. M. Nunes, A. Reis, M. R. M. Domingues and M. A. Coimbra, *J. Agric. Food Chem.*, 2006, **54**, 3428.
27. A. S. P. Moreira, Study of structural changes in models of coffee galactomannans, Masters Thesis, University of Aveiro, 2011.
28. F. M. Nunes and M. A. Coimbra, *J. Agric. Food Chem.*, 2001, **49**, 1773.
29. F. M. Nunes and M. A. Coimbra, *J. Agric. Food Chem.*, 2002, **50**, 7046.
30. A. Oosterveld, J. S. Harmsen, A. G. J. Voragen and H. A. Schols, *Carbohydr. Polym.*, 2003, **52**, 285.
31. A. Oosterveld, A. G. J. Voragen and H. A. Schols, *Carbohydr. Polym.*, 2003, **54**, 183.
32. A. S. P. Moreira, J. Simões, A. T. Pereira, C. P. Passos, F. M. Nunes, M. R. M. Domingues and M. A. Coimbra, *Carbohydr. Polym.*, 2014, **112**, 48.
33. A. S. P. Moreira, J. Simões, F. M. Nunes, D. V. Evtuguin, P. Domingues, M. A. Coimbra and M. R. M. Domingues, *J. Agric. Food Chem.*, 2016, **64**, 1831.
34. F. M. Nunes and M. A. Coimbra, *J. Agric. Food Chem.*, 2002, **50**, 1429.
35. R. J. Redgwell, D. Curti, M. Fischer, P. Nicolas and L. B. Fay, *Carbohydr. Res.*, 2002, **337**, 239.
36. F. M. Nunes, A. Reis, A. M. S. Silva, M. R. M. Domingues and M. A. Coimbra, *Phytochemistry*, 2008, **69**, 1573.
37. P. Capek, M. Matulová, L. Navarini and F. Suggi-Liverani, *Carbohydr. Polym.*, 2010, **80**, 180.
38. M. Matulová, P. Capek, S. Kaneko, L. Navarini and F. S. Liverani, *Carbohydr. Res.*, 2011, **346**, 1029.
39. R. J. Redgwell, V. Trovato, D. Curti and M. Fischer, *Carbohydr. Res.*, 2002, **337**, 421.
40. R. J. Redgwell, C. Schmitt, M. Beaulieu and D. Curti, *Food Hydrocolloids*, 2005, **19**, 1005.
41. F. J. Morales, C. Fernández-Fraguas and S. Jiménez-Pérez, *Food Chem.*, 2005, **90**, 821.
42. M. Takenaka, N. Sato, H. Asakawa, X. Wen, M. Murata and S. Homma, *Biosci. Biotechnol. Biochem.*, 2005, **69**, 26.
43. H. D. Belitz, W. Grosch and P. Schieberle, in *Food Chemistry*, ed. H. D. Belitz, W. Grosch and P. Schieberle, Springer, Berlin, 4th edn, 2009, pp. 938–951.
44. S. Casal, E. Mendes, M. B. P. P. Oliveira and M. A. Ferreira, *Food Chem.*, 2005, **89**, 333.
45. M. N. Clifford, *J. Sci. Food Agric.*, 2000, **80**, 1033.
46. IUPAC, *Biochem. J.*, 1976, **153**, 23.
47. J.-K. Moon and T. Shibamoto, *J. Agric. Food Chem.*, 2009, **57**, 5823.

48. D. Perrone, A. Farah, C. M. Donangelo, T. de Paulis and P. R. Martin, *Food Chem.*, 2008, **106**, 859.
49. A. L. Waterhouse, in *Current Protocols in Food Analytical Chemistry*, ed. R. E. Wrolstad, John Wiley & Sons, Inc., New York, 2002, pp. I1.1.1–I1.1.8.
50. A. Adams, R. C. Borrelli, V. Fogliano and N. De Kimpe, *J. Agric. Food Chem.*, 2005, **53**, 4136.
51. W. Fiddler, W. E. Parker, A. E. Wasserman and R. C. Doerr, *J. Agric. Food Chem.*, 1967, **15**, 757.
52. D. Perrone, A. Farah and C. M. Donangelo, *J. Agric. Food Chem.*, 2012, **60**, 4265.
53. C. Coelho, M. Ribeiro, A. C. S. Cruz, M. R. M. Domingues, M. A. Coimbra, M. Bunzel and F. M. Nunes, *J. Agric. Food Chem.*, 2014, **62**, 7843.
54. H. M. Rawel, S. Rohn and J. Kroll, *Dtsch. Lebensm.-Rundsch.*, 2005, **101**, 148.
55. P. Montavon, A.-F. Mauron and E. Duruz, *J. Agric. Food Chem.*, 2003, **51**, 2335.
56. V. M. Totlani and D. G. Peterson, *J. Agric. Food Chem.*, 2007, **55**, 414.
57. A. S. P. Moreira, M. A. Coimbra, F. M. Nunes, C. P. Passos, S. A. O. Santos, A. J. D. Silvestre, A. M. N. Silva, M. Rangel and M. R. M. Domingues, *Food Chem.*, 2015, **185**, 135.
58. L. C. Maillard, *C. R. Acad. Sci.*, 1912, **154**, 66.
59. J. E. Hodge, *J. Agric. Food Chem.*, 1953, **1**, 928.
60. H.-Y. Wang, H. Qian and W.-R. Yao, *Food Chem.*, 2011, **128**, 573.
61. F. Hayase, T. Usui and H. Watanabe, *Mol. Nutr. Food Res.*, 2006, **50**, 1171.
62. R. Tressl, G. T. Wondrak, L.-A. Garbe, R.-P. Krüger and D. Rewicki, *J. Agric. Food Chem.*, 1998, **46**, 1765.
63. R. Tressl, G. T. Wondrak, R.-P. Krüger and D. Rewicki, *J. Agric. Food Chem.*, 1998, **46**, 104.
64. T. Hofmann, *Z. Lebensm. Unters. Forsch. A*, 1998, **206**, 251.
65. T. Hofmann, *J. Agric. Food Chem.*, 1998, **46**, 3891.
66. T. Hofmann, *J. Agric. Food Chem.*, 1998, **46**, 3896.
67. L. W. Kroh, T. Fiedler and J. Wagner, *Ann. N. Y. Acad. Sci.*, 2008, **1126**, 210.
68. B. Cämmerer and L. W. Kroh, *Food Chem.*, 1995, **53**, 55.
69. B. Cämmerer, W. Jalyschko and L. W. Kroh, *J. Agric. Food Chem.*, 2002, **50**, 2083.
70. T. Hofmann, W. Bors and K. Stettmaier, *J. Agric. Food Chem.*, 1999, **47**, 391.
71. V. Faist and H. F. Erbersdobler, *Ann. Nutr. Metab.*, 2001, **45**, 1.
72. V. Somoza, *Mol. Nutr. Food Res.*, 2005, **49**, 663.
73. J. A. Rufián-Henares and F. J. Morales, *J. Agric. Food Chem.*, 2007, **55**, 10016.
74. D. Tagliazucchi and A. Bellesia, *Amino Acids*, 2015, **47**, 1077.
75. C. Delgado-Andrade and F. J. Morales, *J. Agric. Food Chem.*, 2005, **53**, 1403.
76. D. Tagliazucchi, E. Verzelloni and A. Conte, *J. Agric. Food Chem.*, 2010, **58**, 2513.

77. M. Daglia, A. Papetti, C. Aceti, B. Sordelli, C. Gregotti and G. Gazzani, *J. Agric. Food Chem.*, 2008, **56**, 11653.
78. B. Poljsak and I. Milisav, *Aging, Oxidative Stress and Antioxidants*, 2013.
79. J. A. Rufián-Henares and S. P. de la Cueva, *J. Agric. Food Chem.*, 2009, **57**, 432.
80. J. A. Rufián-Henares and F. J. Morales, *J. Agric. Food Chem.*, 2008, **56**, 2357.
81. J. A. Rufián-Henares and F. J. Morales, *Food Res. Int.*, 2006, **39**, 33.
82. U. Takahama and S. Hirota, *J. Agric. Food Chem.*, 2008, **56**, 4736.
83. M. Stauder, A. Papetti, D. Mascherpa, A. M. Schito, G. Gazzani, C. Pruzzo and M. Daglia, *J. Agric. Food Chem.*, 2010, **58**, 11662.
84. M. Daglia, R. Tarsi, A. Papetti, P. Grisoli, C. Dacarro, C. Pruzzo and G. Gazzani, *J. Agric. Food Chem.*, 2002, **50**, 1225.
85. N. Reichardt, D. Gniechwitz, H. Steinhart, M. Bunzel and M. Blaut, *Mol. Nutr. Food Res.*, 2009, **53**, 287.
86. M. Jaquet, I. Rochat, J. Moulin, C. Cavin and R. Bibiloni, *Int. J. Food Microbiol.*, 2009, **130**, 117.
87. C. Picard, J. Fioramonti, A. Francois, T. Robinson, F. Neant and C. Matuchansky, *Aliment. Pharmacol. Ther.*, 2005, **22**, 495.
88. G. R. Gibson and X. Wang, *J. Appl. Microbiol.*, 1994, **77**, 412.
89. C. Dell'Aquila, J. Ames, G. Gibson and A. Wynne, *Eur. Food Res. Technol.*, 2003, **217**, 382.
90. J. M. Ames, A. Wynne, A. Hofmann, S. Plos and G. R. Gibson, *Br. J. Nutr.*, 1999, **82**, 489.
91. M. A. Forget, R. R. Desrosiers and R. Béliveau, *Can. J. Physiol. Pharmacol.*, 1999, **77**, 465.
92. A. G. Mysliwiec and D. L. Ornstein, *Clin. Colorectal Cancer*, 2002, **1**, 208.
93. J. A. Rufián-Henares and F. J. Morales, *J. Agric. Food Chem.*, 2007, **55**, 1480.
94. J. A. Rufián-Henares and F. J. Morales, *Food Res. Int.*, 2007, **40**, 995.
95. J. Riordan, *Genome Biol.*, 2003, **4**, 225.1.
96. B. Hernández-Ledesma, P. J. Martín-Álvarez and E. Pueyo, *J. Agric. Food Chem.*, 2003, **51**, 4175.
97. P. Vitaglione, F. Morisco, G. Mazzone, D. C. Amoruso, M. T. Ribecco, A. Romano, V. Fogliano, N. Caporaso and G. D'Argenio, *Hepatology*, 2010, **52**, 1652.
98. I. Paur, T. R. Balstad and R. Blomhoff, *Free Radical Biol. Med.*, 2010, **48**, 1218.
99. Y. Fukushima, M. Kasuga, K. Nakao, I. Shimomura and Y. Matsuzawa, *J. Agric. Food Chem.*, 2009, **57**, 11100.
100. C. Guillet, *Nutr. Clin. Metab.*, 2010, **24**, 109.

CHAPTER 30

Acrylamide

JOSÉ O. FERNANDES

University of Porto, Faculty of Pharmacy, LAQV-REQUIMTE, Laboratory of Bromatology and Hydrology, Rua Jorge Viterbo Ferreira, 228, 4050-313, Porto, Portugal
*E-mail: jose.fernandes@ff.up.pt

30.1 Introduction

Coffee roasting is an aggressive process carried out most commonly at temperatures higher than 200 °C, which is essential to confer coffee beans the physical, chemical and sensory features that make them the source of one of the most appreciated beverages all around the world. Over the roasting process, many complex thermally driven chemical reactions take place. These not only yield the formation of hundreds of compounds that are responsible for the exceptional sensory and healthy properties of the product, but are also accountable for the occurrence of undesirable and toxic by-products. In fact, the mechanisms of formation of desirable and undesirable compounds are very similar, which makes the monitoring and control of the whole process in order to favour the production of "good molecules" over the "bad molecules" extremely difficult.

The "acrylamide problem" is a striking example of this kind of setback. Indeed, acrylamide is a toxic compound, classified by IARC[1] as "probably carcinogenic to humans", whose formation takes place from a pathway, until recently unknown, of the Maillard reaction, the same reaction that

Coffee: Production, Quality and Chemistry
Edited by Adriana Farah

Published by the Royal Society of Chemistry, www.rsc.org

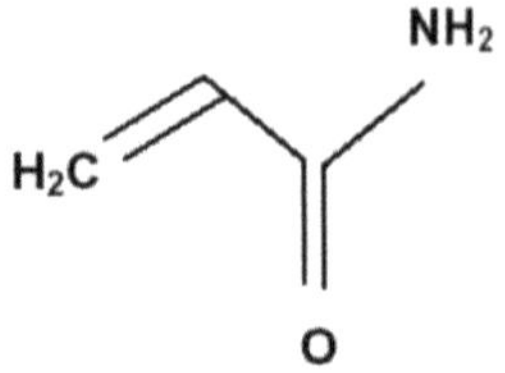

Figure 30.1 Acrylamide structure.

is also the source of a large number of compounds that confer coffee its strong antioxidant ability and most of its pleasant flavour components. Only a thorough knowledge of all the factors involved in the formation/elimination of acrylamide may lead us to realize the best way to keep the pleasant and healthy characteristics of coffee, while achieving the lowest possible toxic effects.

30.2 Chemical Characteristics

Acrylamide, also named 2-propenamide, is a small α,β-unsaturated amide with a molecular mass of 71.08 Da (Figure 30.1). It is an odourless, white crystalline solid, with a melting point of 81–84.5 °C, and a boiling point of 136 °C, highly soluble in water (2.155 g ml^{-1} at 30 °C) and in organic solvents such as methanol, ethanol and acetone, insoluble in heptane and benzene. Acrylamide slowly sublimes at room temperature and polymerizes rapidly and exothermically when melted or exposed to oxidizing agents.[2]

For a long time, acrylamide has been considered a chemical with a wholly industrial origin, synthesized for the first time in 1893, and produced on an industrial scale since the 1950s. It is used as monomer in the production of polyacrylamides and copolymers with different physical and chemical properties.[3] The main uses of polyacrylamides are as flocculants to remove solids from aqueous solutions in mining operations, in the disposal of industrial wastes and in the purification of water supplies. Acrylamide has a large number of other applications including cosmetic additives, soil conditioning agents and in the formulation of grouting agents for the construction of dam foundations, sewers and tunnels.

30.3 Historical and General Occurrence in Foods

The advent of concerns about acrylamide as a food contaminant came around the beginning of this century following a leakage of a sealant product containing acrylamide that occurred in the construction of the Hallandsás tunnel in south-western Sweden, which caused paralysis and even death of cattle and fish fed by water in the neighbourhood.[4] Sweden researchers found the presence of relevant levels of the compound in a range of heat-processed carbohydrates rich food products such as fries and chips, crackers, biscuits and crispbread.[5] This finding was soon confirmed by other food expert researchers,[6] and since then numerous efforts have been developed jointly by official bodies, industry and food researchers in order to evaluate the extent of risk to consumers, find out the mechanisms of formation of the acrylamide and develop strategies to reduce its levels in affected foodstuffs.

Acrylamide has been found in a wide variety of home-cooked and processed foods commonly consumed daily worldwide, including potato crisps, French fries, bread, crispbreads, biscuits, breakfast cereals, coffee and coffee substitutes, which makes it an overall food problem not limited to one or a few particular foods.[7] However, the compound does not occur in such foods subjected to lower temperatures and relatively short process times, *e.g.* boiled potatoes.

A great variability in acrylamide levels can be observed within each food category even in the case of products from the same manufacturer, making the assessment of human dietary exposure difficult. Factors such as variability of acrylamide precursors in the raw material, difference in food composition, difference in process parameters and final heating conditions could be sources of fluctuations.[8] Moreover, final acrylamide contents are often the result of the simultaneous occurrence of formation and elimination reactions, as thoroughly discussed below for roasted coffee beans.

European acrylamide monitoring data compiled by the European Food Safety Authority (EFSA) for the years 2011–2013[9] are summarized in Figure 30.2. As can be seen, the highest acrylamide levels were observed in coffee substitutes (1499 and 4500 μg kg^{-1} at average and 95th percentile, respectively). Acrylamide level in "Coffee (dry)" category was on average 522 μg kg^{-1}, and 1054 μg kg^{-1} at the 95th percentile, being lower in "Roasted coffee (dry)"

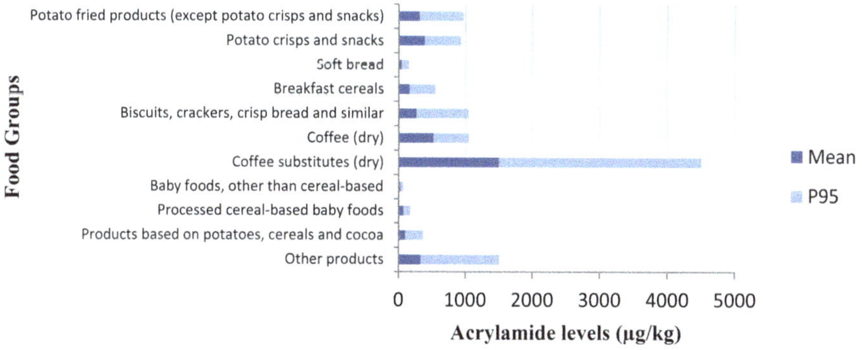

Figure 30.2 Acrylamide levels in selected food groups from European countries: mean values. P95 = percentile 95 (data from ref. 9).

(average: 249 μg kg^{-1}; 95th percentile: 543 μg kg^{-1}) than in "Instant coffee (dry)" (average: 710 μg kg^{-1}; 95th percentile: 1122 μg kg^{-1}).

Because acrylamide is a genotoxic compound, no threshold level exists below which food safety can be disregarded. For that reason the ALARA (as low as reasonably achievable) principle should be followed in what concerns acrylamide levels in food.[10]

Based on the results of the monitoring in the Member States, the EU has set "indicative values" for acrylamide in various foodstuffs.[11] The "indicative values" are neither legal limits nor safety thresholds, but rather intended to indicate the need for a more thorough investigation if the values are exceeded in a particular food product or food category. In respect of coffee products, the "indicative values" are 450 and 900, for roasted coffee and instant coffee, respectively. Accordingly to the higher levels generally found, the highest "indicative values" were set for coffee substitutes, 2000 μg kg^{-1} and 4000 μg kg^{-1}, respectively, for coffee substitutes mainly based on cereals and other coffee substitutes (chicory).

30.4 Mechanisms of Formation in Foods

Soon after the first reports on acrylamide occurrence in certain heated foodstuffs, the Maillard reaction was identified by different researcher groups as the main pathway of acrylamide formation.[12,13] In the presence of hydroxycarbonyl compounds such as reducing sugars or reactive dicarbonyl compounds, asparagine can suffer a thermally driven process of decarboxylation and deamination leading to the formation of acrylamide. Mass spectral studies using stable isotope-labelled compounds have shown that the backbone of the acrylamide molecule – three carbon atoms and the nitrogen of the amide function – is derived from the amino acid asparagine.[14]

Further studies have shown that the main factors affecting the extent of acrylamide formation in most foods are free asparagine and reducing sugars levels in the raw materials, temperature, pH and moisture content. When asparagine and reducing sugars are present, increasingly acrylamide levels are formed at temperatures higher than 120 °C, in a low moisture environment and neutral or slightly basic conditions.

Acrylamide formation involves a cascade of reactions with different highly reactive intermediates and may involve more than one different pathway, as shown in Figure 30.3, where three different pathways are represented: A–the classical pathway starting with the reaction of asparagine with an α-hydroxycarbonyl compound such as glucose; B–an alternative pathway starting with the reaction of asparagine with non-hydroxylated carbonyls; C–the so-called acrolein pathway, which seems to have merely a marginal importance in the bulk of acrylamide formed. One of the key transient intermediates during the thermal degradation process of acrylamide by pathways A and B is 3-aminopropanamide, a compound which can also result from enzymatic decarboxylation of asparagine and as such is already present in the raw materials.[15]

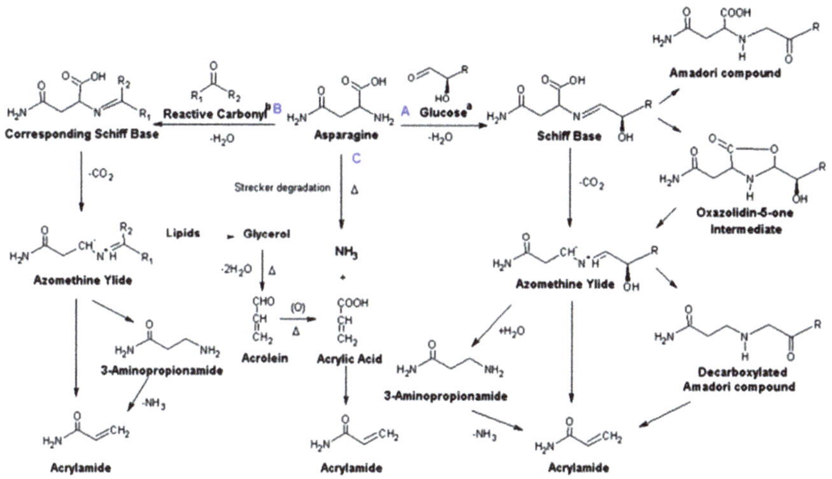

Figure 30.3 Pathways of acrylamide formation from asparagine. Reproduced from ref. 16 with permission from Elsevier. Copyright 2013.

30.4.1 Formation in Coffee

Similarly to what happens with the other foods of concern, the Maillard reaction represents the main route for acrylamide formation in coffee, being initiated by the condensation of asparagine with reducing carbohydrates or reactive carbonyls, when the beans are subjected to the high roasting temperature.[17] However, the dynamics of acrylamide accumulation is somewhat different because coffee roasting takes place at temperatures substantially higher than those normally used in frying and baking processes, usually in the range 220–250 °C. At these temperatures, reactions leading to the formation and elimination of acrylamide are both very active and can coexist in time, which makes their study more difficult.

Acrylamide formation starts rapidly at the beginning of the roasting process and it decreases shortly after reaching a maximum level. Therefore, the degree of roasting will be a key factor in acrylamide content, with light roasted coffees attaining significantly higher amounts than dark roasted ones.[17–21] The decrease observed in acrylamide levels at high degrees of roast certainly occurs when the rate of degradation surpasses the rate of formation, due to the progressive exhaustion of asparagine[18] and reactivity of the compound itself towards reactive species simultaneously present.

Unlike what happens on baking and frying products, there is no apparent source of free carbonyl compounds in green coffee, sucrose being the only sugar present in significant amounts. However, roasting can lead to a pool of neo-formed carbonyls from sucrose decomposition and lipid oxidation, which are likely to play a role in acrylamide formation in coffee.[22] Among others, 5-hydroxymethylfurfural (5-HMF) and 3,4-dideoxyosone originating from sugar decomposition, and 2-octenal, 2,4-decadienal,

2,4-hepatadienal, 4-hydroxynonenal and 4,5-epoxy-2-decenal, arising from fatty acids oxidation were identified in relative high quantities in roasted coffee. 5-HMF in particular could act as an important precursor of acrylamide. It has been demonstrated in model-systems that 5-HMF converts free asparagine in acrylamide more efficiently than glucose during heating,[23] so any ingredient able to stimulate 5-HMF formation would increase acrylamide levels.

Elimination mechanisms of acrylamide have not yet been elucidated satisfactorily. This is not surprising bearing in mind the great reactivity of the compound, thus the many existing possibilities. Indeed, acrylamide possesses two functional groups, the amide group and the electron deficient vinylic double bond, which makes the compound available for a wide range of reactions, including those typically of the amide group, such as hydrolysis dehydration, alcoholysis and condensation with aldehydes, and those characteristic of the vinyl group which can undergo Michael addition type reactions with nucleophilics, including amino and thiol groups of amino acids and proteins.[24,25]

Studies in model-systems showed clearly that coffee melanoidins, formed in large extent during roasting, can react with acrylamide under heating conditions and consequently have a direct influence on the fate of the compound, decreasing the observed levels.[26] Melanoidins are multifunctional negatively charged polymers, and it was hypothesized by the same authors that nucleophilic amino groups arising from the proteinaceous backbone of melanoidins react *via* the Michael addition reaction with acrylamide. The same type of addition reactions was suggested as responsible by the mitigating effect observed by the action of different amino acids with nucleophilic side chains.[27–29]

In contrast to melanoidins, it was shown that chlorogenic acid (5-caffeoylquinic acid, IUPAC nomenclature), a known anti-oxidant present in green coffee in abundant amounts, can contribute to acrylamide formation.[22,30] Although it had no effect on acrylamide formation from asparagine, chlorogenic acid seems to increase acrylamide final levels by three different mechanisms: (i) by increasing the formation of 5-HMF; (ii) by decreasing the activation energy for deamination of 3-aminopropionamide to acrylamide, so easing the respective conversion; (iii) by preventing the attack of produced acrylamide from oxidative free radicals by keeping a high redox potential during the Maillard reaction.[30] These conjugated effects seem to efface the slightly inhibitory action of chlorogenic acid in acrylamide formation by decreasing the amount of neo-formed carbonyl compounds arising from lipid oxidation verified by Kocadağlı and co-workers.[22] In contrast with chlorogenic acid, its quinone derivative inhibits acrylamide formation.[30]

In general, the effect of antioxidants in the levels of acrylamide in roasted coffee or other foods of concern reported in the literature were discordant with some studies claiming mitigation while others had no effect or even an increase, as observed for chlorogenic acid. It was shown in model-systems that stable antioxidants (BHA, BHAT and TBHQ) had no effect or even promoted acrylamide formation, while easily oxidized antioxidants (Ferulic acid, EGCG and Vit. C) inhibited acrylamide formation.[31] The authors suggested that the inhibitory effect could be attributed to the capacity of oxidized forms, once formed, to attack acrylamide, so decreasing the respective levels. In a detailed review of the literature in this area[16] the uneven behaviour of antioxidants was attributed to the ability of antioxidants with different structures or functional groups to react with acrylamide precursors, with intermediates of the reaction or with acrylamide itself, leading to either reducing or promoting effects. The authors proposed that some antioxidants may promote acrylamide formation by acting as donors of carbonyl groups, while others may inhibit acrylamide formation by directly reacting with asparagine or trapping specific Maillard reaction products that cause asparagine precipitation and prevent lipid oxidation. In accordance with these findings, the authors concluded that antioxidant capacity might not be a quality assay to express acrylamide inhibitory ability.

30.5 Occurrence and Factors Affecting the Formation of Acrylamide in Coffees

The main factors that affect acrylamide levels in roasted coffee are coffee species, degree of roasting and storage time. According to the formation/elimination pathways above described it is obvious that some minor changes in the roasting process, such as addition of anti-oxidants, could modulate final acrylamide levels. The amount of the compound effectively ingested by consumers is also dependent on the brew type and respective extension of coffee extraction and dilution factor.

Coffee species may affect greatly acrylamide levels found in roasted beans, with higher amounts found in *Coffea canephora* Pierre (robusta) than in *Coffea arabica* L. (arabica). Several researchers have demonstrated that for identical roasting conditions levels presented by robusta coffees are roughly twice the levels presented by arabica coffees.[18,19,32] The most likely explanation for this differential lies in the fact that free asparagine levels in green coffee beans are slightly higher in robusta beans than in arabica ones as previously reported by Stadler and Scholz[33] and Lantz and co-workers[19] and further confirmed by Bagdonaite and co-workers,[18] who found levels about 60% higher in robusta (797 ± 23 mg kg^{-1}) than in arabica green or unroasted beans (486 ± 96 mg kg^{-1}).

Besides the species, other factors can exert some influence in asparagine levels of the green beans and consequently in acrylamide levels in roasted coffees. As will be described hereafter in the section on mitigation, asparagine accumulation in coffee cherries is influenced by several agronomical factors such as nitrogen supply, micronutrients deficiency and stressing conditions.[34] Any of these factors that lead to increased asparagine levels in coffee beans has the effect of increasing acrylamide formation during roasting. Immature defective beans are usually richer in asparagine, so they should be excluded or subjected to a special treatment consisting in peeling followed by immersion in water.[35] No difference appeared for the two different ways of processing (dry *versus* wet processing) coffee cherries to green coffee beans.[19]

The degree of roast is another key factor in acrylamide levels found in roasted coffee. As explained above, acrylamide formation takes place in the first few moments of the roasting process, quickly reaching a peak from which the contents tend to sharply lower due to the existence of elimination/trapping reactions of the compound. For instance, light roasted coffees present significantly higher amounts than dark roasted ones.[17–21,32] In a recent report provided by the EFSA Panel on Contaminants in the Food Chain regarding acrylamide,[9] the average concentrations of acrylamide in commercial samples were 374, 266 and 187 μg kg^{-1} in light ($n = 45$), medium ($n = 44$) and dark ($n = 15$) roasted coffees, respectively.

The other key factor in final acrylamide levels presented by commercial roasted coffee samples is the marked drop observed in acrylamide levels during storage of roasted coffee.[19,36,37] It was observed for vacuum packed roast and ground coffee that the loss of acrylamide was dependent on the storage temperature.[19] Acrylamide decrease was reported as 33% for instant coffee stored at 25 °C during 12 months.[38] It is likely that the main cause of the observed decrease is the covalent binding of acrylamide to nucleophilic groups of some coffee components, such as ε-NH_2 and SH groups of amino acids which are abundant in the coffee matrix. Supporting this assumption, experiments with radioactive[14] C-labelled acrylamide showed that possible acrylamide adducts remained in the filter residue when coffee ground was brewed, being largely resistant to solvent extraction by different solvents or to sequential polyenzymatic digest, indicative for covalent bond formation.[39]

The amounts of acrylamide in brewed (ready to drink) coffees are highly diverse, from 3 to 68 μg L^{-1} according to the Joint FAO/WHO Expert Committee on Food Additives (JECFA),[40] depending not only on the acrylamide content in the roasted coffee used, but also on the type of preparation and, above all, the respective coffee/water ratio used. Due to its high solubility in water, acrylamide is easily extractable to the coffee brew, usually reaching extraction yields of near 100%, regardless of the preparation process by decoction, infusion or pressure based. The only exception was verified for espresso coffees, the typical coffee brew in southern Europe countries, which are prepared by percolating through a ground coffee cake (usually 6–7 g) a small amount of hot water (usually 30 ml) under high pressure to produce efficiently a very concentrated brew (about 200 g of ground coffee per litre).[41] An extraction efficiency of ~80% was observed for standard espressos, this value being only affected by the brew volume, so by the extension of the contact time between ground coffee and water: in "lungo" espressos (~70 ml) the extraction of acrylamide was almost total, while in "ristretto" espressos (~20 ml) the extraction yield was around 60%.[32] Taking into account the almost total extraction of acrylamide by all the other modes of preparation of coffee brews, acrylamide levels are mainly depending on the amounts of ground coffee per litre of water used for the brewing, which can be very different according to the specific habits of each country, *e.g.* 20 g in the USA,[42] 40 g in Denmark,[43] 50 g in Sweden[44] and 60 g in Norway.[45]

30.6 Contribution of Coffee for the Human Intake of Acrylamide

Accurate data on dietary acrylamide exposure is difficult to obtain owing to the methodological differences in the estimation of dietary acrylamide intake, different ways of describing the different food groups, huge differences between the acrylamide content of the same food according to the home-cooking process and large variations in acrylamide databases for the same product (*e.g.* in an EFSA report,[46] roasted coffee levels varied between 79 and 975 μg kg^{-1}, with a mean of 312 μg kg^{-1}). Furthermore, there are great differences from country to country according to the prevailing dietary patterns and the type of culinary practices. Despite all these difficulties, it is unquestionable that the contribution of coffee to the dietary daily intake of acrylamide can be highly significant, mainly in countries with high coffee consumption.

The first assessment on acrylamide dietary exposure was reported by JECFA in 2006, who estimated a dietary acrylamide intake for the general population, including children, between 1 μg kg^{-1} bw day^{-1} (at the mean) and 4 μg kg^{-1} bw day^{-1} (for a consumer at a high percentile of the distribution).[47] Although subsequent data from several countries presented somewhat lower consumption estimates, these values were further confirmed by JECFA in 2011, for safety evaluation purposes.[40]

Based on mean acrylamide contents of 43 419 food samples from most of the European countries in 2010–2013 and national consumption data, EFSA estimated the mean and 95th percentile dietary acrylamide exposures across surveys and age groups at 0.4 to 1.9 μg kg^{-1} body weight (bw) per day and 0.6 to 3.4 μg kg^{-1} bw per day, respectively (EFSA, 2015). Coffee was the second main contributor to total dietary exposure (immediately following the category "Potato fried products (except potato crisps and snacks)") representing up to 34% of the total acrylamide average exposure. When considering the consumers only, the highest consumption levels are observed in the elderly and very elderly age groups, with median average and 95th percentile consumption levels at respectively 17–21 g dry equivalent per day and 40–45 g dry equivalent per day.[6] If we consider the average acrylamide level on coffee stated in the same report – 522 μg kg^{-1} – daily acrylamide intakes from coffee of 8.9–11.0 μg and 20.9–23.5 μg should be considered.

Within the European Prospective Investigation into Cancer and Nutrition (EPIC) study, a large survey aiming to describe the mean dietary acrylamide intake in 27 regional centres of 10 European countries (Spain, Italy,

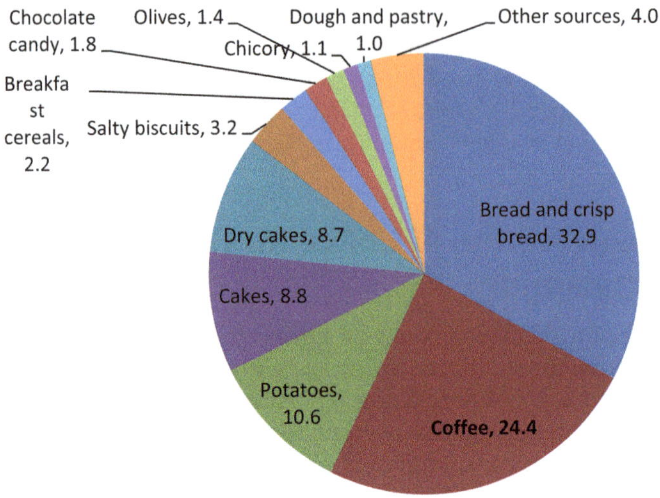

Figure 30.4 Food categories accountable for acrylamide intake (data from ref. 48).

Greece, France, Germany, The Netherlands, UK, Denmark, Sweden and Norway) involving 36 994 men and women, aged 35–74 years, based on a single, standardized 24 hour dietary recall, reported an adjusted acrylamide intake ranging from 13 to 47 μg day^{-1} in men and from 12 to 39 μg day^{-1} in women, with the higher values coming from the northern European countries.[48] Coffee was the second main contributing food group, just after the group "bread and crispbread", accounting for around 25% of acrylamide intake (Figure 30.4).

Besides the differences resulting from different ratios ground coffee/water, the above cited dependence of the roasting degree and coffee variety on acrylamide levels can lead to significant differences in the amount of acrylamide levels ingested by drinking a cup of coffee. As an example, it was reported that the mean levels of a standard espresso cup (6.5 g of ground coffee per 30 ml of water) can range from 0.85 μg (mean of two arabica samples from different geographical origin, submitted to dark roasting) to 9.4 μg (mean of two robusta samples from different geographical origin, submitted to light roasting)[41] (Figure 30.5). By this example it can be concluded that, even among the large consumers of coffee, differences in daily intake of acrylamide can be extremely large, depending on individual consumption habits.

30.7 Mitigation Strategies for the Reduction of Acrylamide in Coffees

Similarly to what happens with other foods of concern regarding the presence of acrylamide, mitigation strategies enforceable in coffee can be divided into three main groups: (i) measures to minimize acrylamide precursors levels *ad initium*, which includes breeding for low asparagine levels, and removal

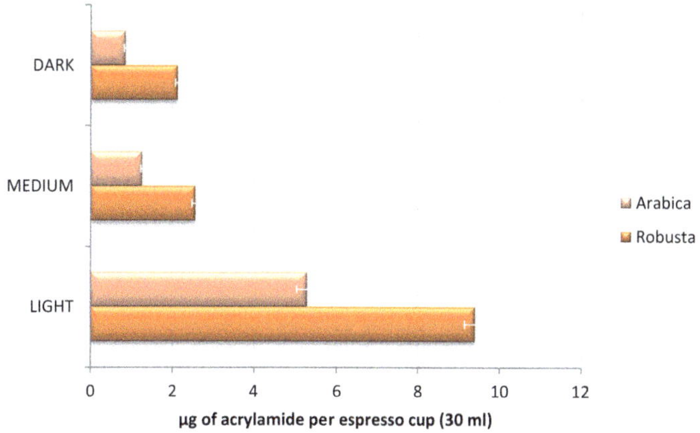

Figure 30.5 Amounts of acrylamide per cup of a standard espresso (data from ref. 41).

of asparagine before processing; (ii) measures to minimize the formation (and increase the elimination) of the compound during processing, which includes formulation and/or processing parameters changes; (iii) measures of removing or trapping acrylamide already formed, along the post-processing steps. The different mitigation approaches may be based on biological, physical or chemical methods.

Given that acrylamide is a by-product of the Maillard reaction, and is essential for the development of important sensorial properties of foods such as colour, taste, flavour and texture, it becomes rather difficult to decrease acrylamide levels without affecting the organoleptic characteristics of the food products. In coffee this challenge has an even greater relevance because minor changes in the standardized roasting process systematically give rise to a major decrease in the quality of the final product. It should also be noted that most of the mitigation measures that have been proposed were only tested at laboratory or pilot scale, it not being possible to know whether the success claimed can be replicated when applying the same measure at industrial scale. Overall, there are few proposals presented so far that are truly suitable to be used by the coffee industry.

30.7.1 Mitigation Strategies Based on Reduction of Asparagine

The most effective way to reduce acrylamide in heated foods is to reduce precursor levels in the raw food materials. Because asparagine has been described as the limiting factor for acrylamide formation in coffee, possible pre-processing mitigation strategies have been focused on asparagine reduction in green coffee beans.

Taking into account that free asparagine levels in raw coffee beans lie within a very narrow range, typically from 0.2 to 1 g kg^{-1}, the opportunity for possible control or reduction by selection of beans with relatively low amounts of free asparagines is small.[49] Halford and co-workers[34] reviewed the factors that influence the accumulation of asparagine in plants, which include an exacerbated nitrogen supply together with lack of nutrients such as potassium, sulfur, phosphorous and magnesium, and stressing conditions namely exposition to toxic metals such as cadmium, pathogen attacks and drought or salt stress. The lack of data on the importance of these factors specifically in the coffee did not allow so far the emergence of proposals in this field aiming to obtain coffee raw materials with decreased levels of asparagine. Genetic modification and other genetic techniques such as identification of quantitative trait loci for low grain asparagine concentration provide another possibility of obtaining cereal and coffee grains with reduced levels of asparagine[34] but again no studies are described regarding coffee.

The enzymatic degradation of asparagine present in raw materials previously to the heat processing by means of asparaginase have been shown to be one of the most promising mitigation strategies in a variety of foods. The use of asparaginase, an enzyme that converts asparagine into aspartic acid and ammonia, was first suggested by Zyzak and co-workers[14] and has been further subject of several patent applications. There are two companies that provide the enzyme commercially. Their use was generally recognized as safe by the US authorities and it has also been given a favourable evaluation as a food additive by JECFA.[50] A very useful review of asparaginase usage in the reduction of acrylamide levels in cooked food was recently provided by Xu and co-workers.[51]

The first application to coffee of the asparaginase pre-treatment in order to reduce asparagine levels in the coffee green beans and consequently attain roasted coffee with reduced acrylamide contents was patented by Dria and co-workers in 2007.[52] Basically, it consisted of subjecting green coffee beans to a water solution containing asparaginase, after a series of preliminary treatments essential to promote a proper contact between the enzyme and the substrate. These treatments could include (i) reduction into fragments (milling, grinding) of the coffee beans, so as to increase the contact surface between the green coffee and the enzyme; (ii) exposure of the coffee beans to the action of cellulases, hemicellulases and/or pectinases, for degrading the cellulose and thus the structure of the beans; (iii) drying the coffee beans, or treating the coffee beans with low pressure or atmospheric pressure steam, so as to open the pores of the beans and facilitate the penetration inside the beans of the water solution containing the enzyme. The complexity of these preliminary treatments that the beans of green coffee have to undergo, and which appear substantially indispensable for obtaining an effective interaction between the enzyme in solution and the asparagine contained in the green coffee beans, make the method particularly complicated to implement and substantially uneconomic.[53]

Better results of the asparaginase pre-treatment of green coffee beans could be achieved by incubating the wetted green beans in a manner similar to that employed in the decaffeination process usually applied by the industry, and infusing asparaginase during the steps of this process. Using this approach at laboratory scale, reductions of 55–74% in the acrylamide levels of roasted coffee were achieved by Hendriksen and co-workers[54] while Xu and co-workers[55] claim losses in acrylamide final levels in the order of 69–86%, making use of slightly higher dosages of the enzyme. Navarini and co-workers[53] have patented recently a method based on a similar approach, which differs from the previous in the additional use, besides asparaginase, of an enzyme aspartase able to degrade aspartic acid. According to the authors, aspartic acid formed from asparagine degradation added to aspartic acid already present in green coffee beans are likely to promote acrylamide formation, so better results are obtained when a simultaneous decrease of asparagine and aspartic acid are achieved.

The use of immobilized enzymes has recently been shown as capable of providing a large reduction of costs by treatment with asparaginase at the industrial level, due to the huge reduction in the amount of enzyme and also water used in the process.[56]

Despite all the advances observed in the use of asparaginase to obtain roasted coffee with lower levels of acrylamide, some more factors will have to be developed for its industrial application. Furthermore, it is necessary to guarantee that major changes in the aromatic profile of the final product do not occur. According to the toolbox provided by Food Drink Europe (FDE)[49] the use of asparaginase is likely to be more successfully applied to robusta green coffees, which are usually steam treated for flavour modification purposes than for arabica green beans, which have been seen severely affected in their "so appreciated" taste and flavour in coffee brews.

30.7.2 Mitigation Strategies Based on Alterations of the Roasting Processing Conditions

As stated above, acrylamide levels in coffee beans decreased with increasing thermal input. At the temperatures applied during coffee roasting, the reaction that leads to the depletion of acrylamide dominates near the end of the cycle. Darker roasting tends to reduce acrylamide levels. Nevertheless, roasting to a darker colour is not a practical mitigation method as it results in deep changes in the flavour profile of the end product, leading to rejection by many consumers. Moreover, dark roasting not only effects some beneficial attributes of coffee, such as its antioxidant activity, but also leads to the formation of undesirable products.[57]

More promising seems to be the proposal of Anese and co-workers (2014)[58] on roasting under vacuum conditions. Using a pressure of 0.15 kPa the authors have obtained a medium roast degree coffee with approximately 50% less acrylamide than conventionally roasted coffee, while maintaining favourable organoleptic properties. The observed reduction was probably due to a stripping effect exerted by the low pressure generated inside the oven during the vacuum process, preventing the accumulation of acrylamide and/or promoting the sublimation of the compound.

Steam roasting at a temperature of 200–240 °C allowed a significant decrease of acrylamide levels when compared with conventional roasting.[59] The reason lies in the longer roasting time needed to reach the desired final colour, so acrylamide elimination reactions have more time to develop. Nevertheless, the results of the comparative sensory analysis have shown that the method led to an undesired alteration of the organoleptic properties of the coffee, namely a higher acidity and less "roasty" notes. When comparing products similar in flavour profile, the steam-roasting process offers only a minor reduction potential of <10%.

It was also shown that decreasing the roasting air velocity from 1 ms^{-1} to 0.5 ms^{-1} results in a decrease of acrylamide level, regardless of the temperature used (190 °C–216 °C), but again at the expense of the antioxidant activity and sensory characteristics of the end product.[60]

30.7.3 Mitigation Strategies Based on Removing or Trapping of Acrylamide Already Formed

The removal of acrylamide by volatilization was studied by Anese and co-workers,[61,62] who demonstrated that the process requires a previous hydration step of the roasted coffee because acrylamide removal is hindered by the glassy state of the coffee matrix. In these conditions, vacuum treatment promotes, simultaneously with the release of acrylamide, loss of many volatile compounds responsible for coffee flavour.

The use of supercritical extraction with CO_2 as elimination technique of acrylamide from roasted coffee beans was also investigated.[63] Temperature and addition of ethanol were the variables that affected the extraction process to the highest extent. The best working conditions were found to be 100 °C, 200 bar and 9.5% w/w ethanol, resulting in a efficiency of acrylamide removal of 45% when the duration of the extraction was below 525 min, and 79% when the extraction time raised up to 1305 min. The caffeine content in the coffee beans was not significantly affected but organoleptic alterations were not studied. The process has the advantage to use equipment already used by industry in the coffee decaffeination process, although it is expected that subjecting the coffee beans to the proposed conditions for nearly 24 h causes significant changes in the aroma profile of coffee.

A first approach to the possibility of removing acrylamide by applying bacterial enzymes able to hydrolyse acrylamide was developed by Cha.[64] The author verified that acrylamidases – enzymes that specifically hydrolyze acrylamide to acrylic acid and ammonia – are able to degrade totally the acrylamide present in a solution of 100 mg coffee/10 ml water, although some inhibitory effect from coffee was observed, requiring the use of amounts of enzyme extracts several times more than is normally required in the absence of coffee. The enzymes were active even at 70 °C, which makes them useful to be added right after brewing from a coffee maker or applied to coffee filters. However, the method needs further investigation regarding the path of neo-formed acrylic acid, a compound known to be corrosive and cause severe skin irritation. Moreover, ammonia released in the reaction could negatively affect flavour profile.

Yeast fermentation was proposed by Akillioglu and Gőkmen (2014)[65] as a promising technique for the simultaneous mitigation of acrylamide and 5-HMF, another heat-induced contaminant of concern, in instant coffees. The authors claimed reductions of about half of the acrylamide present by using baker's yeast (*Saccharomyces cerevisiae* 1–2% w/v) as fermentative agent, during 24 hours, in the presence of 10% sucrose. The presence of sucrose was essential to enhance the activity of yeast cells, hence the degradation of acrylamide. According to the authors the developed method can be easily adapted by the industry because it can be integrated on the fermentation step into the regular production of instant coffee. Again, the main drawback of the proposal is the likely formation of acrylic acid and the possible loss of sensory attributes.

An almost total (about 95%) removal of acrylamide from canned coffees containing levels of 6.2 ± 0.02 μg l^{-1} was recently reported,[66] by means of heat treatment with adding cysteine at 121 °C for 6 minutes. The process can be performed during the manufacturing of canned coffees because the used conditions are the same as the heat sterilization process typically used. However, further studies are needed to verify the possible regeneration of acrylamide during long-term storage at low temperatures and additionally to evaluate the toxicity of cysteine-S-β-propionamide, the compound resulting from Michael addition reaction of acrylamide with cysteine.

As previously reported, an important decrease of acrylamide levels during storage of roasted coffee was usually observed, regardless of any technological intervention. Hence, one can assume that acrylamide levels in commercial coffee samples are dependent on the respective storage time. Because flavour decay is also observed throughout storage time, and flavour corresponds to one of the key parameters that affect coffee quality, storage does not represent therefore a true mitigation opportunity.[57]

30.8 Final Considerations

Exposure assessments show that coffee is undoubtedly one of the food staples of most concern as regards the intake of acrylamide. Up to now, there have been no economically feasible solutions for the production of roasted coffee with significantly lower levels of acrylamide. However, the most recent results from the extensive research efforts carried out by several scientists groups on the complex mechanistic pathways of formation and elimination of acrylamide throughout the roasting process create the firm expectation that such objective will be possible to realize in the near future and thus minimize the amount of acrylamide ingested by consumers whenever they take a cup of this drink so much appreciated all over the world.

References

1. International Agency for Research on Cancer [IARC], *IARC Monogr. Eval. Carcinog. Risks Hum.*, 1994, **60**, 389.
2. K. B. Girma, V. Lorenz, S. Blaurock and F. T. Edelmann, *Coord. Chem. Rev.*, 2005, **249**(11), 1283.
3. E. A. Smith and F. W. Oehme, *Rev. Environ. Health*, 1991, **9**(4), 215.
4. T. Reynolds, *J. Natl. Cancer Inst.*, 2002, **94**(12), 876.
5. E. Tareke, P. Rydberg, P. Karlsson, S. Eriksson and M. Tornqvist, *J. Agric. Food Chem.*, 2002, **50**(17), 4998.
6. J. S. Ahn, L. Castle, D. B. Clarke, A. S. Lloyd, M. R. Philo and D. R. Speck, *Food Addit. Contam.*, 2002, **19**, 1116.
7. D. R. Lineback, J. R. Coughlin and R. H. Stadler, *Annu. Rev. Food Sci. Technol.*, 2012, **3**, 15.
8. R. Medeiros-Vinci, F. Mestdagh and B. Meulenaer, *Food Chem.*, 2012, **133**, 1138.
9. EFSA Panel on Contaminants in the Food Chain (CONTAM), *EFSA J.*, 2015, **13**(6), 4104.
10. F. Pedreschi, M. S. Mariotti and K. Granby, *J. Sci. Food Agric.*, 2014, **94**, 9.
11. Commission Recommendation of 10.1.2011 on investigations into the levels of acrylamide in Food, http://ec.europa.eu/food/food/chemicalsafety/contaminants/recommendation_100 1011_acrylamide_food_pl.pdf. Accessed on 2016-03-15.
12. D. Mottram, B. Wedzicha and A. Dodson, *Nature*, 2002, **419**, 448.
13. R. H. Stadler, I. Blank, N. Varga, F. Robert, J. Hau, P. A. Guy, M.-C. Robert and S. Riediker, *Nature*, 2002, **419**, 449.
14. D. V. Zyzak, R. A. Sanders, M. Stojanovic, D. H. Tallmadge, B. L. Eberhart, D. K. Ewald, D. C. Gruber, T. R. Morsch, M. A. Strothers, G. P. Rizzi and M. D. Villagran, *J. Agric. Food Chem.*, 2003, **51**(16), 4782.
15. M. Granvogl, M. Jezussek, P. Koehler and P. Schieberle, *J. Agric. Food Chem.*, 2004, **52**, 4751.
16. C. Jin, X. Wu and Y. Zhang, *Food Res. Int.*, 2013, **51**(2), 611.
17. H. Guenther, E. Anklam, T. Wenzl and R. H. Stadler, *Food Addit. Contam.*, 2007, **24**(1), 60.
18. K. Bagdonaite, K. Derler and M. Murkovic, *J. Agric. Food Chem.*, 2008, **56**, 6081.
19. I. Lantz, R. Ternite, J. Wilkens, K. Hoenicke, H. Guenther and G. H. D. van der Stegen, *Mol. Nutr. Food Res.*, 2006, **50**, 1039.
20. H. Z. Senyuva and V. Gokmen, *Food Addit. Contam.*, 2005, **22**, 204.
21. D. Taeymans, A. Andersson, P. Ashby, I. Blank, P. Gonde, P. van Eijck, V. Faivre, S. P. Lalljie, H. Lingnert, M. Lindblom, R. Matissek, D. Muller, R. H. Stadler, A. Studer, D. Silvani, D. Tallmadge, G. Thompson, T. Whitmore, J. Wood and D. Zyzak, *J. AOAC Int.*, 2004, **88**(1), 234.
22. T. Kocadağli, N. Gőncűoğlu, A. Hamzahoğlu and V. Gőkmen, *Food Funct.*, 2012, **3**, 970.

23. V. Gokmen, T. Kocadağli, N. Gőncűoğlu and B. A. Mogol, *Food Chem.*, 2012, **132**, 168.
24. E. Capuano and V. Fogliano, *LWT–Food Sci. Technol.*, 2011, **44**, 793.
25. M. Friedman, *J. Agric. Food Chem.*, 2003, **51**(16), 4504.
26. S. Pastoriza, J. A. Rufian-Henares and F. J. Morales, *LWT–Food Sci. Technol.*, 2012, **45**, 198.
27. A. Adams, S. Hamdani, F. Van Lancker, S. Méjri and N. Kimpe, *Food Res. Int.*, 2010, **43**(5), 1517.
28. G. Koutsidis, S. P. Simons, Y. H. Thong, Y. Haldoupis, J. Mojica-Lazaro, B. L. Wedzicha and D. S. Mottram, *J. Agric. Food Chem.*, 2009, **57**(19), 9011.
29. R. Zamora, R. M. Delgado and F. J. Hidalgo, *J. Agric. Food Chem.*, 2010, **58**, 1708.
30. Y. Cai, Z. Zhang, S. Jiang, M. Yu, C. Huang, R. Qiu, Y. Zou, Q. Zhang, S. Ou, H. Zhou, Y. Wang, W. Bai and Y. Li, *J. Hazard. Mater.*, 2014, **268**, 1.
31. S. Ou, J. Shi, C. Huanga, G. Zhanga, J. Tenga, Y. Jianga and B. Yang, *J. Hazard. Mater.*, 2010, **182**, 863.
32. R. Alves, C. Soares, S. Casal, M. B. Oliveira and J. O. Fernandes, *Food Chem.*, 2010, **119**, 929.
33. R. H. Stadler and G. Scholz, *Nutr. Rev.*, 2004, **62**(12), 449.
34. N. G. Halford, T. Y. Curtis, N. Muttucumaru, J. Postles, J. S. Elmore and D. S. Mottram, *J. Exp. Bot.*, 2012, **63**(8), 2851.
35. E. C. Dias, *Perfil de aminoácidos nos frutos verdes do cafeeiro processados por via seca e via úmida*, Dissertation, Universidade Federal de Lavras, Brazil, 2008, p. 1.
36. T. Delatour, A. Périsse, T. Goldmann, S. Riediker and R. H. Stadler, *J. Agric. Food Chem.*, 2004, **52**(15), 4625.
37. K. Hoenicke and R. Gaterman, *J. AOAC Int.*, 2005, **88**, 268.
38. J. Michalak, E. Gujska, M. Czarnowska, J. Klepacka and F. Nowak, *Plant Foods Hum. Nutr.*, 2016, **71**, 115.
39. M. Baum, N. Böhm, J. Görlitz, I. Lantz, K. H. Merz, R. Ternité and G. Eisenbrand, *Mol. Nutr. Food Res.*, 2008, **52**(5), 600.
40. JECFA, Safety evaluation of certain contaminants in food. Acrylamide, 72nd Meeting of the Joint FAO/WHO Expert Committee on Food Additives (JECFA), FAO JECFA Monograph 8, 2011, pp. 1–151, WHO Food Additive Series 63http://whqlibdoc.who.int/publications/2011/9789241660631_eng.pdf.
41. C. Soares, R. Alves, S. Casal, M. B. Oliveira and J. O. Fernandes, *J. Food Sci.*, 2010, **75**(3), 57.
42. D. Andrzejewski, J. Roach, M. Gay and S. Musser, *J. Agric. Food Chem.*, 2004, **52**, 1996.
43. K. Granby and S. Fagt, *Anal. Chim. Acta*, 2004, **520**, 177.
44. K. Svensson, L. Abramsson, W. Becker, A. Glynn, K.-E. Hellenäs, Y. Lind and J. Rosén, *Food Chem. Toxicol.*, 2003, **41**, 1581.
45. E. Dybing and T. Sanner, *Toxicol. Sci.*, 2003, **75**, 7.

46. EFSA Panel on Contaminants in the Food Chain (CONTAM), *EFSA J.*, 2011, **9**(4), 2133.
47. JECFA, Evaluation of certain food contaminants, 64th Report of the Joint FAO/WHO Expert Committee on Food Additives, WHO Technical Report Series, No. 930http://whqlibdoc.who.int/trs/WHO_TRS_930_eng.pdf, 2006, pp. 8–26, 93.
48. H. Freisling, A. Moskal, P. Ferrari, G. Nicolas, V. Knaze, F. Clavel-Chapelon, *et al.*, *Eur. J. Nutr.*, 2013, **52**, 1369.
49. Food Drink Europe (FDE) Acrylamide Toolbox, 2014, http://www.fooddrinkeurope.eu/uploads/publications_documents/AcrylamideToolbox_2013.pdf.
50. JECFA, Compendium of food additive specifications, 68th Meeting of the Joint FAO/WHO Expert Committee on Food Additives. FAO JECFA Monograph 4, 2007, p. 3.
51. F. Xu, M.-J. Oruna-Concha and J. S. Elmore, *Food Chem.*, 2016, **210**, 163.
52. G. Dria, D. Zyzak, R. Gutwein, F. Villagran, H. Young, P. Bunke, P. Lin, J. Howie and R. Schafermeyer, *US Pat.* 7220440 B2, 2007.
53. L. Navarini, L. Terra, S. Colomban, V. Lonzarich and F. Liverani, *US Pat.* 20140193541 A1, 2014.
54. H. V. Hendriksen, G. Budolfsen and M. J. Baumann, *Aspects Appl. Biol.*, 2013, **116**, 41.
55. F. Xu, P. Khalid, M.-J. Oruna-Concha and J. S. Elmore, in *Flavour Science*, ed. A. J. Taylor and D. S. Mottram, Context Products Ltd, Packington, United Kingdom, 1st edn, 2015, pp. 563–566.
56. H. V. Hendriksen, K. Puder and A. G. Olsen, *International Pat.* WO/2014/161935, 2014.
57. C. Summa, B. Calle, M. Brohee, R. H. Stadler and E. Anklam, *LWT*, 2007, **40**, 1849.
58. M. Anese, M. C. Nicoli, G. Verardo, M. Munari, G. Mirolo and R. Bortolomeazzi, *Food Chem.*, 2014, **145**, 168.
59. V. Theurillat, V. Leloup, R. Liardon, R. Heijmans and P. Bussmann, *Proceedings of ASIC 21st International Conference, 11–15 September, Montpellier, France*, 2006.
60. G. Budryn, E. Nebesny and J. Oracz, *Int. J. Food Prop.*, 2015, **18**, 290.
61. M. Anese, L. Manzocco and E. Maltini, *Eur. Food Res. Technol.*, 2005, **221**, 434.
62. M. Anese, M. Suman and M. C. Nicoli, *Food Eng. Rev.*, 2009, **1**, 169.
63. M. Banchero, G. Pellegrino and L. Manna, *J. Food Eng.*, 2013, **115**, 292.
64. M. Cha, *Eur. Food Res. Technol.*, 2013, **236**, 567.
65. H. G. Akillioglu and V. Gőkmen, *Food Res. Int.*, 2014, **61**(7), 252.
66. Y. Narita and K. Inouye, *J. Agric. Food Chem.*, 2014, **62**, 12218.

CHAPTER 31

β-Carbolines

DANIELA A. C. RODRIGUES AND SUSANA CASAL*

LAQV@REQUIMTE, Serviço de Bromatologia e Hidrologia, Faculdade de Farmácia, Universidade do Porto, Rua de Jorge Viterbo ferreira, 228, 4050-313 Porto, Portugal
*E-mail: sucasal@ff.up.pt

31.1 Introduction

In addition to coffee popularity due to its sensory/stimulant properties, the increasing amount of literature attributing potential health effects to regular coffee consumption has triggered research into the quest for potential responsible compounds. This has been the case of β-carbolines, whose presence in coffee is still regarded with caution, due to both the high amounts, making coffee one of the richest dietary sources, as well as their strong biological actions, particularly in the nervous system.

β-Carbolines are mostly known as naturally occurring alkaloids, derived from tryptophan/tryptamine, being present in plants and fruits, cooked/processed foods, beverages/fermented beverages, tobacco smoke, and biological tissues and fluids (such as blood plasma and urine), usually derived from tetrahydro-β-carbolines oxidation.[1–3] The amounts seem to depend on several factors, including precursors availability, ripening, processing method (*e.g.* fermentation, heating, smoking), processing temperature, storage time, presence of oxidants/antioxidants or other types of additives, among others.[2,4,5] Usually, raw and fermented foods are richer in tetrahydro-β-carbolines-3-carboxylic acids, while thermally processed ones have

Coffee: Production, Quality and Chemistry
Edited by Adriana Farah

Published by the Royal Society of Chemistry, www.rsc.org

lower amounts of tetrahydro-β-carbolines and higher contents of the fully aromatic β-carbolines, particularly harman and norharman.[2,4,6] Roasted coffee is, as previously mentioned, one of the richest sources of β-carbolines in the diet.[3,7–9]

The aim of this chapter is to briefly review the current knowledge on β-carbolines analysis and content in coffee, comparing it to other food sources. The physiological aspects are explored in *Coffee: Consumption and Health Implications*.

31.2 Chemical Properties and Formation Routes

β-Carbolines are fully aromatic tricyclic heterocyclic compounds with a pyrido[3,4-*b*]indole ring. Tetrahydro-β-carbolines have a similar structure, but with a totally reduced pyrido ring (Table 31.1). Both might contain several substituents, giving rise to an extended family of derived compounds.[4,10] They are regarded as alkaloids, due to their natural presence in several plant

Table 31.1 Physicochemical properties of simple carbolines: norharman and harman, tetrahydro-β-carboline, and tetrahydro-β-carboline-1-carboxyilic acid. Data from ref. 4 and 13–18.

	Norharman	Harman	Tetrahydro-β-carboline	Tetrahydroharman-3-carboxilic acid
Molecular formula	$C_{11}H_8N_2$	$C_{12}H_{10}N_2$	$C_{11}H_{12}N_2$	$C_{13}H_{14}N_2O_2$
CAS number	244-63-3	486-84-0	16502-01-5	5470-37-1
Molecular weight	168.20	182.22	172.23	230.26
Systematic name	9*H*-Pyrido[3,4-*b*] indole	1-Methyl-9*H*-pyrido[3,4-*b*] indole	2,3,4,9-Tetrahydro-1*H*-pyrido[3,4-*b*] indole	1-Methyl-2,3,4,9-tetrahydro-1*H*-pyrido[3,4-*b*] indole-3-carboxylic acid
Synonymous	9*H*-β-Carboline; carbazoline; 2,9-diazafluorence	1-Methyl-9*H*-β-carboline; Aribine; Passiflorian	Tetrahydronorharman; tetrahydro-9*H*-β-carboline; tryptoline	Tetrahydroharmane-3-carboxylic acid; 1-methyl-1,2,3,4-tetrahydro-β-carboline-3-carboxylic acid
Melting point	199–201°C	237–238°C	206–208°C	243–245°C
Chemical structure	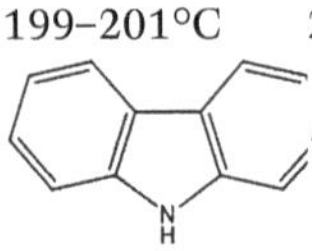	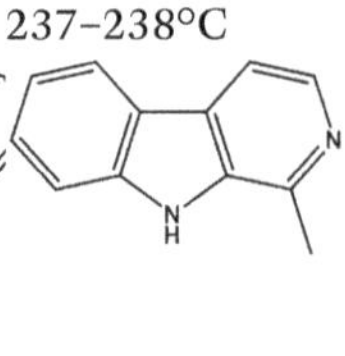	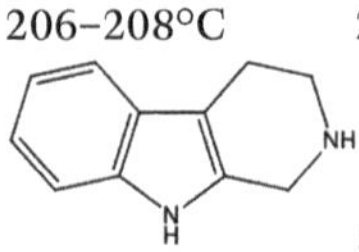	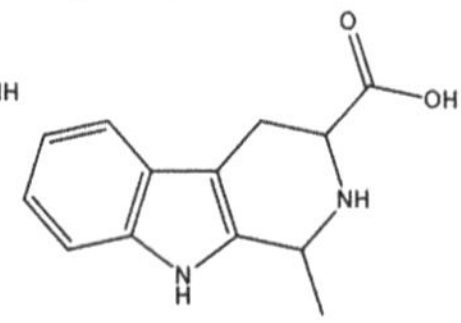

families, but are also formed during food processing, namely cooking, fermentation, and storage.[5]

Simple carbolines biosynthesis seems to occur from indoleethylamine derivatives (such as tryptophan, tryptamine, or serotonine) and aldehyde/α-keto acids (such as formaldehyde or acetaldehyde), through a chemical or enzymatic Pictet–Spengler condensation, resulting in the formation of tetrahydro-β-carboline-3-carboxylic acids. Oxidation of the pyrido ring provides β-carbolines.[4,6,11]

Table 31.1 details some of the most important physicochemical properties of the two main β-carbolines in coffee, norharman and harman, and two of the most common tetrahydro-β-carbolines, tetrahydronorharman and tetrahydroharmane-3-carboxylic acid. Norharman and harman are also frequently named as non-polar heterocyclic amines. Their fluorescence is also among the most interesting properties, being responsible for the native fluorescence of some animals, such as scorpions, while providing a selective and sensitive tool for their detection in analytical methods, as will be detailed later.[4,12,13]

31.3 β-Carbolines and Tetrahydro-β-carbolines in Beverages and Food

Tetrahydro-β-carbolines are the most prevalent carbolines in foods. Norharman and harman are the most common β-carbolines, particularly in coffee.

β-Carbolines and tetrahydro-β-carbolines can be found in some traditional medicinal plants (*Passiflora incarnata*, *Tribulus terrestris*, *Pueraria lobata*, and *Ananas comosus*), where they were first reported, but also in edible plants and fruits (wheat, rice, maize, soy, mushrooms, grapes, pineapple, banana, nuts), fermented beverages (beer, wine, vinegar, whisky, brandy, sake), fermented foods (yogurt, cheese), seasonings, and thermally processed foods (meat, fish, cookies, roasted bread, chocolate, roasted coffee, *etc.*).[2,4,6,8,19]

Table 31.2 summarizes the amounts of β-carbolines in several beverages and foods, with an individual detail of harman and norharman amounts, due to their relevance in coffee. Tetrahydro-β-carbolines were also included as a separate group, with tetrahydro-β-carbolines-carboxylic acid being generally the main compound.

The occurrence of tetrahydro-β-carbolines in fruits and vegetables suggests that the naturally present free aldehydes (acetaldehyde and formaldehyde) might react with tryptophan, serotonine, or tryptamine under mild acidic environments, releasing tetrahydro-β-carbolines.[15,16,20] These compounds have also demonstrated *in vitro* antioxidant activity, with a higher ability to scavenge free radicals than ascorbic acid, suggesting that they might also contribute to the antioxidant activity of several foods.[18]

The presence of tetrahydro-β-carbolines in alcoholic beverages and fermented products results from oxidation of the corresponding

Table 31.2 Norharman, harman, and tetrahydro-β-carbolines values in some beverages and food sources.[a]

Food matrices		Norharman	Harman	Tetrahydro-β-carbolines	References
Vegetables/fruits					
Flour	μg 100 g^{-1}	n.d.	n.d.–1.8	10–250	3 and 4
Toasted bread	μg 100 g^{-1}	4.2–16.4	n.d.	n.d.	3 and 23
Bread	μg 100 g^{-1}	n.d.–6.5	n.d.–0.2	4–251	3, 4 and 15
Cookies/biscuits	μg 100 g^{-1}	0.8–3.4	0.04–0.5	n.d.	3
Breakfast cereals	μg 100 g^{-1}	n.d.–18.7	n.d.–9.2	n.d.	3
Fruit/juices/purees	μg 100 g^{-1}	n.d.–0.55	n.d.–0.18	n.d.–1160	3, 4, 15, and 18
Jams/marmalades	μg 100 g^{-1}	n.d.–0.1	n.d.–0.04	0.6–42	3 and 4
Nuts/dry fruits	μg 100 g^{-1}	n.d.–1.4	n.d.–0.7	n.d.	2 and 3
Chocolate/cocoa	μg 100 g^{-1}	0.4–12.3	0.6–23.4	115–670	3 and 26
Roasted coffee[b]	μg 100 g^{-1}	145–1593	13–375	n.d.	7, 8, and 24
Meat					
Uncooked	μg 100 g^{-1}	n.d.	n.d.	n.d.	3 and 6
Cooked	μg 100 g^{-1}	n.d.–80	n.d.–17	n.d.	2, 3, 4, 17, 23, and 25
Smoked	μg 100 g^{-1}	n.d.	n.d.	n.d.–1480	4, 6, 14, and 22
Fish					
Uncooked	μg 100 g^{-1}	n.d.–2.7	n.d.	n.d.	2, 3, and 6
Cooked	μg 100 g^{-1}	n.d.–20	n.d.–13	n.d.	2–4 and 6
Smoked	μg 100 g^{-1}	n.d.–0.45	n.d.–0.02	3–1220	3, 6, 15, and 22
Fermented products					
Wine	μg 100 ml^{-1}	n.d.–0.31	n.d.–3.9	n.d.–1780	3, 4 and 15
Vinegar	μg 100 ml^{-1}	n.d.–0.35	n.d.–1.43	0.1–0.95	2–4 and 15
Beer	μg 100 ml^{-1}	0.3–2.3	0.07–14	n.d.–1703	2, 4, 14, and 15
Soft drinks	μg 100 ml^{-1}	n.d.	n.d.	n.d.–35	4 and 15
Yoghurt	μg 100 g^{-1}	n.d.	n.d.	0.7–13	3, 4 and 15
Milk	μg 100 ml^{-1}	n.d.	n.d.	n.d.	3 and 15
Cheese	μg 100 g^{-1}	n.d.–47	n.d.–43	n.d.–256	3, 4, 15 and 27
Smoked cheese	μg 100 g^{-1}	n.d.	n.d.	7–610	22
Soy sauce	μg 100 ml^{-1}	3.2–5.2	13–25	10–900	3, 4, 15 and 27

[a]n.d. – not detected.
[b]Further detailed in Table 31.3.

tetrahydro-β-carboline-3-carboxylic acids during alcoholic fermentation, formed from tryptophan and aldehydes released by yeasts.[3,15,21] Tetrahydro-β-carbolines are also the main β-carbolines found in smoked products, probably formed from formaldehyde, a naturally occurring compound in wood smoke.[2,3,22]

Conversely, harman and norharman β-carbolines are more prevalent in cooked products, including meat, fish, roasted cereals, and coffee, with an almost total absence of tetrahydro-β-carbolines. They are present in residual amounts in fruits, vegetables, and dairy products. Differences between cooked foods may be a consequence of the presence of precursors and cooking conditions (temperature and/or time).[2–4,22]

31.4 Norharman and Harman β-Carbolines in Coffee

Taking into consideration the high worldwide coffee consumption, norharman and harman intake through coffee brews could have a relevant impact on the human body. This is a highly discussed field, not totally clarified, as detailed in *Coffee: Consumption and Health Implications*. However, in order to support biological and epidemiological studies, it is mandatory to have a consistent knowledge on effective food intake, and therefore also on the technological factors that might have influence on the β-carbolines amounts in coffee brews. Table 31.3 contains some of the results achieved in diverse published papers.

So far, only harman and norharman have been reported in coffee, probably due to thermal degradation of carboline intermediates.[8] Roasting is clearly the most determinant parameter, with β-carbolines being almost absent in unroasted coffee. However, the amounts achieved after roast are highly variable, being dependent on the roast degrees and coffee species (*C. arabica* L. or arabica and *C. canephora* Pierre or robusta).[8,9] Indeed, robusta roasted coffee has consistently higher amounts of both β-carbolines than arabica. Therefore, we can expect that the presence of robusta in a coffee blend contributes to higher content of β-carbolines in the brew. Knowing that the amount of precursors is one of the factors that influence the amount of β-carbolines in coffee, the consistently higher tryptophan amounts in robusta green coffee beans might be implicated.[8,9]

Published results also demonstrate that β-carboline amounts are correlated to the brewing process. Indeed, the different coffee/water ratios, extractive temperatures and times, characteristic of each method, together with the variable cup sizes, results in a high dispersion of β-carboline intake per standard brew, influencing β-carboline daily ingestion.[7,9] Indeed, although the results for coffee brews are detailed in Table 31.3 on 100 mL basis, one cannot disregard that a common espresso has only around 30 mL, while a drip filter coffee mug can have more than 120 mL. Also, as previously discussed, the presence of robusta coffee in the blend, typical in Mediterranean espresso coffee, will also contribute to increased β-carboline amounts per cup.

Soluble (instant) is also a source of β-carbolines, with highly variable amounts. Roasted cereals, such as chicory, malt, barley and rye, also contain β-carbolines, particulalrly chicory, but in lower amounts than coffee.[28]

Table 31.3 Norharman and harman values in coffee and coffee beverages.

Matrices		Norharman	Harman	Sources
Ground coffee				
Unroasted arabica	μg 100 g^{-1}	2–4	2–4	7 and 29
Unroasted robusta	μg 100 g^{-1}	5–6	5–10	29
Roasted arabica	μg 100 g^{-1}	145–766	13–116	8 and 24
Roasted robusta	μg 100 g^{-1}	296–1593	80–375	8 and 29
Blend	μg 100 g^{-1}	188–311	33–69	7
Decaffeinated	μg 100 g^{-1}	252–550	33–70	7
Coffee brews				
Arabica espresso	μg 100 ml^{-1}	8–21	3–9	8
Robusta espresso	μg 100 ml^{-1}	30–60	10–20	8
Blends espresso	μg 100 ml^{-1}	14–35	5–15	7 and 8
Decaffeinated espresso	μg 100 ml^{-1}	15–32	4–10	8
Mocha	μg 100 ml^{-1}	16–35	5–14	8
Press-pot	μg 100 ml^{-1}	9–18	3–8	8
Filter	μg 100 ml^{-1}	3–20	1–9	7 and 8
Soluble coffees				
Plain coffee	μg 100 g^{-1}	106–753	27–211	3, 7, 9, 23, 30
Decaffeinated coffee	μg 100 g^{-1}	454–934	98–141	7
Blends with coffee	μg 100 g^{-1}	245–589	92–242	9
Plain barley	μg 100 g^{-1}	92–132	27–52	9
Cereals	μg 100 g^{-1}	144–223	63–92	9
Plain chicory	μg 100 g^{-1}	128–200	94–134	9 and 28

31.5 Analysis of β-Carbolines and Tetrahydro-β-carbolines in Foods

Due to their low concentrations, and particularly in complex thermally processed foods, isolation of carbolines, purification, and enrichment are mandatory steps for accurate quantitative analysis, together with the use of adequate internal standards.[2,4,24,27] Although lipid–liquid extraction is usually mentioned, particularly with ethyl acetate under basic pH, solid phase extraction (SPE) provides a faster and cleaner approach, while combining sample concentration under the same procedure. Both reversed-phase SPE, using classical C18 phases, and ion-exchange-SPE have been used, but the latter, with strong cation exchange (SCX) – benzenesulfonic acid, is presently the most common approach for food extracts, providing cleaner extraction than the former.[2,4,23,24,27]

Simultaneously, in order to avoid artifact formations, particularly new tetrahydro-β-carbolines from available precursors, protective measures should be implemented. Together with the previous removal of aldehydes from all reagents and solvents, the addition of aldehyde-trapping agents during sample extraction, such as semicarbazide or fluorescamine, is among the most common measures.[4,15] Knowing that tetrahydro-β-carbolines formation is also pH dependent, being favored under mild acidic conditions, the use of neutral or slightly alkaline conditions during extraction also avoids formation of analytical artifacts.[4] Direct sample analysis is also a possible approach, but this is not compatible with most of the chromatographic techniques.

For quantification, effective separation from interferences and coeluting compounds is necessary to achieve selectivity, requiring advanced chromatographic techniques, such as capillary gas chromatography (GC), high-performance liquid chromatography (HPLC), or capillary electrophoresis.[2,4,19,27] Gas chromatography requires chemical derivatization to convert carbolines into volatile molecules, commonly achieved by acylation with chloroformates or anhydrides, usually followed by liquid–liquid extraction under adequate pH. Despite this disadvantage, the β-carbolines and tetrahydro-β-carbolines molecules provide good mass spectra for detection with mass spectrometry (GC-MS).[4]

Liquid chromatography is usually the most common approach, using reversed-phase interactions, and several detection systems. Fluorescence and mass spectrometry (MS) are preferred for a selective and sensitive detection, although UV, chemiluminescence, or electrochemical detection are also reported. HPLC-MS-MS in combination with soft ionization techniques such as electrospray ionization (ESI) is useful, particularly for identification, because the protonated molecules $[M + H]^+$ are usually the most abundant ions. For quantitative purposes, fluorescence detection is more commonly used.[2,4,27] For more detail on each of these methodologies, the review provided by Herraiz is recommended.[4]

For coffee, sample preparation is dependent on the coffee matrix, with coffee beverages and soluble coffee or congeners requiring direct dilution and filtration, and solid coffee grounds demanding efficient solid–liquid extraction, before SCX-SPE clean-up. Separation and quantification by HPLC with fluorescence detection is the most common approach, with mass spectrometry used for confirmation purposes.[4]

31.6 Conclusion

In conclusion, coffee/coffee products are among the highest sources of β-carbolines in the diet, but the amounts in the brew depend on several technological factors, expected amounts being difficult to define. Tetrahydro-β-carbolines, despite being highly prevalent in fruits and vegetables, including fermented or smoked foods, are apparently absent in roasted coffee, probably due to the high temperatures used for roasting. These data are interesting for coffee bioactivity, because β-carbolines are a recognized class of compounds with intense biological activities, and their presence in coffee might be partially responsible for coffee's health effects.

References

1. H. Tsuchiya, K. Yamada, K. Tajima and T. Hayashi, *Alcohol Alcohol.*, 1996, **31**, 197.
2. W. Pfau and K. Skog, *J. Chromatogr. B: Anal. Technol. Biomed. Life Sci.*, 2004, **802**, 115.
3. T. Herraiz, *Food Addit. Contam.*, 2004, **21**, 1041.
4. T. Herraiz, *J. Chromatogr. A*, 2000, **17**, 483.
5. I. I. Koleva, T. A. Beek, A. E. Soffers, B. Dusemund and I. M. Rietjens, *Mol. Nutr. Food Res.*, 2012, **56**, 30.
6. T. Herraiz, *Food Addit. Contam.*, 2000, **17**, 859.
7. T. Herraiz, *Food Addit. Contam.*, 2002, **19**, 748.
8. R. C. Alves, S. Casal and B. P. Oliveira, *J. Agric. Food Chem.*, 2007, **55**, 1832.
9. R. C. Alves, E. Mendes, B. P. Oliveira and S. Casal, *Food Chem.*, 2010, **120**, 1238.
10. B. Gutsche and M. Herderich, *J. Chromatogr. A*, 1997, **767**, 101.
11. T. Herraiz and J. Galisteo, *Food Chem.*, 2014, **451**, 42.
12. S. J. Stachel, S. A. Stockwell and D. L. Vranken, *Chem. Biol.*, 1999, **6**, 531.
13. S. M. Wabaidur, S. H. Lee, Z. A. Alothman, M. R. Siddiqui and S. M. Alam, *Spectrochim. Acta, Part A*, 2013, **110**, 179.
14. N. P. Sen, S. W. Seaman, B. P. Lau, D. Weber and D. Lewis, *Food Chem.*, 1995, **54**, 327.
15. T. Herraiz, *J. Agric. Food Chem.*, 1996, **44**, 3057.
16. T. Herraiz, *J. Agric. Food Chem.*, 1998, **46**, 3484.
17. M. Jägerstad, K. Skog, P. Arvidsson and A. Solyakov, *Eur. Food Res. Technol.*, 1998, **207**, 419.
18. T. Herraiz and J. Galisteo, *J. Agric. Food Chem.*, 2003, **51**, 7156.
19. E. A. Abourashed, J. Vanderplank and I. A. Khan, *Pharm. Biol*, 2003, **41**, 100.
20. T. Herraiz, *J. Agric. Food Chem.*, 1999, **47**, 4883.
21. T. Herraiz, *Food Chem.*, 1999, **66**, 313.
22. E. Papavergou and T. Herraiz, *Food Res. Int.*, 2003, **36**, 843.
23. M. Murkovic, *Anal. Bioanal. Chem.*, 2007, **389**, 139.
24. S. Casal, E. Mendes, J. Fernandes, M. Oliveira and M. Ferreira, *J. Chromatogr. A*, 2004, **1040**, 105.
25. Y. Totsuka, H. Ushiyama, J. Ishihara, R. Sinha, S. Goto, T. Sugimura and K. Wakabayashi, *Cancer Lett.*, 1999, **143**, 139.
26. T. Herraiz, *J. Agric. Food Chem.*, 2000, **48**, 4900.
27. L. Agui, C. P. Farfal, P. Y. Sedeno and J. M. Pingarron, *Anal. Chim. Acta*, 2007, **585**, 323.
28. E. Wojtowicz, R. Z. Wojtasiak, K. Przygonski and S. M. Szkudlarz, *Food Chem.*, 2015, **175**, 280.
29. A. F. R. Gomes, M.S. thesis, Oporto University, 2006.
30. T. Herraiz and C. Chaparro, *Life Sci.*, 2006, **78**, 795.

CHAPTER 32

Polycyclic Aromatic Hydrocarbons

OLGA VIEGAS*[a,b], OLÍVIA PINHO[a,b] AND ISABEL M. P. L. V. O. FERREIRA[a]

[a]LAQV/REQUIMTE, Laboratório de Bromatologia e Hidrologia, Departamento de Ciências Químicas, Faculdade de Farmácia, Universidade do Porto, 4051-401 Porto, Portugal; [b]Faculdade de Ciências da Nutrição e Alimentação da Universidade do Porto, 4200-465 Porto, Portugal
*E-mail: olgaviegas@fcna.up.pt

32.1 Introduction

Polycyclic Aromatic Hydrocarbons (PAHs) are a ubiquitous class of organic compounds, comprising hundreds of different compounds, each of them containing two or more fused aromatic rings.[1,2] They present low solubility in water and are very lipophilic, persistent, and bioaccumulative compounds.[2]

PAHs can be formed from different combustion and pyrolysis processes and thus their natural or anthropogenic sources are numerous. PAHs comprise the largest class of chemical compounds known to be cancer causing agents. They are toxic, although not all are carcinogenic, but these may act as synergists.[2] PAHs have been implicated as causative agents of breast, lung, and colon cancers and have been associated with immunosuppression, teratogenicity, neuro-, reproductive, and developmental toxicities, and

Coffee: Production, Quality and Chemistry
Edited by Adriana Farah

Published by the Royal Society of Chemistry, www.rsc.org

acute or chronic toxicity in other tissues such as bone morrow and intestinal mucosa.[3,4]

Exposure to PAHs is a major concern for human health and competent organizations established target compounds based on occurrence and toxicity of some of them (US-EPA,[5] EFSA,[1] ICPS,[4] IARC[6]). Food seems to be the major source of PAHs exposition for human population. Contamination of foodstuffs by PAHs can occur due to environmental pollution; however, it mostly arises from intense thermal processing, such as roasting and smoking.

Roasting is a crucial step for the development of coffee characteristics. It is an intense thermal treatment during which coffee beans are heated at high temperatures depending on the desired characteristics of the final product.[7] During roasting chemical and physical properties of green coffee beans change drastically, some compounds are formed and others eliminated, providing flavour, colour, aroma, acidity, and body,[8,9] and the concomitant formation of heat-generated toxic compounds, such as PAHs, also occurs.

This chapter summarizes the present knowledge of the mechanism of PAHs formation, methods of analysis, as well as occurrence in coffee, highlighting the differences of coffee processing from the bean to the brew. A general approach about PAHs chemical properties and formation in foods is described. Studies related to PAHs occurrence in coffee will be depicted and discussed.

32.2 Chemical Structures of PAHs

PAHs represent a major class of organic compounds, comprising hundreds of compounds, each of them containing two or more fused aromatic rings, without other elements than carbon and hydrogen, in a linear or angular configuration.[1,2,10] PAHs containing two or three fused benzene rings are called "light PAHs" and those containing four or more benzene rings are called "heavy PAHs". Heavy PAHs are more stable and more toxic than light ones.[2] Table 32.1 presents structures and chemical and abbreviated names of eight "heavy PAHs", benz(*a*)anthracene, chrysene, benzo(*b*)fluoranthene, benzo(*k*)fluoranthene, benzo(*a*)pyrene, indeno(1,2,3-*cd*)pyrene, benzo(*g*,*h*,*i*)perylene, dibenz(*a*,*h*)anthracene, considered by EFSA[1] as the possible indicators of the carcinogenic potency of PAHs (PAH8) in food. These priority PAHs from UE (PAH8), are the same eight carcinogenic PAHs from US-EPA list.[5]

PAHs can be present in vapour in the atmosphere. PAHs have a high tendency to absorb onto small atmospheric particles that can fall and deposit in sediments, soils, and vegetation. Due to the hydrophobic properties of PAHs, in aqueous environment they are generally found also adsorbed on particulate matter and on humic matter, or dissolved in any oil contaminant. Both vapour pressure and solubility of PAHs in water are inversely proportional to their number of aromatic rings.[1,11]

Table 32.1 Compound names and abbreviations, chemical structure and IARC (International Agency for Research on Cancer) classification of PAH8 EFSA.[1]

Compound name	Abbreviation	Structure	IARC group
Benz(*a*)anthracene	BaA	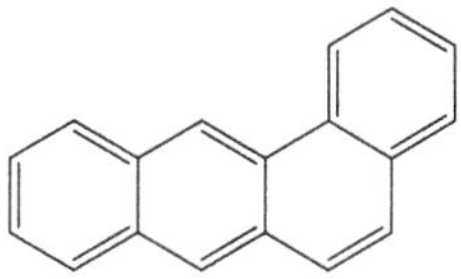	2B
Chrysene	Ch	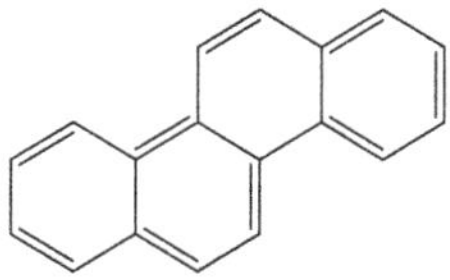	2B
Benzo(*b*)fluoranthene	BbF	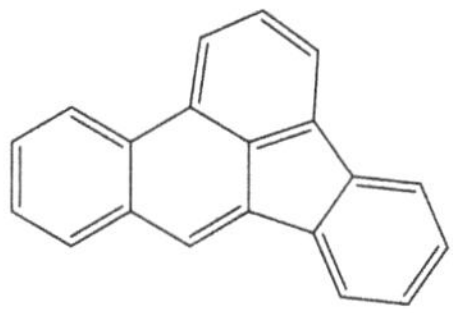	2B
Benzo(*k*)fluoranthene	BkF	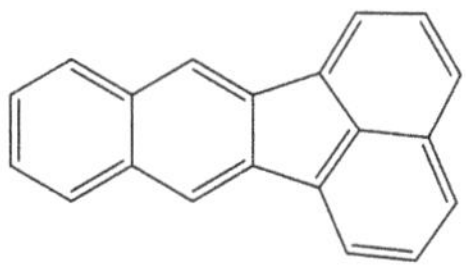	2B
Benzo(*a*)pyrene	BaP	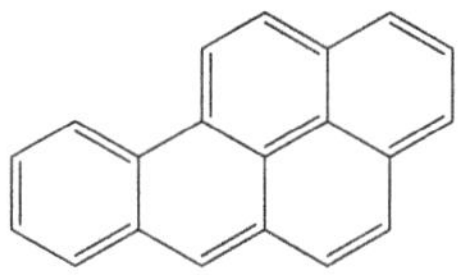	1
Indeno(1,2,3-*cd*)pyrene	IP	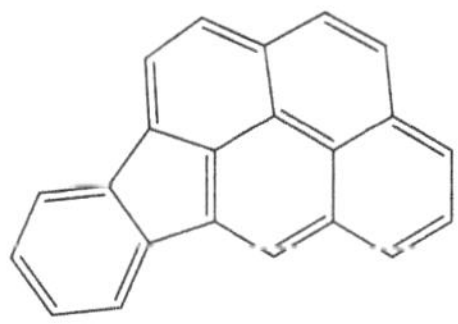	2B
Benzo(*g*,*h*,*i*)perylene	BgP	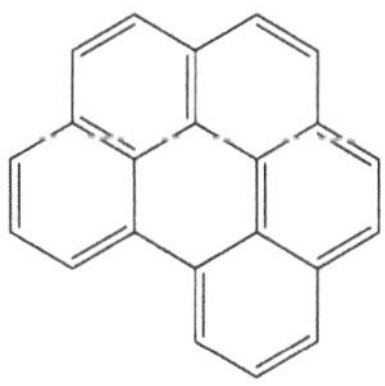	3
Dibenz(*a*,*h*)anthracene	DhA	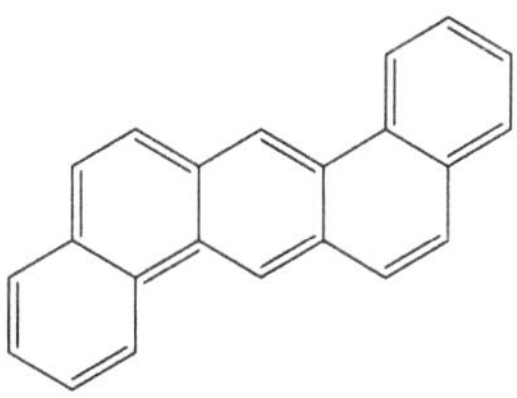	2A

32.3 PAHs Formation Mechanism

Although the exact mechanism of formation of PAHs is not precisely known, it is generally considered that PAHs are formed by condensation of smaller organic compounds in absence of oxygen by either pyrolysis or pyrosynthesis. Pyrolysis (from the Greek-derived elements pyro "fire" and lysis "separating") is a thermochemical decomposition or organic material at elevated temperatures in the absence of oxygen or any halogen. It involves the simultaneous change of chemical composition and physical phase, and is irreversible. Therefore pyrolysis is a type of thermolysis and is most commonly observed in organic materials exposed to high temperatures. In the roasting process, during pyrolysis that starts occurring at temperatures above 200 °C organic compounds are fragmented. At very high temperatures, fragmentation becomes easier and the free radicals produced recombine to form stable polynuclear aromatic compounds (pyrosynthesis). A common reaction may involve Diels–Alder-type rearrangements to yield the PAHs.[11] The content of PAHs formed increased linearly with temperature, which affects both the structure and diversity of PAHs formation.[12]

Smoke generated upon heating lipids at lower temperatures (200 °C) during 2 h generates PAHs, including the carcinogenic benz(*a*)anthracene, benzo(*a*)pyrene, and dibenz(*a*,*h*)anthracene. The mechanisms proposed are the intramolecular cyclization of unsaturated fatty acids (such as oleic acid, linoleic acid, and linolenic acid), and the degradation of saturated fatty acids (such as stearic) generating lower molecular weight compounds with benzene rings that follow Diels–Alder cycloaddition.[12]

32.4 PAHs Formation in Foods

PAHs occur as contaminants in different food categories, drinking water, and beverages. PAHs can reach the food chain in different ways: environmental contamination, through the air (by direct deposition from the atmosphere), soil (by transfer), and/or water (deposition and transfer), *e.g.* fruits, vegetables, and seafood; contamination from packaging materials; and contamination from wood fumes and/or generation from food components during the thermal processing of foods.[1,13,14] The foods with the highest concentrations of PAHs include meat products, smoked fish, smoked cheese, vegetables, tea, and roasted coffee.[1] However, other food groups despite their low PAHs content can contribute significantly to PAHs exposure due to their high daily intake, namely cereals and cereal products. PAHs are typically in tens of micrograms per kilogram in foods.[1,15]

The US-EPA[14] designed 16 PAHs as compounds of interest based on their occurrence, the following eight being carcinogenic: benzo(*a*)anthracene, chrysene, benzo(*b*)fluoranthene, benzo(*a*)pyrene, benzo(*k*)fluoranthene, indeno(1,2,3-*cd*)pyrene, benzo(*g*,*h*,*i*)perylene, dibenzo(*a*,*h*)anthracene. The European Union[1] specified the sum of those eight PAHs, named PAH8, as the suitable indicator of carcinogenic potency of PAHs in food. Nevertheless, the sum of the first four (PAH4) may be enough for this purpose due to their higher frequency in foods.

The median dietary exposure to PAH8 across European countries considering the mean and high dietary consumers was respectively 1.729 and 3.078 µg per day, cereal and seafood products being the greatest sources,[1] whereas the average exposure of the American population was estimated to be between 1 and 5 µg per day, unprocessed grains and cooked meats being the greatest sources.[16] Considering the benzo(*a*)pyrene alone, the EFSA report describes 0.235 and 0.389 µg per day respectively for mean and high dietary consumers, whereas for the American population similar levels were reported ranging from 0.005 to 0.305 µg per day.[17]

The US Food and Drug Administration have not established maximum limits for the PAHs content in foods. The European Union sets different maximum levels for benzo(*a*)pyrene (ranging from 1.0 to 6.0 µg kg^{-1}) and also maximum limits (ranging from 1.0 to 35.0 µg kg^{-1}) for PAH4 in different foods. For example, the limits established for smoked meat or fish were 2 µg kg^{-1} for benzo(*a*)pyrene and 12 µg kg^{-1} for PAH4, whereas for grilled or barbecued meat they were 5 µg kg^{-1} for benzo(*a*)pyrene and 30 µg kg^{-1} for PAH4. The higher limits were for smoked bivalve molluscs. No limits were proposed for coffee.[18] However, in the EFSA report,[1] coffee powder, cacao, and dried tea are the food group that present the highest average content of benzo(*a*) pyrene (around 5 µg kg^{-1}). The average content of PAH8 in powdered coffee is around 17 5 µg kg^{-1}.[1]

32.5 PAHs Formation During Coffee Roasting

The presence of PAHs in coffee can be originated from different sources including atmospheric pollution (contamination of the green beans) and technological processes, such as roasting and drying. Coffee roasting includes an intense thermal process but is a crucial step for the development of the characteristic complex mixture of aroma, flavour, and colour. The chemical reactions essential to develop these key parameters of coffee during the roasting step lead at the same time also to the development of incidental compounds as PAHs, suspected carcinogenic, and mutagenic.[19,20]

Concerning the content of PAHs in green coffee beans, there are reports that indicate low PAHs.[8,21,22] There is no doubt that PAHs in coffee are mainly formed during the roasting of beans. Different roasting conditions mean different expected coffee flavours[9] and at the same time different PAHs content formation. The process is typically performed at temperatures of 160–300 °C (more commonly 210–240 °C) and the beans are roasted for times ranging from a few minutes to about 30 min.[7,11]

The high temperature (160–240 °C) firstly drives off the moisture whilst the undesirable exothermic pyrolysis may occur at the latter roasting stages, because thermal decomposition occurs in addition to chemical and physical changes. The kinetics of these remarkable changes is a function of time and temperature.[7,23] Two types of roasting process are widely used for manufacturing coffee, (1) indirect roasting where the flames or heat source do not make contact with the beans, although the combustion gases from the burner do make contact with the beans, and (2) direct-fired roasting where beans contact directly with the burner flame and the combustion gases. The latter process can lead to a high content of PAHs in coffee.[11,24]

The roasting industry in general uses rotating drum roasters for batch or continuous processes that tumble the green coffee beans in a current of hot combustion gases. Beans are mixed with hot gases in a horizontal rotating drum or in a vertical fixed drum roaster with paddles.[7] The basic disadvantage of these traditional machines is that they require high temperatures/long times in order to roast the beans. That favours the scorching of some beans, oil and char deposits on the cylinder walls, and, regularly, fires when doing a dark roast. Additionally, the machines are difficult to clean after processing, resulting in roasted beans with an acrid, smoky taste.[25] Other kinds and recent roasters such as fluidized bed and spouted bed roasting facilitate roasting for short time periods and better heat transfer (as seen in Chapter 8). The spouted bed needs less energy consumption, but may result in less homogeneous roasting than fluidized bed.[7]

The pyrolysis of coffee lipids, carbohydrates, and amino acids forms PAHs. However, due to the complexity of the coffee matrix it is highly difficult to propose schemes to describe the chemical reactions that occur during roasting.[26] Model systems are often used for elucidation of PAHs formation. Sharma *et al.*[27] demonstrated that the polyphenol chlorogenic acid present in coffee matrix can form several PAHs (pyrene, fluoranthene, phenanthrene, and anthracene) by pyrolysis, and also suggested a possible rearrangement of smaller to larger PAHs. It was also demonstrated that steroids present in coffee beans, such as stigmasterol, may also be considered as a first-order rate for formation of phenanthrene and anthracene, but at 650 °C, in model system.[28]

Also, trigonelline alone or in combination with amino acids after heating at 250 °C has shown mutagenic activity that can result from heterocyclic aromatic amines and/or PAHs.[26,29]

The roasting conditions differ depending on the expected characteristics of the final product (light, medium, or dark) and the type of roaster.[7,26] Thus, roasting conditions (time and temperature) must be optimized to achieve the maximum aroma and flavour, but also should be controlled to avoid the formation of the potential carcinogenic and mutagenic compounds.[30] Experiments should be conducted in order to understand the chemical reaction for PAHs formation in coffee beans during roasting, not only the time and temperature but also the type of roasters. Moreover, being dependent on the chemical composition of coffee the influence of species should be investigated.

Although most PAHs present low solubility, the different types of coffee brewing may affect the real PAHs exposition of these compounds to consumers.[8,9]

32.6 Analytical Methods for PAHs Determination

PAHs occurrence in foodstuffs has been studied for a long time, and different methods have been validated for extraction of PAHs from food depending on the kind of matrix; however, there is still a lack of standardized methods for their analysis.[20] Furthermore, in foods PAHs occur at ppb levels that imply several extraction steps, tedious pre-concentration procedures, and difficulty in achieving repeatability in the concentrations of PAHs with two or three rings. However, it is important to develop rapid and reliable analytical methods for accurate determination of contaminants in foods in order to estimate the actual exposure of humans to PAHs.[31]

Due to PAHs lipophilic properties their extraction is associated with lipid constituents of food, thus, PAHs have to be purified, which in most methods requires the use of large volumes of hazardous solvents and solid phase extraction (SPE) cartridges (silica, alumina, and C_{18}).[32]

The most common methods for the isolation of PAHs from foods involve saponification of lipids by methanolic or ethanolic KOH or NaOH solution followed by extraction procedures to isolate the PAHs-containing fraction. Different procedures are described for PAHs extraction, namely, the traditional liquid–liquid extraction, Soxhlet, sonication, or SPE, and more recently solid-phase microextraction (SPME), supercritical-fluid extraction, gel permeation (GPC), molecularly imprinted polymers (MIPs), and microwave-assisted extraction. Owing to the complexity of the matrices, these methods, in general, present inconstant recoveries, and in some cases, interfering peaks in the chromatograms.[31] Other new techniques have been tested for PAHs analysis, such as QuEChERS (quick, easy, cheap, effective, rugged, and safe) method and dispersive liquid–liquid microextraction (DLLME) that allows lower solvent consumption.[20,32]

Chromatographic analysis has been used as the main instrument for the analytical determination of PAHs in food; mainly liquid chromatography (LC) and gas chromatography (GC), including also comprehensive techniques (GC × GC and LC × LC), with different detectors.[32] GC coupled with MS for determination of PAHs combines better selectivity, sensitivity, and ease of use. On the other hand, HPLC is more appropriate for PAHs isomers separation,[20] for example, concerning the isomers chrysene/triphenylene, the first is a target PAH but the second may interfere with chrysene quantification by GC. Thus, HPLC is particularly selected due to the lack of GC resolving some pairs of peaks.[1,20,32] HPLC with fluorescence (FLD) has been the most common method for PAHs quantification in foods due to its selectivity and sensitivity.[1,20] Diode array detection has often been placed in tandem with FLD.[32] MS detection is also used and presents high resolution and sensitivity, especially MS/MS.[13] No differences were found in the results obtained in a collaborative study on the analysis of PAHs in edible oils, using either HPLC/FLD or GC/MS.[33]

32.7 Analytical Methods for PAHs Determination in Coffee

There is still a gap in the literature regarding validated analytical methods for PAHs determination in coffee samples.[34] Different analytical procedures can be found for ground coffee, and brew, so comparison between results is sometimes difficult.[35] In general, the methods applied to PAHs analysis in coffee follow the same techniques used for other foodstuffs; the main problems are the low detection values and the interferences (*e.g.* lipid content around 15% in ground coffee).

Houessou *et al.*[36] analyzed PAHs coffee brew using a polystyrenedivinylbenzene cartridge and elution by a mixture of methanol/tetrahydrofuran (10 : 90 v/v) and analyzed HPLC-FLD. However, the authors concluded that the identification of traces of PAHs could not be achieved due to matrix effects, and the presence of co-eluting compounds could not be excluded. Later the authors[35] compared different treatment methods, namely, Soxhlet extraction, pressurized liquid extraction (PLE), alkaline saponification, SPE, LLE. The best results were obtained for PLE with hexane/acetone 50 : 50 (v/v) under 150 °C that was combined with different clean-up methods (*in situ* purification, C_{18} SPE, silica SPE, acid digestion, and alkaline saponification). Alkaline saponification followed by cyclohexane extraction and silica was required to eliminate the interfering compounds. The authors used both HPLC-DAD and GC-MS/MS for PAHs identification. In further studies the authors[8,26] used the same extraction technique but opted for the HPLC-FLD detection for quantification and other detectors for confirmation.

García-Falcón *et al.*[37] proposed a rapid analytical method, for seven heavy PAHs with minimal purification, broadly applicable to roasted coffees or their aqueous extracts. The method consists of extraction with hexane, clean-up with a silica cartridge, concentration to dryness, and analysis of residue for HPLC-FLD. The method allowed to confirm the presence of the selected PAHs with recoveries higher than 87% and satisfactory LODs and LOQs, much lower than the restrictions of EU directives.[38] Tfouni *et al.*[9,21] extracted PAHs from coffee powder or brew using Soxhlet and LLE, respectively, followed by rotavapor evaporation and purification by silica cartridge.

Orecchio *et al.*[24] proposed a simple method for determination of 28 PAHs based on saponification and LLE with small volumes of hexane, excluding a further purification step due to the following analysis of the extract by GC-MS (SIM). The method was described as more selective, rapid, and economical than one based on SPE. Recoveries were better than 85% and LODs much lower than the restrictions given for EU directives.[38]

Duedahl-Olesen *et al.*[39] extracted PAHs (PAH4) from coffee powder and brew using *n*-hexane under reflux, after solvent evaporation clean-up with GPC and SPE silica columns were performed and then analyzed by GC-MS.

Sadowska-Rociek *et al.*[40] recently proposed a modified QuEChERS method, including dispersive solid-phase extraction using as sorbents primary secondary amine (PSA) and strong anion exchange (SAX) followed by a final solvent exchange from acetonitrile to hexane. This was followed by gas chromatography selected ion monitoring mass spectroscopy analyses, resulting in satisfactory compound recoveries (70–104%) for the examined compounds in coffee samples.

Recently Pissinatti *et al.*[34] optimized and validated a method for the determination of 10 EPA priority PAHs in roasted coffee analysis. In spite of the evolution of several new fast analytical methodologies, the authors suggested an extraction and clean-up based on PLE followed by LLE and purification by silica columns. The use of isotope dilution GC-MS enabled the absence of matrix effects. The average recoveries were between 87 and 111%. Limits met the criteria of EU regulations.[18] The authors suggested their method as a reliable tool for screening PAHs in roasted coffee samples, including official monitoring programs.

32.8 Occurrence of PAHs in Coffee

Several studies were performed on PAHs content of ground coffee, however they are not representative of human exposition, due to their consumption as a beverage.[1] On the other hand, there is scarce information on PAHs content in brewed coffee and the studies usually involve coffee brews prepared with ground roasted coffee from commercial brands and scarce information was given about the coffee species, cultivar, or the roasting degree of the ground roasted coffee used for brewing.[9]

The occurrence of PAHs in green coffee beans is caused by environmental contamination, since they are ubiquitous compounds in nature. Another possible way of contamination of the green coffee beans may be due to the exhaust gases of cars or the transportation in contaminated jute and sisal bags,[41] however, strict rules are adopted by bag manufactures to avoid contamination.[42] Since the formation of PAHs in coffee occurs mainly during the roasting steps it is expected that green coffee beans are the least contaminated samples.

Stanciu *et al.*[22] reported a mean content of 0.023 μg (PAH8, except chrysene that was not analysed) per kg of green coffee (from a Romanian market without information about origin). Houessou *et al.*[8] analysed PAHs in green coffee beans from Cuba; when considering the sum of the same PAHs studied by Stanciu *et al.*,[22] 0.20 and 0.47 μg kg^{-1} were found in two different lots. Benz(*a*)anthracene and benzo(*b*)fluoranthene were the main contributors for this sum value. Chrysene was also quantified at levels of 0.43 and 0.45 μg kg^{-1}, thus the sum of PAHs at least doubles reaching values of 0.63–0.97 μg kg^{-1}. According to CONTAM panel of EFSA[1] chrysene was the most frequently detected from PAH8; in green coffee beans it is possible to observe the same trend. Houessou *et al.*[8] analyzed also fluoranthene, pyrene, phenanthrene, and anthracene; a total of 15.22 and 20.61 μg kg^{-1} were described for the sum of the 11 PAHs, in green arabica beans.

Tfouni *et al.*[21] analyzed PAHs (benz(*a*)anthracene, benzo(*b*)fluoranthene, benzo(*k*)fluoranthene, and benzo(*a*)pyrene) in green arabica and robusta beans from Brazil. The summed PAHs contents were 0.086 and 0.108 μg kg^{-1} for arabica and robusta, respectively. These values are very low compared with the arabica green coffee studied by Houessou *et al.*,[8] who described 0.20–0.47 μg kg^{-1} for the same PAHs and higher than those quantified by Stanciu *et al.*[22] (0.02 μg kg^{-1}).

32.8.1 PAHs Formation under Controlled Roasting Conditions

Green coffee is a very complex matrix and different reactions occur during roasting. Several coffee constituents may be considered as possible precursors. In order to understand PAHs occurrence in coffee samples and to elucidate their formation in coffee matrix during roasting conditions, it is important to study the roasting parameters under pilot scale. Houessou *et al.*[8] studied the effect of roasting conditions on formation of PAHs (phenanthrene, anthracene, fluoranthene, pyrene, benz(*a*)anthracene, chrysene, benzo(*b*)fluoranthene, benzo(*k*)fluoranthene, benzo(*a*)pyrene dibenz(*a*,*h*)anthrecene, benzo(*g*,*h*,*i*)perylene), using a pilot spouted bed roaster. Studies were performed during 5 and 20 min of roasting at different temperatures (Tables 32.2 and 32.3). In ground coffee, phenanthrene and fluoranthene are the main PAHs with concentrations ranging from (3.12 to 46.87 μg kg^{-1}), followed by pyrene (1.28 to 6.06 μg kg^{-1} except the sample roasted at very strong conditions reaching 53.05 μg kg^{-1}). Stronger roasting conditions (20 min, 260 °C) led to significant content of chrysene and benz(*a*)anthrecene (near 13 μg kg^{-1}). Above 220 °C, phenanthrene (three rings), anthracene (three rings), and benz(*a*)anthracene (four rings) increase with temperature; however, for phenanthrene a decrease is observed at 260 °C. Formation of pyrene (four rings), chrysene (four rings), and benzo(*g*,*h*,*i*)perylene (six rings) (but in lower concentration) requires stronger conditions (around 260 °C, 20 min.). The authors suggested that the lower molecular PAHs with the increase of the roasting degree undergo a partial degradation leading to the formation of PAHs with a high number of rings.

Later Houessou *et al.*[26] predicted the system global behaviour of PAHs formation. Again they observed that above 220 °C the PAHs formation is very low. With this study they reinforce that low molecular weight PAHs appear to be degraded at high temperatures probably to form larger PAHs. However, the kinetic models presented were unable to elucidate the real underlying mechanisms, their reactions remaining unknown. For pyrene, benz(*a*)anthracene, chrysene, and anthracene, satisfactory correlations with experimental and predicted concentrations were achieved. Pilot studies performed during

Table 32.2 Mean PAH contents of ground coffees (μg kg^{-1}) and the respective coffee brews (μg L^{-1}) as well the respective transfer coefficients (C_t). Ground coffee roasted for 5 min under different inlet air temperatures. Reprinted with permission from ref. 8. Copyright 2007 American Chemical Society.

	Roasting time: 5 min															
	Arabica Lot A										Arabica Lot B					
	200 °C			220 °C			250^a °C	260 °C			240 °C			260 °C		
	Ground (μg kg^{-1})	Brew (μg l^{-1})	*Ct* (%)	Ground (μg kg^{-1})	Brew (μg l^{-1})	*Ct* (%)	Ground (μg kg^{-1})	Ground (μg kg^{-1})	Brew (μg l^{-1})	*Ct* (%)	Ground (μg kg^{-1})	Brew (μg l^{-1})	*Ct* (%)	Ground (μg kg^{-1})	Brew (μg l^{-1})	*Ct* (%)
Light PAHs																
Phe	6.05	0.21	13.9	6.51	0.49	30.1	9.26	8.39	0.49	23.4	4.71	0.33	28.6	11.02	0.42	15.3
A	0.68	Tr		0.74	tr		1.99	1.80	tr		0.56	tr		0.89	tr	
Heavy PAHs																
Fl	6.86	0.39	22.7	7.50	0.44	23.5	7.27	10.98	0.76	27.7	3.12	0.17	22.6	4.61	0.35	30.4
P	3.57	0.16	17.9	3.63	0.18	19.8	4.43	6.60	0.29	17.6	1.28	0.062	19.4	1.83	0.13	28.4
BaA	0.75	0.031	16.6	nd	tr		1.33	3.86	0.15	15.5	0.40	0.025	24.3	1.06	0.082	30.9
Ch	1.85	0.087	18.8	1.73	0.14	32.4	1.85	1.58	0.11	27.9	0.33	0.024	29.6	0.63	0.046	29.2
BbF	tr	Tr		0.18	tr		nd	0.43	tr		0.20	tr		0.20	tr	
BkF	nd	Tr		nd	tr		nd	0.17	tr		tr	tr		tr	tr	
BaP	tr	Nd		tr	nd		tr	0.20	tr		tr	nd		nd	nd	
DhA	nd	Nd		nd	nd		nd	0.37	nd		nd	nd		nd	nd	
BgP	nd	nd		nd	nd		nd	0.78	nd		nd	nd		nd	nd	

[a]No information was given concerning brew composition. nd, not detected; tr, traces (compound detected at levels below the limit of detection of the method).

Table 32.3 Mean PAH contents of ground coffees (μg kg^{-1}) and the respective coffee brews (μg l^{-1}) as well the respective transfer coefficients (C_t). Ground coffee roasted for 20 min under different inlet air temperatures. Reprinted with permission from ref. 8. Copyright 2007 American Chemical Society.

	Roasting time: 20 min															
	Arabica Lot A										Arabica Lot B					
	200 °C			220 °C			250[a] °C	260 °C			240 °C			260 °C		
	Ground (μg kg^{-1})	Brew (μg l^{-1})	*Ct* (%)	Ground (μg kg^{-1})	Brew (μg l^{-1})	*Ct* (%)	Ground (μg kg^{-1})	Ground (μg kg^{-1})	Brew (μg l^{-1})	*Ct* (%)	Ground (μg kg^{-1})	Brew (μg l^{-1})	*Ct* (%)	Ground (μg kg^{-1})	Brew (μg l^{-1})	*Ct* (%)
Light PAHs																
Phe	7.45	0.21	11.3	14.6	0.42	11.5	46.87	17.37	0.45	10.4	18.08	0.26	5.7	44.95	0.17	1.5
A	1.01	tr		1.08	tr		1.90	2.17	0.019	3.5	1.91	tr		1.95	tr	
Heavy PAHs																
Fl	4.99	0.22	17.6	4.34	0.16	14.7	11.48	17.54	0.79	18.0	4.69	0.27	23.0	11.5	0.34	11.8
P	3.47	0.12	13.8	2.6	0.093	14.3	5.87	53.05	0.37	2.8	2.29	0.078	13.6	6.06	0.15	9.9
BaA	0.54	tr	5.7	1.4	0.014	4.0	6.52	12.67	0.02	0.6	3.68	tr		6.52	tr	
Ch	2.21	0.048	8.7	1.84	0.027	5.9	1.29	12.82	0.087	2.7	1.47	0.02	5.4	1.54	0.021	5.5
BbF	0.14	0.14	3.5	0.16	tr		0.20	nd	nd		0.2	tr		0.29	tr	
BkF	nd	tr		tr	tr		tr	tr	tr		tr	tr		tr	tr	
BaP	tr	tr		tr	tr		nd	nd	nd		0.2	nd		0.21	tr	
DhA	nd	nd		nd	nd		nd	nd	nd		nd	nd		nd	nd	
BgP	nd	nd		nd	nd		nd	1.71	nd		nd	nd		nd	nd	

[a]No information was given concerning brew composition. nd, not detected; tr, traces (compound detected at levels below the limit of detection of the method).

different time periods indicate that the roaster temperature should be lower than 240 °C for 15 min in order to avoid the formation of carcinogenic PAHs. Higher roasting temperatures led to PAHs formation, but, on other hand, the same higher temperatures may degrade other incidental compounds, such as acrylamide and 5-hydroxymethylfurfural.[26]

Other studies describe the effect of roasting in the formation of PAHs (benz(*a*)anthracene, benzo(*b*)fluoranthene, benzo(*k*)fluoranthene, and benzo(*a*)pyrene),[21] evaluating beans of both important *Coffea* species, *C. arabica* and *C. canephora*. Coffee beans were roasted in a rotating drum at 200 °C for different times and were analyzed before and after roasting. PAHs summed levels ranged from 0.086 to 0.814 μg kg^{-1} (*C. arabica*) and from 0.108 to 0.392 μg kg^{-1} (*C. canephora*), respectively, in green and roasted samples. In *C. canephora* samples the summed PAHs increased with the darker roasting degrees, but roasting arabica beans did not follow the same behaviour. The authors concluded that no correlation was observed between roasting degree, coffee cultivar, and content of PAHs.

32.8.2 PAHs Occurrence in Coffee Samples from Commercial Brands

Usually the commercial brands of ground roasted coffee contain scarce information on the label concerning coffee species, cultivar, or the roasting conditions, although a few contain the roasting degree. Thus, even though a great number of samples from different countries was analyzed, it is difficult to make correlations between these parameters and the PAHs content. Table 32.4 summarizes data from PAH8 found in ground coffee to highlight the variability of results described in the literature.

Jimenez *et al.*[19] searched for the influence of roast degree on the content of 18 PAHs (16 EPA and two methylnaphthalene isomers) in 33 commercial coffees from different origins (available on USA market) comprising light, medium, and dark roasted coffee, including instant and decaffeinated brands. Results are summarized in Table 32.5. The authors concluded that when the concentrations found in the samples were analyzed all together it was possible to observe a trend associating the increase in PAHs concentration with the roasting degree; however, other parameters than the degree of darkness in coffee contributed to higher PAHs concentrations. A few exceptions were observed: one dark coffee ($n = 4$) presented values lower than the light ($n = 3$) or medium dark ($n = 2$) ones. Roasting is crucial for PAHs development in coffee beans, but different brands are produced using different blends, and roasting conditions, namely, type of roaster, and time *vs.* temperature, which may explain the outlier results. Comparison was also performed with the PAHs content of decaffeinated and instant coffees (Table 32.5), which presented lower PAHs contents. In all samples fluorene, pyrene, chrysene, and benzo(*k*)fluoranthene were found. Still in the same study, fewer compounds were quantified in instant coffee and the sum of 18 PAHs was lower compared with other

Table 32.4 PAHs content ($\mu g\ kg^{-1}$) in different commercial roasted coffee samples collected from different works.[a]

Ref.	40	29	39			45	46	28	35		41
Coffee type	Roasted	Roasted	Roasted	Instant	Decaffeinated	Roasted	Roasted	Roasted	Roasted	Instant	Instant
PAHs	$\mu g\ kg^{-1}$	$\mu g\ kg^{-1}$		$\mu g\ kg^{-1}$		$\mu g\ kg^{-1}$	$\mu g\ kg^{-1}$	$\mu g\ kg^{-1}$	$\mu g\ kg^{-1}$	$\mu g\ kg^{-1}$	$\mu g\ kg^{-1}$
BaA	0.78	1.3	0.742	0.473	0.014	0.15–1.57	nd	<0.83–1.06	nd–51.40	nd–18.99	—
Ch	nd	1.6	—	—	—	0.14–1.90	1.8	1.00–1.76	nd–68.15	2.48–14.89	nd
BbF	0.65	0.60	0.219	0.104	0.028	nd–3.90	nd	<0.55–0.67	nd–48.21	2.75–19.55	nd
BkF	0.25	0.31	0.085	0.495	0.10	tr–48.72	2.6	<0.32	nd–41.24	2.41–21.65	nd
BaP	1.23	0.33	0.293	0.857	0.119	tr–0.36	nd	<0.40–0.51	nd–34.84	4.63–12.33	nd–102.9
DhA	0.24	nd	0.040	0.045	0.011	nd–tr	nd	<0.26	nd–18.40	nd–11.18	nd
BgP	1.23	nd	nq	nq	nq	nd	nd	<0.42	nd	nd–10.43	nd–53.4
IP	—	—	0.113	0.142	0.014	—	nd	<0.33	nd–18.74	nd–11.25	nd–38.7

[a]nd, not detected; nq, not quantified.

Table 32.5 PAHs content (μg kg^{-1}) in different types of roasted coffee. Reprinted with permission from ref. 19. Copyright 2014 Taylor & Francis.[a]

	Coffee type				
	Ground coffee			Dark instant	Dark decaffeinated
	Light	Medium	Dark		
Light PAHs					
Na	8.2 ± 13.4	27.9–79.1	27.8–561.4		101.7
Ace	39.3–55.5	56.5–155.0	44.6–512.4		188.2
Ac	6.8–11.4	6.1–17.3	7.4–37.4		20.8
F	24.4–42.1	29.0–75.5	25.3–184.2	11.7–14.2	74.2
Phe	1.6–4.0	3.3–5.6	2.6–14.0		6.5
A	1.7–5.1	5.5–9.0	2.8–44.1		12.7
Heavy PAHs					
Fl	6.1–10.2	12.2–21.7	9.2–85.3		29.8
P	191.5–274.5	141.6–352.8	156.1–458.8	114.1–125.5	59.8
BaA	2.9–4.9	5.2–5.9	3.0–19.1		5.3
Ch	126.1–185.2	84.0–242.0	65.1–371.2	69.9–71.8	258.6
BbF	0.4–1.2	1.0–1.1	0.3–10.5		1.8
BkF	8.8–10.8	15.6–31.6	14.0–53.9	1.4–1.8	7.1
BaP	1.4–3.3	4.9–2.4	3.5–18.5	nd–6.0	13.7
DhA	21.4–27.0	071.4–57.1	12.7–57.4		10.4
BgP	1.8		0.5–16.8		1.7
IP	51.2–70.5	71.4–57.1	26.4–212.0		101.6

[a]nd, not detected.

types of samples. Chrysene and pyrene were the most abundant PAHs in the evaluated instant coffees.

García-Falcón *et al.*[37] studied the presence of PAHs (benzo(*b*)fluoranthene, benzo(*k*)fluoranthene, benzo(*g*,*h*,*i*)pyrene, indeno(1,2,3-*cd*)pyrene, benz(*a*)anthracene, dibenz(*a*,*h*)anthracene) in 12 market samples of instant coffees with different trade names, six with caffeine and six without, three of each six were obtained by natural roasting and the others were torrefacto coffee (in which beans are roasted in the presence of added sugar). In only two torrefacto samples benzo(*b*)fluoranthene, benzo(*b*) fluoranthene, and benzo(*a*)pyrene were found at levels between 0.03 and 0.1 μg kg^{-1} for benzo(*b*)fluoranthene and between 0.01 and 0.04 μg kg^{-1} for the others, far below any potential legal restriction, *e.g.* EU directives for benzo(*a*)pyrene (1 μg kg^{-1}).[18,38] The authors concluded that PAHs were not extracted into water in the preparation of instant coffees from ground roasted coffee and emphasized that instant coffee can be considered as a product with PAHs content between roasted coffee grounds and coffee brews, since in the instant coffee production the brew is dried and concentration occurs.

Duedahl-Olesen *et al.*[39] analyzed 25 PAHs in commercial roasted coffee and focused on the benz(*a*)anthracene, chrysene, benzo(*b*)fluoranthene, and benzo(*a*)pyrene, based on the EFSA[1] statement that the sum of these four (PAH4) is the appropriate indicator for the occurrence and toxicity in foods. The authors found that the content of PAH4 for ground coffee ranged between 2.2 and 5.1 μg kg^{-1} in instant coffee, whereas dark roasted coffee had a mean content of 3.1 ± 0.40 μg kg^{-1} and medium roasted coffee presented lower content, 1.0 ± 0.50 μg kg^{-1}. The sum of the 25 PAHs followed the same order. Instant coffee is produced by preparing an industrial coffee brew from roasted coffee ground followed by freeze drying. The authors concluded that this process could produce higher composition in lower molecular weight PAHs, which explains the higher total sum of the 25 PAHs. However, these PAHs present a lower risk to health. Although PAH4 followed the same trend, no observation was made about it. Increased content of chrysene and benz(*a*)anthracene were observed in dark roasted coffee (0.8–1.2 μg kg^{-1} and 0.2–2.4 μg kg^{-1}) compared with medium roasted (0.1–0.4 μg kg^{-1} and 0.1–1.0 μg kg^{-1}) agreeing with findings of Houessou *et al.*[8] in the pilot scale study described previously. The concentration of benzo(*a*)pyrene was found to range from 0.2–1.0 μg kg^{-1}. Also under controlled concentrations only traces[9,29] or lower[9,21] concentrations were described.

Sadowska-Rociek *et al.*[40] analyzed ground roasted and instant coffees from the Polish market and also observed higher content of PAH4 in ground coffee (7.20–68.15 μg kg^{-1}) than in instant coffee (2.97–19.55 μg kg^{-1}). Light PAHs were the dominating compounds, phenanthrene being the most abundant compound (209.67 μg kg^{-1} in ground roasted coffee).

Pissinati *et al.*[34] analyzed 10 heavy PAHs (fluorene, pyrene, and PAH8 compounds) in roasted coffee samples, and the sample with highest contamination contained 11.29 ± 2.33 μg kg^{-1} of their sum. Benzo(*a*)pyrene contributed with 0.50 ± 0.10 μg kg^{-1}. Considering only the PAH4 of EFSA[1] the content was 3.87 ± 0.61 μg kg^{-1}. From the 24 samples of roasted coffee analyzed, chrysene was the only PAH detected in all samples. Therefore, the authors concluded that monitoring only benzo(*a*)pyrene as indicator of PAHs content in coffee leads to mistaken information, as already reported by EFSA[1] for other foodstuffs. Several studies in which the content of benzo(*a*)pyrene was analyzed in different brands, blends, and coffee types, from different countries, the content of benzo(*a*)pyrene ranged from not detected to extremely high contents as in the case of instant coffee samples analyzed by Grover *et al.*[43] and Iwegbue *et al.*[44] with 98.3 and 102.9 μg kg^{-1}, respectively.

In general, when light PAHs were included in the analysis of ground coffee, quite high values are observed, but also with higher variability of results, due their volatility. However, they have low toxicity with toxic equivalent factors (TEFs)[45] (the relative potency factor to the toxic potential of each compound assigning to the benzo(*a*)pyrene toxicity a value of 1) as 0.001, and their potential risks are low even when higher content is present. Only anthracene, a three-ring PAH, presented a TEF 0.01.

32.8.3 PAHs Transfer to the Coffee Brew

Assessment of the exposure to PAHs from coffee is a difficult task because it depends on the brewing technique used that influences greatly the transfer ratio of PAHs to the brew, and the consumption of the population.[34]

Due to the low solubility of PAHs in water, their transfer from coffee powder to the infusion during the coffee making has been reported as rather limited,[1,8] however some brewing methods extract these compounds more efficiently or they don't use lipophilic paper filter to retain them. Examples are unfiltered, espresso, and French press coffees. This can increase the PAHs content.[46]

As summarized in Tables 32.2 and 32.3 Houessou *et al.*[8] observed similarities in the pattern of PAHs in coffee brew and in ground coffee; independent of the roasting degree; phenanthrene and fluoranthene were the main PAHs with contents ranging from 0.16 to 0.79 $\mu g\ l^{-1}$, followed by pyrene 0.062 to 0.37 $\mu g\ l^{-1}$. The authors estimated that the percentage of PAHs transferred to the brew always remained below 35%.[7] However, no correlations between extractability and octanol–water partition, or the roasting temperature, were observed. Furthermore, for dark-roasted coffee, which usually presents higher PAHs, a slightly lower extractability during brewing was described, probably due to the formation of oil droplets.

Duedahl-Olesen *et al.*[39] observed that 5–14% of PAH4 from ground coffee samples were transferred to the brew. The authors observed that PAH4 from the two dark roasted samples analyzed were transferred to the brew (5–9%). Considering the medium roasted coffees, PAHs were transferred (0–14%) to the brew in four from the five analyzed samples. Regarding instant coffee samples that in powder represent higher content of PAH4, no PAHs were detected in the brews prepared from instant coffees, which is hard to understand since the brewing of instant coffee implies dissolution of instant coffee powder in the hot water, without any filtration.

Considering roasted coffee beans from different species, Tfouni *et al.*[9] also observed that coffee brews prepared with *C. arabica* presented mean sum of PAHs (benz(*a*)anthracene, benzo(*b*)fluoranthene, benzo(*k*)fluoranthene, and benzo(*a*)pyrene) of 0.015 to 0.105 $\mu g\ l^{-1}$, higher than the beverages prepared with *C. canephora* (0.011 to 0.111 $\mu g\ l^{-1}$). It has been hypothesized that since *C. canephora* presents higher content of caffeine, and caffeine is known to form a complex with PAHs,[47,48] this could possibly facilitate PAHs transfer from the ground roasted coffee to the brew. However Tfouni *et al.*[9] observed that caffeine contents in the beverages do not seem to influence the PAHs transfer. The authors concluded that PAHs levels present in the analyzed coffee brews are low considering the maximum permitted levels in the Brazilian and European regulation[18] for different foods.

Orecchio *et al.*[24] prepared brews from 13 commercial coffee samples with a *Moka* coffeemaker and for the same four PAHs analyzed by Tfouni *et al.*[9] verified a wide range of levels (0.001–0.161 $\mu g\ l^{-1}$). The mean value of the sum of PAH8 was 0.148 μg l^{-1}. Phenanthrene, acenaphthylene, and fluoranthene were the most abundant compounds. Phenanthrene was always much more abundant than the isomeric anthracene; this is important from the point of view of toxicity, since anthracene is the only three-ring PAH presenting a higher TEF (0.01). These results are in agreement with controlled roasting conditions performed by Houessou *et al.*[8]

In a study by Kayali-Sayadi *et al.*,[49] PAHs content depended on the sample elaboration. Decaffeinated and green coffees contained 1.99 and 1.65 ng l^{-1} coffee brew, respectively. Roasted coffee contained 2.10 ng l^{-1} and torrefacto coffee presented higher concentration, 2.87 ng l^{-1}. The authors also concluded that the roasting process increased PAHs contents in the samples. All evaluated roasted samples contained benzo(*a*)pyrene while green coffee did not.

32.9 Conclusions

High variability in the content of PAHs of different coffee has been observed in the published studies even comparing samples with similar label description and among samples produced under controlled conditions with the same degree of roasting and same cultivar, which suggests a high influence of analytical method on results.

The most abundant PAHs in coffee samples are phenanthrene, fluoranthene, and pyrene with contents in some cases exceeding largely 100 μg kg^{-1}. These compounds pose lower risk and do not belong to those considered by EFSA as the possible indicators of the carcinogenic potency of PAHs (PAH8) that comprise the eight carcinogenic PAHs from the US-EPA list. However, significant contents of the carcinogenic benz(*a*)anthracene and chrysene have been also found, especially in coffees with higher degree of roast that result from a possible transformation from low molecular to high molecular PAHs with the increasing roasting degree. The most studied PAHs compound, benzo(*a*)pyrene, was reported in commercially ground and instant coffees in the range from <0.01 μg kg^{-1} to near 100 μg kg^{-1}. Due to the low solubility in water, the extractability of compounds from coffee powder to the brew is generally low (less than 35%), especially in filtered coffee. Coffee brews show similar profiles with ground coffee, phenanthrene, fluoranthene, and pyrene still being the main PAHs. Additionally, with the increase of roasting degree a slightly lower extractability seems to occur, but this hypothesis deserves further investigation.

In general, considering the contents of PAHs in coffee and other foods as well as the maximum values above which these compounds represent risks to human health, the contents of PAHs in coffee brew indicate that coffee does not significantly contribute to the daily human intake of carcinogenic PAHs. However, as new roasting and brewing methods are constantly being created, brew concentrations in varied types of coffee brews consumed in different parts of the world still deserve investigation. Implications of human consumption of PAH through coffee will be explored in *Coffee: Consumption and Health Implications*.

References

1. EFSA, *EFSA J.*, 2008, **724**, 1–114.
2. T. Wenzl, R. Simon, E. Anklam and J. Kleiner, *TrAC, Trends Anal. Chem.*, 2006, **25**, 716–725.
3. A. Ramesh, S. A. Walker, D. B. Hood, M. D. Guillen, K. Schneider and E. H. Weyand, *Int. J. Toxicol.*, 2004, **23**, 301–333.
4. WHO – IPCS – International Programme on Chemical Safety, Environmental Health Criteria 202, 1998, (http://www.inchem.org/documents/ehc/ehc/ehc202.htmhttp://www.inchem.org/documents/jecfa/jecmono/v042je03.htm), accessed in 10/07/2015.
5. US-EPA – United States Environmental Protection Agency, (http://www2.epa.gov/sites/production/files/2015-09/documents/priority-pollutant-list-epa.pdf), accessed in 10/07/2015.
6. iIARC Working Group, *IARC Monogr. Eval. Carcinog. Risks Hum.*, 2012, **100F**.
7. A. Fabbri, C. Cevoli, L. Alessandrini and S. Romani, *J. Food Eng.*, 2011, **105**, 264–269.
8. J. K. Houessou, S. Maloug, A. S. Leveque, C. Delteil, B. Heyd and V. Camel, *J. Agric. Food Chem.*, 2007, **55**, 9719–9726.
9. S. A. V. Tfouni, C. S. Serrate, F. M. Leme, M. C. R. Camargo, C. R. A. Teles, K. M. V. A. B. Cipolli and R. P. Z. Furlani, *LWT--Food Sci. Technol.*, 2013, **50**, 526–530.
10. A. Luch and W. M. Baird, in *Comprehensive Toxicology*, ed. C. A. McQueen, Academic Press, Oxford, 2nd edn, 2010, vol. 14, pp. 85–123.
11. J. H. Park and T. M. Penning, in *Process-Induced Food Toxicants: Occurrence, Formation, Mitigation, and Health Risks*, ed. R. H. Stadler and D. R. Lineback, John Wiley & Sons, Inc., New Jersey, 2009, pp. 243–282.
12. B. H. Chen and Y. C. Chen, *J. Agric. Food Chem.*, 2001, **49**, 5238–5243.
13. P. Plaza-Bolanos, A. G. Frenich and J. L. M. Vidal, *J. Chromatogr. A*, 2010, **1217**, 6303–6326.
14. O. Viegas, P. Novo, E. Pinto, O. Pinho and I. M. P. L. V. O. Ferreira, *Food Chem. Toxicol.*, 2012, **50**, 2128–2134.
15. H. Choi, R. Harrison, H. Komulainen and J. M. D. Saborit, in *WHO Guidelines for Indoor Air Quality: Selected Pollutants*, Geneva, 2010.
16. C. A. Menzie, B. B. Potocki and J. Santodonato, *Environ. Sci. Technol.*, 1992, **26**, 1278–1284.
17. K. E. Anderson, F. F. Kadlubar, M. Kulldorff, L. Harnack, M. Gross, N. P. Lang, C. Barber, N. Rothman and R. Sinha, *Cancer Epidemiol., Biomarkers Prev.*, 2005, **14**, 2261–2265.
18. European Commission, *Off. J. Eur. Union L.*, 2011, **215**, 9–16.
19. A. Jimenez, A. Adisa, C. Woodham and M. Saleh, *J. Environ. Sci. Health, Part B*, 2014, **49**, 828–835.
20. Z. Zelinkova and T. Wenzl, *Polycyclic Aromat. Compd.*, 2015, **35**, 248–284.
21. S. A. V. Tfouni, C. S. Serrate, L. B. Carreiro, M. C. R. Camargo, C. R. A. Teles, K. M. V. A. B. Cipolli and R. P. Z. Furlani, *Int. J. Food Sci. Technol.*, 2012, **47**, 406–415.

22. G. Stanciu, S. Dobrinas, S. Birghila and M. Popeseu, *Environ. Eng. Manage. J.*, 2008, 7, 661–666.
23. V. D. Nagaraju and S. Bhattacharya, *J. Food Sci. Technol.*, 2010, **47**, 674–677.
24. S. Orecchio, V. P. Ciotti and L. Culotta, *Food Chem. Toxicol.*, 2009, **47**, 819–826.
25. V. D. Nagaraju, C. T. Murthy, K. Ramalakshmi and P. N. Srinivasa Rao, *J. Food Eng.*, 1997, **31**, 263–270.
26. J. K. Houessou, D. Goujot, B. Heyd and V. Camel, *J. Agric. Food Chem.*, 2008, **56**, 3648–3656.
27. R. K. Sharma and M. R. Hajaligol, *J. Anal. Appl. Pyrolysis*, 2003, **66**, 123–144.
28. P. F. Britt, A. C. Buchanan, M. K. Kidder and C. V. Owens, *J. Anal. Appl. Pyrolysis*, 2003, **66**, 71–95.
29. X. Wu, K. Skog and M. Jagerstad, *Mutat. Res., Genet. Toxicol. Environ. Mutagen.*, 1997, **391**, 171–177.
30. C. Petisca, T. Perez-Palacios, A. Farah, O. Pinho and I. M. P. L. V. O. Ferreira, *Food Bioprod. Process.*, 2013, **91**, 233–241.
31. O. Viegas, P. Novo, O. Pinho and I. M. P. L. V. O. Ferreira, *Talanta*, 2012, **88**, 677–683.
32. G. Purcaro, S. Moret and L. S. Conte, *Talanta*, 2013, **105**, 292–305.
33. R. S. Jose, A. G. Ruiz, C. von Holst, T. Wenzl and E. Anklam, *Anal. Bioanal. Chem.*, 2008, **391**, 1397–1408.
34. R. Pissinatti, C. M. Nunes, A. G. de Souza, R. G. Junqueira and S. V. C. de Souza, *Food Control*, 2015, **51**, 140–148.
35. J. K. Houessou, C. Delteil and V. Camel, *J. Agric. Food Chem.*, 2006, **54**, 7413–7421.
36. J. K. Houessou, C. Benac, C. Delteil and V. Camel, *J. Agric. Food Chem.*, 2005, **53**, 871–879.
37. M. S. García-Falcón, B. Cancho-Grande and J. Simal-Gandara, *Food Chem.*, 2005, **90**, 643–647.
38. European Commission, *Off. J. Eur. Union*, 2006, **L364**, 5–24.
39. L. Duedahl-Olesen, M. A. Navaratnam, J. Jewula and A. H. Jensen, *Polycyclic Aromat. Compd.*, 2015, **35**, 74–90.
40. A. Sadowska-Rociek, M. Surma and E. Cieslik, *Food Anal Method*, 2015, **8**, 109–121.
41. S. Moret, K. Grob and L. S. Conte, *Z. Lebensm.-Unters. -Forsch. A*, 1997, **204**, 241–246.
42. L. Kolling-Speer and K. Speer, in *Espresso Coffee, the Science of Quality*, ed. A. Illy and R. Viani, Elsevier Academic Press, Italy, 2005, pp. 165–166.
43. I. S. Grover, R. Sharma, S. Singh and B. Pal, *Environ. Monit. Assess.*, 2013, **185**, 6459–6463.
44. C. M. A. Iwegbue, H. Agadaga, F. I. Bassey, L. C. Overah, G. O. Tesi and G. E. Nwajei, *Int. J. Food Prop.*, 2015, **18**, 2124–2133.
45. I. C. T. Nisbet and P. K. Lagoy, *Regul. Toxicol. Pharmacol.*, 1992, **16**, 290–300.

46. A. Farah, in *Coffee: Emerging Health Effects and Disease Prevention*, ed. Y.-F. Chu, Wiley-Blackwell, 2012, pp. 21–58.
47. S. Moret and L. S. Conte, *J. Chromatogr. A*, 2000, **882**, 245–253.
48. R. R. Navarro, H. Ichikawa, K. Morimoto and K. Tatsumi, *Chemosphere*, 2009, **76**, 1109–1113.
49. M. N. Kayali-Sayadi, S. Rubio-Barroso, M. P. Cuesta-Jimenez and L. M. Polo-Diez, *J. Liq. Chromatogr. Relat. Technol.*, 1999, **22**, 615–627.

CHAPTER 33

Coffee Volatile and Aroma Compounds – From the Green Bean to the Cup

CHAHAN YERETZIAN*, SEBASTIAN OPITZ, SAMO SMRKE AND MARCO WELLINGER

Zurich University of Applied Sciences, Institute of Chemistry and Biotechnology, 8820 Wädenswil, Switzerland
*E-mail: yere@zhaw.ch

33.1 Introduction

Many hands in many countries on many continents from people who have never met are needed to make a cup of coffee – people who pour their hearts into creating the cup that we enjoy. From the hands of the farmer and farmworkers that plant, cultivate and harvest the cherries to the ones that dry, sort and process them in the origin countries. From the hands that pack and ship the beans to the hands that roast them, arriving in the hands that grind and extract the brew, and finally into the hands of those who consume it. From the seed to the cup, coffee is a product that goes through the hands of many, seemingly unrelated, people. Together, these hands compose this complex and fascinating value chain that crafts one of the world's most consumed and popular beverages. According to the International Coffee Organisation (ICO), in 2011 the coffee sector employed approximately

Coffee: Production, Quality and Chemistry
Edited by Adriana Farah

Published by the Royal Society of Chemistry, www.rsc.org

26 million farmers and coffee workers,[1] and it is estimated that around 125 million people worldwide depend on coffee for their livelihoods.[2] If we include coffee-related jobs such as the truck driver or the port worker who transport the coffee, as well as those serving coffee in cafes and restaurants or working at factories such as paper cup manufacturers, this figure would increase further. At the consumer end, over 2.25 billion cups of coffee are consumed in the world every day.[3] These numbers indicate that coffee is a crop of major economic importance to producing as well as to consuming countries. Coffee is today the most widely commercialized tropical product on the international market.[1]

Coffee is a lifestyle product that has penetrated the daily routine of many people and that is continuing to expand into new and emerging markets. While there are different factors that can be put forward to explain the success story of coffee, such as the mildly stimulating effect of caffeine, the social facilitation over a cup of coffee or the emerging positive health implications, the unique and complex flavour profile is probably the main reason why so many enjoy a cup of coffee. This chapter will specifically address the aroma of coffee.

33.2 Coffee Aroma – From Seed to Cup

The aroma of a cup of coffee is the perceptible result of a series of transformations from the seed to the cup.[4] Figure 33.1 outlines various factors that impact the aroma of coffee. Genetic predispositions, environmental and climatic factors, harvest and post-harvest practices, sorting, grading, storage and transport, roasting, grinding and extraction and finally consumption practices all impact to varying degrees on the final result. This journey can schematically be condensed into three stages: origin → transformation → release. All three play a specific role in the final aroma.

- Origin – the predispositions: the green coffee variety (genetics) with its specific set of chemical precursors in the green bean forms the basis for the aroma compounds that will later develop and are released during consumption to the sensory receptors, which triggers a sensation and leads to a perception of the aroma. The agronomy, climate, harvest practices and post-harvest treatment already at this stage start to modulate the genetic predisposition of the coffee bean and will affect the composition of the green bean as it goes into roasting.
- Transformation – linking the seed to the cup: even the best green coffee will be ruined if not properly processed. The craft and science of creating good coffee is to master every step when transforming the coffee. However, the smell and taste of a green bean provide no clue as to what it might turn into once roasted. The first and most significant step in the transformation process is roasting. This unlocks the potential of a green bean and creates the coffee aroma. However, grinding, degassing and storage are also very important.

Figure 33.1 Every step matters in obtaining a good cup of coffee: the aroma of coffee is the result of a long chain of transformations, which link the seed to the cup. It is influenced by genetic, agricultural, chemical and technical factors. Most importantly, the many people and the care taken by all those involved in these transformations are the most important factors in shaping the final quality in the cup.

- Release – connecting to the consumer: at the end of the long journey of the coffee bean, both in distance and in time, the coffee finally comes into contact with the consumer. The way we extract, prepare and consume our coffee greatly affects the flavour we perceive.

33.3 The Sensory Experience of Coffee

People's motives for drinking coffee vary widely. Some just want the caffeine kick to keep them going. More and more people, though, are currently drinking coffee not for the energy rush, but for the flavour – a trend that was strongly promoted by the advent of the specialty coffee movement in the late 1960s. More recently, reported health benefits from moderate coffee consumption are becoming an additional reason for people to choose to drink coffee – a trend fuelled by insights gained from research in the polyphenols

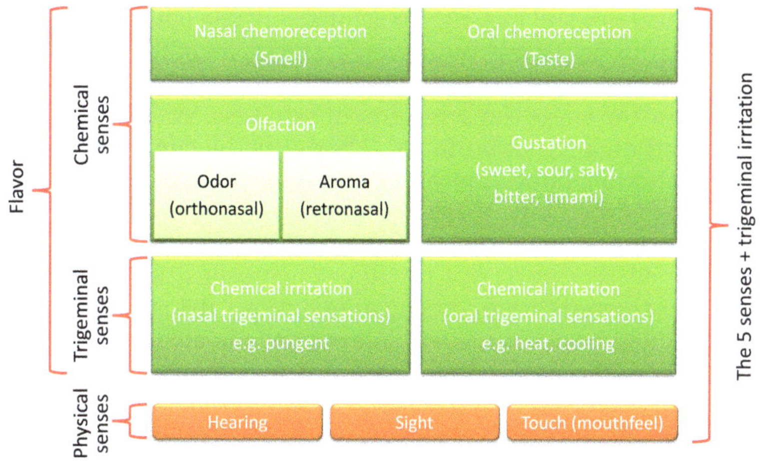

Figure 33.2 Our senses represent our windows to the world and to our cup of coffee. While all are relevant and contribute to the holistic sensory experience while drinking coffee, some are considered more relevant than others. In this chapter, we specifically focus on volatile organic compounds that contribute towards the sense of olfaction.

and antioxidants of coffee.[5] Finally, and something which has always been the case, coffee has been a fabulous facilitator of social interactions – a trend that was propelled by the global rise of coffee shops.

In this chapter, we focus on the aroma of coffee, which is an important facet of the multidimensional sensory experience, as shown in Figure 33.2. Indeed, drinking coffee elicits a multi-sensory experience which involves all our senses, including olfaction, taste, touch, trigeminal sensation, vision and possibly hearing. Aroma is an odour; it is often referred to as smell and is sensed by receptors in the nose. An important distinction is made between the orthonasal and the retronasal aroma. When we have coffee in front of us, we can smell it. We call this the orthonasal, or above-the-food, aroma. Volatile organic compounds (VOCs) released from coffee enter our nasal cavity during inhalations and reach the olfactory epithelium. The olfactory centre resides high along the roof of the nasal cavity, just below and between the eyes. It covers an area of 9 cm^2 (3 × 3 cm) and contains some 10–20 million receptor cells.[6–8] In order for these centres to be stimulated, the odour molecules have to be inhaled into the nose and carried up to the roof of the nose. The second type of aroma of significance to consumers is the aroma released in the mouth when drinking coffee. While coffee is in the mouth and also after swallowing, volatile compounds are released from the drink and transported by various airflows (mouth movement, respiration) from the oral cavity to the pharynx, passing the soft palate (velum palatinum). When exhaling, volatiles are swept by the airflow coming from the oral cavity and the lungs and are released through the nose. During their transport from the oral cavity through the pharynx to the nasal cavity, VOCs pass along the olfactory epithelium and may trigger an olfactory perception. This aroma is termed the retronasal, or in-mouth, aroma, and is part of the flavour of coffee along with the taste aspect, as perceived during eating or drinking.[9,10]

Taste, or gustation, is the sensation of saltiness, sweetness, sourness, bitterness and umami (savouriness). Flavour is the combination of taste and smell. Another important additional sensation to describe the flavour of coffee is the body. It can be light, like a dry light wine, or it can be heavy, like a red wine. The perception of irritants is mediated not by taste and smell receptors, but by other chemosensitive receptors. The perceptual characteristics of chemical irritation, also called chemesthesis, are mediated by non-specific, multimodal somatosensory fibres and are a property of the skin.[11] These chemical senses are complemented by the physical senses of sight or vision, hearing and touch (somatosensation). Figure 33.2 summarizes the multidimensional space of the sensory experience.

Besides the various modalities of a sensory experience, as outlined in Figure 33.2, a complete sensory experience can be decomposed into four hierarchical levels, as outlined in Figure 33.3.

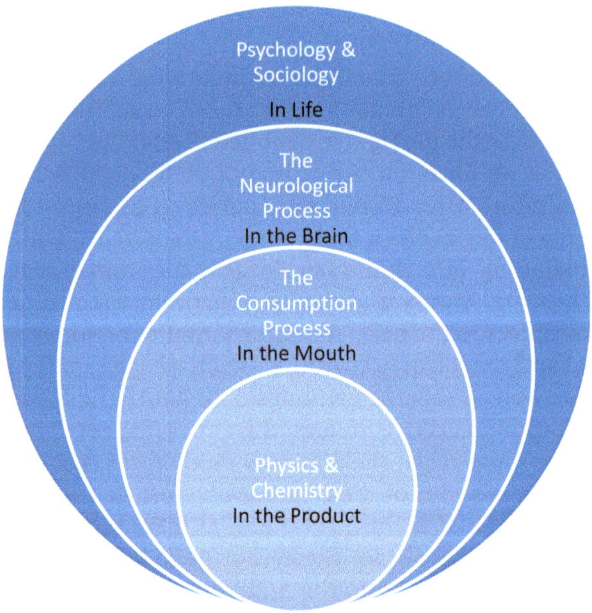

Figure 33.3 The sensory experience of coffee is a truly holistic one where objective and measurable aspects of the product interact with processes occurring at the individual/personal level in the mouth. These are complemented by subjective processes in the brain and are finally modulated by psychological and sociological elements such as mood, memories and social context. All four levels collectively contribute to the overall sensory experience elicited when consuming coffee.

Starting at the level of the product itself, its chemical composition and physical properties, the product can be integrated into an increasingly larger context, including the specific consumer, her/his physiological, neurological and psychological characteristics.

1. Chemical and physical properties: this first and best understood factor of the sensory experience is exclusively related to the product itself, without reference to the specific consumer and mode of consumption. These are properties that can be precisely measured by instrumental/analytical techniques and focus on the concentration of aroma and taste compounds in the product, on compounds that elicit cooling and pungent impressions, on temperature and texture (*e.g.* viscosity, % of total dissolved solids, viscosity) and on colour.
2. The consumption process: a specific cup of coffee might have a precisely measurable composition of aroma compounds, yet these compounds may be released in-mouth and transported to the sensory receptors differently from person to person, leading to inter-individual differences in perception. Individual characteristics in consumption patterns (volume of sip, breathing rhythm, swallowing pattern, movement of the coffee in the mouth, *etc.*) and in-mouth physiognomy and physiology (shape of nasal cavity, amount and composition of saliva) will modulate the sensory experience and lead to inter-individual differences in the sensory experience, even if exactly the same coffee is consumed.[12–20] Indeed, a given coffee will not "taste" the same for different people. This second factor, therefore, concerns the process of drinking and all aspects related to the physiological and physical environment in the mouth that leads to the liberation of the aroma and its transport from the oral to the nasal cavity, where the olfactory receptors are located. By recognizing and exploring such factors involved in a person's perception of coffee aroma, it was possible to develop novel analytical approaches that pave the way for the science of individualized aroma.[9,10,21]

   Gierczynski *et al.* studied the effect of textural modifications of solid milk gels on *in vivo* aroma release. The aroma perception was investigated with a panel of 14 subjects. Great inter-individual differences were observed on aroma-release data, and the consequences of these differences on aroma perception were studied.[22] While this may be related to difference in consumption behaviours and in in-mouth physiognomy and physiology, the fact cannot be excluded that neurological differences may be at play as well.
3. The neurological make-up of individuals: besides individual differences in consumption patterns and in-mouth characteristics, the sensory experience is also modulated by individual differences in the initial sensation at the level of individual receptors, as well as in final and conscious perception. Sensation refers to the process of sensing

our environment through touch, taste, sight, sound and smell.[11,23,24] This information is sent to our brains in raw form where perception comes into play. Perception is the way we interpret these sensations and, therefore, make sense of everything around us. It is affected by a complex transduction process from activation of the multiple sensory receptors to the final response pattern at the level of the central nervous system, where interactions from other senses may be integrated.[25–27] Sensation is hence the process that allows our brains to take in information *via* our five senses, which can then be experienced and interpreted (*i.e.* turned into perception) by the brain. The individual sensory experience is, therefore, affected not only by the composition of sensory active compounds in the coffee and modulated by the individual's in-mouth characteristics and consumption habits. Individuals also differ in their neurological and physiological make-up. In the same way as the in-mouth environment, this make-up will lead to an individually modulated sensory experience. Hence, the same coffee may elicit distinctively different sensory experiences for different people.[28–31] A well-documented example is the perception of bitterness (sensitivity to 6-*n*-propylthiouracyl (PROP) or phenylthiocarbamide (PTC)). It was established that individuals may vary in the extent to which they perceive bitter tasting compounds[32] and that this inheritable trait is related to the density of fungiform papillae on the tongue.[33,34] There are reports that PROP sensitivity affects macronutrient selection.[35] Referring to smell, individual differences in sensitivity to the odour of 4,16-androstadien-3-one (a pheromone) were observed, and a bimodal sensitivity distribution was reported. A smaller group of individuals showed a high sensitivity to androstadienone (supersmellers).[36] Whether this specific sensitivity to androstadienone is due to exposure-induced changes at the receptor level or to genetic factors[37] remains to be determined. Furthermore, the inability of some people to perceive specific odours, termed anosmia, is also well established.[38,39]

4. Psychology and cognition: finally, individual history, past experiences, expectations, product familiarity,[40,41] psycho-social and cognitive factors[42] such as culture, mood, conditioning and social context can all affect the way a person experiences a cup of coffee ([40,43–50]). He or she catches an aroma of a dish and is suddenly immersed in a flurry of vivid memories, often emanating from childhood. What is it about smells that can trigger memories so strong and real it feels like the person has been transported back in time? This is known as "odour-evoked autobiographical memory" or the Proust phenomenon, after the French writer Marcel Proust. In his famous novel *In Search of Lost Time*, the narrator dips a madeleine cookie into a cup of tea and is transported back into time as long-forgotten memories of his childhood come flooding back. Indeed, research shows that odours are especially effective as reminders of past experience, much more so than cues from

other senses, such as sights or sounds.[51,52] This might have to do with the way the brain processes odours and memories. Smells are routed through the olfactory bulb, which is the smell-analysing region in the brain. This is closely connected to the amygdala and hippocampus, brain regions that handle memory and emotion. The close connection may explain why a scent might be tied to vivid memories in the brain, and then come flooding back when one is exposed to that particular odour trigger.

In the following sections, we look in detail at the subject of aroma analysis of coffee and explore the aroma of coffee from the green bean to the cup. In Figure 33.4, PTR-ToF-MS (Proton-Transfer-Reaction Time-of-Flight Mass-Spectrometry) profiles of green coffee, roasted whole bean coffee, roast and ground (R&G) coffee and finally of a coffee brew are shown, all plotted on an identical intensity scale. PTR-ToF-MS is an emerging analytical technique that has first been applied by Yeretzian and co-workers (since 1997) to coffee aroma analysis[53–55] and is today an established technique in the field.[56] While green coffee shows a distinctively different profile of much lower intensities than roasted coffee, grinding the roasted beans leads to a strong increase of the volatile intensities.

In Section 33.4, we will first address the VOC profile of green/raw coffee. While the actual aroma of the coffee beverage is formed during roasting, the VOC profile of green coffee might reveal beans that will negatively impact the quality of roasted coffee.[57] From Section 33.5 on, we will move into discussing the aroma of roasted coffee and "what makes coffee smell so good".[58,59]

33.4 Dynamic Headspace Analysis of Green Bean Volatile Compounds

Volatile fingerprinting of green coffee beans has been used to identify potential markers for defective and low-quality beans.[60,61] Differences in green bean volatile profiles were observed between coffees of different varieties that were harvested from the same farm at different degrees of ripeness.[62] Climate conditions, as well as the terroir, are also reflected in the volatile profiles of the green beans.[63] Hence, the complex composition of green bean VOCs seems to indirectly relate to coffee properties that are important for the aroma in roasted coffee.

We studied the green coffee volatile composition by means of dynamic headspace PTR-ToF-MS. Compared to traditional GC-MS, PTR-ToF-MS can achieve better sensitivity and reproducibility of the analysis. A highly reproducible way of measuring volatiles from coffee (green or roasted) is to perform a dynamic measurement of the headspace composition. In this experiment, the headspace over a sample is constantly flushed by nitrogen gas and the volatiles, dynamically released from the sample, are continuously measured

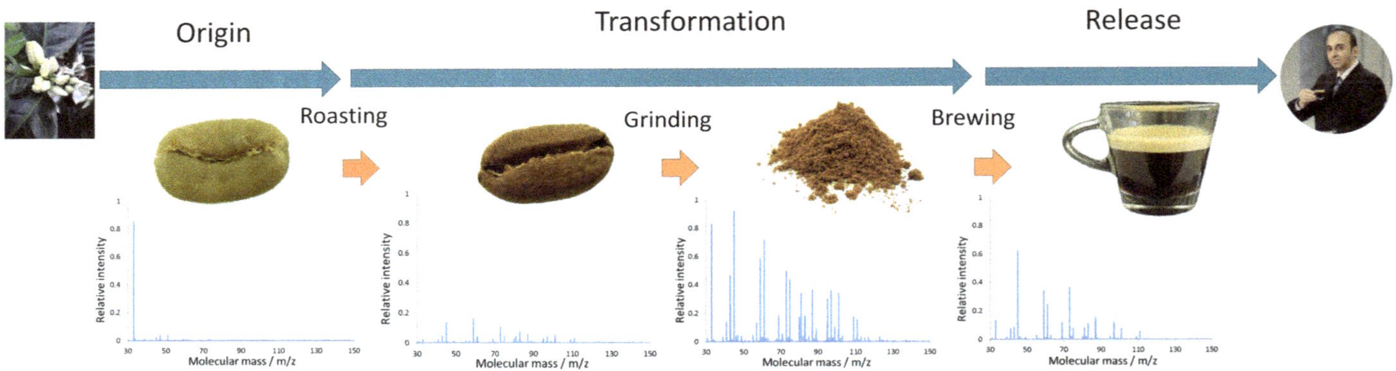

Figure 33.4 PTR-ToF-MS mass spectra profiles of coffee as green whole beans, roasted whole beans, roast and ground beans and coffee brew. The volatiles of ground coffee clearly show the strongest overall intensities.

by a direct PTR-ToF-MS analysis. In the case of green coffee samples, this improves the reliability of the measurements, since volatiles formed by surface oxidation are rapidly depleted and only those from the inside of the bean are measured. We found that measuring volatile composition over ground green coffee leads to irreproducible results, most likely from uncontrollable oxidation on the surface of the particles during sample preparation, as was also observed when analysing ground green coffee with GC-MS.[62]

In contrast, VOC profiles of whole green beans resulted in highly reproducible profiles when using dynamic headspace PTR-ToF-MS. Volatile profiles of green whole beans of *C. arabica* L. (12 samples) and *C. canephora* Pierre, cv. robusta (10 samples) were measured. A very large diversity of volatile profiles was found, ranging from 87 to 170 mass peaks detected in a green coffee sample with intensity over a threshold of 1 ion count per second. The number of molecular mass peaks that could be reliably assigned in all green coffee samples was relatively small – only 17 masses were present in the profiles of every sample. The instrumental analysis was performed in a highly automatized way, and the data were analysed using a non-targeted approach. Successful assignment of peaks in all samples is needed if the quantitative data are statistically analysed, *e.g.* with principal component analysis or building a model to relate the instrumental data to other coffee properties.

The composition of green coffee volatiles is very different from that of roasted coffee. One study has reported 12 alcohols, 6 acids, 6 esters, 5 ketones, 4 pyrazines, 3 furans and 3 aldehydes detected in coffee beans from 3 farms in a single region.[60] Another study reported 68 GC peaks, detected in green coffee samples from a single farm (2 different varieties), with 48 of those identified by comparing to a library of GC-MS spectra.[62] Differences are seen when comparing these compounds to roasted coffee aroma, since green coffee does not contain any of the important coffee aroma compounds, although similar groups of compounds can be observed. Indeed, most of the important coffee aroma compounds are formed during thermal treatment (roasting) of coffee and their content depends critically on the time temperature roast profile and roast degree. The groups that are not present in green coffee, but which are important for roasted coffee aroma, are furanones, phenols and sulfur-containing compounds.

The mass peaks and corresponding tentatively identified compounds in 22 green coffee samples of our study are presented in Table 33.1. The results show that small unsaturated hydrocarbons, alcohols and aldehydes were common to all samples analysed by dynamic headspace PTR-ToF-MS. The detected intensities of dynamic headspace signals were highly reproducible for repetitive analysis of a green coffee sample, but were strongly variable between samples. Using principle component analysis, we were able to differentiate all of the samples, in spite of the fact that the number of mass peaks included in the analysis was low.[22]

The key result of the PTR-ToF-MS profiling of green coffee volatiles shows that there is a very strong diversity and differences in volatile intensities

Table 33.1 List of mass peaks detected in all 22 evaluated coffee samples (12 arabica and 10 robusta) applying dynamic headspace of green whole beans with PTR-ToF-MS for volatile analysis. The mass accuracy in ppm is presented as standard deviation of a total of 100 measurements. Compounds were tentatively identified based on molecular formula and information obtained by GC-MS.[60,62]

Exact molecular mass (protonated)/*m*/*z*	Mass accuracy/ppm	Molecular formula	Compound name (tentative identification)
31.018	3.4	CH_3O^+	Formaldehyde
33.034	2.6	CH_5O^+	Methanol
39.032	6.2	*Not identified*	*Not identified*
41.039	1.0	$C_3H_6^+$	Cyclopropane or propadiene
43.054	12.1	$C_3H_8^+$	Propene or cyclopropane
45.034	0.8	$C_2H_5O^+$	Acetaldehyde
47.050	3.4	$C_2H_7O^+$	Ethanol
51.044	1.6	$C_2H_6O_2^+$	Ethylene glycol
55.054	21.8	$C_4H_7^+$	Unsaturated hydrocarbon
57.035	7.0	$C_3H_5O^+$	Acrolein or cyclopropanone
57.071	1.2	$C_4H_9^+$	Unsaturated hydrocarbon
59.050	1.2	$C_3H_7O^+$	Acetone, propanal
69.071	3.8	$C_5H_9^+$	Unsaturated hydrocarbon
71.086	13.2	$C_5H_{11}^+$	Unsaturated hydrocarbon
83.086	9.3	$C_6H_{11}^+$	Unsaturated hydrocarbon
87.044	17.8	$C_4H_7O_2^+$	Butyrolactone
101.059	19.5	$C_5H_9O_2^+$	*Not identified*

when comparing coffees that do not come from the same farm or region. While the overall intensity of the signals from the dynamic volatile profiles is low for green coffee (compared to roasted coffee), the spectrum itself is highly complex. In Figure 33.5, a mass section of the spectrum from two different arabica coffees is presented, where large differences in the signal intensities are shown for peaks having relatively strong signal intensities.

The relative simplicity of measuring volatiles in green beans (when comparing to the analysis of non-volatile components) opens a wide range of possible applications for rapid analysis of volatile composition of green beans, such as controlling consistency of quality, origin of green beans, monitoring deterioration and changes during storage or during shipment of the green beans. Most importantly, green coffee VOC profiles may reveal specific defects and quality issues that will affect the cup profile.

33.5 Roasted Coffee Aroma Compounds

Among the various sensory modalities, aroma (smell) is of paramount importance to the quality of coffee. In order for coffee to be sensed by our nose, aromatic VOCs are released from *e.g.* the brew and reach the olfactory epithelium, a region in the upper part of the nasal cavity which contains the

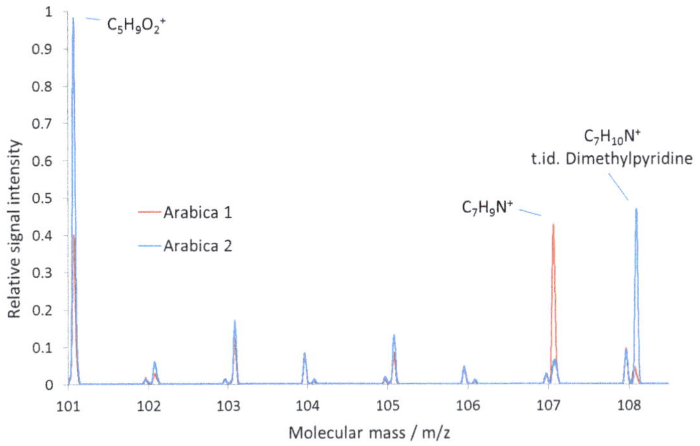

Figure 33.5 Dynamic PTR-ToF-MS mass spectra in the range of 101–108 *m*/*z* for two arabica coffees. For three of the mass peaks shown in the mass window, only the molecular formula could be assigned and, for one, it was possible to tentatively identify the compound as dimethylpyridine, based on identification with GC-MS.

nerve endings that allow us to smell. But what are these volatile aromatic compounds that compose the coffee aroma?

Scientific efforts to elucidate the origin of the rich and distinctive aroma of coffee, and ultimately to understand "What Makes that Coffee Smell so Good",[58,59] can be traced back to 1880 when Bernheimer identified a few volatile compounds in coffee.[64] But the first significant level of progress can most likely be attributed to Reichstein and Staudinger, who, in 1926, identified and patented several important aroma active compounds in coffee.[65,66] Mainly fuelled by progress in analytical techniques, in particular gas chromatography (GC), the number of publications on coffee aroma and the number of identified coffee VOCs has rapidly increased since then. Today, around 1000 VOCs have been reported in coffee, which includes compounds from both green and roasted coffee.[67]

For many years, scientists have concentrated on identifying the VOCs in coffee. However, in the 1970s it had already become clear that only a small fraction of these volatiles – perhaps 5% – are odoriferous and hence relevant to the aroma. As a result, the focus shifted towards these few sensory relevant aroma-active compounds in the headspace of coffee (HS, the air-space above the coffee).

Several instrumental methods have been developed to identify and quantify the odour-relevant volatiles, assess their odour impact and characteristics, and recombine coffee aromas from the identified and quantified main aroma compounds. Given the extensive differences in coffee genetics, geographical origins, cultivation practices and processing techniques, it is not surprising that publications on coffee aroma composition differ with respect to the relative importance given to the different aroma compounds. Consequently, different studies often report slightly different lists of VOCs responsible for the aroma of the particular coffee being studied. The different strategies and analytical techniques used for measuring VOC compositions are an additional source of variability in the ranking of the main aroma compounds. Hence, in order to 'eliminate' variability due to differences between, for example, species, in Table 33.2 we focus on just the *Coffea arabica*.

Studies by Grosch and his co-workers in the mid-1990s concluded that fewer than 30 VOCs are important for the aroma of roasted coffee.[70,72,74,96–99] In follow-up omission experiments, they suggested that the actual number of indispensable coffee aroma compounds could be as small as nine.[75,77,78,100] Based on this detailed work, a condensed list of coffee aroma VOCs has been compiled in Table 33.3. Instead of grouping compounds into chemical families, as in Table 33.2, compounds are grouped in Table 33.3 into sensory families. Besides concentrations in roast and ground coffee, concentrations in the liquid coffee extract and the extraction yields have also been included. The last column in Table 33.3 marks the 12 compounds that were considered by Grosch and his co-workers to be particularly important with their corresponding structural formulas shown in Figure 33.6.[75,77,100] Omitting these compounds from a coffee aroma model (individually or as groups) leads to a significant difference in the coffee aroma profile.

33.6 Analytical Techniques for Coffee Aroma Analysis

33.6.1 Gas Chromatography

The "work-horse" for coffee aroma analysis has been gas chromatography (GC). Over the past 30 years it has led to the elucidation, characterization and quantification of compounds that are relevant to coffee aroma. One potential weakness of such an approach is that it is slow, making it unsuitable for monitoring fast processes in real time. Yet, in spite of these and other drawbacks, GC analysis is still considered an excellent approach for understanding and reconstituting the flavour of coffee. The ultimate proof that we "understand" the aroma of coffee is when we are able to reconstitute it.[77,100]

33.6.2 Olfactometry – When the Human Nose Becomes a Detector

Many aroma active compounds in coffee only appear in the HS at very low concentrations, some of which can hardly be detected by instrumental detectors. The only detector capable of sensing these highly potent coffee aroma compounds is our nose. It is of course also the most relevant detector, when it comes to aroma compounds. Hence, aroma chemists use their nose (or a panel of "sniffers") to detect compounds that are eluted at the end of the GC column. Trained sniffers also provide a sensory description of the compound and rate the perceived intensity. In combination with the chromatographic retention time and mass spectral profile, the aroma note often allows the compound to be unambiguously identified. Techniques which combine separation by gas chromatography and the human nose as a detector are called GC-Olfactometry – GC-O[103] – and are explained in the following section.

Two major GC-O screening techniques have been developed: one by Grosch and his co-workers (called the Aroma Extract Dilution Analysis)[78,104] and the other by Acree and his co-workers (called CHARM analysis).[105–113] These techniques are shown schematically in Figure 33.7. Both evaluate a dilution series of an original aroma extract using GC-O. The occurrence of an aroma in each dilution is noted. As the dilution increases, the compounds with lower odour potency successively cease to be sensed, and only the most potent are detected at higher dilutions. The number of times each odorant was sensed across the dilutions is then summed up. The greater the number of dilutions in which an odorant is sensed, the higher is the odour potency in both AEDA and CHARM analysis. This leads to plots of dilution or CHARM values. Both AEDA and CHARM analysis originally proposed that the larger the dilution or CHARM value, the more important the contribution of the respective aroma compound to the overall aroma. While this interpretation has evolved, both techniques are still widely used to estimate the relative contribution of various VOCs to the aroma of coffee.[4,106,108,109,111,114] An alternative technique, called GC-SNIF, was introduced by Chaintreau and his co-workers.[115] In this method, the intensities of the aromagram peaks were based on the detection frequencies of the odorants perceived by a panel of assessors at the sniffing port of a GC-O. This approach allowed standard deviations to be calculated and hence led to more quantitative data-analysis than previous GC-O methods. For a review of GC-O, with a specific discussion of coffee, please refer to ref. 116.

An interesting application of GC-O was the identification of the chemicals responsible for the mouldy/earthy off-notes found in lots of green Mexican coffees.[4,114] GC-profiles obtained from a reference sample (a sample free of off-notes, assessed by expert tasters) and a defective mouldy sample, showed minor differences and no indication was found as to what the compounds responsible for the off-note could be. The same samples were then subjected to a GC-O sniffing analysis. This led to the recognition of several differences between the extracts, which could be related to the off-flavour and resulted in the identification of a series of compounds that caused the off-flavour (Figure 33.8). In particular, peaks that were related to earthy, green, chemical and mouldy attributes could be identified as 2-methyl isoborneol, 2,4,6-trichloroanisole, geosmin and various pyrazines. The chemical identification allowed the development of a hypothesis concerning the source of the off-flavour and, subsequently, propose mitigation strategies.

Table 33.2 Flavour active VOCs in arabica coffee (adapted from ref. 68). The first column lists 72 VOCs, classified into 15 different groups of chemical compounds. Data compiled from ref. 27, 40 and 69–95.

Key odorant	CAS ID#	Concentration[a] (ppb)	Aroma descriptor	Sensory threshold[b] (ppb)
Aldehyde				
2-Methylbutanal	96-17-3	20 700	Rancid, almond-like, toasty	1.3
2-Methylpropanal	78-84-2	–	Toasty, caprylic, cheesy, dark chocolate, ethereal, fruity, malty, pungent	–
3-Methylbutanal	590-86-3	18 600	Fruity, almond-like, toasty, ethereal, chocolaty, peachy, fatty	0.35
(*E*)-2-nonenal	18829-56-6	19	Fatty, green, cucumber, citrus	0.08
Acetaldehyde	75-07-0	139 000	Pungent, ethereal, fresh, lifting, penetrating, fruity, musty	0.7
4-Methoxybenzaldehyde	123-11-5	–	Sweet, powdery, vanilla, anise, woody, coumarin, creamy	27
Phenylacetaldehyde	122-78-1	–	Sweet, fruity, honey, floral, fermented	–
Propanal	123-38-6	17 400	Ethereal, pungent, earthy, alcoholic	10
Acid				
2-Methylbutyric acid	116-53-0	25 000	Acidic, fruity, dirty, cheese	10
3-Methylbutyric acid	503-74-2	18 060–32 180	Cheesy, dairy, acidic, sour, pungent, fruity, stinky	700
Ester				
Ethyl-2-methylbutyrate	7452-79-1	3.9	Fruity, berry	0.5
Ethyl-3-methylbutyrate	117442-70-3	14	Fruity	0.6
Furan				
Furfural	98-01-1	5880–19 370	Sweet, brown, woody, bready, caramellic	280
2-((Methylthio)methyl) furan/2-furfuryl methyl sulfide	1438-91-1	–	Smoke, roast, onion, garlic, sulfuraceous, pungent, vegetable, horseradish	–
2-Furanmethanol acetate/fur-	623-17-6	24 520–40 040	Onion, garlic, sulfury, pungent, vegetable,	–

5-Methyl-2-furancarboxyalde-hyde/2-methyl furfural	620-02-0	–	Sweet, caramellic, bready, brown, coffee-like	6000
Furfuryl formate	13493-97-5	–	Ethereal	–
Furfuryl methyl ether	13679-46-4	–	Roasted coffee	–
Furfuryl disulfide	4437-20-1	–	Sulfury, coffee, roasted chicken, meaty, onion, cabbage	–
Sulfur-containing compounds				
Dimethyl trisulfide	3658-80-8	28	Sulfurous, cooked onion, savory, meaty, cabbage-like	0.001
Bis(2-methyl-3-furyl)disulfide	28588-75-2	–	Meaty, roasted scallion, onion, sulfury	0.00076
Methional	3268-49-3	213–240	Boiled potato-like, musty, tomato, earthy, vegetable, creamy	0.2
Thiols				
3-Mercapto-3-methyl butyl formate	50746-10-6	130	Green blackcurrant, herbal, fruity, roasted, sweaty	0.0035
2-Furfurylthiol	98-02-2	1080–5080	Roasty (coffee-like), sulfurous	0.01
2-Methyl-3-furanthiol	28588-74-1	60–68	Sulfury, meaty, fishy, metallic, boiled	0.007
3-Mercapto-3-methylbutylacetate	50746-09-3	7.5	Roasty, fruity, sulfurous, sweet	–
3-Methyl-2-butene-1-thiol	5287-45-6	13	Sulfury, smoky, leeky, onion	0.0003
Methanethiol	74-93-1	4550	Rotten eggs, meat or fish, cabbage, garlic, cheesy	0.02
Thiophene				
3-Methylthiophene	616-44-4	–	Fatty, winey	–
Thiazole				
2,4-Dimethyl-5-ethylthiazole	38205-61-7	–	Nutty, roasty, meaty, earthy	–
Furanone				
Dihydro-2-methyl-3(2*H*)-furanone	3188-00-9	7580–30 000	Sweet, bread, buttery, nutty	0.005

(*continued*)

Table 33.2 (*continued*)

Key odorant	CAS ID#	Concentration[a] (ppb)	Aroma descriptor	Sensory threshold[b] (ppb)
2-Ethyl-4-hydroxy-5-methyl-3(2*H*)-furanone (homofuraneol)	27538-10-9	16 800	Sweet, caramel, candy	20
3-Hydroxy-4,5-dimethyl-2(5*H*)-furanone (sotolone)	28664-35-9	1.1–11 470	Extremely sweet, strong caramel, maple, burnt sugar, coffee	20
4-Hydroxy-2,5-dimethyl-3(2*H*)-furanone (furaneol)	3658-77-3	10930–112 000	Sweet, candy, caramel, strawberry, sugar	10
5-Ethyl-3-hydroxy-4-methyl-2(5*H*)-furanone (abhexon)	144831-60-7	104–160	Seasoning-like, caramel-like	7.5
5-Ethyl-4-hydroxy-2-methyl-3(2*H*)-furanone	27538-09-6	17 300	Sweet, caramel, bready, maple, brown sugar, burnt	1.15
Ketone				
1-Octen-3-one	4312-99-6	–	Herbal, mushroom, earthy, musty, dirty	0.0036
2,3-Hexadione	3848-24-6	–	Burnt, buttery, caramel, chocolate cream, creamy, fruity, oily, pear, sweet	–
2,3-Butanedione	431-03-8	48 400–50 800	Buttery, creamy, fatty, oily, sweet, vanilla	0.3
2,3-Pentanedione	600-14-6	3540–39 600	Buttery, caramel, creamy, penetrating, sweet	20
4-(4′-Hydroxyphenyl)-2-butanone	5471-51-2	1	Sweet, fruity, berry, jam, raspberry, ripe, floral (raspberry ketone)	1–10
1-(2-Furanyl)-2-butanone	4208-63-3	–	rummy	–
Norisoprenoid				
(*E*)-β-Damascenone	23726-93-4	195–255	Honey-like, fruity, apple, rose, honey, tobacco, sweet	0.00075
Phenolic compounds				
Guaiacol	90-05-1	2000–17 970	Phenolic, burnt, smoke, spice, vanilla, woody	2.5
4-Ethyl guaiacol	2785-89-9	800–24 800	Spicy, smoky, bacon, phenolic, clove	25
4-Vinyl guaiacol	7786-61-0	8000–64 800	Spicy, dry woody, fresh amber, cedar, roasted peanut	0.75

Pyrazine				
2,3-Dimethylpyrazine	5910-89-4	2580–6100	Nutty, coffee, peanut butter, walnut, caramel, leather	800
2,5-Dimethylpyrazine	123-32-0	4550–11 730	Cocoa, roasted nuts, roast beef, woody, grass, medical	80
2,3-Diethyl-5-methylpyrazine (hazulnut pyrazine)	18138-04-0	73–95	Nutty-roast, musty, meaty, vegetable, roasted hazelnut	0.09
2-Ethyl-3,5-dimethylpyrazine	27043-05-6	55–330	Nutty-roast	0.04
2-Ethyl-3,6-dimethyl-pyrazine (3,6-cocoa pyrazine)	13360-65-1	2570–5980	Potato, cocoa, roasty, nutty	8.6
2-Methoxy-3,5-dimethylpyrazine (3,5-cocoa pyrazine)	13925-07-0	1.1	Earthy, burnt, almonds, roasted nuts, coffee	0.006
2-Methoxy-3,2-methylpropylpyrazine	24683-00-9	–	Green, pea green, bell pepper	–
2-Methoxy-3-isopropylpyrazine	25773-40-4	2.4	Earthy, pea, beany	0.002
3-Ethenyl-2-ethyl-5-methylpyrazine	181589-32-2	–	Earthy	–
6,7-Dihydro-5-methyl-5*H*-cyclopentapyrazine	23747-48-0	–	Nutty-roast, earthy, baked potato, peanut, roasted	–
Ethylpyrazine	13925-00-3	–	Peanut butter, musty, nutty, woody, roasted cocoa	4000
Pyridine				
Pyridine	110-86-1	21280–65520	Fishy	77
Pyrrole	109-97-7	–	Sweet, nutty, ethereal	–
1-Methyl pyrrole	96-54-8	–	Smoky, woody, herbal, negative notes–defective beans	–
Terpene				
Linalool	78-70-6	–	Flowery, citrus, orange, terpy, waxy, rose	0.17
Limonene	138-86-3	–	Citrus, herbal, terpene, camphor	4
Geraniol	106-24-1	–	Sweet, floral, fruity, rose, waxy, citrus	1.1

[a]The second column shows the range of reported concentrations in ppb (μg g^{-1}) in roasted arabica coffee. For some VOCs no information on concentration was reported.
[b]In cases where different sensory detection thresholds were published, the one included in the table corresponds to the lowest one reported.

Table 33.3 Quantitative composition of roasted coffee, ground coffee and coffee extract/beverage, for the 28 most potent coffee aroma compounds, determined for a medium-roasted columbian arabica coffee. (a) Mass-fraction of the aroma compound in roast and ground coffee powder.[101,102] (b) Concentration of aroma compounds in the extract/beverage (54 g of R&G extracted with 1 L of hot water).[74,101] (c) Mass-fraction of aroma compounds in the extract/beverage (calculated from ref. 74 and 101; brew ratio = 18.5 : 1 L water to 54 g R&G). (d) Extraction yield.[101] (e) Compounds that caused a perceived (significant) change of aroma, when omitted from the aroma model.[75] The numbers in brackets behind the compound names correspond to the numbers of the structures shown in Figure 33.6.

Compound name	Powder/μg kg^{-1} (a)	Concentration in brew/(μg L^{-1}) (b)	Yield per mass of coffee used/(μg kg^{-1}) (c)	Extraction-yield/% (d)	Aroma model (e)
Sweet, caramel notes					
2-Methylpropanal (1)	24000	760	14000	59	x
2-Methylbutanal (2)	26000	870	16000	62	x
3-Methylbutanal (3)	17000	570	11000	62	x
2,3-Butandione	49000	2100	39000	79	
2,3-Pentandione	35000	1600	30000	85	
4-Hydroxy-2,5-dimethyl-3(2*H*)-furanone	140 000	7200	130 000	95	
2(5)-Ethyl-4-hydroxy-5(2)-methyl-3(2*H*)-furanone	16000	800	15000	93	
Vanillin	4100	210	3900	95	
Earthy notes					
2-Ethyl-3,5-dimethylpyrazine (4)	400	17	310	79	x
2-Ethenyl-3,5-dimethylpyrazine (5)	53	1	19	35	x
2,3-Diethyl-5-methylpyrazine (6)	100	3.6	67	67	x
2-Ethenyl-3-ethyl-5-methylpyrazine	15	0.2	4	25	
3-Isobutyl-2-methoxypyrazine (7)	120	1.5	28	23	x

Roasty, sulfury notes					
2-Furfurylthiol (8)	1700	17	320	19	x
2-Methyl-3-furanthiol	60	1.1	20	34	
Methional	250	10	185	74	
3-Mercapto-3-methylbutyl format	130	5.7	105	81	
3-Methyl-2-buten-1-thiol	13	0.6	11	85	
Methanthiol	4400	170	3100	72	
Dimethyl trisulfide	28	–	–	–	
Phenolic					
Guaiacol	2400	120	2200	73	x
4-Ethylguaiacol	1800	48	900	49	
4-Vinylguaiacol (9)	45000	740	14000	30	
Fruity, flowery					
Acetaldehyde (10)	120000	4700	87000	73	x
Propanal (11)	17400	–	–	–	x
(*E*)-β-Damascenone	260	1.6	30	11	
Sharp					
3-Hydroxy-4,5-dimethyl-2(5*H*)-furanone/ sotolone	1900	80	1500	78	x
5-Ethyl-3-hydroxy-4-methyl-2(5*H*)-furanone/ furaneol (12)	104	–	–	–	

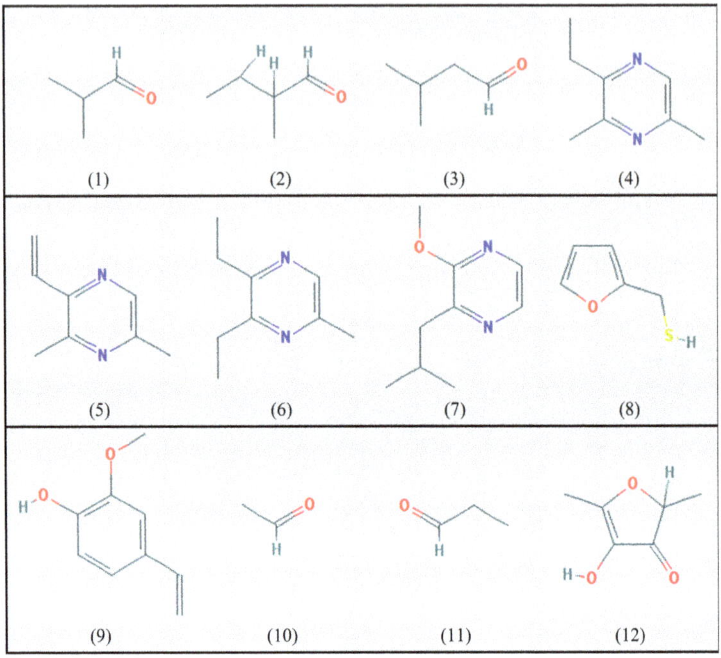

Figure 33.6 Twelve compounds that are considered of particular importance to the aroma of coffee. The numbers in brackets correspond to the numbers behind the names in Table 33.2/first column. Omitting these compounds, individually or as part of compound groups, from a coffee aroma model, leads to a significant difference in the aroma profile. They are therefore considered to be of particular importance to the aroma of coffee.[75]

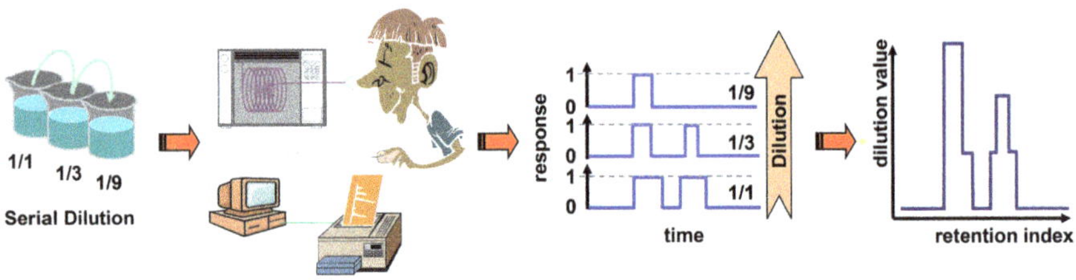

Figure 33.7 Schematic description of Gas Chromatography-Olfactometry, GC-O. Aroma extract dilution analysis (AEDA) and CHARM analysis are the two most common realizations of GC-O.

The example in Figure 33.8 demonstrates some of the strengths of GC-O: the identification of VOCs responsible for olfactory off-notes in a defective sample. It compensates the lack of sensitivity to low concentration flavour active compounds of other detection systems. In this study, it was clear that instrumental detection failed to recognize the defects documented in the sensory profile.[4,114]

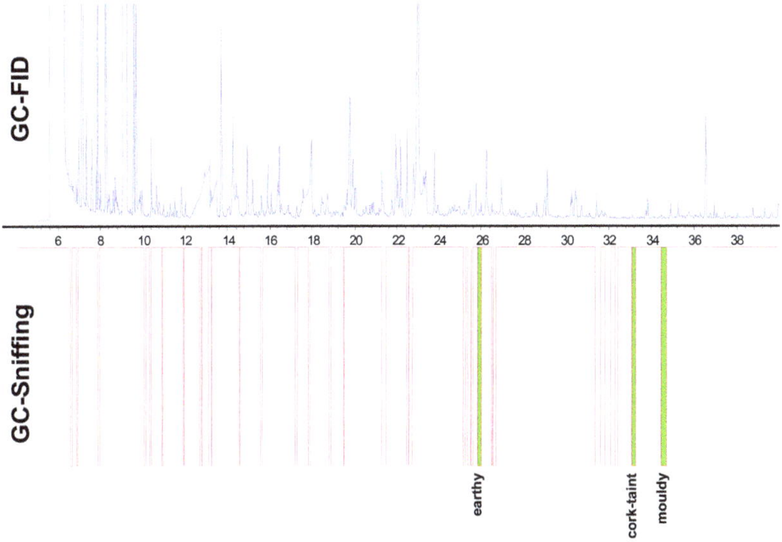

Figure 33.8 Comparison of a GC chromatogram on the top with the GC-O profile analysed by one sniffer on the bottom for a sample with a mouldy/earthy defect. All sniffed signals without sensory descriptors represent signals that are typical to coffee and are not related to the off-note.

33.7 Trends and New Developments in Coffee Aroma Analysis

Over the past 30 years, our understanding of the aroma of coffee has grown steadily. Today, we believe that the list of aroma active compounds is essentially complete. Nevertheless, research into coffee aroma is still in its infancy. Major efforts are currently being made to develop a range of novel analytical technologies and strategies. We would like to highlight three major trends that we believe will shape the future of coffee aroma research. These are (i) development of time-resolved analytical technologies (Section 33.7.1), (ii) progress in individualized aroma science with the development of analytical techniques that allow the capturing of such differences and, consequently, the developing of a better understanding of inter-individual difference in aroma release, sensation and perception (Section 33.7.2) and (iii) mathematical/statistical models to predict sensory profiles from instrumental measurements (Section 33.7.3).

33.7.1 Time-resolved Analytical Techniques

A major new development in analytical technologies represents the time-resolved methods based on direct injection mass spectrometry using optical and laser ionization[117–123] and chemical ionization.[10,56,124–138] Here, we will discuss two applications of time-resolved approaches, one on coffee roasting and one on extraction. Both are dynamic processes requiring a time resolution of approximately one second.

33.7.2 Analysis of Aroma Formation During Roasting

The crucial step in creating the coffee aroma is roasting. One possibility for studying the formation of VOCs during roasting is to take samples at different times during the roasting process and analyse these off-line by gas chromatography.[101,139–143] However, these techniques are not only time-consuming, but often require complex sample preparation processes before analysis, with the additional risk that irregularities in these processes might affect the outcome of the analysis. In contrast, on-line measurements of the roaster off-gas provide a direct insight into the dynamics of VOC formation in real time, are very sensitive and do not require sample preparation (hence avoiding potential distortion of information). An already well-proven technology for on-line analysis of coffee roasting is proton-transfer-reaction mass spectrometry (PTR-MS).[53–55,125,144] Furthermore, alternative techniques based on resonant laser ionization coupled to time-of-flight mass analysis[117,121–123] or ion trap mass spectrometry[145] have also been applied to explore the coffee roasting process. More recently, with single-photon ionization time-of-flight mass spectrometry (SPI-ToF-MS), single bean roasting has been performed on arabica as well as on robusta beans.[123] Combining soft ionization *via* proton–transfer reaction with high mass resolution of a time-of-flight instrument (PTR-ToF-MS)[146] provides the advantage of a fast analytical technique to record information about VOCs formed during roasting in just one single mass spectrum. On-line analysis of coffee roasting with PTR-ToF-MS allows the monitoring of the formation of VOCs in real-time.[136,138,147,148]

Two examples for on-line analysis by PTR-MS of roaster off-gas during coffee roasting are provided below. The first example represents a Colombian arabica coffee which was roasted using different roaster gas temperatures in order to achieve the same dark roast degree (CTN 67 on the Neuhaus scale). The temperature of the roaster gas was set to 228 °C, 238 °C, 248 °C and 258 °C (isothermal roasting) for four roasting trials and more than 50 VOCs were measured simultaneously during each trial. Figure 33.9 shows the experimental setup, while Figure 33.10 shows the time–intensity profiles for two selected compounds – furfural and 5-methylfurfural – during the four different isothermal roasting trials. While the coffee reaches the same dark roast degree of CTN 67 in all four trails, the roasting times differed. At 228 °C (Figure 33.10, green line), the roasting time corresponded to 25 minutes (1500 seconds). In contrast, roasting the coffee at 238 °C, 248 °C and 258 °C (in the black, pink and blue traces) meant that the time taken to reach CTN 67 was reduced to ~14 min, ~10 min and ~7.5 min. respectively. Besides the obvious reduction in roasting time with increased temperature, the time–intensity profiles for the two selected compounds differed in two major aspects: (i) a strong increase in the intensity of the VOCs in the off-gas with increasing

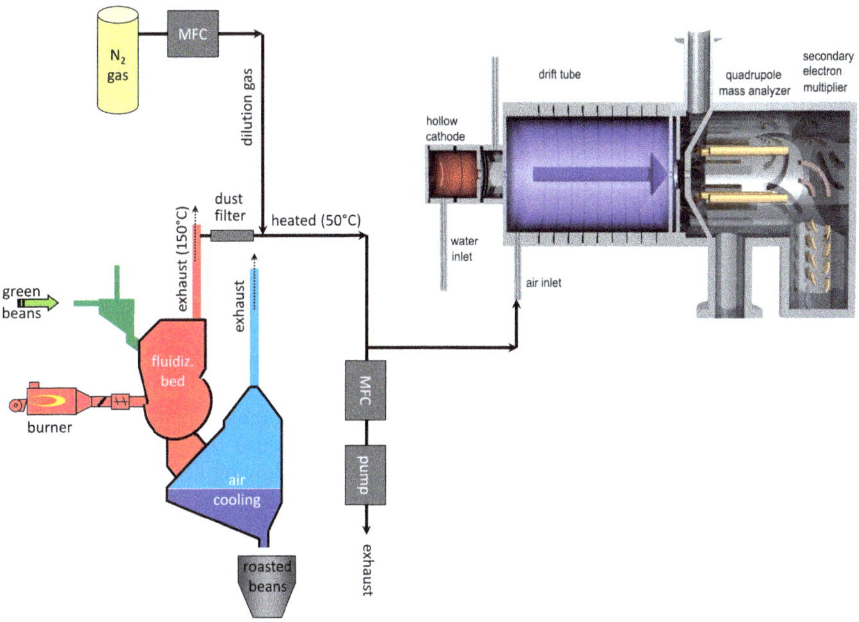

Figure 33.9 Experimental setup for on-line measurements of the roaster off-gas during roasting on a 200 g per batch fluidized bed sample roaster from Neuhaus Neotec. The gas from the roaster off-gas is sampled through a dust filter and diluted with nitrogen gas to reduce the temperature and humidity of the gas. All sampling lines are heated to 50 °C to prevent condensation. While most of the sampled gas is pumped through the Mass Flow Controller (MFC) to the exhaust of the pump, a small fraction is sampled *via* the air inlet into the drift tube of the PTR-MS. In the drift-tube, VOCs are ionized by proton transfer from H_3O^+ and mass analysed by a quadruple mass filter.

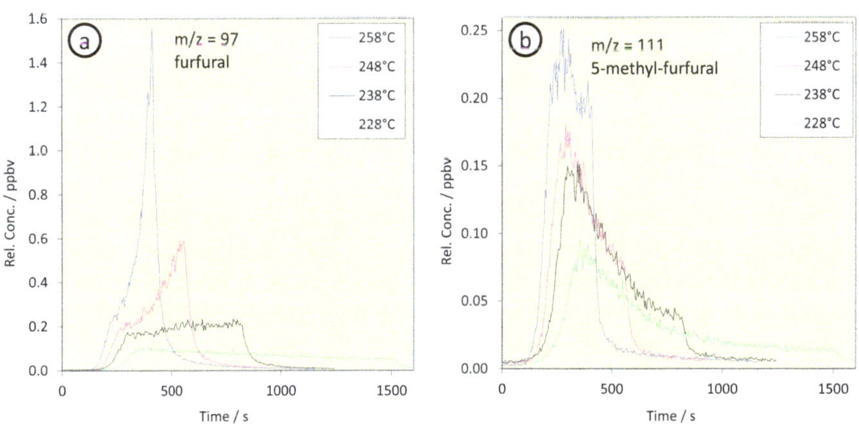

Figure 33.10 The PTR-MS ion traces that are shown were monitored on-line and in real-time in the off-gas of a fluidized bed roaster (200 g batch size). Two compounds are shown (tentative assignment) during four different roasting trials for different roaster gas temperatures; (a) shows the trace for the compound furfural, (b) shows the compound 5-methyl-furfural.

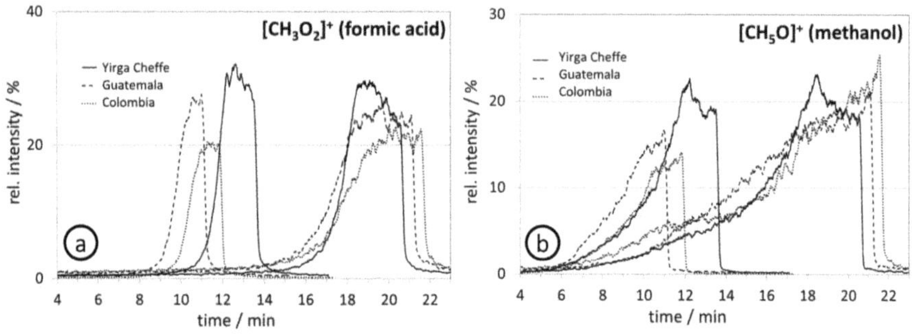

Figure 33.11 Three different arabica coffees were roasted to medium roast degree (103 Pt on the Probat scale). Coffees were roasted either by applying a lower roasting temperature (lower burner intensity), or by applying a higher roasting temperature (medium burner intensity). The roasting process can be visualized by following the formation of the two volatiles, which peak at different time points. On the one hand, there is a difference in roasting time for the three coffees, where coffee from Guatemala is roasted the fastest and followed by the coffee from Colombia and Yirga Cheffe. On the other hand, lower burner capacity prolongs the roasting time for all coffees and the difference in roast time is no longer visible between the three coffees. (a) shows the trace for a compound with mass formula $[CH_3O_2]^+$, which is tentatively assigned to formic acid. (b) on the right shows the compound with mass formula $[CH_5O]^+$, which is tentatively assigned to methanol.

temperature, which corresponds to a concomitant increase in the compound's concentration in the roasted coffee.[138] Clearly, short-time high-temperature (STHT) roasting generates coffee that exhibits significantly higher aroma intensity, in comparison to long-time low-temperature (LTLT) roasting to the same roast degree. The impact of the time–temperature profiles (for identical roast degree) has been confirmed in gas chromatographic[138] and sensory (unpublished) studies. (ii) Modifying the time–temperature roasting profile may alter the dynamics of the compound formation and hence modulate the formation of intermediates. By exploring the formation of a range of phenolic compounds on-line, it was possible to demonstrate the sequential formation of compounds, from precursors *via* various intermediate stages to the final VOC.[123]

The second example examines the roasting of three distinct single-origin *arabica* coffees. Figure 33.11 shows the time–temperature profiles of two exemplary VOCs: formic acid as an example of a volatile organic acid formed during roasting, and methanol as an example of a VOC with alcohol functionality. All coffees were roasted to the same medium roast degree (103 Pt Colorette 3b, Probat), using either a medium or a low burner intensity/temperature. The left frame shows the PTR-ToF-MS ion traces of formic acid. Roasting at medium burner intensity, the first to reach the target roast degree was the Central American coffee from Guatemala, where the peak in formic acid intensity already happens after 10 minutes. The second fastest coffee was the one from South America (Colombia), while the Yirga Cheffe took the longest time with formic acid peaking only after 12 minutes. In a subsequent set of experiments, the same coffees were roasted at low burner intensity. Besides extending the roasting times in general, the first to reach the target roast degree of 103 Pt was now the Yirga Cheffe. These examples demonstrate, for instance, that the aroma development strongly depends on the time–temperature roasting and differs among arabica and robusta coffee.[147] In a recent publication, the implication of such observations on split *versus* mixed roasting was discussed.[138]

33.7.3 Extraction Kinetics of Coffee Aroma Compounds

Using proton-transfer-reaction time-of-flight mass-spectrometry (PTR-ToF-MS), we investigated the extraction dynamic of 95 ion traces in real time (time resolution: 1 second) during espresso coffee preparation.[149] Fifty-two of these ions were tentatively identified. This was achieved by on-line sampling of the VOCs close to the coffee extract flow, at the exit of the extraction hose of the espresso machine (single serve capsules). The results show considerable differences in the extraction kinetics for different compounds, which led to an evolution of the volatile profiles in the extract flow and consequently to an evolution of the aroma profile in the cup.

The time–intensity profiles in Figure 33.12 show different extraction dynamics for the analyzed VOCs (left frame). Four example compounds are shown: methylpropanal, pyridine, methyl furan and guaiacol. The time to reach the maximum intensity ranged from 2 to 20 seconds. Once the maximum had been reached, the intensity fell at different rates, depending on the compound. This decrease in intensity provides information on

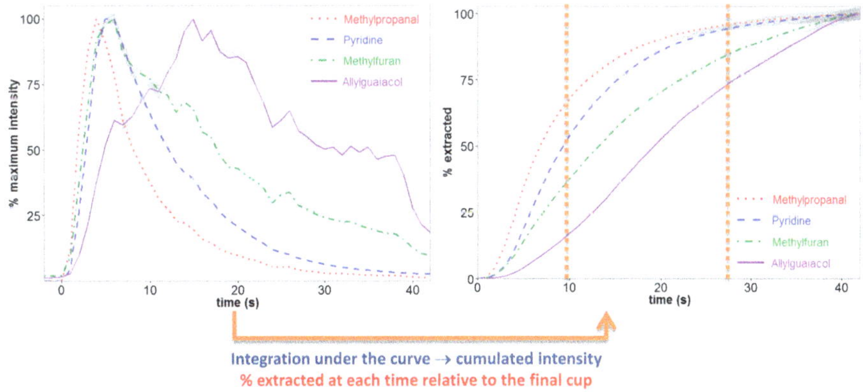

Figure 33.12 Time intensity profiles for one specific capsule (lungo, 42 second extraction time, 110 mL extraction volume) showing differences in extraction kinetics. Left frame: data normalized to the maximum intensity of each of the four VOCs. Right frame: integration of the area under the curve at each time point as a percentage of the total area at the end of the extraction. The shaded ribbons show the 95% confidence interval.

the extraction kinetics of the compounds. A fast decrease implies that the compound is already extracted over a relatively short time period, whereas a slow decrease implies that the compound is extracted over a longer period. The implication of such observations is that the aromatic composition varies as the extraction progresses, with some compounds being fully extracted within the first few seconds of the extraction while other compounds extract much slower and are starting to dominate the extract at prolonged extractions. Using this analytical approach, we are currently exploring the impact of extraction pressure, the temperature and the mineral content of the extraction water.

This technique was also applied to study the time-resolved extraction dynamics with a professional machine (with porta filter) and is currently being applied to a fully automatic machine extraction. Such studies allow both a novel and detailed insight to be obtained into the extraction process and an understanding of how the various instrumental (*e.g.* temperature and mineral composition of the extraction water; pressure of extraction; shape of porta filter, pre-infusion, ...) and coffee (variety, roast degree, grind, ...) parameters and the brew ratio affect the extraction process.

33.7.4 Moving Towards an Individualized Aroma Science – In-mouth Coffee Aroma

By looking at the actual flavour experience in more detail that occurs whilst coffee is being drunk, it becomes clear that it is a highly dynamic experience, quickly changing and evolving in the mouth. In order to develop a better understanding of the aroma perceived by a consumer, it is important to develop techniques that capture the temporal evolution of the aroma during the actual process of consumption.[10,124,150–154]

Coffee aroma evolves in the mouth during drinking and leaves a typical after-odour, or finish, in the mouth for several minutes after swallowing – we have termed this dynamic evolution of the in-mouth aroma "the melody of coffee".[10,21] The nose-space technique allows these dynamic processes to be visualized, and provides a vivid insight into aroma release and its temporal evolution in the mouth.

Figure 33.13 shows the nose-space profiles from four different assessors (four male assessors aged 33–43 and labelled A to D) when drinking the same coffee.[10] The nose-space profile is the time-resolved representation of the intensity of volatile aroma compounds in the air exhaled through the nose. Maxima represent exhalation while minima are periods of inhalation. The time axis (in seconds) runs through all four experiments without interruption and shows the relative time of analysis for the four assessors. Five different mass signals are shown, differentiated by the colour. All experiments were conducted following a strict protocol. The assessor introduced the two ends of the nose-piece into his two nostrils. Initially, and during three complete breathing cycles, the assessor had no coffee in his mouth; the PTR-MS signals during this phase represent VOC that naturally occur in the assessor's

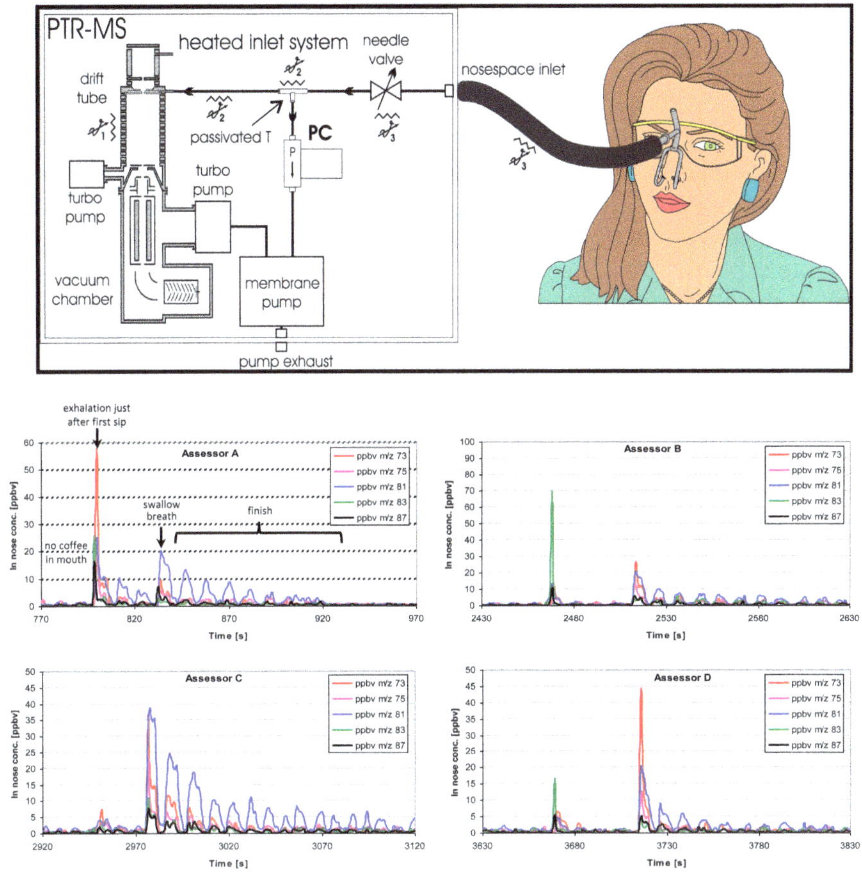

Figure 33.13 Top frame: schematic of the nosepiece (exhaled nose air sampler) for sampling breath-by-breath the air exhaled through the nose during drinking coffee. The nosepiece is connected to a PTR-MS, for on-line monitoring of the concentration of volatile organic compounds in the exhaled breath air. Bottom four frames: nose-space profile of four different assessors (A, B, C, D) while drinking an espresso coffee. Five different PTR-MS ion-mass trace intensity in the exhaled air are shown. All four assessors drank the same coffee following the identical breathing and drinking protocol, while their breath-air exhaled through the nostrils was measured on-line by PTR-MS. Four characteristic phases of the consumption process are outlined. (1) The three first exhalations were done with the nosepiece connected to the two nostrils while the assessors had no coffee in their mouths. Very low signal intensities are observed in the exhaled air. Before the fourth exhalation, the assessors take a sip of coffee. (2) "Exhalation just after first sip" corresponds to the nose-space profiles of the exhalation just after taking coffee into the mouth. During the subsequent two exhalations, the assessors keep the coffee in their mouths. (3) "Swallow-breath" corresponds to the nose-space exhalation following just after swallowing of the coffee. "Finish" corresponds to the sequence of nose-space profiles after swallowing when the assessors no longer have any coffee left in their mouths. Five *m*/*z* ion traces are shown and qualified by their *m*/*z* values. They have tentatively been assigned as: *m*/*z* 73: 2-butanone; *m*/*z* 75: methyl acetate; *m*/*z* 81: pyrazine or/and a fragment of fururyl alcohol; *m*/*z* 83: 2-methylfuran; *m*/*z* 87: 2-,3-methylbutanal (57%); diacetyl (43%).[10] The time resolution used for recording the nose-space spectra was 0.5 s.

breathed air (→background). After the third exhalation, assessors take a sip of approximately 10 mL coffee. The first exhalation just after the first sip showed high intensities for assessors A and B, while assessor C had very low intensities. This indicated that significant inter-individual differences occur in the transfer of VOC from the oral cavity to the nose space (nasopharynx). Possible explanations for such variations are difference (i) in the release of VOC in the mouth due to differences in mouth flora or temperature, (ii) in the opening of the velum (uvula) which connects the mouth cavity to the nasopharynx or (iii) in air flows and breathing patterns. After three breathing cycles with coffee in the mouth, assessors swallow the coffee. The first exhalation just following the swallowing is termed the swallow breath and in general shows relatively high intensities. This is explained by the wetting of the trachea with coffee which releases VOC into the airflow of the first exhalation. For some assessors (*e.g.* C and D), the swallow breath shows the highest intensities in VOC of all exhalations. After the swallow breath, assessors have no coffee in their mouths but still show significant intensities in the exhaled air that slowly decreases in intensity over time. This is often called the after-odour or finish.

Assessor A drank the coffee first, from seconds 770 to 970; assessor B drank the coffee between seconds 2430 to 2630, and so forth. It is obvious that the four nose-space profiles are different. The main differences pertain to: (i) the absolute intensities of the three characteristic moments during the in-mouth consumption experience – the first sip, the swallow-breath and the after-odour or finish – and (ii) the relative intensities of the individual VOCs (the balance or VOC profile) for each of these characteristic moments. For assessor C, the coffee aroma is essentially composed of the swallow-breath aroma and the persisting after-odour. Comparing assessors A and C with B and D, we notice that their first sip and swallow-breath aromas are different in terms of aroma balance.

The nose-space experiment demonstrates that the coffee aroma that reaches the olfactory receptors can vary over a large range between assessors, even if exactly the same coffee is drunk. The fraction and composition of VOCs that are actually released from a substance in the mouth and transported to the olfactory receptors depends not just on the composition of the substance, but is also strongly modulated by anatomic and physiologic characteristics of the person, and may be further modulated by the person's consumption and breathing habits. This is a demonstration of what we may all already know – a given coffee or given food does not taste the same to everybody.

Analyses of mouth- and nose-spaces have been used abundantly in research over the past few decades for medical purposes as well as for sensory analysis. All of these sampling methods have been using the analysis of the respiratory flow to determine the content of volatiles present in the lungs, the mouth or the nasal cavity. With regard to sensory analysis, this means that the compounds of interest that are accumulating in the mouth and nasal

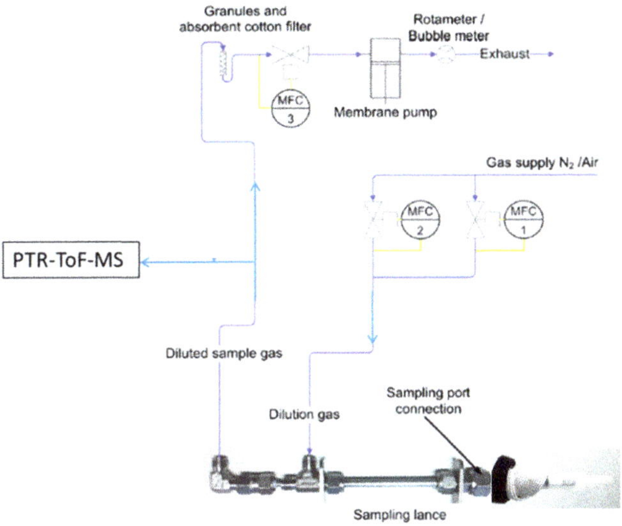

Figure 33.14 Sampling setup with direct mouth-space sampling setup.

cavity become diluted by the respiratory flow, leading to low signal intensities and that fact that only a limited number of flavour active compounds can be monitored. Hence, one aspect of the research in our laboratory is to increase the signal intensity from VOCs present in the mouth cavity.

To this end we have tested, compared and optimized three different sampling methods: (i) nose-space sampling of the exhalation stream (with nose-piece). (ii) Mouth-space sampling of the exhalation stream, using a Buffered End-Tidal breath sampling inlet (BET) by Ionicon. (iii) Direct mouth-space: drawing sample gas directly from the mouth cavity while breathing through the nose – mouthpiece from BET (non-re-breathing mouthpieces) coupled to a sampling and dilution lance[155] – a schematic of the setup is shown in Figure 33.14.

The measurements of seven different coffees have revealed the same order of magnitude in differences between the three sampling methods. Figure 33.15 shows two series of graphs. The upper series depicts the intensities for *m*/*z* 153.055 and *m*/*z* 153.091, tentatively identified as vanillin and ethyl-guaiacol respectively, which represent signals with the lowest intensity that were still distinguishable from the background. The lower series shows the intensities of *m*/*z* 73.065, *m*/*z* 87.080 and *m*/*z* 101.060, tentatively identified as 2-methylpropanal, 2-/3-methylbutanal and 2,3-pentadione, respectively, which represent signals with high intensities.

In summary, the signal area of our new direct mouth-space sampling method was at least a factor of five higher in intensity, compared to the indirect mouth-space sampling and a factor of at least 20 higher compared to nose-space sampling.

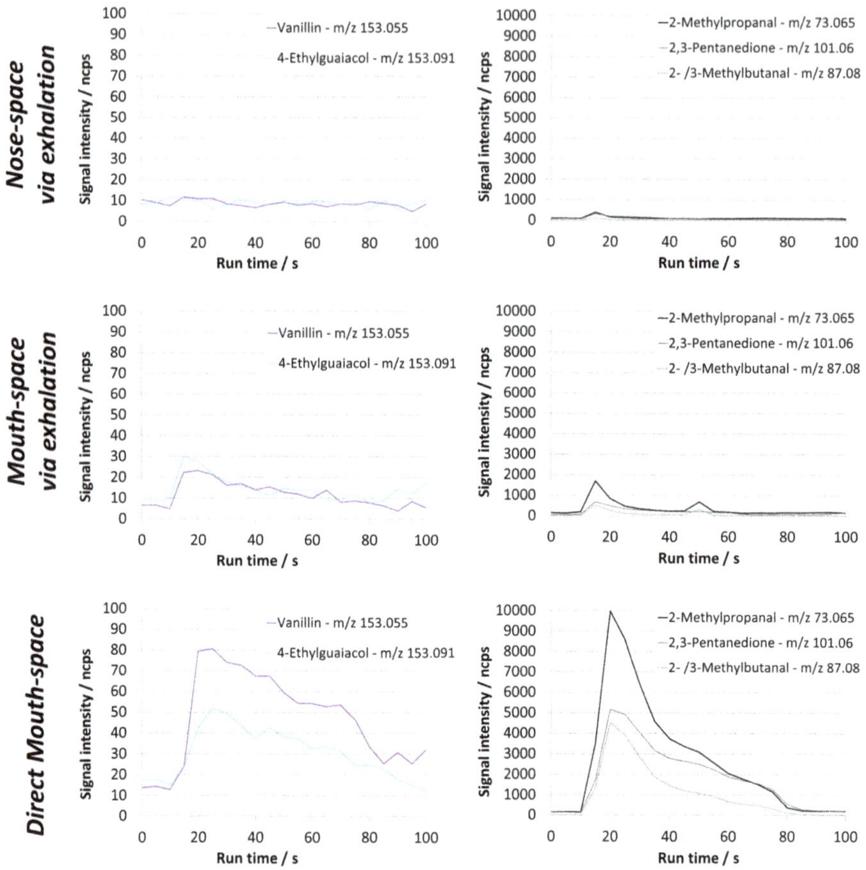

Figure 33.15 Comparison of signal intensities of low intensity compounds (to the left) and high intensity compounds (to the right) for three different sampling methods: upmost panels: nose-space; middle panels: BET mouth-space; lowest panels: direct mouth-space.

The results from the comparison of the three different sampling methods have shown a massive improvement in signal intensity of the newly developed direct mouth-space sampling compared to the other two (conventional) sampling methods. The improvement in signal intensity can be directly linked to the fact that this approach uses a direct sampling of the volatiles present in the mouth-cavity instead of indirectly sampling of these in the exhalation stream. It therefore enables the sensitivity of the measurement to be increased under otherwise equal conditions such as time resolution and a given measurement setup. Thereby, the direct mouth-space method enables a more sensitive dynamic measurement of volatiles during the aftertaste or lingering phase following the ingestion of food or drinks.

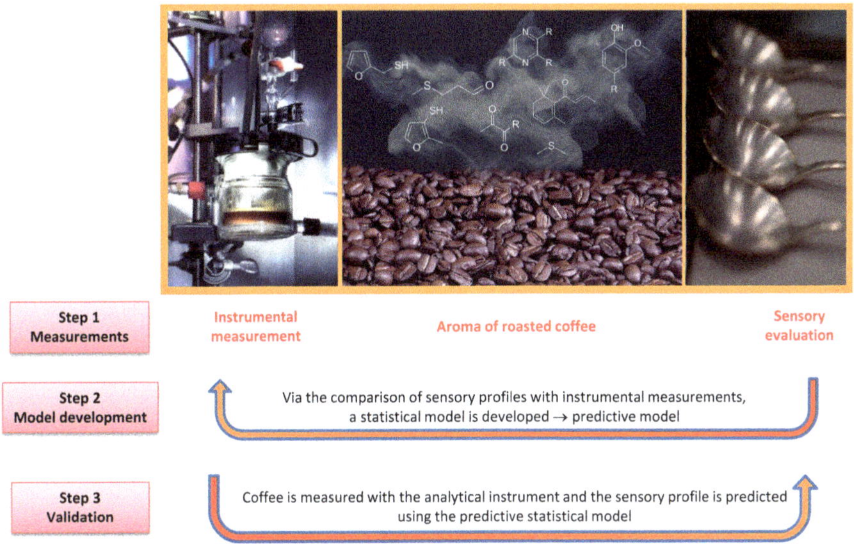

Figure 33.16 Predicting sensory profiles from instrumental measurements: a three-step approach.

33.7.5 Predicting Sensory Profile From Instrumental Measurements

Improved strategies and methods for the correlation of sensory and instrumental analysis are being developed with the ultimate objective of predicting the sensory profile from instrumental measurements. While this represents a truly challenging endeavour, it is often considered the holy grail of flavour science.[132–134]

The strategy presented in ref. 132 can be schematically described as a three-step process, as outlined in Figure 33.16. In the first step, a range of coffees are analysed by instrumental techniques (*e.g.* PTR-MS and/or GC-MS) and are assessed by a trained sensory panel. The second step concerns the development of a mathematical/statistical model that predicts the sensory profiles based on measured instrumental data. Thirdly, in order to validate the predictive model, a series of coffees were measured by instrumental methods, and the sensory profiles predicted based on the model were developed in step 2. Subsequently, the same coffees were profiled by the sensory panel and the profiles compared to the predicted ones. If the match was considered satisfactory, the model was successfully validated and can be applied to predict the sensory profiles of coffee, based on measured instrumental data.

In Figure 33.17, the application of a model was applied to two capsules that was developed specifically for single-serve coffees.[132,133] Predicted sensory profiles were generated using a formerly established predictive model based on PTR-ToF-MS measurements (coloured area) and then superimposed on

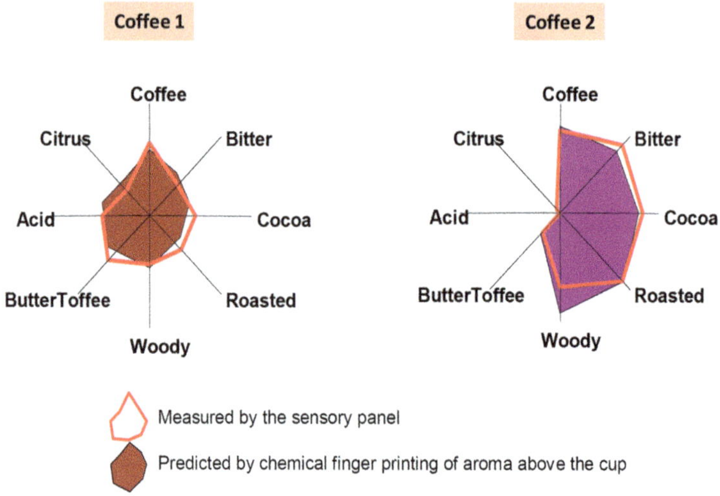

Figure 33.17 Comparison of predicted sensory profiles, based on instrumental measurements (PTR-MS) and sensory profiles for two selected coffees. The coffees analysed here were two single-serve coffee capsules.

the sensory profiles of the same capsules created by a sensory panel (red line). Clearly, a very good match was achieved.

33.8 What Next?

In the past, research in coffee aroma tended to focus on the identification, quantification and qualification of main coffee aroma compounds, and it is believed that essentially all relevant compounds have been identified. Consequently, the focus is shifting towards new fields. We see three major trends (among others) that we believe will dominate research in coffee aroma in the years to come.

Technological and analytical progress in instrumentation, and on-line techniques with high time resolution and very high sensitivity will certainly be one of the most prominent and relevant instrumental developments.

Understanding individual coffee flavour perception and preferences is a second major field of research that will attract significant attention. Novel tools and strategies will be developed to measure the volatile aroma compounds delivered breath-by-breath to the nose at an individual level. Understanding the basis of the differences in aroma delivery during coffee drinking and sensation/perception will contribute to the development of individualized aroma science.

Predicting the sensory profile of coffee from instrumental measurements is possibly the most significant challenge in flavour science and will certainly attract major attention and much effort for many more years to come.

Today, flavour science is moving into a discipline which is truly multidisciplinary and which requires a new breed of scientists.[21] What was once the playground of food and flavour scientists and analytical chemists is today a complex scientific platform where experts from biology, psychophysics, psychology, organic chemistry, analytics, material sciences, physics, mathematics and health meet collaboratively with food and flavour scientists.

Acknowledgements

This work would not have been possible without the help and support of many colleagues and friends with whom we have been able to share our fascination for our work on the science of coffee aroma. They are: Werner Lindinger, Alfons Jordan, Martin Graus, Imre Blank, Philippe Pollien, Santo Ali, Christian Lindinger, Ralf Zimmermann, Ralph Dorfner, Alexia N. Glöss, Barbara Schönbächler, Flurin Wieland and many more.

References

1. O. Monteiro da Silva and C. A. Moreira Leite, *Income Content of the World Coffee Exports*, AD-minister, 2013, pp. 69–82.
2. B. Lewin, D. Giovannucci and P. Varangis, *Coffee Markets: New Paradigms in Global Supply and Coffee Markets Supply and Demand*, 2004.
3. S. Ponte, The `Latte Revolution'? Regulation, Markets and Consumption in the Global Coffee Chain, *World Development*, 2002, **30**, 1099–1122.
4. E. Cantergiani, H. Brevard, Y. Krebs, A. Feria-Morales, R. Amado and C. Yeretzian, Characterisation of the aroma of green Mexican coffee and identification of mouldy/earthy defect, *Eur. Food Res. Technol.*, 2001, **212**, 648–657.
5. B. A. Goodman, *Coffee Consumption and Health*, Nova Science Publishers, Inc., Hauppauge, New York, 2012.
6. S. Firestein, A code in the nose, *Sci. STKE*, 2004, **2004**, pe15.
7. S. Firestein, How the olfactory system makes sense of scents, *Nature*, 2001, **413**, 211–218.
8. R. C. Araneda, A. D. Kini and S. Firestein, The molecular receptive range of an odorant receptor, *Nat. Neurosci.*, 2000, **3**, 1248–1255.
9. J. B. German, C. Yeretzian and V. B. Tolstoguzov, Olfaction, where Nutrition, Memory and Immunity Intersect, in *Flavours: Chemistry, Technology and Resources*, ed. R. G. Berger, Springer Verlag, Berlin, Germany, 2007, pp. 25–41.
10. C. Yeretzian, P. Pollien, C. Lindinger and S. Ali, Individualization of Flavor Preferences: Toward a Consumer-centric and Individualized Aroma Science, *Compr. Rev. Food Sci. Food Saf.*, 2004, **3**, 152–159.
11. J. Delwiche, The impact of perceptual interactions on perceived flavor, *Food Quality and Preference*, 2004, **15**, 137–146.

12. A. Buettner, Influence of human salivary enzymes on odorant concentration changes occurring in vivo. 1. Esters and thiols, *J. Agric. Food Chem.*, 2002, **50**, 3283–3289.
13. A. Buettner and J. Beauchamp, Chemical input - Sensory output: Diverse modes of physiology-flavour interaction, *Food Quality and Preference*, 2010, **21**, 915–924.
14. A. Buettner, A. Beer, C. Hannig and M. Settles, Observation of the swallowing process by application of videofluoroscopy and real-time magnetic resonance imaging-consequences for retronasal aroma stimulation, *Chem. Senses*, 2001, **26**, 1211–1219.
15. A. Buettner, A. Beer, C. Hannig, M. Settles and P. Schieberle, Physiological and analytical studies on flavor perception dynamics as induced by the eating and swallowing process, *Food Quality and Preference*, 2002, **13**, 497–504.
16. A. Buettner, S. Otto, A. Beer, M. Mestres, P. Schieberle and T. Hummel, Dynamics of retronasal aroma perception during consumption: Cross-linking on-line breath analysis with medico-analytical tools to elucidate a complex process, *Food Chem.*, 2008, **108**, 1234–1246.
17. A. Buettner and P. Schieberle, in *Changes in the Concentrations of Key Fruit Odorants Induced by Mastication*, ed. D. D. Roberts and A. J. Taylor, ACS, Washington, DC, 2000, pp. 87–98.
18. A. Buettner and P. Schieberle, Exhaled Odorant Measurement (EXOM) – A New Approach to Quantify the Degree of In-Mouth Release of Food Aroma Compounds, *Lebensm.-Wiss. Technol.*, 2000, **33**, 553–559.
19. A. Buettner and P. Schieberle, Influence of mastication on the concentrations of aroma volatiles – some aspects of flavour release and flavour perception, *Food Chem.*, 2000, **71**, 347–354.
20. M. Mestres, N. Moran, A. Jordan and A. Buettner, Aroma Release and Retronasal Perception during and after Consumption of Flavored Whey Protein Gels with Different Textures. 1. in Vivo Release Analysis, *J. Agric. Food Chem.*, 2004, **53**, 403–409.
21. I. Blank, M. Wust and C. Yeretzian, Expression of multidisciplinary flavor science: research highlights from the 12th Weurman Symposium, *J. Agric. Food Chem.*, 2009, **57**, 9857–9859.
22. I. Gierczynski, H. Laboure and E. Guichard, In Vivo Aroma Release of Milk Gels of Different Hardnesses: Inter-individual Differences and Their Consequences on Aroma Perception, *J. Agric. Food Chem.*, 2008, **56**, 1697–1703.
23. P. Mombaerts, F. Wang, C. Dulac, S. K. Chao, A. Nemes, M. Mendelsohn, J. Edmondson and R. Axel, Visualizing an olfactory sensory map, *Cell*, 1996, **87**, 675–686.
24. P. Mombaerts, Molecular biology of odorant receptors in vertebrates, *Annu. Rev. Neurosci.*, 1999, **22**, 487–509.
25. R. J. Stevenson, J. Prescott and R. A. Boakes, Confusing tastes and smells: How odours can influence the perception of sweet and sour tastes, *Chem. Senses*, 1999, **24**, 627–635.

26. J. Prescott, V. Johnstone and J. Francis, Odour-taste interactions: effects of attentional strategies during exposure, *Chem. Senses*, 2004, **29**, 331–340.
27. J. S. Ribeiro, R. F. Teófilo, F. Augusto and M. M. C. Ferreira, Simultaneous optimization of the microextraction of coffee volatiles using response surface methodology and principal component analysis, *Chemom. Intell. Lab. Syst.*, 2010, **102**, 45–52.
28. M. Damm, J. Vent, M. Schmidt, P. Theissen, H. E. Eckel, J. Lotsch and T. Hummel, Intranasal volume and olfactory function, *Chem. Senses*, 2002, **27**, 831–839.
29. K. Zhao, P. W. Scherer, S. A. Hajiloo and P. Dalton, Effect of Anatomy on Human Nasal Air Flow and Odorant Transport Patterns: Implications for Olfaction, *Chem. Senses*, 2003, **29**, 365–379.
30. J. T. Hornung, D. J. Smith, D. B. Kurtz, T. White and D. A. Leopold, Effect of nasal dilaors on nasal structures, sniffing strategies, and olfactory ability, *Rhinology*, 2001, **39**, 84–87.
31. D. A. Leopold, The Relationship between Nasal Anatomy and Human Olfaction, *Laryngoscope*, 1988, **98**, 1232–1238.
32. L. M. Bartoshuk, V. B. Duffy and I. J. Miller, PTC/PROP tasting: anatomy, pschophysics, and sex effects, *Physiol. Behav.*, 1992, **56**, 1165–1171.
33. L. H. Snyder, Inherited Taste Deficiency, *Science*, 1931, **74**, 151–152.
34. U. K. Kim, E. Jorgenson, H. Coon, M. Leppert, N. Risch and D. Drayna, Positional cloning of the human quantitative trait locus underlying taste sensitivity to phenylthiocarbamide, *Science*, 2003, **299**, 1221–1225.
35. M. M. Kamphuis and M. S. Westerterp-Plantenga, PROP sensitivity affects macronutrient selection, *Physiol. Behav.*, 2003, **79**, 167–172.
36. J. N. Lundström, T. Hummel and M. J. Olsson, Individual Differences in Sensitivity to the Odor of 4,16-Androstadien-3-one, *Chem. Senses*, 2003, **28**, 643–650.
37. K. K. Yee and C. J. Wysocki, Odorant exposure increases olfactory sensitivity: olfactory epithelium is implicated, *Physiol. Behav.*, 2001, **72**, 705–711.
38. S. Assouline, M. I. Shevell, R. J. Zatorre, M. Jones-Gotman, M. D. Schloss and K. Oudjhane, Children Who Can't Smell the Coffee: Isolated Congenital Anosmia, *J. Child Neurol.*, 1998, **13**, 168–172.
39. N. D. Abolmaali, V. Hietschold, T. J. Vogl, K.-B. Hüttenbrink and T. Hummel, MR Evaluation in Patients with Isolated Anosmia Since Birth or Early Childhood, *Am. J. Neuroradiol.*, 2002, **23**, 157–164.
40. W. Grosch, Specificity of the human nose in perceiving food odorants, in *Weurman Flavour Research Symposium, Freising*, ed. P. Schieberle and K.-H. Engel, Deutsche Forschungsanstalt für Lebensmittelchemie, Garching, Freising, 2000, pp. 213–219.
41. C. Reverdy, P. Schlich, E. P. Köster, E. Ginon and C. Lange, Effect of sensory education on food preferences in children, *Food Quality and Preference*, 2010, **21**, 794–804.

42. F. Grabenhorst, E. T. Rolls and A. Bilderbeck, How Cognition Modulates Affective Responses to Taste and Flavor: Top-down Influences on the Orbitofrontal and Pregenual Cingulate Cortices, *Cereb. Cortex*, 2008, **18**, 1549–1559.
43. E. D. Capaldi, Conditioned food preferences, in *Why We Eat what We Eat: The Psychology of Eating*, ed. E. D. Capaldi, American Psychological Association, Washington, DC, 1996.
44. P. Rozin, *Towards a Psychology of Food Choice*, Institut Danone, Bruxelles, 1998.
45. F. Lucas and A. Sclafani, Flavor preferences conditioned by high-fat versus high-carbohydrate diets vary as a function of session length, *Physiol. Behav.*, 1999, **66**, 389–395.
46. C. Perez, F. Lucas and A. Sclafani, Increased flavor acceptance and preference conditioned by the postingestive actions of glucose, *Physiol. Behav.*, 1998, **64**, 483–492.
47. J. A. Mennella and G. K. Beauchamp, Flavor experiences during formula feeding are related to preferences during childhood, *Early Hum. Dev.*, 2002, **68**, 71–82.
48. A. M. D. Hirsch, S. G. Harris, J. Fawcett, and A. Hirsch, *What Flavor Is Your Personality? Discover Who You Are by Looking at what You Eat*, Sourcebooks Trade Naperville, Illinois, 2001.
49. J. Mojet and E. P. Koster, Texture and flavour memory in foods: an incidental learning experiment, *Appetite*, 2002, **38**, 110–117.
50. C. Sulmont, S. Issanchou and E. P. Koster, Selection of odorants for memory tests on the basis of familiarity, perceived complexity, pleasantness, similarity and identification, *Chem. Senses*, 2002, **27**, 307–317.
51. S. Chu and J. Downes, Proust nose best: odors are better cues of autobiographical memory, *Memory and Cognition*, 2002, **30**, 511–518.
52. J. Willander and M. Larsson, Smell your way back to childhood: autobiographical odor memory, *Bull. Psychon. Soc.*, 2006, **13**, 240–244.
53. C. Yeretzian, A. Jordan, H. Brevard, and W. Lindinger, in *Time-Resolved Headspace Analysis by Proton-Transfer-Reaction Mass-Spectrometry Flavour Release, Washington, DC, 2000*, ed. D. D. Roberts and A. J. Taylor, ACS, Washington, DC, 2000, pp. 58–72.
54. C. Yeretzian, A. Jordan, H. Brevard, and W. Lindinger, in *On-line Monitoring of Coffee Roasting by Proton-Transfer-Reaction Mass-Spectrometry, Flavour Release, Washington, DC, 2000*, ed. D. D. Roberts and A. J. Taylor, ACS, Washington, DC, 2000, pp. 112–123.
55. C. Yeretzian, H. Brevard, A. Jordan, A. Hansel, and W. Lindinger, On-line analysis of coffee roasting by proton-transfer-reaction/mass-spectrometry (PTR/MS), ed. N. M. ACS, in *218th ACS National Meeting,* New Orleans, 1999.
56. F. Biasioli, C. Yeretzian, F. Gasperi and T. D. Märk, PTR-MS monitoring of VOCs and BVOCs in food science and technology, *Trends Anal. Chem.*, 2011, **30**, 968–977.

57. C. R. Eloy Dias and C. Yeretzian, Investigating Coffee Samples by Raman Spectroscopy for Quality Control - Preliminary Study, *Int. J. Exp. Spectroscopic Tech.*, 2016, **1**, 1–5.
58. W. Grosch, Warum Riecht Kaffee so Gut?, *Chemie in unserer Zeit*, 1996, **30**, 126–133.
59. T. H. Parliment and H. D. Stahl, What Makes That Coffee Smell So Good, *CHEMTECH*, 1995, **25**, 38–47.
60. A. T. Toci and A. Farah, Volatile fingerprint of Brazilian defective coffee seeds: corroboration of potential marker compounds and identification of new low quality indicators, *Food Chem.*, 2014, **153**, 298–314.
61. A. T. Toci and A. Farah, Volatile compounds as potential defective coffee beans' markers, *Food Chem.*, 2008, **108**, 1133–1141.
62. S. Smrke, I. Kroslakova, A. N. Gloess and C. Yeretzian, Differentiation of degrees of ripeness of Catuai and Tipica green coffee by chromatographical and statistical techniques, *Food Chem.*, 2015, **174**, 637–642.
63. B. Bertrand, R. Boulanger, S. Dussert, F. Ribeyre, L. Berthiot, F. Descroix and T. Joet, Climatic factors directly impact the volatile organic compound fingerprint in green Arabica coffee bean as well as coffee beverage quality, *Food Chem.*, 2012, **135**, 2575–2583.
64. O. Bernheimer, Zur Kenntniss der Röstproducte des Caffees, *Monatshefte für Chemie*, 1880, **1**, 456–467.
65. T. Reichstein and H. Staudinger, Method for isolating the aromatic principle contained in roasted coffee, *Br. Pat.*, 246454, 1926.
66. H. Staudinger and T. Reichstein, Producing artificial coffee oil, *Br. Pat.* 260960, 1928.
67. L. M. Nijssen, C. A. Visscher, H. Maarse, L. C. Willemsense, and M. H. Boelens, *Volatile Compounds in Food. TNO Nutrition and Food Research Institute*, Zeist, The Netherlands, 1996, vol. 7th.
68. W. B. Sunarharum, D. J. Williams and H. E. Smyth, Complexity of coffee flavor: A compositional and sensory perspective, *Food Res. Int.*, 2014, **62**, 315–325.
69. I. Blank and W. Grosch, Evaluation of potent odorants in dill seed & dill herb (*Anethum graveolens* L.) by aroma extract dilution analysis, *J. Food Sci.*, 1991, **56**, 63–67.
70. I. Blank, A. Sen and W. Grosch, Potent odorants of the roasted powder and brew of arabica coffee, *Z. Lebensm.-Unters. Forsch.*, 1992, **195**, 239–245.
71. C. Cerny and W. Grosch, Quantification of character-impact odour compounds of roasted beef, *Zeitschrift fuer Lebensmittel-Untersuchung und-Forschung*, 1993, **196**, 417–422.
72. P. Semmelroch and W. Grosch, Analysis of Roasted Coffee Powders and Brews by Gas Chromatography-Olfactometry of Headspace Samples, *Lebensm.-Wiss. Technol.*, 1995, **28**, 310–313.

73. M. Czerny, R. Wagner and W. Grosch, Detection of odor-active ethenylalkylpyrazines in roasted coffee, *J. Agric. Food Chem.*, 1996, **44**, 3268–3272.
74. P. Semmelroch and W. Grosch, Studies on Character Impact Odorants of Coffee Brews, *J. Agric. Food Chem.*, 1996, **44**, 537–543.
75. M. Czerny, F. Mayer and W. Grosch, Sensory study on the character impact odorants of roasted arabica coffee, *J. Agric. Food Chem.*, 1999, **47**, 695–699.
76. M. Czerny and W. Grosch, Potent odorants of raw Arabica coffee. Their changes during roasting, *J. Agric. Food Chem.*, 2000, **48**, 868–872.
77. F. Mayer, M. Czerny and W. Grosch, Sensory study of the character impact aroma compounds of a coffee beverage, *Eur. Food Res. Technol.*, 2000, **211**, 272–276.
78. W. Grosch., Chemistry III: Volatile compounds, in *Coffee: Recent Developments*, ed. R. J. Clarke and O. G. Vitzthum, Blackwell Science, London, 2001, pp 68–89.
79. H. D. Belitz, W. Grosch, and P. Schieberle, *Food Chemistry*, Springer, Berlin Heidelberg, 4th edn, 2009.
80. M. Czerny, M. Christlbauer, M. Christlbauer, A. Fischer, M. Granvogl, M. Hammer, C. Hartl, N. Hernandez and P. Schieberle, Re-investigation on odour thresholds of key food aroma compounds and development of an aroma language based on odour qualities of defined aqueous odorant solutions, *Eur. Food Res. Technol.*, 2008, **228**, 265–273.
81. M. Akiyama, K. Murakami, N. Ohtani, K. Iwatsuki, K. Sotoyama, A. Wada, K. Tokuno, H. Iwabuchi and K. Tanaka, Analysis of volatile compounds released during the grinding of roasted coffee beans using solid-phase microextraction, *J. Agric. Food Chem.*, 2003, **51**, 1961–1969.
82. M. Akiyama, K. Murakami, M. Ikeda, K. Iwatsuki, A. Wada, K. Tokuno, M. Onishi and H. Iwabuchi, Analysis of the Headspace Volatiles of Freshly Brewed Arabica Coffee Using Solid-Phase Microextraction, *J. Food Sci.*, 2007, **72**, C388–C396.
83. M. Akiyama, K. Murakami, Y. Hirano, M. Ikeda, K. Iwatsuki, A. Wada, K. Tokuno, M. Onishi and H. Iwabuchi, Characterization of Headspace Aroma Compounds of Freshly Brewed Arabica Coffees and Studies on a Characteristic Aroma Compound of Ethiopian Coffee, *J. Food Sci.*, 2008, **73**, C335–C346.
84. T. Michishita, M. Akiyama, Y. Hirano, M. Ikeda, Y. Sagara and T. Araki, Gas Chromatography/Olfactometry and Electronic Nose Analyses of Retronasal Aroma of Espresso and Correlation with Sensory Evaluation by an Artificial Neural Network, *J. Food Sci.*, 2010, **75**, S477–S489.
85. W. Holscher and H. Steinhart, Investigation of roasted coffee freshness with an improved headspace technique, *Zeitschrift fuer Lebensmittel-Untersuchung und-Forschung*, 1992, **195**, 33–38.

86. J. S. Ribeiro, F. Augusto, T. J. Salva and M. M. Ferreira, Prediction models for Arabica coffee beverage quality based on aroma analyses and chemometrics, *Talanta*, 2012, **101**, 253–260.
87. K. Kumazawa and H. Masuda, Identification of Odor-Active 3-Mercapto-3-methylbutyl Acetate in Volatile Fraction of Roasted Coffee Brew Isolated by Steam Distillation under Reduced Pressure, *J. Agric. Food Chem.*, 2003, **51**, 3079–3082.
88. R. J. Clarke and O. G. Vitzthum, *Coffee: Recent Developments*, Blackwell Sciences Ltd., 2001.
89. M. A. Gianturco, A. S. Giammarino and P. Friedel, The volatiles constituents of coffee – III The structure of two heterocyclic compounds and the synthesis of tetrahydrofuranones, *Tetrahedron*, 1964, **20**, 1763–1772.
90. R. Silwar, H. Kamperschroer and R. Tressl, Gaschromatographic-mass spectrometric investigation of coffee aroma – quantitative determination of steam-volatile aroma constituents, *Chem., Mikrobiol., Technol. Lebensm.*, 1987, **10**, 176–187.
91. G. A. Burdock., *Fenaroli's Handbook of Flavor Ingredients*, CRC Press/Taylor & Francis, Boca Raton, 6th edn, 2010.
92. P. Salo, Determining the Odor Thresholds for Some Compounds in Alcoholic Beverages, *J. Food Sci.*, 1970, **35**, 95–99.
93. J. Barrett, S. A. Halsey and T. L. Peppard, Flavour Stability Of Ales Brewed Using Hydrolysed Maize Syrup—Sensory And Analytical Aspects, *J I Brewing*, 1983, **89**, 356–360.
94. M. Larsen and L. Poll, Odour thresholds of some important aroma compounds in strawberries, *Zeitschrift fuer Lebensmittel-Untersuchung und-Forschung*, 1992, **195**, 120–123.
95. R. G. Buttery and L. C. Ling, 2-ethyl-3,5-dimethylpyrazine and 2-ethyl-3,6-dimethylpyrazine: Odor thresholds in water solution, *Food Science and Technology (LEB)*, 1997, **30**, 109–110.
96. I. Blank, A. Sen and W. Grosch, in *Aroma impact compounds of arabica and robusta coffees. Qualitative and quantitative investigations,* ASIC-14eme Colloque Scientifique International sur le Café, Paris, 1992, ASIC, Paris, 1991, pp. 117–129.
97. A. Scn, G. Laskawy, P. Schieberle and W. Grosch, Quantitative Determination of á-Damascenone in Foods Using a Stable Isotope Dilution Assay, *J. Agric. Food Chem.*, 1991, **39**, 757–759.
98. W. Grosch, Determination of Potent Odourants in Food by Aroma Extract Dilution Analysis (AEDA) and Calculation of Odour Activity Values (OAVs), *Flavour Fragrance J.*, 1994, **9**, 147–158.
99. P. Semmelroch, G. Laskawy, I. Blank and W. Grosch, Determination of potent odourants in roasted coffee by stable isotope dilution assays, *Flavour Fragrance J.*, 1995, **10**, 1–7.
100. F. Mayer and W. Grosch, Aroma simulation on the basis of the odourant composition of roasted coffee headspace, *Flavour Fragrance J.*, 2001, **16**, 180–190.

101. F. Mayer, M. Czerny and W. Grosch, Influence of provenance and roast degree on the composition of potent odorants in Arabica coffees, *Eur. Food Res. Technol.*, 1999, **209**, 242–250.
102. D. S. Mottram, B. L. Wedzicha and A. T. Dodson, Acrylamide is formed in the Maillard reaction, *Nature*, 2002, **419**, 448–449.
103. G. H. Fuller, R. Steltenkamp and G. A. Tisserand, The Gas Chromatograph with Human Sensor: Perfumer Model, *Ann. N. Y. Acad. Sci.*, 1964, **116**, 711–724.
104. F. Ullrich and W. Grosch, Identification of the Most Intense Volatile Flavor Compounds Formed during Autoxidation of Linoleic-Acid, *Z. Lebensm. Unters. Forsch.*, 1987, **184**, 277–282.
105. T. E. Acree in, *Gas Chromatography-Olfactometry*, ed. C. T. Ho and C. H. Manley, Markel Dekker, Inc., Institute of Food Technology, New York, 1993, pp. 77–94.
106. T. E. Acree and J. Barnard, Gas Chromatography-Olfactometry and CharmAnalysisTM, in *Trends in Flavour Research*, Elsevier Science B.V., 1994, pp. 211–220.
107. D. D. Roberts and T. E. Acree, Simulation of Retronasal Aroma Using a Modified Headspace Tewchnique: Investigating the Effects of Saliva, Temperature, Shearing, and Oil on Flavor Release, *J. Agric. Food Chem.*, 1995, **43**, 2179–2186.
108. T. E. Acree, GC/olfactometry, *Anal. Chem.*, 1997, **69**, 170A–175A.
109. K. D. Deibler, T. E. Acree, and E. H. Lavin, in *Aroma analysis of coffee brew by gas chromatography-olfactometry*, ed. E. T. Contis, 1998, pp. 69–78.
110. K. D. Deibler, T. E. Acree and E. H. Lavin, Solid phase microextraction application in gas chromatography/olfactometry dilution analysis, *J. Agric. Food Chem.*, 1999, **47**, 1616–1618.
111. Y. W. G. Feng and T. E. Acree, Gas Chromatography Olfactometry in Aroma Analysis, *Foods Food Ingredients J. Jpn.*, 1999, **179**, 57–66.
112. K. D. Deibler, E. H. Lavin, R. S. Linforth, A. J. Taylor and T. E. Acree, Verification of a mouth simulator by in vivo measurements, *J. Agric. Food Chem.*, 2001, **49**, 1388–1393.
113. T. E. Acree, J. Barnard and D. G. Cunningham, A Procedure for the Sensory Analysis of Gas Chromatography Effluents, *Food Chem.*, 1984, **14**, 273–286.
114. E. Cantergiani, H. Brevard, R. Amado, Y. Krebs, A. Feria-Morales, and C. Yeretzian, Characterisation of mouldy/earthy defect in green Mexican coffee, in *ASIC-18eme Colloque Scientifique International sur le Café, Paris, 2000*, ASIC, Paris, 2000, pp. 43–49.
115. P. Pollien, A. Ott, F. Montigon, M. Baumgartner, R. MunozBox and A. Chaintreau, Hyphenated headspace gas chromatography sniffing technique: Screening of impact odorants and quantitative aromagram comparisons, *J. Agric. Food Chem.*, 1997, **45**, 2630–2637.
116. B. d'Acampora Zellner, P. Dugo, G. Dugo and L. Mondello, Gas chromatography–olfactometry in food flavour analysis, *J. Chromatogr. A*, 2008, **1186**, 123–143.

117. R. Zimmermann, H. J. Heger, C. Yeretzian, H. Nagel and U. Boesl, Application of Laser Ionization Mass Spectrometry for On-line Monitoring of Volatiles in the Headspace of Food Products: Roasting and Brewing of Coffee, *Rapid Commun. Mass Spectrom.*, 1996, **10**, 1975–1979.
118. R. Dorfner, R. Zimmermann, A. Kettrup, C. Yeretzian, A. Jordan and W. Lindinger, Vergleich zweier massenspektrometrischer Verfahren zur Direktanalyse in der Lebensmittelchemie, *Lebensmittelchemie*, 1999, **53**, 32–34.
119. R. Dorfner, R. Zimmermann, C. Yeretzian, and A. Kettrup, in *On-Line Analysis of Food Processing Gases by Resonance Laser Mass Spectrometry (REMPI-TOFMS): Coffee Roasting and Related Applications, ASIC-18eme Colloque Scientifique International sur le Café, Paris, 2000*, ASIC, Paris, 2000, pp. 136–142.
120. R. Dorfner, T. Ferge, T. Uchimura, C. Yeretzian, R. Zimmermann, and A. Kettrup, in *Laser/Chemical Ionisation – Mass Spectrometry as an On-Line Analysis Technique for Monitoring the Coffee Roasting Process, ASIC-19eme Colloque Scientifique International sur le Café, Paris, 2002*, ASIC, Paris, 2002.
121. R. Dorfner, T. Ferge, A. Kettrup, R. Zimmermann and C. Yeretzian, Real-time monitoring of 4-vinylguaiacol, guaiacol, and phenol during coffee roasting by resonant laser ionization time-of-flight mass spectrometry, *J. Agric. Food Chem.*, 2003, **51**, 5768–5773.
122. R. Dorfner, T. Ferge, C. Yeretzian, A. Kettrup and R. Zimmermann, Laser mass spectrometry as on-line sensor for industrial process analysis: Process control of coffee roasting, *Anal. Chem.*, 2004, **76**, 1386–1402.
123. R. Hertz-Schönemann, R. Dorfner, C. Yeretzian, T. Streibel and R. Zimmermann, On-line process monitoring of coffee roasting by resonant laser ionisation time-of-flight mass spectrometry: bridging the gap from industrial batch roasting to flavour formation inside an individual coffee bean, *J. Mass Spectrom.*, 2013, **48**, 1253–1265.
124. S. Ali, P. Pollien, C. Lindinger and C. Yeretzian, *In vivo* analysis of aroma release while eating food: a novel set-up for monitoring on-line nosespace air, *1st International Conference on Proton Transfer Reaction Mass Spectrometry and its Applications*, ed. A. Hansel and T. Märk, Innsbruck, 2003, pp. 161–164.
125. C. Yeretzian, A. Jordan, R. Badoud and W. Lindinger, From the green bean to the cup of coffee: investigating coffee roasting by on-line monitoring of volatiles, *Eur. Food Res. Technol.*, 2002, **214**, 92–104.
126. T. Karl, C. Yeretzian, A. Jordan and W. Lindinger, Dynamic measurements of partition coefficients using proton-transfer-reaction mass spectrometry (PTR–MS), *Int. J. Mass Spectrom.*, 2003, **223–224**, 383–395.
127. P. Pollien, A. Jordan, W. Lindinger and C. Yeretzian, Liquid-air partitioning of volatile compounds in coffee: dynamic measurements using proton-transfer-reaction mass spectrometry, *Int. J. Mass Spectrom.*, 2003, **228**, 69–80.

128. P. Pollien, A. Jordan, W. Lindinger and C. Yeretzian, Liquid–air partitioning of volatile compounds in coffee: dynamic measurements using proton-transfer-reaction mass spectrometry, *Int. J. Mass Spectrom.*, 2003, **228**, 69–80.
129. P. Pollien, C. Lindinger, C. Yeretzian and I. Blank, Proton transfer reaction mass spectrometry is a suitable tool for monitoring on-line the formation of acrylamide in food and maillard systems, *Anal. Chem.*, 2003, **75**, 5488–5494.
130. C. Yeretzian, A. Jordan and W. Lindinger, Analysing the headspace of coffee by proton-transfer-reaction mass-spectrometry, *Int. J. Mass Spectrom.*, 2003, **223–224**, 115–139.
131. C. Lindinger, P. Pollien, S. Ali, C. Yeretzian, I. Blank and T. Märk, Unambiguous identification of volatile organic compounds by proton-transfer reaction mass spectrometry coupled with GC/MS, *Anal. Chem.*, 2005, **77**, 4117–4124.
132. C. Lindinger, D. Labbe, P. Pollien, A. Rytz, M. A. Juillerat, C. Yeretzian and I. Blank, When machine tastes coffee: instrumental approach to predict the sensory profile of espresso coffee, *Anal. Chem.*, 2008, **80**, 1574–1581.
133. C. Lindinger, C. Yeretzian and I. Blank, When Machine Tastes Coffee: Successful Prediction of Coffee Sensory Profiles by Instrumental Methods Based on On-line PTR-MS, *Chimia*, 2009, **63**, 292.
134. C. Lindinger, R. C. H. de Vos, C. Lambot, P. Pollien, A. Rytz, E. Voirol-Baliguet, R. Fumeaux, F. Robert, C. Yeretzian, and I. Blank, Coffee Chemometrics as a New Concept: Untargeted Metabolic Profiling of Coffee, in *Expression of Multidisciplinary Flavour Science – 12th Weurman Symposium*, ed. I. Blank, M. Wust, and C. Yeretzian, Zurich University of Applied Science, Wintherthur, 2010, pp. 581–584.
135. F. Biasioli, C. Yeretzian, T. D. Märk, J. Dewulf and H. Van Langenhove, Direct-injection mass spectrometry adds the time dimension to (B)VOC analysis, *TrAC, Trends Anal. Chem.*, 2011, **30**, 1003–1017.
136. F. Wieland, A. N. Gloess, M. Keller, A. Wetzel, S. Schenker and C. Yeretzian, Online monitoring of coffee roasting by proton transfer reaction time-of-flight mass spectrometry (PTR-ToF-MS): towards a real-time process control for a consistent roast profile, *Anal. Bioanal. Chem.*, 2012, **402**, 2531–2543.
137. F. Wieland, A. N. Gloess, M. Keller, A. Wetzel, S. Schenker and C. Yeretzian, On-line process control of the roast degree of coffee, *Chimia (Aarau)*, 2012, **66**, 443.
138. A. N. Gloess, A. Vietri, F. Wieland, S. Smrke, B. Schönbächler, J. A. Sánchez-López, S. Petrozzi, S. Bongers, T. Koziorowski and C. Yeretzian, Evidence of different flavour formation dynamics by roasting coffee from different origins: On-line analysis with PTR-ToF-MS, *Int. J. Mass Spectrom.*, 2014, **365–366**, 324–337.
139. S. Schenker, C. Heinemann, M. Huber, R. Pompizzi, R. Perren and F. Escher, Impact of roasting conditions on the formation of aroma compounds in coffee beans, *J. Food Sci.*, 2002, **67**, 60–66.

140. J. Baggenstoss, L. Poisson, R. Kaegi, R. Perren and F. Escher, Roasting and aroma formation: effect of initial moisture content and steam treatment, *J. Agric. Food Chem.*, 2008, **56**, 5847–5851.
141. J. Baggenstoss, L. Poisson, R. Kaegi, R. Perren and F. Escher, Coffee roasting and aroma formation: application of different time-temperature conditions, *J. Agric. Food Chem.*, 2008, **56**, 5836–5846.
142. A. S. Franca, L. S. Oliveira, R. C. S. Oliveira, P. C. M. Agresti and R. Augusti, A preliminary evaluation of the effect of processing temperature on coffee roasting degree assessment, *J. Food Eng.*, 2009, **92**, 345–352.
143. J. K. Moon and T. Shibamoto, Role of roasting conditions in the profile of volatile flavor chemicals formed from coffee beans, *J. Agric. Food Chem.*, 2009, **57**, 5823–5831.
144. C. Lindinger, P. Pollien, S. Ali, T. Märk and C. Yeretzian, in *Coupling GC-MS with PTR-MS for unambiguous chemical characterisation of on-line PTR-MS spectra, 1st International Conference on Proton Transfer Reaction Mass Spectrometry and its applications, Innsbruck, 2003*, ed. A. Hansel and T. Märk, Innsbruck, 2002, pp. 157–160.
145. E. Schramm, A. Kurten, J. Holzer, S. Mitschke, F. Muhlberger, M. Sklorz, J. Wieser, A. Ulrich, M. Putz, R. Schulte-Ladbeck, R. Schultze, J. Curtius, S. Borrmann and R. Zimmermann, Trace detection of organic compounds in complex sample matrixes by single photon ionization ion trap mass spectrometry: real-time detection of security-relevant compounds and online analysis of the coffee-roasting process, *Anal. Chem.*, 2009, **81**, 4456–4467.
146. A. Jordan, S. Haidacher, G. Hanel, E. Hartungen, L. Maerk, H. Seehauser, R. Schottkowsky, P. Sulzer and T. D. Maerk, A high resolution and high sensitivity proton-transfer-reaction time-of-flight mass spectrometer (PTR-TOF-MS), *Int. J. Mass Spectrom.*, 2009, **286**, 122–128.
147. A. N. Gloess, A. Vietri, S. Bongers, T. Koziorowski, and C. Yeretzian, in *On-line analysis of the coffee roasting process with PTR-ToF-MS: changes in flavor formation for different coffee varieties, 24th Food and Feed Chemistry (Section 17)*, ed. Association Scientifique Internationale du, C., Association Scientifique Internationale du Cafe, 2012.
148. A. N. Gloess, B. Schonbachler, B. Klopprogge, L. D'Ambrosio, K. Chatelain, A. Bongartz, A. Strittmatter, M. Rast and C. Yeretzian, Comparison of nine common coffee extraction methods: instrumental and sensory analysis, *Eur. Food Res. Technol.*, 2013, **236**, 607–627.
149. J. A. Sánchez-López, R. Zimmermann and C. Yeretzian, Insight into the Time-Resolved Extraction of Aroma Compounds during Espresso Coffee Preparation: Online Monitoring by PTR-ToF-MS, *Anal. Chem.*, 2014, **86**, 11696–11704.
150. M. Graus, C. Yeretzian, A. Jordan, and W. Lindinger, in *In-mouth coffee aroma: breath-by-breath analysis of nose-space while drinking coffee, ASIC-19eme Colloque Scientifique International sur le Café, Paris, 2002*, ASIC, Paris, 2002.
151. S. Ali, P. Pollien, C. Lindinger, and C. Yeretzian, in *In-vivo analysis of aroma release while eating food: a novel set-up for monitoring on-line*

nosespace air, 1st International Conference on Proton Transfer Reaction Mass Spectrometry and its applications, Innsbruck, 2003, ed. A. Hansel and T. Märk, Innsbruck, 2003, pp. 161–164.

152. D. Mayr, T. Maerk, W. Lindinger, H. Brevard, and C. Yeretzian, In-vivo analysis of banana aroma by Proton Transfer Reaction-Mass Spectrometry, in *Flavour Research at the Dawn of the twenty-first century*, ed. J. L. Le Quere and P. X. Etiévant, Lavoisier/Intercept, TEC & DOC, 2003, pp. 256–259.
153. D. Mayr, T. Maerk, W. Lindinger, H. Brevard and C. Yeretzian, Breath-by-breath analysis of banana aroma by proton transfer reaction mass spectrometry, *Int. J. Mass Spectrom.*, 2003, **223–224**, 743–756.
154. D. D. Roberts, P. Pollien, C. Lindinger and C. Yeretzian, Nosespace Analysis with Proton-Transfer Reaction Mass Spectrometry: Intra- and Interpersonal Variability, in *Handbook of Flavor Characterization: Sensory Analysis, Chemistry, and Physiology*, ed. K. Deibler, Dekker, New York, 2003, pp. 151–162.
155. M. Wellinger, S. Biollaz, J. Wochele and C. Ludwig, Sampling and Online Analysis of Alkalis in Thermal Process Gases with a Novel Surface Ionization Detector, *Energy Fuels*, 2011, **25**, 4163–4171.

CHAPTER 34

Phytochemicals From Coffea Leaves

MARIA TERESA SALLES TREVISAN*[a], RICARDO FARIAS DE ALMEIDA[a], ANDREA BREUER[b] AND ROBERT W. OWEN[b]

[a]Departamento de Química Orgânica e Inorgânica, Universidade Federal do Ceará, CP 12200, 60451-970-Fortaleza, Ceará, Brazil; [b]Division of Preventive Oncology, National Center for Tumor Diseases, and German Cancer Research Center (DKFZ), Im Neuenheimer Feld 460, Im Neuenheimer Feld 581, Heidelberg, Germany
*E-mail: mariattre@hotmail.com

34.1 Introduction

Consumed for centuries, coffee and tea are the most popular beverages after water. Tea from the *Cammelia sinensis* plant has a long history of use as a medicinal and healthy drink.[1,2] It is produced by infusion of shredded dried leaves (tea grounds), or also nowadays by infusing tea-bags in near boiling water. The current global annual production of *C. sinensis* tea is about 5.07 million tonnes, comprising 20% green, 2% oolong and the remainder black.[3]

Coffee (*Coffea*) is an even more important commodity, with annual production of about 9.0 Mt.[4] Each year, it involves about 500 million people from cultivation to final consumption.[5] In addition to the seeds, coffee leaves have been used for tea preparation and consumption, particularly in Africa and Asia, but not in Europe and America. It is prepared from dried green leaves, in a similar manner to *C. sinensis* tea.

Coffee: Production, Quality and Chemistry
Edited by Adriana Farah

Published by the Royal Society of Chemistry, www.rsc.org

The use of *Coffea* leaf tea has been common for three to four centuries in Africa, where it is a native plant. The leaf of arabica coffee has been consumed in the continent to treat diarrhoea and intestinal pain. It is still used to manage HIV/AIDS in Kamuli, Sembabule, Kabale, and Gulu districts of Uganda.[6] Dried leaves are used for preparation of a tea named "Quti" in Ethiopia.[7] The salty tea made with the leaf of Liberian coffee, on the other hand, has been described as a laxative drug.[8]

In contrast to the considerable amount of research performed on green coffee seeds, there are relatively few studies concerning the chemical constituents of the other parts of the coffee plant, including the leaves.[9–11] This is also true in relation to the use of coffee leaf as a medicinal plant.[6,7]

Coffee leaf is a rich source of various polyphenolic compounds. A number of studies have shown a relation between the consumption of either tea or coffee and disease prevention, attributed mainly to the presence of polyphenolic compounds in the beverages.[12] Among these phenolic compounds, mangiferin is the least studied and known compound, although it holds high potential as a cancer chemopreventive agent. In addition to phenolic compounds, caffeine is also present in coffee leaves, as in *C. sinensis* tea.[13]

The consumption of coffee leaf tea as a natural source of mangiferins and other bioactive compounds may contribute significantly to elevate the intake of these potentially health-promoting compounds which are not easily available from other sources. This chapter will briefly explore the phytochemical composition of coffee leaves and the potential of coffee leaf tea to promote human health.

34.2 Phytochemical Composition of Coffee Leaves

As cited before, coffee leaf is a rich source of various polyphenolic compounds. About fifteen phenolic compounds have been identified in methanolic extracts of *Coffea* leaves: six chlorogenic acids (3-*O*-caffeoylquinic acid, 4-*O*-caffeoylquinic acid, 5-*O*-caffeoylquinic acid, 3,4-*O*-dicaffeoylquinic acid, 3,5-*O*-dicaffeoylquinic acid and 4,5-*O*-dicaffeoylquinic acid), two xanthones (mangiferin and isomangiferin) and the flavonoids rutin glucoside, quercetin diglucoside, kaempferol diglucoside rhamnoside, quercetin-3-*O*-rhamnoglucoside (rutin), quercetin-3-*O*-rhamnogalactoside, quercetin-3-*O*-glucoside and kaempferol-3-*O*-rhamnoglucoside (Almeida *et al.*, unpublished (Figure 34.1)).

Additional phenolic compounds such as hydroxycinamic acids, epicatechin and isoquercitin, procyanidin B1, and isorhamnetin glucoside, as well as various secondary metabolites in the essential oils (aromatic compounds, monoterpene hydrocarbons, oxygenated monoterpenes, sesquiterpene hydrocarbons, oxygenated sesquiterpenes, among others) have been reported in methanolic and ethanolic extracts of *Coffea* leaves, especially in young leaves, which normally contain higher contents of these compounds.[15–17] Among the constituents identified in the essential oil of *C. arabica* leaves are 3-methylbutanal, 1-penten-3-ol, (*Z*)-2-pentenol, hexanal, (*E*)-2-hexenal, (*Z*)-3-hexenol, 1-hexanol, heptanal, 2-heptanol, (*E*,*E*)-2,4-hexadienal, α-thujene, α-pinene,

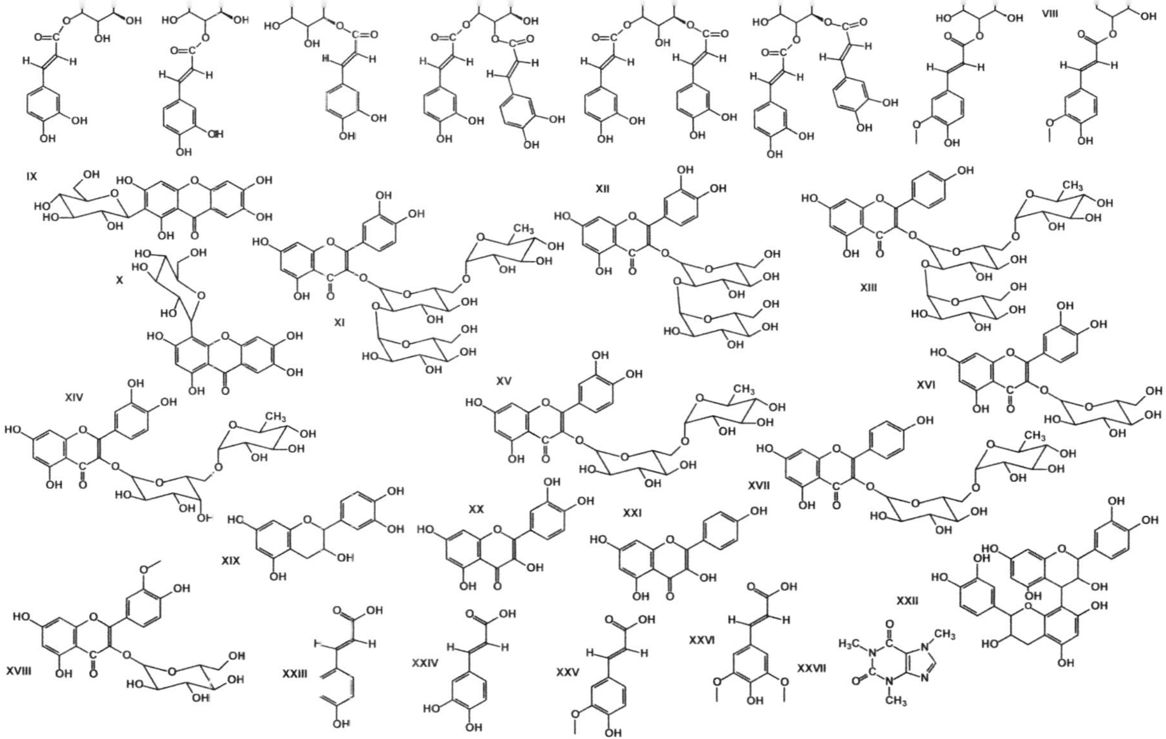

Figure 34.1 Chemical structures of secondary metabolites identified in the leaves of *Coffea arabica*: **I** 3-*O*-caffeoylquinic acid, **II** 4-*O*-caffeoylquinic acid, **III** 5-*O*-caffeoylquinic acid (chlorogenic acid), **IV** 3,4-*O*-dicaffeoylquinic acid, **V** 3,5-*O*-dicaffeoylquinic acid, **VI** 4,5-di-*O*-caffeoylquinic acid, **VII** 4-feruloylquinic acid, **VIII** 5-feruloylquinic acid, **IX** mangiferin, **X** isomangiferin, **XI** rutin glucoside, **XII** quercetin diglucoside, **XIII** kaempferol diglucoside rhamnoside, **XIV** quercetin-3-*O*-rhamnogalactoside, **XV** quercetin-3-*O*-rhamnoglucoside (rutin), **XVI** quercetin-3-*O*-glucoside, **XVII** kaempferol-3-*O*-rhamnoglucoside, **XVIII** isorhamnetin glucoside, **XIX** epicatechin, **XX** quercetin, **XXI** kaempferol, **XXII** procyanidin B1, **XXIII** *p*-coumaric acid, **XXIV** caffeic acid, **XXV** ferulic acid, **XXVI** sinapic acid, **XXVII** caffeine. Data from ref. 10, 16, 17, 19 and 38 and Almeida *et al.*, private commun.

5-ethyl-2(5*H*)-furanone, benzaldehyde, sabinene, 1-octen-3-ol, myrcene, (*E*,*Z*)-2,4 heptadienal, α-terpinene, *p*-cymene, limonene, 1,8-cineole, (*Z*)-β-ocimene, phenylacetaldehyde, (*E*)-β-ocimene, γ-terpinene, *cis*-linalool oxide (furanoid), *trans*-linalool oxide (furanoid), linalool, (*Z*,*Z*)-3,6-nonadienal, terpinen-4-ol, α-terpineol, methyl salicylate, safrole, α-gurjunene, β-caryophyllene, α-(*E*)-ionone, (*E*)-β-farnesene, *ar*-curcumene, γ-curcumene, β-(*E*)-ionone, (*Z*,*E*)-α-farnesene, γ-cadinene, tetradecanal, *cis*-cadin-4-en-7-ol, benzyl salicylate, (*E*)-phytol.[17]

A compilation of data on the content of bioactive compounds in coffee leaves is presented in Table 34.1. Unfortunately, most reported values in the literature result from methanolic and ethanolic extractions. For mangiferin, however, it has been reported that despite its low solubility, 1 minute infusion in boiling water was enough to release 53% of the compound, compared to methanol extraction.[18]

As in coffee seeds, phenolic compounds are the main secondary metabolites and bioactive compounds in the leaves. They derive from multiple branches of the phenylpropanoid biosynthetic pathways (Figure 34.2), one of the most-characterized secondary metabolic routes in plant systems. These compounds are involved in plant response to biotic and abiotic stresses,[19] mainly due to their antioxidant properties.[20] In this way, they may play a role in adaptation to environmental changes[19,20] and in co-evolution with pests and diseases.[19] In humans, there are a number of potential beneficial effects depending on their bioavailability.

Leaves from coffee plants grown under full sun exposure and according to traditional practices presented the highest concentrations of practically all the evaluated phenolic compounds when compared with the leaves obtained from coffee trees using protective tools to avoid climate change stress.[14,21] Wind protection, soil cover and green manure decrease plant stress leading to lower secondary metabolites production in relation to coffee grown in traditional cultivation.[14,15,18] Shade treatment can effectively improve the quality of tea beverage, by decreasing the phenolic content. UV radiation increases phenylpropanoid biosynthesis as demonstrated in tea plants grown in shade or directly exposed to sunlight.[22] Leaves collected from the upper part of the coffee tree contain higher concentrations of all 15 phenolic compounds evaluated than leaves next to the ground. The upper parts of the plant are more exposed to sun and consequently under more stressful conditions, thereby necessitating greater production of secondary metabolites for plant protection.[15,21]

Below, the three main classes of coffee leaves secondary metabolites in coffee leaves are briefly presented.

34.2.1 Chlorogenic Acids

In the *Coffea* genus, chlorogenic acids accumulate to high levels in seeds and leaves and are of great economic importance in coffee production *via* two major species: *Coffea arabica* L. (60–70% of the worldwide coffee production) and *C. canephora* Pierre ex Froehner.[9,23–25]

Among the chlorogenic acids, under current International Union of Pure and Applied Chemistry (IUPAC) recommendations, 3-*O*-caffeoylquinic acid (3-CQA) is now designated 5-*O*-caffeoylquinic acid (5-CQA). 5-CQA, also called chlorogenic acid, was for a long time the only chlorogenic acid commercially available and has been extensively studied due to its antioxidant capacity in addition to other biological effects.[26] Structurally, 5-CQA is the ester formed between caffeic acid and the 5-hydroxyl position of L-quinic acid. Isomers of chlorogenic acid in *Coffea* leaves include the caffeoyl ester at other hydroxyl sites on the quinic acid ring: 4-*O*-caffeoylquinic acid (4-CQA) and 3-*O*-caffeoylquinic acid (3-CQA). At least three major esters between two molecules of caffeic acid and L-quinic acid have been identified in coffee leaves, as in the coffee seeds, 3,4-di-*O*-caffeoylquinic acid, 3,5-di-*O*-caffeoylquinic acid and 4,5-di-*O*-caffeoylquinic acid (Figure 34.1).

Additional hydroxycinnamic acids, such as *p*-coumaric, ferulic, and sinapic acid, which are also precursors of chlorogenic acids in coffee seeds, have been identified in the leaves of *C. arabica* and *C. benghalensis.*[16]

Chlorogenic acids act as antioxidants in plants and protect animals against degenerative, age-related diseases when supplied in their diet. In plants, they can also modulate the feeding behaviour of insects.[27–29] Among many promising effects, caffeic and chlorogenic acids inhibited the activity of M. SssI and Dnmt1 and decreased the methylation level at the RAR beta promoter gene in breast cancer cell lines.[30]

For more information on chlorogenic acids see Chapters 24 and 25 and *Coffee: Consumption and Health Implications*.

34.2.2 Mangiferins

Mangiferin (1,3,6,7-tetrahydroxyxanthone-*C*2-β-glucoside) is a *C*-glycosylated xanthone, which is highly stable and widely distributed in the plant kingdom as reviewed by Vyas *et al.*[31] Isomangiferin is a structural isomer of mangiferin differing only in the position of the C-glucosidic moiety.[32] These compounds are found in many plant species, in particular those from the Anacardiaceae and Gentianaceae families, for example: *Cyclopia genistoides* and *Cyclopia subternata*, popular for the preparation of a herbal tea named "green honeybush". The leaf of *Cyclopia* plants is the most well-known viable natural source of mangiferin, in which amounts as high as 11.8 g 100 g^{-1} dried extract (n.b. not g 100 g^{-1} plant material) have been reported.[33] Mangiferin was first observed in mango fruit in 1922, but it was only isolated and identified in 1957.[34,35] In coffee leaves, it is a relatively newly discovered compound, being only detected in the leaves in 2008.[36]

The first indication for the existence of mangiferin in coffee leaves was reported by Campa *et al.*,[9] who examined samples from seven African species belonging to the *Coffea* taxa (including *C. arabica* and wild species such as *C. salvatrix*) and found a wide range of contents (from 0.03 to 16.36 g 100 g^{-1}, dry weight). Although usual amounts are much lower, some of the amounts reported by Campa *et al.* for coffee leaves can be comparable to the amounts in *Cyclopia* leaves. Trevisan *et al.*,[18] however, reported contents in *C. arabica*

Table 34.1 Compilation of reported data on the content of bioactive compounds in coffee leaves.

Bioactive compound	*C. arabica*	*C. heterocalyx cf.*	*C. eugenioides*	*C. anthonyi*	*C. salvatrx*	*C. pseudozanguebariae*	*C. sessiliflora*
g 100 g^{-1} dry leaves							
3-*O*-Caffeoylquinic acid	0.10[10]	0.13[10]	0 × 10 [10]	0.04[10]	0.28[10]	0.06[10]	0.17[10]
4-*O*-Caffeoylquinic acid	0.26[10]	0.12[10]	0.16[10]	0.09[10]	0.50[10]	0.09[10]	0.30[10]
5-*O*-Caffeoylquinic acid	3.23[10]						
	0.04–0.31 (ref. 38)	0.99[10]	2.74[10]	1.70[10]	1.07[10]	0.85[10]	0.35[10]
3,4-Di-*O*-caffeoylquinic acid	0.05[10]	0.01[10]	0.02[10]	0.00[10]	0.01[10]	0.002[10]	0.01[10]
3,5-Di-*O*-caffeoylquinic acid	0.17[10]	0.02[10]	0.01[10]	0.001[10]	0.05[10]	0.02[10]	0.01[10]
4,5-Di-*O*-caffeoylquinic acid	0.04[10]	0.06[10]	0.04[10]	0.00[10]	0.04[10]	0.03[10]	0.02[10]
Total feruloylquinic acid	0.12[10]	0.00[10]	0.00[10]	0.00[10]	0.00[10]	0.00[10]	0.00[10]
Total chlorogenic acids	3.96[10]						
	0.24–1.86[a]	1.33[10]	3.07[10]	1.84[10]	1.94[10]	1.05[10]	0.85[10]
Mangiferin	0.05–0.42[19]						
	0.57[a]						
	0.02–0.08 (ref. 38)	4.52[10]	5.68[10]	12.06[10]	16.36[10]	4.91[10]	0.03[10]
Isomangiferin	0.02–0.08[19]						
Total mangiferins	0.07–0.50[19]						
	0.15–0.88[a]						
Total flavonoids[b]	0.12–0.66[a]						
Caffeine	0.59–0.97[a]						
	0.15–0.65 (ref. 38)						
	C. arabica	*C. benghalensis*					
μg mL^{-1} non-hydrolysed or hydrolysed extract							
Caffeic acid	1.82–86.52[17]	2.26–125.55[17]					
5-*O*-Caffeoylquinic acid	605.65[17]	428.73[17]					
p-Coumaric acid	0.86–1.10[17]	0.80[17]					
Ferulic acid	103.90[17]	8.50[17]					
Sinapic acid	8.21[17]	7.34[17]					
Quercetin-3-*O*-glucoside	6.36[17]	7.44[17]					
Rutin	1.60[17]	12.88[17]					
Quercetin	11.96[17]	42.73[17]					
Kaempferol	1.54[17]	5.72[17]					

[a](Almeida *et al.*, unpublished).

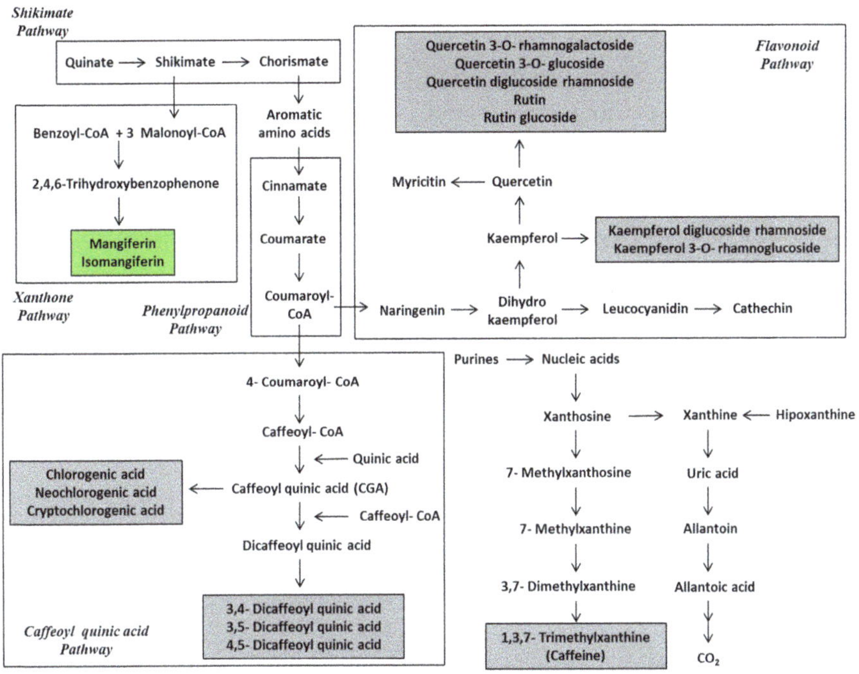

Figure 34.2 The biosynthetic pathways of the phenolic compounds identified in coffee leaves. Data from ref. 128.

leaves from Brazil and Costa Rica ranging from 0.07 g to 0.50 g 100 g^{-1} dry weight, lower than the amount reported by Campa *et al.*[9] for *C. arabica* leaves from Africa (0.57 g 100 g^{-1}).

Mangiferins have shown to play a role in increasing the resistance to coffee leaf rust, a disease caused by *Hemileia vastatrix* Berk. et Br., which causes defoliation and drying of branches, notwithstanding a relation with the biochemical mechanism of plant defence against pathogens and UV irradiation.[37,38] A better understanding of the biosynthetic pathway of xanthones and their glycosylated derivatives is essential to determine the level of involvement of mangiferin in plant stress response.

Mangiferins, other xantones and phenolic compounds in general are potential natural anti-oxidative and chemopreventive agents. It is well known that the addition of mango by-products extract (a good source of mangiferins) to broiler and laying chicken diets can be a simple and convenient strategy to introduce these substances into meats and eggs to prevent or delay lipid oxidation and improve the oxidative stability of these foods.[39–43]

The potential nutraceutical and pharmaceutical significance of mangiferins has also been extensively demonstrated and continues to attract much attention, especially concerning its potential to fight degenerative diseases, mainly as antioxidants, a property that enables protection against damage to human cells, *via* inhibition of lipid peroxidation[44,45] resulting in reduced oxidative stress, DNA damage and therefore degenerative diseases.[46–49] A number of reports have been published on mechanisms of anti-oxidative and related effects,[50–56] including metal chelation[51,57] and free radical scavenging.[58–61] *In vitro*, mangiferins can also exert anti-inflammatory, immunomodulatory, cell cycle arrest, anti-proliferative, anti-apoptotic, anti-genotoxic and anti-viral effects which cumulatively result in anti-tumour activity.[46] The main effect of mangiferin on key enzyme systems is the inhibition of aldose reductase, catalase, COXs, α-glucosidase, isomaltase, hypoxanthine-xanthine oxidase, creatine kinase, IκB kinase, nitric oxide synthase and reductase. The effects in functional systems such as lymphocytes occur by activation/inhibition of proliferation or specific cytolytic T-killers formation, but in functional systems such as macrophages they occur by phagocytic activity inhibition, controlling the expression of genes for primary inflammation mediators and inhibition of reactive oxygen species (ROS) and nitric oxide (NO) production.[48]

Despite the potential of mangiferin and isomangiferin to improve health, they are not produced under pharmacologic license, and are not yet authorized by regulatory agencies. The use of pure isolated compounds has been found to be less effective than the use of crude mixtures suggesting that the synergism of various polyphenols is important for maximum antioxidative activity.[49] A natural alternative source of mangiferins is coffee leaf tea or extract that can be consumed in the diet. The addition of dried coffee leaf extract as a partial substitution for instant coffee, for example, which contains chlorogenic acids but no mangiferin, could be a useful way of increasing mangiferin intake.

According to reports from the International Agency for Research on Cancer (IARC), the incidence of cancer is increasing in the world. There were 14.1 million cases and over 8.2 million mortalities due to cancer in 2012 compared to 10 million cases and 6 million mortalities in 2000 (ref. 46) in a baseline population of 7.1 billion in 2012 and 6.1 billion in 2000. Such statistics have increased the interest in recent years in the study of potential cancer chemopreventive agents. The effect of natural products such as polyphenols in the prevention and co-treatment against an array of different cancers has been demonstrated *in vitro* and *in vivo*.[62–76] Mangiferins and coffee leaf tea appear to be promising products in this sense and their bioavailability and effect in humans should be thoroughly studied.

In addition to cancer preventive properties, mangiferins are reported to exert anti-allergic,[77,78] anthelminthic,[78] antiviral[79] and gastro-protective properties,[80,81] and potent hepatoprotective effects on CCl_4-induced liver damage in mice.[82] They have also shown to improve long-term cholinergic memory deficits by acetylcholinesterase (AChE) inhibition or cholinergic receptor stimulation and inhibition of NF-*k*B activation *in vivo*.[83] Atropine, a cholinergic antagonist, blocks the stimulatory effects of mangiferin as well, suggesting that these natural xanthones could induce Ca^{2+} mobilization in the submucosal neurons, thereby stimulating gut motility and secretion being an alternative for the treatment of functional gastrointestinal disturbances such as dyspepsia and postoperative ileus.[84]

Fortification with mangiferin has already been used in foodstuffs to treat diabetes[85] and to co-treat and help prevent neurodegenerative diseases and related ageing symptoms, at least in part, due to its antioxidant and anti-inflammatory effects. Inflammation is known to play an important role in the development of schizophrenia[86–94] and co-treatment with a mangiferin extract could possibly have positive ameliorative effects. Mangiferin also decreases glutamate-induced neurotoxicity and memory loss induced by scopolamine.[83,92,95,96]

The bioavailability of a phytochemical is dependent on different factors including its bioaccessibility (quantity of compound released from the food matrix in the digestive tract), solubility in gastrointestinal fluids, cellular uptake, compound metabolism and efficiency of the circulatory system.[46] The optimal health benefits of mangiferin could not be fully evaluated due to its poor water solubility (1.57 mg dm^{-3}),[18,97] which is a limiting factor for its nutraceutical potential, and its bioavailability in rats was very low.[98]

Micro or nano-encapsulation of a number of compounds has improved solubility in general, and this is particularly suitable for compounds such as mangiferin. Spray-drying formulations can impact on retention of mangiferin in the particle as demonstrated by the comparison of a pectin formulation *versus* a chitosan polysaccharide, with pectin being found to have a better retention of mangiferin in the particles than a chitosan formulation.[99] Mangiferin also showed strong interactions with copolymer micelles allowing enhanced solubility in water, providing a higher bioavaibility.[100] Galactomannan from *D. gardneriana* (GLMDg) and its sulfated derivative (GLMDgS) were also used as polymeric matrices for mangiferin encapsulation by a spray-drying process, showing potential as a bio-based drug delivery system for controlled release.[97]

The mer-[$RuCl_3$(dppb)(H_2O)] complex, in the presence of a reducing agent, was used to prepare new carbonyl ruthenium(II) complexes to solubilize mangiferin. The oxidizing property of this aqua complex allowed a stabilization of the compound *trans* [$RuCl_2$(dppb)(mang)], containing the natural pharmaceutical product, mangiferin.[101]

When assessing the health potential of a compound, it is important to also consider its toxicity. Both mangiferin and its aglycone alone did not show cytotoxic effect.[102] The toxic dose of mangiferin in mice was reported to be 400 mg kg^{-1}, being generally regarded as non-toxic.[103,104] Thorough investigation on the absorption, metabolism and safety of biotransformed products such as homomangiferin and norathyriol (Figure 34.3) is also required. Both compounds were found to be products of human faecal bacteria *in vitro*[105] and it is worth noting that it has been shown that norathyriol was equally cytotoxic (IC_{50} = 51 μM) against both colon and lung cancer cell lines, whereas mangiferin was ineffective.[105]

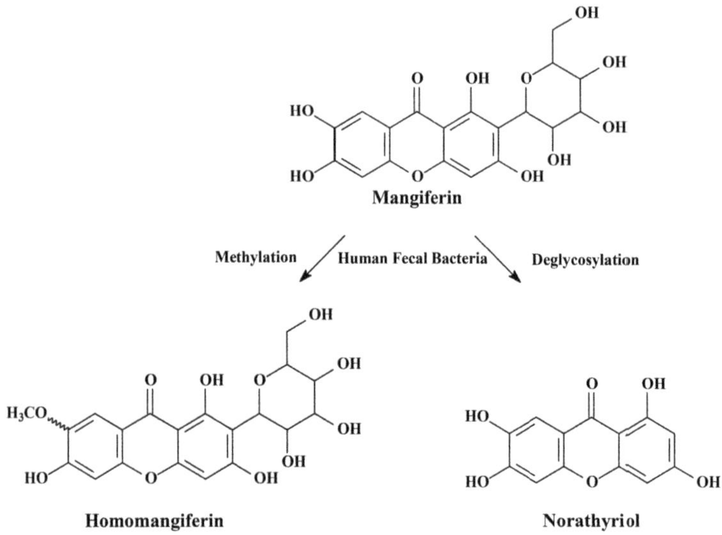

Figure 34.3 Homomangiferin and norathyriol have been identified as intestinal metabolites of mangiferin. Data from ref. 106, 129 and 130.

Despite the lack of studies on mangiferin toxicity in humans, evidence on non-toxicity exists based on evidence from occasions when mangiferin has been taken orally. For example, in the Center of Pharmaceutical Chemistry, in Cuba, mango bark extract containing mangiferin as a major component has been administrated orally to more than 7000 patients for 10 years.[47] Additionally, as previously mentioned, coffee leaf tea has been consumed in Africa for centuries.

34.2.3 Rutin

The main flavonoid compound in coffee leaves is rutin. Rutin plays a role in plant UV defence and its content increases rapidly after UV-B radiation.[106] The content depends on many factors, but 811 mg 100 g^{-1} (dry weight) has been detected in leaves of *Coffea arabica* Caturra Vermelho cv. IAC-477 (Almeida *et al.*, submitted). Values of 826 mg 100 g^{-1} and 195 mg 100 g^{-1} were obtained for husk and pulp, respectively, considered to be by-products that can be incorporated to coffee leaf tea.[107] These contents of rutin are much higher than those found in *Camellia sinensis* (L.) Kuntze, for example (120 mg 100 g^{-1} dry weight). The leaves and stems of *Ilex paraguariensis* used for maté tea preparation also contain a considerable amount of rutin (up to 50 mg 100 g^{-1})[108].

Rutin, also known as vitamin P, is widely consumed in the diet. It has exhibited important pharmacologic activities *in vitro* and *in vivo*, including anti-inflammatory, anti-diabetes, anti-adipogenic, neuroprotective,[109] gastroprotective[110,111] and anti-glycation[112,113] effects. These effects are mainly attributed to the potent antioxidant property of rutin as a free radical scavenger.[114–116] The lethal dose of rutin in mice was reported to be 950 mg kg^{-1}.[117]

34.2.4 Caffeine

Caffeine (1,3,7-trimethylxanthine) is a purine alkaloid. Purine alkaloids are secondary metabolites derived from purine nucleotides, and they are present in nearly 100 species in 13 orders of the plant kingdom,[118] including the flowers of several citrus species and fungi.

Together with chlorogenic acids, caffeine is the most abundant secondary metabolite in coffee leaves.[119] It has been proposed that caffeine is sequestered in the vacuoles of coffee leaves as a chlorogenic acid complex.[1] Caffeine is produced in young *Coffea* leaves and immature fruits, and continues to accumulate gradually during the maturation of these organs. The caffeine content in young leaves of *Coffea arabica* is higher compared to old leaves.[120,121] Its physiological function is chemical defence against pathogens and herbivores, and allelopathic effects against competing plant species. An example is the chemical defence against the coffee berry borer *Hypothenemus hampei*.[122] Since nearly 40% of the world agricultural production is lost by herbivores and diseases production of low-caffeine agriculturally important crops, the use of caffeine from coffee leaves could be generally effective to control both pest insects and pathogens.[123,124]

Caffeine contents are similar in all *Camelia sinensis* tea types, ranging from 2.3 g 100 g^{-1} in green tea, to 1.87 g 100 g^{-1} in oolong tea).[125,126] In coffee leaf, the content is reported to be half of this amount[13,37,125] (Table 34.1). It is well known that caffeine is not altered or degraded during the roasting process of coffee seeds and it contributes to the perceived bitterness of coffee brews, decreasing its acceptability when in excess.[127] The lower caffeine content in coffee leaves could benefit the acceptance of this new kind of tea, in comparison with both *C. sinensis* tea and coffee beverage.

34.3 Conclusions

The coffee leaf is a potential low-cost source of polyphenolic compounds with high antioxidative capacity that may potentially help protect the body against damage linked to oxidative stress, among other types of damage. In addition to other compounds, mangiferin and rutin are mainly concentrated in the leaves of the *Coffea* plant. Coffee leaf extracts could be explored in different ways, for tea preparation and for enrichment of human foods, animal chow, instant coffee, among other uses. Whole coffee leaf extracts are probably more potent than pure isolated mangiferin or rutin highlighting the synergism between mangiferin, rutin and other coffee polyphenols (*e.g.*, chlorogenic acids) for enhanced activity.

Improved technology and marketing of green coffee leaves needs to be developed, which would benefit both farmers and consumers.

References

1. M. Friedman, *Mol. Nutr. Food Res.*, 2007, **51**, 116.
2. J. V. Higdon and B. Frei, *Crit. Rev. Food Sci. Nutr.*, 2003, **43**, 89.
3. K. Chang, in *World tea production and trade. Current and future development*, FAO, 2015, pp. 1–13.
4. Data from https://www.fas.usda.gov/data/coffee-world-markets-and-trade, obtained December 2016.
5. K. D. Batista, W. L. Araújo, W. C. Antunes, P. C. Cavatte, G. A. B. K. Moraes and S. C. V. Martins, *et al.*, *Trees*, 2012, **26**, 459.
6. M. Lamorde, J. R. S. Tabuti, C. Obua, C. K. Byobona, H. Lanyero, P. B. Kibwika, *et al.*, *J. Ethnopharmacol.*, 2010, **130**, 43.
7. M. Giday, Z. Asfaw and Z. Woldu, *J. Ethnopharmacol.*, 2010, **132**, 75.
8. A. Belayneh and N. F. Bussa, *J. Ethnobiol. Ethnomed.*, 2014, **10**, 18.
9. C. Campa, L. Mondolot, A. Rakotondravao, L. P. Bidel, A. Gargadennec, E. Couturon, P. La Fisca, J. J. Rakotomalala, C. Jay-Allemand and A. P. Davis, *Ann. Bot.*, 2012, **110**, 595.
10. L. Mondolot, P. La Fisca, B. Buatois, E. Talansier, A. de Kochko and C. Campa, *Ann. Bot.*, 2006, **98**, 33.
11. M. N. Clifford, J. Kirkpatrick, N. Kuhnert, H. Roorendaal and P. R. Salgado, *Food Chem.*, 2008, **106**, 379.
12. Y. Wang and C.-T. Ho, *J. Agric. Food Chem.*, 2009, **57**, 8109.
13. R. Fang, S. P. Redfern, D. Kirkup, E. A. Porter, G. C. Kite, L. A. Terry, M. J. Berry and M. S. J. Simmonds, *Food Chem.*, 2017, **220**, 517.
14. L. Gobbo-Neto and L. G. Lopes, *Quim. Nova*, 2007, **30**, 374.
15. R. F. Almeida, C. M. Ulrich, R. W. Owen, M. T. S. Trevisan, *Proceedings of ASIC Conferences*, 2014, vol. 1, p. 42.
16. É. B. Patay, T. Németh, T. S. Németh, R. Filep, L. Vlase, N. Papp, *et al.*, *Farmacia*, 2016, **64**, 1.
17. C. Quijano-Célis, D. Piedrahita and J. A. Pino, *J. Essent. Oil-Bear. Plants*, 2015, **18**, 486.
18. M. T. S. Trevisan, R. F. Almeida, G. Soto, E. M. Virginio, C. M. Ulrich and R. W. Owen, *Food Anal. Methods*, 2016, **9**, 2649.
19. V. Cheynier, G. Comte, K. M. Davies, V. Lattanzio and S. Martens, *Plant Physiol. Biochem.*, 2013, **72**, 1.
20. S. Caretto, V. Linsalata, G. Colella, G. Mita and V. Lattanzio, *Int. J. Mol. Sci.*, 2015, **16**, 26378.
21. F. Delaroza, M. Rakocevic, G. B. Malta, R. E. Brunsc and I. S. Scarminio, *J. Braz. Chem. Soc.*, 2014, **25**, 1929.
22. Q. Zhang, Y. Shi, L. Ma, X. Yi and J. Ruan, *PLoS One*, 2014, **9**, e112572.
23. M. N. Clifford, T. Williams and D. M. Bridson, *Phytochemistry*, 1989, **28**, 829.

24. C. L. Ky, J. Louarn, S. Dussert, B. Guyot, S. Hamon and M. Noirot, *Food Chem.*, 2001, **75**, 223.
25. International Coffee Organization (ICO), *2011, ICO Annual Review 2011/12*, https://www.ico.org/documents/cy2012-13/annual-review-2011-12e.pdf, accessed 23 april 2017.
26. D. Kremr, T. Bajer, P. Bajerová, S. Surmová and K. Ventura, *Quim. Nova*, 2016, **vol. 39**(4), 530–533.
27. W. P. Chen and L. D. Wu, *Int. J. Clin. Exp. Pathol.*, 2014, **7**, 8797.
28. Y. Hamamura, in *Control of insect behavior by natural products*, ed. D. L. Wood, R. M. Silverstein, and M. Nakajima, Academic Press, New York, 1970, pp. 72–73.
29. A. Mubarak, C. P. Bondonno, A. H. Liu, M. J. Considine, L. Rich and E. Mas, *et al.*, *J. Agric. Food Chem.*, 2012, **60**, 9130.
30. W. J. Lee and B. T. Zhu, *Carcinogenesis*, 2006, **27**, 269.
31. A. Vyas, K. Syeda, A. Ahmad, S. Padhye and F. H. Sarkar, *Mini-Rev. Med. Chem.*, 2012, **12**, 412.
32. D. de Beer, G. Jerz, E. Joubert, V. Wray and P. Winterhalter, *J. Chromatogr. A*, 2009, **1216**, 4282.
33. D. De Beer and E. Joubert, *J. Food Compos. Anal.*, 2010, **23**, 289.
34. S. Iseda, *Bull. Chem. Soc. Jpn.*, 1957, **30**, 625.
35. S. Iseda, *Bull. Chem. Soc. Jpn.*, 1957, **30**, 629.
36. P. Talamond, L. Mondolot and A. Gargadennec, *et al.*, *Acta Bot. Gallica*, 2008, **155**, 513.
37. M. T. S. Trevisan, J. R. Paiva, R. F. Almeida, R. W. Owen, V. L. Carvalho and R. L. Cunha, *The 26th International Conference on Coffee Science*, 2016, p. 215.
38. G. Franklin, L. F. R. Conceição, E. Kombrink and A. C. P. Dias, *Phytochemistry*, 2009, **70**, 65.
39. E. R. Freitas, A. S. Borges, A. L. F. Pereira, V. K. G. Abreu, M. T. S. Trevisan and P. H. Watanabe, *Poult. Sci.*, 2015, **94**, 295.
40. A. L. F. Pereira, T. F. Vidal, M. C. Teixeira, P. F. Oliveira, M. M. M. Vieira, J. F. F. Zapata, R. C. F. F. Pompeu and E. F. Freitas, *Braz. J. Food Technol.*, 2010, **13**, 293.
41. A. L. F. Percira, T. F. Vidal, M. C. Teixeira, P. F. Oliveira, R. C. F. F. Pompeu, M. M. M. Vicira and J. F. F. Zapata, *Cienc. Tecnol. Aliment.*, 2011, **31**, 135.
42. E. R. Freitas, A. S. Borges, M. T. S. Trevisan, P. H. Watanabe, A. L. Cunha, A. L. F. Pereira, V. K. Abreu and G. A. J. Nascimento, *Pesqui. Agropecu. Bras.*, 2012, **47**, 1025.
43. E. R. Freitas, A. S. Borges, M. T. S. Trevisan, A. L. Cunha, N. M. Braz, P. H. Watanabe and G. A. J. Nascimento, *Pesqui. Agropecu. Bras.*, 2013, **48**, 714.
44. G. M. Sánchez, M. A. Rodríguez, A. Giuliani, A. J. Núñez Sellés, N. P. Rodríguez, H. Fernández, O. S. Leon and L. Re, *Phytother. Res.*, 2003, **17**, 197.
45. F. R. Ferreira, I. B. Valentim, E. L. C. Ramones, M. T. S. Trevisan, C. Olea-Azar, F. Perez-Cruz, F. C. De Abreu and M. O. F. Goulart, *LWT--Food Sci. Technol.*, 2013, **51**, 129.

46. F. Gold-Smith, A. Fernandez and K. Bishop, *Nutrients*, 2016, **8**, 396.
47. A. J. Nunez-Selles, R. Delgado-Hernandez, G. Garrido-Garrido, D. Garcia-Rivera, M. Guevara-Garcia and G. L. Pardo-Andreu, *Pharmacol. Res.*, 2007, **55**, 351.
48. M. M. M. Pinto, M. E. Sousa and M. S. J. Nascimento, *Curr. Med. Chem.*, 2005, **12**, 2517.
49. M. Masibo and Q. He, *Compr. Rev. Food Sci. Food Saf.*, 2008, **7**, 309.
50. T. Sato, A. Kawamoto, A. Tamura, Y. Tatsumi and T. Fujii, *Chem. Pharm. Bull.*, 1992, **40**, 721.
51. S. Ghosal, G. Rao, V. Saravana, N. Misra and D. Rana, *Indian J. Chem., Sect. B: Org. Chem. Incl. Med. Chem.*, 1996, **35**, 561.
52. P. Scartezzini and E. Speroni, *J. Ethnopharmacol.*, 2000, **71**, 23.
53. M. Born, P. A. Carrupt, R. Zini, F. Brée, J. P. Tillement, K. Hostettmann and B. Testa, *Helv. Chim. Acta*, 1996, **79**, 1147.
54. G. M. Sanchez, L. Re, A. Giuliani, A. J. Nunez-Selles, G. P. Davison and O. S. Leon-Fernandez, *Pharmacol. Res.*, 2000, **42**, 565.
55. S. Muruganandan, S. Gupta, M. Kataria, J. Lal and P. K. Gupta, *Toxicology*, 2002, **176**, 165.
56. Y.-Y. Soong and P. J. Barlow, *Food Chem.*, 2004, **88**, 411.
57. S. Y. Tang, M. Whiteman, A. Jenner, Z. F. Peng and B. Halliwell, *Free Radical Biol. Med.*, 2004, **36**, 1588.
58. C. G. Sampaio, L. S. Frota, H. S. Magalhães, L. M. U. Dutra, D. C. Queiroz, R. S. Araújo, H. Becker, J. R. R. Souza, N. M. P. S. Ricardo and M. T. S. Trevisan, *Int. J. Biol. Macromol.*, 2015, **78**, 273.
59. D. Garcia, R. Delgado, F. M. Ubeira and J. Leiro, *Int. Immunopharmacol.*, 2002, **2**, 797.
60. J. M. Leiro, E. Álvarez, J. A. Arranz, I. G. Siso and F. Orallo, *Biochem. Pharmacol.*, 2003, **65**, 1361.
61. P. M. Pauletti, I. C. Gamboa, D. H. S. Silva, M. C. M. Young, D. M. Tomazela, M. N. Eberlin and V. S. Bolzani, *J. Nat. Prod.*, 2003, **66**, 1384.
62. P. Rajendran, T. Rengarajan, I. Nishigaki, G. Ekambaram and D. Sakthisekaran, *J. Cancer Res. Ther.*, 2014, **10**, 1033.
63. J. Lv, Z. Wang, L. Zhang, H. L. Wang, Y. Liu, C. Li, J. Deng, W. Yi and J. K. Bao, *J. Anim. Vet. Adv.*, 2013, **12**, 352.
64. P. Rajendran, T. Jayakumar, I. Nishigaki, G. Ekambaram, Y. Nishigaki, J. Vetriselvi and D. Sakthisekaran, *Int. J. Biomed. Sci.*, 2013, **9**, 68.
65. X. Y. Hu, J. G. Deng, L. Wang and Y. F. Yuan, *Med. Chem.*, 2013, **9**, 1058.
66. I. Rodeiro, R. Delgado and G. Garrido, *Cell Proliferation*, 2014, **47**, 48.
67. B.-P. Zhang, J. Zhao, S.-S. Li, L.-J. Yang, L.-L. Zeng, Y. Chen, J. Fang, APS 2014, **35**, 257.
68. H. Li, J. Huang, B. Yang, T. Xiang, X. Yin, W. Peng, W. Cheng, J. Wan, F. Luo and H. Li, *Toxicol. Appl. Pharmacol.*, 2013, **272**, 180.
69. S. Das, B. Nageshwar Rao and B. S. Satish Rao, *Chem.-Biol. Interact.*, 2011, **193**, 129.
70. J. Xiao, L. Liu, Z. Zhong, C. Xiao and J. Zhang, *Oncol. Rep.*, 2015, **33**, 2815.

71. B. Zhang, J. Zhao, S. Li, L. Zeng, Y. Chen and J. Fang, *Pharm. Biol.*, 2015, **53**, 503.
72. J. Zhao, B. Zhang, S. Li, L. Zeng, Y. Chen and J. Fang, *Int. J. Mol. Med.*, 2014, **33**, 1348.
73. J. S. Jung, K. Jung, D. H. Kim and H. S. Kim, *Pharmacol. Res.*, 2012, **66**, 95.
74. D. García-Rivera, R. Delgado, N. Bougarne, G. Haegeman and W. Vanden Berghe, *Cancer Lett.*, 2011, **305**, 21.
75. B. Zhang, J. Fang and Y. Chen, *Lett. Drug Des. Discovery*, 2013, **10**, 239.
76. L. Torre, F. Bray, R. Siegel, J. Ferlay, J. Lortet-Tieulent and A. Jemal, *Ca-Cancer J. Clin.*, 2015, **65**, 87.
77. D. G. Rivera, I. H. Balmaseda, A. A. Leon, B. C. Hernandez, L. M. Montiel, G. G. Garrido, S. Cuzzocrea and R. D. Hernandez, *J. Pharm. Pharmacol.*, 2006, **58**, 385.
78. D. Garcia, M. Escalante, R. Delgado, F. M. Ubeira and J. Leiro, *Phytother. Res.*, 2003, **17**, 1203.
79. C. Yoosook, N. Bunyapraphatsara, Y. Boonyakiat and C. Kantasuk, *Phytomedicine*, 2000, **6**, 411.
80. A. C. Carvalho, M. M. Guedes, A. L. de Souza, M. T. S. Trevisan, A. F. Lima, F. A. Santos and V. S. Rao, *Planta Med.*, 2007, **73**, 1372.
81. T. C. Morais, S. C. Lopes, K. M. M. B. Carvalho, B. R. Arruda, F. T. C. Sousa, M. T. S. Trevisan, V. S. N. Rao and F. A. Santos, *World J. Gastroenterol.*, 2012, **18**, 3207.
82. M. Rasool, E. P. Sabina, P. S. Mahinda and B. Clara-Gnanaselvi, *Comp. Clin. Pathol.*, 2012, **21**, 865.
83. Y. Lemus-Molina, M. V. Sánchez-Gómez, R. Delgado-Hernández and C. Matute, *NeuroToxicology*, 2009, **30**, 1053.
84. T. C. Morais, B. R. Arruda, H. S. Magalhães, M. T. S. Trevisan, D. A. Viana, V. S. Rao and F. A. Santos, *Naunyn-Schmiedeberg's Arch. Pharmacol.*, 2015, **2**, 1.
85. M. Wada, , Japanese Patent, JP2007204462-A, 2007.
86. P. S. Sellamuthu, P. Arulselvan, B. P. Muniappan, S. Fakurazi and M. Kandasamy, *J. Med. Food*, 2013, **16**, 719.
87. L. Amazzal, A. Lapôtre, F. Quignon and D. Bagrel, *Neurosci. Lett.*, 2007, **418**, 159.
88. G. L. Andreu, R. Delgado, J. Velho, C. Curti and A. E. Vercesi, *Eur. J. Pharmacol.*, 2005, **513**, 47.
89. J. C. Barreto, M. T. S. Trevisan, W. E. Hull, G. Erben, E. S. de Brito, B. Pfundstein and G. Würtele, *et al.*, *J. Agric. Food Chem.*, 2008, **56**, 5599.
90. L. Márquez, B. García-Bueno, J. L. Madrigal and J. C. Leza, *Eur. J. Nutr.*, 2012, **51**, 729.
91. G. Martínez Sánchez, E. Candelario-Jalil, A. Giuliani, O. S. León, S. Sam, R. Delgado and A. J. Núñez Sellés, *Free Radical Res.*, 2001, **35**, 465.
92. V. S. Rao, A. C. Carvalho, M. T. S. Trevisan, G. M. Andrade, H. V. Nobre Júnior, M. O. Moraes, H. I. Magalhães, H. Iury, T. C. Morais and F. A. Santos, *Pharmacol. Rep.*, 2012, **64**, 848.

93. H. Rahmoune, L. W. Harris, P. C. Guest and S. Bahn, (n.d.), *Rev. Psiq. Clín.*, 2013, **40**, 28.
94. C. C. T. Aguiar, C. D. Alves, F. A. R. Rodrigues, F. W. A. Barros, F. C. F. de Sousa, S. M. M. Vasconcelos and D. S. Macedo, *J. Bras. Psiquiatr.*, 2010, **59**, 52.
95. K. Jung, B. Lee, S. J. Han, J. H. Ryu and D. H. Kim, *Biol. Pharm. Bull.*, 2009, **32**, 242.
96. G. L. Pardo Andreu, N. Maurmann, G. K. Reolon, C. B. de Farias, G. Schwartsmann, R. Delgado and R. Roesler, *Eur. J. Pharmacol.*, 2010, **635**, 124.
97. L. Dutra, I. Cavalcante, D. de Brito, I. Vieira, M. T. S. Trevisan, M. E. Ribeiro, S. Yeates and N. M. P. S. Ricardo, *J. Braz. Chem. Soc.*, 2017, **28**, 1341.
98. D. Han, C. Chen, C. Zhang, Y. Zhang and X. Tang, *J. Pharm. Biomed. Anal.*, 2010, **51**, 260.
99. J. R. R. De Souza, J. P. A. Feitosa, N. M. P. S. Ricardo, M. T. S. Trevisan, H. C. B. de Paula, C. M. Ulrich and R. W. Owen, *Food Hydrocolloids*, 2013, **33**, 10.
100. R. R. De Almeida, H. S. Magalhães, J. R. R. De Souza, M. T. S. Trevisan, I. G. P. Vieira, J. P. A. Feitosa, T. G. Araújo and N. M. P. S. Ricardo, *Ind. Crops Prod.*, 2015, **69**, 284.
101. M. I. F. Barbosa, E. R. Santos, A. E. Graminha, A. L. Bogado, L. R. Teixeira, H. Beraldo, M. T. S. Trevisan, J. Ellena, E. E. Castellano and B. L. Rodrigues, *Polyhedron*, 2011, **30**, 41.
102. W. Dou, J. Zhang, G. Ren, L. Ding, A. Sun, C. Deng, X. Wu, X. Wei, S. Mani and Z. Wang, *Int. Immunopharmacol.*, 2014, **23**, 170.
103. J. Lei, C. Zhou, H. Hu, L. Hu, M. Zhao, Y. Yang, Y. Chuai, J. Ni and J. Cai, *J. Cell. Biochem.*, 2012, **113**, 2633.
104. G. C. Jagetia and M. S. Baiga, *Phytomedicine*, 2005, **12**, 209.
105. J. R. R. Souza, J. P. A. Feitosa, M. T. S. Trevisan, N. Ricardo, E. Frei, C. Ulrich, and R. Owen, *Cytotoxicity Screen of Mangiferin and its Major Metabolite Norathyriol in Human Tumor Cell Lines*, 2012, https://doi.org/10.13140/2.1.3547.7768.
106. J. B. Harbone and C. A. Williams, *Phytochemistry*, 2000, **55**, 581.
107. L. R. P. García, C. R. Biasetto, A. R. Araujo and V. L. Del Bianchi, *Food Sci. Technol.*, 2015, **35**, 704.
108. R. Filip, P. López, G. Giberti, J. Coussio and G. Ferraro, *Fitoterapia*, 2001, **72**, 774.
109. A. R. Gulpinar, I. E. Orhan, A. Kan, F. S. Senol, S. A. Celik and M. Kartal, *Food Res. Int.*, 2012, **46**, 536.
110. I. T. Abdel-Raheem, *Basic Clin. Pharmacol. Toxicol.*, 2010, **107**, 742.
111. C. La Casa, I. Villegas, C. La de Alarcon Lastra, V. Motilva and M. J. Martin Calero, *J. Ethnopharmacol.*, 2000, **71**, 45.
112. S. Pashikanti, D. R. de Alba, G. A. Boissonneault and D. Cervantes-Laurean, *Free Radical Biol. Med.*, 2010, **48**, 656.

113. T. Nagasawa, N. Tabata, Y. Ito, N. Nishizawa, Y. Aiba and D. D. Kitts, *Mol. Cell. Biochem.*, 2003, **249**, 3.
114. A. M. Mahmoud, *Exp. Toxicol. Pathol.*, 2012, **64**, 783.
115. L. C. N. Abraham, T. Masakuni, H. Isao and T. Hajime, *Food Chem.*, 2008, **109**, 415.
116. J. Yang, J. Guo and J. Yuan, *LWT-Food Sci. Technol.*, 2008, **41**, 1060.
117. T. Guardia, A. E. Rotelli, A. O. Juarez and L. E. Pelzer, *Farmaco*, 2001, **56**, 68.
118. H. Ashihara, H. Sano and A. Crozier, *Phytochemistry*, 2008, **69**, 841–856.
119. F. Wei, K. Furihata, M. Koda, F. Hu, R. Kato, T. Miyakawa, *et al.*, *J. Agric. Food Chem.*, 2012, **60**, 10118.
120. J. C. D. Chaves, M. Miyazawa, M. F. M. Bloch and J. K. Yamaki, *Acta Sci., Agron.*, 2004, **26**, 287.
121. K. I. Raju and N. H. Gopal, *J. Coffee Res.*, 1979, **9**, 83.
122. O. Guerreiro Filho and P. Mazzafera, *J. Agric. Food Chem.*, 2003, **51**, 6987–6991.
123. Y.-S. Kim and H. Sano, *Phytochemistry*, 2008, **69**, 882–888.
124. M.-L. Nurminen, *et al.*, *Eur. J. Clin. Nutr.*, 1999, **53**, 831–839.
125. J.-E. Lee, B.-J. Lee, J.-O. Chung, H.-N. Kim, E.-H. Kim, S. Jung, H. Lee, S.-J. Lee and Y.-S. Hong, *Food Chem.*, 2015, **174**, 452.
126. Y. Zuo, H. Chen and Y. Deng, *Talanta*, 2002, **57**, 307.
127. S. Casal, M. B. Oliveira and M. A. Ferreira, *Food Chem.*, 2000, **68**, 481.
128. P. M. Dewick, *Medicinal Natural Products: A Biosynthetic Approach*, Wiley, New York, 3rd edn, 2009.
129. Y. Niu, J. Liu, H. Liu, L. Gao, G. Feng, X. Liu and L. Li, *Pharm. Biol.*, 2016, **54**, 1680.
130. K. Sanugul, T. Akao, Y. Li, N. Kakiuchi, N. Nakamura and M. Hattori, *Biol. Pharm. Bull.*, 2005, **28**, 1672.

Section II

Incidental Contaminants

CHAPTER 35

Mycotoxins

REBECA CRUZ AND SUSANA CASAL*

LAQV@REQUIMTE, Serviço de Bromatologia e Hidrologia, Faculdade de Farmácia, Universidade do Porto, Rua de Jorge Viterbo ferreira, 228, 4050-313 Porto, Portugal
*E-mail: sucasal@ff.up.pt

35.1 Introduction

In a world where consumers' trust is imperative, industries and competent authorities have come together to ensure high levels of food safety. Therefore, every stage of the food chain, from the farm to the table, must be duly controlled in order to identify potential hazards and to apply effective preventive measures.

A "hazard" can be defined as any physical, chemical, or biological agent that, when present in food, can potentially bring harmful health consequences. As regards to chemical hazards, in the past few years, mycotoxin contamination of food products became a frequent concern. In fact, among all hazard categories defined by the "Rapid alert system for food and feed" (RASFF), mycotoxins were targeted as a main hazard in border rejection notifications during 2013.[1]

Among all susceptible crops, coffee has been a prevalent contaminated commodity. The first scientific study regarding the presence of these harmful fungal metabolites in coffee was published in the early 1970s, and aimed at the assessment of sterigmatocystin in green coffee beans.[2] From this point forward, several studies have been dedicated to the development of increasingly sensitive analytical methodologies, and further applied on green and

Coffee: Production, Quality and Chemistry
Edited by Adriana Farah

Published by the Royal Society of Chemistry, www.rsc.org

roasted coffee beans and coffee products, particularly concerning ochratoxin (OTA) content.[3–5]

Hence, the present chapter provides an update on the major mycotoxins in coffee products, analytical methods advantages and drawbacks, preceded by a summary of their individual amounts detected in these foodstuffs.

35.2 Major Mycotoxins in Coffee

35.2.1 General Features

Mycotoxins – from the Greek μύκης (*mykes*, *mukos*) "fungus" and τοξικόν (*toxikon*) "poison" – are small toxic chemicals (MW ~300) produced by filamentous fungi, belonging mostly to *Aspergillus*, *Penicillium* and *Fusarium* genera, as a consequence of their metabolism under favourable environmental conditions. Mycotoxicology was born in 1960, when more than 100 000 young turkeys died from the so-called "Turkey-X disease" after eating peanut meal contaminated with aflatoxins.[6] Since then, thousands of these microbiological metabolites have been discovered and classified in distinct chemical classes (Table 35.1). For instance, common *Aspergillus* mycotoxins – aflatoxins – and their biosynthetical precursor – sterigmatocystin, are difuranocoumarin derivatives, while fumonisins, often produced by *Fusarium* species, consist of sphinganine derivatives.[7] In addition to such chemical diversity, the same mycotoxin can be produced by different fungal species (such as OTA, which is produced by *Aspergillus ochraceus* and *Penicillium verrucosum*), and the same fungal species can produce more than one mycotoxin (such as deoxynivalenol and zearalenone, produced by *Fusarium graminearum*).[8] This variability hardens the identification of the fungi responsible for a mycotoxin contamination, and the analysis of mycotoxins as well.

Toxigenic fungi may infect agricultural crops at both pre- and post-harvest stages. Fungal contamination by *Fusarium*, as an example, generally occurs before or immediately after harvesting, while *Penicillium* and *Aspergillus* are more likely to proliferate during drying and subsequent storage steps.[7] By infecting agricultural commodities, they may enter the food and feed chains, causing a myriad of acute and/or chronic detrimental health effects, such as immunosuppressive, carcinogenic, estrogenic, gastrointestinal and kidney events.[8] In addition to these health concerns, mycotoxins have major economic impacts in numerous crops, particularly in maize, nuts, dry fruits and coffee, due to crop yield and value reduction, human health costs, together with costs regarding preventive measures.[1] Therefore, several countries have adopted regulations to limit mycotoxin exposure through food, and narrow their levels to as low as possible.[9,10]

Coffee is mostly traded as green coffee beans, obtained from coffee cherries after a complex processing chain, involving different ways of eliminating the outer shells (as fresh pulp or dry husks), while drying and releasing the green coffee beans. Although the final product's quality is known to be highly influenced by coffee species and varieties, edaphoclimatic features, as well as

Table 35.1 Major mycotoxins identified in coffee products.

Common name	Abbreviation	CAS number	Structural formula	Molecular formula	Molecular weight	IARC class
Ochratoxin A	OTA	303-47-9	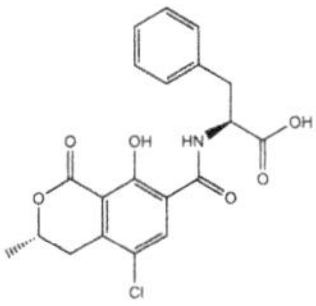	$C_{20}H_{18}ClNO_6$	403.81	2B
Aflatoxin B1	AFB1	1162-65-8	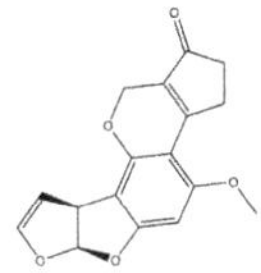	$C_{17}H_{12}O_6$	312.27	1
Aflatoxin B2	AFB2	7220-81-7	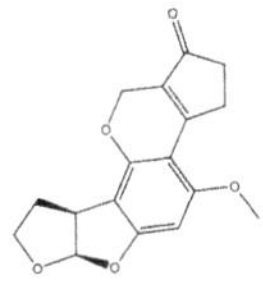	$C_{17}H_{14}O_6$	314.29	1
Aflatoxin G1	AFG1	1165-39-5	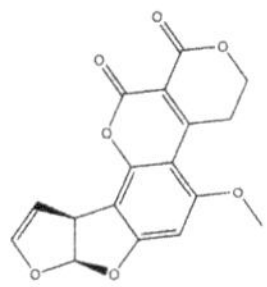	$C_{17}H_{12}O_7$	328.27	1
Aflatoxin G2	AFG2	7241-98-7	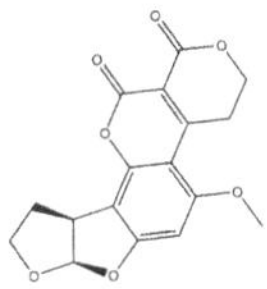	$C_{17}H_{14}O_7$	330.29	1
Sterigmatocystin	STC	10 048-13-2	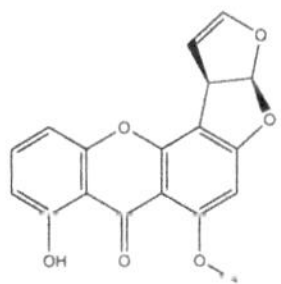	$C_{18}H_{12}O_6$	324.28	2B

the processing method, the sensorial burst of this food product occurs only after roasting, where the last inner shell remains might be released (coffee silverskin, commonly representing 1% of total initial weight). In addition, the brew preparation method (instant coffee, espresso, moka, Turkish, *etc.*) will offer very distinct experiences to consumers, as well as different chemical compositions to both beverage and spent coffee grounds (which often comprise 45% of total initial weight).[11]

Despite usually being subjected to meticulous quality control processes, coffee is susceptible to several plagues and diseases during both pre- and post-harvest steps. For instance, mycotoxin contamination has been a major issue, causing economic losses in producing countries as well as having an impact on consumers' well-being.

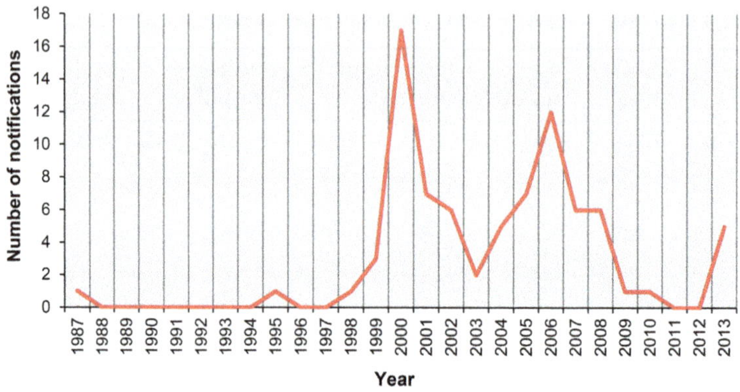

Figure 35.1 Notifications on mycotoxins presence in coffee products from European markets (1997–2013). Data from ref. 1.

Since 1987, coffee products have received 81 hazard notifications in several European countries (Figure 35.1), reduced to only five notifications in 2013. The majority of these notifications concerned green coffee beans contamination with OTA, while other mycotoxins seem to be less prevalent.[1] However, the apparent increase in notification events in the last two decades cannot be regarded as an increased contamination, but rather associated with an enlarged awareness for mycotoxin presence, a consequence of both the implementation of more restrictive legislation and development of increasingly sensitive analytical methods, as will be further discussed.

35.2.2 Ochratoxin A

Ochratoxins are pentaketide metabolites of *Aspergillus* and *Penicillium* species, consisting of a dihydroisocoumarin core linked to a β-phenylalanine residue (Table 35.1). Due to its toxicity and prevalence in food, OTA is the most important congener, though other related structures are known to exist (as ochratoxins B, C, α and β). OTA (2′S-ochratoxin A, (2S)-2-[[(3R)-5-chloro-8-hydroxy-3-methyl-1-oxo-3,4-dihydroisochromene-7-carbonyl]amino]-3-phenylpropanoic acid) was named after *A. ochraceus*, from which it was first isolated in 1965.[12] Its health consequences are mainly related to nephrotoxic events, but it was also proven to be hepatotoxic, teratogenic, immunotoxic and carcinogenic.[7] Furthermore, the International Agency for Research on Cancer (IARC) classifies OTA as a class 2A hazard, *i.e.* as a potential carcinogenic substance to humans.[13] As a consequence, the European Union (EU) set limits for OTA in roasted coffee (5.0 $\mu g\ kg^{-1}$) and instant coffee (10.0 $\mu g\ kg^{-1}$).[9]

Coffee husks are high in OTA, but these layers are removed during processing. However, soil-contaminated beans were also reported to have higher OTA content and therefore should be avoided, this being a general recommendation in good manufacturing practices.[14] No limits are defined in the

EU for green coffee, because the roasting step is usually regarded as a safety barrier in the production chain.[13] However, while mycotoxins are generally stable under usual cooking temperatures, the intensity of coffee roasting, surpassing 200 °C, can destroy OTA in a great extent (8–98%), with a dependence on the roasting conditions and initial contamination levels.[15] Therefore, only some European countries have settled statutory legal limits for green coffee (Finland and Switzerland, 5 μg kg^{-1}; Italy, Portugal and Spain, 8 μg kg^{-1}; Czech Republic, 10 μg kg^{-1}; Hungary, 15 μg kg^{-1}; Greece, 20 μg kg^{-1}).[16] Nevertheless, recent evidences on the formation of nephrotoxic OTA degradation products (2′R-OTA and 2′-decarboxy-OTA), during coffee roasting,[47] may question the safety of roasted coffee, since these derivatives are not regulated. Moreover, new information on UV photo-degradation of OTA will definitely open new paths of research in terms of identifying the resulting derivatives and their presumed toxicity.[48]

In the last decade, several scientific studies have confirmed the presence of OTA in green, roasted and instant coffee, as well as in coffee brews, wastes and food supplements (Table 35.2). The majority of the samples were originated or acquired from Latin American countries, though the highest degree of contamination was observed in insect damaged green coffee beans from Kenya (499 μg kg^{-1}).[24] The ratio of positive samples among analysed samples is usually high (around 65%), both in green and in roasted coffee, but most samples are within the legal range. No differences between caffeinated and non-caffeinated coffees are perceived.[18] Regarding instant coffee, the prevalence of positive samples is higher (>90%), but most samples are also below the limit defined. However, many surveys found some coffee products readily available to consumers that surpassed the aforementioned regulated limits.[5,30–32,37,40]

Presently, due to the increasing consumption of coffee-based food supplements, a new wave of concern has emerged. In particular, green coffee extracts are advertised as promising to be natural promoters of weight loss, due to their caffeine and chlorogenic acids content.[49] Vaclavik and co-workers[44] quantified up to 136 μg kg^{-1} of OTA in dietary supplements containing green coffee (Table 35.2). Since there is no implemented legislation regarding OTA content in these products, the lack of surveys and the rising of their consumption invoke the development of scientific studies in order to provide more data on this specific matter.

Also, in times when words like "sustainability", "waste management" and "reuse" are becoming integrated in our quotidian, some foods are potentially cost-effective sources of valuable compounds that can be returned to the food chain (or other industries), adding value to regular food products. Therefore, the main by-products of the coffee agro-industry are not an exception (Figure 35.1), being mostly used to recover bioactive phenolic constituents.[50] However, its sustainable reuse may not be contaminant-free and thus, when it returns to the food chain, safety control evaluations should be performed. Unfortunately, in the past ten years, only two studies have been conducted in coffee by-products aiming at the assessment of OTA.[45,46] In particular, coffee

Table 35.2 Summary of reported studies on Ochratoxin A (OTA) prevalence in coffee products and by-products from 2005 to 2014.[a]

Product	Sampling period	Origin/retail country	n	% Positive	Content (μg kg^{-1})	Analytical methodology	Reference
Green coffee	n.s.	Vietnam, India, Uganda, Costa Rica, Colombia, Brazil	36	69	n.d.–8.05	HPLC-FLD	17
	2003	Brazil	136	22	0.84–4.70	HPLC-FLD	18
	2003	Brazil	136	15	n.d.–7.30	Indirect ELISA	18
	n.s.	Brazil	25	100	0.73–37.22	HPLC-FLD	19
	2006–2007	India	80	74	0.2–13.5	HPLC-FLD	20
	2005	Japan	11	27	1.4–7.6	HPLC	21
	2006–2007	Thailand	64	94	0.6–27	Direct ELISA	22
	2007	India, Vietnam, Cameroon, Ethiopia, Brazil, Salvador	14	100	0.567–2.553	HPLC-FLD	23
	n.s.	Brazil	289	44	n.d.–100	HPLC-FLD	24
	n.s.	Kenya	229	90	n.d.–499.3	HPLC-FLD	25
	2008–2009	Jordan	11	63	2.19–6.57	Direct ELISA	26
	2004–2009	Japan	21	29	n.s.–0.76	HPLC-FLD	27
	n.s.	Panama	21	19	4.90–37.73	ELISA	28
	n.s.	Brazil	10	0–100	n.d.–25.7	HPLC-FLD	29
	n.s.	Denmark, Kenya, Tanzania, Uganda	18	40	n.s.–2.8	UPLC-MS/MS	5
Roasted coffee	n.s.	Vietnam, India, Uganda, Costa Rica, Colombia, Brazil	9	100	0.19–1.20	HPLC-FLD	17
	n.s.	France	24	46	n.d.–20.8	HPLC-FLD	30
	2005	Japan	9	33	1.1–3.3	HPLC	21
	2007	India, Vietnam, Cameroon, Ethiopia, Brazil, Salvador	14	n.s.	n.d.–0.930	HPLC-FLD	23
	n.s.	France	30	83	n.d.–15.08	HPLC-FLD	31
	2008	Spain	45	96	1.30–5.24	HPLC-FLD	32
	n.s.	Malaysia	8	38	n.d.–4.43	HPLC-DAD	33
	n.s.	Italy	30	27	1.03 ± 0.17	HPLC-FLD	34
	2004–2009	Japan	84	55	n.s.–2.75	HPLC-FLD	27
	n.s.	Czech Republic	48	n.s.	n.d.–1.37	HPLC-FLD	35
	n.s.	Chile	24	100	0.47 ± 0.20	HPLC-FLD	36
	2013–2014	Spain	103	89	1.56–32.40	LC-MS/MS	37
	n.s.	Denmark, Kenya, Tanzania, Uganda	57	46	n.s.–8.3	UPLC-MS/MS	5

Instant coffee	2004	Brazil	82	99	0.17–6.29	HPLC-FLD	38
	2007	Italy	9	100	0.3–2.2	HPLC-FLD	23
	n.s.	Malaysia	5	100	1.58–9.82	HPLC-DAD	33
	2011	Italy	44	95	0.32–6.40	HPLC-FLD	39
	2004–2009	Japan	126	98	n.s.–4.23	HPLC-FLD	27
	2012	Portugal	23	100	0.15–11.8	HPLC-FLD	40
	n.s.	Chile	39	100	1.80 ± 1.81	HPLC-FLD	36
	n.s.	Denmark, Kenya, Tanzania, Uganda	25	56	n.s.–21	UPLC-MS/MS	5
	2011–2012	Czech Republic	12	92	0.1–4.91	HPLC-FLD	41
Coffee brews							
Espresso/ Moka/ Auto-drip	n.s.	Vietnam, India, Uganda, Costa Rica, Colombia, Brazil	27	7	n.d.–0.112[b]	HPLC-FLD	16
Canned/ Bottled	2007	Japan	30	57	n.d.–0.037[b]	LC-MS/MS	42
Filtration/ Piston	n.s.	France	30	93	n.d.–0.882	HPLC-FLD	31
Canned	2004–2009	Japan	76	58	n.s.–0.04	HPLC-FLD	26
	n.s.	Lebanon	30	n.s.	0.508	HPLC-FLD	43
Coffee supplements							
Green coffee extract	n.s.	n.s.	50	36	2.7–136	UPLC-MS/MS	44
Coffee by-products							
Silverskin	n.s.	Italy	n.s.	n.s.	<0.04	HPLC-FLD	45
Silverskin	n.s.	Italy	30	n.s.	17.8–36.1	Direct ELISA	46

[a]DAD, diode array detector; ELISA, enzyme-linked immunosorbent assay; FLD, fluorescence detector; HPLC, high performance liquid chromatography; MS, mass spectrometry; n.d., not detected; n.s., not specified; OTA, ochratoxin A; UHPLC, ultra pressure liquid chromatography.
[b]Results are expressed in micrograms per liter of sample.

silverskin is being advertised as a "new functional ingredient", but the OTA amounts found by Toschi *et al.* (2014)[46] are disturbing (Table 35.2), justifying further studies.

Regardless of OTA being more prevalent in coffee products, and the only one with legislated limits, other mycotoxins such as aflatoxins and fumonisins, and sterigmatocystin, have also been reported in many scientific studies, as will be subsequently detailed.

35.2.3 Aflatoxins

Similarly to OTA, aflatoxins are difuranocoumarin mycotoxins (Table 35.1), and *Aspergillus* sp. metabolites, in this case namely of *A. flavus* and *A. parasiticus*.[6] Aflatoxins, including aflatoxin B_1, B_2, G_1, G_2 and M_1, are known as genotoxic and hepatocarcinogenic, and may be responsible for acute and chronic toxicity in humans, being confirmed as a group 1 agent in IARC classification.[51] Maximum total aflatoxin content in food (sum of B_1, B_2, G_1 and G_2), as well as aflatoxin B_1 content alone, due to its major toxic features, are well defined in European legislation for many food products, though coffee is not included.[6] On the other hand, the US Food and Drug Administration (FDA) recommended a legal action for aflatoxin content in foods above 20 μg kg^{-1}.[52]

Temperature and moisture are primary factors for mould growth control. *Aspergillus flavus* is favoured by temperatures in the 29–35 °C range, which is slightly higher than ochratoxin A, and above 18.5% of relative humidity, being able to tolerate easily a wide range of pH, though 5.5 and 7.5 are more suitable for its development.[53] Similarly to other mycotoxins, aflatoxins are heat-stable contaminants, being able to moderately resist temperature processing, thus reaching the human food chain.

However, unlike OTA, only a single RASFF report of aflatoxin contamination in coffee has been reported since 1987, and it was regarding green coffee beans from the Democratic Republic of the Congo.[1] Likewise, there is a lack of scientific studies confirming the presence and amounts of this mycotoxin in coffee products. Nevertheless, Soliman[54] reported, in 2002, low total concentrations from 0.76 to 8.92 μg kg^{-1} for green coffee and roasted coffee. More recent data from Franco and co-workers[28] confirmed their presence in 3 out of 21 samples of green coffee from Panama, with a total content varying from 1.51 to 1.93 μg kg^{-1}. Also, García-Moraleja and colleagues[37] verified that aflatoxins have high occurrence (>50%) but low degrees of contamination (average of 1.95 μg kg^{-1}; range of 0.25 to 13.12 μg kg^{-1}) in roasted coffees, aflatoxins B_2 and G_2 being the ones present in higher amounts, with no significant differences among different coffee types (caffeinated *vs.* non-decaffeinated), or roasting procedures (natural *vs.* torrefacto).[37] Despite representing apparently a lower concern then OTA, their increased toxicity should not be neglected. Also, according to Paterson and co-workers,[55] climate changes might induce changes in mycotoxins contamination patterns, and aflatoxins might become more prevalent than

OTA in the future, with serious implications on coffee production and consumers' health. The same might be applied to fumonisins, presently of no concern in coffee products.

35.2.4 Sterigmatocystin

Sterigmatocystin is also an *Aspergillus* sp. secondary metabolite, produced at 23–29 °C mainly by *A. versicolor*, being regarded as a biogenic precursor of aflatoxin B_1 (Table 35.1) and thus recognized as a group 2B agent according to IARC.[51,56] Again, the presence of sterigmatocystin is not regulated so far in the European Union.

Food contamination by this mycotoxin is frequently observed in grains and grain-based products, mostly at post-harvest stages. Nevertheless, its prevalence in coffee products is low. It was first detected in green coffee beans in high amounts (1143 $\mu g\ kg^{-1}$) back in 1973.[2] Just recently, García-Moraleja and colleagues[37] found two positive roasted samples with 23.8 and 36.5 $\mu g\ kg^{-1}$, but there is a lack of studies concerning its quantification or detection in coffee products.

35.3 Analysis of Mycotoxins in Coffee Products

In the past, many studies associated with mycotoxins in food focused on the ascertainment of both presence and identification of the contamination source. However, the positive recognition of the presence of mycotoxigenic fungus is not directly related to mycotoxins contamination, as well as mycotoxin-contaminated samples may not have active mould growth. Furthermore, the toxic effects of mycotoxins can easily be observed at low concentrations and, thus, their assessment methodologies demand to be very sensitive and reproducible. In order to provide faster, cost-effective and more sustainable analysis, multi-residue techniques are usually chosen for mycotoxins evaluation, trying to bypass their distinct physicochemical features.

While for other foodstuffs diverse techniques have been used for mycotoxin analysis, such as electrochemical, colorimetric or even bioassays, two major techniques are usually described for coffee: immunochemical (mostly restricted to Enzyme-Linked Immunosorbent Assay, ELISA) and chromatographic methods, as Table 35.2 corroborates and as will be further detailed.

Regardless of the accuracy of the analytical method, sampling represents a crucial step for the assessment of food contaminants. Therefore, Regulation (EC) no. 401/2006[10] lays down the methods of sampling and analysis for the official control of mycotoxins in foodstuffs, including roasted coffee beans, ground roasted coffee and soluble coffee.[10] In summary, depending on the coffee lot weight, the product is divided in sublots from where up to 100 samples (≈100 g each) are taken and further gathered in representative samples of up to 10 kg, this latter being subjected to the analysis. Since mycotoxins formation is not homogeneous in the whole food product, if an adequate sampling is not assured, misleading results are easily reported.

Prior to extraction, coffee beans are usually dried, ground and sieved to achieve a better homogenization. According to several references mentioned in Table 35.2, OTA extraction is commonly performed with sodium bicarbonate solutions (1% or 3%, w/v), or with mixtures of sodium bicarbonate and methanol, in different proportions (mainly 70 : 30), with a possible addition of polyethylene glycol 8000 to reduce foam formation.

35.3.1 Immunoassays

Competitive ELISA methods have been applied to the analysis of OTA in green coffee and coffee silverskin.[18,22,28,46] This immunochemical approach has proven to be highly selective, avoiding complex preliminary clean-up steps, thus resulting in simple and time effective analysis. However, despite allowing the simultaneous analysis of several samples, ELISA detection is often less sensitive, and the outcomes less accurate, than the ones from chromatographic analysis. Indeed, Fujii and co-workers[18] compared an indirect ELISA method with High Performance Liquid Chromatography (HPLC) analysis for OTA quantification in green coffee samples, corroborating the previous statements. Hence, immunochemical methods should be mostly regarded as qualitative rather than quantitative procedures.

Most studies in coffee matrices use direct competitive ELISA, thus avoiding the possible cross reactivity of secondary antibody leading to a less specific signal, more frequent at indirect ELISA.

Currently, the use of immunoaffinity columns for selective clean-up and concentration purposes before HPLC determinations is the most common technique, described in most of the works detailed in Table 35.2.

35.3.2 Chromatographic Analysis

Chromatographic methods for OTA assessment in coffee matrices are also summarized in Table 35.2. Following extraction, samples are often subjected to clean-up procedures, such as liquid–liquid extraction with chloroform, solid-phase extraction (using mixed-mode, reversed-phase/strong anion-exchange columns), the previously mentioned immunoaffinity columns and, more recently, molecularly imprinted polymers (MIP). These last two methods present increased selectivity, being usually applied in solid-phase extraction sorbents and used when mass spectrometry (MS) detection is not available, otherwise the Quick, Easy, Cheap, Effective, Rugged and Safe (QuEChERS) extraction methods are a more economical approach.

After being purified and concentrated, chromatographic separation is commonly performed using HPLC on C_{18} reverse-phase columns and methanol (or water) mixtures with acetonitrile and acetic acid, in isocratic mode. Subsequently, detection is frequently conducted by fluorescence (FLD, $\lambda_{exc} \approx$

333 nm and $\lambda_{em} \approx 460$ nm), and further confirmed by mass detection. OTA analysis by HPLC-FLD results in average limits of detection from 0.03 to 3.0 μg kg^{-1}, mostly depending on the clean-up strategy.[57]

Modern LC-MS/MS equipment provides highly selective and sensitive determinations, resulting in precise and multicomponent outcomes within a short period of time.[5,42,44] For mycotoxins, MS with electrospray ionization is usually operated in positive mode in the dynamic multiple reaction monitoring, using two transitions (quantification and confirmation). For OTA analysis, the precursor ion is *m*/*z* 404, while product ions are, generally, *m*/*z* 239 and *m*/*z* 358, leading to suitable detection limits (0.3–1.1 μg kg^{-1}).

Gas chromatography methods in the assessment of coffee mycotoxins are being discontinued, due to the need for prior derivatization, and higher detection limits (5–50 μg kg^{-1}), while TLC methods are used mostly for qualitative or semi-quantitative purposes when advanced chromatographic equipment is unavailable.

35.4 Conclusions and Future Perspectives

Regardless of being subjected to a high-temperature processing (roasting), coffee contamination by mycotoxins is greatly prevalent, although increasingly at lower levels due to the successful quality measures taken during the coffee processing chain. However, the expected climatic changes might induce important alterations in mycotoxins qualitative and quantitative patterns, requiring increased attention. The analytical methods available, particularly the most recent multiparametric ones and the increased sensitivity achieved, represent important advances in mycotoxin analysis, but also show that mycotoxin contamination is more frequent than expected. Furthermore, in times when food residues are increasingly being tentatively reintegrated in the food chain, particular attention should be devoted to the safety of using coffee by-products with this purpose, because the outer shells eliminated represent the core of mycotoxin contamination.

Acknowledgements

This work received financial support from the European Union (FEDER funds through COMPETE) and National Funds (FCT, Fundação para a Ciência e Tecnologia) through project UID/QUI/50006/2013.

References

1. *The Rapid Alert System for Food and Feed – 2013 Annual Report*, Publications Office of the European Union, Luxembourg, 2014.
2. I. F. H. Purchase and M. E. Pretorius, *J. AOAC Int.*, 1973, **56**(1), 225.
3. M. H. Taniwaki, J. I. Pitt, A. A. Teixeira and B. T. Iamanaka, *Int. J. Food Microbiol.*, 2003, **82**(2), 173.

4. B.-H. Liu, Z.-J. Tsao, J.-J. Wang and F.-Y. Yu, *Anal. Chem.*, 2008, **80**(18), 7029.
5. K. F. Nielsen, A. F. Ngemela, L. B. Jensen, L. S. de Medeiros and P. H. Rasmussen, *J. Agric. Food Chem.*, 2015, **63**(3), 1029.
6. W. P. Blount, *J. Br. Turk. Fed.*, 1961, **9**, 52.
7. M. J. Sweeney and A. D. W. Dobson, *Int. J. Food Microbiol.*, 1998, **43**, 141.
8. S. Marin, A. J. Ramos, G. Cano-Sancho and V. Sanchis, *Food Chem. Toxicol.*, 2013, **60**, 218.
9. Regulation (EC) 1881/2006, *Off. J. Eur. Communities: Legis.*, 2006, **L364**, 5.
10. Regulation (EC) 401/2006, *Off. J. Eur. Communities: Legis.*, 2006, **L70**, 12.
11. P. Esquivel and V. M. Jiménez, *Food Res. Int.*, 2012, **46**, 488–495.
12. K. J. Van Der Merwe, P. S. Steyn, L. Fourie, De B. Scott and J. J. Theron, *Nature*, 1965, **205**, 1112.
13. Commission Regulation (EU) No 105/2010, *Off. J. Eur. Communities: Legis.*, 2010, **L35**, 7.
14. R. R. M. Paterson, N. Lima and M. H. Taniwaki, *Food Res. Int.*, 2014, **61**, 1.
15. M. B. M. Ferraz, A. Farah, B. T. Iamanaka, D. Perrone, M. V. Copetti, V. X. Marques, A. A. Vitali and M. H. Taniwaki, *Food Control*, 2010, **21**, 872.
16. *OTA risk management: guidelines for green coffee buying*, International Coffee Organization, 2005, http://www.ico.org/documents/ed1939e.pdf.
17. A. Pérez De Obanos, E. González-Peñas and A. López De Cerain, *Food Addit. Contam.*, 2005, **22**(5), 463.
18. S. Fujii, R. M. R. Ribeiro, M. B. S. Scholz, E. Y. S. Ono, C. E. C. Prete, E. N. Itano, Y. Ueno, O. Kawamura and E. Y. Hirooka, *Food Addit. Contam.*, 2006, **23**(9), 902.
19. E. A. Vargas, T. B. Whitaker, E. A. Dos Santos, A. B. Slate, F. B. Lima and R. C. A. Franca, *Food Addit. Contam.*, 2006, **23**(1), 62.
20. T. N. Gopinandhan, G. S. Kannan, P. Panneerselvam, K. Velmourougane, Y. Raghuramulu and J. Jayarama, *Food Addit. Contam., Part B*, 2008, **1**(1), 51.
21. S. Kumagai, M. Nakajima, S. Tabata, E. Ishikuro, T. Tanaka, H. Norizuki, Y. Itoh, K. Aoyama, K. Fujita, S. Kai, T. Sato, S. Saito, N. Yoshiike and Y. Sugita-Konishi, *Food Addit. Contam., Part A*, 2008, **25**(9), 1101.
22. P. Noonim, W. Mahakarnchanakul, K. F. Nielsen, J. C. Frisvad and R. A. Samson, *Int. J. Food Microbiol.*, 2008, **128**, 197.
23. L. La Pera, G. Avellone, V. Lo Turco, G. Di Bella, P. Agozzino and G. Dugo, *Food Addit. Contam.*, 2008, **25**(10), 1257.
24. L. R. Batista, S. M. Chalfoun, C. F. Silva, M. Cirillo, E. A. Varga and R. F. Schwan, *Food Control*, 2009, **20**, 784.
25. D. Duris, J. K. Mburu, N. Durand, R. Clarke, J. M. Frank and B. Guyot, *Food Addit. Contam.*, 2010, **27**(6), 836.
26. N. M. Salem and R. Ahmad, *Food Control*, 2010, **21**(8), 1099.
27. Y. Sugita-Konishi, Y. Kamata, T. Sato, T. Yoshinari and S. Saito, *Food Addit. Contam., Part A*, 2013, **30**(8), 1392.
28. H. Franco, A. Vega, S. Reyes, J. De Léon and A. Bonilla, *Arch. Latinoam. Nutr.*, 2014, **64**(1), 42.

29. M. H. Taniwaki, A. A. Teixeira, A. R. R. Teixeira, M. V. Copetti and B. T. Iamanaka, *Food Res. Int.*, 2014, **61**, 161.
30. P. Mounjouenpou, N. Durand, B. Guyot and J. P. Guiraud, *Food Addit. Contam.*, 2007, **24**(7), 730.
31. M. Tozlovanu and A. Pfohl-Leszkowicz, *Toxins*, 2010, **2**, 1928.
32. M. B. Coronel, S. Marin, G. Cano, A. J. Ramos and V. Sanchis, *Food Control*, 2011, **22**, 414.
33. T. P. Lee, B. Saad, W. S. Khayoon and B. Salleh, *Talanta*, 2012, **88**, 129.
34. A. Prelle, D. Spadaro, A. Denca, A. Garibaldi and M. L. Gullino, *Toxins*, 2013, **5**, 1827.
35. F. Malir, V. Ostry, A. Pfohl-Leszkowicz, J. Toman, I. Bazin and T. Roubal, *Toxins*, 2014, **6**, 3438.
36. O. Galarce-Bustos, M. Alvarado, M. Vega and M. Aranda, *Food Control*, 2014, **46**, 102.
37. A. García-Moraleja, G. Font, J. Mañes and E. Ferrer, *Food Control*, 2015, **57**, 282.
38. A. P. De Almeida, J. Alaburda, L. Shundo, V. Ruvieri, S. A. Navas, L. C. A. Lamardo and M. Sabino, *Braz. J. Microbiol.*, 2007, **38**(2), 300.
39. A. Vecchio, V. Mineo and D. Planeta, *Food Control*, 2012, **28**, 220.
40. S. Casal, T. Vieira, R. Cruz and S. Cunha, *Food Res. Int.*, 2014, **61**, 56.
41. J. Skarkova, V. Ostry, F. Malir and T. Roubal, *Anal. Lett.*, 2015, **46**, 1495.
42. S. Noba, A. Uyama and N. Mochizuki, *J. Agric. Food Chem.*, 2009, **57**(14), 6036.
43. F. Raad, L. Nasreddine, C. Hilan, M. Bartosik and D. Parent-Massin, *Food Chem. Toxicol.*, 2014, **73**, 35.
44. L. Vaclavik, M. Vaclavikova, T. H. Begley, A. J. Krynitsky and J. I. Rader, *J. Agric. Food Chem.*, 2013, **61**, 4822.
45. R. C. Borrelli, F. Esposito, A. Napolitano, A. Ritieni and V. Fogliano, *J. Agric. Food Chem.*, 2004, **52**, 1338.
46. T. G. Toschi, V. Cardenia, G. Bonaga, M. Mandrioli and M. T. Rodriguez-Estrada, *J. Agric. Food Chem.*, 2014, **62**, 10836.
47. B. Cramer, M. Königs and H.-U. Humpf, *J. Agric. Food Chem.*, 2008, **56**(14), 5673–5681.
48. R. Ibarz, A. Garvín, S. Rojas, E. Azuara and A. Ibarz, *LWT--Food Sci. Technol.*, 2016, **73**, 147.
49. K. Tanaka, S. Nishizono, S. Tamaru, M. Kondo, H. Shimoda, J. Tanaka and T. Okada, *Food Sci. Technol. Res.*, 2009, **15**, 147.
50. P. Esquivel and V. M. Jiménez, *Food Res. Int.*, 2012, **46**(2), 488.
51. IARC monographs on the evaluation of carcinogenic risks to humans, 100F, 2012, 229, Aflatoxins.
52. FDA. CPG, Sec. 555.400: Foods–adulteration with aflatoxins, 1995, http://www.fda.gov/, accessed May 19, 2015.
53. G. U. Holmquist, H. W. Walker and H. M. Stahr, *J. Food Sci.*, 1983, **48**(3), 778.
54. K. Soliman, *J. Agric. Food Chem.*, 2002, **50**, 7477.

55. R. R. M. Paterson, N. Lima and M. H. Taniwaki, *Food Res. Int.*, 2014, **61**, 1.
56. V. Betina, *Mycotoxins: Chemical, Biological and Environmental Aspects (Bioactive Molecules)*, Elsevier Science, 1989, vol. 9, p. 437.
57. A. Prelle, D. Spadaro, A. Denca, A. Garibaldi and M. L. Gullino, *Toxins*, 2013, **5**, 1827.

CHAPTER 36

Pesticide Residues

SARA C. CUNHA* AND JOSÉ O. FERNANDES

LAQV-REQUIMTE – Laboratório de Bromatologia e Hidrologia, Faculdade de Farmácia, Universidade do Porto, Rua de Jorge de Viterbo Ferreira 228, 4050-313 Porto, Portugal
*E-mail: sara.cunha@ff.up.pt

36.1 Introduction

According to the International Coffee Organization, there has been steady growth in world production of coffee over the last 50 years (1963–2013). The average growth rate since 1963 was 2.4%, with 2.8% annual growth in the market-controlled period, and 2% since 1990. In crop year 2012/13, world production reached 145.1 million bags, the largest ever recorded.[1] This extensive coffee production was developed at the expense of new agricultural lands (contributing to the deforestation) and due to intensification of agrochemicals use.

Pesticides have been used in coffee culture since the 1920s, when for the first time "Bordeaux mixture" was successfully used in India for control of "coffee rust" disease. However, only in the last years has the development of effective insecticides and organic fungicides enabled a wide range of products to be used against coffee pests and diseases.

Over the 1990s, the global pesticide sale remained relatively constant, between 270 and 300 billion dollars, of which 44% were herbicides, 29% were insecticides, 21% were fungicides/bactericides and 6% others. Over the period 2007 to 2008, herbicides ranked first in three major categories

Coffee: Production, Quality and Chemistry
Edited by Adriana Farah

Published by the Royal Society of Chemistry, www.rsc.org

of pesticides (insecticides, fungicides/bactericides, herbicides).[2] Fungicides/bactericides increased rapidly and ranked second. Europe is now the largest pesticide consumer in the world, followed by Asia. As for countries, China, United States, France, Brazil and Japan are the largest pesticide producers, consumers or traders in the world.[2] In coffee, crop insecticides are the main class of pesticide used followed by fungicides and herbicides.

The aim of this chapter is to present the pesticides used in coffee, their chemical characteristics and classification. Additionally, we will present a review on the analysis of pesticide residues in coffee beans and coffee products, with an emphasis on the extraction and clean-up procedures employed before the chromatographic determination. Also, we discuss pesticides used for coffee crops and their residue levels found in coffee products and relate it to consumer exposure.

36.2 Pesticide Definition, Classification and Pesticide Use

The Food and Agriculture Organization of the United Nations defines pesticides as any substance or mixture of substances used to prevent, destroy or control any pest, including vectors of human or animal disease, unwanted species of plants or animals causing harm during or otherwise interfering with the production, processing, storage, transport or marketing of food, agricultural commodities, wood and wood products or animal feedstuffs, or substances which may be administered to animals for the control of insects, arachnids or other pests in or on their bodies. The term includes substances intended for use as a plant growth regulator, defoliant, desiccant or agent for thinning fruit or preventing the premature fall of fruit, and substances applied to crops either before or after harvest to protect the commodity from deterioration during storage and transport.[3]

The classification of pesticides in its many aspects – chemistry, biochemistry, toxicology, *etc.* – is an extremely complex task, given the diversity of chemicals and subjects to consider. Pesticides can be classified according to the organism attacked (insecticide – insect; fungicide – fungi; herbicide – plants, acaricide – mites, nematicides – nematodes, avicides – birds, *etc.*), according to the chemical structure (organophosphates, carbamates, organochlorines, ureas, pyrithroids, azoles, phenols, *etc.*) or according to the mode of action (anticholinesterase, chitin synthesis inhibitor, anticoagulant, glutamine synthesis inhibitor, protein synthesis inhibitor, mitochondrial respiration inhibitor, *etc.*).[4]

Regarding coffee, the Common Code for the Coffee Community (4C), a joint initiative of coffee producers, trade and industry, and social or environmental unions, has classified the pesticide used in different coffee-producing countries according to their toxicological characteristics in red, yellow or green.[5]

- Red: highly toxic (WHO classes Ia and Ib) and not compatible with sustainable coffee production.
- Yellow: moderate hazardous (WHO Class II) and should be used only as part of a monitoring system.
- Green: might be used as part of an integrated pest management strategy.

Overall, insecticides, fungicides and herbicides are the most common pesticides used in coffee crop and coffee storage. Insects are the most serious pest of coffee; over 900 species are known to infest the crop.[6] Several diseases caused by pathogenic microfungi and occasionally by bacteria can also cause serious damage in different plant organs.[6] The literature on all these pests is extensive and in a chapter of this length it is impossible to give a detailed account of all the pesticides used to treat them. Instead, the major pesticide classes used in coffee crop are surveyed, with appropriate examples and references. The main chemical characteristics of 21 pesticides listed for coffee by the *Codex Alimentarius* are described in Table 36.1.

36.2.1 Insecticides

Insecticides are used to control or kill insects. The major chemical classes of insecticides are: organochlorine compounds, organophosphate, carbamates, pyrethroids and benzoylureas.[7] Organochloride pesticides are organic compounds with five or more chlorine atoms. Organophosphorous insecticides contain a phosphate group and their basic structure is described by Schrader's formula shown in Figure 36.1.

The structural diversity of organophosphorus is due to different substituents at the phosphorus atom. Substituents at R^1 and R^2 are alkyl, alkoxy, alkylthio or amino groups, and the substituent at X is a labile acyl residue (halide, cyano, phenol or thio group), known as the leaving group.

Carbamates are organic pesticides derived from carbamic acid with the general formula shown in Figure 36.2.

Pyrethroids insecticides on the other hand are synthetic analogues of the naturally occurring pyrethrins; a product of flowers from the pyrethrum plant (*Chrysanthemum cinerariaefolium*).

Benzoylureas are chemical derivatives of *N*-benzoyl-*N'*-phenylurea (Figure 36.3).

Organochloride insecticides such as endosulfan are commonly used to control coffee berry borer, *Hypothenemus hampei* Ferrari. Coffee berry borer is the most serious insect pest of coffee worldwide that can cause direct damage in the coffee bean.

Organophosphate and pyrethroid insecticides can be used for suppression of coffee leaf miner, *Leucoptera Coffeela*, which causes serious damage in leaves (Barrera, 2008).[8] In some areas of tropical America, the coffee leaf miner is considered to be the principal insect pest of coffee.[8] Many other insect pests that attack coffee including stem borers, aphids, green scales (*e.g. Coccus viridis*), mealybugs (*e.g. Planococcus citri*), thrips, caterpillars and nematodes (*e.g. Meloidogyne incognita, Pratylencus coffeae*) are treated with insecticides.

Table 36.1 Pesticides in coffee listed by *Codex Alimentarius*, classification, chemical proprieties, stability and use (https://www.ncbi.nlm.nih.gov accessed February 2016; Tomlin (2003)).

Pesticide (IUPAC name)	Chemistry	Classification	MM, log K_{ow}, solubility	Stability	Use
Aldicarb ([(*E*)-(2-methyl-2-methylsulfanylpropylidene)amino] *N*-methylcarbamate)	Carbamate	Insecticide, acaricide, nematicide	190.26326 g mol^{-1}, log K_{ow} = 1.13, sparingly soluble in certain organic solvents, most soluble in chloroform and acetone	Stable under normal storage conditions and in acidic media but decomposes rapidly in alkaline media and at temperatures above 100 °C	Used against leaf miner
Boscalid (2-chloro-*N*-[2-(4-chlorophenyl)phenyl] pyridine-3-carboxamide)	Carboxamide	Fungicide	343.20668 g mol^{-1}, log K_{ow} = 2.96, solubility in water, 4.6 mg L^{-1} at 20 °C	Stable to hydrolysis at pH 4.5, 7	Used against white mould
Carbendazim and benomyl (methyl *N*-(1*H*-benzimidazol-2-yl) carbamate)	Benzimidazole	Fungicide	191.18666 g mol^{-1}, log K_{ow} = 1.52, solubility in water at 24 °C: 29 mg L^{-1} at pH 4; 8 mg L^{-1} at pH 7; 1.49 mg L^{-1} at pH 8	Stable for at least 2 years below 50 °C. Stable in acids, forming water-soluble salts	Used against coffee berry disease
Carbofuran ((2,2-dimethyl-3H-1-benzofuran-7-yl) *N*-methylcarbamate)	Carbamate	Insecticide, nematicide	221.2524 g mol^{-1}, log K_{ow} = 2.32, solubility in water, 351 mg L^{-1} at 25 °C	Stable under neutral or acid conditions, unstable in alkaline media	Used against coffee root-knot nematode
Chlorpyrifos (*O*,*O*-diethyl-*O*-3,5,6-trichloro-2-pyridyl phosphorothioate)	Organophosphate	Insecticide	350.586302 g mol^{-1}, log K_{ow} = 4.96, solubility (g 100 g^{-1}): acetone 650, benzene 790, carbon disulfide 590, carbon tetrachloride 310, chloroform 630, diethyl ether 510, ethanol 63, ethyl acetate >200, isooctane 79. Readily soluble in other organic solvents	Rate of hydrolysis increases with pH	Used against coffee cherry borer and coffee leaf miner

Clothianidin (1-[(2-chloro-1,3-thiazol-5-yl)methyl]-2-methyl-3-nitroguanidine)	Neonicotinoid	Insecticide	249.67802 g mol^{-1}, $\log K_{ow} = 0.7$ at 25 °C, solubility in water: 0.304 (pH 4), 0.340 (pH 10) (both g L^{-1}, 20 °C). In heptane less than 0.00104, xylene 0.0128, dichloromethane 1.32, methanol 6.26, octanol 0.938, acetone 15.2, ethyl acetate 2.03 (all in g L^{-1}, 25 °C)	Stable to zinc, zinc ions, iron, iron ions, aluminum and aluminum ions at 25 °C and 54 °C for 14 days	
Cypermethrin ([cyano-(3-phenoxyphenyl)methyl] 3-(2,2-dichloroethenyl)-2,2-dimethylcyclopropane-1-carboxylate)	Pyrethroid	Insecticide	416.29716 g mol^{-1}, $\log K_{ow} = 6.60$, soluble in methanol, acetone, xylene, dichloromethane	Relatively stable in neutral and weakly acidic media, with optimum stability at pH 4. Hydrolyzed in alkaline media. Relatively stable to light in field situations. Thermally stable up to 220 °C	Used against coffee cherry borer
Disulfoton (*O*,*O*-diethyl *S*2-ethylthioethyl phosphorodithioate)	Organophosphate	Insecticide, acaricide	274.404022 g mol^{-1}, $\log K_{ow} =$ 4.02, miscible in *n*-hexane, dichloromethane, 2-propanol, toluene	Relatively stable in acidic and neutral media. Hydrolyzed in alkaline media	Used against leaf miner
Endosulfan	Cyclodiene	Insecticide, acaricide	406.92514 g mol^{-1}, $\log K_{ow} = 3.83$ (alpha); $\log K_{ow} = 3.62$ (beta), solubility in ethyl acetate 200, dichloromethane 200, toluene 200, ethanol 65, hexane 24 (all in g L^{-1}, 20 °C)	Slowly hydrolyzed in aqueous acids and alkalis, with the formation of the diol and sulfur dioxide	Used against coffee cherry borer
Haloxyfop (2-[4-[3-chloro-5-(trifluoromethyl)pyridin-2-yl]oxyphenoxy]propanoic acid)	Aryloxyphenoxypropionate	Herbicide	361.70035 g mol^{-1}, $\log K_{ow} = 1.34$ (20 °C, unsated pH), solubility in water 43.4 mg L^{-1} (pH 2.6, 25 °C)	Time for 50% (DT_{50}) loss in water 75 days (pH 5)	Control weeds

(*continued*)

Table 36.1 (*continued*)

Pesticide (IUPAC name)	Chemistry	Classification	MM, log K_{ow}, solubility	Stability	Use
Imidacloprid (*N*-[1-[(6-chloropyridin-3-yl)methyl]-4,5-dihydroimidazol-2-yl] nitramide)	Neonicotinoid	Insecticide	255.661 g mol^{-1}, log K_{ow} = 0.57 at 21 °C, solubility in water, 6.1 × 10^{2+} mg L^{-1} at 20 °C	Stable to hydrolysis at pH 5–11	Used against coffee leaf miner
Permethrin ((3-phenoxyphenyl)methyl 3-(2,2-dichloroethenyl)-2,2-dimethylcyclopropane-1-carboxylate)	Pyrethroid	Insecticide	391.2877 g mol^{-1}, log K_{ow} = 6.50, soluble in most organic solvents except ethylene glycol. Soluble in acetone, ethanol, ether, and xylene	Stable to heat (> or = 2 years at 50 °C), more stable in acid than alkaline media with optimum stability *ca.* pH 4	
Phorate (*O*,*O* diethyl *S*-ethylthiomethyl phosphorodithioate)	Organophosphate	Insecticide, acaricide, nematicide	260.377442 g mol^{-1}, log K_{ow} = 3.56, soluble in water, 50 mg L^{-1} 25 °C	Stable under normal storage conditions for at least 2 years. Aqueous solutions degraded by light (DT_{50} 1.1 day); stability to hydrolysis optimum at pH 5–7	Used against sucking and chewing insects, leafhoppers, leaf miner, mites
Propiconazole (1-[[2-(2,4-dichlorophenyl)-4-propyl-1,3-dioxolan-2-yl] methyl]-1,2,4-triazole)	Triazole	Fungicide	342.22038 g mol^{-1}, log K_{ow} = 3.72, soluble in water, 100 mg L^{-1} at 25 °C	Stable up to 320 °C; no significant hydrolysis	Used against coffee leaf rust
Pyraclostrobin (methyl *N*-[2-[[1-(4-chlorophenyl)pyrazol-3-yl]oxymethyl] phenyl]-*N*-methoxycarbamate)	Strobilurin	Fungicide	387.81692 g mol^{-1}, log K_{ow} = 3.99 at 22 °C, Soluble in water, 1.9 mg L^{-1} at 20 °C	Stable more 30 days (pH 5–7, 25 °C)	Used against coffee leaf rust

Spirodiclofen ([3-(2,4-dichlorophenyl)-2-oxo-1-oxaspiro[4.5]dec-3-en-4-yl] 2,2-dimethylbutanoate)		Acaricide	411.31886 g mol^{-1}, $\log K_{ow} = 5.8$ (pH 4); 5.1 (pH 7); both ambient temperature, soluble in water, 50 (pH 4); 190 (pH 7); both in μg L^{-1} at 20 °C	Hydrolysis DT50 119.6 days (pH 4), 51.2 days (pH 7), 2.5 days (pH 9) (all 20 °C)	Used against red spider mite and coffee ringspot mite
Tebuconazole (1-(4-chlorophenyl)-4,4-dimethyl-3-(1,2,4-triazol-1-ylmethyl) pentan-3-ol)	Diacylhydrazine	Insecticide	307.81838 g mol^{-1}, $\log K_{ow} = 3.7$, soluble in water, 36 mg L^{-1} at pH 5–9, 20 °C	Stable to elevated temperatures, and to photolysis and hydrolysis in pure water	Used against berry spot disease and American leaf disease
Terbufo (*S-tert*-butylthiomethyl *O*,*O*-diethyl phosphorodithioate)	Organophosphate	Insecticide, nematicide	288.430602 g mol^{-1}, $\log K_{ow} =$ 4.48, readily soluble in most organic solvents: aromatic hydrocarbons, chlorinated hydrocarbons, alcohols, acetone	Stable for at least 2 years at room temperature	
Thiamethoxam ((NE)-*N*-[3-[(2-chloro-1,3-thiazol-5-yl)methyl]-5-methyl-1,3,5-oxadiazinan-4-ylidene] nitramide)	Neonicotinoid	Insecticide	291.7147 g mol^{-1}, $\log K_{ow} = -0.13$ at 25 °C, soluble in water 4.1, acetone 48, ethyl acetate 7.0, dichloromethane 110, toluene 0.680, methanol 13, *n*-octanol 0.620 (all in g L^{-1})		Used against coffee leaf miner
Triadimefon (1-(4-chlorophenoxy)-3,3-dimethyl-1-(1,2,4-triazol-1-yl) butan-2-one)	Triazole	Fungicide	293.74874 g mol^{-1}, $\log K_{ow} =$ 2.77, moderately soluble in most organic solvents except aliphatics	Stable to hydrolysis at 22 °C	Used against coffee rust
Triadimenol (1-(4-chlorophenoxy)-3,3-dimethyl-1-(1,2,4-triazol-1-yl) butan-2-ol)	Triazole	Fungicide	295.76462 g mol^{-1}, $\log K_{ow} = 2.90$, soluble in water, 120 mg L^{-1} at 20 °C	Both diasteroisomers are stable to hydrolysis DT_{50}(20 °C) more than 1 year	Used against coffee rust

Figure 36.1 Schrader's formula for organophosphorus compounds.

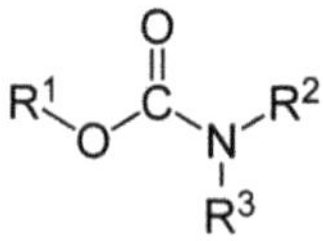

Figure 36.2 Chemical structure of carbamate compounds. R^1 is an alcohol group, R^2 is a methyl group and R^3 is usually hydrogen.

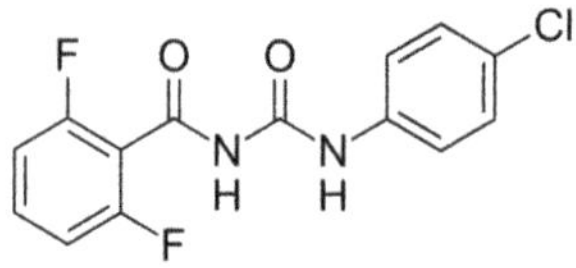

Figure 36.3 Chemical structure of diflubenzuron, a commonly used benzoylurea insecticide.

36.2.2 Fungicides

Fungicides are used to control molds and fungi. According to the chemical structure, two groups of fungicides can be identified: inorganic and synthetic organic. The most widely used synthetic fungicides belong to the chemical classes of triazoles, dithiocarbamates, anilinopyrimidines, strobilurins and benzimidazoles.[7]

Triazoles fungicides refers to any of the heterocyclic compounds with molecular formula $C_2H_3N_3$, having a five-membered ring of two carbon atoms and three nitrogen atoms. Triazoles fungicides include epoxiconazole, triadimenol, propiconazole, prothioconazole, metconazole, cyproconazole, tebuconazole, flusilazole and paclobutrazol.

Dithiocarbamates are organosulfur compounds represented by a general structure shown in Figure 36.4.

Strobilurin fungicides are derived from a group of natural fungicidal derivatives of β-methoxyacrylic acid. The simplest are strobilurin A and oudemansin A (Figure 36.5).

Anilinopyrimidines include any fungicides whose structure is based on a pyrimidine ring with an aniline substituent, for example cyprodinil (Figure 36.6).

Fungicides can be contact, translaminar or systemic. Contact fungicides are not taken up into the plant tissue and protect only the plant

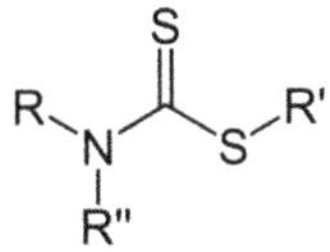

Figure 36.4 Chemical structure of dithiocarbamate compounds. *R′* can be substituted by an alkyl, alkylene, aryl or other similar group, and *R″* usually by a metal ion.

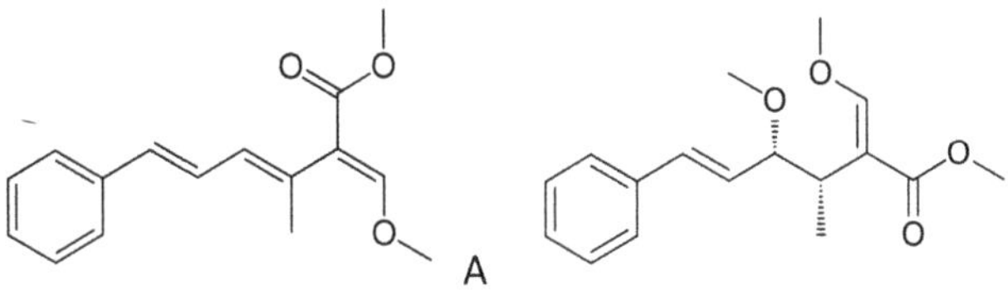

Figure 36.5 Chemical structure of (A) strobilurin A and (B) oudemansin A.

where the spray is deposited. Translaminar fungicides redistribute the fungicide from the upper, sprayed leaf surface to the lower, unsprayed surface. Systemic fungicides are taken up and redistributed through the xylem vessels.

Copper fungicides are some of the most effective and economical methods of controlling the rust pathogen in susceptible coffee varieties. They are used in control of coffee leaf rust, *Hemileia vastatri*, which is an obligate parasite and is transmitted when urediniospores (spores produced from the brown-red rust pustules) disperse from one part of the plant to another, or to a new uninfected plant. Systemic fungicides that can also be used in coffee to treat this disease include triazoles such as triadimefon and propiconazole and strobilurins such as azoxystrobin. No systemic fungicide such as captafol and copper-based fungicides have been effectively used in control of coffee berry disease, *Colletotrichum kahawae*. This is a very serious disease that can destroy up to 80% of crop, causing damage in coffee berries.[8] Other diseases in coffee including cercospora spot (*Cercospora coffeicola*) and bacterial blight (*Pseudomonas syringae pv. garcae*) can also be controlled with fungicides.

36.2.3 Herbicides

Herbicides are used to control weeds. They can be classified according to their chemical class, selectivity, nature of action and application characteristics (plant region and time). Herbicides that destroy all vegetation are classified as total or non-selective, while those that control some weeds without damaging other agricultural cultures are defined as selective.[7] Non-selective pesticides, *e.g.* paraquat, play a key role in the quest for sustainable coffee production by controlling weeds that would otherwise seriously reduce productivity (Figure 36.7).

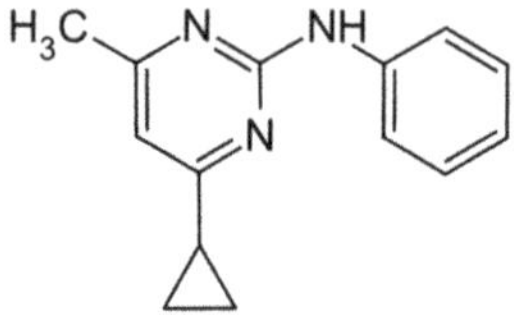

Figure 36.6 Chemical structure of cyprodinil.

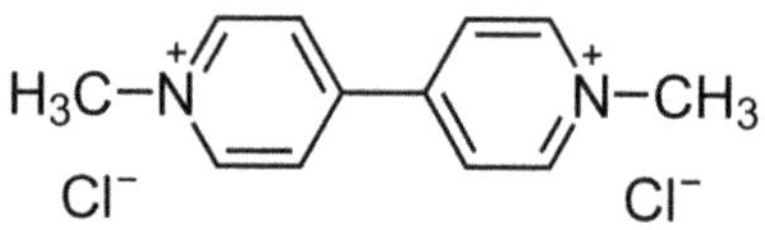

Figure 36.7 Chemical structure of paraquat.

Contact herbicides exert their action only on the part of the plant where they have been deposited (*e.g.* paraquat). Systemic herbicides, on the other hand, penetrate the plant and reach regions that are far from the point of application. Translocation within the plant occurs *via* the phloem and the xylem.[7] A non-selective systemic herbicide that is absorbed through plant leaves and green stem tissues is for example glyphosate. It is effective in killing a wide variety of weeds, including grasses, broad-leafed and woody plants in the coffee fields.

Herbicides can be applied in the regime of pre-sowing, pre-emergence and post-emergence. Although several herbicides are registered to be used in coffee plantations, only a few of them are selective enough to be applied directly over the crop, at post-emergence of weeds like fluazifop-*P*-butyl, clethodim or fomesafen.[9]

36.3 Physicochemical Proprieties

To understand the behavior of a specific compound in the environment and along the food chain, specifically its persistency and wide spreading potential, several physicochemical features can be determined, particularly in such a huge and complex family of compounds as pesticides. Hence, some of the most important physicochemical characteristics will be detailed.

Physical properties of pesticides are known to contribute to the mode of action, dosage, mode of application and its capability of entering into a suitable biological reaction. The physical properties of pesticides vary greatly according to their chemical nature and formulation.[10]

Water solubility is a key feature that affects the primary transformations processes namely hydrolysis, photolysis, oxidation, reduction and biodegradation reactions. Pesticides that are very soluble in water will tend not to accumulate in soil or biota because of its strong polar nature. However, they can transport away from the application site by runoff or irrigation water to reach the surface waters.

The octanol–water partition coefficient (K_{ow}) is characteristic of lipophilicity of the molecule and gives an indication of the pesticide tendency to accumulate in the environment, living organisms and foods. It is usually considered that substances with a log K_{ow} value higher than 3 can show accumulation persistence. The polarity of the molecules is also strong correlated with logK_{ow}. Non-polar pesticides are characterized by logK_{ow} values higher than 4–5 whereas polar pesticides have logK_{ow} values lower than 1–1.5 and moderated pesticides have values between these two values.[11]

Another crucial property in terms of physicochemical characterization is the vapor pressure (Vp), which allows predicting the behavior of organic contaminants in the environment and calculating Henry's law constants (H). The Vp of a pesticide is a potential indicator of its volatility, permitting a prediction of pesticide prone to evaporate from leaf and soil surface after application. The tendency of pesticide to volatize from water solution into air is largely defined by their H value, a high value favoring volatilization while a pesticide with an H value $<10^5$ Pa m^3 mol^{-1} shows little tendency to volatize.[11]

The soil partition coefficient (K_d) is used to describe a chemical's affinity of a pesticide for adsorption by solids and its potential for leaching though the soil. The values for K_d variy greatly because the organic matter content of the soil is not considered in the equation. The preferred parameter to determine the soil's ability to adsorb pesticides is the soil adsorption coefficient (K_{oc}) since it considers the organic matter content of the soil. K_{oc} is the ratio (at equilibrium) of the mass of a substance, adsorbed onto a unit mass of soil, relative to the mass of the substance remaining in water solution. K_{oc} is also a unitless parameter and its value is dependent on the organic matter content of the soil, polarity of the chemical and soil pH.[10] Pesticides with K_{oc} below 50 are considered to be highly mobile compounds.[11]

Degradation of pesticides is the breaking down or chemical transformation of a pesticide into other compounds that are not necessarily simpler and less toxic compared to the parent molecule.[10] The rate of degradation of pesticides is usually measured in terms of half-life ($t_{1/2}$), which is the time required for the depletion of half (or 50%) of the amount of pesticide present initially. The degradation process or breakdown of pesticides can occur in plants, animals and in the soil and water; or it can take place during the processing upon exposure to temperature or ultra-violet (UV) radiation. The degradation involves oxidation, reduction, hydrolysis, photolysis and conjugation reactions. The type of the reaction which a pesticide undergoes is largely determined by the pesticide inherent physicochemical properties and the environmental compartment (water, soil, air, biota) in which it is hosted.[10]

36.4 Legislation

In order to protect human safety and health resulting from pesticides application in coffee production, many countries have these chemicals under strict legislation and surveillance. A detailed review related to the global pesticide legislation has been recently published by Handford *et al.*[12] In this section some of the most important legislation aspects of four key regions – Europe, United States, Japan and Brazil – will be presented.

During the past decade, a series of legislative actions on pesticide were applied by the European Union (EU) in order to harmonize the maximum residues limits (MRL) in all member states. Currently, the MRLs are applied to 315 fresh products and to the same products after processing including coffee beans and coffee (Regulation EC 178/2006). A major focus of current EU legislation, along with harmonizing the pesticide registration process in Europe, is the future sustainable use of pesticides and adoption of integrated pest management.[12]

In the United States of America the US Food Drug Administration establishes tolerances for most products, which are the maximum amount of a pesticide allowed to remain in or on a food, as part of the process of regulating pesticides. Presently, for coffee about 43 pesticides are listed with tolerance limits.

In 2006, the Ministry of Health, Labour and Welfare of Japan established new MRLs for some food commodities and comprehensively reviewed existing MRLs. Codex MRLs have been adopted as Japanese MRLs, where in existence. Since its application, all Japanese agriculture products and imports have had to comply with the positive list. For substances with no established MRL a uniform default level of 0.01 ppm is set as the maximum limit.[12]

In Brazil, the leading coffee exporter in the world, there are two monitoring programs to ensure compliance with National MRLs (more than 80 listed for coffee); the Program on Pesticide Residue Analysis, which is coordinated by the Ministry of Health, and the National Residue and Contaminant Control Program, coordinated by the Ministry of Agriculture.[12]

Most countries in the world use MRLs to regulate pesticides, however, the levels established differ greatly between countries. Despite pesticide MRLs not being harmonized on a global scale, international instruments such as Codex have been established to assist in international trade. There are 21 MRLs listed for coffee as of December 2012 (Table 36.1).

36.5 Analytical Methods for Pesticide Residues Determination

Consumer concerns associated with the stringent regulations and legislation have been responsible, in large part, for the development of new analytical methods to determine pesticide residues in food. Due to the wide scope of pesticide analytes and large number of different commodities, associated to the low levels involved, pesticide residues analysis is one of the most challenging and complex areas of analytical chemistry.[13] Coffee beans and coffee are generally hard to analyze by conventional chromatographic techniques due to their extremely complex composition. This probably explains the small number of studies published on pesticide residue analysis in coffee.

Conventionally, pesticide analysis involves the use of a water-miscible solvent (*e.g.* methanol, acetonitrile) followed by liquid–liquid partitioning with an organic solvent (*e.g.* *n*-hexane with acetonitrile). This approach is further followed by a clean-up using cartridge solid-phase (SPE) or gel-permeation chromatography.

Lodevico *et al.*[14] extracted a neonicotinoid insecticide (imidacloprid and its metabolites) from green coffee beans, roasted beans and freeze-dried coffee with a mixture of methanol and 1% H_2SO_4 (1/3 v/v) followed by a clean-up involving hexane partitioning and XAD-4 resin before gas-chromatography mass spectrometry (GC-MS) analysis.

In another study,[15] 69 pesticide residues were extracted from coffee beans with ethyl acetate followed by gel-permeation chromatography (GPC) and combined double cartridge, solid-phase (SPE) clean-up before GC-MS analysis. Despite the quality of the separation of 69 pesticide residues from the matrix this procedure is time-consuming and quite laborious.

To overcome these drawbacks, techniques such as solid-phase microextraction (SPME) and QuEChERS (quick, easy, cheap, effective, rugged and safe) have been employed in the analysis of pesticide residues in coffee beans and coffee. These approaches require any or very small solvent volumes and often combine extraction and clean-up into a single step.

SPME involves a sorbent coating placed on a glass fiber device, and analytes are adsorbed onto or absorbed into the sorbent through an equilibrium process.[16,17] The fiber can be placed into a liquid sample or into the headspace above it (closed system). The extraction step is influenced by parameters that affect the distribution between the analytes and the sorbent coating, such as physicochemical properties (*e.g.* partitioning coefficients), volumes (of both the sample and coating), temperature, time, sample agitation, pH, ionic strength and other matrix effects. After extraction, a desorption step is needed for analysis by GC or liquid chromatography (LC).

Copobiango *et al.*[18] applied the polydimethylsiloxane (PDMS) fiber, in the extraction of nine organophosphorus pesticides (dichlorvos, phorate, guthion, phosdrin, guthion, malathion, methyl-malathion, di-syton and co-ral) from coffee grains and leaves. The fiber was immersed directly into the sample (0.1805 to 0.1842 g with 16 mL of Milli-Q water) for 40 min at 30 °C, before GC-MS analysis.

A pronounced advantage in comparison to the previous methods is that SPME can be automated and has advantages of generating no waste, being easy to perform and using little space or labware. Despite being a more environmentally friendly technique, SPME is more expensive than others such as QuEChERS that makes use of common laboratory material and encompasses a small number of steps.

QuEChERS was introduced by Anastassiades *et al.*[19] for the multiclass, multiresidue analysis of pesticides in fruits and vegetables. It is a very flexible approach that has been adapted for the analysis of many pesticide residues in food and animal tissues.[13] The main concept entails the use of an organic solvent, typically acetonitrile, for extraction of the pre-homogenized sample, which already contains water (or water is added), followed by addition of anhydrous $MgSO_4$ and NaCl or sodium acetate to induce partitioning of the extract (upper phase) from the water (lower phase). The extraction is typically conducted by shaking, but blending is also an option. After centrifugation, an aliquot of the extract is mixed with sorbent(s) in a centrifuge tube in an approach known as dispersive-SPE clean-up. The final extract is analyzed for pesticides by chromatographic methods, with detection ideally using MS techniques.

QuEChERS is a very versatile method that allows working at different pHs and several modifications can be performed at the clean-up step, although this technique cannot easily be automated and the enrichment factor is very poor.

QuEChERS techniques were applied in coffee beans for analysis of 51, 123 and 10 pesticide residues by Pizzutti *et al.*,[20] Dias *et al.*[21] and Mekonen *et al.*,[22] respectively. The main difference from the original method was the use of acetonitrile acidified with 1% acetic acid as extraction solvent, in order to avoid the degradation of some pesticides (captan, captafol, folpet and dicofol), instead of acetonitrile alone.

Due to the promising results shown by SPME and QuEChERS their application will likely expand in the near future. Encouraging the use of sustainable analytical methods not only reduces waste generation in the laboratory, but it also leads to a global environmental awareness with subsequent major health impact.

The determination of pesticide residues in coffee has traditionally been performed by mass spectrometry coupled to gas chromatography (GC-MS). These techniques allow: (i) the detection of a wide range of analytes independently of their elemental composition; (ii) the simultaneous quantification and identification of the detected compounds; (iii) the spectrometric resolution of co-eluting peaks; and (iv) potentially faster analysis times than other chromatographic techniques.[15,18,20]

Despite the advantages of GC-MS systems, the analysis of polar, non-volatile or/and thermally labile pesticides by this technique is limited, usually requiring chemical derivatization. Liquid chromatography tandem-mass spectrometry (LC-MS/MS) has become a standard approach to expand the range of pesticides quantified and identified in complex matrices; this approach was used by Dias *et al.*[21] applied in the analysis of 123 pesticide residues in coffee beans.

36.6 Pesticide Residues in Coffee Beans and Beverage

Coffee is typically cultivated in tropical and subtropical areas at high elevations, often in rugged mountainous zones, and naturally grows under a shaded canopy of trees. Traditionally coffee is often integrated in agro-forestry systems in which tree species are cultivated together with coffee and other agricultural commodities. In these systems the use of pesticides is lower than those of industrial plantations.

Intensive pesticide use within industrial coffee production often employs pesticides that present serious health and ecological concerns. Sampling of imported green coffee beans conducted by the US Food and Drug Administration (FDA) in the late 1970s and early 1980s revealed frequently the presence of dichlorodiphenyltrichloroethane (DDT), benzine hexachloride (BHC) and other pesticides banned in the US because of long-term persistence in the environment and possible carcinogenicity.[23] Later, in 1993, the FDA verified from a survey of 60 imported green coffee bean samples from 21 countries that some of them contained different pesticide residues, such as chlorpyrifos at concentrations ranging from 0.01 to 0.04 mg kg^{-1} and pirimiphos methyl at 0.01 mg kg^{-1}.[24]

In 2009 the Food Safety Department of Japan detected the presence of pyraclostrobin and flutriafol pesticides residue in coffee containers imported from Brazil. The residual level detected was 0.02 and 0.09 mg kg^{-1}, while the limit established by Japanese standards corresponded to 0.01 ppm for both chemicals.

In 2011 the Canadian Food Inspection Agency detected pesticide residues in only two of the 297 coffee samples analyzed, in levels compliant with maximum residue limits (MRLs) established.[25]

A study in Brazil showed that residues of different pesticides such as endosulfan, chlorpyrifos, cypermethrin and captafol have been detected in coffee beans from either registered or illegal sources.[20] Another study in Brazil showed that 7 out of 15 coffee samples were contaminated with pesticides, with carbendazim, thiamethoxam and triadimenol above the MRL allowed for these pesticides.[21] DDT, cypermethrin, permethrin, deltamethrin, chlorpyrifos ethyl and endosulfan (α and β) were detected in Ethiopian green coffee beans at concentrations ranging from 0.0111 to 1.115 mg kg^{-1}.[22] DDT and endosulfan are not authorized for use in coffee; this result reinforces the need for more monitoring in a wide range of pesticides to safeguard consumer health.

Additionally, some researchers have studied the persistence of pesticides applied in coffee crops and the subsequent residue levels in coffee beans and, in a few cases, processed coffees. Mekonen *et al.*[26] have studied the effect of household coffee processing (washing, roasting and brewing) on pesticide residues, namely endosulfan α, endosulfan β, cypermethrin, permethrin, deltamethrin, chlorpyrifos ethyl, heptachlor epoxide, hexachlorobenzene, *p'p*-DDE, *p'p*-DDD, *o'p*-DDT and *p'p*-DDT. They observed that washing of coffee beans led to a 14.63–57.69% reduction, while the roasting process reduced up to 99.8%. Chlorpyrifos ethyl, permethrin, cypermethrin, endosulfan α and β in roasted coffee and all of the 12 pesticides in the coffee brewing processes were not detected.

Oliveira *et al.*[27] verified that the preparation of coffee beverage caused a decrease of chlorpyrifos and dichlorvos levels, but did not change the concentration of methyl parathion. In this study, powder coffee free of pesticides was spiked with a mixture of pesticides, being the coffee beverage prepared with hot water as commonly performed in Brazil.

Although there has been increasingly more information regarding pesticide residues in coffee beans, studies on the evaluation of coffee processing conditions on the pesticide amounts in the final roasted product are, as far as concerned, almost absent. Additionally, few studies have reported the amounts of pesticide residues in coffee beverage and evaluated the risk consumption.

Due to the high coffee consumption worldwide it is of great importance to determine the presence of pesticide residues in such products, to assess whether the amount of residues is within the maximum residue limits (MRLs) values recommended for each pesticide and if the acceptable daily intake of pesticide residues in beverage coffee is safe.

36.7 Final Considerations

The control of pests and diseases in coffee crops requires the application of a wide range of pesticides such as insecticides, fungicides or herbicides. The use of pesticides, however, should be selective, having the maximum effect on the pest and the minimum effect on beneficial organisms and also must minimize the risk of toxic residues for humans.

To meet regulatory requirements and protect the consumer and the environment, many analytical methods have been reported to monitor and control residue levels in coffee. The use of classical extraction techniques requiring large volumes of harmful solvent has been reduced by new techniques such as SPME and QuEChERS. These methods are ideally suited for the powerful analytical features of GC-MS and LC-MS/MS and can provide rapid, reliable and sensitive analysis of pesticides in coffee matrices.

In general, the achievement of low limits of quantification for pesticide residues analysis is easily attained by MS methods. Therefore, research should focus on providing useful information for risk assessment studies, namely the effect of processing on residue amounts in coffee and the importance of different consumption patterns.

References

1. ICO, 214-World coffee trade (1963–2013): a review of the markets, challenges and opportunities facing the sector.
2. W. J. Zhang, F. B. Jiang and J. F. Ou, Global pesticide consumption and pollution: with China as a focus, *Proc. Int. Acad. Ecol. Environ. Sci.*, 2011, **1**, 125–144.

3. FAO, *International Code of Conduct on the Distribution and Use of Pesticides, (Revised Version)*, Food and Agriculture Organization of the United Nations, Roma, Itália, p. 6.
4. B. Ballantyne and T. C. Marrs, Pesticides: an overview of fundamentals, in *Pesticide Toxicology and International Regulation*, ed. T. C. Marrs and B. Ballantyne, John Wiley and Sons, Chichester, West Sussex, UK, 2004, ch. 1.
5. J. M. Waller, M. Bigger and R. J. Hillocks, *Coffee Pests, Diseases and Their Management*, ed. J. M. Waller, M. Bigger and R. J. Hillocks, UK, 2007, ch. 15, pp. 336–361.
6. R. Bardner, *"Pest Control" in Coffee Botany, Biochemistry, and Production of Beans and Beverage*, ed. M. N. Clifford and K. C. Willson, Croom Helm, Bristish Library, 1985, ch. 8, pp. 208–218.
7. P. Cabras, Pesticides: Toxicology and Residues in Food, in *Food Safety-Contaminants and Toxins*, ed. J. P. F. D'Mello, CABI Publishing in CAB International, Oxon, UK, 2003, pp. 91–124.
8. J. F. Barrera, Coffee Pests and their Management, in *Encyclopedia of Entomology*, ed. J. L. Capinera, Springer, 2008, pp. 961–998.
9. E. N. Alcântara, Avaliação de herbicidas para cafeeiros em formação, in *Congresso Brasileiro da Ciência das Plantas Daninhas*, Foz do Iguaçu, Resumos. Londrina: SBCPD, 2000, vol. 22, p. 344.
10. J. T. Zacharia, *Identity, Physical and Chemical Properties of Pesticides, Pesticides in the Modern World – Trends in Pesticides Analysis*, ed. M. Stoytcheva, InTech, 2011, DOI: 10.5772/17513, available from: http://www.intechopen.com/books/pesticides-in-the-modern-world-trends-in-pesticides-analysis/identity-physical-and-chemical-properties-of-pesticides.
11. D. Tsipi, H. Botitsi and A. Economou, *"Pesticide Chemistry and Risk Assessment" at the book Mass Spectrometry for the Analysis of Pesticide Residues and their Metabolites*, ed. TsipiD., BotitsiH. and EconomouA., John Wiley & Sons, Inc, 2015, pp. 1–35.
12. C. E. Handford, C. T. Elliott and K. Campbell, A review of the global pesticide legislation and the scale of challenge in reaching the global harmonization of food safety standards, *Integr. Environ. Assess. Manage.*, 2015, **11**, 525–536.
13. S. C. Cunha, S. J. Lehotay, K. Mastovska, J. O. Fernandes and M. B. P. P. Oliveira, Sample preparation approaches for the analysis of pesticide residues in olives and olive oils at the book Olives and Olive Oil, in *Health and Disease Prevention*, ed. V. R. Preedy and R. R. Watson, Oxford:Academic Press, 2010, pp. 653–666.
14. R. G. Lodevico and Q. X. Li, Determination of total Imidacloprid residues in coffee by gas chromatography-mass spectrometry, *Anal. Lett.*, 2002, **35**, 315–326.
15. X. Yang, J. Wang, D. C. Xu, J. W. Qiu, M. Ying and J. Cui, Simultaneous determination of 69 Pesticide Residues in Coffee by Gas Chromatography-Mass Spectrometry, *Food Anal. Methods*, 2010, **4**, 186–195.

16. C. L. Arthur, L. M. Killam, K. D. Buchholz, J. Pawliszyn and J. R. Berg, Automation and optimization of solid-phase microextraction, *Anal. Chem.*, 1992, **64**, 1960–1966.
17. J. Pawliszyn, *Solid Phase Microextraction Theory and Practice*, Wiley-VCH, New York, NY, 1997.
18. H. L. V. Copobiango and Z. L. Cardela, A solid-phase microextraction method for the chromatographic determination of organophosphorus pesticides in fish, water, potatoes, guava and coffee, *J. Braz. Chem. Soc.*, 2005, **16**, 907–9014.
19. M. Anastassiades, S. J. Lehotay, D. Štajnbaher and F. J. Schenck, Fast and easy multiresidue method employing acetonitrile extraction/partitioning and "dispersive solid-phase extraction" for the determination of pesticide residues in produce, *J. AOAC Int.*, 2003, **86**, 412–431.
20. I. R. Pizzutti, A. de Kok, C. Dickow Cardoso, B. Reichert, M. de Kroon, W. Wind, L. Weber Righi and R. Caiel da Silva, A multi-residue method for pesticides analysis in green coffee beans using gas chromatography-negative chemical ionization mass spectrometry in selective ion monitoring mode, *J. Chromatogr. A*, 2012, **1251**, 16–26.
21. C. M. Dias, F. A. Oliveira, F. D. Madureira, G. Silva, W. R. Souza and Z. L. Cardeal, Multi-residue method for the analysis of pesticides in Arabica coffee using liquid chromatography/tandem mass spectrometry, *Food Addit. Contam., Part A*, 2013, **30**, 1308–1315.
22. S. Mekonen, A. Ambelu and P. Spanoghe, Pesticide residue evaluation in major food items of Ethiopia using the QuerChERS method: a case study from the Jimma zone, *Environ. Toxicol. Chem.*, 2014, **33**, 1294–1302.
23. S. A. Hearne, *Harvest of Unknowns: Pesticide Contamination in Imported Foods*, New York, Natural Resources Defense Council, 1984, pp. 16–17.
24. R. M. Jacobs and N. J. Yess, Survey of imported green coffee beans for pesticide residues, *Food Addit. Contam.*, 1993, **10**, 575–577.
25. FSAP – Food Safety Action Plan (2011), 2010–2011 Pesticides in Coffee, Fruit Juice and Tea, http://www.inspection.gc.ca/food/chemical-residues-microbiology/chemical-residues/pesticides/eng/1351913846907/1351913943956.
26. S. Mekonen, A. Ambelu and P. Spanoghe, Effect of household coffee processing on pesticide residues as a means of ensuring consumers safety, *J. Agric. Food Chem.*, 2015, **63**, 8568–8573.
27. J. J. V. Oliveira, I. S. D. Loredo, J. Rodrigues and M. F. P. M. Castro, Avaliação dos níveis residuais de clorpirifós, diclorvós e paration metílico na borra e na bebida de café, *Rev. Bras. Toxicol.*, 2003, **15**, 75–78.

Subject Index

www.ingramcontent.com/pod-product-compliance
Ingram Content Group UK Ltd.
Pitfield, Milton Keynes, MK11 3LW, UK
UKHW020731190926
8624IPUK00035B/17

9 781782 62004